Sensors and Biosensors, MEMS Technologies and its Applications

Sergey Y. Yurish
Editor

Sensors and Biosensors, MEMS Technologies and its Applications

Book Series: Advances in Sensors: Reviews, Vol. 2

International Frequency Sensor Association Publishing

Editor
Sergey Y. Yurish
Sensors and Biosensors, MEMS Technologies and its Applications
Advances in Sensors: Reviews, Vol.2

E-mail (for orders and customer service enquires): ifsa.books@sensorsportal.com
Visit our Home Page on http://www.sensorsportal.com

ISBN-10: 84-616-4154-X
ISBN-13: 978-84-616-4154-3
BN-20130415-XX
BIC: TJFC

Acknowledgments

As Editor I would like to express my undying gratitude to all authors, editorial staff, reviewers and others who actively participated in this book. We want also to express our gratitude to all their families, friends and colleagues for their help and understanding.

Contents

Contributors

Azian Azamimi Abdullah Biomedical Electronic Engineering Department, School of Mechatronic Engineering, Universiti Malaysia Perlis, 02600 Arau, Perlis, Malaysia

Hepsiba K. Anga Digital Systems Group, CSIR-Central Electronics Engineering Research Institute, Pilani-333031, India

Arjun Ashok School of Electrical Engineering, VIT University, Vellore - 632014, India

David Assan Civil and Environmental Engineering Department, School of Engineering, Idaho State University, Pocatello, ID 83209, USA

Alireza Bahrampour Department of Physics, Sharif University of Technology, Tehran, Iran

Wamadeva Balachadran Centre for Electronic Systems Research (CESR), Electronics and Computer Engineering, School of Engineering and Design, Brunel University, London, UK

Nikolaos V. Boulgouris Centre for Electronic Systems Research (CESR), Electronics and Computer Engineering, School of Engineering and Design, Brunel University, London, UK

Scott Cambron Department of Bioengineering University of Louisville, Louisville, KY 40292, USA

Yaoyao Cao School of Agriculture and Food Engineering, Shandong University of Technology, No.12, Zhangzhou Road, Zibo 255049, Shandong Province, P. R. China

Yaw-Jen Chang Mechanical Engineering Department, Chung Yuan Christian University, Zhong Li, Taiwan

Kosum Chansiri Department of Pathology, Faculty of Medicine, Srinakharinwirot University, Sukhumvit 23, Bangkok, 10110, Thailand

T. Y. Chin Bioscience Technology Department, Chung Yuan Christian University, Zhong Li, Taiwan

Wen-Yaw Chung Electronic Engineering Department, Chung Yuan Christian University, Zhong Li, Taiwan

Andre M. Gobin Department of Bioengineering University of Louisville, Louisville, KY 40292, USA

Poonam Goel School of Electrical and Electronics Engineering, Nanyang Technological University Singapore, 639798

Navneet Gupta Department of Electrical and Electronics Engineering Birla Institute of Technology and Science Pilani, Rajasthan, India, 333 031

Kuang-Pin Hsiung Taiwan Unison Biotech Inc., Hsin Chu City, Taiwan, R.O.C.

Xiaofeng Hu Oil Crops Research Institute of the Chinese Academy of Agricultural Sciences, Wuhan 430062, China

Neda Jahangiri Department of photonics, Kerman graduate University of Technology, Kerman, Iran

Thongchai Kaewphinit Innovative Learning Center, Srinakharinwirot University, Sukhumvit 23, Bangkok 10110, Thailand

Robert S. Keynton Department of Bioengineering University of Louisville, Louisville, KY 40292, USA

K. A. Koparkar Department of Physics, Sant Gadge Baba Amravati University, Amravati 444 602, India

Alexander Kukla Institute of semiconductor physics of NASU, Kiev, pr-t Nauki 41, 03028, Ukraine

Santosh Kumar Digital Systems Group, CSIR-Central Electronics Engineering Research Institute, Pilani-333031, India

James C. K. Lai Civil and Environmental Engineering Department, School of Engineering, Idaho State University, Pocatello, ID 83209, USA

Pick Yern Lee Biomedical Electronic Engineering Department, School of Mechatronic Engineering, Universiti Malaysia Perlis, 02600 Arau, Perlis, Malaysia

Solomon W. Leung Civil and Environmental Engineering Department, School of Engineering, Idaho State University, Pocatello, ID 83209, USA

Jing Li Oil Crops Research Institute of the Chinese Academy of Agricultural Sciences, Wuhan 430062, China

Peiwu Li Oil Crops Research Institute of the Chinese Academy of Agricultural Sciences, Wuhan 430062, China

Yuxin Liu Division of Bioengineering, School of Chemical & Biomedical Engineering, Nanyang Technology University, Division Bioengineering, Singapore 637457, Singapore

Kochuthomman Joseph Mampilly School of Electrical Engineering, VIT University, Vellore - 632014, India

Nadarajah Manivannan Centre for Electronic Systems Research (CESR), Electronics and Computer Engineering, School of Engineering and Design, Brunel University, London, UK

Muhammad Saiful Badri Mansor Process Tomography Research Group, Control & Instrumentation Engineering Department, Faculty of Electrical Engineering, Universiti Teknologi Malaysia, 81310 Skudai, Johor Bahru, Malaysia

Shahzad Memon Centre for Electronic Systems Research (CESR), Electronics and Computer Engineering, School of Engineering and Design, Brunel University, London, UK

Siti Zarina Mohd. Muji Department of Computer Engineering, Faculty of Electrical and Electronic Engineering, Universiti Tun Hussein Onn Malaysia, 86400, Parit Raja, Batu Pahat, Johor, Malaysia

Azad Noor Centre for Electronic Systems Research (CESR), Electronics and Computer Engineering, School of Engineering and Design, Brunel University, London, UK

Martin G. O'Toole Department of Bioengineering University of Louisville, Louisville, KY 40292, USA

P. C. Panchriya Digital Systems Group, CSIR-Central Electronics Engineering Research Institute, Pilani-333031, India

Alexey Pavluchenko Institute of semiconductor physics of NASU, Kiev, pr-t Nauki 41, 03028, Ukraine

Dhruvinkumar N. Patel Department of Bioengineering University of Louisville, Louisville, KY 40292, USA

Dorota G. Pijanowska Institute of Biocybernetics and Biomedical Engineering, Polish Academy of Sciences, Warsaw, Poland

P. Bhanu Prasad Digital Systems Group, CSIR-Central Electronics Engineering Research Institute, Pilani-333031, India

Lu Qiao School of Agriculture and Food Engineering, Shandong University of Technology, No.12, Zhangzhou Road, Zibo 255049, Shandong Province, P.R. China

Ruzairi Abdul Rahim Process Tomography Research Group, Control & Instrumentation Engineering Department, Faculty of Electrical Engineering, Universiti Teknologi Malaysia, 81310 Skudai, Johor Bahru, Malaysia

Prabhu Ramanathan School of Electrical Engineering, VIT University, Vellore - 632014, India

Sudha Ramasamy School of Electrical Engineering, VIT University, Vellore - 632014, India

M. Saliminasab Department of photonics, Kerman graduate University of Technology, Kerman, Iran

Somchai Santiwatanakul Department of Biochemistry, Faculty of Medicine, Srinakharinwirot University, Sukhumvit 23, Bangkok, 10110, Thailand

A. L. Sharma School of Instrumentation, Devi Ahilya University, Takshila Campus, Khandwa Road, Indore-452001, India

Ashish Kumar Sharma Department of Electrical and Electronics Engineering Birla Institute of Technology and Science Pilani, Rajasthan, India, 333 031

Xia Sun School of Agriculture and Food Engineering, Shandong University of Technology, No.12, Zhangzhou Road, Zibo 255049, Shandong Province, P. R. China

Xinghua Sun Department of Bioengineering University of Louisville, Louisville, KY 40292, USA

Majid Taraz Department of Physics, Shahid Bahonar University of Kerman, Kerman, Iran

Kimberly Jane Uy Electronic Engineering Department, Chung Yuan Christian University, Zhong Li, Taiwan

Xiangyou Wang School of Agriculture and Food Engineering, Shandong University of Technology, No.12, Zhangzhou Road, Zibo 255049, Shandong Province, P.R. China

Sazali Yaacob Biomedical Electronic Engineering Department, School of Mechatronic Engineering, Universiti Malaysia Perlis, 02600 Arau, Perlis, Malaysia

R. D. S. Yadava Sensor & Signal Processing Laboratory Department of Physics, Faculty of Science Banaras Hindu University, Varanasi 221005, India

Hao Chun Yang Electronic Engineering Department, Chung Yuan Christian University, Zhong Li, Taiwan

Ting Ya Yang Electronic Engineering Department, Chung Yuan Christian University, Zhong Li, Taiwan

Yi Ying Yeh Electronic Engineering Department, Chung Yuan Christian University, Zhong Li, Taiwan

Li Yu Oil Crops Research Institute of the Chinese Academy of Agricultural Sciences, Wuhan 430062, China

Sergey Y. Yurish Research & Development Department, Technology Assistance BCNA 2010, S. L., Parc UPC-PMT, Edificio RDIT-K2M C/ Esteve Terradas, 1, 08860, Castelldefels, Barcelona, Spain

Zulkarnay Zakaria Biomedical Electronic Engineering Department, School of Mechatronic Engineering, Universiti Malaysia Perlis, 02600 Arau, Perlis, Malaysia

M. H. Zandi Department of Physics, Shahid Bahonar University of Kerman, Kerman, Iran

Chen Zhai School of Agriculture and Food Engineering, Shandong University of Technology, No. 12, Zhangzhou Road, Zibo 255049, Shandong Province, P. R. China

Guandong Zhang Department of Bioengineering University of Louisville, Louisville, KY 40292, USA

Zhaowei Zhang Oil Crops Research Institute of the Chinese Academy of Agricultural Sciences, Wuhan 430062, China

Ying Zhu School of Agriculture and Food Engineering, Shandong University of Technology, No.12, Zhangzhou Road, Zibo 255049, Shandong Province, P.R. China

Preface

It is my great pleasure to present the second volume from the *Advances in Sensors: Reviews* book Series started by the IFSA Publishing in 2012.

The second volume titled '*Sensors and Biosensors, MEMS Technologies and its Applications*' contains eighteen chapters with sensor related state-of-the-art reviews and descriptions of latest achievements written by authors from academia and industry from 12 countries: China, India, Iran, Malaysia, Poland, Singapore, Spain, Taiwan, Thailand, UK, Ukraine and USA.

This book ensures that our readers will stay at the cutting edge of the field and get the right and effective start point and road map for the further researches and developments. By this way, they will be able to save more time for productive research activity and eliminate routine work.

Built upon the series *Advances in Sensors: Reviews* - a premier sensor review source, it presents an overview of highlights in the field and becomes. Coverage includes current developments in physical sensors, biosensors, immunosensors, data acquisition systems, MEMS technologies and its applications.

This volume is divided into three main parts: physical sensors, biosensors, nanoparticles, MEMS technologies and applications. With this unique combination of information in each volume, the Advances in Sensors book Series will be of value for scientists and engineers in industry and at universities, to sensors developers, distributors, and users.

Like the first volume of this book Series, the second volume also has been organized by topics of high interest. In order to offer a fast and easy reading of the state of the art of each topic, every chapter in this book is independent and self-contained. The eighteen chapters have the similar structure: first an introduction to specific topic under study; second particular field description including sensing applications.

Chapter 1 reviews various physical and chemical sensors and sensor systems for smartphones. Coming technological limitations and challenges are outlined. Two design approaches traditional and novel, advanced approach for such sensors are described in details and illustrated by examples. The advanced design approach is based on the smart system integration and allows to eliminate existing technological limitations. It lets to create smart sensor systems faster and easier, which make smartphones smarter and intelligent.

Chapter 2 presents an analysis of synchronization between two coupled nonlinear surface acoustic wave (SAW) delay line oscillators where coupling is provided via another linear phase SAW delay line. The analysis is aimed at determining the sensitivity of synchronization frequency for perturbations in the coupling SAW delay line, and then to explore whether the coupling SAW device can be used for making a better chemical sensor in comparison to the usual polymer-coated SAW delay line oscillator sensor. The dynamics of coupled system is analyzed for small perturbations in limit cycles under phase approximation. Considering the noise suppression characteristics and high sensitivity regions of synchronization it is found that the coupled SAW oscillators in synchronized states have potential for making high performance SAW sensors.

The main focus of Chapter 3 is to review the existing fingerprint sensing technologies in terms of liveness detection, and discusses hardware based 'liveness detection' techniques reported in the literature for automatic fingerprint biometrics.

Chapter 4 confined about synthesis of conjugated polymer through plasma polymerization process. The merits and demerits associated with plasma polymerization are discussed in comparison with chemical rout polymerization and electrochemical polymerization various aspects regarding the chemical and physical properties of plasma polymer film is different from conventional methods of synthesis of conjugated polymer. Plasma-enhanced chemical vapour deposition (PECVD) is also reviewed within their respective sections. This different structure of plasma polymerized thin film leads to potential application in various fields. The plasma polymerization processes of conjugated polymers are strappingly affected by various parameters such as pressure, separation between two electrodes. These parameters discussed on the basis of Pascen's law. This review shortlisted and discuss some advantages and disadvantages associated with plasma polymerization.

Chapter 5 describes substrate materials its characteristics, fabrication process and application for non-Silicon fabrication micromachining technologies as such as printed circuit board (PCB), low temperature co-fired ceramic (LTCC) and liquid crystal polymers (LCP) in addition to new upcoming technologies such as polymer core conductor and polydimethylsiloxane (PDMS).

Chapter 6 presents the rapidly emerging field of MEMS based instruments used in medical field and discuss its present and future applications.

Chapter 7 is about the development of MEMS switch for RF application in past two decades. Perhaps the most important RF MEMS component is RF MEMS Switch. RF MEMS switches are the fundamental part of the RF systems because these switches provide automatic redirection of RF signals in RF Communication devices. RF MEMS switch design parameters like actuation voltage, insertion loss, isolation, return loss, switch lifetime, switching speed, and temperature sensitivity are discussed in detail inside this chapter. These design parameters show significant impact RF MEMS switch development towards better performance. The motive of the chapter is to provide an overall device picture, current status and the research efforts that carried out to maturing this technology.

Chapter 8 highlights the research carried out during the last 5 years on amperometric acetylcholinesterase (AChE) biosensors for determination of organophosphorous and carbamate pesticides in a wide range of samples. Various immobilization protocols and modified electrode methods used for constructing AChE biosensors are also described in details. Future prospects toward the development of selective, sensitive biosensing systems are discussed.

Chapter 9 reviews quartz crystal microbalance DNA based biosensor for clinical diagnosis and detection.

Chapter 10 presents an overview of electrochemical immunosensors for the detection of pesticides residues and various immobilization protocols of Ab or hapten, such as physical adsorption, covalent coupling, entrapment, oriented immobilization, avidin–biotin affinity reaction, self-assembled monolayer, nanoparticles. Future prospects toward the immobilization protocols for the development of electrochemical immunosensor are discussed.

The objective of Chapter 11 is to have a look on the interaction of electromagnetic field on biological tissue from the view of physiological and electrical engineering. The understanding of the phenomena in the interaction may give clear picture on the fundamental concept of dielectric dispersion in term of suitable frequency application and the reaction of biological tissue to it. The other factors such as thermal effect also will be gone through. This review may provide some useful information to the research of magnetic induction tomography on the biological tissue imaging and electromagnetic therapy which apply electromagnetic field at certain frequency in their application.

The use of miniaturized sensing platform for the analysis of biotoxins is the subject of Chapter 12. Materials for sensing platform and detection methods for biotoxins are discussed in details. Applications to the analysis of phytotoxins, animal toxins, marine toxins, microbial toxins are also described. Finally, future challenges and opportunities are discussed.

Chapter 13 examines the usage of gold nanoshells (GNS), with near infrared (nIR) resonant properties, to improve the detection of antigen/biomarkers/viral vectors using immunoassays that rely upon an adventitious change in the surface plasmon resonance when multiple particles interact with antigens. Methods of producing GNS, conjugation of biomolecules to the nanoparticle surface, quantification of the number of bound antibodies on the gold surface, and two immunoassays for detection of antigen/biomarkers/viral vectors at various concentrations will be presented. The demonstrated techniques allow one to produce GNS with specific nIR resonant properties, allow specific targeting to various antigens, analytes, biomarkers, viral particles, and cells via antibodies or ligands, and have the potential to be used to various applications including photothermal ablation therapy.

Chapter 14 describes the performance of ultra-high performance sensors fabricated with identical biocomposite materials and procedures, except the anchoring conductive materials. The anchoring materials were glassy carbon, Pt, Au and Ag; the biocomposite layer consisted of polymer/Au nanoparticles/enzyme. The enzymes in the biocomposite layers are specific for the target analytes which enable the coupling (detection) reactions to occur. The enzymes used in this study were lactate dehydrogenase (LDH), glutamate dehydrogenase (GDH), and hemoglobin. The specific target analytes for the detection included lactate, NH_4^+, NO_2^-, and peroxide. The biosensors developed here were tested with solution concentrations ranged from 10^{-4} to 10^{-16} M; however, except for NO2-, the lower concentration limit could be orders of magnitude lower. In all, the anchoring materials have drastic effect to the performance of this ultra-high performance biosensor platform; in addition, nature of the enzymes can alter the stability of these sensors.

Chapter 15 includes an overview of Electronic nose and Electronic Tongue based Biomimetic systems for classification and authentication of beverages. Biomimetic systems, in particular electronic-nose (e-Nose) and electronic-tongue (e-Tongue), show promising utilization and advantages in various fields and for this reason emphasis here is made on electronic-nose and electronic-tongue, the recently added systems in the field of biomimetics. In brief, Electronic-Nose and Electronic-Tongue are the systems which mimic the human olfaction and gustation respectively. They are used for automatic analysis and recognition (classification) of liquids and gases (e-Nose for gases and E-tongue for liquids). The e-Nose can determine the fingerprint of a complex volatile or dissolved compound mixture by an array of non-specific/semi-specific sensors coupled to a pattern recognition system. Similar to the human nose, the electronic nose operates by recognizing the overall pattern of components constituting the given sample stored in a database, in a manner analogous to the recollection of olfaction perception in the human brain. e-Tongue is analogous to e-Nose but related to sense of taste. Much emphasis and importance is given regarding the selection of sensors and sensing methods employed for these biomimetic systems. An optimized selection is made regarding the sensors and sensing methods based upon the application involved, accuracy level to be met, cost and complexity of the system involved. Not only the sensors and the sensing methods involved in these systems but also the pattern recognition algorithms developed for these specific systems are of importance. Despite the optimized selection of sensors and sensing methodology, the performance of these systems is greatly affected by the quality of functioning of its pattern recognition block. Various techniques and methods have been developed and used which serve the purpose. Besides an overview of Electronic Nose and Electronic Tongue systems, basic sensing principles of these systems and the different types of sensors used in these systems are included. Two different case studies have been presented wherein the Electronic Nose system was used for the classification and authentication of Indian Teas while the Electronic Tongue system was used for the classification and authentication of Indian wines. The chapter also deals with Artificial neural networks, and in particular with the probabilistic neural network (PNN) and other data analysis methods like principal component analysis (PCA) and linear discriminant analysis (LDA) which have been used as data analysis methods in the two case studies presented.

Chapter 16 is the compiled description of the design considerations taken while developing the magnetic bead based biosensor systems. The work will show experimental proof of this determination theory. This work also presents a system that will make use of magnetic particle assay (MPA) as a means of implementing this diagnosis. The work also includes a discussion on the sensor that is used to quantify the number magnetic particles after performing MPA on the sample as well as a prototype developed as a proof of concept for this work.

Chapter 17 discusses a novel sensor for non-invasive determination of body analytes by using ultralow-threshold stimulated Raman lasing in microspheres. We theoretically employed this sensor for non-invasive measuring of blood glucose levels. Also a composite sensing system of an optical microsphere resonator and silver nanoparticles based on surface enhanced Raman scattering and stimulated Raman scattering techniques towards a point of care diagnostic system for acute myocardial infarction using the Troponin I biomarker in HEPES buffered solution is proposed.

Chapter 18 presents a simple and robust multipoint data acquisition bus built on top of the standard RS232 interface with the minimum additional non-standard hardware components. The presented network structure and associated protocol can be used to form a general purpose distributed data acquisition bus. Basic methods of data flow control and failure recovery suitable for the described architecture are described.

Each chapter has been written by different contributors. Many of contributors are members of the editorials board of different journals related to the field and some of them are IFSA members.

We hope that readers enjoy this book and that can be a valuable tool for those who involved in research and development of various MEMS, biosensors, physical, chemical sensors and sensor systems.

Sergey Y. Yurish

Editor
IFSA Publishing

Barcelona, Spain

Chapter 1
Smart Sensors for Smartphones: How to Make it Smarter ?

Sergey Y. Yurish

A smartphone without sensors is nothing more than an annoyingly small laptop.

(Bruce Kasanoff)

1.1. MEMS and Sensor Markets for Smartphones and Tablets

The modern smartphone market is a high growth and potentially huge market. *IHS* predicts global smartphone shipments will rise by 28 percent in 2013 to 836 million units, up from 654 million in 2012 (Fig. 1.1). A semiconductor marketing & consulting research company *Semico Research Corporation* (USA) projects that this market has a Compound Annual Growth Rate (CAGR) of 21.9 % approaching 1.1 billion units by 2015 [1]. In this year 44 % of all mobile phones will be smartphones. However, according to *IHS* already 2013 will mark the first year that smartphones will account for more than 50 % of global cellphone shipments.

A *Consumer Electronics Association* (CEA) representative at a press conference on 2013 Consumer Electronics Show (CES) in Las Vegas (USA) said that the technology industry has entered the "post-smartphone era" when communication functions such as calls and texting are no longer the main focus for smartphones. The representative noted that smartphones now mainly are used for non-communication types of functions.

CEA also said that consumer electronics products are becoming "sensorized"- i.e., devices like smartphones are making increasing use of sensors that allow the digitization of everyday things. Specific types of applications mentioned included infrared, near-field communications and moisture sensors that tell users when their plants need to be watered.

S. Y. Yurish
Research & Development Department, Technology Assistance BCNA 2010, S. L.
Parc UPC-PMT, Edificio RDIT-K2M, C/ Esteve Terradas, 1, 08860 Castelldefels, Barcelona, Spain

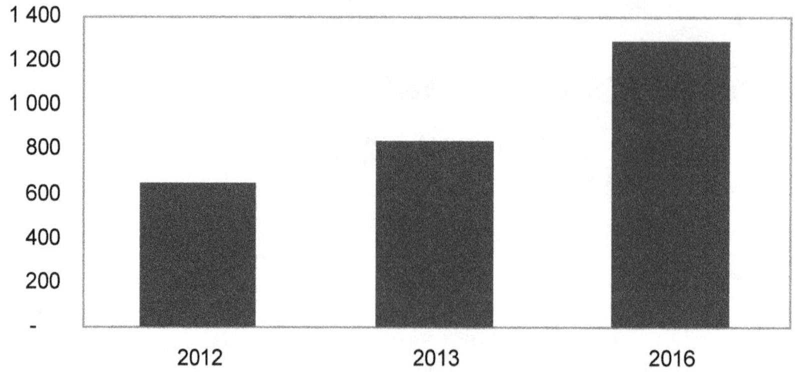

Fig. 1.1. Worldwide smartphone shipment forecast (Millions of units) (Source: IHS iSuppli + Screen Digest Research, January 2013).

CEA specifically mentioned micro-electro-mechanical systems (MEMS), which commonly serve as sensors in smartphone applications, including gyroscopes, accelerometers and pressure sensors.

Global revenue for MEMS in consumer and mobile applications is anticipated to reach $ 3.2 billion in 2013, up 22 percent from $ 2.6 billion in 2012, as presented in Fig. 1.2. Revenue will soar to $ 4.9 billion in 2016, according to the IHS iSuppli MEMS & Sensors Service at *IHS*.

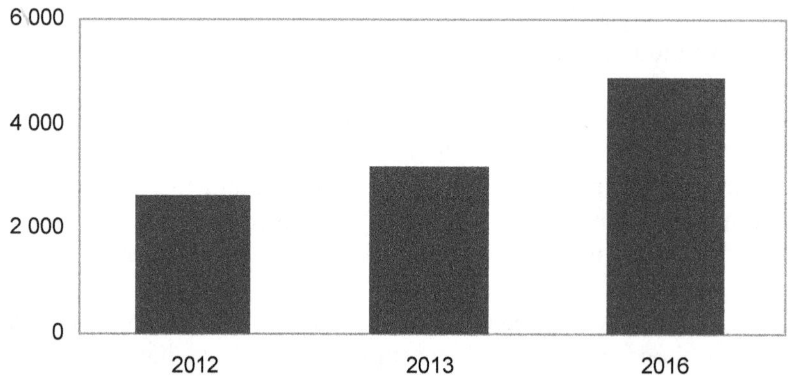

Fig. 1.2. Global forecast of MEMS for Consumer and Mobile Electronics (in $ M) (Source: IHS Screen Digest Research, January 2013).

In terms of functionality, sensors continue to improve significantly. Future smart devices (smartphones and tablets) will include new sensor types, such as biometric, pressure, and environmental sensors, along with ones currently in most smartphones [2]. More sensors and the great penetration of mobile devices will drive global sensor shipments to 6 billion units by 2015, 1/3 of which will represent new sensor types (Fig. 1.3) [3].

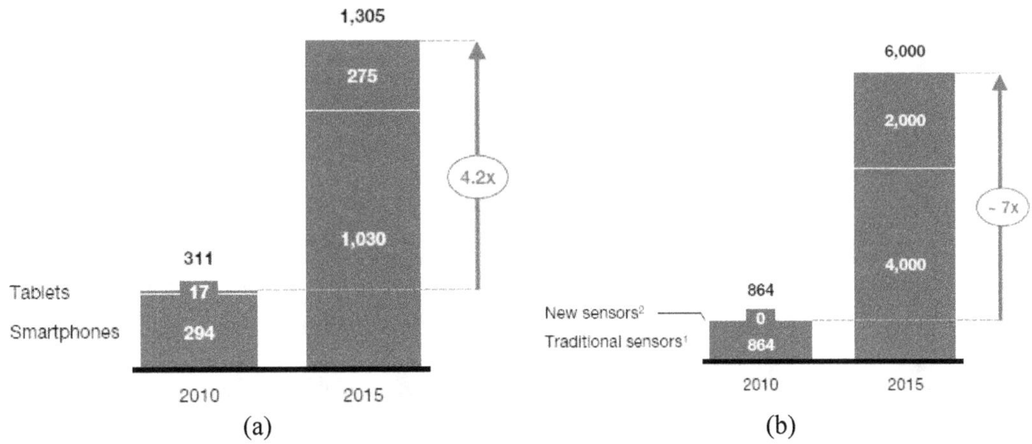

Fig. 1.3. Global smart devices (a), and sensors (b) shipments (Million units) [3]. (Source: IHS iSuppli, McKinsey analysis).

The *Juniper Research* declares, that smartphone sensors (MEMS and others) to be $ 6 bn business by 2016.

MEMS and sensors are very important and extremely popular components that enable many of the new features on smartphones and tablets. In the new report "MEMS for Cell Phones and Tablets" *Yole Developpement* predicts 19.8 % annual growth to reach $5.4B in 2017 [4]. First of them it is motion sensors such as accelerometers, magnetometers and more recently gyroscopes (it is the hottest market segment, which is still growing quickly); combo sensors, MEMS microphones and BAW filters and duplexers. Many other MEMS products are under development. New MEMS devices will benefit from the mobile device growth that is predicted for the coming years: phones and tablet will represent a 2.9 B units in 2017 and most of them will integrate 5 to 10 MEMS devices [4]. Ten new MEMS applications will to be worth more than $100 million in 2017, versus 3 categories of MEMS devices in high-volume production today. The global MEMS market growth diagram for cell phones and tablets is shown in Fig. 1.4.

New market intelligence from *ABI Research* projects the smartphones are the largest market for MEMS sensors in mobile devices. The total annual revenues for sensors will be $ 1.3 billion by the end of 2012. MEMS sensor revenues for smartphones are forecast to be worth $ 4.4 billion in 2017. Media tablets were the second largest market for MEMS sensors in mobile devices and are predicted to be almost $ 835 million in 2017 [5].

Global revenue from motion sensor technology in smartphones and tablets will expand to $ 2.1 billion in 2015, up from $ 1.1 billion in 2011 [6]. The motion sensor category consists of products including MEMS accelerometers; MEMS gyroscopes; electronic compasses (3-axis magnetometers) and MEMS pressure sensors.

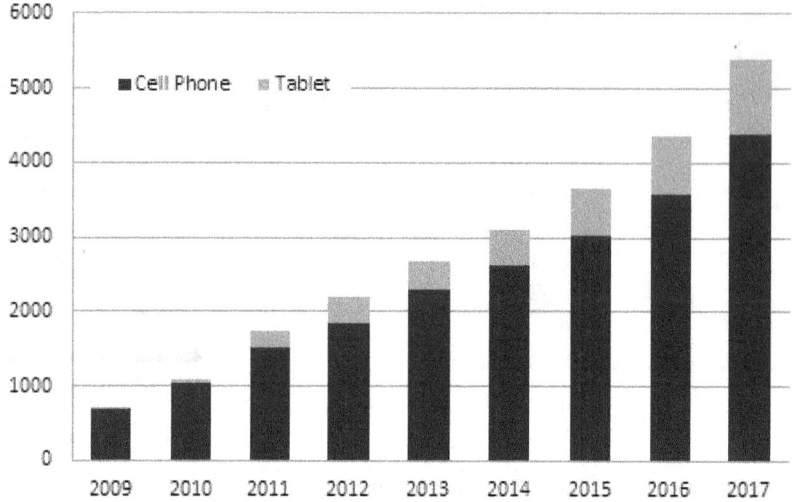

Fig. 1.4. Global MEMS market for cell phones and tablets (in $ M) [4].

The global market revenue forecast of MEMS motion sensors in smartphones and tablets is shown in Fig. 1.5.

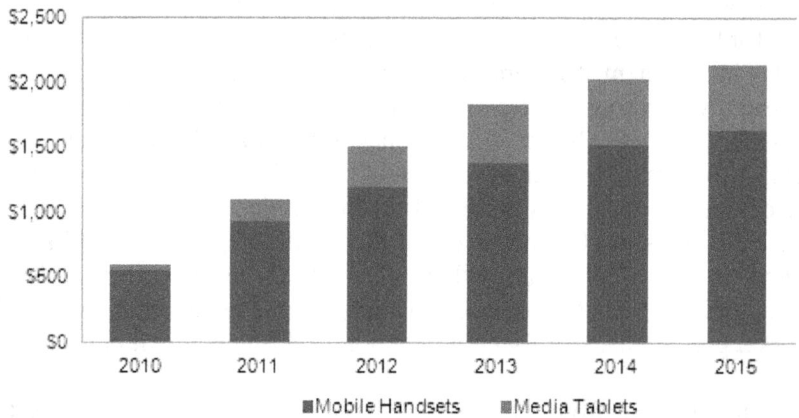

Fig. 1.5. Global market revenue for MEMS motion sensors in smartphones and tablets (in $ M) [6].

Only few years ago the first accelerometers were integrated into mobile phones, at a price level similar to gyroscopes today. Now accelerometers are viewed as commodity products in some platforms and their penetration was above one third of cell phones in 2010. According to *ABI Research* it estimates 32 % of MEMS sensor shipments will be accelerometers in mobile devices. Nevertheless, gyroscopes were the largest portion of the MEMS sensor market in revenue terms and have reached $ 350 million in 2012.

According to the 'IHS iSuppli Microelectromechanical Systems (MEMS) Market Brief' report from information and analytics provider *IHS*, by 2016, combo sensor revenue will amount to $ 1.4 billion accounting for 71 percent of the overall motion sensor space in consumer and mobile applications. Discrete accelerometers, gyroscopes and compasses will make up the remainder of the market, or 29 percent, based on 2016 revenue of $ 561.9 million, in view of steadily shrinking usage [7].

Speaking about modern smartphone applications market, a healthcare applications market segment is very dynamic and promised (Fig. 1.6).

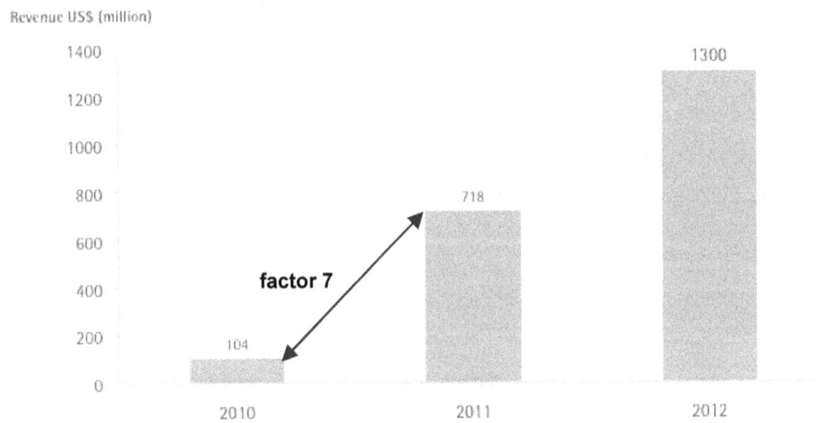

Fig. 1.6. Global revenue for mobile healthcare applications in 2012 (Source: research2guidance) [8].

30 % of total smartphone uses (500 million people) will have used mHealth applications by 2015. Many of these applications are sensor based applications: 76 % of total mHealth application market revenue will come from related services and products such as sensors.

Modern smartphones have a limited number of built-in sensors. As rule, it is so-called "must have" sensors, mentioned above (accelerometer, gyroscope and magnetic sensors), and in several models, there are also light (proximity) and pressure sensors. But based on the *International Frequency Sensor Association (IFSA)* Survey 2012 "What Sensors are Needed in a Mobile Phone ?" 36 % of responders from 265 would like to have much more sensors in their mobile devices (Fig. 1.7). First of all it is various gases detectors, air quality detectors, temperature, humidity, altitude and ultra-violet sensors.

Among other suggestions, responders mentioned also radiation and glucose sensors, fire alarms, alcohol detectors and breath analyzers. Several other emerging sensors have been also identified, with the potential to become new killer applications in 5 to 10 years: environmental sensors, fingerprint sensors, etc. …

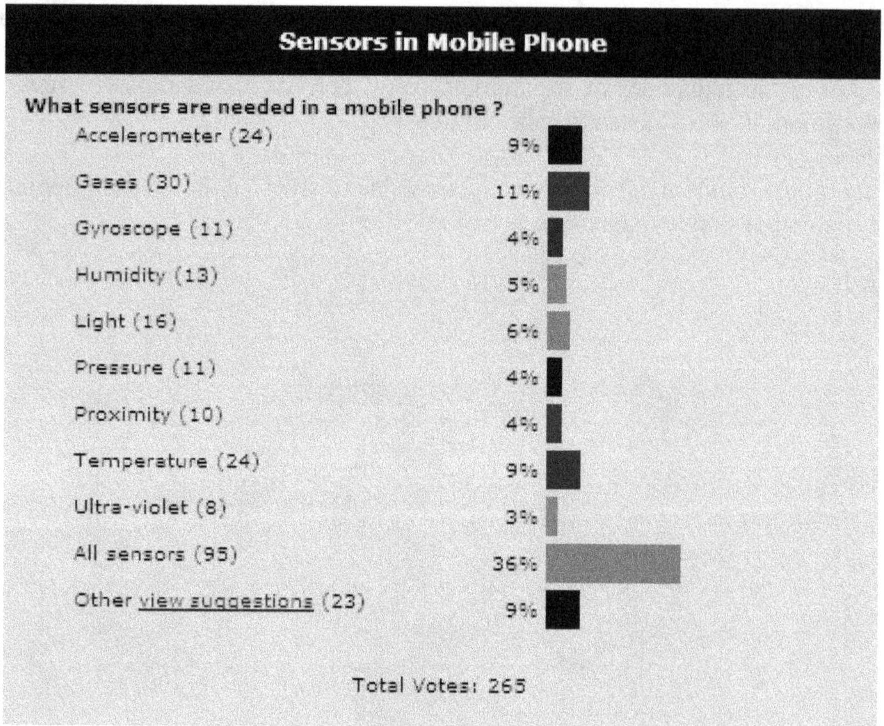

Fig. 1.7. IFSA Survey 2012 results (Source: IFSA).

Smartphone users expect their device to be intuitive and capable of interacting without pressing a button. Sensors make this possible.

1.2. Sensors for Smartphones: Sate-of-the-Art

1.2.1. Physical Sensors

Based on design and communication features all sensors for smartphones can be divided into three main groups: built-in (embedded, internal) sensors and external sensors (Fig. 1.8). In turn, external sensors can be divided into two main subgroups: wire sensors (detachable units, physically connected to a USB port or dock-connector) and wireless sensors (connected by Bluetooth with a wearable sensing module, for example).

Let consider some interesting examples from each of sensor groups in details. In this review, we will not consider image sensors (for smartphones' photo and video cameras) and microphones. Such devices exist in all smartphones and have good performances. The future development is expected in direction of megapixel resolution improvement for cameras. We will not also consider a GPS, because of it is a 'system/application', rather than a separate sensor device.

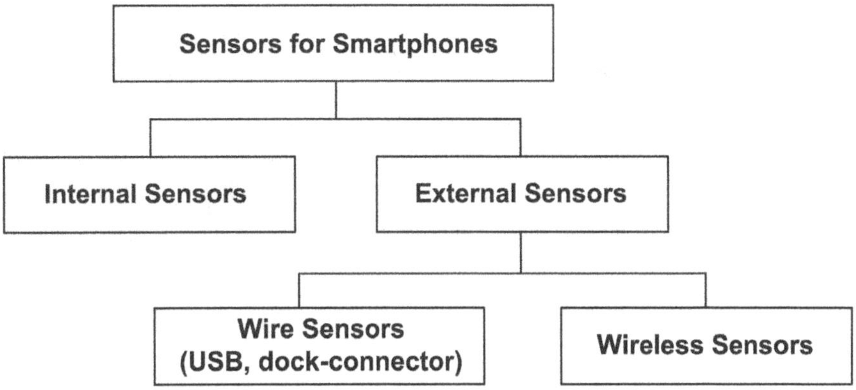

Fig. 1.8. Smartphone sensors classification.

Aginova, Inc. (USA) announced the launch of iCelsius and iCelsius Pro, designed to extend iPads (iCelsius Pro) by employing a temperature probe to convert iPhone, iPad or iPod touch into a digital thermometer for various applications [9]. The temperature sensors should be connected to a smartphone's dock-connector and can provide accurate temperature readings in the ranges of -30 ^{0}C to +70 ^{0}C (iCelsius) and in the ranges of -30 ^{0}C to +150 ^{0}C (iCelsius Pro). The temperature probes has absolute error ±0.2 ^{0}C (±0.4 ^{0}F) at 25 ^{0}C, and ±1^{0}C over whole range (iCelsius), and ±2.7 ^{0}C (iCelsius Pro). The iCelsius Pro uses a 4" stainless steel tip to report temperature readings through the iPad when the probe is in thermal contact with matter. Digital thermometers based on iPhone and iPad, and temperature probes are shown in Fig. 1.9 (a-c) accordingly. There are some different probes for various applications.

iCelsius RH - Temperature & Humidity Sensor can measure temperature (-40 ^{0}C to +120 ^{0}C) and relative humidity (0-100 % RH) [10].

MIT Mobile Experience Laboratory has developed a ultra-violet sensor for Android, which provides real-time and useful information to the user through a user-friendly interface in order to limit overexposure to a known human carcinogen to address UV radiation and skin cancer prevention [11]. The architecture selected uses the powerful features of a smartphone to interact with the user and a wireless sensor to monitor the UV radiation. UV index data, measured by the handheld sensor, are sent to the mobile phone by using a low-energy wireless protocol. The incoming data, postprocessed by the mobile application, are displayed to the user with relevant and useful information to help the user to make an appropriate decision regarding the sun protection. By continuously monitoring and providing meanful history graphs, this tool allows people to have a long term view of their UV exposition to modify their behavior towards the sun.

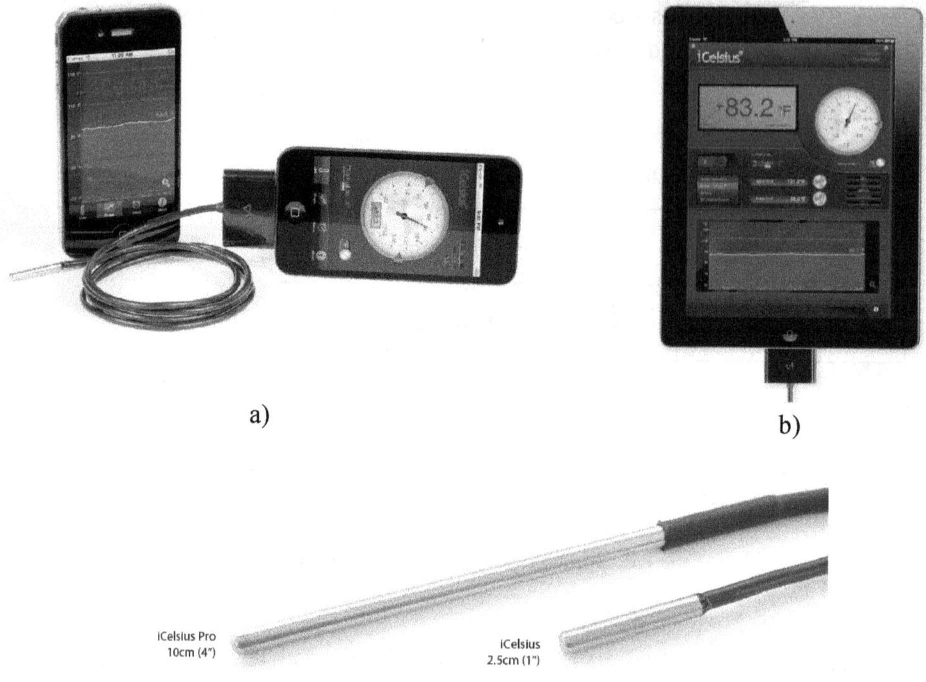

c)

Fig. 1.9. Digital thermometers based on iPhone (a), iPad (b), and temperature probes iCelcius Pro and iCelcius (c).

The first smartphone-compatible prototype of dosimeter-radiometer DO-RA (*Intersoft Eurasia*, Russia), will measure the equivalent dose of radiation exposure over a set period of time [12]. Whenever the phone has the program running in dosimeter mode, the dose will be displayed on the screen. In radiometer mode, the DO-RA measures equivalent dose rate of radiation emitted by ground, water, foodstuffs and other objects.

The unique, world's smallest all-purpose dosimeter for measuring radiation can either be integrated into the design of new mobile phones at a production stage or used as a detachable unit for multiple electronic devices (Fig. 1.10). The unique α-, β- and γ-radiation sensor is based on a semiconductor detector, the patent for the graphene sensor has already been issued. The detachable DO-RA radiation sensor can also be connected via a standard USB port. The remote sensor communicates with a mobile phone/smartphone and the elements of the DO-RA via such data transmission technologies as Bluetooth, Wi-Fi, etc.

Using Bluetooth connection, the data collected by the DO-RA can be transmitted to other electronic devices positioned within 10 meters.

The prototype of dosimeter-radiometer has already been adapted to work with practically all mobile operational platforms such as iOS, Android, WP7, BalckBerry, JavaME, as well as Windows, MacOS, Linux, etc.

a) b)

Fig. 1.10. Detachable dosimeter-radiometer DO-RA for smartphone: a) sensor's print board; b) unit with a smartphone [12].

If a mobile phone has a built-in GPS/GLONASS capability, the DO-RA will automatically produce radiological condition reports (equivalent dose rate levels) for the area where the customer is located. The collected data may be transmitted in real time via mobile communication providers or other data transmission networks.

According to the field tests, the dosimeter-radiometer device consumes no more than 1 % of the regular smartphone battery capacity. The test was conducted on iPhone-4 and 4 s.

In the future new generations of the DO-RA device will be produced with the help of the radiation sensors built with Silicium (Si) and graphene nanostructure. It will be a measuring device of high accuracy and great compactness. The evolution of DO-RA devices is shown in Fig. 1.11. The DO-RA device can become a one standard microchip with a sensor inside the aforementioned devices. The radiation sensor will look like a Smartphone camera, i.e. its linear dimensions will be similar.

A smartphone that can measure radiation levels was also demonstrated by the Japanese mobile phone giant *NTT DoCoMo* [13]. The smartphone has several changeable "jackets" fitted with sensors, to monitor body mass as well as level of skin-damaging ultraviolet light. But the shell that measures radiation levels is likely to prove the most popular.

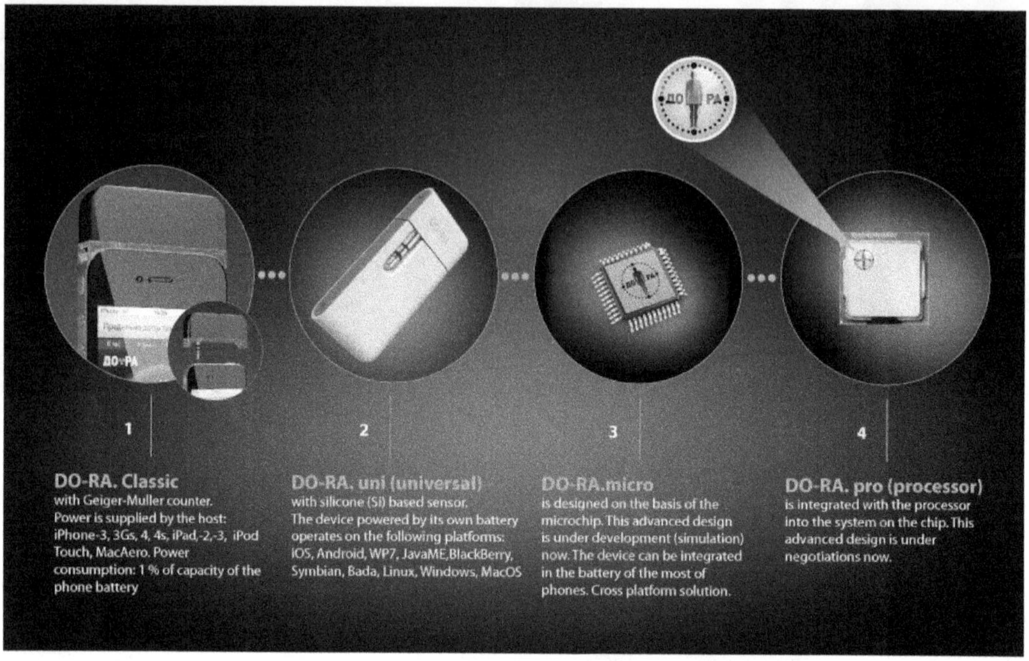

Fig. 1.11. The evolution of DO-RA devices [12].

1.2.2. Chemical Sensors and Biosensors

Various chemical and biosensors essentially extend smartphone applications. Some interesting examples are described below.

Sensorex (USA) has introduced PH-1 pH meter for iPhone, iPad and iPod to measure and record pH values in the lab or field (Fig. 1.12) [14]. The pH sensor is connected to the smartphone' dock-connector. The supplied app provides real time display of pH, millivolts, ambient temperature and solution temperature. When used with a GPS enabled device, the pH Meter application will record where and when measurements were taken. The absolute error of measurement is ± 0.01 pH after calibration. The pH measuring range is 0 ... 14; the solution temperature range - 0 ... 100 ^{0}C, and ambient temperature range - 0 to 40 ^{0}C.

Suitable for any process where pH measurements are routinely performed. By recording the pH value with a time stamp and location, this app eliminates any chance of mistakes when recording readings. Works with location aware devices to log the coordinates where readings were taken for analysis and auditing purposes.

Sanofi-Aventis (USA) manufactures the first available smallest innovative blood glucose meter iBGStar that seamlessly connects to the Apple iPhone and iPod touch allowing to view and analyze accurate, reliable information in 'real time' [15]. The meter requires no battery changes and is instead recharged similarly to a phone (Fig. 1.13).

Fig. 1.12. PH-1 pH meter for iPhone, iPad and iPod (Source: *Sensorex's* web site [14]).

a) b)

Fig. 1.13. Blood glucose meter iBGStar: (a) separate unit; (b) embedded into a smartphone.

The iBGStar meets today's stringent industry standards for accuracy and features plasma equivalent calibration for assurance with each and every reading. The main iBGStar specification is the following:

Blood sample size: 0.5 microliters;
Test strips: BGStar test strips;
Test time: 6 seconds;
Memory: 300 test results;
Operating Temperature: 10 ^{0}C to 40 ^{0}C (50 ^{0}F to 104 ^{0}F);
Blood calibration: Plasma equivalent;

Result Range: 1.1 to 33.3 mmol/L;
Size: L = 56 mm, W = 24 mm, H = 10 mm.

Sensorcon, Inc. (USA) has developed Sensordrone - an external, small key fob sized device, which can be connected to Android based smartphones via Bluetooth (2.1 and 4.0) [16]. In addition, the device includes also digital (RS232, I^2C) and analog signals (3.3 V) inputs for communication with external hardware (such as other sensors or any other peripheral that can communicate via the supported protocols). The USB port is used only for charging. The Sensordrone includes the following chemical and physical sensors:

Precision Electrochemical Gas Sensor – Calibrated for Carbon Monoxide (also can be used for precision measurements of Alcohol, Hydrogen, and others);
Gas Sensor for Oxidizing Gases – MOS type for Chlorine, Ozone, Nitrogen Dioxide, etc.;
Gas Sensor for Reducing Gases – (MOS type for methane, Propane, alcohols, other hydrocarbons, etc.);
Temperature – Simple resistance temperature sensor type;
Humidity;
Pressure – can be used for Barometer, Altimeter, Blood Pressure, etc.;
Non-Contact Thermometer – Infrared sensor for scanning object temperature;
Proximity Capacitance – fluid level, intrusion detection, stud finder, etc.;
RGB Color Intensity.

Application examples are:
- Air quality;
- Breath alcohol;
- Color matching;
- Non-contact thermometer;
- Weather monitoring;
- Crowd sourced Sensor Data;
- Games (totally new concepts);
- Flatulence (use gas sensors).

Let consider Sensordrone's sensors general technical specifications and metrological performances in details.

Precision Gas Sensor. The precision gas sensor is a professional grade electrochemical sensor (fuel cell sensor) that is very sensitive to a variety of toxic gases, alcohol, and hydrogen. It comes pre-calibrated for Carbon Monoxide (CO). The output from the precision gas sensor is given in ppm of CO with the following specifications:

- Resolution: 1 ppm;
- Range: 0-2 000 ppm;
- Accuracy: ± 10 % of reading;
- Response Time: 10-20 seconds to reach 90 % of final signal.

Although the precision gas sensor comes pre-calibrated for CO, it will respond to a variety of gases. For example, the sensitivity to Hydrogen is about 10-20 % of CO. What this means in practice is that if the sensor is exposed to 1,000 ppm of Hydrogen, the CO output would give a reading of 100 to 200 ppm. Apps can be made for various gases such as hydrogen, breath alcohol (ethanol) and various others by simply multiplying the CO output by a suitable sensitivity factor. A selection of sensitivity factors will be provided in follow on data-sheets.

Oxidizing and Reducing Gas Sensors. The Oxidizing and Reducing gas sensors are general purpose semiconductor sensors that respond to a broad range of gases. They are not as sensitive or accurate as the precision gas sensor, but respond to more gases, so are particularly suitable for finding gas leaks or gas emitting sources. One sensor is tuned to respond to oxidizing gases, a second sensor is tuned to respond to reducing gases. General specifications for CO and NO_2 gas sensors are shown in Table 1.1.

Table 1.1. General specifications for CO and NO_2.gas sensors.

	Reducing Gas Sensor: when sensing CO	**Oxidizing Gas Sensor: when sensing NO_2**
Min. Detectable Concentration, ppm	5	0.050
Range, ppm	1 000	0-5
Response Time: T90/ time to reach 90 % of final signal, s	30-60 seconds	30-90 seconds

Temperature Sensor. A silicon bandgap type temperature sensor has the following metrological performance:

- Accuracy: ± 0.5 ^{0}C (± 1 ^{0}C near lower/upper ends of range);
- Range: -20 ^{0}C to + 60 ^{0}C (limit of overall system);
- Response Time: 20-60 seconds to reach 90 % of final signal/change in signal.

It may take several minutes when transitioning from hot to cold environments. The worst case will happen when Sensordrone is heated during the initial phases of charging, and then quickly moved to a cold environment. If fast temperature measurements are required, the infrared temperature sensor may be a better choice.

Humidity Sensor. A capacitive type humidity sensor has the following metrological performance:
- Accuracy: ± 2 % RH from 20-80 % RH (± 4 % outside this range);
- Range: 0 to 100 % RH;
- Response Time: 10-180 seconds depending on temperature change.

The %RH is calculated based on air moisture content and temperature. If the temperature remains constant and the moisture content changes, %RH response will be fast. If both temperature and humidity change dramatically (e.g. moving from indoors to outdoors during winter months in a cold climate), it may take a several minutes for both the temperature sensor and humidity to stabilize.

Pressure Sensor. The pressure sensor is a MEMS type pressure sensor that is used for absolute (as opposed to differential) pressure measurements. It is most suited for measuring changes in ambient pressure. The pressure sensor has the following metrological performance:

- Accuracy: ± 0.1 kPa;
- Resolution: 1.5 Pa (pressure) / 0.3 m (altitude);
- Range: 30 to 110 kPa;
- Response Time: ~1 second.

Infrared Temperature (non-contact temperature sensor). The infrared temperature sensor uses a thermopile to absorb infrared energy being emitted from an object, and uses the corresponding voltage change to determine the temperature of the object. The infrared temperature sensor can be used to make non-contact temperature measurements.

- Accuracy: ± 1 to 3 ^{0}C (depending on ambient vs. measured object temperature),
  ± 0.2 to 0.3 ^{0}C (with suitable averaging and stable temperate);
- Normal Range: -20 ^{0}C to + 60 ^{0}C (-40 ^{0}C to +125 ^{0}C with increasing error);
- Response Time: 1-5 seconds.

Proximity Capacitance. The proximity capacitance sensor consists of a set of electrodes located underneath the bottom cover of the Sensordrone, and a capacitance to digital converter circuit. Any material within close proximity to the electrodes will cause a change in the capacitance between the electrodes.

- Resolution: 0.5 fF (in lowest range, 12 bit of full scale in all other ranges);
- Range: 4 selectable ranges (0 to 0.5 pF, 0 to 1 pF, 0 to 2 pF, 0 to 4 pF);
- Response Time: < 1 second.

Color and Light Sensors. A photodiode array with filters for Red, Green, Blue and Clear (visible light) are used for these sensed parameters. The response time is < 1 s.

Similar to the Sensordrone, the *Texas Instruments* (USA) proposes the Bluetooth SensorTag, which lets to connect accelerometer, gyroscope, compass, humidity/temperature, pressure and IR temperature sensors to smartphones via Bluetooth.

Designed by NASA device contains several chemical sensors to connect to a smartphone, and may not be too far away from commercialization [17]. It's housed in a small case that attaches to a smartphone. (Fig. 1.14). A postage-sized chip with 16 nanosensors, each made up of a different nano-structure material that can respond to different chemicals in different ways. The chip requires about 5 mW of power so it can run for 8 continuous hours on a single charge when connected to a smartphone.

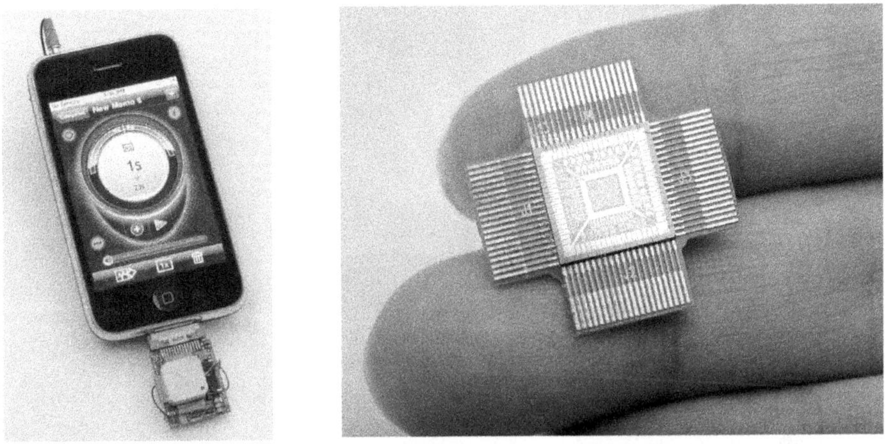

Fig. 1.14. Chemical sensors array for smartphone [17].

The device is chiefly designed to monitor carbon monoxide in the user's home along with real-time tracking of levels of chlorine, ammonia and methane in the immediate environment. A version of the device has been installed on the International Space Station in 2008. It monitors fuel leaks in launch vehicles and also monitors air quality - especially formaldehyde - in the air inside the Space Station.

LifeWatch V Android smartphone, manufactured by *LifeWatch AG*, is comes equipped with built-in medical sensors that are able to keep track of a number of health records including blood glucose, body temperature, one-lead ECG, heart rate, blood oxygen saturation, body fat percentage and stress levels (based on heart rate variability). The mHealth smartphone based sensor system is shown in Fig. 1.15. Each sensor is easy to operate which just needs to place finger on it only for collecting the medical measurements. The collected data will automatically be saved to a remote server securely and is retrievable from the cloud services for follow-up anytime, anywhere. The device could give innovative clinical laboratories a new opportunity to continuously monitor an individual's incoming measurements and provide clinical pathology consultations. The medical specification of the health care smartphone is shown in Table 1.2.

The device will run Android 2.3 Gingerbread with a fast Qualcomm MSM7227 Turbo chip. It comes with 512 Mb of RAM, 1 Gb of internal storage, and two cameras. The front-facing camera has low resolution though the rear camera is 5 MP. While Bluetooth and Wi-Fi are standard, the phone currently only supports 3G.

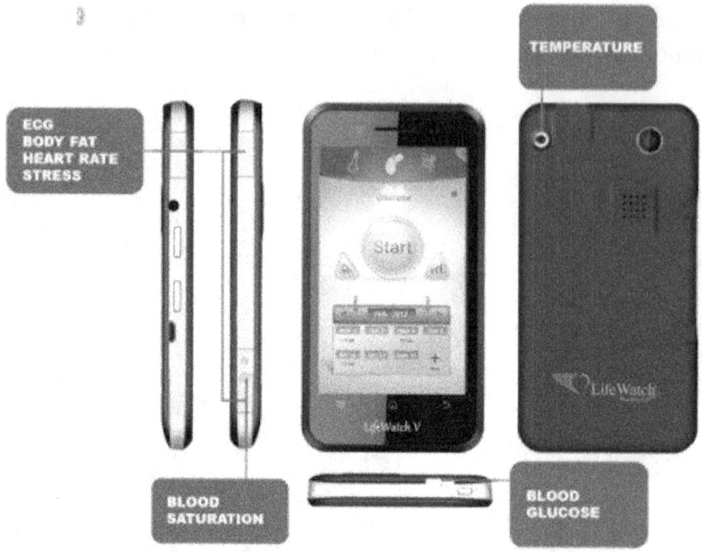

Fig. 1.15. mHealth smartphone-based sensor system (Source: *LifeWatch AG's* press release).

Table 1.2. Medical specification of LifeWatch V smartphone-based sensor system.

Medical Specifications	
EKG (1-Lead)	EKG dynamic range 0.15-5 mV
	Frequency response 0.5-40 Hz
	Sampling 12 bit, 250 samples /sec
	Test time: average of 35 sec
Heart rate	HF measurement range 30 to 250 bpm
	QRS detection sensitivity > 98 %
	QRS detection predictability > 98 %
	Result time: average of 10 sec
SpO_2	Method: Reflective R & IR emission
	Oxygen saturation range: 70 to 100 %
	Accuracy: Standard deviation ± 1 digits
	Test time: Average of 20 sec
Glucose	Range: 10-650 mg/dl
	Sample size: 0.4-0.7 L
	Accuracy: ± 10 mg/dL for glucose level 75 mg/dL
	Test time: 3-5 sec
IR-Thermometer	Range: 34 to 41 ^{0}C; Accuracy: ± 0.2 ^{0}C
Body fat percentage	Method: Bioelectrical Impedance
Operating temperature	10 to 40 ^{0}C (40 to 104 ^{0}F)

Some mobile instrumentation platforms (so called 'labs-on-a-smartphone') and sensor systems based on smartphones have been recently described in journals and presented on conferences; many R&D projects are now also under development. So, a sensor system for colorimetric measurements is described in [18]. This article presents a simple mobile instrumentation platform based on a smartphone using its built-in functions for colorimetric diagnosis. The color change as a result of detection is taken as a picture through a CCD camera built in the smartphone, and is evaluated in the form of the hue value to give the well-defined relationship between the color and the concentration. To prove the concept in the present work, proton concentration measurements were conducted on pH paper coupled with a smartphone for demonstration. This article is believed to show the possibility of adapting a smartphone to a mobile analytical transducer, and more applications for bioanalysis are expected to be developed using other built-in functions of the smartphone [18].

The chemical sensor-enabled smartphone provides a unique (bio)sensing platform for monitoring airborne or waterborne hazardous chemicals or microorganisms for both single user and crowdsourcing security applications [19]. Chemical sensoring of GaN-based blue LED chips with those indicators has also been achieved by plasma treatment of their surface, and the micrometer-sized devices have been tested to monitor O_2 in the gas phase to show their full functionality [19].

A Korean research team has developed a transparent sensor that can be attached on a mobile phone or a window of a car [20]. It is a low power and high sensitivity transparent chemical sensor that can operate without a heater. Unlike the existing chemical sensors, the new sensor has a distinctive feature of detecting chemical molecules through self-activation. Its electric power consumption is 0.2 microwatt, equivalent to one one-thousandth of that of earlier products. If the transparent sensor is attached on a mobile phone with an ordinary battery, it can operate for more than six months without being recharged. Its sensitivity is more than 1,000 times higher than the existing thin film sensors. It can detect toxic gases such as nitrogen dioxide, sulfurous acid gas, and acetone at levels below 1 ppb (1 nanogram per gram). In particular, the sensor's board is transparent because its composing materials are glass and oxide electrodes [20].

The study published in [21] shows a valuable impact of the complementary metal oxide semiconductor (CMOS) smartphone's image sensor and indium nanoparticle (InNP) substrates used for the detection of antigen-antibody interactions. Dielectric layer of antigen-antibodies bind on the conducting metal surface, which is helpful for the scattering of electro magnetic radiation. Different types of antigens such as Interferon Gamma, C-reactive protein and Troponin I were adsorbed on the InNP substrate and interact with specific antibodies. The antigen-antibody detection was based on the photon count measurements before and after protein binding observed by CMOS image sensor, which is converted into digital form with the help of analog to digital converter (ADC). Various thickness of InNP substrates were analyzed for antigen-antibody interaction. The photon counts gradually decrease when subsequent protein layers adsorbed on the substrates. This photon number is directly proportional to the digital number observed with the CMOS sensor for detecting antigen-antibody interactions [21].

Alongside with existing built-in (internal, embedded) and external (wire) sensors there are many physical and chemical external, wearable, wireless sensors. Several examples are shown in Fig. 1.16.

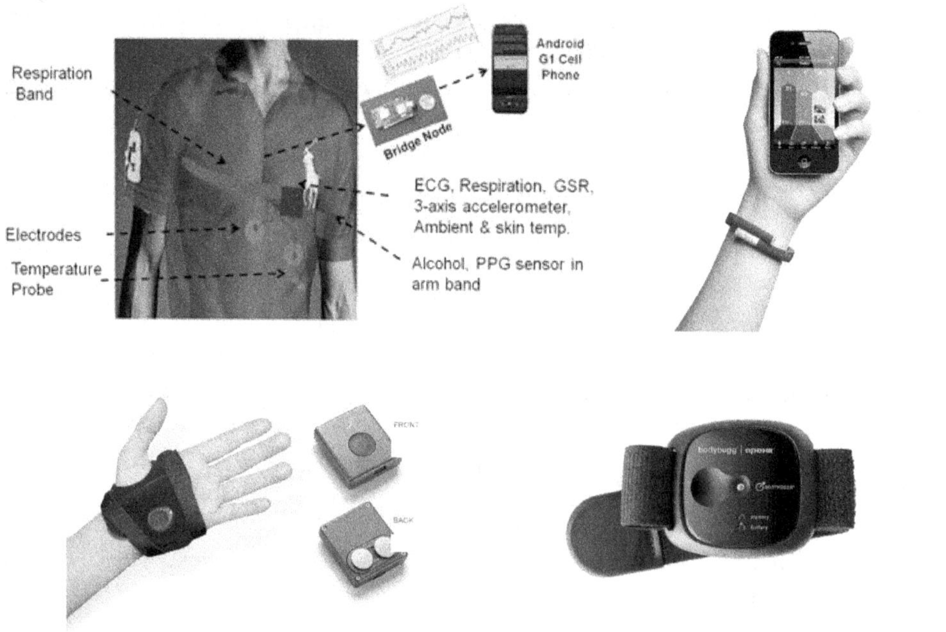

Fig. 1.16. Examples of wearable wireless sensors.

It is clear, that the best solution will be to have a lot of various internal, embedded sensors, which are built-in inside the standard smartphone. It is predictable, that in 2017 smartphone will have 18 different MEMS-based sensors as minimum. But such promised design approach must be based on the latest achievements in smart system integration, and there are some imitations. It is so called technological limitation, connected with microelectronic and nanoelectronic developments and smart system integration. How to overcome this limitation and design different cost-effective multisensor systems for modern smartphones are described below.

1.3. Design Approach: Gadget or Measuring Instrument ?

1.3.1. Technology Limitation

Dependent on type of sensors output, only two types of outputs are used till now in modern smartphones: analog output (voltage) and digital output (SPI and/or I^2C sensor buses). This fact strongly limits possibilities of smart system integration for smartphones.

Sensors usually are analog output devices. To input information from such sensors to system microcontroller or application processor in smartphones it is necessary to convert analog output signal to digital with the help of analog-to-digital converter (ADC). Taking into account low power analog signal on sensors' outputs, it is necessary to use also an analog amplifier and filter. But below the 100 nm technology, the design of analog and mixed-signal circuit becomes perceptibly more difficult [22, 23]. Such analog components are not "process compatible". This is particularly true for low supply voltage near 1 V or below. The result is not only increased design effort, long development time, high risk, cost and the need for very high volumes, but also growing power consumption, lost performance and flexibility.

Digital circuits, however, become faster, smaller, and less power hungry. They scale very well with scaling CMOS technologies. Their power consumption and speed performance improves significantly due to reduced parasitic capacitance. On the other hand, this results in challenges for the design of interface electronics: matching and noise become a serious issue, while the dynamic range is reduced due to the supply voltage reduction. The fast switching transitions reduce the susceptibility to noise, e.g. flicker noise in the transistors. There are also a few drawbacks, such as the generation of power supply noise or the lack of power supply rejection. Because of these drawbacks the analog circuits do not become much smaller in area in smaller technologies. Still, the advantages are overwhelming and suggest implementing as many system components as possible in the digital or quasi-digital domain. However, regarding sensors and transducers, the number of physical phenomenon, on the basis of which direct conversion sensors with digital outputs can be designed, is essentially limited. There is no natural phenomenon with discrete performances changing under pressure, temperature, etc. In this case, an ADC and analog multiplexers (for sensor arrays, multisensing and multi-parametric systems) are used. However, since scaling technology results in higher device speeds, the improved timing and frequency resolution at ADC conversion can be achieved when the signals are transformed into a frequency/time signal domain. When going to smaller technologies reduced parasitic gate capacitances result in smaller gate delays. This improved frequency/time resolution makes frequency (time)-to-digital conversion an interesting and promising alternative for ADC converters. Based on all the above, and existing standard CMOS technological 90 nm, 45 nm, 40 nm, 32 nm and 28 nm processes, a main challenge has arisen to changing a traditional analog (voltage) informative signal to a quasi-digital (frequency-time) informative signal in order to eliminate difficulties and limitations of analog and mixed-signal circuits design.

Today, more and more manufacturers propose digital output sensors for smartphones with SPI or I^2C output, but they are based on a standard ADC, and all mentioned design problems including the smart sensor system integration can not be eliminate by this way.

How to solve these design problems and create sensor systems for smartphones with high metrological performances and process compatibility ? First of all it is necessary to create a core (a new component) for smart sensor systems: programmable, universal, high-performance frequency (time parameters)-to-digital converters (FDC) instead of traditional ADC. This unit directly influences such sensor metrological characteristics, such as accuracy and the conversion time, as well as power consumption. It is clear that such FDC must be based on novel, advanced methods of measurement for frequency-time parameters of signals, and two different design approaches at the same time: technological and structural-algorithmic approaches that open a way towards radically new forms and uses of nano- and microelectronic technologies for bridging smart sensors and system integration. Such FDC will circumvent analog impairments in nanometer-scale CMOS technologies.

The challenges arise also when high resolution, linearity, low power consumption, high dynamic range, reliability and robustness come into the play. Hence, the only promising FDC concepts are those which have the ability to exploit the advantages of digital circuits. The FDCs have the potential to become trend-setting technology for recently and future scaled CMOS technologies for both the monolithic approach (System-on-Chip, System-in-Package) and multi-chip approach to drastically increase the integration level. It opens great perspectives for application of such technology in modern and future smartphones.

Smart sensor systems go beyond micro systems for single (or multi-) physical, chemical or biological, electrical or non-electrical parameter measurements combined with advanced signal processing and parameter-to-digital conversion. Smart sensor systems integration addresses the demand for miniaturized multifunctional devices and specialized connected and interfacing solutions.

In alternative design approach, frequency-time parameters of electrical signal must be used as an informative parameter on a sensor's output. First of all it is frequency, period, duty-cycle, pulse-width modulated signs, phase-shift, pulse number, etc. In such so-called quasi-digital sensors these informative parameters are proportional to measurand, and a frequency-to-digital converter should be used to obtain a digital output. According to the International Frequency Sensor Association's (IFSA) 2012 study, quasi-digital sensors share approximately 20 % of the modern global sensors market and among them frequency output sensors share 70 % [24]. Today there are quasi-digital sensors and transducers for any physical and chemical, electrical or non-electrical quantities on the modern market [25]. In addition, any voltage signal can be easy converted to frequency signal with the help of integrated, low cost voltage-to-frequency converter [26].

Frequency-time domain (or quasi-digital) sensors are rather interesting from a technological and fabrication compatibility point of view, due to the simplifications of the signal conditioning circuitry and measurand-to-digital converter, as well as metrology performances and the hardware for realization. The last one essentially influences the chip area. Moving from the traditional analog (voltage and current) signal domain to the frequency-time signal domain lets us achieve many well known and validated benefits due to properties of frequency, such as the informative output signal for sensors and transducers, namely:

- High noise immunity;
- High output signal power;
- Wide dynamic range;
- High accuracy of frequency standards;
- Simplicity of signal switching and interfacing;
- Simplicity of signal integration and coding;
- Multiparametricity;
- Modern process compatibility.

More details about each of the mentioned advantages of frequency output are available in [22].

Using frequency as the output signal for sensors is extremely useful alternative to the conventional analog voltage output signal and is easily accomplished with relatively few components. By eliminating the need for an ADC, frequency output sensor schemes reduce the cost of sensor systems.

No output standardization is necessary as in the case of analog signal domain. Many types of sensing elements and read-out circuitry can be merged in this way on a single chip, in System-on-Chip (SoC) or Systems-in-a-Package (SiP). In addition, such approaches will provide a great opportunity to create new self-adaptive smart sensors and sensor systems for the future r smartphones.

A duty-cycle output signal is also widely used as an information-carrying parameter for different quasi-digital sensors. It is quite immune to interfering signals, such as voltage spikes, and the ratio between time intervals does not depend on the absolute value of any component.

The state-of-the-art review of modern quasi-digital sensors has shown the obvious tendency of the relative error decreasing up to 0.003 % and below [22, 25]. These devices are working in broad frequency ranges: from several hundredth parts of Hz up to several MHz. The extension of their "intelligent" capabilities including intelligent signal processing is traced. The process of miniaturization boosts the creation of multichannel, multifunction (multiparameter, miltivariable) one-chip smart sensors and sensors arrays.

There are two possible design approaches for implementing digital smart sensors and sensor systems for smartphones. Let consider it in derails below.

1.3.2. Traditional Design Approach and Sensor Interfacing

The importance of microsystems is continuously growing because of the combination of two trends: the progress in silicon sensor technology and the introduction of new techniques for interface circuits. The cost of measuring systems has been greatly reduced due to the batch fabrication of both the sensors and the interface circuits. Interface circuits are the most critical part of the signal processing chain, which means that the overall performance of the system strongly depends on the quality of the parameter-to-digital converter.

Two motion sensors processing solutions for consumer applications are described in [27]. In the standard non-integrated 6-axis motion sensor processing solution (Fig. 1.17) a dedicated microcontroller may be required. The drawback is the additional cost of a microcontroller, which precludes their use in cost-sensitive applications.

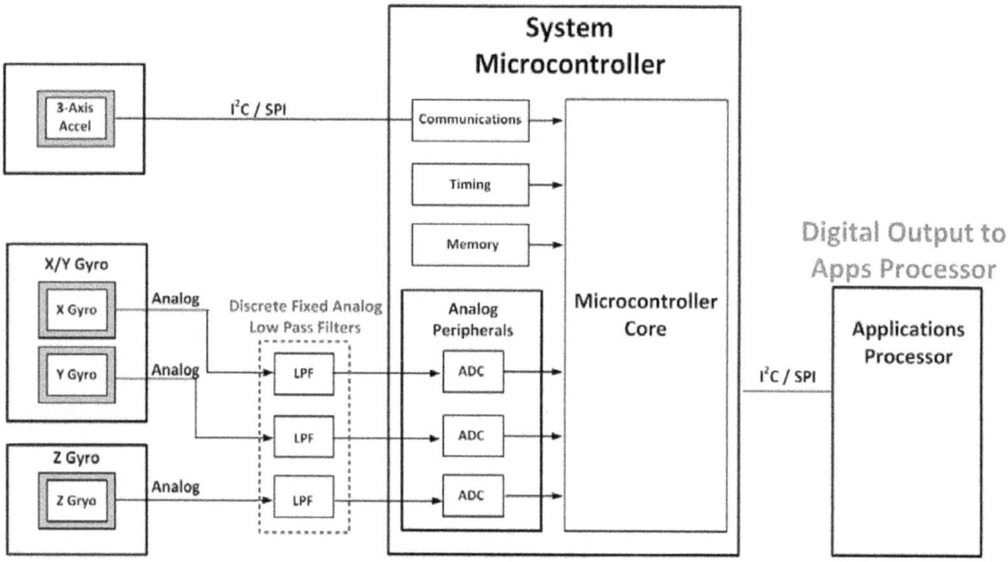

Fig. 1.17. Non-integrated 6-axis motion sensor procession solution [27].

As part of its signal processing, the fully integrated 6-axis motion processing solution proposed by *InvenSense, Inc.* (Fig. 1.18) has fixed-frequency anti-aliasing filters as part of its ADC block, followed by programmable digital low-pass filter, which negate the need for external signal conditioning and microcontrollers. Integrated analog gyroscopes IDG-650, ISZ-650 and IC IME-3000 including accelerometers and temperature sensor are show in Fig. 1.19.

Nevertheless that second solution uses smart sensors, it has the same disadvantage like the first one: analog components as ADCs and analog multiplexers. These components mean mixed design. The analog multiplexer introduces additional error of measurements, and ADC as rule has a limited by power supply and noises resolution and low conversion speed. In addition, both solutions are not compatible with the modern and future technological processes and effective system integration approach. In order to achieve acceptable metrological performance, complex sensor fusion algorithms must be used [27, 28].

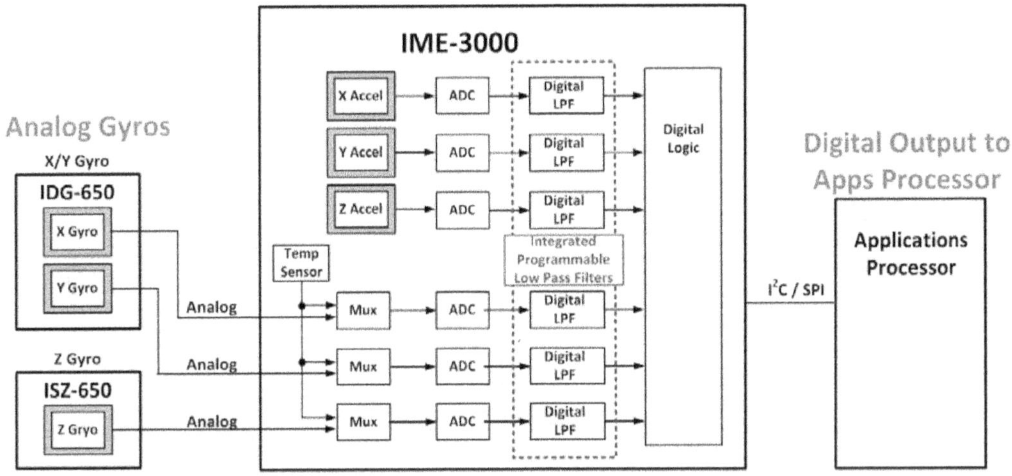

Fig. 1.18. Integrated 6-axis motion sensors processing solution [27].

Fig. 1.19. *InvenSense* integrated circuits for 6-axis motion sensors processing solution [27].

A Unified MEMS sensor interface development is described in [29]. A standards development group is exploring the feasibility of a new interface standard that would enable even multiple degree-of-freedom arrays of MEMS sensors to use a simple common interface. Such an interface—which would be used by accelerometers, magnetometers, gyroscopes, altimeters, compasses, proximity sensors and non-MEMS sensors.

A two-wire interface standard is shown in Fig. 1.20 as an example of how up to 23-pins from a SoC (Fig. 1.20 a) could be eliminated with a universal multi-drop interface (Fig. 1.20 b) - a kind of network-for-sensors, instead of using I^2C or SPDIF [29].

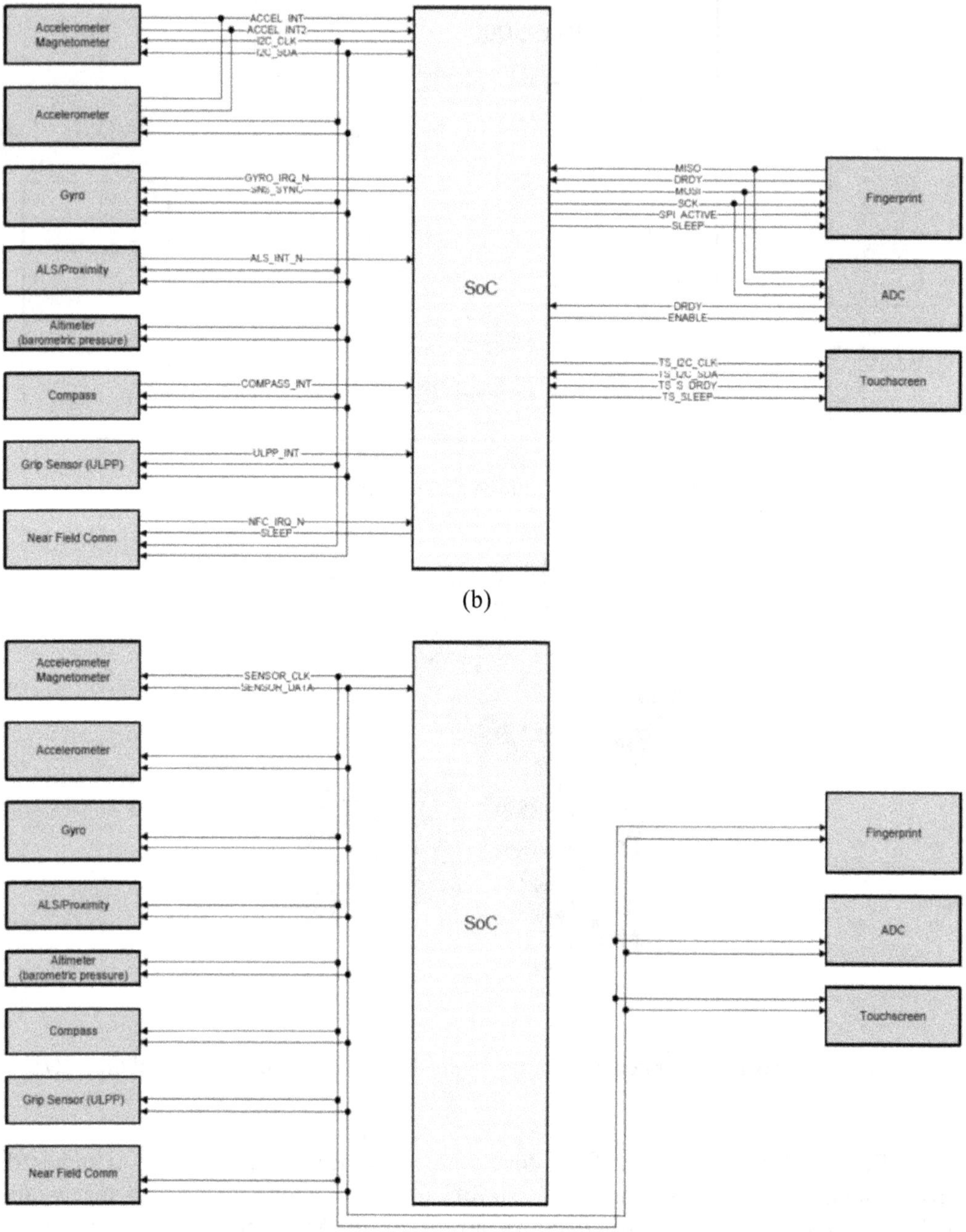

Fig. 1.20. MEMS sensor system: 23 pins on a SoC to manage all the sensors (a), and required pins number due to a universal interface (b).

This would greatly simplify the design and cut the bill-of-materials cost of mobile devices with many of sensors. Of course, such solution will be possible if sensors vendors will support a common standard. In addition, both approaches use ADC (Fig. 1.20), and are in the front of coming technological limitations, discussed above.

Design solution for digital sensors can be based on the ultra-low power, compact sensor hub microcontroller for smartphones ML610Q792 (Fig. 1.21) - the industry's smallest microcontroller from *LAPIS Semiconductor* (formerly *OKI Semiconductor*) a ROHM Group company, designed for integrated, low-power control of multiple sensors in smartphones [30].

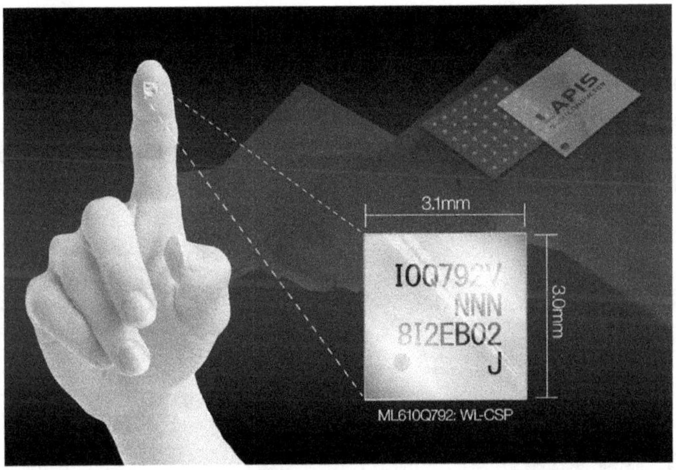

(a)

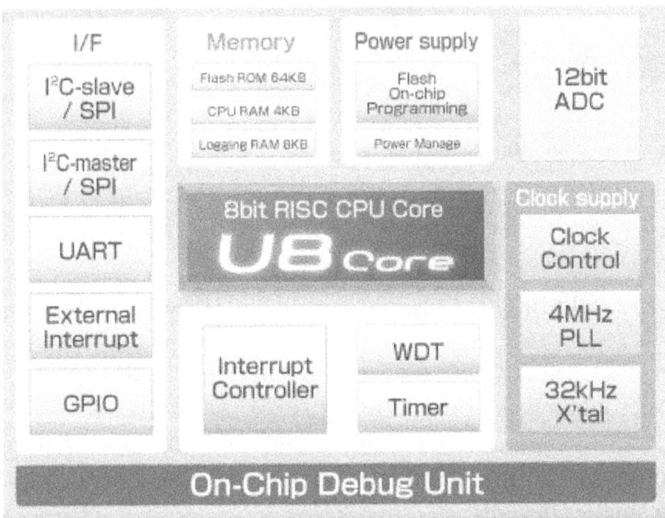

(b)

Fig. 1.21. Sensor Hub Microcontroller ML610Q792 for Smartphones: small package (a), and block diagram (b) [30].

Remarkably low power consumption not only prolongs battery life, but enables support for wireless communication in compact products like pedometers and smartphone accessories. The microcontroller incorporates an original high-performance 8-bit RISC CPU Core (U8) and 16-bit coprocessor for arithmetic calculations. Dual interfaces allow for both sensor connection and connecting to the main chipset, while the ultra-compact WL-CSP form factor (3.1 mm × 3.0 mm) supports high-density designs. In addition, a development board containing multiple sensors and drivers is provided along with a Software Development Kit (SDK) that includes sample source code and libraries for data logging, calorie calculation, pedometer function, and device drivers for interfacing with the host processor.

One of the future growth strategies is to offer total sensor solutions that provide greater added-value through product synergy combining ROHM sensors (proximity, ambient light, Hall ICs, temperature) and *Kionix* accelerometers and gyroscopes with *Lapis Semiconductor* low-power microcontrollers [30]. ML610Q792 sensors compatibility is shown in Fig. 1.22.

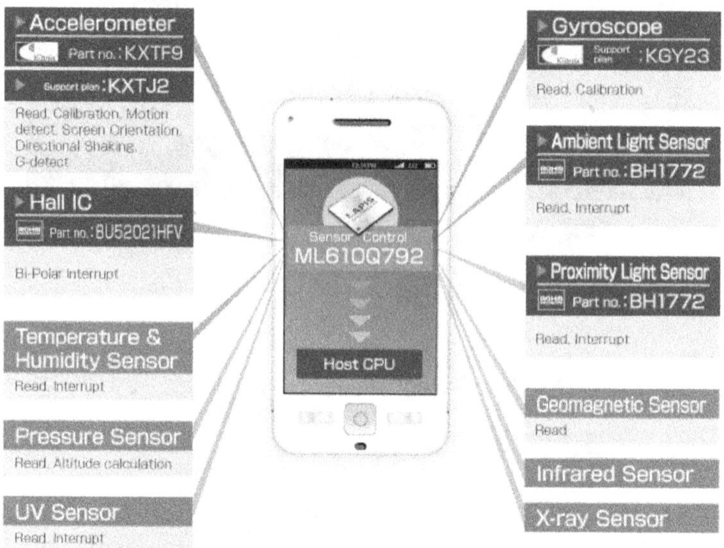

Fig. 1.22. ML610Q792 Sensors Compatibility (Drivers and Firmware).

The microcontroller features are the following:

- 8-bit RISC Core;
- Ultra-compact 48-pin 3.1 mm × 3.0 mm WL-CSP (Wafer-Level Chip-Scale Package) contributes to smaller designs;
- Comprehensive software support includes sensor drivers and key algorithms;
- Controls multiple sensors installed on smartphones;
- Low power consumption: 0.6 μA or less in HALT mode;
- Built-in 64 KB Flash ROM; onboard write supported.

The sensor hub microcontroller is working with analog and digital (RS232, SPI and I^2C) sensors. However, it is difficult to use it with quasi-digital sensors. In turn, digital sensors are based on standard ADCs, and therefore have many disadvantages mentioned above.

Atmel also has announced a range of microcontroller-based solutions that integrate touch and sensor hub functionality to enable superior user experience for a variety of mobile devices including smartphones, tablets, Ultrabooks, and convertible PCs.

1.3.3. Combo Sensors

Following the requirements of smart system integration (miniaturization challenges) for smartphones applications some companies, for example, *Kionix* and *STMicroelectronics*, start to manufacture so-called combo sensors.

According to *Yole Developpement* [31] one clear motivation for combo sensors is the possibility to reduce both cost and footprint by combining some sensors into one package with a single ASIC. However this cost benefit is not yet obvious with the dynamism of the evolution of each individual sensor. Such benefit will be very application dependent. Lower cost units combining multiple MEMS sensors are poised for healthy growth, starting with ESCs (Electronic Stability Control System), bringing opportunity for new players and demands for sensor management solutions. This trend is showing up first in the more mature automotive MEMS sector, where the price of the sensor unit for the ESC can now be significantly reduced, by combining the accelerometer and the gyroscope in one package with one ASIC. Adoption is a little slower on the consumer side, where the fast changing technology means discrete device prices are still falling rapidly, so products from even six months before have less of a cost advantage [31].

The second key motivation from combo sensors relies on sensor fusion. New functionality can now be offered using multiple sensing elements. This is an opportunity for players in the MEMS industry to compensate for the drastic decrease in price by selling high value solutions that include more software content. There is likely room for multiple alternatives, with the sensor makers supplying the algorithms to combine and cross-calibrate the sensor data and do some standard applications, while the software and chipset makers supply the higher level, specialty functions. Combo sensors require more complex software for the sensor fusion calculations, and those will likely need to be done on an MCU, not just the usual ASIC. This is driving changes in the supply chain, as makers of microcontrollers, software, and subsystems start to take over more of the sensor management [31].

The new solutions build on *Atmel's* leading maXTouch controllers and incorporate the company's patented maXFusion sensor fusion technology to address the growing motion sensor market. Atmel maXFusion combines inputs from multiple sensors that detect motion, including accelerometers, magnetometers and gyroscopes, to provide real-time direction, orientation and inclination data bringing visibly superior performance to a range of applications including gaming, navigation, and virtual reality.

Yole Developpement states that combination sensors, will jump from very tiny volumes currently to penetrate some 40 % of the $ 2.7 billion consumer inertial market and more than 12 % of the $1.1 billion automotive inertial market by 2016 [31] and $ 1.7 billion opportunity by 2017. As it was mentioned above, by 2016, *IHS* forecasts that combo sensor revenue will amount to $1.4 billion, accounting for 71 % f the total motion sensor market in consumer and mobile applications.

Existing on the modern market combo sensors as rule includes 2-3 motion sensors into one package. For example, an iNEMO Inertial Module LSM333D - 9 degrees of freedom sensing solution from *STMicroelectronics,* contains a 3D accelerometer, a 3D gyroscope and a 3D magnetometer [32]. The LSM333D includes an I^2C serial bus interface supporting standard and Fast mode 100 kHz and 400 kHz, and SPI serial standard interface. All built-in sensors can be enabled or set in power-down mode separately for smart power management.

Two new 6-axis accelerometer-magnetometer solutions KMX61G and KMX61 available from *Kionix* [33]. The KMX61G accelerometer/magnetometer, with integrated Sensor Fusion software is the industry's first smart and intelligent magnetic gyro solution. The KMX61 is a high performance accelerometer/magnetometer E-compass solution. Available with either I^2C or SPI output, the 4 × 4 × 0.9 mm parts also include an auxiliary input to form the basis for a 9-axis IMU solution.

Other companies as *Freescale* and *InvenSense* also develop smart and intelligent solutions for combo sensors. So, *Freescale* manufactures Xtrinsic FXOS8700CQ sensor, which includes 3-axis accelerometer and 3-axis magnetometer [34]. The integrated 6-axis motion sensors IME-3000 from *InvenSense* was described above (see Fig. 1.18). Another integrated 6-axis solution (MPU-6500 model) includes 3-axis accelerometer and 3-axis gyroscope [35].

Nevertheless on many advantages (smaller die, lower parasitic connections, power saving, intelligent functions, etc.), which brings such sensor solutions to mobile devices applications, many digital combo sensors as usually are based on tradition ADC with a limited resolution, and technological limitations from smart system integration point of view are still a bottleneck for the next generation of combo sensors, which will include much more various sensors in one package.

1.3.4. Advanced Design Approach

The first 10 years of 21st century smartphones became "smarter", got more powerful processors and operating systems. As it was written in [36], "following the massive proliferation of sensors in smartphones, many have predicted that mobile devices would soon see new generations of smart sensors". Over the last few years, while sensors in smartphones have gotten smaller, now consume less power, and feature better performance, they haven't gotten much intelligence; while the performance of individual sensors has increased, their functionality has not expanded [36]. The motivation for using smart sensors lies in reducing the power consumption. A smart sensor could also choose to optimize its performance for a specific set of contexts or a segment of use cases [36]. Developers are encouraged to make applications that utilize the data from multiple sensors. But there are key limitations of existing smart, intelligent sensors and sensor systems for smartphones [34]:

- Level of complexity to implement more different sensors in the same device;
- Design of software adapting to each new sensor (interface, data sampling rate, etc.).

In order to eliminate these limitations *Technology Assistance BCNA 2010, S. L.* (Barcelona, Spain) has developed a novel integrated circuit - Universal Sensors and Transducers Interface (USTI-MOB), which lets to build easily various smart sensor systems. The USTI-MOB can work practically with any quasi-digital sensors, available on the modern sensor market, and especially designed for smartphone applications. In addition, the USTI-MOB can be used with resistance, capacitance and resistive bridge sensing elements. The prototype of USTI-MOB IC is shown in Fig. 1.23.

Fig. 1.23. USTI-MOB prototype in 5 × 5 mm MLF package.

The IC USTI-MOB will have the same measurement modes as the early developed IC USTI [22, 37], but different metrological performance and electrical characteristics (see Table 1.3), connected with the specific application in smartphones (reduced clock frequency, supply voltage and power consumption).

Each of ICs will have one generating mode (for calibration purposes) and 26 measuring modes to measure: frequency, period, its ratio and difference, duty-cycle, duty-off factor, time interval between start- and stop- pulses, pulse width, pulse space, phase shift, frequency deviation (absolute and relative), rotational speed and pulse number.

Each of ICs will have three popular serial communication interfaces and buses: RS232, SPI and I^2C. Several practical examples (cases study) of multisensor systems for various applications will be demonstrated and tested.

Table 1.3. Main technical performances of USTI and USTI-MOB ICs.

No.	Performance	IC	
		USTI	**USTI-MOB**
1.	Frequency range of measurement, Hz	0.05 ... 7.5×10^6	0.05 ... 2×10^6
2.	Programmable relative error, %	1 ... 0.0005	1 ... 0.0005
3.	Supply voltage, V	5.0	2.7
4.	Current consumption (active mode), mA	11	< 3
5.	Current consumption (sleep mode), μA	450	< 125
6.	Operation temperature range, ^{0}C	- 40 ... +85	- 40 ... +85
7.	Miniaturized package type	MLF, 28-pad	MLF, 28-pad
8.	Overall dimensions, mm	5×5×1	4×4×1 and 5×5×1

The USTI-MOB can be connected directly to an application processor or to a sensor hub microcontroller in a smartphone, for example, ML610Q792 (*Lapis Semiconductor*). The multisensor system for smartphones based on the USTI-MOB IC and various quasi-digital sensors is shown in Fig. 1.24. The system also contains a digital multiplexer (MX) controlled by the USTI-MOB IC (the MX can be also controlled by the sensor hub microcontroller or application processor). The MX's output is connected to the first USTI's channel. The second channel can be used for a direct sensor connection.

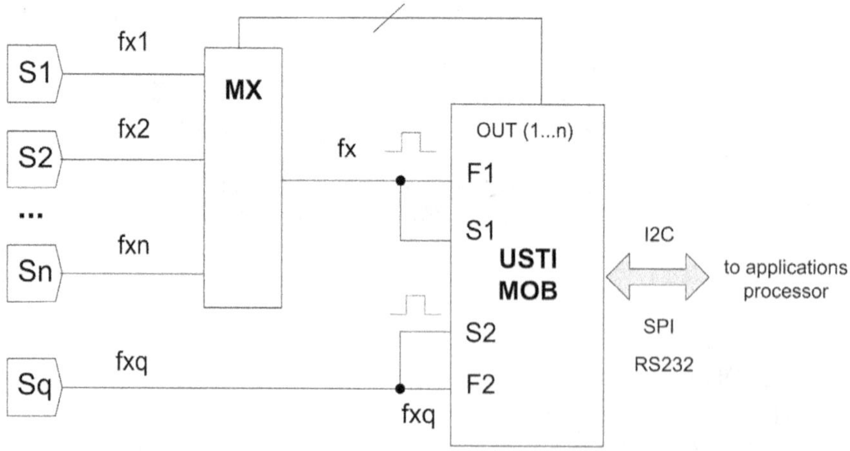

Fig. 1.24. Multisensor system for smartphones with quasi-digital sensors.

In case of analog sensors, an intermediate voltage-to-frequency conversion should be used with the help of integrated voltage-to-frequency converter [26]. A sensor system for analog output sensors is shown in Fig. 1.25. In both cases time-division, space-division and combining channeling can be used [38].

Taking into account that many quasi-digital sensors, integrated voltage-to-frequency converters, digital multiplexers and IC USTI-MOB are manufactured in standard CMOS processes it opens a wide horizon for the future smart sensor system integration.

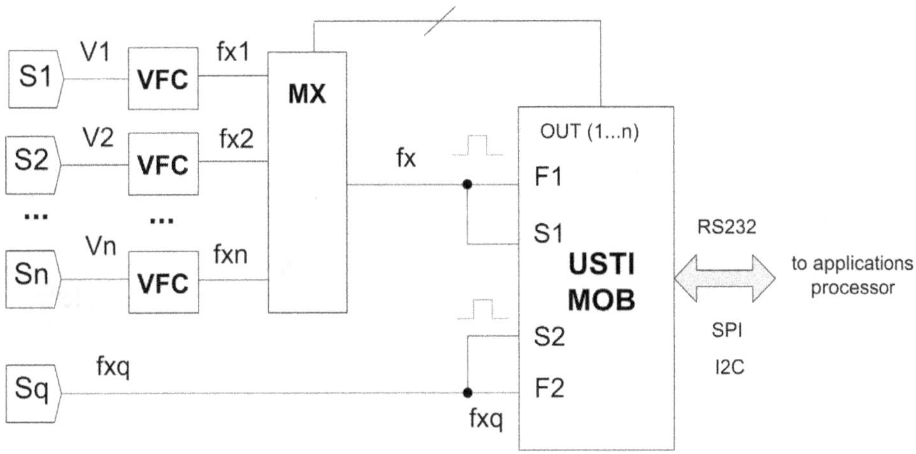

Fig. 1.25. Multisensor system for smartphones with analog sensors.

According to *InvenSense*, the hierarchy for mobile sensor integration contains three levels dependent on involved hardware [35]:

I. Individual sensors:

- Lowest level, but can provide built in intelligence to lower system power and embed sensor expertise;
- Low power consumption, so many sensors can stay active for long periods of time.

II. Sensor Hub:

- Aggregates sensor inputs, can provide some information processing to determine if the AP should be woken up;
- Uses less power than the AP, signal processing can be shared across all sensors.

III. Application Processor:

- Used for the "heavy lifting" in processing the sensor data and enabling applications;
- Uses the most power when active. Needs to be able to sleep for long periods of time when the system is idle to save power.

Due to advanced technologies, which where used at the USTI-MOB design, this component can be used on all these three levels of integration. On the lowest level of individual sensors, the IC can be used for three difference sensors: two quasi-digital and one parametric sensing element (resistive, capacitive or resistive bridge). On the 2nd level of integration the USTI-MOB IC can aggregate many sensors output via a digital multiplexer, and on the 3rd level of integration the USTI-MOB can be embedded into an application processor as a firmware IP. For example, it will be compatible with a wide range of solutions offered by *Atmel*, namely, the UC3L sensor hub solutions with customizable firmware.

The CMOS digital IC USTI-MOB also compatible with the SoC design approach, when sensors, hub and USTI-MOB itself can be integrated into a single product.

Let's consider a multisensor system example - a smartphone based weather station (Fig. 1.26). It contains tree sensors: a quasi-digital, temperature, duty-cycle output sensor SMT160-30 in from *Smartec* (The Netherlands) in small-size HEMP or SOIC-8 housing [39]; a frequency output module for humidity measurement, and barometric pressure, bridge-type, miniature SMD sensor MS54XX [40]. All of these sensors can be connected directly to one USTI-MOB IC at the same time without any multiplexers.

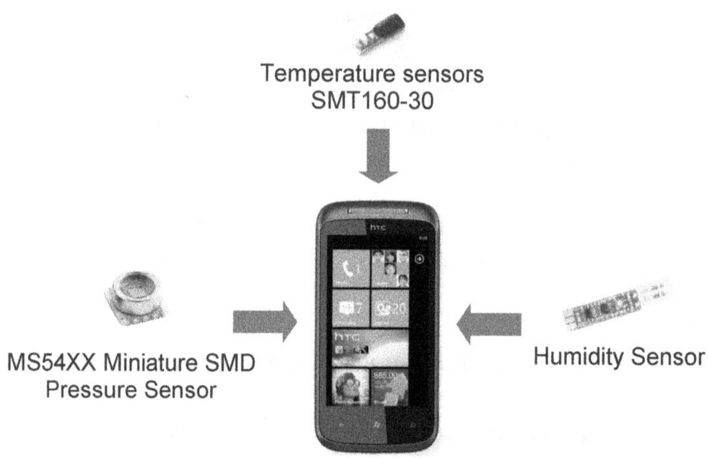

Fig. 1.26. Smartphone based weather station.

Commands for USTI-MOB working in I^2C communication mode are shown in Fig. 1.27. The format of result of measurement in BCD is shown in Fig. 1.28. The binary format also possible. In this case instead of <07> command the <08> command must be used [36].

The humidity module is connected to the first IC's cannel, and temperature sensor - to the second channel. Resistance passive bridge measurements are described in details in [22, 37, 41]. The programmable relative error of USTI-MOB for frequency measurements (humidity) must be selected in one order less (or at least in 5 times less), than the sensor error in order to be neglected [22, 42].

<06><00> ; Frequency measurement initialization in the 1st channel (humidity)
<02><02> ; Set up the conversion error 0.25 %
<09> ; Start a measurement
<03> ; Check result status: returns '0' if ready or not '0' if busy
<07> ; Get result in BCD format

<06><14> ; Duty-cycle measurement initialization in the 2nd channel (temperature)
<09> ; Start a measurement
<03> ; Check result status: returns '0' if ready or not '0' if busy
<07> ; Get result in BCD format

<06><12> ; Set up a resistance-bridge B_x measurement mode (pressure)
<10><13> ; Set the charging time 20 ms
<09> ; Start measurement
<03> ; Check result status: returns '0' if ready or not '0' if busy
<07> ; Read conversion result

Fig. 1.27. Commands for USTI-MOB (I^2C slave communication mode).

<Sign><I5><I4><I3><I2><I1><I0><F0><F1><F2><F3><F4><F5>

<Sign> - sign of result: 0x20 (space char) for positive result; 0x2D (minus char) for negative result;
<In> - byte n of integer part of result
<Fn> - byte n of fractional part of result

Fig. 1.28. BCD format for result of measurement.

So, frequency output humidity sensing modules have 2-3 % error [43], therefore the programmable relative error for the USTI-MOB should be 0.25 %. The digital multiplexer does not introduce any error.

The duty-cycle is measured by the IC with a maxim possible accuracy, which is determined by time interval measurement error, and it is not necessary to set up the programmable relative error for this measuring mode.

The resistive bridge measurement requires two external components: capacitor and resistor R=220 Ohm. The considerations about value calculation for capacitor and time constant (commands <10>) are described in details in [22, 37, 41].

Let's consider another example: design of absolute digital output barometric pressure sensor for smartphones.

According to the IHS iSuppli MEMS &Sensors Service at information and analytics provider *IHS*, the global shipments of microelectromechanical system MEMS pressure sensors in cellphones are set to rise to 681 million units in 2016, up more than eightfold from 82 million in 2012 [44]. Shipments this year are expected to double to 162 million units, as presented in the attached figure, primarily due to Samsung usage of pressure sensors in the Galaxy S4 and other smartphone models. Although pressure sensors are very useful currently in smartphones, they hold strong potential for the future. The most interesting application now is the fast Global Positioning System (GPS) lock, wherein the GPS chipset can lock on to a satellite signal and calculate positions more quickly by using the pressure sensor to determine the smartphone's altitude. However, the most exciting use for pressure sensors in the future will be indoor navigation, an area with massive potential growth in retail and travel applications. Pressure sensors will provide the floor accuracy required to determine which level a user is on within a structure [44].

A block diagram of modern barometric pressure sensor based on a traditional design approach is shown in Fig.1.29. It contains many analog components such as analog multiplexer, analog front end, ADC and digital filter. The DSP is used for temperature compensation.

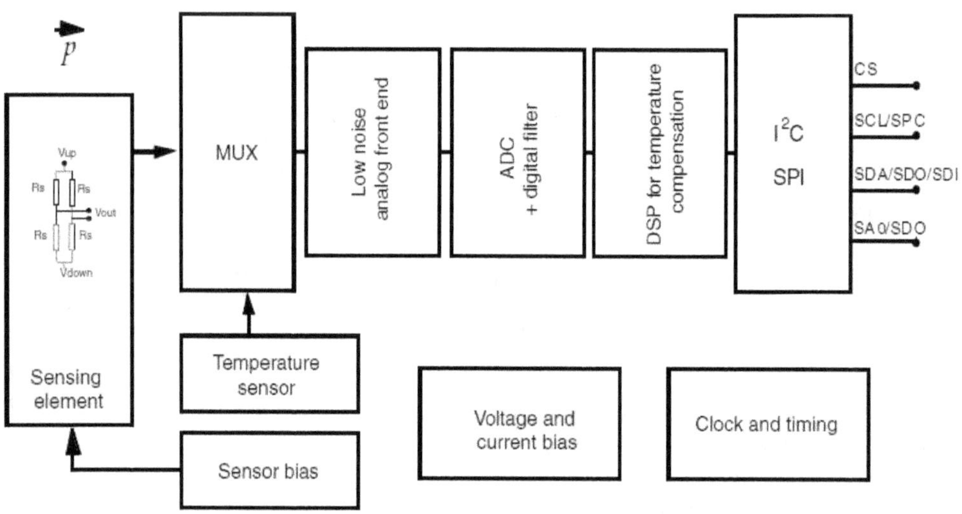

Fig. 1.29. Barometric pressure sensor based on traditional design approach.

A block diagram of barometric pressure sensor based on advanced design approach described above is shown in Fig. 1.30. It contains the USTI-MOB, which convert pressure and temperature to digital according to the I^2C and SPI interfaces, and performs the temperature compensation. It does not need any mentioned analog components neither the DSP processor for temperature compensation. It is expediently to use a CMOS temperature sensor with frequency, period, duty-cycle or PWM output in this design. One addition quasi-digital sensor can be connected directly to the 2-channel USTI-MOB IC.

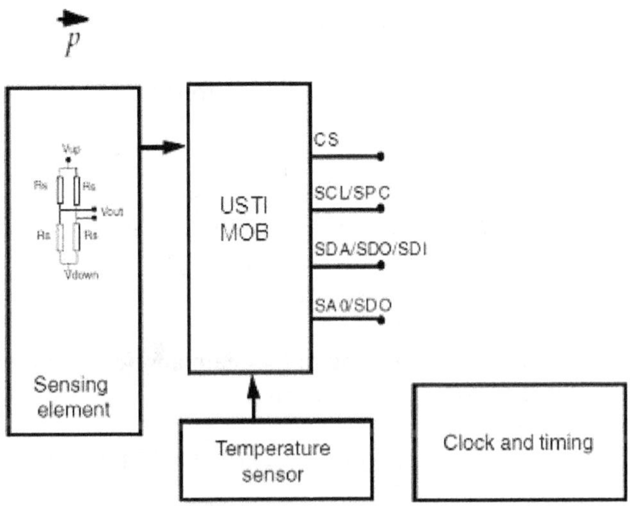

Fig. 1.30. Barometric pressure sensor based on advanced design approach.

Such integrated sensor can be manufactured using a standard CMOS process that allows a high level of integration to design a dedicated circuit, which is trimmed to better match the sensing element characteristics and metrological performance. All functions of the USTI-MOB can be also realized in firmware IP form and to be implemented inside a sensor hub microcontroller, applications or system microprocessors in smartphones.

The USTI-MOB can work also with any quasi-digital sensors available on the modern market, including temperature, RGB, IR and light sensors from *ams-TAOS*, *Analog Devices*, *Maxim*, *Melexis*, *Hamamatsu* and other sensor manufacturers. Any existing voltage output sensors can be also used with the USTI-MOB with the help of any intermediate voltage-to-frequency converters.

1.4. Summary and the Future Trends

In order to implement more different sensors in smartphones by easy way and overcome existing technological limitations due to complex analog and mixed design, developers must use the alternative, advanced design approach based on precision frequency-to-digital conversion for both: sensor systems and digital sensors for smartphones. This will drastically increase the number of sensors, which can be embedded into modern smartphones and tablets. As a result, the number of software applications for such devices will be increased in geometrical progression.

The extension of sensors "intelligent" capabilities including self-adaptation will be available. The process of miniaturization and smart system integration based on the novel design approach boost the creation of multichannel, multifunction (multiparameter) one-chip smart sensors and combo sensors arrays.

Advantages of the proposed design approach are the following:

1) Optimized, process compatible architecture for the future smart system integration;
2) Much more sensors can be connected to a smartphone;
3) Much better metrological performances of various sensors and sensor systems;
4) Wide functionalities, including such new intelligent function as self-adaptation through the dedicated interface and software;
5) Simplified sensor fusion algorithms;
6) Reduction of the data communication with the main applications processor;
7) Reduced implementation cost and time-to-market.

The core component of novel design approach - IC of USTI-MOB will be available on the modern market in 2013 from *Technology Assistance BCNA 2010, S. L.* [45].

To be successful on the market of smart devices, companies also need to be smart. I hope this chapter will help them.

References

[1]. MEMS in Smartphones: Growth at What Price? Press Release, Semico Research, Phoenix, USA, 17 May 2011 (http://www.semico.com/press/press.asp?id=288).
[2]. Venkat Atluri, Umit Cakmak, Richard Lee, Shekhar Varanasi, Makinng Smartphones Brilliant: Ten Trends, Recall No. 20, A new era of Personalized Computing, *McKinsey & Company*, pp. 8-13.
[3]. Venkat Atluri, Umit Cakmak, Richard Lee, Shekhar Varanasi, Six Sensing: The Future of Mobile Device Sensors, Recall No. 20, A new era of Personalized Computing, *McKinsey & Company*, pp. 45-49.
[4]. MEMS for Cell Phones & Tablets, *Yole Developpement*, France, 5 July 2012.
[5]. Mobile Device Sensor Market Worth $1.2 billion at End of 2012, *ABI Research*, London, United Kingdom, 21 September 2012.
[6]. Motion Sensor Market for Smartphones and Tablets to Double by 2015, *IHS iSuppli Research*, September 2011.
[7]. Jérémie Bouchaud, Tablets and Smartphones Drive 700 Percent Surge in Motion Combo Sensor Market in 2012, *IHS iSuppli Research*, 8 October 2012.
[8]. Mobile Health Market Report 2011-2016, research2guidance, 2011.
[9]. iCelsius Thermometer for iPad / iPhone / iPod touch, *Aginova, Inc.*, USA.
[10]. iCelsius User Guide, *Aginova, Inc.*, USA, 2012.
[11]. http://mobile. mit.edu/research/new-sensors-smartphones
[12]. http://intersofteurasia.ru/eng/

[13]. Smartphone to measure radiation, *BBC News Technology*, 22 September 2011.
[14]. PH-1- pH Meter for iPhone, iPod & iPad, Product Specification Sheet, *Sensorex Corporation*, USA, 2012 (http://www.sensorex.com/products/more/ph1_iphone_pH_meter).
[15]. http://www.ibgstar.us/
[16]. Sensordrone: Specifications & User Guide, *Sensorcon, Inc. (USA)*, November 2012.
[17]. Brian Dolan, NASA developing noninvasive smartphone health sensor, *Mobihelathnews*, February 2012.
[18]. Byoung-Yong Chang, Smartphone-based Chemistry Instrumentation: Digitization of Colorimetric Measurements, *Bull. Korean Chem. Soc.*, Vol. 33, No. 2, 2012, pp. 549-552.
[19]. Guillermo Orellana, Elias Muoz, Luz K. Gil-Herrera, Pablo Muoz, Juan Lopez-Gejo Carlos Palacio, Integrated luminescent chemical microsensors based on GaN LEDs for security applications using smartphones, in *SPIE Proceedings of the Optical Materials and Biomaterials in Security and Defence Systems Technology IX Conference*, Vol. 8545, Edinburgh, United Kingdom, 24 September 2012.
[20]. Yoon Dae-Won, Smartphone Comes with Gas Sensor, *Korea IT News*, 3 September 2012. (http://english.etnews.com/device/2640348_1304.html).
[21]. Devadhasan J. P., Marimuthu M., and Sanghyo Kim, in *Proceedings of the International Conference on Convergence (ICTC' 2011)*, 28-30 September 2011, pp. 383-387.
[22]. Yurish S. Y., Digital Sensors and Sensor Systems: Practical Design, *IFSA Publishing*, 2011.
[23]. Stephan Henzler, Time-to-Digital Converters, *Springer*, 2010.
[24]. International Frequency Sensor Association (IFSA), http://www.sensorsportal.com
[25]. Sensors Web Portal: http://www.sensorsportal.com/HTML/Sensor.htm
[26]. Voltage-to-Frequency Converters (VFC) and Current-to-Frequency Converters (CFC): (http://www.sensorsportal.com/HTML/SENSORS/Voltage_to_Frequency_Converters.htm).
[27]. Steve Nasiri, David Sachs and Michael Maia, Selection and Integration of MEMS-based Motion Processing in Consumer Apps, *DSP Design Line*, 8 July 2009.
[28]. William Wong, Sensor Fusion or Sensor Confusion ? *Electronic Design*, 28 June 2012.
[29]. R. Colin Johnson, Unified MEMS sensor interface sought, *EETimes*, 13 November 2012.
[30]. Ultra-Low Power, Ultra-Compact Sensor Hub Microcontroller for Smartphones, Press Release, *Sensors & Transducers Magazine-(e-Digest)*, Vol. 142, Issue 7, July 2012.
[31]. Inertial Combo Sensors for Consumer & Automotive Market Report, *Yole Developpement,* France, November 2011.
[32]. LSM333D iNEMO Inertial Module: 9 degrees of freedom sensing solution, Preliminary Datasheet, *STMicroelectronics*, March 2012.
[33]. Kionix web site: http://www.kionix.com/6-axis-combo-parts
[34]. Stéphane Gervais-Ducouret, Smart & Intelligent Sensors, Invited Presentation at *Workshop on the Intelligent Sensor Hub (SENSORDEVICES-WISH' 2012)*, 19-24 August 2012, Rome, Italy, 2012.
[35]. Stephen Lloyd, CMOS-MEMS for the next generation of "Combo" Sensors, *InvenSense*, 29 September 2012.
[36]. Ian Chen, A Case for Smart Sensors, *Sensors Magazine*, 14 December 2012.
[37]. Universal Sensors and Transducers Interface (USTI), Specification and Application Note, *Technology Assistance BCNA 2010, S. L.*, 2010.
[38]. Kirianaki N. V., Yurish S. Y., Shpak N. O., Deynega V. P., Data Acquisition and Signal Processing for Smart Sensors, *John Wiley & Sons*, Chichester, UK, 2002.
[39]. SMT160-30 Digital Temperature Sensor, Datasheet, *Smartec*, 14 November 2010.
[40]. MS54XX Miniature SMD Pressure Sensor, Datasheet, *Measurement Specialties*, 20 November 2012.
[41]. Yurish S. Y., Universal Interfacing Circuit for Resistive-Bridge Sensors, in *Proceedings of the 1st International Conference on Sensor Device Technologies and Applications (SENSORDEVICES' 2010),* 18 – 25 July 2010, Venice/Mestre, pp. 211-216.

[42]. Novitsliy P. V., Zograf I. A., Errors Estimation for Measuring Results, *Energoatomizdat*, Leningrad, 1991 (in Russian).
[43]. List of Humidity and Moisture Sensors Manufacturers at Sensors Web Portal: http://www.sensorsportal.com/HTML/SENSORS/HumiditySens_Manufacturers.htm
[44]. Jessie Shen, Smartphones to drive MEMS pressure sensor market through 2016, says IHS, Press Release, *DIGITIMES*, 21 March 2013.
[45]. Technology Assistance BCNA 2010, S. L., http://www.techassist2010.com/

Chapter 2
Phase Dynamics, Synchronization and Sensing by SAW Delay Line Coupling Between Nonlinear SAW Oscillators

R. D. S. Yadava

2.1. Introduction

Phase dynamics and synchronization of delay coupled nonlinear dynamical systems have been studied mostly via mathematical models of delay differential equations of various types. In these studies, no explicit expression is given to model parameters (which are often in normalized form) in terms of real physical system characteristics. Therefore, it becomes difficult for a system designer to take advantage of these mathematical models in nonlinear science for creating novel system designs, or for improving system performance. The primary focus in mathematical model studies is to capture generalities in nonlinear dynamics that generate various control, synchronization or chaotic behavior in biological, physical and chemical systems [1-11]. From the perspective of system innovation, it is desirable to model the dynamics of practically realizable physical systems and analyze their behavior so that insight gained could be used for system design and development. Recently, several novel engineering applications have been identified by understanding nonlinear dynamics of autonomous oscillators and coupled nonlinear oscillators system. Some examples are: synchronization and security of communication networks [12-14], synchronization of chaotic lasers and chemical oscillators [15], optoelectronics [16], phase locking of relativistic magnetrons [17], data mining [18] and robotics control [19].

Synchronization of coupled nonlinear oscillators is a particular state in which all the oscillators adjust their rhythms (frequency or phase) to a common mode of oscillation. Individual oscillators in the uncoupled state are autonomous stable self-sustained systems. This means they have independent source of energy, are in continuous periodic motion on a fixed closed trajectory in state space, and if perturbed by an external

---

R. D. S. Yadava
Sensor & Signal Processing Laboratory, Department of Physics, Faculty of Science
Banaras Hindu University, Varanasi - 221005, India

transient force they return to the stable state by drawing or dissipating energy. Nonlinearity in dynamics is essential for the system to remain autonomous and stable. By coupling two such self-sustained autonomous oscillators the mutual interaction allows energy flow from each other that influences their motion. The dynamics in coupled state depends on the type of nonlinearity (whether delayed or non-delayed) and the strength of interaction [4, 11]. A coupling is described as being weak if the amplitudes of oscillations remain constant in time but the phases can flow freely between the coupled oscillators. A weak coupling facilitates adjustment of phase dynamics of the interacting oscillators, and synchronization occurs. The limit cycle oscillators are stable autonomous systems which are in motion with a fixed angular velocity. The phenomenon of synchronization in coupled limit cycle oscillators is referred to as frequency entrainment or phase locking in which both oscillators move with common frequency maintaining some constant phase shift [11]. The dynamics of interacting nonlinear oscillators has been extensively modeled and analyzed for understanding variety synchronization phenomena in nature such as circadian rhythm in living organisms, cardiac cycle, animal locomotion etc [3, 4, 11].

In this chapter, we present theoretical models for the autonomous and the coupled dynamics of an important class of oscillators that have found wide applications in the domain of physical, chemical and biological sensing. The mentioned oscillator class is of the surface acoustic wave (SAW) device delayed self-feedback radio frequency oscillators. In particular, we consider the coupling of SAW delay line oscillators where the coupling is also defined by a SAW delay line. We analyze conditions for phase synchronization, and calculate dependence of synchronization frequency on delay perturbations. We determine the sensitivity of the system in synchronized state against perturbations in the coupling delay, and discuss the possibilities for using synchronization modes for advanced SAW sensor development.

A SAW oscillator is converted into a sensor by making the measurand of interest interact in some way with the surface acoustic waves travelling between the input and the output ports of the SAW device. The interaction must produce a change in the oscillator frequency related to the quantity of measurand. The most basic SAW parameter that undergoes a change is the SAW velocity. Consequently, the feedback delay time and the oscillation frequency are changed. Surface acoustic wave energy is confined within about one wavelength of the surface region; therefore, any perturbation in the surface condition that affects the SAW velocity can, in principle, be sensed. The changes in SAW velocity due to surface strain, temperature, pressure and conductivity makes the basis for physical sensors; and that due to sorption and desorption of chemical compounds provide the basis for chemical and biological sensors [20-26]. The change in oscillator frequency in response to measurand defines the sensor output (or the signal). In chemical and biological sensors, the SAW propagation path needs selective sensitization for the chemical analytes of interest. This is usually done by coating a thin polymer film onto the surface wave propagation path that selectively sorbes analyte molecules. The analyte loadings produce changes in SAW velocity. The chemical sensors are made by tuning weak chemical interaction processes between target molecules and polymer coatings in order that the sensor operates in reversible manner.

The biological sensors also operate in a similar manner where the volatile organic compounds originated in biological processes (biogenic amines or metabolites) are sensed [25, 26]. The SAW oscillator sensors make an important class of chemical vapor sensors. In most common form they are used as a sensor array for making an electronic nose where the individual sensors are functionalized for selectively capturing different chemicals of interest [20, 27-31].

The dynamic response and nonlinearity of polymer coated SAW chemical sensors with respect to sorption and desorption kinetics has been modeled [32-34] and used for chemical identification purposes in some recent studies [34-39]. These studies are based on SAW delay line oscillators in stable limit cycle operation. Most of the studies on SAW chemical sensors focus primarily on the selection of polymer coatings [27-30, 40, 41], the selection of device type and the frequency of operation for achieving high chemical discrimination ability, high sensitivity and high signal-to-noise ratio [20]. Some studies are also focused on the device design aspects seeking enhancement of sensitivity of the sensing platform [21, 42, 43]. Only a few studies are aimed at understanding the dynamics of SAW oscillators and coupled SAW oscillators system [44-47].

2.2. SAW Delayed Self-Feedback Oscillator

A self-sustained oscillator is a nonlinear oscillatory system which maintains its stable rhythmic cycles solely due to its internal properties. It has an internal source of energy and dissipative nonlinear dynamics. Stability means the oscillations exist with certain nondecaying amplitude, and the internal source exactly compensates for the energy dissipation in cyclic motions. Any perturbation in the amplitude is restored to the stable state either by additional dissipation or by drawing energy from the internal source. This essentially needs the system dynamics to be nonlinear. If perturbed from equilibrium, a linear system can not maintain stable amplitude oscillations [11, p.37]. The self-sustained oscillators are also referred to as *autonomous* as they do not need support from an external energy source for maintaining their dynamics. In the phase space description of the dynamical system the self-sustained oscillators are represented by closed loop trajectories called the *limit cycles*. The phase space (state space) is the geometrical space defined by all the variables of the system and their time derivatives. The state of the system at some instant is represented by a phase point (that is, the set of values of all coordinates in the phase space). The time evolution of the system is described by the trace of phase points called trajectories. The trajectories of periodic oscillations are closed loop curves in phase space called cycles. The trajectory corresponding to a self-sustained oscillator is called a limit cycle as the nearby trajectories are attracted to it via dissipative nonlinear dynamics of the system. The limit cycle of a two variable oscillator system is 2π- periodic. The limit cycle of a harmonic oscillator is a circle of radius equal to the amplitude of oscillations and the phase point rotating with a constant angular speed equal to the frequency of oscillations. The stability of a limit cycle oscillator means that its amplitude is stable against perturbations in amplitude (that is, perturbations in transverse directions to the limit cycle). It perturbed, the amplitude decays to the stable equilibrium state along the limit cycle. However, the perturbations in phase along the limit cycles neither decay nor grow. This is stated as the phase point being in state of *neutral equilibrium*. In other words, if the phase point is shifted by a small force it remains in that shifted state. The main implication of this characteristic is that the phases can be easily adjusted by external action. This provides the basis for synchronization in coupled oscillators system [11, Ch. 2].

2.2.1. SAW Delay Line

A SAW delay line is defined by a pair of interdigital transducers (IDTs) fabricated on a piezoelectric crystal surface. The two transducers are positioned in-line at centre-to-centre separation of length L as shown in Fig. 2.1. A radio frequency voltage source connected across the input IDT applies oscillatory electric fields between adjacent electrodes of opposite polarities. That generates oscillatory strain fields because of the piezoelectricity, and the surface acoustic waves (SAW) are launched in both directions away from the transducer. For the transducers with regular electrodes (that is, uniform pitch and period) the surface acoustic waves are seen to be effectively arising from the center of the IDT [48]. The waves reaching the output IDT induce current in its electrodes due to reciprocal piezoelectric effects. Consequently, an output voltage appears across the load connected to the output IDT. If the output IDT also has regular electrode geometry the incoming surface waves are seen to be effectively received at the center of the IDT. The surface acoustic wave components approaching towards the edges of the piezo-crystal substrate from both the IDTs are absorbed by acoustic absorbers deposited as shown in the figure. The absorption of outgoing waves is important for avoiding complications in the functioning of SAW device that occur due to reflections from edges reaching the transducers with multiple delays.

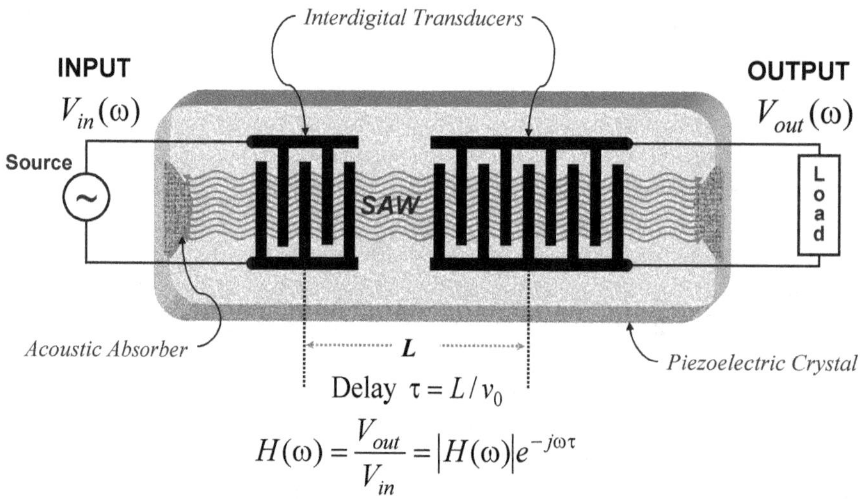

Fig. 2.1. SAW delay line.

The SAW device defined thus is visualized as a simple two-port delay line— an electrical network with input to output delay given by $\tau = L / v_0$ where v_0 is the propagation velocity of the surface acoustic waves. The SAW delay line with both transducers having regular electrode periodicity is nondispersive; that is, all the frequency components are delayed by the same amount τ. The harmonic voltage transfer function of the delay line SAW device can therefore be written as $H(\omega) = V_{out}(\omega) / V_{in}(\omega) = |H(\omega)| e^{-j\omega\tau}$ where $\omega = 2\pi f$ denotes the angular frequency of the input signal, $|H(\omega)|$ represents the amplitude transfer function and $\varphi_{SAW} = -\omega\tau$ represents the linear phase transfer.

2.2.2. SAW Delay Line Oscillator

A SAW delay line oscillator is a feedback oscillator circuit in which the SAW delay line has been used as feedback network connecting the output of a nonlinear amplifier to its input as shown in Fig. 2.2. The open-loop voltage transfer function of the circuit is defined as the ratio of voltages at the output of SAW delay line to the input of amplifier if the connectivity between them at the input end is cut open. It is given by $G_{OL}(\omega) = A(\omega)H(\omega) = |A(\omega)||H(\omega)| e^{j(\varphi_{SAW} + \varphi_A)}$ where $A(\omega) = |A(\omega)| e^{j\varphi_A}$ represents the amplifier voltage transfer function with $|A(\omega)|$ and φ_A being the gain and the phase shift respectively. In close-loop operation, the voltage transfer function (ratio of amplifier output to input) becomes $G_{CL}(\omega) = A(\omega)/[1 - A(\omega)H(\omega)]$. The circuit becomes unstable under the condition $A(\omega)H(\omega) \geq 1$ which means the open loop gain $|A(\omega)||H(\omega)| \geq 1$ and the phase shift $\varphi_{SAW} + \varphi_A = -2n\pi$. At the amplifier output, the signal appears repeatedly in phase after passing through the feedback path and getting amplified. Therefore, the voltage at the amplifier output grows by adding up constructively. A state of stable amplitude self-sustained oscillations can emerge if $|A(\omega)||H(\omega)| = 1$. Under this condition the amplifier gain just compensates for the losses in the feedback path for the signals arriving in phase. These are the well known Berkhausen criteria for setting up stable harmonic oscillations in feedback circuits.

A SAW delay line oscillator is an autonomous oscillator system. It contains all the essential elements of an autonomous system— energy source in the form of dc voltage supply, dissipation in the form of losses through SAW delay line and nonlinear dynamics as provided by the nonlinear voltage transfer function of the amplifier. The system does not need any external input for triggering or sustaining the oscillations. The inherent noises in constituent circuit elements of the loop trigger oscillations. The final voltage amplitude level is limited either by predefined amplifier nonlinearity or by dc supply driving the amplifier into saturation. In the above we used symbol φ_A to denote the phase shift across the amplifier. In fact, any other constant phase shift contribution due to non-SAW circuit elements in the loop can be absorbed into φ_A. The oscillator output voltage can be expressed in complex notation as $v_{out}(t) = |v_{out}(0)|e^{j(\omega t+\varphi_0)}$ where $|v_{out}(0)|$ denotes the stable amplitude level, ω is the angular frequency, and φ_0 is the constant initial phase. The initial phase can be interpreted to be the phase accumulated during the transient phase of the oscillations build up until the stable state is established.

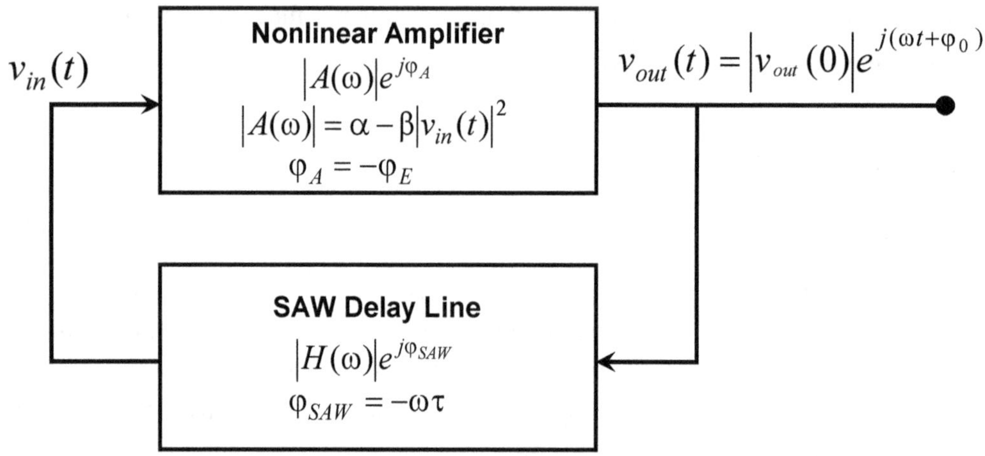

Fig. 2.2. Nonlinear self-sustained limit cycle SAW delay line oscillator.

2.2.3. Amplifier Transfer Function

The SAW delay line oscillator is a delayed self-feedback nonlinear dynamical system. The feedback comes from the transmission of the output signal back to the input via SAW delay line. The nonlinearity which controls the stability of amplitude is provided by the amplifier. The time delay in SAW device controls the frequency of oscillations. The start up comes from the ever present noisy signals or the transients at the time of power switching. In order to formulate the dynamics of this system we must assume some specific form for the amplifier nonlinearity, and define the signal transfer characteristics across the SAW delay line. Here, we assume that the amplifier transfer function to be of the following form

$$A = |A(\omega)|\exp(j\varphi_A) \tag{2.1a}$$

with

$$|A(\omega)| = \alpha - \beta|v_{in}(t)|^2 \text{ and } \varphi_A = -\varphi_E, \tag{2.1b}$$

where α and β are the frequency independent real constants, $|v_{in}(t)|$ is the amplitude of the signal at the input, and φ_A is the constant phase shift across the amplifier. The symbol φ_E with –ve sign has been used for φ_A in (2.1 b) for two reasons: (i) to signify that it may include contributions from other non-SAW components in the loop circuit such as connectors, phase shifter etc., and (ii) the delay in signal transmission across the amplifier, though negligibly small compared to the SAW delay time, is finite. Therefore, the sign of phase should be consistent with the definition of delay as $-d\varphi/d\omega$, which also means that φ_E is taken to be a positive quantity. With increasing amplitude level the amplifier gain is reduced in proportion to $|v_{in}(t)|^2$. The amplifications of various frequency components at $v_{in}(t)$ occur in accordance with to their amplitudes.

2.2.4. SAW Delay Line Transfer Response

Next, we consider the signal transfer characteristics of a simple SAW delay line defined by two in-line uniform aperture (unapodized) interdigital transducers as shown in Fig. 2.3. The description presented here is largely adapted from [49, Ch. 4]. Transducer A is connected to a voltage source V_A and transducer B is short circuited. Both the transducers are bidirectional. The surface potential waves launched by Transducer A from its Port 1 are received by the Transducer B at its Port 1. The received potential waves induce current in the electrodes of Transducer B. Assuming that the electrodes are non-reflecting this current is shown as short circuited current I_B. Also, the voltage source at A has been assumed to have zero impedance. Further, it is assumed that the surface waves going away from the transducers from the respective ports 2 are effectively absorbed at the substrate edges.

The short circuit current under these conditions has been calculated to be [49, Sec. 4.7.1]

$$I_B = V_A H_A(\omega) H_B(\omega) e^{-jk_0 d}, \tag{2.2}$$

where $H_A(\omega)$ and $H_B(\omega)$ are defined as the frequency responses of unapodized transducers, $k_0 = \omega / v_0$ is the wave vector of surface potential waves on the free surface and d is the separation between the launching port of transducer A (port 1) and the receiving port of transducer B (port 1). The position of acoustic ports on either side of the transducers is symmetrically taken near the edges of the outermost electrodes. Their precise location is not important. The x-axis is taken along the surface wave propagation direction. The acoustic aperture W of the transducers is assumed to be large enough so that surface potential in the lateral direction is uniform. The frequency response of a transducer is defined in terms of the surface wave potential it generates at a distance x from its center for unit voltage applied (V_A or $V_B = 1$). In Eq. (2.2), $H_A(\omega)$ and $H_B(\omega)$ refer to the launching port 1 of the two transducers with respect to the origin of x-axis taken at the centre of each transducer separately and the positive x-axis pointing away from the transducers as shown in Fig. 2.3.

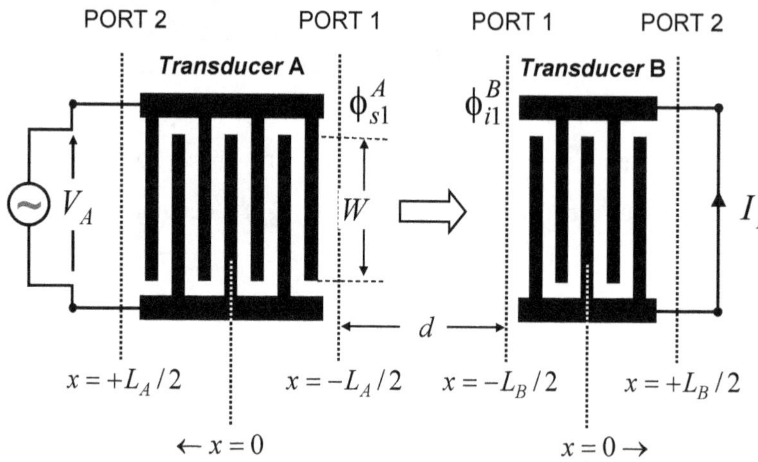

$$I_B = V_A H_A(\omega) H_B(\omega) e^{-jkd}$$

Fig. 2.3. Signal transfer across a simple SAW delay line. The input transducer A is connected to voltage source V_A and the output transducer B is short-circuited. I_B is the induced short-circuit current.

The surface potential wave at port 1 of the transducer A is given as [49, Eq. 4.50]

$$\phi_{s1}^{A}(\omega) = j\Gamma_s V_A \overline{\rho}_e^{A}(k_0)\exp(-jk_o L_A/2), \tag{2.3}$$

where $\overline{\rho}_e^{A}(k)$ is the Fourier transform of $\rho_e^{A}(x)$ which is electrostatic charge density on transducer A for unit applied voltage, and Γ_s is the constant of the piezoelectric substrate material defined as follows [49, Eq. 4.21]:

$$\Gamma_s \approx \frac{1}{\varepsilon_0 + \varepsilon_p^T}\frac{v_0 - v_m}{v_0} \quad \text{with,} \quad \varepsilon_p^T = \left(\varepsilon_{11}\varepsilon_{33} - \varepsilon_{13}^2\right)^{1/2}, \tag{2.4}$$

where v_0 and v_m are the SAW velocities for a free surface and for metalized surface respectively, ε_0 is the permittivity of vacuum, ε_p^T is the permittivity constant obtained by using permittivity elements ε_{ij} measured at constant stress.

The surface potential wave launched at port 2 is essentially the conjugate of ϕ_{s1}^{A}. By defining

$$H_A(\omega) = (\omega W \Gamma_s)^{1/2} \overline{\rho}_e^A(k_0)\exp(-jk_0 L_A/2) \tag{2.5}$$

Eq. (2.3) is rewritten as

$$\phi_{s1}^{A}(\omega) = jV_A H_A(\omega)\left[\frac{\Gamma_s}{\omega W}\right]^{1/2}. \tag{2.6}$$

This wave is incident on port 1 of transducer B as

$$\phi_{i1}^{B}(\omega) = \phi_{s1}^{A}\exp(-jk_0 d). \tag{2.7}$$

The induced short-circuited current in transducer B is given as

$$\begin{aligned} I_B &= -j\omega W \phi_{i1}^{B} \overline{\rho}_e^B(k_0)\exp(-jk_0 L_B/2) \\ &= -j\phi_{i1}^{B} H_B(\omega)\left[\frac{\omega W}{\Gamma_s}\right]^{1/2} \end{aligned} \tag{2.8}$$

where

$$H_B(\omega) = (\omega W \Gamma_s)^{1/2} \overline{\rho}_e^B(k_0)\exp(-jk_0 L_B/2) \tag{2.9}$$

Thus, by using Eqs. (2.6) and (2.7) in Eq. (2.8) one obtains Eq. (2.2) for the short circuit current in the transducer B due to voltage applied across transducer A. It should be mentioned here that the same relation also holds for a two transducer delay line if one transducer is apodized and other unapodized. However, the apodized transducer response now contains an additional multiplication factor that accounts for the positioning and apodization of electrodes [49, Eqs. 4.126, 4.127].

If we consider that all the electrodes in transducers A and B have same width and periodicity (*regular electrodes*) then in view of the chosen origin for the x-axis the charge density functions $\rho_e^A(x)$ and $\rho_e^B(x)$ are obviously symmetric (even). Therefore, their Fourier transforms $\rho_e^A(k_0)$ are $\rho_e^B(k_0)$ are also real and even. By using Eqs. (2.5) and (2.9) in Eq. (2.2) the relation for I_B can thus be written as

$$I_B = V_A \omega W \Gamma_s \overline{\rho}_e^A(k_0)\overline{\rho}_e^B(k_0)\exp(-jk_0 L). \tag{2.10}$$

where $L = L_A/2 + L_B/2 + d$ is the center-to-center separation between transducer A and transducer B. We can manipulate this relation further by noting that the acoustic conductance for a uniform transducer is $G(\omega) = \omega W \Gamma_s |\overline{\rho}_e(k_0)|^2$ [49, eq. 4.60], and $k_0 L = \omega\tau$, hence:

$$I_B = V_A \sqrt{G_A(\omega) G_B(\omega)} \exp(-j\omega\tau). \tag{2.11}$$

This relation describes the *short-circuit response* of a simple SAW delay line composed of two interdigital transducers having uniform aperture and regular electrode geometry; the input transducer is connected to zero impedance voltage source and the output transducer is shorted. In practical situations however both the input (transducer A) and output (transducer B) transducers will be terminated into finite impedances. Therefore, proper corrections must be made in $G_A(\omega)$ and $G_B(\omega)$ by calculating the actual voltage appearing across the input transducer and the actual current flowing through the output transducer under loaded conditions [49, Sec. 4.8.1, 8.4.2]. It has been shown that circuit effects appear as properly calculated correction factors $F_A(\omega)$ and $F_B(\omega)$ multiplied to $G_A(\omega)$ and $G_B(\omega)$ respectively [49, Eq. 4.137]. That is, in Eq. (2.11), $G_A(\omega) \rightarrow G_A(\omega)F_A(\omega)$ and $G_B(\omega) \rightarrow G_B(\omega)F_B(\omega)$.

In general, it is difficult to accurately calculate the circuit effects due to several uncertainties associated with the operating conditions and parasitic effects. A more pragmatic approach would be to minimize the circuit effects by using an amplifier with small input impedance and large output impedance, and by using a SAW delay line with large input impedance and small output impedance. Note that in the feedback oscillator circuit (Fig. 2.2), the amplifier output is the voltage source which is connected to the input transducer of the SAW delay line, and the output transducer is connected to the amplifier input. Thus, with this combination of impedances the connectivity of SAW feedback path would remain close to the zero impedance voltage source and the short-circuit output current as described by Eq. (2.11). Moreover, it is not difficult to achieve these combinations as the transducer admittances can be easily adjusted by controlling the number of electrode pairs and acoustic aperture, and extensive expertise is available for amplifier designs.

2.2.5. Delay Differential Equation for SAW Delay Line Oscillator

Let us consider the SAW delay line that connects the output of the amplifier to its input in the feedback oscillator circuit as shown in Fig. 2.4 (a). We assume that the amplifier transfer characteristics are as defined by Eqs. (2.1 a, b). The time delay across the amplifier is insignificant compared to the delay across the SAW delay line. The time evolution of the oscillator output therefore depends almost wholly on the voltage transfer characteristics of the SAW delay line. In order to obtain the delay differential equation for this system we can use the short-circuit response of the SAW delay line as expressed by Eq. (2.11). However, we need to change the notations for current and voltages to make it consistent with the amplifier input and output. We shall use $i_{sc}(t)$ for I_B, $v_{out}(t)$ for V_A as shown in the figure.

Recall that the short-circuit current $i_{sc}(t)$ arises due to the surface acoustic wave potential ϕ_{i1}^{B} incident on the transducer B, which is the delayed version of the surface wave potential ϕ_{s1}^{A} launched by the transducer A under the action of the amplifier output voltage $v_{out}(t)$. If the transducer B were not short-circuited then $i_{sc}(t)$ can be seen as the charging current of the electrostatic capacitance associated with this transducer under the action of surface acoustic wave potential ϕ_{i1}^{B}. Hence, the voltage appearing at the amplifier input $v_{in}(t)$ can be seen as the voltage developed across the transducer B due to charging of its capacitance by the surface acoustic wave traveling under it. The $i_{sc}(t)$ and $v_{in}(t)$ can be related by using the equivalent circuit representation of the transducer B. Fig. 2.4(b) and (c) respectively show the parallel and the series equivalent circuit representations of a uniform unapodized transducer [49, Sec.7.1.1]. In these circuits, C_T denotes the electrostatic capacitance of the transducer, and $G_a(\omega)$ and $B_a(\omega)$ denote respectively the acoustic conductance and susceptance parts; the transducer admittance $Y_T = G_a(\omega) + jB_a(\omega) + j\omega C_T$.

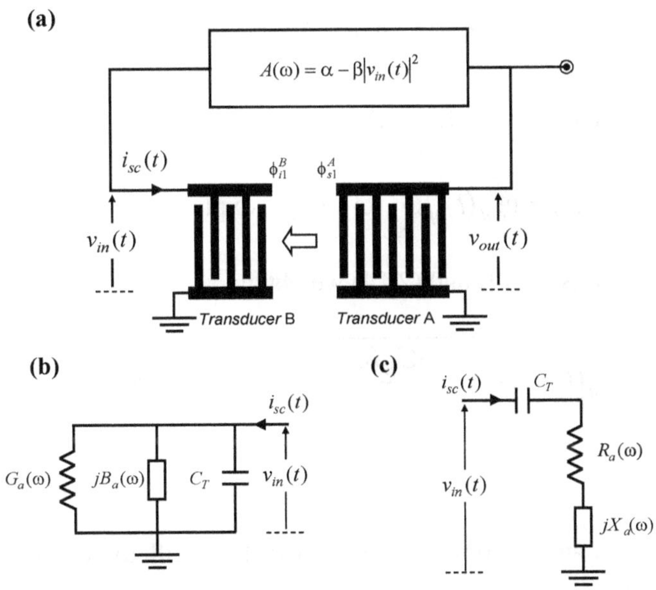

Fig. 2.4. (a) SAW delay line feedback oscillator, (b) and (c) parallel and series equivalent circuit representation of an interdigital SAW transducer respectively.

In general, C_T part dominates over the $G_a(\omega)$ and $B_a(\omega)$ contributions to the admittance in most common simple SAW delay lines. In series equivalent representation, $R_a(\omega)$ and $X_a(\omega)$ are known as radiation resistance and acoustic reactance respectively. The value of $R_a(\omega)$ measures the amount of electrical energy converted into the acoustic energy by the transducer. The transducer impedance is given as $Z_B = R_a(\omega) + jX_a(\omega) + 1/j\omega C_T$ with

$$R_a(\omega) = \frac{G_a(\omega)}{\{G_a(\omega)\}^2 + \{B_a(\omega) + \omega C_T\}^2} \tag{2.12a}$$

$$X_a(\omega) = \frac{1}{\omega C_T} - \frac{B_a(\omega) + \omega C_T}{\{G_a(\omega)\}^2 + \{B_a(\omega) + \omega C_T\}^2}. \tag{2.12b}$$

For most practical transducers, $\omega C_T >> G_a(\omega) >> B_a(\omega)$. That means, $R_a(\omega) \approx G_a(\omega)/(\omega C_T)^2$ and $X_a(\omega) \approx 0$. Therefore, $v_{in}(t)$ can be seen to arise primarily from the charging of receiving transducer capacitance C_T through $R_a(\omega) - C_T$ network by current $i_{sc}(t)$. Denoting the receiving transducer capacitance by C_B we can therefore write

$$i_{sc}(t) = C_B \frac{dv_{in}(t)}{dt}. \tag{2.13}$$

In terms of the new notations the Eq. (2.11) is rewritten as

$$i_{sc}(t) = v_{out}(t)\sqrt{G_A(\omega)G_B(\omega)}\exp(-j\omega\tau). \tag{2.14}$$

Thus, by combining Eqs. (2.13) and (2.14) we obtain

$$\dot{v}_{in}(t) = \frac{\sqrt{G_A(\omega)G_B(\omega)}}{C_B} v_{out}(t)\exp(-j\omega\tau). \tag{2.15}$$

where $"\cdot" \equiv d/dt$.

In order to arrive at an equation of motion for the oscillator output we can use the amplifier input-output relationship

$$v_{out}(t) = Av_{in}(t) \tag{2.16}$$

and write,

$$\dot{v}_{out}(t) = A\dot{v}_{in}(t) + \dot{A}v_{in}(t) \tag{2.17}$$

Now, if we use $\dot{v}_{in}(t)$ from Eq. (2.15) and $v_{in}(t)$ from Eq. (2.16) into Eq. (2.17) we obtain

$$\dot{v}_{out}(t) = \left(AX + \dot{A}/A\right)v_{out}(t)\text{,} \tag{2.18}$$

where

$$X = \frac{\sqrt{G_A(\omega)G_B(\omega)}}{C_B}\exp(-j\omega\tau)\text{.} \tag{2.19}$$

Equation (2.18) represents the general equation of motion for the SAW feedback oscillator circuit in which SAW delay line specifics appears through the factor X as defined by Eq. (2.19). In the expression for X the subscripts A and B respectively refer to the launching transducer (connected to amplifier output) and the receiving transducer (connected to amplifier input). Also, it has been assumed that the launching transducer is driven by a voltage source (amplifier output with small resistance compared to transducer input impedance) and almost total short-circuit current is used up in charging the receiving transducer capacitance. The latter effectively means that the amplifier input terminal is a virtual current source of infinite impedance.

The oscillator dynamics close to *steady state condition* can be simplified by noting that the amplifier gain reaches nearly a constant level, hence, the time rate of its variation $\dot{A} \approx 0$, and Eq. (2.18) can be written as

$$\dot{v}_{out}(t) = \frac{A(\omega)\sqrt{G_A(\omega)G_B(\omega)}}{C_B}v_{out}(t)\exp(-j\omega\tau)$$

or

$$\dot{v}_{out}(t) = \frac{A(\omega)\sqrt{G_A(\omega)G_B(\omega)}}{C_B}v_{out}(t-\tau) \tag{2.20}$$

where

$$v_{out}(t-\tau) = v_{out}(t)e^{-j\omega\tau}$$

is the amplifier output delayed by the SAW delay line by time τ. In these equations, it may be convenient to express the SAW specific factor X in terms of the more commonly used voltage transfer function (or the frequency response) of the SAW delay line. The voltage transfer function, denoted as $H(\omega)$, is defined as the ratio of the output voltage to the input voltage applied at frequency ω; that is, $H(\omega) = V_B(\omega)/V_A(\omega)$ where $V_A(\omega)$ and $V_B(\omega)$ may in general be complex valued.

By writing $v_{in}(t) = V_B(\omega)\exp(j\omega t)$ and $v_{out}(t) = V_A(\omega)\exp(j\omega t)$ in Eq. (2.15) results in

$$X = j\omega H(\omega). \tag{2.21}$$

with X defined by Eq. (2.19). This gives the harmonic voltage transfer function

$$H(\omega) = \frac{X}{j\omega} = \frac{\sqrt{G_A(\omega)G_B(\omega)}}{\omega C_B}\exp(-j\omega\tau)\exp(j3\pi/2). \tag{2.22}$$

Thus, the voltage transfer function across the SAW delay line can be written as

$$H(\omega) = |H(\omega)|\exp(-j\omega\tau)\exp(j3\pi/2) \tag{2.23a}$$

with

$$|H(\omega)| = \frac{\sqrt{G_A(\omega)G_B(\omega)}}{\omega C_B}. \tag{2.23b}$$

It may be remarked here that the voltage transfer function of a SAW delay line driven by a voltage source with internal resistance R_s and terminated into a load R_L is given by [48, Eq. 4.29]

$$H(\omega) = \frac{Y_{AB}R_L}{(1+Y_{AA}R_s)(1+Y_{BB}R_L) - Y_{AB}^2 R_s R_L}, \tag{2.24}$$

where Y_{AA} and Y_{BB} are the admittances of the input and the output transducers respectively, and Y_{AB} is the short-circuit transfer admittance defined as $Y_{AB} = (I_A/V_B)_{V_A=0}$. The $Y_{AB}^2 R_s R_L$ term in the denominator represents the triple-transit interference (TTI) effect due to electrode reflections. It may be reminded that in the present derivation of equations of motion we ignored TTI effect, and assumed small voltage source and load resistances compared to the input and the output transducers. These simplifications are valid or they can be made valid by proper designing of transducers and circuitry.

In view of these considerations we finally write the equation of motion for the SAW delay line oscillator near steady state condition, Eq. (2.20), in terms of the voltage transfer function $H \equiv H(\omega)$ as

$$\dot{v}_{out}(t) = j\omega A H v_{out}(t) \tag{2.25}$$

where $A \equiv A(\omega)$; and, for a general condition, Eq. (2.18), as

$$\dot{v}_{out}(t) = \left(j\omega AH + \dot{A}/A\right)v_{out}(t) \tag{2.26}$$

where $A \equiv A(\omega,t)$. By writing $j = e^{j\pi/2}$ and using Eq. (2.23a) for H we can write $jH = |H(\omega)|e^{-j\omega\tau}$ in these equations. These are linear time-delay differential equations with time-dependent coefficient $P(t) = j\omega AH + \dot{A}/A$ in Eq. (2.26) and time-invariant coefficient $Q = j\omega AH$ in Eq. (2.25). The time dependence is contained in the nonlinear amplifier gain $A \equiv A(\omega,t)$ through $|v_{in}(t)|$ and the time-delay component in contained in the SAW delay line voltage transfer function $H \equiv H(\omega)$.

2.2.6. Delay Differential Equation by Feng and Chicone

To our knowledge, the only other work reported on the delay differential equation for SAW delay line oscillator is that by Feng and Chicone [44] which was subsequently used in [45, 46] to study synchronization phenomena in coupled delay line oscillators and sensors. With some differences in the notation they obtained a voltage transfer relation across the SAW delay line similar to that given by Eq. (2.15) above. In [44] the notations $v_{in}(t)$ and $v_{out}(t)$ refer to the delay line input and output transducers respectively, whereas in the present work these refer to the amplifier input and output respectively. Therefore, for making a comparison of with our results the subscripts '*in*' and '*out*' in [44] associated with the voltages must be exchanged. The voltage transfer relation in [44, eq. 6] is analogous to Eq. (2.15) above, and is written as

$$\dot{v}_{in}(t) = h\sum_{n=0}^{N-1}\sum_{m=0}^{M-1} v_{out}(t - \frac{L + ma + na}{V_R}). \tag{2.27}$$

In deriving this relation a SAW delay line with two uniform interdigital transducers having M and N electrode pairs (fingers) in the input and the output respectively were assumed. The other parameters are: V_R SAW velocity, a electrode period, L gap length between transducers referred to $a/2$ distances away from the centers of innermost electrodes, and h a constant which accounts for the effect of electro-acoustic transduction processes at the SAW substrate. For sinusoidal signals the series summation in Eq. (2.24) was evaluated to yield the frequency response for the delay line as [44, Eqs. 8, 14, 15]

$$G(\omega) = \frac{h}{\omega} g(\omega) \exp(-j\omega\tau) \tag{2.28a}$$

where

$$\tau = \frac{L + Ma + Na}{2V_R} \tag{2.28b}$$

and

$$g(\omega) = \frac{\sin(\pi M\omega/\omega_c)}{\sin(\pi\omega/\omega_c)} \frac{\sin(\pi N\omega/\omega_c)}{\sin(\pi\omega/\omega_c)}. \tag{2.28c}$$

It can be seen that τ in Eq. (2.29) represents the center-to-center time delay between the input and output transducers (same as τ in our treatment), and the factor $hg(\omega)/\omega = |G(\omega)|$ is the voltage amplitude transfer function of the SAW delay line. For this expression of the SAW delay line transfer function to be equal to that given by Eq. (2.23) one must have $|G(\omega)| = |H(\omega)|$, or

$$h = \frac{\omega|H(\omega)|}{g(\omega)} = \frac{\sqrt{G_A(\omega)G_B(\omega)}}{C_B g(\omega)}. \tag{2.29}$$

The theory of SAW transducers based on the impulse response models describes the transducer frequency response as a product of two factors: an *element factor* $E(\omega)$ associated with charging of electrodes and an *array factor* $g'(\omega)$ due to the periodic arrangement of electrodes; that is, for a given transducer $|H_T(\omega)| = E_T(\omega)g'_T(\omega)$ [48, Sec. 4.6], [49, Sec. 4.1, 4.7.3, 8.1.1]. From the field theoretical analysis of SAW interdigital transducers it is shown that the transducer frequency response is related to its acoustic conductance as $G_a(\omega) = |H_T(\omega)|^2$ [49, Eq. 4.124]. Therefore,

$$\begin{aligned} \sqrt{G_A(\omega)G_B(\omega)} &= |H_A(\omega)||H_B(\omega)| \\ &= E_A(\omega)E_B(\omega)[g'_A(\omega)g'_B(\omega)] \\ &= E_A(\omega)E_B(\omega)g'(\omega) \end{aligned} \tag{2.30}$$

where $g'(\omega)$ is the product of the array factors for the two transducers. Actually, the array factor $g'(\omega)$ and $g(\omega)$ in (2.28c) are the same. Therefore,

$$h = \frac{E_A(\omega)E_B(\omega)}{C_B}. \tag{2.31}$$

In [44], h was parameterized as product of three factors K_1, K_2 and K_3 through the relation $h = K_1K_2K_3/2V_R$ where K_1 is proportionality constant between mechanical force and voltage at an electrode, K_2 is proportionality constant between charge produced and mechanical displacement at an electrode, and K_3 is proportionality constant between total charge on a transducer and voltage across it. It may be noted that K_3 is equivalent to $1/C_T$ in our case. We can go a step further by using the known expression for the element factor for a uniform transducer [49, Eqs. 4.60, 8.4]

$$E_{A,B}(\omega) = (\omega W\Gamma_s)^{1/2}\overline{\rho}_e^{A,B}(k_0). \tag{2.32}$$

By taking into account the contributions from for both the transducers in the SAW delay line response we obtain

$$E(\omega) = E_A(\omega)E_B(\omega) = \omega W\Gamma_s\overline{\rho}_e^{A}(k_0)\overline{\rho}_e^{B}(k_0). \tag{2.33}$$

Thus,

$$h = \frac{\omega W\Gamma_s\overline{\rho}_e^{A}(k_0)\overline{\rho}_e^{B}(k_0)}{C_B}. \tag{2.34}$$

With these clarifications for the definition of h factor in the paper of Feng and Chicone [44] their delay line transfer response given by Eq. (2.28) becomes equivalent to that used in the present work defined through Eq. (2.23).

Finally, in the close-loop operation with amplifier gain A the delay differential equation for the amplifier output analogous to Eqs. (2.20) and (2.25) was obtained as [44, Eq. 17]

$$\dot{v}_{out}(t) = Ah\sum_{n=0}^{N-1}\sum_{m=0}^{M-1} v_{out}\left(t - \frac{L + ma + na}{V_R}\right). \tag{2.35}$$

Based on Eq. (2.35), the authors in their later studies proposed a simplified form [45]

$$\dot{v}_{out}(t) = Ahf[Bv_{out}(t-\tau)], \tag{2.36}$$

where the function $f(x)$ is mathematically introduced to account for series summation in Eq. (2.35) and amplifier nonlinearity and profile scaling through the feedback path. In this equation the parameters A and B were introduced to specify the amplifier gain and the profile scaling.

2.2.7. Self-Sustained Limit Cycle Oscillations

Equations of motion (2.25) and (2.26) can alternately be viewed as ordinary differential equations with complex coefficients P and Q. That means, we can expect that the system supports complex solutions of the form

$$v_{out}(t) = |v_{out}(t)| e^{j\varphi(t)} \tag{2.37}$$

with $|v_{out}(t)|$ and $\varphi(t)$ respectively being the instantaneous amplitude and phase. Substituting this into these equations, and equating the real and the imaginary parts on both sides we obtain separate equations for the motion for the amplitude and the phase as follows:

from Eq. (2.25),
$$(|v_{out}(t)|)^{\dot{}} = \omega |A(\omega)||H(\omega)| \cos(\omega\tau + \varphi_E) |v_{out}(t)| \tag{2.38}$$

$$\dot{\varphi}(t) = -\omega |A(\omega)||H(\omega)| \sin(\omega\tau + \varphi_E) \tag{2.39}$$

from Eq. (2.26),
$$(|v_{out}(t)|)^{\dot{}} = [\omega |A(\omega,t)||H(\omega)| \cos(\omega\tau + \varphi_E) + \dot{A}(\omega,t) / A(\omega,t)] |v_{out}(t)| \tag{2.40}$$

$$\dot{\varphi}(t) = -\omega |A(\omega,t)||H(\omega)| \sin(\omega\tau + \varphi_E)\,. \tag{2.41}$$

In order to seek the conditions for limit cycle oscillations we assume that the circuit parameters are defined in such a way that the system asymptotically approaches the steady state at some frequency. The limit cycle oscillations are closed-loop trajectories in phase plane. They are characterized by the time-invariant amplitude and angular frequency conditions, that is, $(|v_{out}(t)|)^{\dot{}} = 0$ and $\dot{\varphi}(t) = \text{const}$ respectively. Applying the amplitude stability condition with the general equation of motion (2.40) we obtain

$$(|A(\omega,t)|)^{\dot{}} = -\omega |A(\omega,t)|^2 |H(\omega)| \cos(\omega\tau + \varphi_E)\,. \tag{2.42}$$

By integrating this equation and assuming some arbitrary initial condition defined as $|A(\omega,t)| = |A(\omega,0)|$ at $t = 0$ we obtain

$$A(\omega,t) = |A(\omega,0)| \left[1 - \omega t |A(\omega,t)||H(\omega)| \cos(\omega\tau + \varphi_E)\right]. \tag{2.43}$$

Similarly, integrating Eq. (2.41), and assuming for the initial phase value $\varphi(t) = \varphi(0)$ at $t = 0$ we obtain

$$\varphi(t) = \varphi(0) - \omega t|A(\omega,t)||H(\omega)|\sin(\omega\tau + \varphi_E). \qquad (2.44)$$

For a possible limit cycle state, both Eq. (2.43) and Eq. (2.44) must be satisfied simultaneously for some ω. Eq. (2.44) tells that $\varphi(t)$ can be interpreted to be the linear phase on the limit cycle with period $2\pi/\omega$ if $A(\omega,t)$ reaches some constant level in the steady state. That is, if $A(\omega,t) \to A(\omega,\infty)$ (say). In that case, $\dot{\varphi}(t) = -\omega|A(\omega,\infty)||H(\omega)|\sin(\omega\tau + \varphi_E)$ becomes the angular velocity on the limit cycle. Note that even if $A(\omega,t) \to A(\omega,\infty)$ the r.h.s. of Eq. (2.43) still has a linear dependence on time. This obviously contradicts the time-invariance of $A(\omega,\infty)$ imposed on the steady state phase characteristics of Eq. (2.44). This contradiction can be sorted out only if we find a 'ω' such that $\cos(\omega\tau + \varphi_E) = 0$; this in turn implies $\sin(\omega\tau + \varphi_E) = \pm 1$. Under this condition, $\dot{\varphi}(t) = \omega|A(\omega,\infty)||H(\omega)|$ where we have taken –ve sign so that phase is always interpreted as increasing quantity. Since by definition $\dot{\varphi}(t)$ must be equal to ω on the limit cycle we find that $|A(\omega)||H(\omega)| = 1$ on the limit cycle. In conclusion, Eqs. (2.38)-(2.41) admit limit cycle solutions only if

$$|A(\omega)||H(\omega)| = 1 \text{ and } \sin(\omega\tau + \varphi_E) = -1 \qquad (2.45a)$$

or equivalently

$$\operatorname{Re} A(\omega)H(\omega,\tau) = 1 \text{ and } jA(\omega)H(\omega,\tau) = 1. \qquad (2.45b)$$

By integrating Eq. (2.38) directly we can examine the behavior of the system close to steady state conditions. We can easily see that

$$|v_{out}(t)| = |v_{out}(\infty)|e^{\lambda t} \qquad (2.46)$$

with,

$$\lambda = \omega|A(\omega)||H(\omega)|\cos(\omega\tau + \varphi_E), \qquad (2.47)$$

where $|v_{out}(\infty)|$ denotes the steady state amplitude level. On the limit cycle, $\cos(\omega\tau + \varphi_E) = 0$, therefore, $\lambda = 0$ and $|v_{out}(t)| = |v_{out}(\infty)|$. Deviation from the limit cycle can be represented by assigning small perturbation to either $|A(\omega)||H(\omega)| = 1 \pm \delta$ or $\cos(\omega\tau + \varphi_E) = \pm\delta$ where δ denotes a small positive real number. If the value of $|A(\omega)||H(\omega)| > 1$ the frequency will shift to make $\cos(\omega\tau + \varphi_E) < 0$ so that $\lambda < 0$ and amplitude decays to stable level. If the value of $|A(\omega)||H(\omega)| < 1$ the frequency will shift to make $\cos(\omega\tau + \varphi_E) > 0$ so that $\lambda > 0$ and amplitude grows to stable level.

In view of the phase and amplitude expressions given Eqs. (2.44) and (2.46) we can express mathematically the stable limit cycle oscillations as

$$v_{out}(t) = |v_{out}(\infty)| e^{j(\omega t + \varphi_0)}. \tag{2.48}$$

To understand the nature of amplitude variation, and how an arbitrary solution is attracted to the limit cycle, we can consider the solution of Eq. (2.40) by direct integration. It can be easily seen that

$$v_{out}(t) = |v_{out}(\infty)| \frac{|A(\omega,t)|}{|A(\omega,\infty|)} e^{j(\omega t + \varphi_0)}, \tag{2.49}$$

where $A(\omega,\infty)$ denotes the amplifier gain in the steady state condition and we have applied Eq.(2.45) for motions near limit cycle. The maximum gain of the amplifier occurs at the time of start up with arbitrarily small input, $|A(\omega)| = \alpha - \beta |v_{in}(t)|^2 \Rightarrow A_{\max} \approx \alpha$. As the signal in loop grows towards steady state, $A(\omega,t) \rightarrow A(\omega,\infty)$.

2.3. SAW Delayed Coupling of SAW Delay Line Oscillators

2.3.1. Two SAW Delay Line Oscillators Coupled by On-Chip Acoustic Field

Earlier, Chicone and Feng based on their delay differential equation [45, Eq. (36)] presented an analysis of the nonlinear dynamics of two coupled SAW delay line oscillators. In the system considered, two *adjacent* SAW delay lines fabricated on the same substrate but supported by separate amplifier circuitry defined the two SAW delay line oscillators. The coupling was assumed to occur via surface acoustic fields on the SAW substrate. Without specifying the details of interaction a delayed coupling was assumed between the outputs of the two oscillators [denoted as $v_{out}(t)$ and $w_{out}(t)$] in the following form:

$$\dot{v}_{out}(t) = Ahf\left(Bv_{out}(t-\tau)\right) + Cf\left(Bw_{out}(t-(\tau+\iota))\right) \tag{2.50}$$

$$\dot{w}_{out}(t) = Dhf\left(Bw_{out}(t-(\tau+\kappa))\right) + Ef\left(Bv_{out}(t-(\tau+\upsilon))\right), \tag{2.51}$$

where A and D are the amplifier gains, C and E represent the cross coupling strengths, τ and $\tau+\kappa$ are the SAW time delays in the feedback paths of the two oscillators, $\tau+\iota$ and $\tau+\upsilon$ are the acoustic cross coupling time delays. The delay lines were assumed to be constructed of only one electrode pair (finger) in the input and the output transducers. The multi-finger transducers represented by the delay differential [45, Eq. (35)] were considered in a later study through simulation [46]. For the analysis, the coupled equations were transformed to dimensionless form by normalization of the system parameters (amplifier gains, time delays and coupling strengths). The time-averaged equations (averaged over fast variations across electrode periods) were analyzed for bifurcation points, synchronization and stability through linearization near equilibrium points (steady states obtained after setting the time rate of variation of averaged outputs zero). The mathematical analysis presented is quite complicated due to involvement of several parameters (some defined and some arbitrarily introduced), and also due to several stages of parameters/variables transformation or representation. Even though the paper [45] presents a detailed bifurcation analysis of the parametric space, it is difficult to identify and associate some of these parameters quantitatively with practically adjustable system characteristics (particularly the cross-coupling).

However, some interesting observations were made by considering simpler cases of nearly identical oscillators. It was found that for each oscillator to be self-excited the amplifier gain and coupling must exceed a threshold value.

If the two oscillators differ only in feedback delays, and if the delay in one oscillator is below threshold in the uncoupled state, then after coupling it is slaved to the second self-excited oscillator. That is, the slave oscillator starts oscillating with the same amplitude and frequency as the master oscillator.

If both the oscillators are self-excited, and if the delay difference is small and coupling is weak, their oscillations are synchronized— both oscillating with the same frequency maintaining a constant phase difference which is in proportion to the difference in the delays. The synchronized state solutions in this case are defined by the condition [45, Eq. 16]

$$\delta_2 + \left(\frac{\gamma_1 r_2^2 + \gamma_2 r_1^2}{r_1 r_2} \right) \sin(\theta_1 - \theta_2) = 0, \tag{2.52}$$

where δ_2 represents the delay difference, (r_1, r_2) represent the amplitudes, (γ_1, γ_2) represent cross-coupling strengths and (θ_1, θ_2) represent the phases with subscripts $(1, 2)$ referring to the two oscillators; all these quantities are in the normalized form. For identically built oscillators with symmetric coupling, $r_1 = r_2$ and $\gamma_1 = \gamma_2 = \gamma$ (say). Denoting the phase difference $\theta_1 - \theta_2$ by ϕ_1, the parametric condition for synchronization to occur becomes

$$\delta_2 + 2\gamma \sin\phi_1 = 0 \text{ or } \sin\phi_1 = -\frac{\delta_2}{2\gamma} \tag{2.53}$$

which has solutions if and only if

$$|\delta_2| \le 2\gamma . \tag{2.54}$$

It is found that the equality in this condition (that is, $\delta_2 = \pm 2\gamma$) corresponds to the saddle-node bifurcation where two synchronization branches (one stable and other unstable) meet. This is independent of the normalized loop gain parameter α (assumed to be the same for the identically built oscillators, $\alpha_1 = \alpha_2 = \alpha$). It is shown that in this case, under the condition

$$\alpha = -\frac{\delta_2}{\pi} \pm \frac{1}{2}\sqrt{4\gamma^2 - \delta_2^2} \tag{2.55}$$

Hopf bifurcation (that leads to limit cycles) occurs. For conditions deviated from these special points no closed form solutions are found, but simulation based analyses have revealed rich bifurcation structure in parameter space.

The synchronization analysis with respect to the variation of parameter δ_2 for fixed values of $\gamma = 0.001$ and $\alpha = 0.002$ reveals that perfect synchronization occurs at $\delta_2 = -0.008$, and then in the interval $-0.002 \le \delta_2 \le 0.002$; at $\delta_2 = -0.005$ torus bifurcation occurs, and for small variations beyond $\delta_2 > 0.002$ modulated output appears; and for δ_2 significantly beyond 0.002 the two oscillators vibrate independently and the synchronization is lost.

As an impact of this synchronization behavior on dual delay line SAW sensors (where one oscillator serves as reference and the other as sensor) it is pointed out that if the dissimilarity between the two oscillators as represented by the parameter δ_2 is less than the threshold $\delta_2 \le 2\gamma$ the oscillators remain in synchronized state. As a consequence any change in sensing oscillator frequency due to presence of stimuli can not be detected. This situation must be avoided for functioning of sensors. On the other hand, it has been suggested that in the synchronization state the phase difference (instead of frequency difference) between the two oscillators can be used as satisfactory measure of sensor signal. In the extended later study [46] the authors analyzed the impact of cross-coupling on the conventional difference frequency output of the dual SAW oscillator sensors. It was found that the coupling reduces the sensitivity and also makes the sensor response nonlinear [46, Fig. 9].

2.3.2. Two SAW Delay Line Oscillators Coupled by a SAW Delay Line

The dynamics of a single SAW delay line oscillator is formulated in Section 2.2. It has possibilities for generating varied outputs by proper control of the amplifier nonlinearity (through α and β) and the SAW delay transfer response (through $|H(\omega)|$ and τ). In Section 2.2, we sought conditions for stable limit cycle harmonic oscillations. Such oscillators have been most commonly used for sensing volatile organic chemicals since the first report by Wholtjen and Dessy [50]. For this purpose, SAW delay path in the feedback loop is sensitized by depositing a thin polymer film that can selectively sorb the target chemical analytes. The chemical sorption in the film produces a change in delay time which shifts the equilibrium (limit cycle) point for stable oscillations. The measurement of this shift is the basis for SAW chemical sensing. Usually two nearly identical oscillators (one sensitized by polymer coating and other unsensitized or uncoated) are used for making a sensor. The difference in frequencies of oscillations (low pass filtered after mixing) is measured as the sensor signal. The uncoated polymer serves as reference so that any spurious frequency shift due to non-chemical factors such as temperature, humidity etc. is cancelled out in the differential mode of signaling.

The synchronization analysis in [44-46] had considered this type of dual delay line oscillator sensors. The analyses presented in these papers clearly showed a link between the coupling strength, loop gain and the range of delay differences over which the oscillators remain synchronized [45, Eqs. (52) through (55)]. The relations are however generic; and, it is not clear how to use them as control parameters in the sensor design or performance analysis. Motivated by this, recently, we considered the coupling between two nonlinear SAW delay line oscillators by a third SAW delay line with fully specified characteristics [47]. Restricting our analysis to the case of weak coupling which could be described by stable-amplitude phase dynamics we analyzed conditions for synchronization and looked for the possibility of making *coupling SAW delay line as sensing platform*. We assumed that the individual SAW delay line oscillators are in the stable limit cycle states, and coupled their outputs via a SAW delay line having the same synchronous frequency as that of the coupled oscillators. The formulation of internal dynamics of autonomous SAW oscillators in Section 2.2 and the coupling by a SAW device with specified characteristics will provide flexibility and parametric control for designing and optimizing coupled SAW oscillators system for sensing applications.

2.3.2.1. Coupling Configuration

Fig. 2.5 depicts a coupling scheme between two SAW delay line oscillators. The SAW oscillators are the same as that described in Section 2.2.2 and 2.2.7. The coupling is defined by connecting their outputs through a simple SAW delay line. The coupling SAW delay line delays the output from one oscillator and feeds it as input to the feedback network of the other oscillator. The coupling is taken to be reciprocal so that a similar feedback occurs from the second oscillator to the first. Thus, the feedback signal in each oscillator is derived from its current output and the delayed output from the other oscillator. The combined signal is then processed by the feedback SAW network and applied to the amplifier input. The coupling interactions are completely specified in terms of the stable states of the uncoupled oscillators and the transfer response of the coupling delay line. The system behavior is now controlled by three delay times associated with the three delay lines, besides their amplitude transfer responses and those of the two amplifiers. It may be noted that this configuration, once formulated for its dynamics, can easily be reduced to two simpler configurations where (i) two non-SAW autonomous oscillators are coupled by a SAW delay line, and (ii) two SAW delay line oscillators are coupled by a non-SAW electrical network. A comparative analysis of these configurations may provide insight about which SAW device (that used in the self-feedback or in the cross-coupling) would make a better sensing platform in the synchronized state.

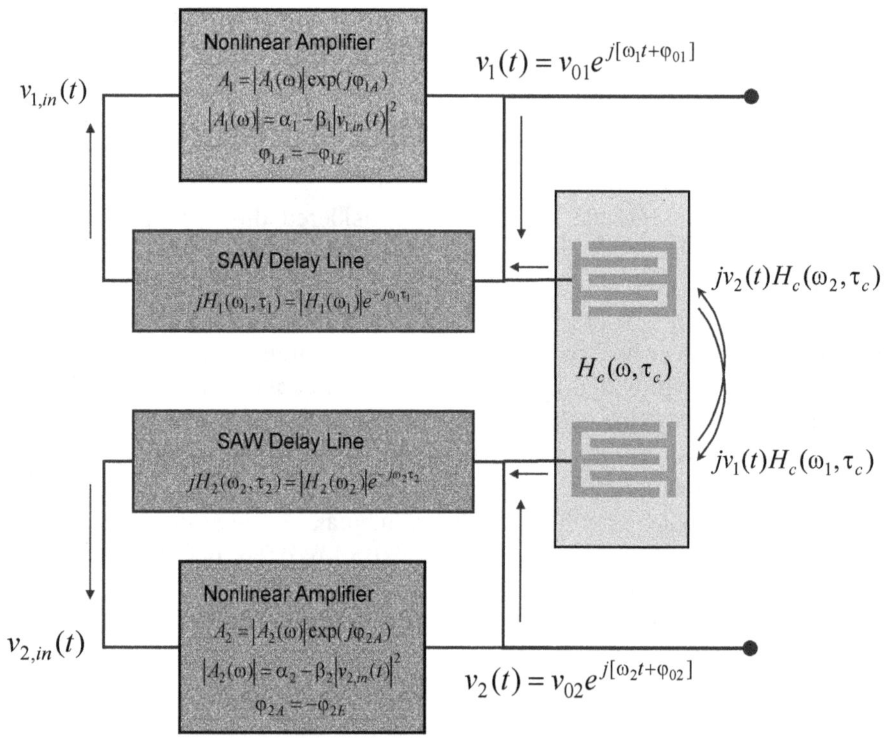

Fig. 2.5. Coupling configuration for mutual interaction between two nonlinear SAW delay line oscillators. The coupling is provided by the SAW delay line on the right.

All the SAW delay lines are assumed to have identical electrode periodicity in their input and output transducers. That is, the centre frequencies of all the SAW components are the same. However, their amplitude transfer responses and delay times could in general be different. In particular, the pass bandwidth and insertion loss of the coupling delay line is assumed to be large compared to the feedback delay lines in order to ensure that it provides constant amplitude transfer and weak coupling over frequency range of oscillations supported by the two coupled oscillators. The SAW delay line transfer characteristics can easily be tuned to the desired values by adjusting the number electrode pairs and apertures of the interdigital transducers and the separation between their phase centres [48]. The feedback delay lines can be designed to have narrower pass band and arbitrary delays to suit specific needs.

2.3.2.2. Coupled Delay Differential Phase Equations

Assuming that the two SAW oscillators in the uncoupled states are in their respective limit cycles and the coupling is weak, we set up equations of motion for pure phase coupling. That means, the amplitudes of the oscillators are not changed by coupling, but their phases can easily flow from one to other. Incorporating amplitude perturbations and arbitrary coupling strength for a generalized treatment will not only add to mathematical complications, but one would have to deal with instabilities and chaotic dynamics. Our primary motivation here is to study the coupled synchronized system for sensing applications. Therefore, it is convenient to avoid strong coupling situations, and focus mainly on the phase dynamics under weak coupling approximation.

Prior to coupling each oscillator is assumed to be in the stable limit cycle state with sinusoidal oscillations of the form $v(t) = v_0 e^{j(\omega t + \varphi_0)}$ where $v_0 \equiv |v_{out}(\infty)|$ is the constant amplitude, ω is the angular frequency of the oscillations referred to as the *natural frequency*, and φ_0 is the initial phase. Irrespective of how these oscillations were triggered to reach self-sustained limit cycles the initial phase can be taken to be any convenient reference phase point after the steady state conditions are established. It may represent the total phase accumulated from the start of triggering event to the full scale amplitude build up. Since each oscillator has independent triggering point and path of temporal evolution, the initial phase values at the reference point will, in general, be different for different oscillators. The influence of coupling on the phases of the two oscillators can be expressed by assigning them a phase function of the form

$$\varphi(t) = \omega' t + \varphi_0 , \qquad (2.56)$$

where ω' denotes the instantaneous frequency. In doing so, basically we are assuming that the phase perturbation due to coupling does not disturb the sinusoidal character of the limit cycle oscillations. In what follows we adapt the theory of weakly coupled phase oscillators as presented in [11, Ch. 7, 8]. However, the notable difference is that we are dealing with phase coupling of the autonomous SAW delay line oscillators by a SAW delay line.

In the following we shall use subscripts 1, 2 and c to designate quantities that pertain to oscillator 1, oscillator 2 and coupling delay line respectively. The voltage transfer functions $H_i(\omega_i, \tau_i), i = 1,\ 2,\ c$ for the SAW delay lines are defined by Eq. (2.23). The delay differential equation for the oscillators close to limit cycles in the uncoupled state is defined by Eq. (2.20) or (2.25). The coupling of their outputs through the SAW delay line allows a fraction of output from the oscillator 2 to be fed into the input of feedback SAW delay line in oscillator 1, and vice versa, as shown in Fig. 2.5. In view of the weak coupling assumption the effects of this signal flow on the amplitudes are assumed to be negligible. The primary effect is on the oscillation phases. The limit cycle of an autonomous dynamical system is invariant under time shift $t \rightarrow t - \Delta t$, which is equivalent to the limit cycle invariance under phase shifts $\Delta\phi = \omega\Delta t$ [11, Sec. 7.1.1]. Therefore, the phase shift of oscillator 2 due to signal flow from oscillator 1, and the phase shift of oscillator 1 due to signal flow from oscillator 2 do not perturb the limit cycles of the respective oscillators.

In presence of coupling, the feedback signal in the close loop of each oscillator has a component from the other oscillator that has been transferred by the coupling delay line. Thus, the voltages at the input transducer of feedback SAW delay lines are

$$v_{1,S}(t) = v_1(t) + jH_c(\omega_2, \tau_c)v_2(t) \tag{2.57a}$$

$$v_{2,S}(t) = v_2(t) + jH_c(\omega_1, \tau_c)v_1(t), \tag{2.57b}$$

where

$$v_1(t) = v_{01}e^{j(\omega_1 t + \varphi_{01})} \tag{2.57c}$$

$$v_2(t) = v_{02}e^{j(\omega_2 t + \varphi_{02})} \tag{2.57d}$$

$$jH_c(\omega_2, \tau_c) = |H_c(\omega_2)|e^{-j\omega_2\tau_c} \tag{2.57e}$$

$$jH_c(\omega_1, \tau_c) = |H_c(\omega_1)|e^{-j\omega_1\tau_c}. \tag{2.57f}$$

In this equation $v_1(t)$ and $v_2(t)$ respectively denote the outputs of the oscillator 1 and the oscillator 2, and $jH_c(\omega, \tau_c)$ is the voltage transfer function of the coupling delay line, see Eqs. (2.22) and (2.23). The rate of voltage change at the output of each oscillator as given by Eq. (2.20) or (2.25) in the uncoupled state will therefore be modified due to the presence of second terms on the r.h.s. of Eqs. (2.57a, b). Proceeding in the same way as before while deriving Eqs. (2.20) and (2.25) we obtain the modified delay differential equations as

$$\dot{v}_1(t) = j\omega_1 A_1(\omega_1)H_1(\omega_1, \tau_1)v_1(t) + j^2\omega_2 A_1(\omega_2)H_1(\omega_2, \tau_1)H_c(\omega_2, \tau_c)v_2(t) \tag{2.58a}$$

$$\dot{v}_2(t) = j\omega_2 A_2(\omega_2)H_2(\omega_2, \tau_2)v_2(t) + j^2\omega_1 A_2(\omega_1)H_2(\omega_1, \tau_2)H_c(\omega_1, \tau_c)v_1(t), \tag{2.58b}$$

where

$$jH_1(\omega_1,\tau_1)=\left|H_1(\omega_1)\right|e^{-j\omega_1\tau_1} \tag{2.58c}$$

$$jH_1(\omega_2,\tau_1)=\left|H_1(\omega_2)\right|e^{-j\omega_2\tau_1} \tag{2.58d}$$

$$jH_2(\omega_2,\tau_2)=\left|H_2(\omega_2)\right|e^{-j\omega_2\tau_2} \tag{2.58e}$$

$$jH_2(\omega_1,\tau_2)=\left|H_2(\omega_1)\right|e^{-j\omega_1\tau_2}\,. \tag{2.58f}$$

In order to determine the effect of coupling on phases of the two oscillators we denote their modified values by $\varphi_1(t)$ and $\varphi_2(t)$. The weak coupling ensures that the limit cycle amplitudes remain unperturbed, and the flow of phases from each other retains their harmonic characteristic. Thus, we write

$$v_1(t)=v_{01}e^{j\varphi_1(t)} \tag{2.59a}$$

$$v_2(t)=v_{02}e^{j\varphi_2(t)} \tag{2.59b}$$

which yields

$$\dot{v}_1(t)=j\dot{\varphi}_1(t)v_1(t) \tag{2.60a}$$

$$\dot{v}_2(t)=j\dot{\varphi}_2(t)v_2(t)\,. \tag{2.60b}$$

By combining Eqs. (2.58) and (2.60) we obtain:

$$\dot{\varphi}_1(t)=\mathrm{Re}\left[\omega_1A_1(\omega_1)H_1(\omega_1,\tau_1)+j\omega_2A_1(\omega_2)H_1(\omega_2,\tau_1)H_c(\omega_2,\tau_c)\frac{v_2(t)}{v_1(t)}\right] \tag{2.61a}$$

$$\dot{\varphi}_2(t)=\mathrm{Re}\left[\omega_2A_2(\omega_2)H_2(\omega_2,\tau_2)+j\omega_1A_2(\omega_1)H_2(\omega_1,\tau_2)H_c(\omega_1,\tau_c)\frac{v_1(t)}{v_2(t)}\right]. \tag{2.61b}$$

Note that $\dot{\varphi}_1(t)$ and $\dot{\varphi}_2(t)$ in Eqs. (2.61a) and (2.61b) are equal to the respective real parts on the r.h.s. Since,

$$H(\omega,\tau)=\left|H(\omega)\right|\exp(-j\omega\tau)\exp(j3\pi/2)=-j\left|H(\omega)\right|\exp(-j\omega\tau)$$

as defined through Eq. (2.23), we note that for the two oscillators in their limit cycle motion:

$$\mathrm{Re}[A(\omega)H(\omega)] = \mathrm{Im}[|A(\omega)||H(\omega)|\exp(-j\omega\tau)] = -|A(\omega)||H(\omega)|\sin(\omega\tau + \varphi_E) = 1 \tag{2.62a}$$

$$\mathrm{Im}[A(\omega)H(\omega)] = \mathrm{Re}[|A(\omega)||H(\omega)|\exp(-j\omega\tau)] = |A(\omega)||H(\omega)|\cos(\omega\tau + \varphi_E) = 0\,. \tag{2.62b}$$

In order to obtain the r.h.s. real parts in Eqs. (2.61 a, b) we rearrange, expand, use Eqs. (2.62 a, b) and proceed as follows.

Eq. (2.61a):

$$\begin{aligned}
\dot{\varphi}_1(t) &= \mathrm{Re}\left[\left(\omega_1 + j\omega_2 H_c(\omega_2,\tau_c)\frac{A_1(\omega_2)H_1(\omega_2,\tau_1)}{A_1(\omega_1)H_1(\omega_1,\tau_1)}\frac{v_2(t)}{v_1(t)}\right)A_1(\omega_1)H_1(\omega_1,\tau_1)\right] \\
&= \omega_1\,\mathrm{Re}[A_1(\omega_1)H_1(\omega_1,\tau_1)] + \omega_2\,\mathrm{Re}[A_1(\omega_1)H_1(\omega_1,\tau_1)]\mathrm{Re}\left[jH_c(\omega_2,\tau_c)\frac{A_1(\omega_2)H_1(\omega_2,\tau_1)}{A_1(\omega_1)H_1(\omega_1,\tau_1)}\frac{v_2(t)}{v_1(t)}\right] \\
&= \omega_1 + \omega_2\,\mathrm{Re}\left[j(-j)|H_c(\omega_2)|e^{-j\omega_2\tau_c}\frac{|A_1(\omega_2|e^{-j\varphi_{E2}}}{|A_1(\omega_1|e^{-j\varphi_{E1}}}\frac{(-j)|H_1(\omega_2)|e^{-j\omega_2\tau_1}}{(-j)|H_1(\omega_1)|e^{-j\omega_1\tau_1}}\frac{v_{02}e^{j\varphi_2(t)}}{v_{01}e^{j\varphi_1(t)}}\right] \\
&= \omega_1 + \omega_2|H_c(\omega_2)|\frac{|A_1(\omega_2|}{|A_1(\omega_1|}\frac{|H_1(\omega_2)|}{|H_1(\omega_1)|}\frac{v_{02}}{v_{01}}\,\mathrm{Re}\,e^{j[\{\varphi_2(t)-\omega_2(\tau_c+\tau_1)\}-\{\varphi_1(t)-\omega_1\tau_1\}-\{\varphi_{E2}-\varphi_{E1}\}]} \\
&= \omega_1 + K_1\,\mathrm{Re}\left[e^{j[\{\varphi_2(t)-\omega_2(\tau_c+\tau_1)\}-\{\varphi_1(t)-\omega_1\tau_1\}-\{\varphi_{E2}-\varphi_{E1}\}]}\right]
\end{aligned} \tag{2.63a}$$

with

$$K_1 = \omega_2|H_c(\omega_2)|\frac{|A_1(\omega_2|}{|A_1(\omega_1|}\frac{|H_1(\omega_2)|}{|H_1(\omega_1)|}\frac{v_{02}}{v_{01}} \tag{2.63b}$$

Similarly from Eq. (2.61b):

$$\dot{\varphi}_2(t) = \omega_2 + K_2\,\mathrm{Re}\left[e^{j[\{\varphi_1(t)-\omega_1(\tau_c+\tau_2)\}-\{\varphi_2(t)-\omega_2\tau_2\}-\{\varphi_{E1}-\varphi_{E2}\}]}\right] \tag{2.63c}$$

with

$$K_1 = \omega_1|H_c(\omega_1)|\frac{|A_2(\omega_1)|}{|A_2(\omega_2)|}\frac{|H_2(\omega_1)|}{|H_2(\omega_2)|}\frac{v_{01}}{v_{02}} \tag{2.63d}$$

These expressions can be greatly simplified if we take both oscillators to be nominally identical except for the feedback delays. That is, if we assume that the differences in their amplitudes, frequencies and SAW delay line amplitude responses are small enough so that we can make approximations: $v_{01} \approx v_{02}$, $|H_1(\omega_2)| \approx |H_1(\omega_1)|$, $|H_2(\omega_1)| \approx |H_2(\omega_2)|$, $|A_1(\omega_2)| \approx |A_2(\omega_1)|$ and $\varphi_{E1} \approx \varphi_{E2}$. Further we assume that the coupling SAW delay line has flat amplitude response over the frequency range of interest so that $|H_c(\omega_2)| \approx |H_c(\omega_1)| = H_c$ (say), hence, $K_1 = \omega_2 H_c$, $K_2 = \omega_1 H_c$. Note that these conditions can easily be fulfilled by proper designing of the SAW delay lines and circuitry. The delay times, however, we are keeping different because the phase dynamics is most sensitive to these delays, and we intend to explore these delay dependencies for various possible applications. In view of these simplifications we write the coupled phase equations as

$$\dot{\varphi}_1(t) = \omega_1 + K_1 \cos[(\varphi_2(t) - \omega_2(\tau_c + \tau_1)) - (\varphi_1(t) - \omega_1\tau_1)] \tag{2.64a}$$

$$\dot{\varphi}_2(t) = \omega_2 + K_2 \cos[(\varphi_1(t) - \omega_1(\tau_c + \tau_2)) - (\varphi_2(t) - \omega_2\tau_2)]. \tag{2.64b}$$

The factors K_1 and K_2 define the strength of coupling. In view of weak coupling, harmonicity of limit cycles and continuous phase flow we denote $\varphi_2(t) - \omega_2(\tau_1 + \tau_c) = \varphi_2(t - \tau_1 - \tau_c)$, $\varphi_1(t) - \omega_1(\tau_2 + \tau_c) = \varphi_1(t - \tau_2 - \tau_c)$, $\varphi_1(t) - \omega_1\tau_1 = \varphi_1(t - \tau_1)$ and $\varphi_2(t) - \omega_2\tau_2 = \varphi_2(t - \tau_2)$. A more elaborate justification is given in [47]. Thus, finally, we write the coupling phase equations for the two nearly identical SAW delay line oscillators coupled by a SAW delay line as

$$\dot{\varphi}_1(t) = \omega_1 + K_1 \cos[\varphi_2(t - \tau_c - \tau_1) - (\varphi_1(t - \tau_1))] \tag{2.65a}$$

$$\dot{\varphi}_2(t) = \omega_2 + K_2 \cos[\varphi_1(t - \tau_c - \tau_2) - (\varphi_2(t - \tau_2))] \tag{2.65b}$$

with

$$K_1 = \omega_2 H_c \tag{2.65c}$$

$$K_2 = \omega_1 H_c. \tag{2.65d}$$

Eqs. (2.65 a-d) define an asymmetric coupling because in general $K_1 \neq K_2$ and $\tau_1 \neq \tau_2$. However, note that $\omega_1 >> \omega_2 H_c$, $\omega_2 >> \omega_1 H_c$ because $H_c << 1$, and $\omega_1 \approx \omega_2$. Therefore, in order to simplify the analysis of coupled phase dynamics we can represent the coupling strength by some mean value $K = \omega_m H_c$ where mean frequency can be taken to either arithmetic mean $\omega_m = (\omega_1 + \omega_2)/2$ or geometric mean $\omega_m = \sqrt{\omega_1\omega_2}$. Therefore, for further analysis here, we rewrite the phase equations as

$$\dot{\varphi}_1(t) = \omega_1 + K \cos[\varphi_2(t - \tau_1 - \tau_c) - \varphi_1(t - \tau_1)] \tag{2.66a}$$

$$\dot{\varphi}_2(t) = \omega_2 + K \cos[\varphi_1(t - \tau_2 - \tau_c) - \varphi_2(t - \tau_2)], \tag{2.66b}$$

where

$$K = \omega_m H_c. \tag{2.66c}$$

It is to be noted that these equations are derived by considering the coupling between oscillators via their feedback paths. The first term in the argument of cosine tells that the phase of one oscillator is transferred to the other with delay equal to the sum of delays through the coupling delay line and the self-feedback time, while the second term represents the self-feedback. In this we did not include the direct phase transfer from the coupling delay line. However, if we desire to take that also into account it is obvious from the structure of these equations that an additional term would appear in cosine argument: $\varphi_2(t-\tau_c)$ in the equation for $\dot{\varphi}_1(t)$, and $\varphi_1(t-\tau_c)$ in the equation for $\dot{\varphi}_2(t)$. That is, the equations with complete phase transfer are

$$\dot{\varphi}_1(t) = \omega_1 + K\cos[\varphi_2(t-\tau_c) + \varphi_2(t-\tau_1-\tau_c) - \varphi_1(t-\tau_1)] \quad (2.67a)$$

$$\dot{\varphi}_2(t) = \omega_2 + K\cos[\varphi_1(t-\tau_c) + \varphi_1(t-\tau_2-\tau_c) - \varphi_2(t-\tau_2)], \quad (2.67b)$$

where

$$K = \omega_m H_c .$$

Similarly, it is easy to se that if the coupling is desired to be direct (that is, without the signal from the second oscillator going through the feedback path of the first) then $\varphi_2(t-\tau_1-\tau_c)$ term from Eq. (2.67 a) and $\varphi_1(t-\tau_2-\tau_c)$ term from Eq. (2.67 b) must be absent. The coupled phase equations in this become

$$\dot{\varphi}_1(t) = \omega_1 + K\cos[\varphi_2(t-\tau_c) - \varphi_1(t-\tau_1)] \quad (2.68a)$$

$$\dot{\varphi}_2(t) = \omega_2 + K\cos[\varphi_1(t-\tau_c) - \varphi_2(t-\tau_2)], \quad (2.68b)$$

where

$$K = \omega_m H_c .$$

Next, we analyze the phase dynamics of these coupling configurations to seek the existence of stable synchronization states, and to explore whether their dependencies on coupling parameters (H_c, τ_c) are appropriate for making better SAW sensors.

2.4. Synchronization

In order to analyze the synchronization behavior of the coupled system it is convenient to transform the coupled phase equations in terms of new variables defined as the sum and difference of the oscillator phases. That is, we define the new variables as

$$\psi_1(t) = \varphi_2(t) - \varphi_1(t) \quad (2.69a)$$

$$\psi_2(t) = \frac{\varphi_2(t) + \varphi_1(t)}{2} \tag{2.69b}$$

By taking time derivatives and making substitutions from the coupled phase equations (that is, by using Eqs. 66-68) we obtain

$$\dot{\psi}_1(t) = (\omega_2 - \omega_1) + K[\cos A - \cos B] \tag{2.70a}$$

$$\dot{\psi}_2(t) = \frac{\omega_2 + \omega_1}{2} + \frac{K}{2}[\cos A + \cos B] \tag{2.70b}$$

where

$$\left.\begin{array}{l} A = \varphi_1(t-\tau_c) + \varphi_1(t-\tau_2-\tau_c) - \varphi_2(t-\tau_2) \\ B = \varphi_2(t-\tau_c) + \varphi_2(t-\tau_1-\tau_c) - \varphi_1(t-\tau_1) \end{array}\right] \text{in fully coupled Eqs. (67a, b)}$$
$$\left.\begin{array}{l} A = \varphi_1(t-\tau_2-\tau_c) - \varphi_2(t-\tau_2) \\ B = \varphi_2(t-\tau_1-\tau_c) - \varphi_1(t-\tau_1) \end{array}\right] \text{in feedback coupled Eqs. (66a, b)} \tag{2.71}$$
$$\left.\begin{array}{l} A = \varphi_1(t-\tau_c) - \varphi_2(t-\tau_2) \\ B = \varphi_2(t-\tau_c) - \varphi_1(t-\tau_1) \end{array}\right] \text{in dierctly coupled Eqs. (68a, b)}$$

These equations can be further manipulated by using trigonometric identities $\cos A - \cos B = -2\sin\frac{A+B}{2}\sin\frac{A-B}{2}$ and $\cos A + \cos B = 2\cos\frac{A+B}{2}\cos\frac{A-B}{2}$, and by assuming that the oscillators maintain their harmonic character during adjustment of phases for synchronization. In view of the latter, we can express oscillator phases as $\varphi(t) = \omega' t + \varphi_0$ where ω' denotes instantaneous frequency and φ_0 initial phase.

Synchronization occurs when the phase difference between the two oscillators becomes time-invariant, that is, $\psi_1(t) = \varphi_2(t) - \varphi_1(t) = \text{const}$, $\dot{\psi}_1(t) = 0$. This is possible only if their phase variations adapt to a common time dependence maintaining some constant phase difference. The common mode frequency (denoted by Ω) is called the synchronization frequency. In synchronization state, both the oscillators pull their frequencies to the same level, $\dot{\varphi}_1(t) = \dot{\varphi}_2(t) = \Omega$. As the coupled system approaches towards synchronization the instantaneous frequency mismatch $\omega'_d \to 0$, mean instantaneous frequency $\omega'_m \to \Omega$. Note that in the synchronized state, $\dot{\psi}_2(t) = \Omega$, hence, $\psi_2(t) = \Omega t + \text{const}$.

2.4.1. Fully Coupled Configuration

Eqs. (2.67a, b) in this case transform to

$$\dot{\psi}_1(t) = (\omega_2 - \omega_1) + 2K\sin\left[\frac{3\psi_1(t)}{2} - (\omega_2' - \omega_1')(\tau_c + \frac{\tau_2+\tau_1}{2})\right]\sin\left[\frac{\psi_2(t)}{2} - (\omega_2' + \omega_1')\tau_c + \frac{(\omega_2' - \omega_1')(\tau_2 - \tau_1)}{2}\right] \quad (2.72a)$$

$$\dot{\psi}_2(t) = \frac{\omega_2 + \omega_1}{2} + K\cos\left[\frac{3\psi_1(t)}{2} - (\omega_2' - \omega_1')(\tau_c + \frac{\tau_2+\tau_1}{2})\right]\cos\left[\frac{\psi_2(t)}{2} - (\omega_2' + \omega_1')\tau_c + \frac{(\omega_2' - \omega_1')(\tau_2 - \tau_1)}{2}\right] \quad (2.72b)$$

Introducing symbols: $\omega_2 - \omega_1 = \omega_d$ for natural frequency mismatch (detuning), $\tau_2 - \tau_1 = \tau_d$ for self-feedback delay mismatch, $\frac{1}{2}(\omega_2 + \omega_1) = \omega_m$ for mean natural frequency, $\frac{1}{2}(\tau_2 + \tau_1) = \tau_m$ for mean self-feedback delay, $\omega_2' - \omega_1' = \omega_d'$ for instantaneous frequency difference, and $\frac{1}{2}(\omega_2' + \omega_1') = \omega_m'$ for instantaneous mean frequency we can rewrite Eqs. (2.72 a, b) as

$$\dot{\psi}_1(t) = \omega_d + 2K\sin\left[\frac{3\psi_1(t)}{2} - \omega_d'(\tau_c + \tau_m)\right]\sin\left[\frac{\psi_2(t)}{2} - 2\omega_m'\tau_c + \frac{\omega_d'\tau_d}{2}\right] \quad (2.73a)$$

$$\dot{\psi}_2(t) = \omega_m + K\cos\left[\frac{3\psi_1(t)}{2} - \omega_d'(\tau_c + \tau_m)\right]\cos\left[\frac{\psi_2(t)}{2} - 2\omega_m'\tau_c + \frac{\omega_d'\tau_d)}{2}\right]. \quad (2.73b)$$

Applying synchronization conditions $\dot{\psi}_1(t) = 0$, $\dot{\psi}_2(t) = \Omega$, $\omega_d' = 0$ and $\omega_m' = \Omega$ with Eqs. (2.72 a, b) we obtain the following relations where superscript (*) has been used to denote that they refer to synchronized state

$$\sin\frac{3\psi_1^*}{2}\sin\left(\frac{\psi_2^*(t)}{2} - 2\Omega\tau_c\right) = -\frac{\omega_d}{2K} \quad (2.74a)$$

$$\cos\frac{3\psi_1^*}{2}\cos\left(\frac{\psi_2^*(t)}{2} - 2\Omega\tau_c\right) = \frac{\Omega - \omega_m}{K}. \quad (2.74b)$$

These are the coupled equations between synchronization phase difference ψ_1^* and frequency Ω. Note that Ω is defined through time rate of variation of $\psi_2^*(t)$. We can eliminate $\psi_2^*(t)$ from these equations by using the identity $\cos x = \pm\sqrt{1 - \sin^2 x}$ and making substitution from Eq. (2.74 a) into (2.74 b). We obtain

$$\Omega = \omega_m \pm K\cos\frac{3\psi_1^*}{2}\sqrt{1-\frac{\omega_d^2}{4K^2\sin^2\frac{3\psi_1^*}{2}}} \tag{2.75}$$

This relation tells that the system allows synchronization for a given frequency detuning ω_d and coupling strength K only over a range of phase differences ψ_1^* because a real solution for Ω can exist only if the quantity within the square root is positive. That is, the synchronization can occur if

$$\psi_1^* \geq \frac{2}{3}\sin^{-1}\left(\frac{\omega_d}{2K}\right). \tag{2.76}$$

If the condition of equality is satisfied in this equation the synchronization occurs at the mean frequency ω_m. Otherwise, the synchronization frequencies are symmetrically positioned around the mean frequency ω_m. Eq. (2.75) however does not give complete solution. We need to know ψ_1^* to determine Ω, or vice versa. By eliminating ψ_1^* from Eqs. (2.74 a, b) we can obtain a relation between $\psi_2^*(t)$ and Ω. The result is

$$\Omega = \omega_m \pm K\cos\left(\frac{\psi_2^*(t)}{2} - 2\Omega\tau_c\right)\sqrt{1-\frac{\omega_d^2}{4K^2\sin^2\left(\frac{\psi_2^*(t)}{2} - 2\Omega\tau_c\right)}}. \tag{2.77}$$

A comparison of Eqs. (2.75) and (2.77) tells that $\frac{3}{2}\psi_1^* = \frac{1}{2}\psi_2^*(t) - 2\Omega\tau_c$ which by using $\psi_2^*(t) = \Omega t + \text{const}$ becomes

$$3\psi_1^* = \Omega(t - 4\tau_c) + \text{const}. \tag{2.78}$$

Eq. (2.78) obviously can not be satisfied because the l.h.s. is fixed-valued whereas the r.h.s. is a linear function of time. It appears therefore that a fully coupled configuration where coupling occurs through delayed direct signal as well as delayed feedback signal may not exhibit synchronization.

2.4.2. Feedback Coupled Configuration

Eqs. (2.66 a, b) in this case transform to

$$\dot{\psi}_1(t) = (\omega_2 - \omega_1) - 2K \sin\left[\psi_1(t) - \frac{1}{2}(\omega_2' - \omega_1')(\tau_2 + \tau_1 + \tau_c)\right] \sin\left[\frac{1}{2}(\omega_2' + \omega_1')\tau_c - \frac{1}{2}(\omega_2' - \omega_1')(\tau_2 - \tau_1)\right] \quad (2.79a)$$

$$\dot{\psi}_2(t) = \frac{\omega_2 + \omega_1}{2} + K \cos\left[\psi_1(t) - \frac{1}{2}(\omega_2' - \omega_1')(\tau_2 + \tau_1 + \tau_c)\right] \cos\left[\frac{1}{2}(\omega_2' + \omega_1')\tau_c - \frac{1}{2}(\omega_2' - \omega_1')(\tau_2 - \tau_1)\right] \quad (2.79b)$$

which can be rewritten as

$$\dot{\psi}_1(t) = \omega_d - 2K \sin(\omega_m' \tau_c - \tfrac{1}{2}\omega_d' \tau_d) \sin(\psi_1(t) - \tfrac{1}{2}\omega_d' \tau_s) \quad (2.80a)$$

$$\dot{\psi}_2(t) = \omega_m + K \cos(\omega_m' \tau_c - \tfrac{1}{2}\omega_d' \tau_d) \cos(\psi_1(t) - \tfrac{1}{2}\omega_d' \tau_s) \quad (2.80b)$$

where $\tau_s = \tau_1 + \tau_2 + \tau_c$. In synchronization state we get

$$\sin \psi_1^* = \frac{\omega_d}{2K \sin \Omega \tau_c} \quad (2.81a)$$

$$\cos \psi_1^* = \frac{\Omega - \omega_m}{K \cos \Omega \tau_c}. \quad (2.81b)$$

Eqs. (2.81 a, b) relates the synchronization frequency Ω and locked phase difference ψ_1^* in terms of frequency detuning, coupling delay time τ_c and coupling strength K. Note that, $K = \omega_m H_c$, in which H_c denotes the voltage amplitude transfer response of the coupling SAW delay line. From these equations ψ_1^* will be given by

$$\left.\begin{aligned} \psi_1^* &= \sin^{-1}\left(\frac{\omega_d}{2K \sin \Omega \tau_c}\right) && \text{for in-phase} \\ &= \pi - \sin^{-1}\left(\frac{\omega_d}{2K \sin \Omega \tau_c}\right) && \text{for anti-phase} \end{aligned}\right] \text{synchronized oscillations} \quad (2.82)$$

where the synchronization frequency is determined from the transcendental equation

$$\Omega = \omega_m \pm K \cos \Omega \tau_c \sqrt{1 - \frac{\omega_d^2}{4K^2 \sin^2 \Omega \tau_c}} \quad (2.83)$$

under the condition that

$$\frac{\omega_d^2}{4K^2 \sin^2 \Omega\tau_c} < 1.$$

It is interesting to note that the similar synchronization results were obtained by Schuster and Wagner [51] where two limit cycle oscillators interacted after a fix time delay, and by Popovych, Krachkovskyi and Tass [52] where interaction between two nonlinear oscillators were defined by a combination of self-delayed plus direct output from the other oscillator. Fig. 2.6 shows schematically the coupling configurations in these studies and in the present work. The common point to be noted from these analyses is that the synchronization behavior depends on the difference in delay time between the self-feedback and the cross-feed irrespective where it is located.

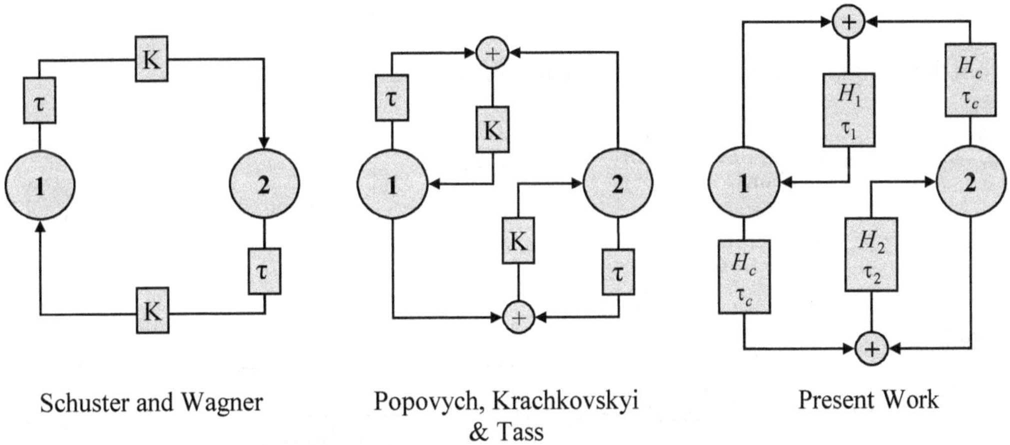

Fig. 2.6. Schematic representation of coupling configurations of two limit cycle oscillators studied as indicated. The circles 1 and 2 denote two oscillators. K coupling strength, τ delay time, and (H_1, H_2, H_c) amplitude transfer responses and (τ_1, τ_2, τ_c) delay times of SAW delay lines.

2.4.3. Direct Coupled Configuration

Eqs. (2.68 a, b) in this case transform to

$$\dot{\psi}_1(t) = (\omega_2 - \omega_1) - 2K \sin\left[\psi_1(t) - \frac{1}{2}(\omega_2' - \omega_1')\tau_c - \frac{1}{2}(\omega_2'\tau_2 - \omega_1'\tau_1)\right] \sin\left[\frac{1}{2}(\omega_2' + \omega_1')\tau_c - \frac{1}{2}(\omega_2'\tau_2 + \omega_1'\tau_1)\right] \quad (2.84a)$$

$$\dot{\psi}_2(t) = \frac{\omega_2 + \omega_1}{2} + K \cos\left[\psi_1(t) - \frac{1}{2}(\omega_2' - \omega_1')\tau_c - \frac{1}{2}(\omega_2'\tau_2 - \omega_1'\tau_1)\right] \cos\left[\frac{1}{2}(\omega_2' + \omega_1')\tau_c - \frac{1}{2}(\omega_2'\tau_2 + \omega_1'\tau_1)\right] \quad (2.84b)$$

which can be rewritten as

$$\dot{\psi}_1(t) = \omega_d - 2K \sin\left[\psi_1(t) - \frac{1}{2}\omega_d'\tau_c - \frac{1}{2}(\omega_2'\tau_2 - \omega_1'\tau_1)\right] \sin\left[\omega_m'\tau_c - \frac{1}{2}(\omega_2'\tau_2 + \omega_1'\tau_1)\right] \quad (2.85a)$$

$$\dot{\psi}_2(t) = \omega_m + K\cos\left[\psi_1(t) - \frac{1}{2}\omega'_d\tau_c - \frac{1}{2}(\omega'_2\tau_2 - \omega'_1\tau_1)\right]\cos\left[\frac{1}{2}\omega'_m\tau_c - \frac{1}{2}(\omega'_2\tau_2 + \omega'_1\tau_1)\right]. \quad (2.85b)$$

In synchronization state we obtain

$$\sin\left(\psi_1^* - \frac{1}{2}\Omega\tau_d\right) = \frac{\omega_d}{2K\sin\Omega(\tau_c - \tau_m)} \quad (2.86a)$$

$$\cos\left(\psi_1^* - \frac{1}{2}\Omega\tau_d\right) = \frac{\Omega - \omega_m}{K\sin\Omega(\tau_c - \tau_m)} \quad (2.86b)$$

where, $\tau_d = \tau_2 - \tau_1$ and $\tau_m = \dfrac{\tau_2 + \tau_1}{2}$. (2.86c)

From these equations we obtain the synchronization phase and frequency as

$$\left.\begin{aligned} \psi_1^* &= \frac{1}{2}\Omega\tau_d + \sin^{-1}\left(\frac{\omega_d}{2K\sin\Omega(\tau_c - \tau_m)}\right) && \text{for in - phase} \\ &= \pi - \left[\frac{1}{2}\Omega\tau_d + \sin^{-1}\left(\frac{\omega_d}{2K\sin\Omega(\tau_c - \tau_m)}\right)\right] && \text{for anti - phase} \end{aligned}\right] \text{synchronized oscillations} \quad (2.87)$$

and

$$\Omega = \omega_m \pm K\cos\Omega(\tau_c - \tau_m)\sqrt{1 - \frac{\omega_d^2}{4K^2\sin^2\Omega(\tau_c - \tau_m)}} \quad (2.88)$$

under the condition that $\dfrac{\omega_d^2}{4K^2\sin^2\Omega(\tau_c - \tau_m)} < 1$.

2.4.4. Synchronization Examples

In this section we present simulation results for three systems with feedback coupled configuration whose synchronization behavior is described by Eqs. (2.82) and (2.83). These systems were defined to have the same mean frequency ω_m, but differed in coupling delay time by an order of magnitude from each other extending from low to high value. The synchronization frequencies were determined by solving the transcendental Eq. (2.83) and the phase difference in synchronized state were obtained by using Eq. (2.82). In order that the systems achieve synchronized condition the coupling strengths were varied and correspondingly detuning frequencies were adjusted. The results are shown in Figs. 2.7-2.9.

The results in Fig. 2.7 (a) show that system goes into synchronized state first for $K = 6$ with synchronization frequency close to the higher frequency oscillator. As the coupling strength increases the synchronization frequency decreases towards the mean frequency. However, at $K \approx 10$ the system bifurcates into two additional states which split into two synchronization branches for higher K. It can be seen that the first synchronization mode merges with one these branches. From Fig. 2.7(b) it is worth noting that the first synchronization mode is locked with phase difference of $\pi/2$, \and it remains in that condition even though the synchronization frequency is changing. It is also noteworthy that the amplitude transfer response where the first time synchronization occurs is $H_c = 0.005$ which corresponds to 46 dB loss delay line, where the bifurcation occurs is $H_c = 0.008$ which corresponds to 42 dB loss.

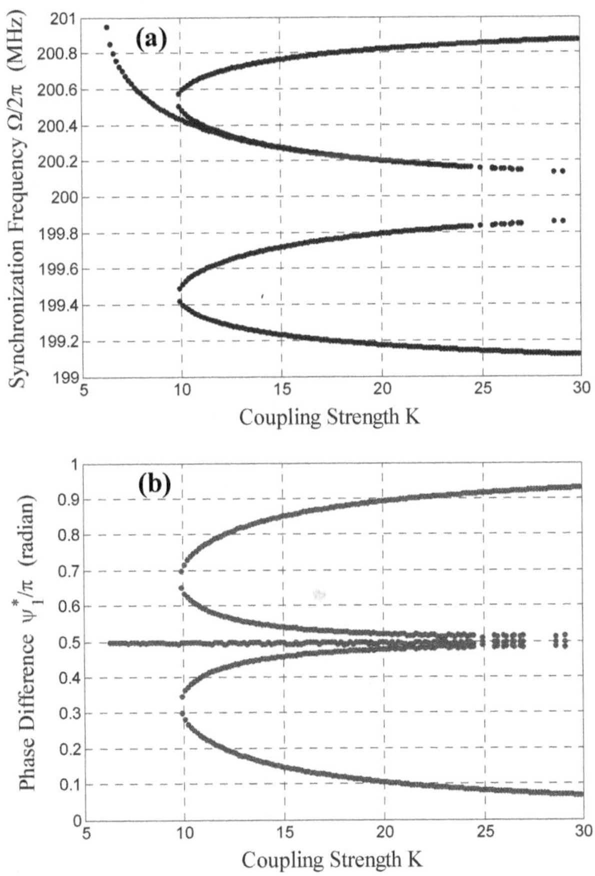

Fig. 2.7. (a) Synchronization frequency and (b) phase difference of SAW delay line coupled two SAW delay line oscillators having natural frequencies $f_1 = \omega_1 / 2\pi = 199\,MHz$ and $f_2 = \omega_2 / 2\pi = 201\,MHz$. The coupling delay time $\tau_c = 0.25\,\mu sec$. The relation $K = \omega_m H_c$ yields $H_c = 0.005$ for $K = 6$, $H_c = 0.008$ for $K = 10$, and $H_c = 0.024$ for $K = 30$.

The results in Fig. 2.8 correspond to 10 times lower coupling delay time ($\tau_c = 0.025\,\mu\text{sec}$) than that corresponding to Fig. 2.7. The first time synchronization occurs at $K = 40$ which correspond to $H_c = 0.03$ or 30 dB loss. The bifurcation point almost coincides at nearly the same K value. The first synchronization mode occurs with $\pi/2$ phase difference as before. Fig. 2.9 shows the results for higher coupling delay time ($\tau_c = 2.5\,\mu\text{sec}$). Note that in this case the first time synchronization begins to occur at $K = 3.2$ corresponding to $H_c = 0.0025$ or 52 dB insertion loss delay line. That is, we note that for longer coupling delays the synchronization begins to occur at weaker coupling strengths. The first synchronization mode in this case also occurs with $\pi/2$ phase difference. Another point to note is that as the coupling delay increases the number of synchronization modes also increase.

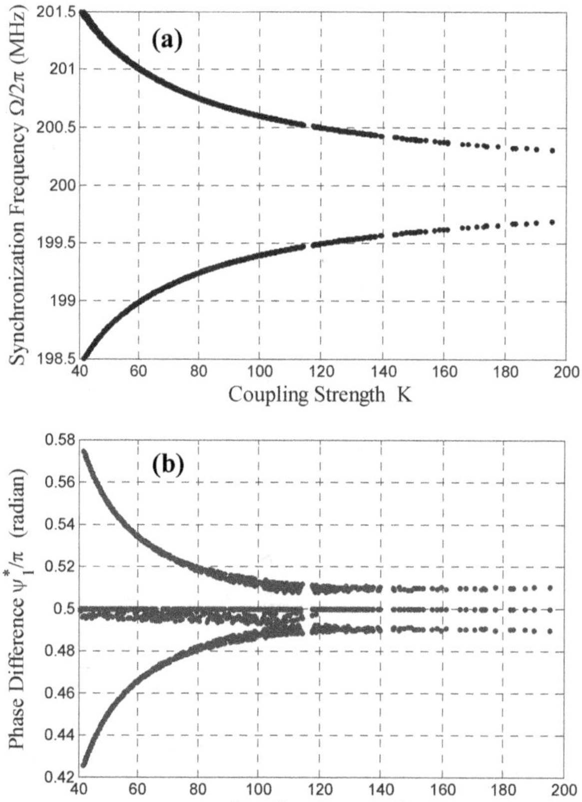

Fig. 2.8. (a) Synchronization frequency and (b) phase difference of SAW delay line coupled two SAW delay line oscillators having natural frequencies $f_1 = \omega_1 / 2\pi = 198.5\ MHz$, $f_2 = \omega_2 / 2\pi = 201.5\ MHz$, and coupling delay time $\tau_c = 0.025\ \mu sec$. $H_c = 0.03$ for $K = 40$, $H_c = 0.08$ for $K = 100$.

2.5. Basis for Synchronization Mode Sensing

Measurement of synchronization frequency of coupled oscillator system can be utilized for making physical or chemical sensors by exploiting parametric dependencies of synchronization. In the above analysis of dual SAW oscillator configuration coupled by SAW delay line we found that the synchronization frequency is primarily determined by the coupling SAW device characteristics (delay time τ_c and amplitude transfer function H_c), see Eqs. (2.83) and (2.88). The parameters of SAW delay line oscillators also appear through the difference and mean of their natural frequencies and feedback delays (ω_d, τ_d, ω_m, τ_m). The influence of τ_c is however most significant as it appears through arguments of sine and cosine functions. In the following we analyze the dependence of Ω on τ_c to explore the possibility of making the coupling SAW delay line as sensing platform.

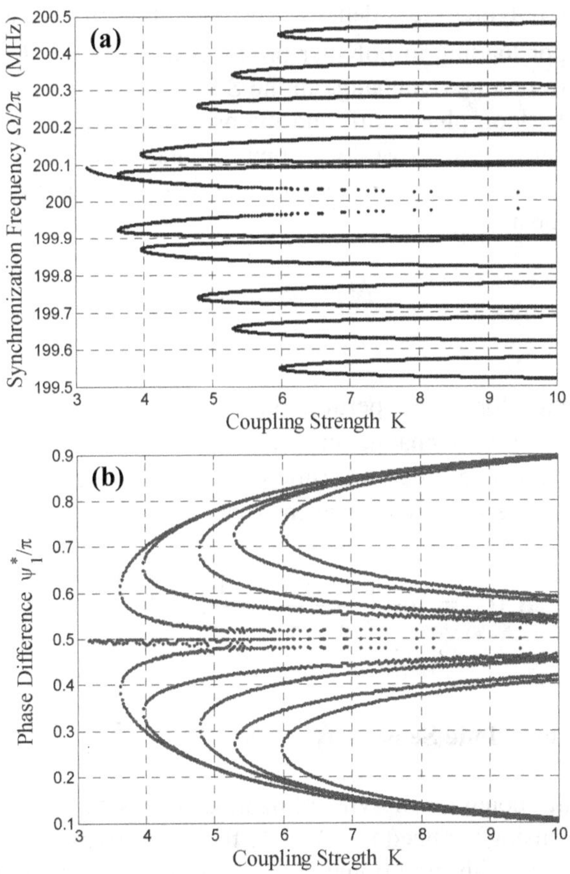

Fig. 2.9. (a) Synchronization frequency and (b) phase difference of SAW delay line coupled two SAW delay line oscillators having natural frequencies $f_1 = \omega_1 / 2\pi = 199.5\ MHz$, $f_2 = \omega_2 / 2\pi = 200.5\ MHz$, and coupling delay time $\tau_c = 2.5\ \mu sec$; $H_c = 0.0027$ for $K = 3.2$ where synchronization occurs first.

SAW chemical sensors in dual oscillator configuration utilize most commonly the perturbation in feedback delay time for frequency change, where the shift in frequency referred to the reference oscillator is read out as sensor signal [20]. For making a chemical sensor the surface acoustic wave propagation path in one feedback SAW delay line is sensitized by depositing a thin polymer coating that selectively sorbs chemical analytes, consequently, perturbs SAW propagation delay. The SAW delay line in the reference oscillator is not sensitized. It helps frequency down conversion for easy and accurate frequency measurement and also in cancellation of spurious influences. The frequency change of oscillator due to chemical mass loading is described the well known Sauerbrey`s relation [53, 20, Ch.5]

$$\frac{\Delta\omega}{\omega} = -\kappa\omega h\Delta\rho\,, \tag{2.89}$$

where κ is the constant depending on the SAW substrate material and propagation mode, ω denotes the SAW oscillator frequency, h denotes the polymer film thickness and $\Delta\rho$ denotes the polymer mass density. The quantity $h\Delta\rho$ is the areal mass density of the film. The vapor loading in polymer produces a change in the areal surface mass density which in turn produces a change in the SAW velocity. As a result, the feedback delay time, hence the loop phase shift and the oscillation frequency are changed. This change occurs because the close-loop phase condition for stable self-sustained oscillations. This condition also relates the change in oscillator frequency to the change in feedback delay time as

$$\frac{\Delta\omega}{\omega} = \frac{\Delta\tau}{\tau}. \tag{2.90}$$

The sensor sensitivity for delay perturbation is measured by the change in SAW oscillator frequency for unit change in delay time, that is, by $\Delta\omega/\Delta\tau$. Usually, a normalized frequency change measure (normalized with respect to unperturbed SAW oscillator frequency) is used for defining sensitivity as

$$S_\tau(\omega) = \lim_{\Delta\tau\to 0}\left(\frac{1}{\omega}.\frac{\Delta\omega}{\Delta\tau}\right) = \frac{1}{\omega}.\frac{\partial\omega}{\partial\tau} = \frac{1}{\tau}. \tag{2.91}$$

2.5.1. Synchronization Mode Sensitivity

In order to assess the potential of synchronization modes in coupled SAW oscillators system for analyte sensing we need to calculate their sensitivities for delay perturbation, and make comparison with traditional SAW oscillator mode sensing. An analytic calculation of synchronization mode sensitivity is difficult because of transcendental nature of relations, Eqs. (2.83) or (2.88). By differentiating Eq. (2.83) with respect to τ_c (keeping in mind the implicit dependency of Ω on τ_c) we obtain

$$S_{\tau_c}(\Omega) = \frac{1}{\Omega}.\frac{\partial\Omega}{\partial\tau_c} = \frac{Z}{1 - Z\tau_c} \tag{2.92a}$$

where

$$Z = \mp K \sin\Omega\tau_c \left[1 - \frac{\omega_d^2}{4K^2 \sin^2\Omega\tau_c} \frac{(\Omega - \omega_m)}{K} \cot^2\Omega\tau_c\right]. \tag{2.92b}$$

The -ve sign in Eq. (2.92 b) corresponds to the +ve sign in Eq. (2.83), and vice versa. One can see that Z is the controlling factor in determining synchronization mode sensitivity. In the limit, $Z\tau_c >> 1$, $S_{\tau_c}(\Omega) \to 1/\tau_c$ which is identical to the delay line SAW oscillator sensitivity as defined by Eq. (2.91). In view of this we define a sensitivity factor

$$\beta = \frac{Z\tau_c}{1 - Z\tau_c} \tag{2.93}$$

such that

$$S_{\tau_c}(\Omega) = \beta S_{\tau_c}(\omega) \tag{2.94}$$

where $S_{\tau_c}(\omega) = 1/\tau_c$ represents the sensitivity of a SAW feedback oscillator had it used the coupling SAW delay line for its frequency control. The factor β serves as a measure for sensitivity comparison between synchronization mode and oscillator mode sensing. The condition $\beta > 1$ signifies superior synchronization mode sensitivity which will occur only if $|Z|\tau_c < 1$. Eq. (2.93) further predicts that if $|Z|\tau_c \to 0$, $\beta \to 0$, and if $|Z|\tau_c \to 1$, $\beta \to \infty$. That is, at least in mathematical terms, there exists possibility for achieving any level of sensitivity from 0 to ∞ by adjusting $|Z|\tau_c$ over [0, 1]. It should however be remembered that for the existence of synchronization mode the constraint $\omega_d^2 / 4K^2 \sin^2\Omega\tau_c < 1$ must be satisfied, see Eqs. (2.83) and (2.88). Let us examine the expression

$$|Z|\tau_c = K\tau_c \sin\Omega\tau_c \left[1 - \frac{\omega_d^2}{4K^2 \sin^2\Omega\tau_c} \frac{(\Omega - \omega_m)}{K} \cot^2\Omega\tau_c\right]. \tag{2.95}$$

The condition $\omega_d^2 / 4K^2 \sin^2\Omega\tau_c < 1$ means

$$\sin^{-1}\left(\frac{\omega_d}{2K}\right) < \Omega\tau_c < \pi - \sin^{-1}\left(\frac{\omega_d}{2K}\right). \tag{2.96}$$

Note that: $K = \omega_m H_c$, $\omega_d << \omega_m$ by several orders of magnitude, and $H_c < 1$. Typically, for 100 MHz dual SAW oscillator $\omega_d / \omega_m \sim$ kHz/MHz $\sim 10^{-3}$, and H_c may lie over 0.25-0.01 corresponding to SAW delay lines with 6-40 dB loss. That means, most practical SAW cases correspond to $\omega_d / 2K < 1$. The permissible range of $\Omega\tau_c$ as defined by Eq. (2.96), therefore, lies symmetrically around $\pi/2$. In the middle at $\Omega\tau_c = \pi/2$, note that from Eq. (2.83) $\Omega = \omega_m$, and from Eq. (2.95) $|Z|\tau_c = K\tau_c = \omega_m \tau_c H_c$. This is the maximum value of $|Z|\tau_c$ in the permissible range of $\Omega\tau_c$. That means the maximum value of sensitivity factor corresponds to $\Omega\tau_c = \pi/2$, and is given as

$$\beta_{\max} = \frac{\omega_m \tau_c H_c}{1 - \omega_m \tau_c H_c}. \tag{2.97}$$

Note that at the limits of permissible $\Omega\tau_c$ range, again we have $\Omega = \omega_m$ which results in $|Z|\tau_c = \frac{1}{2}\omega_d \tau_c$. In this condition the sensitivity factor is minimum, and is obtained as

$$\beta_{\min} = \frac{\omega_d \tau_c}{1 - \frac{1}{2}\omega_d \tau_c}. \tag{2.98}$$

Eqs. (2.97) and (2.98) tell that the sensitivity is maximum when $\Omega\tau_c = \pi/2$ and its decreases continuously as one moves away from this condition. The parameters that determine this condition are ω_d, ω_m, τ_c, and H_c which can be optimum to achieve maximum sensitivity.

Near $\Omega\tau_c = \pi/2$, we can simplify the expression for $|Z|\tau_c$ by approximating $\cot \Omega\tau_c \approx \pi/2 - \Omega\tau_c$, and obtain a general expression for the sensitivity as

$$\beta = \frac{\Omega \tau_c H_c}{1 - \Omega \tau_c H_c}. \tag{2.99}$$

It can be noted that similar results will be obtained for the directly coupled case described by Eq. (2.88). In this case however $(\tau_c - \tau_m)$ instead of τ_c appears. All the above results are therefore valid for this case as well by substituting $(\tau_c - \tau_m)$ in place of τ_c.

2.5.2. Simulation Examples

In Figs. 2.10-2.13 are shown some numerical results for the synchronization frequency Ω and sensitivity factor β and the phase difference ψ_1^* calculated directly by using Eq. (2.92). Three sets of system parameters are selected to represent cases for low and high sensitivity systems. For calculations of sensitivity factor β the synchronization frequencies Ω were obtained first by solving the transcendental Eq. (2.83) for a given value of τ_c, and then variations in τ_c were introduced for calculating sensitivity. The ranges of τ_c perturbations were taken small for representing typical perturbation situations in SAW chemical sensing.

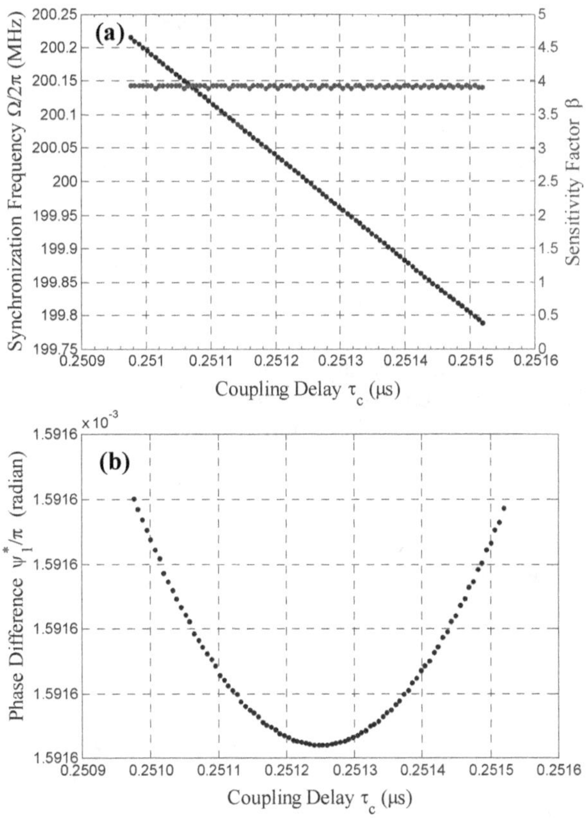

Fig. 2.10. (a) Synchronization frequency and sensitivity factor for coupling SAW delay line as sensor, (b) phase difference between two SAW delay line oscillators at synchronization. The system parameters are: $f_1 = \omega_1 / 2\pi = 199.75$ MHz, $f_2 = \omega_2 / 2\pi = 200.25$ MHz and $H_c = 0.20$.

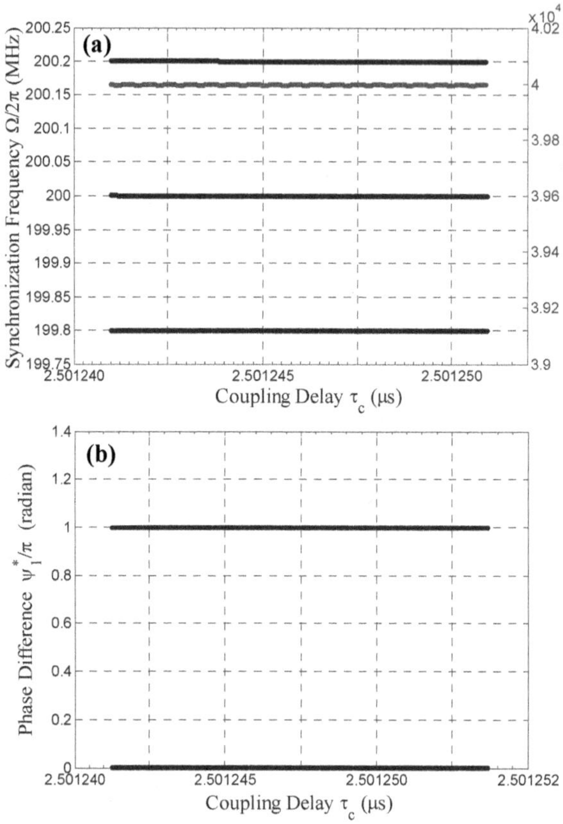

Fig. 2.11. (a) Synchronization frequency and sensitivity factor for coupling SAW delay line as sensor, (b) phase difference between two SAW delay line oscillators at synchronization. The system parameters are: $f_1 = \omega_1 / 2\pi = 199.78$ MHz, $f_2 = \omega_2 / 2\pi = 200.22$ MHz, $H_c = 0.22$.

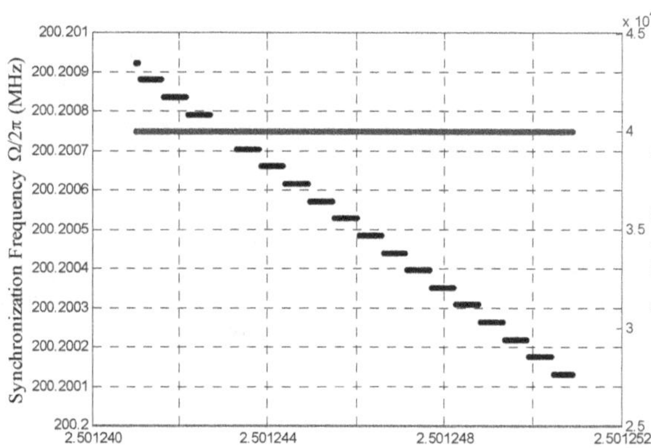

Fig. 2.12. Synchronization frequency and sensitivity for one mode in Fig. 2.11 on magnified scale.

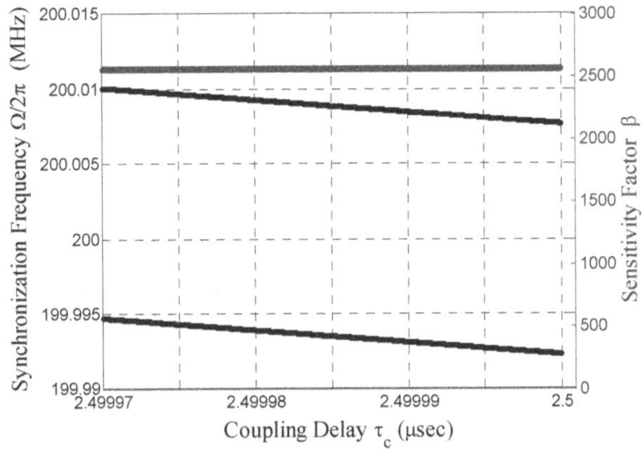

Fig. 2.13. Synchronization frequency and sensitivity factor for coupling SAW delay line as sensor. system parameters are: $f_1 = \omega_1 / 2\pi = 199.90$ MHz, $f_2 = \omega_2 / 2\pi = 200.02$ MHz, $H_c = 0.0025$.

The combinations of parameters for these cases are defined as follows. **Case 1** (results shown in Fig. 2.10): $f_1 = \omega_1 / 2\pi = 199.75$ MHz, $f_2 = \omega_2 / 2\pi = 200.25$ MHz, $H_c = 0.20$ and $\tau_c = [0.2509, 0.2519]$ μsec. The associated values for $K = 251.3$ and $\Delta\tau_c / \tau_c \approx 3$ parts per thousand. The system was found to synchronize in a single mode with very little phase adjustment, see Fig. 2.10 (a, b). The sensitivity enhancement occurs by the factor $\beta \approx 4$. **Case 2** (results shown in Fig. 2.11): $f_1 = \omega_1 / 2\pi = 199.78$ MHz, $f_2 = \omega_2 / 2\pi = 200.22$ MHz, $H_c = 0.22$, $\tau_c = [2.5012, 2.5013]$ μsec. The associated values for $K = 276.4$ and $\Delta\tau_c / \tau_c \approx 4$ parts per million. Because of large τ_c in this case, at least three modes of synchronization consisted of one mode at the mean frequency, and two modes closer to the natural frequencies of oscillators with phase difference of π are observed. In Fig. 2.11 (a) the synchronization frequency does not seem to vary with changes in τ_c. This appearance is however only due the scale of variation used in the figure. In Fig. 2.12 is shown one isolated mode from the Fig. 2.11(a) at the magnified scale where almost linear variation of Ω and constant sensitivity ($\beta \approx 40000$) with regard to changes in coupling delay time are evident. **Case 3** (results shown in Fig. 2.13): $f_1 = \omega_1 / 2\pi = 199.90$ MHz, $f_2 = \omega_2 / 2\pi = 200.02$ MHz, $H_c = 0.0025$, $\tau_c = [(2.5 - 3 \times 10^{-5}), 2.5]$ μsec represents the weakest coupling of the three. The associated values for $K = 3.14$ and $\Delta\tau_c / \tau_c \approx 12$ parts per million. In this case only two synchronization branches are seen. The sensitivity of both the branches are nearly the same having $\beta \approx 2500$.

We noted from the results in Figs. 2.7 through 2.9 that the number of synchronization modes depend on the frequency detuning $\omega_d = (\omega_2 - \omega_1)$ and coupling delay τ_c. For small values for ω_d and τ_c the number of synchronization modes can be restricted to a small number, but this requires strong coupling (high H_c). That is, the onset of synchronization occurs at higher K, see Fig. 2.8 in comparison to Figs. 2.7 and 2.9. By confining to the beginning of these synchronization curves, even single mode synchronization can be achieved. The results in Fig. 2.10 correspond to single mode condition. The sensitivity enhancement is however not so good compared to the three mode condition in Fig. 2.11. It is noticeable that the sensitivity of these modes are nearly identical, Fig. 2.12.

The results in Fig. 2.13 correspond to the coupling delay and amplitude transfer that can be easily implemented in SAW delay lines. The sensitivity enhancement factor is also quite high. The simultaneous presence of two modes is however problematic from the point of view of making a practical system. Besides, the frequency detuning and delay perturbation range are small. That means one needs to handle oscillator stability and interferences more stringently in practical system. It should be noted that the results and arguments presented here are no prescription for practical system design. These only highlight the possibilities for further research and practical exploitation.

2.6. Noise Reduction in Synchronization Mode Sensing

The random frequency fluctuation at the output of an oscillator sensor defines the ultimate limit of readable signal that sensor can produce. In other words, the noise defines the minimum detection limit of the quantity being measured. The frequency fluctuation in a SAW oscillator is usually dominated by loop phase fluctuations. The latter are invariably dominated by contributions from SAW delay fluctuations. The standard deviation of fluctuation is often taken as the measure of uncertainty in the sensor output which depends on the bandwidth of measuring system (or equivalently the measurement time). In order to specify the noise figure of merit of the sensor, that is independent of the measurement condition, usually a normalized measure of the noise power defined per unit bandwidth is used. The most commonly used measure for this purpose is the noise power spectral density specified in $(\mathrm{Hz})^2/\mathrm{Hz}$ for the frequency fluctuation or in $(\mathrm{radian})^2/\mathrm{Hz}$ for the phase fluctuation as a function of position from the carrier frequency (called the offset frequency f). The spectral power density is simply the mean square frequency fluctuation in 1-Hz bandwidth at the offset frequency. The spectral power density for frequency fluctuation $S_F(f)$ and for phase fluctuation $S_\Phi(f)$ are related as [54]

$$S_\Phi(f) = S_F(f)/f^2. \tag{2.100}$$

It is known that the phase locking or frequency synchronization reduces noise in oscillators of both electronic and non-electronic types [55-58]. It has been demonstrated both theoretically and experimentally that when N identical oscillators interact on reciprocal basis the phase noise power spectral density in the synchronized state reduces to $1/N$ of individual oscillators [56, 57]. In a detailed analysis of coupled neuronal (relaxation) oscillators it has been shown recently that the synchronization reduces the internal noise and makes the system more robust against external perturbations [58]. In [11, Sec. 9.2.4] the effect of synchronization on phase noise in two weakly coupled nonlinear limit cycle oscillators has been described. The phase fluctuation of a self sustained oscillator is described as a random walk or diffusion process. The noise power spectral density for a Gaussian frequency fluctuation has been determined to be equal to the diffusion constant $D = 2\sigma^2$ where σ denotes the standard deviation of the frequency random walk. Mutual interaction of the phase oscillators leading to synchronization allows randomness in their phases also to flow as easily through the coupling network and loop as their unperturbed phases. The noise in the coupled state is calculated by defining a new variable that represents the sum of phases of the individual oscillators, $\vartheta = \varepsilon(\varphi_1 + \varphi_2)$. The diffusion coefficient for ϑ is found to be $D_\vartheta = 2\varepsilon^2(\sigma_1^2 + \sigma_2^2)$ where $D_1^0 = 2\sigma_1^2$ and $D_2^0 = 2\sigma_2^2$ are the phase diffusion coefficients of the noninteracting oscillators. The diffusion constant of the individual oscillators in coupled state is obtained as

$$D_1 = D_2 = D = \frac{\sigma_1^2 + \sigma_2^2}{2}. \tag{2.101}$$

If both the oscillators in their free running states have the same amount of noise power spectral density, that is, if $\sigma_1 = \sigma_2 = \sigma$, then from Eq. (2.100) one can see that in the coupled state $D = \sigma^2$. This is ½ the noise power spectral density of the individual oscillators in their free running states. This result is the same as a special case of the above mentioned result on noise reduction by $1/N$ for *N*-coupled oscillators in synchronized state. If the oscillators are not identical the noise in coupled state is always lower than the noisiest oscillator in the free running state. This suggests that the synchronization mode sensors have the potential not only for enhanced sensitivity but also for reduced noise. That is, one can expect that a suitably designed synchronization mode sensor may exhibit much enhanced signal-to-noise ratio.

2.7. Discussion

We took up this study with the purpose of taking advantage of mathematical studies on delay differential equations for designing novel sensing platforms. The conversion of a mathematical model into a practically realizable physical system needs visualization of the system architecture, basic understanding of the system subparts and accurate mathematical description of these parts such as transducer, sensor, and circuit etc. A nonlinear SAW feedback oscillator was provides a mature platform for visualization and implementation of nonlinear dynamical models because individual oscillators are stable autonomous nonlinear systems which can be coupled in various ways to define a higher level dynamical system. The target applications may range from phase locking, clock synchronization, noise suppression, noise generation, multiple frequency generation, chaotic code generation, and wired or wireless sensors etc. Also, a SAW system because of inherent delay (hence memory) and adjustable nonlinear characteristics is suitable for modelling artificial intelligence systems mimicking human brain, swarm intelligence, neuronal firing etc. It looks attractive to draw from the advances in SAW physics and technology and the mathematics of nonlinear delay differential equations to visualize new systems with enhanced performance or novel applications.

This work particularly focused on SAW delay line controlled self-feedback oscillator system. This system makes clean representation of the autonomous delay differential dynamical system. Besides, the SAW science and technology is fairly advanced. The polymer coated SAW oscillator sensors are among some most advanced chemical sensor technologies today. Eq. (2.20) obtained here by using the SAW delay line physics and transfer response model and the delayed self-feedback mechanism through a nonlinear amplifier defines the basic delay differential equation for SAW delay line oscillators. The system attains stable amplitude self-sustained limit cycle (harmonic) oscillations under conditions defined by Eqs. (2.45). A SAW delay line coupling between two such oscillators defines a delay coupled delay differential system whose dynamics is defined by Eqs. (2.58). This system, under the assumption of weak coupling, is amplitude stable and its coupled phase dynamics is defined by phase flows via the coupling SAW delay line. The coupled phase equations are obtained as Eqs. (2.66). These equations are similar to those used earlier in mathematical models such as in [51, 52] with one difference that in our case it is cosine function in the coupling term instead of sine function. This difference does no alter the basic dynamical character of the system but it affects one synchronization branch at the mean frequency as discussed below.

Applying the condition of phase locking we obtain the phase difference relation given by Eq. (2.82) and the synchronization frequency relation given by Eq. (2.83). These relations are expressed in terms of the SAW device parameters which are amenable to design and manipulation for desired system characteristics. The coupling strength is determined by the mean frequency of the oscillators and the amplitude transfer response of the coupling SAW delay line, $K = \omega_m H_c$. The synchronization frequency is determined by K in combination with two additional parameters, namely, frequency detuning ω_d and coupling delay τ_c. The number synchronization branches depend on the combination of these parameters. For arbitrarily large values of all these parameters a large number of branches appear. Such a combination is not suitable for sensor applications where we need a well defined stable mode having sensitivity for perturbation of ω_d or τ_c. The number of synchronization modes can be limited to a few, or to one, for small values of K, ω_d and τ_c as illustrated by Figs. 2.7 through 2.9

and by Figs. 2.10 through 2.13. In latter figures we discussed only the effect of perturbations of coupling delay. Similar conclusions can be for perturbations of ω_d. In that case, the sensor will operate similar to the traditional dual oscillator SAW sensor; however, readout will be through synchronization frequency rather than the difference frequency. The sensitivities as defined by Eqs. (2.92) through (2.94) are almost independent of the synchronization mode at least for combinations shown in Figs. 2.10-2.13. That means, one needs only to isolate one branch for making a sensor. This, perhaps, can be achieved by applying either a frequency selective filter or by working out proper combination of parameters that generate only one synchronization branch, for example, as in Fig. 2.7 for $K = [6, 10]$ and in Fig. 2.9 for $K = [3.2, 3.7]$. The results shown here demonstrate that a trade-off is needed between the achievable sensitivity and limiting the synchronization branch. For practical implementations, it may be necessary to seek alternate solutions in the form of inserting, for example, an adjustable phase shifters in feedback and coupling paths or phase difference read out.

Even though one settles for not too high sensitivity the synchronization mode sensing seems to offer the major advantage of noise suppression. A compromise between sensitivity and noise can still facilitate high performance sensor. The synchronization mode sensing has the potential of integrating the features of stochastic resonance for improving the signal-to-noise as the stochastic resonance also needs nonlinear interaction between the signal and the externally added noise source [58-61], and both perhaps can be integrated into a single nonlinear element. In fact, some studies have analyzed stochastic resonance seen as synchronization phenomena [11, section 2.3.6.3].

Besides, there are options to choose from a variety of SAW devices to choose for feedback or coupling. Most important among these seems to the use of SAW resonators for coupling and SAW dispersive delay lines for feedback. The former may be helpful in defining a synchronization branch because of its narrow passband, and the latter may provide flexibility in adjusting frequency detuning and feedback delay. However, these require further extensive analyses. The present analysis is more in nature of visualizing a possibility rather than specifying a solution.

2.8. Conclusion

Two limit cycle SAW delay line feedback oscillators with nonlinear amplifiers in loops can be phase-locked by coupling their feedback paths through a simple linear phase SAW delay line. In phase-locked state the coupled system oscillates synchronously at a common frequency. However, there exist several synchronization modes depending on the frequency detuning between the coupled oscillators, the coupling strength and the coupling delay. These synchronization modes exhibit high sensitivity conditions for perturbations in the coupling delay, and can be used for sensor applications. From the sensor perspective at least three important features are noticeable: the synchronization frequency varies linearly with perturbations in the coupling delay over the range of interest for most chemical sensing (ppm to sub-ppm) applications; the sensitivity is nearly constant for several synchronization modes; and, the sensitivity can be tuned to any desired level by proper combination of detuning, coupling strength and delay parameters. The simultaneous occurrence of more than one synchronization mode is however disturbing from the point of view of system fabrication, and need both system design and engineering solutions. Nevertheless, the results presented here underline the flexibility and potentiality of synchronization mode sensing for achieving high sensitivity sensors. The equations of motion derived here for autonomous SAW oscillators and for coupled oscillators system are based on developed SAW physics and device characteristics. Therefore, these can be taken to be the basis for further exploration.

References

[1]. T. Erneux, Applied Delay Differential Equations, *Springer Science+Business Media, LLC,* New York, 2009.

[2]. I. Györi, G. Ladas, Oscillation Theory of Delay Differential Equations with Applications, *Clarendon Press*, Oxford, 1991.

[3]. H. Smith, An Introduction to Delay Differential Equations with Applications to Life Sciences, *Springer Science+Business Media, LLC,* New York, 2011.

[4]. S. H. Strogatz, Nonlinear Dynamics and Chaos: with Application to Physics, Biology, Chemistry, and Engineering, *Perseus Books Publishing, LLC*, 1994.

[5]. Y. Kuramoto, Chemical Oscillations, Waves, and Turbulence, *Springer*, Berlin, 1984.

[6]. A. Katok, B. Hassleblatt, Introduction to Modern Theory of Dynamical Systems, *Cambridge University Press*, 1995.

[7]. Visarath In, P. Longhini, A. Palacios (Editors), Applications of Nonlinear Dynamics: Model and Design of Complex Systems, *Springer-Verlag*, Berlin Heidelberg, 2009.

[8]. S. Boccaletti, J. Kurth, G. Osipov, D. L. Valladares, C. S. Zhou, The synchronization of chaotic systems, *Physics Reports*, Vol. 366, 2002, pp. 1–101.

[9]. D. Chawla, E. D. Lumer, K. J. Friston, The relationship between synchronization among neuronal populations and their mean activity levels, *Neural Computation*, Vol. 11, 1999, pp. 1389–1411.

[10]. M. S. Dutra, Armando C. de Pina Filho, Vitor F. Romano, Modeling of a bipedal locomotor using coupled nonlinear oscillators of Van der Pol, *Biological Cybernetics*, Vol. 88, 2003, pp. 286-292.

[11]. A. Pikovsky, M. Rosenblum, J. Kurths, Synchronization: A Universal Concept in Nonlinear Sciences, *Cambridge University Press*, Cambridge, 2003.

[12]. W. C. Lindsey, J.-H. Chen, Mutual clock synchronization in global digital communication networks, *Eur. Trans. Telecomm.,* Vol. 7, No. 1, 1996, pp. 25-37.

[13]. L. E. Larson, J. M. Liu, L. S. Tsimring, Editors, Digital Communications Using Chaos and Nonlinear Dynamics, *Springer Science +Business Media,* New York, 2006.

[14]. P. Stavroulakis, Chaos Applications in Telecommunications, *Taylor & Francis*, Boca Raton, 2006.

[15]. M. Rosenblum, A. Pikovsky, Synchronization: from pendulum clocks to chaotic lasers and chemical oscillators, *Contemporary Physics,* Vol. 44, No. 5, 2003, pp. 401-416.

[16]. T. E. Murphy, A. B. Cohen, B. Ravoori, K. R. B. Schmitt, A. V. Setty, F. Sorrentino, C. R. S. Williams, E. Ott, R. Roy, Complex dynamics and synchronization of delayed-feedback nonlinear oscillators, *Phil. Trans. R. Soc.,* Vol. A368, 2010, pp. 343–366.

[17]. J. Benford, H. Sze, W. Woo, R. R. Smith, B. Harteneck, Phase locking of relativistic magnetrons, *Phys. Rev. Lett.,* Vol. 62, 1989, pp. 969-971.

[18]. T. Miyano, T. Tsutsui, Data synchronization in a network of coupled phase oscillators, *Phys. Rev. Lett.,* Vol. 98, 2007, 024102 (4pp).

[19]. Y.-C. Liu, N. Chopra, Controlled synchronization of heterogeneous robotic manipulators in the task space, *IEEE Trans. Robotics*, Vol. 28, No. 1, 2012, pp. 268-275.

[20]. D. S. Ballantine, R. M. White, S. J. Martin, A. J. Ricco, G. C. Frye, E. T. Zellers, H. Wohltjen, Acoustic Wave Sensors (Theory, Design and Physico-Chemical Applications), *Academic*, San Diego, 1997.

[21]. M. Hoummady, A. Campitelli, W. Wlodarski, Acoustic wave sensors: design, sensing mechanisms and applications, *Smart Mater. Struct*., Vol. 6, 1997, pp. 647–657.

[22]. A. Afzal, F. L. Dickert, Surface acoustic wave sensors for chemical applications, Ch. 10, in: Chemical Sensors: Comprehensive Sensor Technologies, Vol. 4, Solid State Devices (Ed. G. Korotcenkov), *Momentum Press, LLC*, 2011.

[23]. M. Thomson, D. C. Stone, Surface-Launched Acoustic Wave Sensors (Chemical Sensing and Thin Film Charcaterization), *John Wiley & Sons*, New York, 1997.

[24]. T. M. A. Gronewold, Surface acoustic wave sensors in bioanalytical field: Recent trends and challenges, *Anal. Chim. Acta*, Vol. 603, 2007, pp. 119-128.

[25]. K. Länge, B. E. Rapp, M. Rapp, Surface acoustic wave biosensors: a review, *Anal. Bioanal. Chem,* Vol. 391, 2008, pp. 1509–1519.

[26]. M.-I. R. Gaso, C. M. Iborra, Á. M. Baides, A. A. Vives, Surface generated acoustic wave biosensors for the detection of pathogens: a review, *Sensors*, Vol. 9, 2009, pp. 5740-5769.

[27]. L. M. Dorozhkin, I. A. Rozanov, Acoustic wave chemical sensors for gases, *J. Anal. Chem.,* Vol. 56, No. 5, 2001, pp. 399–416.

[28]. J. W. Grate, Acoustic wave microsensor arrays for vapor sensing, *Chem. Rev.,* Vol. 100, 2000, pp. 2627–2648.

[29]. C. K. Ho, E. R. Lindgren, K. S. Rawlinson, L. K. McGrath, J. L. Wright, Development of a surface acoustic wave sensor for in-situ monitoring of volatile organic compounds, *Sensors*, Vol. 3, 2003, pp. 236-247.

[30]. E. T. Zellers, S. A. Batterman, M. Han, S. J. Patrash, Optimal coating selection for the analysis of organic vapor mixtures with polymer-coated surface acoustic wave sensor arrays, *Anal. Chem.,* Vol. 67, No. 6, 1995, pp. 1092-1106.

[31]. R. D. S. Yadava, R. Chaudhary, Solvation, transduction and independent component analysis for pattern recognition in SAW electronic nose, *Sens. Actuators B: Chem.,* Vol. 113, 2006, pp. 1-21.

[32]. S. J. Martin, G. C. Frye, S. D. Senturia, Dynamics and response of polymer-coated surface acoustic wave devices: effect of viscoelastic properties and film resonance, *Anal. Chem.,* Vol. 66, 1994, pp. 2201-2219.

[33]. R. D. S. Yadava, R. Kshetrimayum, M. Khaneja, Multifrequency characterization of viscoelastic polymers and vapor sensing based on SAW oscillators, *Ultrasonics*, Vol. 49, No. 8, 2009, pp. 638–645.

[34]. P. Singh, R. D. S. Yadava, Effect of film thickness and viscoelasticity on separability of vapour classes by wavelet and principal component analyses of polymer-coated surface acoustic wave sensor transients, *Meas. Sci. Technol.,* Vol. 22, 2011, 025202 (15pp.).

[35]. P. Singh, R. D. S. Yadava, Using parametric nonlinearity in SAW sensor transients and Information Fusion for improving electronic nose intelligence, *Int. J. Computational Intelligence Research*, Vol. 6, No. 4, 2010, pp. 919-927.

[36]. P. Singh, R. D. S. Yadava, A fusion approach to feature extraction by wavelet decomposition and principal component analysis in transient signal processing of SAW odor sensor array, *Sensors & Transducers,* Vol. 126, Issue 3, March 2011, pp. 64-73.

[37]. P. Singh, R. D. S. Yadava, Transient feature extraction based on phase space fusion by partial-least-square regression analysis of sensor array signals, *IEEE Conf. Proc. Emerging Trends in Electrical and Computer Technology, IEEE ICETECT 2011*, pp. 676-680.
[38]. P. Singh, R. D. S. Yadava, Discrete wavelet transform and principal component analysis based vapor discrimination by optimizing sense-and-purge cycle duration of SAW chemical sensor transients, *IEEE Conf. Proc. Computational Intelligence and Signal Processing, IEEE CISP 2012,* pp. 71-75.
[39]. P. Singh, R. D. S. Yadava, Wavelet based fuzzy inference system for simultaneous identification and quantitation of volatile organic compounds based on SAW sensor transients, in: Swarm, Evolutionary and Memetic Computing, *Springer LNCS Series*, 7077, 2011, pp. 319-327.
[40]. S. K. Jha, R. D. S. Yadava, Designing optimal model SAW sensor array electronic nose for body odor discrimination. *Sensor Lett.*, Vol. 9, No. 5, 2011, pp. 1612-1622.
[41]. S. K. Jha, R. D. S. Yadava, Data mining approach to polymer selection for making SAW sensor array based electronic nose, *Sensors & Transducers,* Vol. 147, Issue 12, December 2012, pp. 108-128.
[42]. R. D. S. Yadava, Enhancing mass sensitivity of SAW delay line sensors by chirping transducers, *Sens. Actuat. B: Chem.,* Vol. 114, 2006, pp. 127-131.
[43]. R. Kshetrimayum, R. D. S. Yadava, R. P. Tandon, Mass sensitivity analysis and designing of surface acoustic wave resonators for chemical sensors, *Meas. Sci. Technol.,* Vol. 20, 2009, 055201 (10pp).
[44]. Z. C. Feng, C. Chicone, A delay differential equation model for surface acoustic wave sensors, *Sens. Actuat. A,* Vol. 3654, 2003, pp. 1-8.
[45]. C. Chicone, Z. C. Feng, Synchronization phenomena for coupled delay-line oscillators, *Physica D,* Vol. 198, 2004, pp. 212-213.
[46]. D. Seo, C. Chicone, Z. C. Feng, Synchronization problem in delay-line oscillator SAW sensors, *Sens. Actuat. A,* Vol. 121, 2005, pp. 44-51.
[47]. S. S. Jha, R. D. S. Yadava, Synchronization based SAW sensor using delay line coupled dual oscillator phase dynamics, *Sensors & Transducers*, Vol. 141, Issue 6, June 2012, pp. 71-91.
[48]. C. Campbell, Surface Acoustic Wave Devices for Mobile and Wireless Communications, *Academic Press*, 1998.
[49]. D. P. Morgan, Surface-Wave Devices for Signal Processing, *Elsevier*, Amsterdam, 1985.
[50]. H. Wohltjen, R. Dessy, Surface acoustic wave probe for chemical analysis. I. Introduction and instrument description, *Anal. Chem.,* Vol. 51, No. 9, 1979, pp. 1458–1464.
[51]. H. G. Schuster, P. Wagner, Mutual entrainment of two limit cycle oscillators with time delayed coupling, *Prog. Theor. Phys.,* Vol. 81, 1989, pp. 939–945.
[52]. O. V. Popovych, V. Krachkovskyi, P. A. Tass, Twofold impact of delayed feedback on coupled oscillators, *Int. J. Bifurcation and Chaos,* Vol. 17, No. 7, 2007, pp. 2517–2530.
[53]. G. Sauerbrey, Verwendung von schwingquarzen zur wägung dünner schichten und zur mikrowägung, *Z. Phys.,* Vol. 155, 1959, pp. 206-212.
[54]. T. E. Parker, G. K. Montress, Precision surface-acoustic-wave (SAW) oscillators, *IEEE Trans. Ultrason. Ferroelec. Freq. Control,* Vol. 35, No. 3, 1988, pp. 342-364.
[55]. K. Kurokawa, Noise in synchronized oscillators, *IEEE Trans. Microwave Theory Techniques,* Vol. MTT-16, No. 4, 1968, pp. 234-240.
[56]. H. -C., X. Cao, U. K. Mishra, R. A. York, Phase noise in coupled oscillators: theory and experiment, *IEEE Trans. Microwave Theory Techniques,* Vol. MTT-5, No. 5, 1997, pp. 604-615.
[57]. D. J. Needleman, P. H. E. Tiesinga, T. J. Sejnowski, Collective enhancement of precision in networks of coupled oscillators, *Physica D*, Vol. 155, 2001, pp. 324-336.
[58]. N. Tabareau, J.-J. Slotine, Q.-C. Pham, How synchronization protects from noise, *PLoS Computational Biology,* Vol. 6, No. 1, 2010, pp. 1-9.

[59]. S. Kim, S. H. Park, H.-B. Pyo, Stochastic resonance in coupled oscillator systems with time delay, *Phys. Rev. Lett.,* Vol. 82, No. 8, 1999, pp. 1620-1623.

[60]. F. C. Blondeau, X. Godivier, Theory of stochastic resonance in signal transmission by static nonlinear systems, *Phys. Rev.*, Vol. 55, No. 2, 1997, pp. 1478-1495.

[61]. F. C. Blondeau, D. Rousseau, Noise improvement in stochastic resonance: from signal amplification to optimal detection, *Fluctuation and Noise Letters*, Vol. 2, No. 3, 2002, pp. L221-233.

Chapter 3
Fingerprint Sensors: Liveness Detection and Hardware Solutions

Shahzad Memon, Nadarajah Manivannan, Azad Noor, Wamadeva Balachadran, Nikolaos V. Boulgouris

3.1. Introduction

The use of fingerprints for biometric identification is one of the most prevalent authentication methods used today in a number of commercial, civil and forensic applications. In the last decade, research in fingerprint based identification systems has reached a high degree of accuracy [1-4]. The academic and industrial research in fingerprint biometrics has achieved more maturity than other biometric systems, such that it has almost become synonymous. Automatic Fingerprint Biometrics is now an accepted system for securing commercial and homeland security systems. In the past few years, many hardware and software technologies have been tested for capturing, storing, processing and matching fingerprint data. The main reasons for the success and proliferation of this technology are: wider acceptance of fingerprints as a biometric means, and availability of low-cost fingerprint acquisition devices. In spite of their numerous advantages, recent research has pointed out that the current fingerprint sensing technology can be deceived by a fake or artificial fingerprint made of gelatine or similar material [5, 7]. Many believe that fingerprint biometric systems can detect liveness in biometric samples; however, artificially replicated fingerprints have become more convincing, and it has been proved that about 80 %-85 % of these systems are easily fooled by different kinds of fake fingers with stamped patterns. Therefore, the issue of 'Liveness Detection' has emerged in biometrics research. Liveness detection refers to the inspection of the characteristics of a finger in order to check whether the input finger is live or artificial. Currently, a number of fingerprint identification systems have been implemented at various important places such as airport, border and immigration controls. However, in most cases, the ability to detect liveness does not appear to be a major concern of the manufacturers of these systems. The possible measures to detect liveness are only proposed in patents and published literature. In January 2009, a woman

Shahzad Memon
Centre for Electronic Systems Research (CESR), Electronics and Computer Engineering, School of Engineering and Design, Brunel University, London, UK

at Tokyo airport managed to fool a fingerprint reading machine by using a piece of tape stuck on her fingers. Japanese officials said in a report that they suspected many others had been doing the same thing, demonstrating that the biometric systems they installed in 30 airports in 2007 at a cost of $45 million are insufficient [8]. This incident has made it clear that there is no known fingerprint biometric hardware or software solution that is 100 % fool-proof. Various liveness detection methods proposed in literature could be possibly be implemented either at the acquisition stage or at the processing stage. It is extremely difficult to identify these fake fingers simply via image processing algorithms. New sensing technology/or some kind of extended hardware is required, that can differentiate living tissue from artificially created materials resembling it.

This chapter is divided into five sections. Section 3.1 introduces the spoofing issue with fingerprint biometrics; section 3.2 discusses the details of fingerprint sensor technologies. Fingerprint sensing technologies are explained in section 3.3. Proposed hardware based liveness detection techniques are described in section 3.4. The conclusions are drawn in section 3.5.

3.2. Spoofing Techniques in Fingerprint Sensors

Before describing spoofing in more detail, it may be helpful to discuss how the False Acceptance Ratio (FAR), a typical efficiency measure of biometric devices, is related to spoofing. A false-accept occurs when a submitted template is incorrectly matched to a template enrolled by another user [1, 9]. This only refers to a zero effort attempt, i.e., an unauthorized user making an attempt with his/her own biometric to gain access to a system. If the FAR is kept low, then the probability of specific user with criminal intent matching another template is very low. The FAR does not give information on the vulnerability of a system to spoof attacks.

The fraudulent entry of an unauthorized person into a fingerprint recognition system by using a fake fingerprint sample is termed spoofing. The spoofing of fingerprint sensors was first revealed by Network Computing in 1998 [5]. They addressed the vulnerability of fingerprint scanning devices by testing with fake/lifted fingerprints resulting in four out of six devices were found to be susceptible to fake finger attacks. Further research was undertaken by Tsutomu Matsumoto to prepare fake fingerprint stamps using low-cost, easily obtainable tools and materials such as silicon and Gelatine (Fig. 3.1. Method-1& 2). The images produced by these fake fingers were accepted as a real finger tips by 11 types of fingerprint systems (Fig. 3.1: Fingerprint) [10].

Many tests have shown a high FAR with optical/ capacitive fingerprint sensors. In these tests, fake finger- print enrolment was achieved with 68–100 % acceptance [6,7,10,13].

In August 2003, two German hackers claimed to have developed a technique using latent prints on a fingerprint scanner and convert them to a latex fingerprint [14]. They used graphite powder and adhesive tape to recover the latent prints that were then digitally photographed and enhanced by using graphics software.

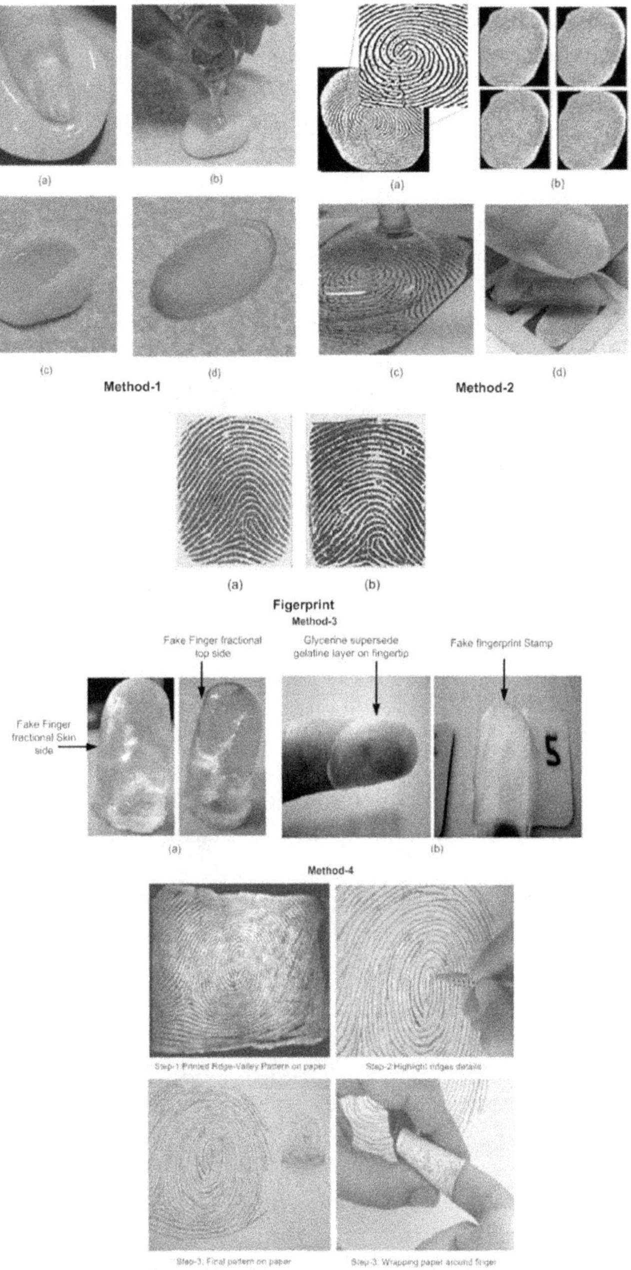

Fig. 3.1. Method-1: Process of making fake fingerprint from plastic/ silicon mould a) Pressing of finger on soft plastic silicon/rubber mould b) Dripping of liquid gelatine/rubber over mould c) Solidification d) Artificial fingerprint stamp [10]. Method-2: a) Imaging of fingerprint from residual fingerprint b) Making and printing of fingerprint c) Masking and printing of fingerprint d) Detaching of fake fingerprint stamp [6, 10], Method-3: GSG based fake finger and fingerprint (a) Glycerine based 3-D fake finger (b) Finger sample [11] Method-4: Four step fake fingerprint preparation to spoof the touchless surrounded biometric system.[12]. Fingerprint: a) Live Finger b) Fake/Gummy Finger [6].

A recent publication [11] has explained some new techniques of making 3D fake fingers (Fig. 3.1 Method-3 a) and fingerprint using materials, such as Glycerine Supersede Gelatine (GSG) (shown in Fig. 3.1 Method-3 b). The GSG based fake fingers have been tested on capacitive, optical and thermal sensors and they have been successfully enrolled and matched.

A touchless surrounded imaging based fingerprint technique was also spoofed using a piece of paper printed with the ridges and valley pattern of a finger. The four step procedure of preparing a fake fingerprint sample for touchless surrounded imaging is illustrated in Fig. 3.1 Method-4.

Much of the activity in spoofing biometric systems has, up until now, been confined to researchers. Moreover, as biometric systems become more widespread, the incentives to misuse or hack biometric systems will grow. Understanding the nature and risk of such attacks will become increasingly important to systems architects, administrators and systems security managers.

3.3. Fingerprint Sensing Technologies

Many sensing technologies have been developed over the past few years. Fingerprint sensors are the most important part of a fingerprint scanner where the image is formed. They can be categorised as optical, electro-optical, capacitive, piezoelectric, thermal, ultrasound and radio frequency. In the early days, most of the sensors were optical or electro-optical, but most of the sensors in the market today are either capacitive or RF-resistive/modulation.

3.3.1. Optical Sensors

Using optical fingerprint sensors, various fingerprint capture techniques have been introduced in last few years in a variety of security applications. In the following section major optical fingerpint sensing techniques are discussed in detail.

3.3.1.1. Frustrated Total Internal Reflection (FTIR)

This is the oldest and most used live scan technique. The finger is placed on top of the prism (as shown in Fig. 3.2 (a) and it is illuminated by a light source from one side. The light rays entering the prism are reflected at the valleys and randomly scattered at the ridges of the finger as they make contact with the prism surface. The lack of reflection of light from the ridges (which appear dark in the image) contrast with the valleys which appear bright. The reflected rays are focused on to a CCD (Charge Coupled Device) by an optical lens to form the image [15-16]. Fig. 3.1 (b) is showing such a optical fingerprint sensor manufactuerd by MAXIS biometrics company in 2009. The

False Acceptance Rate (FAR) of this optical sensor is < 0.0001 and False Rejection Rate (FRR) is < 0.01 % and it captures an image with 500 DPI resolution.

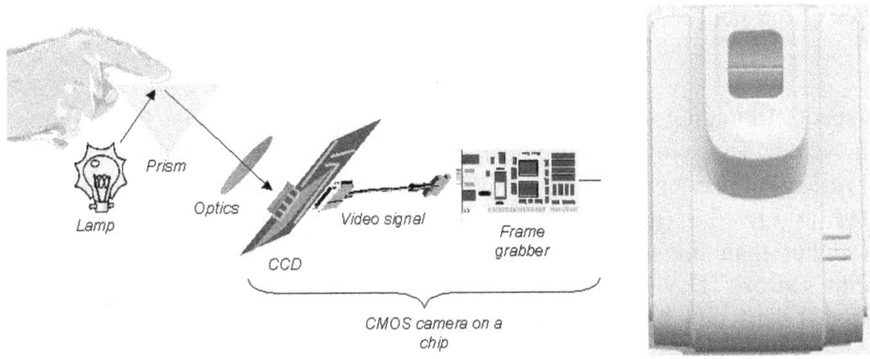

(a) (b)

Fig. 3.2. (a) FTIR Mechanism [17], (b) MIAXIS FPR-620Optical fingerprint reader [18].

As compared to other fingerprint technologies optical fingerprint sensors are robust and much less sensitive to adverse environmental effects such as mechanical shocks or Electro-Static Discharge (ESD). However, they are higly suspectable to non-ideal skin conditions, in particular, if the skin is too dry or not in good contact with the sensor, the image quality will be degraded.

High resolution (500-1200 DPI) optical fingerprint sensors are commercially available and they are in use with various static access control systems. However, optical fingerprint sensors tend to be larger due to its sensing mechanisim, so they are not appropriate for mobile applications that require small sensors in cell phones or Tablet PCs. Another drawback with these sensors is latent fingerprints which are left by previous users of the system, that can be copied and possibly used to prepare fake fingerprints to gain access to a system.

In addition, FTIR based optical fingerprint sensors failed in many liveness detection tests [10-19].This technology can not distinguish between fake and original fingerprints, because it only captures an image from the reflection of light from the surface of the finger skin without measuring the properties of the skin layer.

3.3.1.2. Multispectral Imaging

In optical fingerprint sensors, multispectral imaging technology has recently been developed and introduced by Lumidigm inc USA. It is based upon the principle that different wavelengths of light penetrate into human skin to different depths and are absorbed differently by various chemical components of the skin [20]. Fingerprint sensor based on Multispectral imaging technology collect multiple images of the surface and subsurface of the fingertip skin under a variety of optical conditions and combine them to yield high-quality and complete fingerprint images[21]. The group of raw images captured by multispectral method is also analyzed to ensure that the optical properties of the sample being measured match with those expected from a living finger [22].

A multispectral imaging system has two main modules: a multiple light source and an imaging system. These modules are designed and configured expressly to avoid the total internal reflection phenomena. The imaging mechanism of the multispectral technique is illustrated in Fig. 3.3 (a). This technique utilizes multiple wavelength illumination sources rather than the monochromatic illumination commonly used in FTIR based fingerprint sensors. The orthogonal configuration of linear polarizers emphasizes this multispectral light, which penetrates the surface of the skin [23].

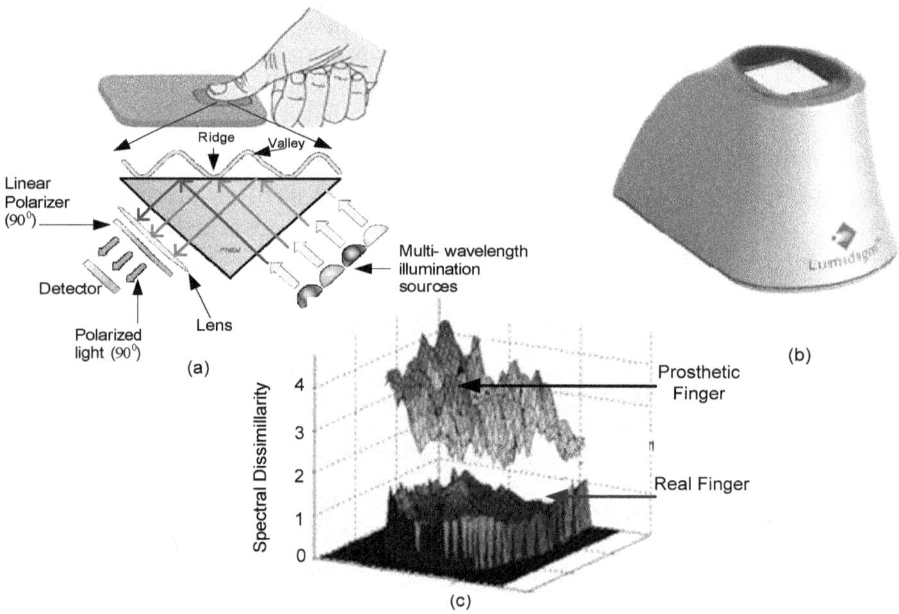

Fig. 3.3. (a) Principle of Multispectral Fingerprint Imaging, (b) Lumidigm Mercury Series M301multispectral imaging technology based Fingerprint Reader [24], (c) Spectral dissimilarity test result [25, 26]

The light undergoes multiple scattering events before emerging from the skin towards the imaging array. In avoiding the optical phenomenon of FTIR, the multispectral imaging based fingerprint sensor is able to collect more identifying data from the finger than the FTIR based finegrprints sensor. Fig. 3.3 (b) shows a commercially available multispectral imaging technology based fingerprint reader Mercury Series M301 manufactured by Lumidigm inc USA in 2010.

The main advantage of this technique is that the combination of surface and subsurface imaging ensures that usable biometric data can be taken across a wide range of environmental and physiological conditions such as: wetness, bright ambient lighting, dry skin, various topical contaminants present on the surface of the finger and poor contact between finger and sensor [21, 22]. In addition, they are very robust and relatively insensitive to adverse environmental effects such as mechanical shocks or Elelctro Static Discharge (ESD) as compared to semiconductor fingerprint sensors. Moreover, this technique is claimed to be able to detect living tissues from non-living tissues or other organic or synthetic materials. An analysis of the surface and subsurface spectral charcterstic differences between a living finger and a prosthetic is illustrated in Fig. 3.3 (c). These differences can be used to detetct prosthetics. However, in the future, the rapid developments in material technology may lead to the engineering of the materials with the same optical properties as living skin and fake fingers. Experts might use these materials to prepare spoofs and it could therefore became difficult to distinguish between live and dead tissues by using multispectral imaging techniques.

3.3.1.3. 3D Touchless Imaging

Compared to the flat touch-based fingerprint sensing systems, contactless or touchless fingerprinting is basically a remote sensing technique used to capture the ridge-valley patterns with no contact between the skin of the finger and the sensing area [27, 12]. The fingerprint representation using the touchless technology is different from the representation of the FTIR approach. While the latter method uses the optical principle of the reflection of the light (see Fig. 3.4 (a)), the touchless approach produces mainly a photograph of the fingerprint [28]. Fig. 3.4 (b) shows the touchless system with five cameras surrounding a finger. Knowing the position and orientation of each camera within a refence co-ordinate system, the five shapes or images are then projected into the 3D space and interpolated together to obtain the 3D shape of the finger [12, 29].The use of multiple views enables the capture of a full nail-to nail fingerpint that is faster than the traditional rolling procedures. Different views can be obtained by either different cameras surrounding the finger or one camera and a set of small mirrors.

Touchless fingerprint devices are already available on the market from Touchless Biometrics Systems (TBS) and FlashScan shown in Fig. 3.4 (c) and (d). However, higher costs compared with other fingerprint sensing technologies, has not encouraged them to become widespread. This fingerprint technique tries to overcome the problem that comes with the legacy optical capture technology, because of the lack of contact between the finger and many rigid surface [29, 27].

The skin does not deform during the capture and the repeatablility of the measure is ensured. With some advantages, this new fingeprint scanning technology introduces some new challenges.The capture approach intrinsically provides a big system vulnerability and increases the risk of impostor intrusion making it insecure and useless for access control or unattended applications in general. Lower image contrast, illumination, correct finger positioning and user convenience must be addressed.

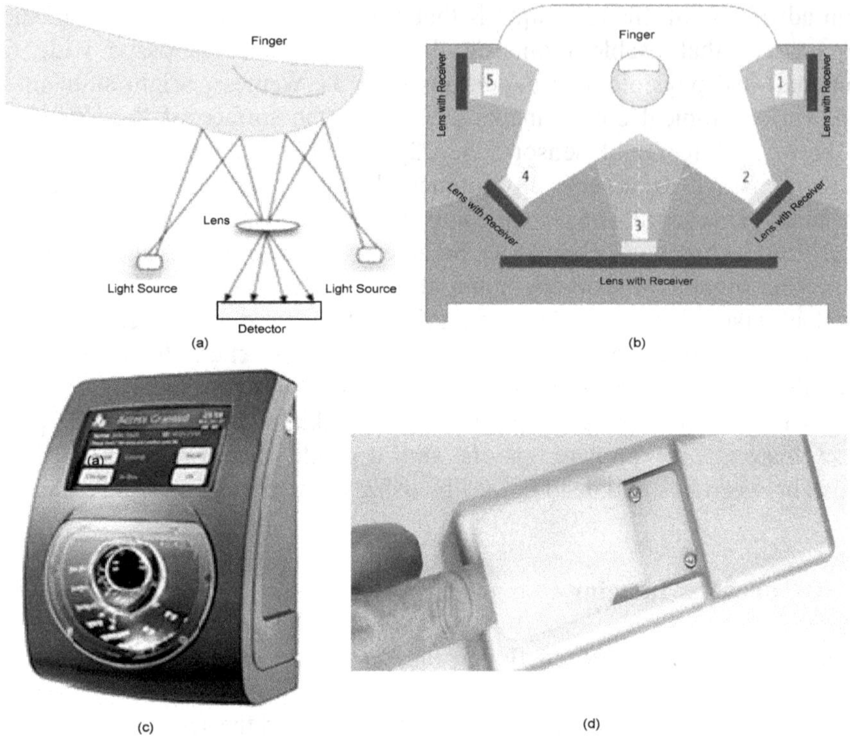

Fig. 3.4. (a) Principle of Toucless Fingerprint Imaging [30], (b) Touchless finegrprint scanner with five optical sources and detect [12], (c) TBS 3DGuard Touchless fingerprint scanning Terminals [31], (d) FlashScan 3D touchless fingerprint reader [32].

Like the other optical imaging technologies, touchless fingerprint technology is not free from liveness detection issue. In fact, it may easier to fool than touch based optical fingerprint technology [12].

3.3.1.4. TFT (Thin Film Transistor) Optical

Generally, TFT based optical detection techniques are used in facsimile and digital copying machines. In recent years, a TFT type optical detection sensor has been used as a fingerprint sensor. The TFT changes its electrical characteristics depending upon the presence of light incident on the device as illustrated in Fig. 3.5. (a).

An optical detecting sensor comprises a window, through which light generated from a light source passes, and a TFT sensor, for generating optical current by detecting the light that is transmitted through the window and then reflected off an object disposed on the optical detecting sensor [17]. Optical current generated from the TFT sensor is stored in a storage capacitor as charges, and the charges stored in the storage capacitor are transmitted by a switching TFT to an external driving circuit. In addition, a light-shielding layer for blocking light is formed over the semiconductor layer where the switching TFT is located.

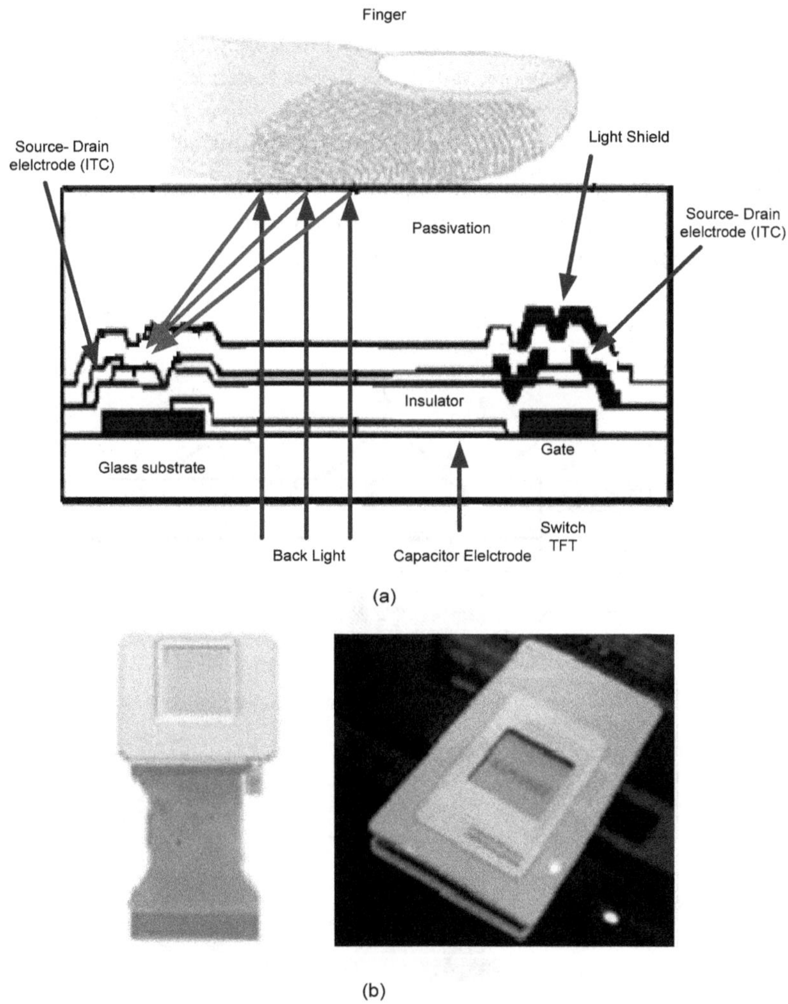

Fig. 3.5. (a) Schemetic of TFT Optical type fingerprint sensor, (b) Two optical TFT prototype from CASIO [17].

TFT based fingerprint sensors could be a useful solution to integrate within touch display based applications because of their compactness and low power consumption; however, it is not possible to detect liveness from finger placed on a TFT sensor because the sensor just scans or copies the fingertip exterior skin layer, which is not sufficient to detect the liveness of the finger.

3.3.1.5. Electro-optical

Some polymers such as Polyphenylene Vinylene (PPV) and Ploytiophene (PT) are able to emit light when properly excited with the proper voltage (usually a high voltage is required). This polymer is directly connected to a CMOS camera, which is necessarily the size of the finger (See Fig. 3.6). Generally, the finger acts as the ground, and the polymer emits light where the ridges touch. As ridges touch the polymer and the valleys do not, the potential is not the same across the surface when a finger is placed on it and the amount of light emitted varies, thus allowing a luminous representation of the fingerprint pattern to be generated and acquired by the imaging layer. This technology has not been tested, but as the basic principle of the technology is similar to that of TFT it can be assumed that it will not be of use liveness detection.

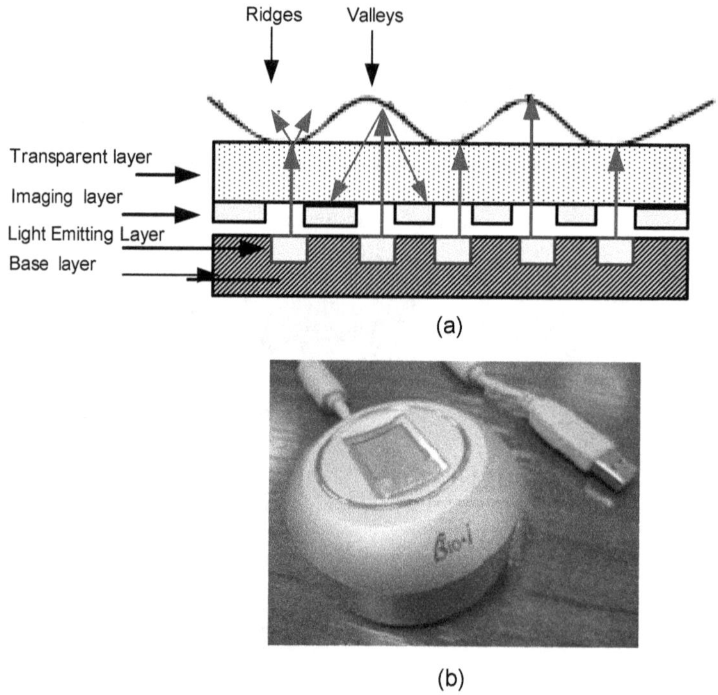

Fig. 3.6. (a) Sensing principle of Electro-optical fingerprint [33], (b) Bio-i CYTE fingerprint sensor manufactured by Testech, Inc. [34].

3.3.2. Capacitive

This sensor is made up of one or more semiconductor chips containing an array of tiny cells, each cell includes two conductor plates covered with an insulating layer. The cells are slightly smaller than the width of one ridge on a finger (The average ridge width of a male is 0.48 mm and a female 0.43 mm). The surface of the finger acts as a third capacitor plate, separated by the insulating layers in the cell structure and, in the case of the fingerprint valleys, a pocket of air. Varying the distance between the capacitor plates (by moving the finger closer or further away from the conducting plates) changes the total capacitance (ability to store charge) of the capacitor. Because of this feature, the capacitor in a cell under a ridge will have a greater capacitance than the capacitor in a cell under a valley [35] (See Fig. 3.7).

The array of pixels is used to map the fingerprint image based on the ridge and valley structure. The distance between the array of pixels and the finger should be very low to provide enough sensitivity, i.e., the coating must be as thin as possible (a few microns).

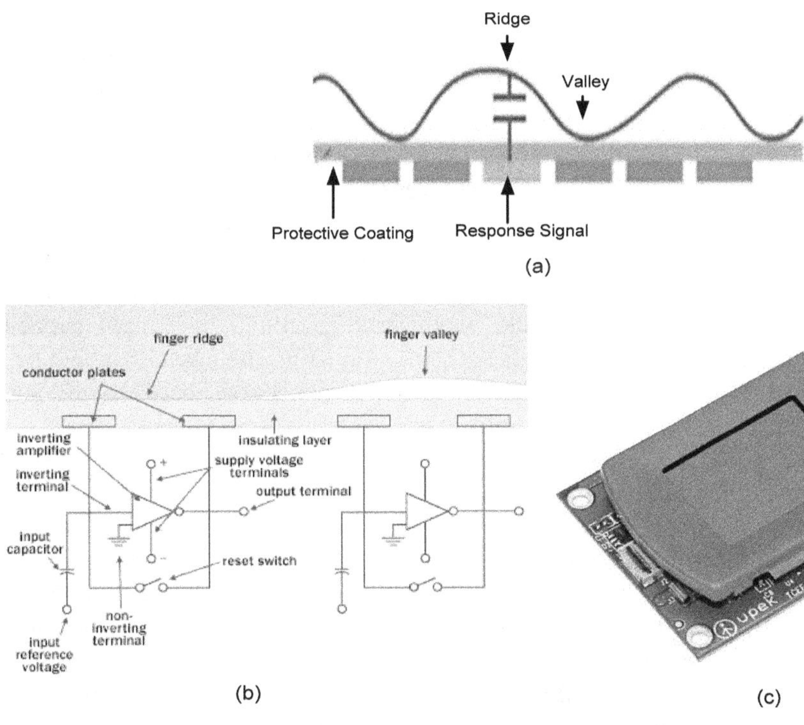

Fig. 3.7. (a) Capacitive Fingerprint Sensing mechanism, (b) Capacitance detection circuit [35], (c) UPEK's TCS1 TouchChip Fingerprint Sensor [36].

The surface layer of fingertip skin, which the capacitive fingerprint sensor detects is prone to damage and contamination in the course of everyday activities. Since the dielectric constant of the surface layer is mainly due to moisture in the dead cells, ridges in dry fingers will have dielectric constants very close to air, resulting in very faded images for capacitive fingerprint sensing. In addition, the capacitive sensing is vulnerable to strong external electrical fields, the most dangerous being ESD.

The simple measurement of the surface characteristics of the skin causes capacitive sensors to be highly vulnerable to spoofing .They scan only the surface of the fingerprint, using dielectric measurements to distinguish between the ridges and valleys of the outer dead skin layer of the fingertip. Capacitive fields do not penetrate very far into the skin and can only image the surface of the finger tip which is not sufficient to detect liveness. Capacitive fingerprint technology has been failed in many liveness detection tests and can be easily spoofed with fake finger stamps [19]. The main advantages of capacitive sensors are their compact size and low cost. Veridicom, Fujitsu, Infineon, Sony, Upek, Hitachi, LighTuning, Melfas, Atrua, NTT and Symwave, etc. are the major manufacturers of capacitive fingerprint sensors are commercially available in the market.

3.3.3. Radio Frequency (RF)

In the radio frequency based fingerprint sensing method, a radio frequency (RF) signal is injected into the finger from one side. A field will be created by RF signal between the finger and the adjacent semiconductor, that mimics the shape of the ridges and valleys of the finger‘s epidemal layer, then the signal is received by the pixel array on the silicon. These pixels array act like antennas (see Fig. 3.8 (a)). These antenna measure the skin’s subsurface (the live layer of skin) features by generating and detecting linear field geometries of the live layer of skin cells originated beneath the skins surface. This is in contrast to the spherical or tubular field geometries generated by simple capacitive sensor, which read only the very top surface of the skin [37]. The signal strength on the receiver pixel will depend on the capacitive/resistive connections between the source and the pixel receiver. An under-pixel amplifier is used to measure the signals.

Images obtained by RF technique, that accurately corresponds to the pattern of the fingerprint are clearer in quality compared with resulting images produced by optical or DC capacitive techniques. This technique is highly resistant to Electro-Static-Discharge (ESD) and it allows the sensor to acquire images from a finger surface with dirt, oil, scars, cuts, or other impurities that can effect other technologies, thus improving the accuracy and reliability of the sensor. Despite being the newest form of fingerprint biometrics, it is also the most popular with more than 8 million RF fingerprint sensors in use today. Fig. 3.8 (a) illustrates authentic TruePrint® RF technology based AES 4000 fingerprint sensor. Validity‘s live flex® RF technology fingerprint sensor based on RF is depicted in Fig. 3.8 (c), it uses 18-24 MHz frequency to capture image of fingertip. The False Acceptance Rate (FAR) of Live flex technology is <1:100,000, and lowest False Rejection Rate (FRR) < 1:500.

With many advantages, RF sensors fail when the sensor surface is wet; because they measure the fingertip features very close (tens of microns) to the surface of the skin the technology is almost as susceptible as capacitive sensors to worn or missing skin features.

Manufacturers claim that RF based fingerprints sensors are very hard to spoof with fake fingerprints. Even though spoofing a RF fingerprint sensor takes a little more sophistication, spoofing is easy when proper materials are used [40]. RF fingerprint scanning technology utilizes the conductive or live layer of skin beneath fingertip surface; however, a gummy finger can still successfully fool the scanner because the conductance of a properly made gummy finger is similar to that of a live finger.

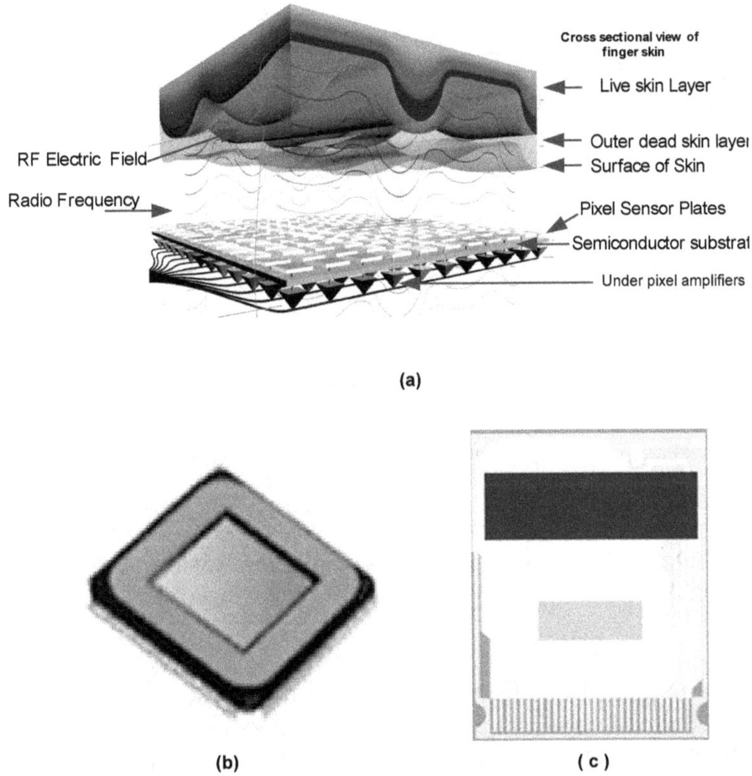

Fig. 3.8. (a) Principal of operation of RF field sensing Fingerprint Sensor [38], (b) Authentic AES 4000 RF based finegrprint [37], (c) Valadity Liveflex VFS 201 Sensor [39].

3.3.4. Thermal

A thermal fingerprint sensor consists of an array of micro heaters and pyro-electric material based sensing elements. The pyro-electric materials (tourmaline, lithium sulphate monohydrate etc) generate voltages on the temperature differentials. Fig. 3.9 (a) illustrates the cross sectional view of a thermal fingerprint sensor. It scans only the surface skin of the fingertip by measuring the heat transferred from the array to the fingertip. When the sensor surface heats up, it creates a temperature difference between itself and the finger. The ridges of the fingertip touching the sensor draw heat away from the sensor faster than the valleys, which are separated from the sensor by insulating air (See Fig. 3.9 (b); therefore, the sensor detects the ridge-valley pattern or image of the fingertip [41].

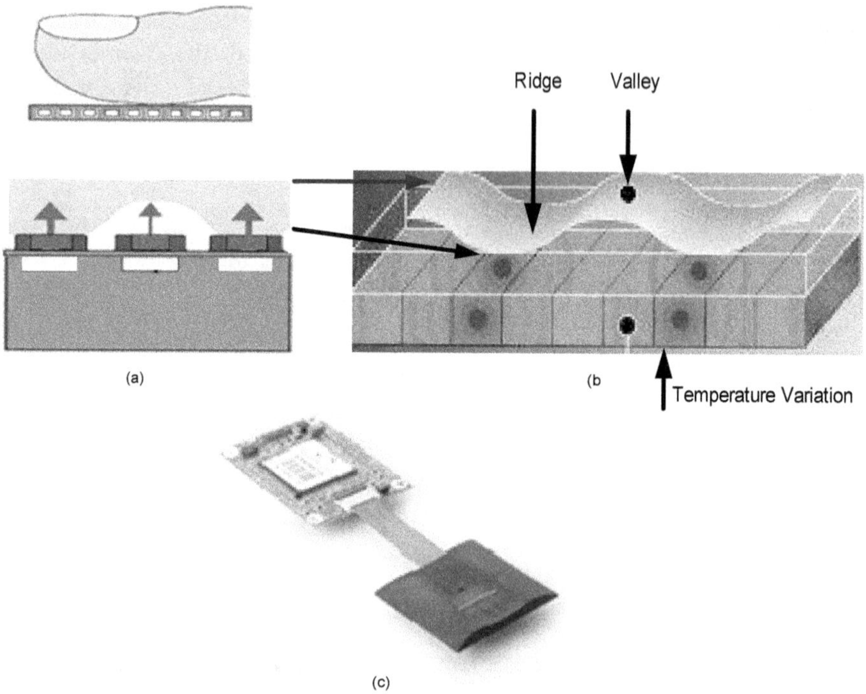

Fig. 3.9. (a) Thermal Fingerprint Sensor [42], (b) Sensing mechanism [17], (c) Atmel AT77C104B FingerChip™ Swipe fingerprint sensor[43].

When a finger is placed on the thermal fingerprint sensor, there is a big change of temperature, and therefore signal, which varies after a short period (less than a tenth of a second). If the finger and the sensor chip have reached thermal equilibrium, and there is no change in temperature, then there is no signal [41]. It is necessary therefore to avoid the possibility of thermal equilibrium between the sensor and fingertip surface.

The main advantages of the thermal sensing based fingerprint sensor is that, it operates well under extreme environmental conditions, such as extreme temperatures, high humidity, oil and dirt; but the heating of the sensor array increases the power consumption which, for portable devices, can significantly shorten battery life.

Thermal sensing method is commercially less common and it exists only in the Atmel FingerChip™ thermal fingerprint sensor which is illustrated in Fig. 3.9 (c). It is based on the swiping method that has a benefit of self cleaning and no latent fingerprint. However, the image quality depends on the user's skill in using the sensor, and is currently the only reasonable option for portable consumer electronics such as cell phones and laptop computers.

Thermal fingerprint sensing technology is also incapable to differentiating between a fake and a live finger tip [19, 14] because it can scan only the surface of the fingertip. The variation in temperature does not differentiate between live ridges-valleys or fake ridges-valleys. In liveness detection tests, thermal sensor was easily spoofed by silicone rubber fingers [19].

3.3.5. Ultrasound

Ultrasound sensors employ acoustic signals transmitted towards the fingertip surface. Acoustic waves travel at different speeds though ridges and air lodged under the skin. The reflected acoustic signal (echo) is captured by a receiver, which generates the fingerprint image [44] (See Fig. 3.10).

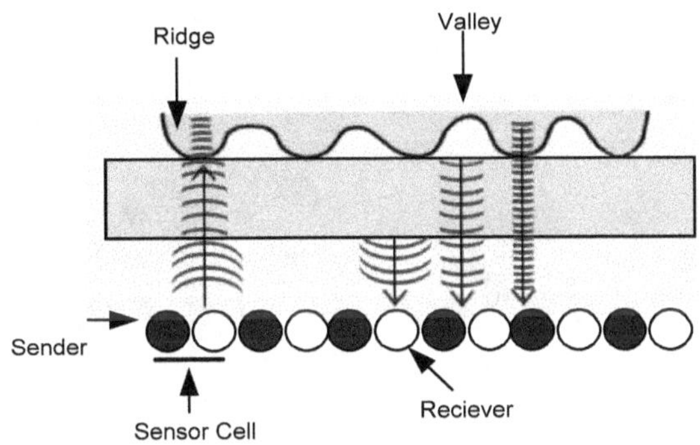

Fig. 3.10. Principles of ultrasound fingerprint sensing [33].

The main advantage of this technology is that it can read the sub-surface of the skin rather than the surface only. Ultrasound fingerprint sensing is not very common in the market because of its size and higher cost [1]. It also comes with some mechanical parts and takes a couple of seconds to generate the image; therefore this technology is not suitable for large scale production. However, it has demonstrated much more tolerance to external conditions that cause poor biometric image quality in optical or capacitive systems, such as ambient light, humidity, extreme temperatures and ESD.

There are still some issues with ultrasonic fingerprint scanning, such as poor image quality and the susceptibility to spoofing [44]. Ultrasonic fingerprint scanners can be compromised by soft artificial material such as gelatine that has the same echoing characteristics as fingers.

3.3.6. Micro-Electro-Mechanical System (MEMS)

Micro Electro-Mechanical System (MEMS) technology based devices have seen much recent interest in fingerprint acquisition systems as it is now possible to design and fabricate extremely tiny silicon switches[45]. A Fingerprint sensor based on MEMS is actually an array of tiny silicon switches [46, 47]. Fig. 3.11 (a). illustrates a 3D model of a MEMS technology based fingerprint sensor and Fig. 3.11 (b) depicts the SEM image of micro switches. The basic sensing principle is, when the ridge touches two adjacent tabs, the switch closes and it remains open when they are under a valley. The sensing method is illustrated in Fig. 3.11 (c). The ridge of a finger surface pushes the protrusion down, and the protrusion deflects the upper electrode. The deflection of the upper electrode increases the capacitance between itself and the lower electrode. Then, the capacitance is detected by the sensing circuit just under the lower electrode [48]. On the other hand, the valley of a finger surface does not push the protrusion, and the capacitance is kept small. Therefore, the capacitance under a ridge Cr is larger than that under a valley Cv. This relationship Cr> Cv is translated into digitized signal levels. With this sensing, the detected signals from all the pixels generate one fingerprint image.

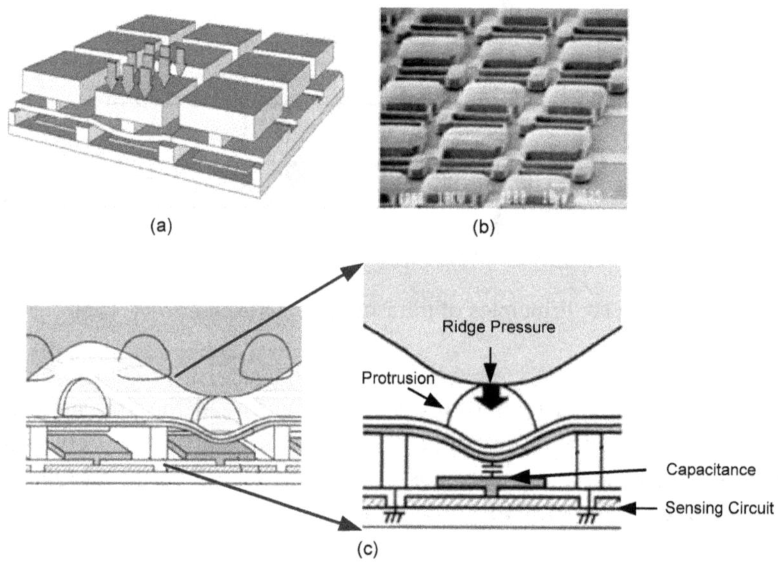

Fig. 3.11. (a) 3D view of MEMS fingerprint sensor [46], (b) SEM Images of MEMS fingerprint sensor [17], (c) Principle of MEMS fingerprint sensors [47, 48].

The significant advantages of this system are: durability, low power consumption, being resistant to ESD, non bulky in shape, and a direct binary output leading to minimal information [48]. However, one significant issue with this technology is that it is not possible to coat the surface of the switches to protect from dust or other environmental effects.

It is not possible to differentiate between fake and real fingertips placed on the sensor, because the sensor cannot detect the difference between a real finger and a fake gummy finger applied with the same pressure, and comprising the same shape, size and weight as a real finger. No liveness tests on MEMS based fingerprint sensors have been reported in literature as yet, and no further development has been done with this technique beyond laboratory. NTT Microsystem Integration Laboratories in Japan and Michigan University (Wise) are working with this technology but it is still under research.

Among the commercial sensors available in the market, some vendors offer liveness detection features as an option, although no fingerprint system currently available is 100 % fool proof. Before describing spoofing in more detail, it may be helpful to discuss how the false accept ratio (FAR), a typical efficiency measure of biometric devices, is related to spoofing. A false-accept occurs when a submitted template is incorrectly matched to a template enrolled by another user [1, 9]. This only refers to a zero effort attempt, i.e., an unauthorized user making an attempt with his/her own biometric to gain access to a system. If the false accept ratio is kept low, then the probability of a specific user with criminal intent matching another template is very low. The false accept ratio does not give information on the vulnerability of a system to spoof attacks.

3.4. Proposed Hardware based Liveness Detection Methods

Liveness detection for fingerprint scanners can be achieved by adding extra hardware to acquire liveness signs. The human body offers a large variety of characteristics, but not all of them comply with foregoing requirements and not all of them can be used with fingerprint technology. For liveness detection systems, the usable properties of the human body can be split into three categories: intrinsic physiological properties, biological or involuntarily generated signals, and responses to a stimulus. The reported hardware based liveness detection techniques are classified and illustrated in Fig. 3.12.

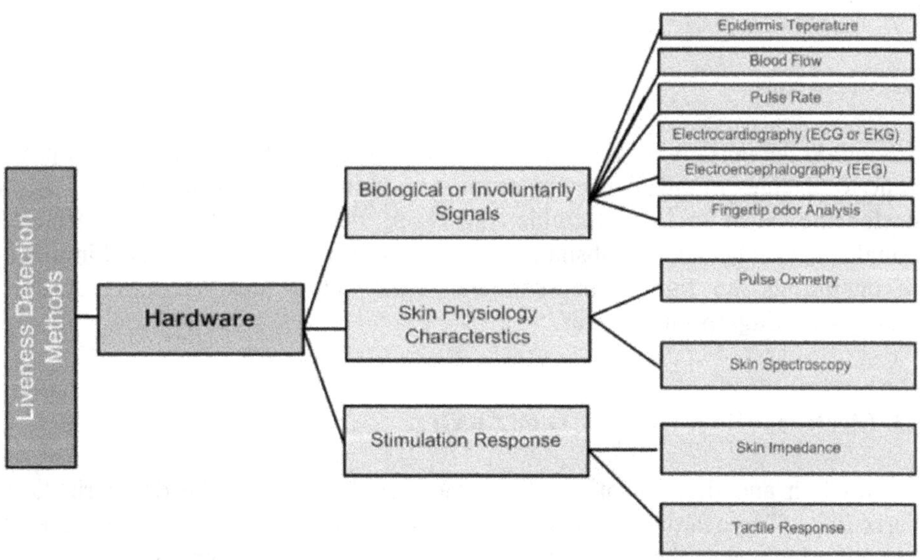

Fig. 3.12. Classification of Hardware based Liveness Detection Methods.

Measurement of biological signals to detect liveness of the fingertip placed on the surface of the sensor is obviously an expensive approach as it requires additional hardware and can be highly invasive: for example, measuring a person's blood pressure can be used for other reasons than for simply detecting the liveness of his/her fingertip.

3.4.1. Biological Signals

A living human body generates many biological signals that can be observed from the many measuring points available on the surface of the body. Some of the biological signals can be detected and measured from the tip of the finger of a person. Most of the hardware based liveness detection techniques recommended in literature use the biomedical sensors to measure the continuation of biological signals to verify the liveness. The following techniques were suggested from different researchers to overcome the liveness issue in fingerprint sensors.

3.4.1.1. Blood Flow

In 1998, a US patent entitled "Anti-Fraud Biometric Sensor" that accurately detects blood flow by Smart Touch LLC [49] describes the method of blood flow detection in the finger. This method uses two light emitting diodes (LEDs) and a photo-detector to determine whether blood is flowing through the finger. Furthermore, this patent claims to have solved these problems by checking if the background light level is above a threshold and by detecting movement of the finger. However, it has been possible to fool similar solutions by simulating blood flow through the use of a flashing light or by moving the imposters' finger.

3.4.1.2. Pulse Rate

In this method the underlying finger's pulse will be sensed. However, practical problems arise due to changes in pulse. The finger must be held for at least a few seconds on the sensor for the pulse to be detectable. However, the pulse rate is also affected by emotional states and can vary substantially if the subject becomes exercised immediately before or during the fingerprint scanning. This would need an extra sensor and processing with fingerprint hardware.

3.4.1.3. Electrocardiography (ECG or EKG)

The contraction and the relaxation of cardiac muscle result in the depolarization and depolarization of myocardial cells. These electrical changes are recorded via electrodes placed on the limbs and chest wall and are transcribed on to a graph paper to produce an electrocardiogram (commonly known as an ECG) [50, 51]. However, the measurement of ECG pulses is easily fooled with a small signal generator, and is an unrealistic option with access to only one fingertip. In addition, the implementation of this system with a fingerprint sensor requires expensive hardware and high processing costs.

3.4.1.4. Electroencephalography (EEG)

An EEG is a painless test that records brain activity. When the brain cells send messages to each other they produce tiny electrical signals [52]. The EEG signals are recorded from the subject while being exposed to a stimulus, which consist of drawings of objects chosen from Snodgrass and Vanderwart picture set [53]. However, the measurement of EEG-pulse is again an unrealistic to use EEG with only access to one fingertip.

3.4.1.5. Finger Skin Odour Analysis

This method is based on the acquisition of the odour by means of an electronic nose. The response in the presence of human skin differs from that in the presence of other materials [54], usually found in fake fingerprints. An odour sensor is used to sample the odour signal and an algorithm allows discrimination of the finger skin odour from that of other materials such as latex, silicone or gelatine [39-40, 54-55, 59-60]. The method of acquisition of an odour pattern consists of sampling the data coming from sensor during a given time interval, usually few seconds. The limit of this method is the time necessary to restore the sensor response, depending upon the sensor characteristic and environmental conditions. In addition, experimental results confirm that this method, which is able to effectively discriminate real fingerprints from fake, can be forged using a wide range of materials.

3.4.1.6. Temperature of Finger Tip Epidermis

Temperature is an involuntarily generated signal under the finger tip. It is quite easy to measure it, but not sufficient to detect liveness. Average temperature on fingertips ranges between 26 °C and 30 °C [56, 57]. However, the temperature depends on the health condition of the user (fever or poor blood circulation could influence the result of the liveness detection). This could make an impostor with a thin artificial fingerprint attached on his real finger be accepted and, on the other hand, a user with poor blood circulation or cold be rejected. If a thin silicon artificial fingerprint is patched on to a real finger, the temperature can be decreased by a maximum of 2 °C, which is well inside the working margins of the sensor. Sensors that are used outdoors often have a broader working margin, giving the intruder even more chance.

3.4.2. Skin Physiology Characteristics

Human fingers are known to display friction ridge skin that consists of a series of ridges and furrows, generally referred to as fingerprints (Fig. 3.13 a, b & c). The friction ridge skin is made of two major layers: dermis (inner layer) and epidermis (outer layer), as shown in Fig. 3.13 d & e.

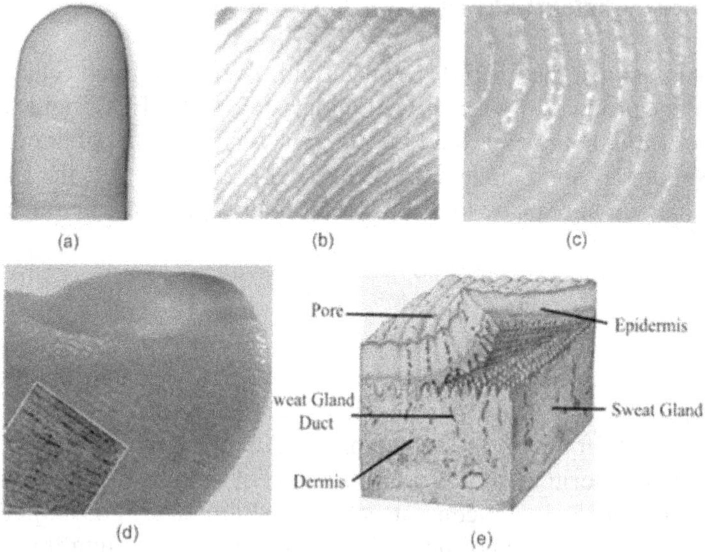

Fig. 3.13. (a) Finger Tip, (b) friction ridge skin, (c) Sweat pores on ridges, (d) The epidermis is partly lifted from the dermis, (e) A three-dimensional representation of the structure of ridged skin [58].

A typical young male has, on an average, 20.7 ridges per centimeter while a female has 23.4 ridges per centimeter [58]. It is suggested that friction ridges are composed of small "ridge units," each with a pore, and the number of ridge units and their locations on the ridge are randomly established. As a result, the shape, size, alignment of ridge units, and their fusion with an adjacent ridge unit are unique for each person (Fig. 3.13 a & c). Although, some cases exist where ridge units fail to compose a ridge, also known as dysplasia, independent ridge units still exist on the skin.

3.4.2.1. Pulse Oximetry

Pulse oximetry is used in the medical field to measure the oxygen saturation of hemoglobin in a patient's arterial blood [59]. In this method the pulse and blood oxygenation are measured by shining the beams of light through the finger tissue (Fig. 3.14).

The oxygenated hemoglobin allows red light to transmit through and absorbs more infrared light while the deoxygenated hemoglobin allows infrared to transmit through and absorbs more red light. Usually, a finger is placed between the source (LEDs) and the receiver (photodiode) acts as a translucent site with good blood flow. Once these absorption levels are detected from the finger, the ratio of absorption at different wavelengths can be obtained.

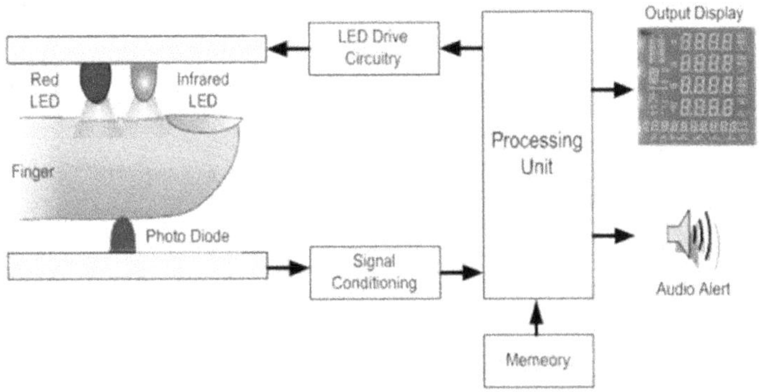

Fig. 3.14. Pulse Oximetry Technique.

The advantage of this method, inherent in its origin, is the well known principle of pulse oximetry, which is especially used in medicine; however, it takes a long time for scanning pulses and possibly the measurements will fail in cold weather because of the lack of microcirculation in the fingertip. In addition, it can easily be circumvented with the use of a thin fake finger layer made using gelatine.

3.4.2.2. Skin Spectroscopy

In skin spectroscopy, as optical technique is used to measure the absorption of light by tissue, fat, blood and melanin pigment. This measurement system is based on an optical source that illuminates a small area of fingertip skin with multiple wavelengths of visible and infrared light as shown in Fig. 3.15.

Light undergoes scattering and absorption at different layers of skin. Scattering is caused by the structural characteristics such as the arrangement of the collagen fibers, and absorption is caused by chromophores in the layers. The depth of light penetration depends on the wavelength of the light and the level of pigmentation. The light is reflected back after being scattered in the skin and is then measured for each of the wavelengths. The system analyzes the reflectance variability of the various wavelengths such as, 400–700 nm as they pass through the skin. Because the optical signal is affected by chemicals and other changes to the skin, skin spectroscopy also provides a sensitive and relatively easy way to confirm that a sample is living tissue [60, 61]. Furthermore, the reflectance spectrum of skin provides information regarding the distribution and concentration of various chromophores present in the skin and is highly dependent on the person's physical characteristics. Thus spectroscopic measurements can be successfully used as a biometric. However, the system needs a moderate environmental conditions and it might slow the access control process.

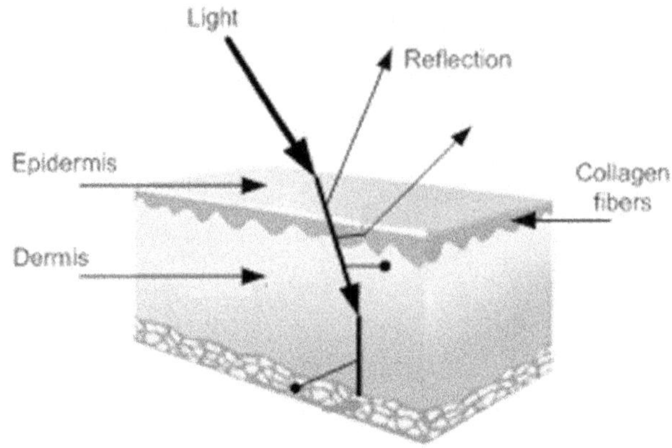

Fig. 3.15. Skin Spectrometry.

3.4.3. Stimulus Response

The stimulus response of the skin can be obtained by applying small external electrical signals and measuring the variation.

3.4.3.1. Skin Impedance

The skin impedance method is based on simultaneous measurements of the electrical bio impedance of different skin layers. The measurements are sensitive to skin properties like stratum corneum impedance and viable skin impedance. Dispersive behaviour of these layers can be detected in the measured frequency range and anisotropy in the stratum corneum [62].

The electrical impedance measurement system is illustrated in Fig. 3.16. This system uses an electrode array with three alternative current injecting electrode sets and one set of voltage pick-up electrodes. The two middle electrodes are connected to the differential voltage input of the frequency response analyzer, and the other three electrode pairs (denoted as i_1, i_2, i_3 in Fig. 3.16) were in turn connected to the internal oscillator. The frequency response analyzer consequently performed three successive four-electrode measurements, each time using the same voltage pick-up electrodes, but different current injecting electrode pairs (denoted as inner, middle and outer in Fig. 3.16). In each measurement, a live or fake finger was placed on the top of the electrode system with a light pressure and then the three frequencies scan was performed using an applied voltage on the current injecting electrodes with discrete frequencies.

The sensitivity field of a four-electrode system is found by taking the dot product of the current density vectors resulting from driving a unity current through the current injecting electrodes and voltage pick-up electrodes, respectively [62]. Up till now, this system is tested at prototype level and proposed as a liveness detection solution for fingerprints sensor modules. However, it is not tested or implemented with actual fingerprint modules.

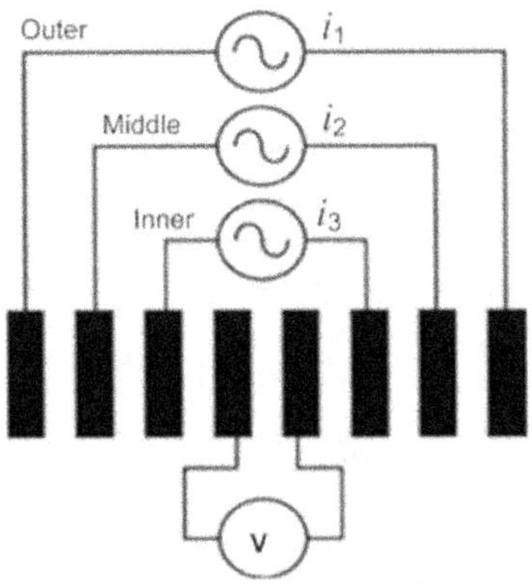

Fig. 3.16. Electrical Impedance Measurement System [62].

3.4.3.2. Electrotactile

The theory behind the electrotactile stimulation is the understanding of human touch sensation. The live skin is sensitive to temperature, vibrations, pressure, electrical voltage and current, with different receptors located at different depth of the skin for each sensation. Such tactile perception capability is absent in fake or dead skin. The eletcrotactile method uses an electrical means to directly activate the nerve to stimulate the sense of touch. Such a tactile system typically involves a matrix of surface electrodes that pervade very small, controlled electric currents into skin (Fig. 3.17).

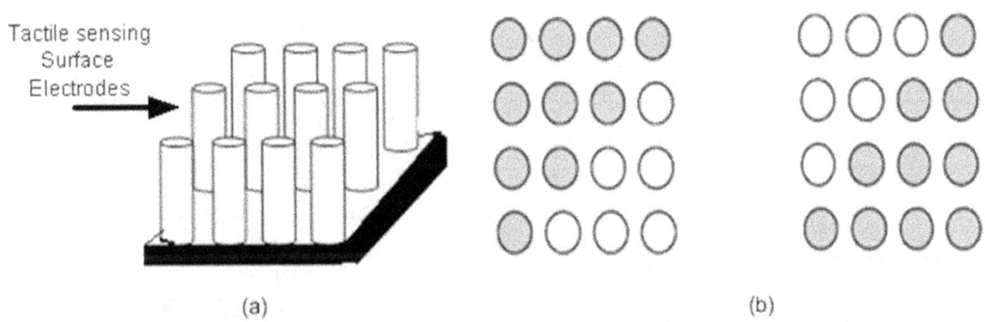

Fig. 3.17. (a) 3D representation of Electrotactile Electrode Array, (b) A 4-by-4 electrode pad showing an upper triangular (left) and lower triangular (right) tactile pattern. Each circle is an electrode. The dark circle represents an active electrode (having current flow) while the light circle represents an inactive electrode [63].

Liveness detection method based on electrical based tactile sensation is proposed in [63]. This system is capable of presenting a tactile pattern of fake and live finger tip skin. The initial results from prototype shows that the proposed approach is indeed able to detect gelatine fake fingers worn over live fingers, even when the gelatine is only 1 mm thick. The only uncertainty is whether a finger with a thin fake layer such as those made using gelatine, can still provide the person with sufficient touch perception to circumvent the system.

The following Table 3.1 summarizes the requirements and limitations of hardware based liveness detection methods discussed in the above sections.

Table 3.1. Summary of Hardware based Liveness Detection Techniques.

Methods	Liveness Detection Technique	Requirements	Limitations
Biological Signals	Epidermis Temperature	Additional Temperature sensing and signal processing hardware	Measurement Time and cost
	Blood Flow	Biomedical sensors and signal processing hardware	• Setup for measuring the noise free bio signals • Interface with fingerprint sensor Cost and size issues
	Electrocardiogram (ECG)		
	Electroencephalogram (EEG)	Detection Electrodes	• Easily fooled (with a small signal generator) • Impractical with access to only one fingertip
	Pulse Rate Detection	Optical Sensors and measurement systems	Environmental conditions Size and cost of hardware Integration with fingerprint scanners
	Odor Analysis	Sensors to detect different type of odor	• Extra sensors • Detection timing and Processing • Cost and system size
Skin Physiological Characteristics	Pulse oximetry	Optical sensors	• Biomedical hardware requirement • Complex signal processing
	Skin spectroscopy	Different Optical sources and sensors Additional measurement Systems	• Size of system • Additional signal processing and software requirements • Not feasible in mobile applications
	Skin Impedance	Electrodes Array External Dc Voltage	Skin condition can change resistance Additional power
	Eleltrotactile	Array of Electrodes with additional measurement system	Deign of electrode array and size Addition Signal processing hardware Integration of fingerprint sensors

The design, cost, user acceptance and implementation are common issues with the proposed hardware based liveness detection solutions. It needs more research and investigation to find a suitable solution that can bring the balance among price, user-friendliness and security of the system.

3.5. Conclusion

To discourage possible attackers from presenting a fake finger using artificial methods, it is important to ensure that the finger presented to the sensor is genuine and it is not fake/artificial or from a cadaver. Consequently, a live finger detection mechanism is a crucial part of any fingerprint identification system used for security reasons. In any practical fingerprint acquisition system, the live finger detection should be performed simultaneously with the capture of the fingerprint which seems unlikely with currently available fingerprint technologies without additional hardware. The only possibility is to redesign the fingerprint sensing technology that will be able to detect liveness from a placed finger without the addition of extra hardware.

For liveness detection in fingerprint sensors, it is necessary to choose a property of fingertip, which is difficult or impossible to imitate. The chosen liveness detection method should be easy to implement as a hardware or software solution, so that the final price of such a security solution would not be too high. For liveness detection from fingertips, an elegant method from among the many possible techniques senses the ionic activities in sweat pores present on fingertip ridges. The authors of this paper are investigating the novel liveness detection method on a fingertip using nanotubes/nanowires based sensor by detecting ionic activities from sweat pores. The findings of this research will be used to characterize a novel nanotechnology based fingerprint sensor which might help to reduce the FAR and FRR and lead to the design of such a sensor with inherent features of liveness detection.

References

[1]. Maltoni, D., Maio, D., Jain, A. K., Prabhakar, S., Handbook of Fingerprint Recognition, 2nd ed., *Springer-Verlag,* London, UK, 2009.

[2]. D. Maltoni and R. Cappelli, Advances in fingerprint modeling, *Image Vision Comput.,* Vol. 27, pp. 258-268, 2/2, 2009.

[3]. S. S. Adebisi, Fingerprint Studies - The Recent Challenges And Advancements: A Literary View, *The Internet Journal of Biological Anthropology,* Vol. 2, 2009, pp. 1-15.

[4]. A. Abhyankar and S. Schuckers, Integrating a wavelet based perspiration liveness check with fingerprint recognition, *Pattern Recognit,* Vol. 42, 3, 2009, pp. 452-464.

[5]. W. David, L. Mike, Six biometric devices point the finger at security, [Online], 2011(10/11), pp. 2, 2010, Available: http://www.networkcomputing.com/910/910r1.html

[6]. T. Matsumoto, H. Matsumoto, K. Yamada and S. Hoshino, Impact of artificial gummy fingers on fingerprint systems, *Proceedings of SPIE,* Vol. 4677, Optical Security and Counterfeit Deterrence Techniques IV, 2002, pp. 275-289.

[7]. S. A. C. Schuckers, Spoofing and Anti-Spoofing Measures, *Information Security Technical Report,* Vol. 7, 12, 2002, pp. 56-62.

[8]. Y. Flink, Million dollar border security machines fooled with ten cent tape, Find Biometrics-Global Identity Management [Online], 2010 (03/13), 2009. Available: http://www.findbiometrics.com/articles/i/6090/

[9]. D. Petrovska-Delacrétaz, G. Chollet, K. Anil and B. Dorizzi, Guide to Biometric Reference Systems and Performance Evaluation, *Springer-Verlag,* New York Inc, 2009.

[10]. T. Matsumoto. Gummy and conductive silicone rubber fingers importance of vulnerability analysis, in *Advances in Cryptology (ASIACRYPT 2002)*, Vol. 2501, pp. 574-575.

[11]. C. Barral and A. Tria, Fake Fingers in Fingerprint Recognition: Glycerin Supersedes Gelatin, *Formal to Practical Security*, 2009, pp. 57-69.

[12]. M. Tistarelli, S. Z. Li and R. Chellappa, Handbook of Remote Biometrics: for Surveillance and Security (Advances in Pattern Recognition), *Springer*, 2009.

[13]. U. Uludag and A. K. Jain, Attacks on biometric systems: A case study in fingerprints, in *Proc. SPIE-EI 2004, Security, Seganography and Watermarking of Multimedia Contents VI,* 2004, pp. 622-633.

[14]. C. Roberts, Biometric attack vectors and defences, *Comput. Secur.,* Vol. 26, 2, 2007, pp. 14-25.

[15]. Henry C. Lee, Robert E. Gaensslen, Advances in Fingerprint Technology, *CRC Press*, 2001.

[16]. C. Roberts. Biometric technologies – fingerprints, *National Cyber Security Centre*, New Zealand, 2006, Available: www.ccip.govt.nz/notes/biometrics-technologies-fingerprints.pdf

[17]. Jean-François Mainguet, Fingerprint sensors, [Online]. 2011 (08/12), 2009, Available: http://pagesperso-orange.fr/fingerchip/biometrics/types/fingerprint_sensors_productsi.htm

[18]. B. MIAXIS, FPR-620 Optical Fingerprint Reader, [online]. 2011 (8/02), Available: http://www.miaxis.net

[19]. H. Kang, B. Lee, H. Kim, D. Shin and J. Kim, A Study on Performance Evaluation of the Liveness Detection for Various Fingerprint Sensor Modules, *Knowledge-Based Intelligent Information and Engineering Systems*, 2003, pp. 1245-1253.

[20]. C. D. Tran, Principles, instrumentation, and applications of infrared multispectral imaging, an overview, *Anal. Lett.,* Vol. 38, 2005, pp. 735-752.

[21]. M. Ennis, R. Rowe, S. Corcoran and K. Nixon, Multispectral sensing for high-performance fingerprint biometric imaging, *Lumidigm, Inc.,* Available: http://www.Lumidigm.com/PDFs/Multispectral_Fingerprint_Imaging.pdf, 2005.

[22]. R. Rowe, K. Nixon and P. Butler, Multispectral Fingerprint Image Acquisition, *Advances in Biometrics*, 2008, pp. 3-23.

[23]. R. K. Rowe, K. Nixon and S. Corcoran, Multispectral fingerprint biometrics, in *Proceedings from the Sixth Annual IEEE SMC Information Assurance Workshop (IAW' 05),* 2005, pp. 14-20.

[24]. I. Lumidigm, Mercury desktop sensor, [Online], 2011(8/10), Available: http://www.lumidigm.com

[25]. Lumidigm Inc., Liveness detection, [Online], 2011(10/12), 2010. Available: http://www.lumidigm.com/liveness-detection/

[26]. K. Nixon and R. Rowe, Spoof detection using multispectral fingerprint imaging without enrollment, in *Proceedings of Biometrics Symposium (BSYM2005*), Arlington, VA, 2005.

[27]. G. Parziale, Touchless Fingerprinting Technology, *Advances in Biometrics*, 2008, pp. 25-48.

[28]. C. Lee, S. Lee and J. Kim, A study of touchless fingerprint recognition system, Structural, Syntactic, and Statistical Pattern Recognition, *Lecture Notes in Computer Science*, Vol. 4109, 2006, pp. 358-365.

[29]. Y. Chen, G. Parziale, E. Diaz-Santana and A. K. Jain, 3D touchless fingerprints: Compatibility with legacy rolled images, in *Proceedings of the Biometrics Symposium: 2006*, 19-21 August 2006, pp. 1-6.
[30]. iFingersys, Fingerprint Biometrics, Vol. 2011, 2011, pp. 08.
[31]. TBS inc. Touchless terminals- the TBS 3D guard series, [Online], 2011(9/13), Available: http://www.tbsinc.com
[32]. Flash Scan3D, 3D Touchless Fingerprint Biometrics, [Online], 2011(09/18), 2011, Available: http://www.flashscan3d.com
[33]. S. Memon, M. Sepasian and W. Balachandran, Review of finger print sensing technologies, in *Proceedings of the IEEE International Multitopic Conference (INMIC' 2008)*, pp. 226-231.
[34]. NeuroTechnology. Bio-i CYTE fingerprint sensor, [Online], 2010 (08/15), 2010. Available: http://www.neurotechnology.com/fingerprint-scanner-testech-bio-i.html.
[35]. T. Harris, Capacitive scanner, [Online], 2011(10/01), pp. 01, 2010. Available: http://computer.howstuffworks.com/fingerprint-scanner3.htm
[36]. i.UPEK, TCS1 TouchChip fingerprint sensor, [Online], 2011 (8/20), Available: http://www.upek.com
[37]. Bergdata Biometrics GmbH, Fingerprint-E field sensors, [Online], 2011 (11/03), pp. 01, 2010, Available: http://www.authentec.com/
[38]. D. R. Setlak, Advances in fingerprint sensors using RF imaging techniques, Automatic Fingerprint Recognition Systems, N. Ratha and R. Bolle, *Springer-Verlag*, New York, 2004.
[39]. Validity Inc., VFS201 Fingerprint Sensor Product Brief, [Online], 2010 (10/01), pp. 01, 2010, Available: http://www.validityinc.com
[40]. K. M. Chan, A. Pop, S. Safarkhah and G. Virdi, A Security Analysis of RF Biometric Fingerprint Scanners, [Online], Available: http://scholar.google.co.uk/scholar?cluster=14072940149125339727&hl=en&as_sdt=1,5
[41]. Hiro Han and Yasuhiro Koshimoto, Characteristics of thermal-type fingerprint sensor, *Biometric Technology for Human Identification, Proc. of SPIE 6944*, Vol. 6944 2008, pp. 1-12.
[42]. J. Han, Z. Tan, K. Sato and M. Shikida, Thermal characterization of micro heater arrays on a polyimide film substrate for fingerprint sensing applications, *J Micromech Microengineering,* Vol. 15, 2005, pp. 282.
[43]. G. MSC Vertriebs GmbH, FingerChip™, Atmel‘s Biometric Sensor, [Online], 2010 (10/01), pp. 01, 2010, Available: http://www.msc-ge.com/en/news/pressroom/newsletter/atmel/nl/2632-www.html
[44]. Y. Saijo, K. Kobayashi, N. Okada, N. Hozumi, Y. Hagiwara, A. Tanaka and T. Iwamoto, High frequency ultrasound imaging of surface and subsurface structures of fingerprints, in *Proceedings of the 30th IEEE Annual International Conference of the Engineering in Medicine and Biology Society, EMBS 2008,* pp. 2173-2176.
[45]. H. Morimura, S. Mutoh, H. Ishii and K. Machida, Integrated CMOS-MEMS technology and its applications, in *Proceedings of the 9th International Conference on Solid-State and Integrated-Circuit Technology (ICSICT' 2008),* pp. 2460-2463.
[46]. N. Sato, K. Machida, H. Morimura, S. Shigematsu, K. Kudou, M. Yano and H. Kyuragi, MEMS fingerprint sensor immune to various finger surface conditions, *IEEE Transactions on Electron Devices,* Vol. 50, 2003, pp. 1109-1116.
[47]. N. Sato, S. Shigematsu, H. Morimura, M. Yano, K. Kudou, T. Kamei and K. Machida, Novel surface structure and its fabrication process for MEMS fingerprint sensor, *IEEE Transactions on Electron Devices,* 52, 5, 2005, pp. 1026-1032.

[48]. M. Damghanian and B. Y. Majlis, Novel Design and Fabrication of High Sensitivity MEMS Capacitive Sensor Array for Fingerprint Imaging, *Advanced Materials Research,* Vol. 74, 2009, pp. 239-242.
[49]. Lapsley, Philip Dean Lee, Jonathan Alexander, Pare, Jr., David Ferrin Hoffman, Ned, Anti-fraud biometric scanner that accurately detects blood flow, *US Patent 5737439,* April 7, 1998.
[50]. K. N. Plataniotis, D. Hatzinakos and J. K. M. Lee, ECG biometric recognition without fiducial detection, in *Proceedings of the Biometrics Symposium 2006*, 19-21 August 2006, pp. 1-6.
[51]. G. Wübbeler, M. Stavridis, D. Kreiseler, R. D. Bousseljot and C. Elster, Verification of humans using the electrocardiogram, *Pattern Recog. Lett.,* Vol. 28, 2007, pp. 1172-1175.
[52]. S. Marcel and J. D. R. Millan, Person Authentication Using Brainwaves (EEG) and Maximum A Posteriori Model Adaptation, *IEEE Transactions on Pattern Analysis and Machine Intelligence,* Vol. 29, 2007, pp. 743-752.
[53]. R. Palaniappan and D. Mandic, EEG Based Biometric Framework for Automatic Identity Verification, *The Journal of VLSI Signal Processing,* Vol. 49, 11/01/, 2007, pp. 243-250.
[54]. P. E. Keller, Electronic noses and their applications, in *Proceedings of the IEEE Technical Applications Conference and Workshops (Northcon' 95),* 1995, pp. 116.
[55]. D. Baldisserra, A. Franco, D. Maio and D. Maltoni, Fake Fingerprint Detection by Odor Analysis, *Advances in Biometrics*, 2005, pp. 265-272.
[56]. Jim Edmond Riviere, Structure and function of skin, *CRC Press*, 2005, pp. 1-18.
[57]. M. Chaberski, Level 3 friction ridge research, *Biometric Technology Today,* Vol. 16, 12, 2008, pp. 9-12.
[58]. Anil K. Jain, Yi Chen and Meltem Demirkus, Pores and ridges: High-resolution fingerprint matching using level 3 features, *IEEE Transactions on Pattern Analysis and Machine Intelligence,* 29, 1, 2007, pp. 15-27.
[59]. R. Derakhshani, S. A. C. Schuckers, L. A. Hornak and L. O'Gorman, Determination of vitality from a non-invasive biomedical measurement for use in fingerprint scanners, *Pattern Recognit,* Vol. 36, 2, 2003, pp. 383-396.
[60]. Centre for Unified Biometrics and Sensors. Skin spectroscopy. *University of Buffalo, the State University of New York* [Online], 2010 (4/3), pp. 01. Available: http://www.cubs.buffalo.edu/skin.shtml
[61]. Ryan R. Emerging biometric technologies, *Secuity infoWatch* [Online], 2010 (03/12), 2009, Available: www.securityinfowatch.com/Access+Control/emerging-biometric-technologies
[62]. O. Martinsen, S. Clausen, J. B. Nysæther and S. Grimnes, Utilizing Characteristic Electrical Properties of the Epidermal Skin Layers to Detect Fake Fingers in Biometric Fingerprint Systems—A Pilot Study, *IEEE Transactions on Biomedical Engineering,* Vol. 54, 2007, pp. 891–894.
[63]. Wei-Yun Yau, Hai-Linh Tran and Eam-Khwang Teoh, Fake finger detection using an electrotactile display system, in *Proceedings of the 10th International Conference on Control, Automation, Robotics and Vision(ICARCV' 2008),* pp. 962-966.

Chapter 4
Plasma Polymerized Thin Film Sensors

K. A. Koparkar

4.1. Introduction

Plasma polymerization is gaining importance for last several years as a tool to modify material surfaces. Organic vapors can be polymerized at low temperatures using plasma enhancement. Plasma polymerization can also be used to produce polymer films of organic compounds that do not polymerize under normal chemical polymerization conditions because such processes involve electron impact dissociation and ionization for chemical reactions.

4.1.1. Historical Prospective

Plasma is not discovery of mankind [1]. The word plasma is derived from the ancient Greek language, where it meant 'that what is built', or 'that what is formed'. In modern language the term plasma means a more or less ionized gas, also called as 'glow discharge'. In 1879 Crookes first described this so-called state of ionized gas as 'a world where matter may exist in a fourth state'. Fifty years later, first Irving Langmuir in 1928 introduced the term 'plasma' which now a day known as fourth state of matter when he in his studies of electrified gases in vacuum tubes [2]. Lightning for instance, an electric discharge in air can be considered plasma. Also the stars or the polar light belong to the better-known plasma phenomena. In general, when a molecule is subjected to 'severe' conditions such as intense heat, the molecule will ionize. The sun, and other stars in our universe, have temperatures ranging from 5000 to 70,000 K or more, and consist entirely of plasma. In the laboratory, plasma can be generated by number of method such as combustion, flames, lasers, controlled nuclear reactions and shocks, but most experimental work carried out by means of an electrical discharge. Plasma polymerization can be defined as 'the formation of polymeric materials under the influence of plasma [3]. Solid deposits from organic compounds formed in a plasma, generated by some kind of electrical discharge, were already described as early as in

K. A. Koparkar
Department of Physics, Sant Gadge Baba Amravati University,
Amravati 444 602, India

1874 [4]. Systematic investigation regarding the synthesis of polymer via plasma polymerization started only in the 1960's. However, at that time very little was known about polymers, and these deposits were considered to be nothing more than undesirable by-products of phenomena associated with electrical discharge [5], following the rapid development in polymer science [3, 6]. Only over the past two decades, the advantages of plasma polymerization have been fully accepted.

4.1.2. Plasma State

In general, plasma can be referred to as the fourth state of matter, which is macroscopic composition of ions, electron, radicals and photons, highly excited, with a neutral electrical charge [15]. Plasma is a highly unusual and reactive environment. It is typically obtained by exciting a gas or vapor into high energetic states by radio frequency (RF), microwave (MW), alternating current (AC) or direct current (DC) current glow discharges, or as well as by the electrons of a hot filament discharge. The high density of ionized and excited species in the plasma can change the surface of normally inert materials [16]. To generate the plasma, it is necessary to ionize atoms or molecules in the gas phase. When an atom or molecules gains enough energy from an external excitation source or through collisions with another molecule, ionization occurs [16]. This happens usually when the molecules are under specific conditions, like extreme heat, generating the so-called hot plasmas, or under electrical glow discharge, for the cold plasmas [17]. The plasmas loose energy to their surroundings through collision and radiation processes; therefore energy must be supplied continuously to the system to maintain the plasma state. The easiest way to supply energy to a system in a continuous manner is with an electrical source. Therefore, electrical glow discharges are the most common plasmas [15].

In cold plasmas, the transfer of power from an electric field to electrons can generate a glow discharge. Different kinds of sources can be used, such as DC or RF glow discharge, or electron cyclotron resonance, as an example, to generate uniform plasma in a relative big area with a controlled electronic density [16]. In general, plasma is referred to as a partially ionized gas that contains positively and negatively charged particles [18]. The plasma state is more highly activated than the solid, liquid or gas state. Consequently, it found plenty application in modern era [19]. This whole transition process is shown in Fig. 4.1.

4.2. Plasma Polymerization

Thin film of materials can be prepared by so many approaches like spin coating, chemical bath deposition etc., but plasma polymerization technique of preparation of thin film is another approach, in which monomer can be converted in polymer in presence of glow discharge [20-22]. It refers to the deposition of polymer films through plasma dissociation, the excitation of an organic monomer gas and subsequent deposition and polymerization of the excited species on the surface of a substrate.

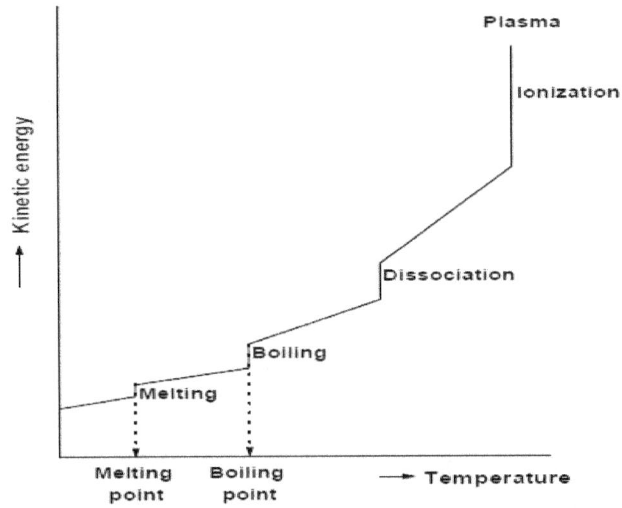

Fig. 4.1. Schematic diagram of the state transition processes (adopted from [19]).

Plasma polymerization is used to deposit films with thicknesses from several 10 to several 1000 of Angstroms. The deposited films are termed as plasma polymerized film, which are typically different from conventional polymers in chemical and physical manner. Whereas conventional polymerization is based on molecular processes during which rearrangements of the atoms within the monomer seldom occur, plasma polymerization is essentially an atomic process. Plasma polymers are, in most cases, highly branched and highly cross-linked [20]. Plasma polymerization is characterized by several features [22]:

1) Plasma polymers are not characterized by repeating units, as is typical for conventional polymers;
2) The properties of the plasma polymer are not determined by the monomer being used but rather by the plasma parameters. For example, an ethylene plasma does not simply give rise to polyethylene, but to a variety of products (including unsaturated groups, aromatic groups and side branches), depending on the plasma conditions;
3) The monomer used for plasma polymerization does not have to contain a functional group, such as a double bond.

Today, Plasma polymerization includes the fact that pinhole free, conformal thin films can be deposited on most substrates, using a relatively simple single-step coating procedure [7]. Plasma polymerization is accepted as an important process for the formation of entirely new materials, and as a valuable technique to modify the surfaces of polymers or of other materials for various applications. Additionally, a wide range of compounds can be chosen as a monomer for plasma polymerization, even saturated hydrocarbons, providing a great diversity of possible surface modifications. These advantages have resulted in the rapid development of plasma technology during the past decades, for applications ranging from adhesion to composite materials, protective coatings, printing, membranes, biomedical etc. Despite the efforts that have been made however, plasma polymerization remains a very complex process that is not well understood [8]. The structure of plasma deposited films is not as well defined as that of conventional polymers, and depends on many different factors such as design of the reactor [9], input power [10], monomer flow rate [11], substrate temperature [12], power level [23], and structure, pressure and flow rate of monomer [26-28]. Initially, most of the studies on plasma polymerization focused on the production of highly cross-linked films. Reactions were generally carried out under high energy plasma conditions, leading to a wide variety of ionization and chemical bond dissociation processes [13]. More recently, interest has shifted towards the synthesis of films that contain high concentrations of specific functional groups at the surface. Different techniques to improve the control over the film chemistry of plasma films have been studied, most of them based on decreasing the input energy during plasma polymerizations. One approach in particular has been very successful in achieving this enhanced film chemistry controllability. By employing pulsed radio frequency plasma instead of the traditional continuous wave plasma, Timmons et al. reported a high control over the film chemistry by variation of the plasma-on or plasma-off times. For many monomers, it was found that a more selective chemistry occurs during the plasma-off relaxation periods, leading to less cross-linked and more 'polymer-like' structures at relatively long off-times [14].

4.3. Structure of Plasma Polymerized Material

In plasma polymer reaction that take place in a plasma environment process are far more complex than those occurring in a conventional polymerization process because plasma polymers are not characterized by repeating units, short chains of monomers characterized by a high degree of cross-linking as compared to typical conventional polymers but many of low-temperature plasma surface modification researches are devoted to the polymer surface property, such as cross-linking, chain scission, creation of functional groups, and their relationship with the surface properties obtained after treatment.

There two products polymerization techniques as shown in Figs. 4.2 and 4.3 represent the difference between a conventional polymer film and plasma polymer film.

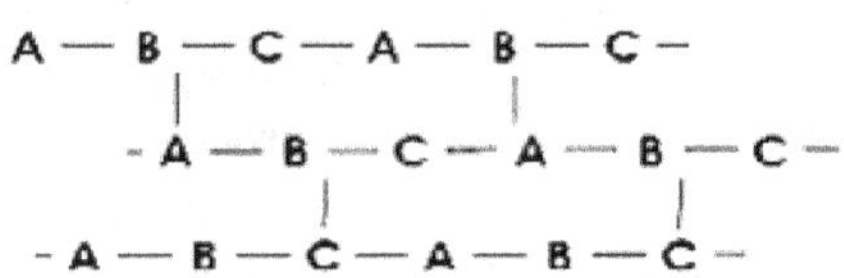

Fig. 4.2. Conventional polymer.

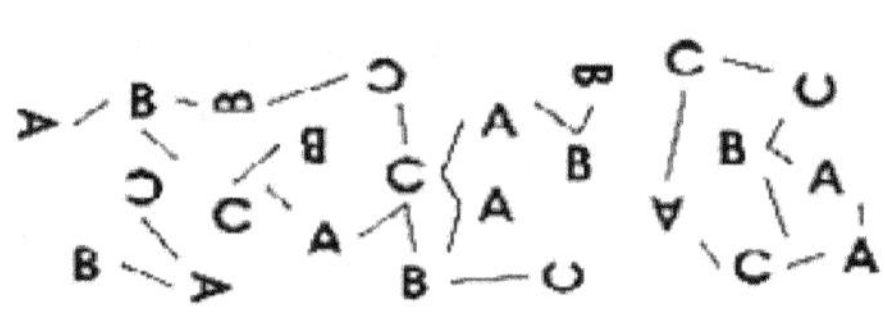

Fig. 4.3. Plasma polymer.

Conventional polymers are made of small unit bounded in long chains and few cross-linking .A, B and C can be the same "mer". The long range of order of conventional polymers, i.e. its degree of polymerization can be controlled by the polymerization parameters. On the other hand, Plasma polymer with short chains of building units A, B and C, and a high degree of cross-linking. A, B and C, are not necessarily identical. Plasma polymers are usually amorphous in nature, i.e. made of short chains of monomers characterized by a high degree of cross-linking and dangling bonds. In the case of hydrogenated carbonic films, it was shown that most carbon atoms are in the sp^3 and sp^2 configurations [91]. Depending on the glow discharge conditions (e.g. on the composite parameter W/FM), the ratio between those types of sites as well as the hydrogen content can be varied [84], in turn influencing [85] among others, the density, the hardness, the band gap, the conductivity, the refractive index and the functionalization. Indeed, a saturated sp^3 film will present less double bond making the polymer film less reactive to its environment.

4.4. Mechanism of Plasma Polymerization

Polymerization mechanism in chemical route is well described by various researchers. Generally, synthesis of conjugated polymers is done in presence of oxidant which provides oxidation potential to attach monomer unit to each other. The amount of energy required to removed the electron from active site of monomer and attach there another monomer unit was supplied by oxidant through oxidation process. Similarly in case of plasma polymerization the amount of energy required to attach monomer is supplied by plasma environment. The concept of this polymerization points out that elemental reactions occurring in plasma polymerization are the fragmentation of monomer molecules, the formation of active sites (radicals) from the monomer, and recombination of the activated fragments to form a polymer. H. Yasuda described this process as the rapid step growth of polymers due to fragmentation of monomers, which occur in a glow discharge region [29-31]. This process is attributed to the various types of collisions occurring simultaneously or separately in the reaction chamber. If fragmentation and recombination operate in plasma, the starting molecules for plasma polymerization will not be restricted to unsaturated compounds such as vinyl compounds; saturated compounds can also deposit plasma polymers. The fragmentation of starting molecules in plasma is represented by two types of reactions, namely the elimination of the weak hydrogen atom and the scission of the C–C bond (Fig. 4.4).

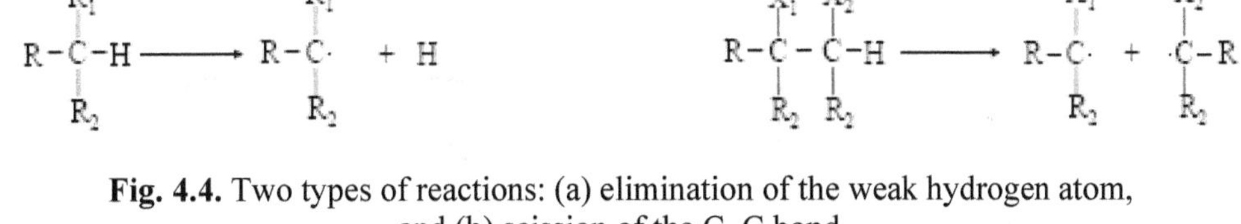

Fig. 4.4. Two types of reactions: (a) elimination of the weak hydrogen atom, and (b) scission of the C–C bond.

Hydrogen elimination is considered to contribute greatly to the polymer-forming process in plasma polymerization. The gas phase of a closed system after plasma polymerization of hydrocarbons is mainly composed of hydrogen as reported by Hillman and co-workers [32]. Fig. 4.5 shows an essential polymer forming process as proposed by H. Yasuda, which can be related to the elimination of hydrogen and scission of C–C bonds. When hydrogen atoms are eliminated from monomer molecules by plasma to form mono-radicals Mi, and bi-radicals Mk, and thereafter, the addition of the radicals to monomer and the recombination between two radicals proceeds to make large molecules with or without radical. Mono-radicals Mi add to the monomer to form a new radical Mj –M. Mono-radical Mi and may recombine with Mj to build a neutral or stable molecule Mi –Mj. A bi-radical can also attack the monomer to form a new bi-radical Mk –M, or recombine with another bi-radical to form a new bi-radical Mk – Mj. The new monomer Mi –Mj is activated by plasma to form mono- or bi-functional activated species. This cycle I is the repeated activation of the reaction products from mono-functional activated species, cycle II describes the recombination of the mono and bi-radical to form larger radicals.

The fragmentation to yield these species depends on how much the electric energy (RF power) and the amount of monomer injected into the system to maintain the plasma. The polymer deposition rate has been related to the Yasuda parameter, (W/FM) which describes the ratio of input power, W to the flow rate, F and monomer molecular weight, M [29]. As soon as plasma is created, the gas phase becomes a mixture of the original monomer and species created. This implies that any material that interacts with plasma becomes a source of monomer for plasma polymerization. Fig. 4.6 depicts the overall scheme of plasma polymerization, which encompasses the principle of the competitive ablation and polymerization (CAP) mechanism [32, 33].

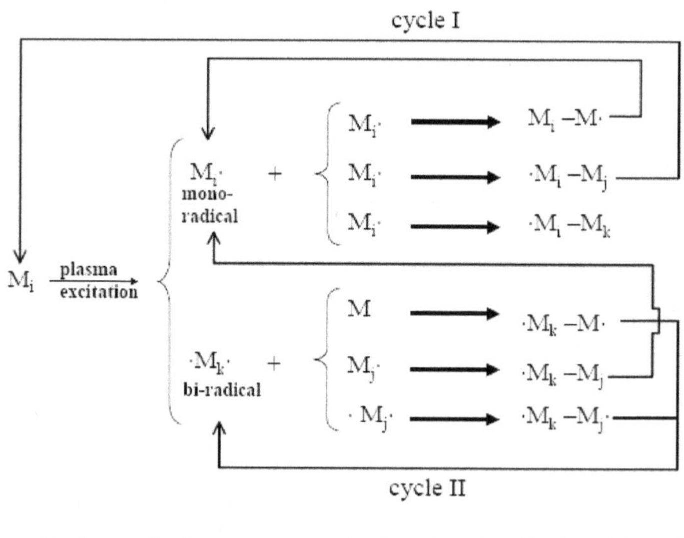

Fig. 4.5. The mechanism of plasma polymerization is visualized a bi-cyclic step-growth. Subscripts i, j, k, merely indicate the difference in the size of species involved (i = j is possible, if i = j = 1 corresponds to original monomer) (adapted from [29]).

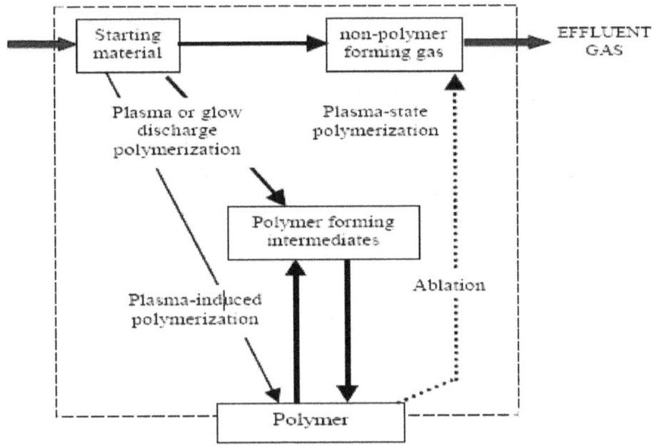

Fig. 4.6. Schematic representation of plasma polymerization mechanism (adapted from [29]).

Thus, the two types of plasma-mechanisms can be differentiated, the plasma-state and plasma-induced polymerization. The plasma-state polymerization describes the main route of polymer formation under the influence of plasma; whereas, plasma-induced polymerization represents the polymerization of condensed monomer on the surface by plasma interaction. In contrast to conventional polymerization, plasma polymerization forms an irregular network-like material which varies from a low to a high cross-linked density [33]. The plasma polymerization of acetylene, ethylene, ethane, styrene and other hydrocarbons have been investigated, which showed different chemical characteristics as compared to conventional method of polymerization. Investigations were based on various operational parameters such input power, deposition pressure and monomers flow. In a conventional polymerization, monomers are linked together through chemical reactions without any alteration of chemical structure of the monomer or, in some cases, with small alteration by loss of fragments from two monomers, e.g. condensation reaction. Therefore, the chemical structure of the formed polymer chains is well predicted from the chemical structure of the monomer. In plasma polymerization, the chemical structure of polymers formed, if the same monomer was used, is never predicted from the monomer structure. In some cases, due to atmospheric reactions, nitrogen, or oxygen are incorporated into the polymers formed by radical combination reactions. Thus, using the conventional method to prepare polymer films involves steps such as monomer synthesis, polymerization, and preparation of coating solution, cleaning or conditioning of substrate, drying and curing. This can be replaced by a "single–step process" in the plasma polymerization, which can be credited to the complication of the chemical reactions in such a process.

4.5. Plasma Deposition Technique

Recently, synthesis of thin polymer films by use of plasma polymerization has emerged as an interesting new approach. Plasma polymerization is a 'dry' technique that doesn't require the use of solvents. The plasma processes are an emergent technology in industrial processes. It has lot of technological applications, developed in the latest decennia; this technique could only be made possible with plasma assisted methods. Examples of industrial plasma processes are surface modification, pollution control, sterilization, cleaning and etching. Plasma can also be found in a lot of well known applications, such as lamps and plasma screens [34, 35]. From an ecological and economical point of view, the absence of solvents and solvent waste offers a lot of advantages. Whereas plasma polymerization was first used for the formation of thin passive layers, it is now used for many different applications.

4.5.1. Low Pressure Plasma Deposition

In this review, literature survey of conjugated polymers at low pressure has been studied. The plasma depositions were performed at low pressure, which was especially successful for deposition of wear resistant coatings and dense gas barrier layers. These reactive vacuum plasmas often result in a significant fragmentation of the monomer structure. The suggestions regarding physical properties of the low pressure plasma polymerized film is discussed by many research groups. The conjugated polymer is deposited in a glow discharge at low pressure after flash evaporating an atomized monomer into the reactor chamber. Vacuum plasma deposition of polyaniline is studied by Cruz et al [36]. In plasma, the aromatic rings of aniline had partially degraded and the resulting deposition was cross-linked. The conductivity of the plasma deposited polyaniline was low and dependent on the humidity. Aromatic rings were fragmented with formation of aliphatic structures is reported by Paterno et al [37] and Zaharias et al [38]. The aromatic ring of aniline was partially retained after plasma polymerization, resulting in a considerably low dielectric constant. Iodine doping decreased the band gap. Plasma polyaniline doped in an iodine chamber was compared to in situ doped plasma polyaniline, which was more stable it showed that Mathai et al [39-41]. When they doped plasma polyaniline with iodine has similar results of Bhat et al [42] and Olayo et al [43, 44]. Plasma deposited polyaniline at low pressure with simultaneous injection of the dopant toluene sulphonic acid, Nastase et al found that in an increased conductivity [45]. Wang et al report that low pressure plasma deposited polyaniline is different compared to chemical and electrochemical synthesis methods. Oligomeric depositions with an unconventional benzene substitution pattern were obtained. The conductivity of plasma polyaniline was significantly lower [46]. Gong et al also deposited a polyaniline like structure with low pressure plasmas [47]. Low pressure plasma polymerization of thiophene is studied by Silverstein et al [48]. The molecular structure of the obtained plasma polymer contained opened thiophene rings and oxygen and nitrogen atoms. A conducting iodine percolation network seemed to have formed upon doping with iodine. Kim et al. shows that plasma deposited polythiophene like structures with reduced band gap [49-51]. Vasquez et al obtained in situ chlorine doped plasma polythiophene by simultaneous injection of chloroform. The conductivity was dependent on the humidity [52]. Groenewoud et al reported the influence of substitutions on the plasma polymerization of thiophene. Iodine doping resulted in charge transfer complexes and carbon iodine chemical bonds. Since the chemical structure of plasma polythiophene changed after exposure to air before doping, the conductivity decreased. Higher conductivities could be reached when the pressure during plasma polymerization was slightly increased, because this resulted in less fragmentation [53-55]. Bae et al reported that plasma deposited polythiophene layers can be used for corrosion protection of metals [56]. Plasma depositions of polythiophene at low pressure were also obtained by other authors [57-62]. The lifetime of chemically produced PEDOT coatings by mild oxygen plasma treatment is improved Zhou et al [63]. Plasma deposited polypyrrole like coatings at low pressure which is a much lower conductivity than conventional polypyrrole is found by Morales et al [64]. Zhang et al reports the formation of a dense yellow film when plasma polymerizing pyrrole. Although these plasma polypyrrole coatings still contained a certain amount of intact pyrrole rings, structural degradation by oxygen contamination, was also observed [65]. Martin et al deposited polypyrrole and polythiophene thin films with a vacuum pulsed RF plasma. The use of a pulsed plasma, which has a lower ionization degree, resulted in slower deposition rates but the chemical structure was better retained and the coating roughness extremely low [66]. Cruz et al reports vacuum plasma deposition of in situ iodine doped polypyrrole. These polypyrrole films were hygroscopic and absorbed water from ambient air, which increased chain mobility of the polymer. This induced ordering of the polymer chains and increased conductivity. Undoped polypyrrole seemed to have more capacity to store water than doped polypyrrole. [67]. Wang et al [68]. The conductivity after iodine doping was significantly lower and less stable. Due to the low conductivity of plasma polymerized polypyrrole, reported by most authors, Kumar et al. [69] even consider using it as a dielectric material. He succeeded in plasma polymerizing polypyrrole like films with a high dielectric constant. Hosono et al show that plasma polymerized polypyrrole coatings, doped with 4-ethylbenzenesulfonic acid can be used to sensor volatile organic compounds. Sensitivity of the sensor was dependent on humidity [70]. Plasma polymerization of polypyrrole coatings at low pressure and Plasma copolymerization can easily be done by simultaneously injecting two monomers is studied by John et al [71] and Eufinger et al [72]. Morales et al reported that plasma copolymererization at low pressure of pyrrole and aniline. The monomer mass ratio influenced the structural evolution and the physical characteristics of the copolymers [73]. Plasma copolymerized pyrrole and silicon containing compounds (hexamethyldisiloxane, hexamethyldisilane and hexamethyldisilazane). Copolymerization had an effect on the physical properties of the film and the hydrophobicity of the surface is reported that Hirotsu et al [74]. Plasma deposited polyaniline-silicon dioxide and polythiophene-silicon dioxide nanocomposites by spraying dispersions of nano sized silica particles in monomer liquids, into the plasma zone is studied by Nastase et al [75]. Synthesis of vacuum plasma polymerized bi-layers is done by Morales et al [76]. Plasma polyaniline/polypyrrole bi-layers were chemically bonded at the interface and had a higher conductivity after doping compared to the separated homopolymers.

4.5.2. Atmospheric Pressure Plasma Deposition

The plasma coating at atmospheric pressure is viable and low cost technique due to no need of expensive vacuum equipment and higher deposition rates. A typical application of plasma technology at atmospheric pressure is the deposition of siloxanes. Some authors show that polymer functionalities can be retained after plasma deposition at atmospheric pressure, which is less aggressive than deposition at low pressure. Klages et also showed that polymerization of glucidyl methacrylate in a pulsed plasma reactor at atmospheric pressure resulted in retention of 90 % of the epoxy groups of glucidyl methacrylate [77-82]. Moreover, Ward et al report that ultrasonic atomization of acrylic acid in an atmospheric pressure helium glow discharge results in deposition of structurally well-defined polymer films. Polymerization proceeded predominantly via the carbon double bond with high retention of the carboxylic acid group. Despite good results with plasma deposition at atmospheric pressure of other polymers, plasma polymerization of conjugated polymers has never been documented before [83].

4.6. Different Techniques to Produce Plasma Discharge

The first experiment with electrical discharges were carried out in 1796, The invention of other types electrical discharges like high voltage radio frequency and microwave discharges, lasers induced plasma polymerization, came up and were explored as well. It was the availability of radio frequency and microwave generators that focused attention on electrode less discharges.

There are two types of plasma can be produced by electrical discharges namely, hot or equilibrium plasma and cold or non-equilibrium plasma. The plasma produced in silent as well as various types of glow discharges belongs to cold plasma which is characterized by a low gas temperature and a high electron temperature. Hot plasma characterized by a high gas temperature and an approximate equality between the gas and electron temperature belongs to that produced in arcs and plasma torches. They are Silent discharge, Direct current (DC) discharge, Alternating current (AC) discharge, Radio frequency (RF) discharge, and microwave (MW) discharge.

The Silent discharge is easy to produce and can operate at pressure up to 1 atmosphere. The discharge tube is usually made of glass or quartz and consists of a co-axial section, the inner and the outer surfaces of which are connected to high voltage power supply (10-20 kV). The discharge can be sustained at line frequency (50 Hz) as well as higher frequencies. The use of higher frequencies has the advantage of higher power transfer at lower voltages. Microwave generators with output between a few watts and a kilowatt can be employed for plasma generation. From the generator (Magnetron or Klystron) the micro wave power to the resonant cavity enclosing the reactor is fed by a co-axial cable or wave guide. With inorganic compounds microwave discharge has been used successfully [86], but with organic substance it is often difficult to avoid complete destruction of the material. Though the discharge equipment can be easily set up, it is difficult to initiate and sustain the discharge at low pressure (~1 Torr). Hence the experiments are usually carried out at pressure of several Torr. This is disadvantageous to certain extent, because as a result of the higher pressure, the gas temperature can rise to several hundreds of degrees, leading to pyrolitic decomposition of the reactants.

In the DC discharge, two metal electrodes are placed inside the plasma reactor and a high voltage is applied to produce plasma [87, 88]. First the reactor cell is evacuated to Torr and pure monomer vapor is fed into the cell raising the pressure ranging from 10^{-4} to 1 Torr. When DC field is applied to the electrodes, the monomer molecules are polymerized and deposited mostly on the cathode. It is observed that if the formed polymer is an insulator, afire forming a few layers of the film, the glow discharge ceases. It is found that when the applied voltage is increased, arcing is developed which disturbs the film. Hence this method cannot be used for the preparation of insulator polymer films of sufficient thickness.

In A.C. discharge, in typical arrangement, two metal electrodes are placed inside the reaction chamber and after evacuating the chamber to about 10' Torr, a low frequency (50 Hz) with high AC voltage (500-1500 V) is applied to obtain the discharge. A detailed description of the reactor chamber is given elsewhere [89]. In this apparatus instead of AC voltage, DC voltage ranging from 500-2000 V can also be employed to obtain thc glow discharge [89].

In RF discharge usually we use above 1 MHz. In this method direct contact between the electrodes and the plasma is not necessary; the electrical power can be fed to the plasma tube either by inductive coupling or by capacitive coupling. In the inductive coupling method, the discharge tube is located at the axis of a coupling coil and radio frequency power is fed to the discharge tube through this coil (Fig. 4.7).

In some cases polymerization as well as deposition takes place in the tail of the glow-discharge [90]. In the capacitive coupling arrangement [91], the power is fed to the discharge tube by means of two copper foil tubes situated few centimeters apart (Fig. 4.8). The working frequencies in radio frequency plasma experiments are limited to below 200 MHz because, with higher and higher working frequencies. Since plasma reactions do not depend appreciably on the operating frequencies, their choice is mainly determined by considerations of handling or shielding. When working in normal unshielded laboratories, care should be taken not to **interfere** with radio frequency signals.

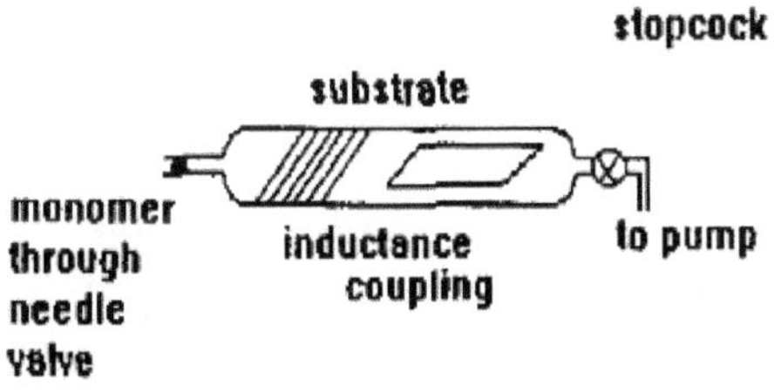

Fig. 4.7. Schematic diagram of inductive coupled RF discharges set up.

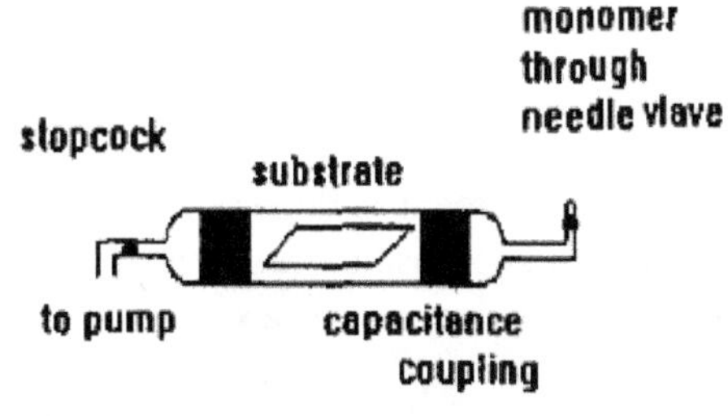

Fig. 4.8. Schematic diagram of capacitive coupled RF discharge set up [91]).

4.7. Principle of Paschen's Law

Paschen's law was discovered by Friedrich Paschen in 1889. This law states that the breakdown voltage between parallel plates immersed in gas can be described as a function of the product of the operating pressure, p, and the gap between the electrodes. Townsend explaining mechanism of the first model, the breakdown gaseous phenomenon in DC discharges in a reactor with two plane parallel electrodes in 1910 [92] and further developed it in 1915 [93]. This can be summed up as follows.

Some electrons (primary electrons) produced by natural sources (e.g. cosmic rays) or technical sources (e.g. UV light) are present in the gas phase. Accelerated by the applied electric field, the primary electrons gain energy and may reach the ionization energy of the neutral gas; the probability for an electron to ionize neutral molecules per unit length, is defined as the first Townsend coefficient α. The ionizing collisions with the gas produce further electrons and ions, which accelerate along the electric field lines: the electrons towards the anode, the positive ions towards the cathode. The ions that hit the cathode with sufficient energy release "secondary" electrons with second Townsend coefficient γ. Based on this model [94], Paschen's law reflects the Townsend breakdown mechanism and is derived from the discharge current expression:

$$i = [i_0 \exp(\alpha \times d)] / [1 - \gamma \times (\exp(\alpha \times d) - 1)] \quad (4.1)$$

Townsend derived an equation for the current i between the electrodes with a separating distance d [95]; the current was measured for three different inter-electrode distances leading to three independent equations. Then, by fitting the part of the current growth characteristics satisfying Equation 4.1, the values of the pre-breakdown current i_0, which is the dark current flowing between the electrode before the inception of the glow and of the two Townsend coefficients [93].

As early as 1889 [97], Paschen's law is the combined effect of the pressure (p) and the inter-electrode distance (d) on the breakdown voltage was experimentally studied. On other hand, breakdown voltage is a function of the product of the pressure and the inter-electrode distance,

$$V_b = f(pd) \quad (4.2)$$

Such a measurement is depicted in Fig. 4.9. Paschen's law is one of the most important relations in gas discharge physics and pioneered the mathematical formulation of dielectric breakdown of gases.

It displays a sharp decrease, a minimum and slow increase of the breakdown voltage as the product pd increases:

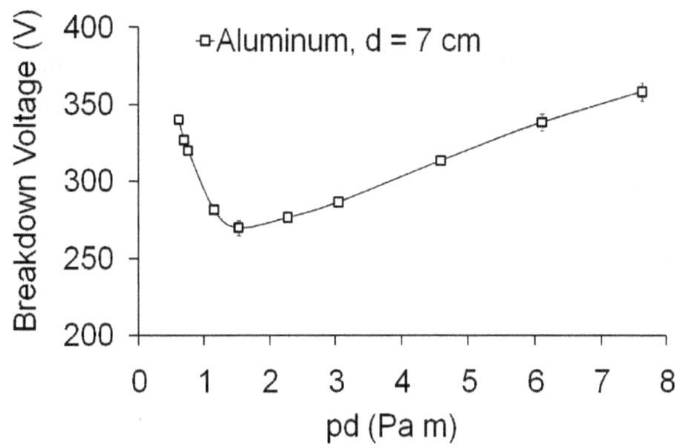

Fig. 4.9. A Paschen curve in an argon discharge with an inter-electrode distance of 7 cm with aluminum electrodes (20 cm^2) (adapted from [97]).

$$V_b = [B_{pd}] / [C + \ln (pd)] \tag{4.3}$$

Above equation described the shape of the Paschen curve.

The sharp increase left of the minimum is explained by the number of gas molecule becoming scarce and the consequent decrease of the ionization probability, This slows that the increase right of the minimum with decrease of the mean free path of the electrons and the consequent need to increase the applied electric field for the electrons to reach the ionization potential of the gas molecules.

4.8. Selection of Gases with Liquids

In plasma polymerization method, we required medium between two electrodes to enhanced glow discharge. This results in higher efficiency of plasma polymerization set up. The gaseous medium in plasma polymerization process is crucial, consequently selection of medium play very important role. The selection specific combination of gases and liquids can be used independently or in combination with other gases to create specific modifications (Table 4.1). The challenge is to choose gases and/or liquids that will yield desired properties for the customer's specific application.

4.9. Surface Modification of Plasma Materials

In the recent years, there has been an enormous research work reported on the use of low-temperature plasmas for the surface process to avoid the chemical waste problems related with wet processing. These types of plasma reactors which use chemically reactive gases might be toxic or corrosive and must be carefully processed, but the much smaller quantities compared to wet processing make this problem much less severe in plasma surface processing. Low-temperature plasma surface modification is the process that alters the physical and chemical characteristics of a surface of material without changing its bulk material properties [99]. This behavior of plasma polymerized surface becoming attractive and can be done on metals and alloys as well as ceramics and polymers. Some of the surface properties that can be modified by low-temperature plasma process include in hardness, adhesion, corrosion, fatigue, friction, oxidation, resistivity, toughness, and wear properties. The examples of low-temperature plasma surface modification processes are plasma oxidation, plasma nitrating, and polymer surface modification.

Table 4.1. Sample gases and liquids used to create plasmas [98].

Gas	Liquid
Oxygen	Methanol
Argon	Water
Helium	Allyl Amine
Nitrogen	Ethylenediamine
Ammonia	Acrylic Acid
Hydrogen	Acetone
Nitrous Oxide	Hydroxyethylmethacrylate
Carbon Dioxide	Ethanol
Air	Toluene
Methane	Diaminopropane
Ethane	Butylamine
Ethylene	Gluteraldehyde
Acetylene	Hexamethyldisiloxane
Tetrafluoromethane	Tetramethylsilane
Hexafluoroethane	Polyethylene glycols
Hexafluoropropylene	Diglymes

4.9.1. Plasma Enhanced Chemical Vapor Deposition (PECVD)

In this method, gas or vapor are inputted into a relatively low power density plasma system. The gas molecules are dissociated by plasma, creating reactive radicals or atoms that condense on the substrate [99]. Low temperature plasma processes are usually preferred to deposit materials on surfaces. Plasma chemical vapor deposition (PECVD) is the technique of forming solid deposition by initiating chemical reactions in plasma state [99]. Its major advantage over thermal chemical vapor deposition is the ability to deposit films at relatively low substrate temperatures [101, 103]. Plasma chemical vapor deposition (PECVD) is especially useful for materials that might vaporize, flow, diffuse, or undergo a chemical reaction at the higher temperatures. Usually, the substrates are heated to improve the film quality, but the substrate temperatures which are required are usually significantly lower than those required for thermal chemical vapor deposition [105]. As was the case for the previously discussed deposition methods, energetic ion bombardment of the growing film has a large influence on films deposited by plasma chemical vapor deposition, but again the mechanistic understanding is missing. In spite of the lack of mechanistic understanding of this process, plasma chemical vapor deposition has many applications in industry. One of the most important applications is the deposition of amorphous hydrogenated silicon films by the plasma decomposition of silane [101]. Another area in plasma chemical vapor deposition is widely used is in the final passivation of integrated circuits, which involves the deposition of silicon nitride (from silane and ammonia or nitrogen), silicon oxide (from silane and nitrous oxide), and silicon oxy-nitride (from silane, ammonia, and nitrous oxide) [105]. The other currently very popular area is the deposition of diamond films (from methane and excess hydrogen) [100]. Plasma polymerization is very similar to plasma chemical vapor deposition, the major difference being that in plasma chemical vapor deposition is concerned with depositing inorganic films, whereas in plasma polymerization the resulting film is an organic polymer [101, 106]. Plasma polymerization is the formation of polymeric materials under the influence of plasma [102]. It is an atomic polymerization different from conventional polymerization [99]. Most of organic gases and vapors, even gases that cannot produce polymer by conventional polymerization methods such as methane, would produce polymeric material (plasma polymers) in the glow discharge. In the system, polymeric materials are deposited on the solid surfaces forming ultra-thin solid films called plasma polymers. Plasma polymers have several advantageous properties such as being pinhole free, highly branched, and highly crossed linked [102]. Although they can form powdery and oily products in certain cases, the formation of such products can be prevented by controlling the system conditions and the reactor design. Due to the reaction, in which high energetic electrons break the chemical bond in plasma polymerization system, plasma polymers have no manifestly repeating units as conventional polymers. Therefore, it is hard to determine the properties of plasma polymers from used monomers in contrast to conventional polymerization which are linked together with a mere alternation of the chemical structure of monomer. Plasma polymerization is the formation of an anti-corrosion protecting interlayer by introducing silicon carbon monomer gas into a low power density glow discharge.

In recent research, the direct current (DC) glow discharge has an ideal method to coat a metallic substrate by plasma polymerization because the polymer nearly entirely deposits on the surface of cathode [107].

4.10. Plasma Polymerized Thin Film Sensor

This work made use of cleaner technologies and environmental friendly reagents to produce polymer films with adsorption properties to be used as sensors in the environmental field. The gas sensors fabricated by using conducting polymers. A gas sensor exhibits the property of changing the resistivity of the sensing material when it is exposed to different gas atmospheres. Today's modern day society has brought numerous luxury items but with them series of problems like air pollution and emission of toxic gasses have also been introduced to our society. The necessity to constantly monitor and control these gases emitted sprouted the need for gas sensors. The various uses of gas sensors vary across a wide range and industries and applications; for example monitoring air pollution, chemical processes and exhaust from combustion engines. Literature survey of gas sensor reflects many approaches of preparation of gas sensing material. But according to present review discussion is restricted to gas sensing material prepared by plasma polymerization method.

Hellegouarch et al reported that, Transparent conducting tin oxide materials have been used in many important applications, such as liquid crystal displays, solar cell, protective coatings and more recently as solid-state sensors to reducing gases. Today, tin oxide sensitive films can be elaborated by many different processes, such as PVD and CVD. SnO_2 thin films are elaborated by a PECVD process in a RF glow discharge reactor. The films have been studied diode reactor according to their properties and their response to specific gases such as ethanol. The detection results show best sensitivities for unbiased films to ethanol at 450 °C at 45 parts-per-billion (ppm) [136].

A. Chaturvedi et al shows that, in this paper deals with selectivity and sensitivity studies on plasma treated tin oxide thick film gas sensors. Tin dioxide thin films were deposited in a low pressure reactor with capacitively coupled electrodes. The electrical power was supplied by an RF 13.56 MHz generator. Response of oxygen, hydrogen, nitrogen and argon plasma treated sensors has been studied upon exposure to CCl_4, C_3H_7OH, CO, LPG, N_2O and CH_4. He shows that the sensitivity of sensors treated in various gaseous. The selectivity of these sensors also improves upon plasma treatment, specifically; hydrogen plasma treated sensor becomes sensitive to CO. This sensor based on two basic parameters, namely the temperature and the concentration. We have demonstrated that the unbiased films presented better responses with higher sensitivities [137].

Shi Liqin et al studies by the induction plasma deposition process is possible to fabricate a gas sensitive film consisting of nanostructured particles supported by rather large particles by controlling several processing parameters of plasma deposition. The development of highly sensitive NO_2 gas sensor, SnO_2 based film is prepared by using radio-frequency induction plasma deposition (RFIPD) method. In this study, the effect of processing parameters and annealing temperatures on the physical properties of thin films was examined by variance characterization method. The sensor films were used to detect gas at a concentration range from 20 to 200 ppb, and the results show that the gas sensors have achieved high sensitivity at extremely low NO_2 concentration down to 20 ppb with the response and recovery time of the SnO_2 gas sensor for NO_2 gas is less than several minutes [138].

F. Kraus et al shows that, the capacitive humidity sensor are fabricated employing a dielectric made from HMDSM at different condition of plasma polymerization setup. The silicon organic thin films of hexamethyldisilazane (HMDSM) prepared by plasma polymerization method. The designing of humanity sensor including comprises interdigital electrodes, the plasma polymer, and a porous top electrode. Capacitive humidity are fabricated employing a plasma polymerization of HMDSM thin film as a moisture sensitive material The structure of humanity sensor is small and characterized exhibiting a response time below good reversibility, and sensitivity [139].

R.V. Dabhade et al reported that, the polymers are used for humidity sensing for quite some time. Today, Chemical modification has also been carried to impart the desired characteristics to the polymers; surface functionalization of polymer surface by oleum has been carried for improving the humidity sensing characteristics of the polymer. Surface modification Plasma polymers has been carried out for a long time namely for adhesion improvement and biomedical applications, he report to the best of our knowledge for the first time humidity sensing by argon plasma-treated polymer namely PMMA with encouraging results. A capacitive RH sensor was fabricated using the argon plasma-treated PMMA as the dielectric for relative humidity sensing. the sensitivity is not linear throughout the RH range measured but as the order of nF are observed in the 0–40% RH range indicating usable range as humidity sensor. In the present work, we are reporting a novel way to synthesize humidity sensing polymeric material namely PMMA by argon plasma treatment. The possible use of plasma surface treatment of polymers for humidity sensing. The plasma polymerized films can be easily tailored by varying the plasma parameters [140].

4.11. Characterization of Plasma Polymers

According to the type of monomer used several types of plasma polymers are distinguished [109,110]. The plasma polymerized materials is characterized by various techniques regarding their application. A number of characterization techniques are used that include: FTIR, XPS, AES, SIMS, RHEED, NMR, ESR, ERDA, RBS, NRA, DSC, TGA, XRD (SAXS), SEM, and AFM (STM). Most frequently used are FTIR, XPS, RBS, ERDA, SEM, and AFM.

Electrical properties are studied by conventional DC techniques, AC properties by impedance spectroscopy.

Optical properties in the visible region use transmission and absorption and ellipsometric measurements.

Surface properties such as wettability (contact angle) [111], adhesion and mechanical properties such as internal stress, hardness, abrasion and friction are measured by routine thin film methods [112].

4.12. Difficulties in Plasma Polymerization Method

The low pressure plasma has been come to a great value in fundamental research as well as plasma technology. This technique lacks in some points such as the necessity of the vacuum system which leads to the costly operation and the batch processes that are unsuitable for the continuous sequence of manufacturing because the samples needed to be transferred into and out of a vacuum system. In low temperature plasma processing is carried out by transferring electrical energy in between two electrodes under reduced pressure in the range of 10 to 1000 Pa. Therefore, in order to overcome such problems, recent trends have been focusing on the development of new plasma sources operated at atmospheric pressure with still keeping the characteristic of low temperature [113]. Several numbers of atmospheric pressure plasma sources have been developed for generating non-equilibrium plasma such as conventional dielectric barrier discharge (DBD) [114-116], planar DBD [117], various types of coronas [118, 119], micro-hollow cathode [120], DC discharge [121], MW discharge [122], and RF glow discharges [123-127]. Some of them have been proved to be utilized for PECVD to effectively deposit thin solid films with characteristics comparable to those films deposited using low pressure modes [127-129].

4.12.1. Application of Plasma Polymerized Thin Films

The polymer films are widely used for various industrial purposes. The main advantages of polymer thin films are that they can be prepared easily and cheaply, that they are stable and flexible and can be mounded into any form. The demands of various kinds of protective and anticorrosion coatings using polymers like Polytetrafluoroethylene (Teflon) and Polypropylene have not diminished over the years. Its chemical inertness and durability of many polymers still make them applicable in various fields. The major applications of plasma techniques are in improving the wettability and adhesion of polymers for surface coatings, inks and dyes [130, 131]. Plasma techniques are widely used in the electronics industry [132] particularly for microelectronics fabrication. Polymer films are very well in use in electronics as capacitors. Polystyrene capacitors are very popular in electronic industry. After the development of n-type and p-type polyacetylene, diodes and transistors are fabricated with this polymer [135], Due to its very good photosensitivity; it can be used for solar cell applications. Although most of the materials involved in these applications are inorganic, they are of interest to polymer chemists because polymers can be involved as resists, insulators or semiconductors. Operations carried out by plasma techniques include photo-resist removal [133] etching silicon compounds and deposition of polymer films [134].

4.13. Conclusion

Traditional methods of polymerization viz. chemical root polymerization and electrochemical polymerization are time consuming process where as plasma polymerization is single step procedure and it is completely dry process. The surface modification of plasma polymerized film is very easy and efficient for various application than other synthesis methods like chemical route polymerization, electrochemical polymerization method etc. It is not easy to achieve a good control over the chemical composition of the surface after modification in plasma polymerization method. Plasma polymerization must require a vacuum system which increases the cost of the operation.

The DC glow discharge process is widely used for plasma polymerization process, where as RF generator, AC field or microwave power etc. have been developed to increase the plasma density and the efficiency of power absorption. Recently plasma polymers films found great scope in the field of chemical sensors especially as gas sensors and humidity sensors.

Acknowledgement

Authors are thankful to Head, Department of Physics Sant Gadge Baba Amravati University, Amravati for providing necessary facilities.

References

[1]. B. G. Yacobi, S. Martin, K. Davis, A. Hudson, M. Hubert, Adhesive bonding in microelectronics and photonics, *J. Appl. Phys.,* Vol. 91, Issue 10, 2002, pp. 6227.

[2]. M. Pascu, C. Vasile, M. Gheorghiu, Modification of polymer blend properties by argon plasma/electronbeam treatment: surface properties, *Materials Chemistry and Physics*, Vol. 80, Issue 2, 2003, pp. 548–554.

[3]. D. Briggs, V. J. I. Zichy, D. M. Brewis, J. Comyn, R. H. Dahm, M. A. Green, M. B. Konieczko, X-rayphotoelectron spectroscopy studies of polymer surfaces, 4-further studies of the chromic acid etching of lowdensity polyethylene, *Surface and Interface Analysis*, Vol. 2, Issue 3, 1980, pp. 107–114.

[4]. R. Mix, J. Friedrich, A. Rau, Polymer Surface Modification by Aerosol Based DBD Treatment of Foils, *Plasma Processes and Polymers*, Vol. 6, Issue 9, 2009, pp. 566–574.

[5]. M. K. Shia, J. Christouda, Y. Holla, F. Cloueta, Functionalization by Cold Plasmas of Polymer Model Surfaces (Hexatriacontane and Octadecyloctadecanoate) Studied by Contact Angle Measurements, XPS, and FTIR Spectroscopy, *Journal of Macromolecular Science, Part A: Pure and Applied Chemistry,* Vol. 30, Issue 2-3, 1993.

[6]. B. Lee, Y. Kusano, N. Kato, K. Naito, T. Horiuchi, H, Koinuma, Oxygen Plasma Treatment of Rubber Surface by the Atmospheric Pressure Cold Plasma Torch, *Jpn. J. Appl. Phys.,* Vol. 36, 1997, pp. 2888-2891.

[7]. W. C. Wang, Y. Zhang, E. T. Kang, K. G. Neoh, Electroless Deposition of Copper on Poly(tetrafluoroethylene) Films Modified by Plasma-Induced Surface Grafting of Poly(4 vinylpyridine), *Plasmas and Polymers*, Vol. 7, 2002, pp. 207-225.

[8]. F. P. M. Mercx, Improved adhesive properties of high-modulus polyethylene structures: 3. Air- and Ammonia, plasma treatment, *Polymer*, Vol. 35, Issue 10, 1994, pp. 2098–2107.

[9]. S. R. Gaboury, M. W. Urban, Quantitative analysis of the Si-H groups formed on poly(dimethylsiloxane) surfaces: an ATR FTi.r. approach, *Polymer*, Vol. 33, Issue 23, 1992, pp. 5085–5089.

[10]. C. F. Amstein, P. A. Hartman, Adaptation of plastic surfaces for tissue culture by glow discharge, *Journal of Clinical Microbiology*, Vol. 2, Issue. 1, 1975, pp. 46-54.

[11]. C. M. Cepeda-Jiménez, R. Torregrosa-Maciá, J. M. Martín-Martínez, Surface modifications of EVA copolymers induced by low pressure RF plasmas from different gases and the irrelation to adhesion properties, *J. Adhesion Sci. Technol.,* Vol. 17, Issue 8, 2003, pp. 1145–1159.

[12]. H. J Busscher, A. W. J Van Pelt, H. P De Jong, J Arends, Effect of spreading pressure on surface free energy determinations by means of contact angle measurements, *Journal of Colloid and Interface Science*, Vol. 95, Issue 1, 1983, pp. 23–27.

[13]. C. Decker, S. Biry, Light stabilization of polymers by radiation-cured acrylic coatings, *Progress in Organic Coatings*, Vol. 29, Issues 1–4, 1996, pp. 81–87.

[14]. M. Gugliotti, Response to Historical Scientific Literature and the Spreading of Oil on Water, *J. Chem. Educ.*, Vol. 86, Issues 7, 2009, pp. 808.

[15]. S. Alvarez, S. Manolache, F. Denes, Synthesis of Polyaniline Using Horseradish Peroxidase Immobilizedon Plasma-Functionalized Polyethylene Surfaces as Initiator, *Journal of Applied Polymer Science*, Vol. 88, Issue 2, 2002, pp. 369–379.

[16]. P. K. Chu, J. Y. Chen, L. P. Wang, N. Huang, Plasma-surface modification of biomaterials B, *Materials Science and Engineering R*, Vol. 36, 2002, pp. 143–206.

[17]. E. M. Liston, L. Martinu, M. R. Wertheimer, Plasma surface modification of polymers for improved adhesion: a critical review, *Journal of Adhesion Science and Technology*, Vol. 7, Issue 10, 1993, pp. 1091-1127.

[18]. E. E. Johnston, B. D. Ratner, Surface characterization of plasma deposited organic thin films, *Journal of Electron Spectroscopy and Related Phenomena*, Vol. 81, Issue 3, 1996, pp. 303–317.

[19]. D. Thiry, N. Britun, S. Konstantinidis, J. Dauchot, L. Denis, R. Snyders, Altering the sulphur content in the propanethiol plasma polymers using the capacitive-to-inductive mode transition in inductively coupled plasma discharge, *Applied Physics Letters*, Vol. 100, Issue 7, 2012, pp. 071-604.

[20]. A. Bogaerts, R. Gijbels, Numerical modelling of gas discharge plasmas for various application, *Vacuum,* Vol. 69, 2003, pp. 37–52.

[21]. M. D. Duca, C. L. Plosceanua, T. Popb, Surface modifications of polyvinylidene fluoride (PVDF) under rfAr plasma, *Polymer Degradation and Stability*, Vol. 61, Issue 1, 1998, pp. 65–72.

[22]. B. Winther-Jensen, K. Norrman, P. Kingshott, K. West, Characterization of Plasma-Polymerized Fused Polycyclic Compounds for Binding Conducting Polymers, *Plasma Processes and Polymers*, Vol. 2, Issue 4, 2005, pp. 319–327.

[23]. G. Clarottia, 1, F. Schuea, J. Sledza, K. E. Geckelerb, W. Göpelc, A. Orsettid, Plasma deposition of thin fluorocarbon films for increased membrane hemocompatibility, *Journal of Membrane Science,* Vol. 61, 1991, pp. 289–301.

[24]. C. M. Chan, T. M. Koa, H. Hiraokab, Polymer surface modification by plasmas and photons, *Surface Science Reports,* Vol. 24, Issues 1–2, 1996, pp. 1–5.

[25]. M. Vincent, Donnelly, L. Daniel, Flamm, H. Richard Bruce, Effects of frequency on optical emission, electrical, ion, and etching characteristics of a radio frequency chlorine plasma, *J. Appl. Phys*., Vol. 58, 1985, pp. 2135-2145.

[26]. C. M. Chan, T. M. Ko, H. Hiraoka, Polymer surface modification by plasmas and photons, Surface Science Reports, Vol. 24, Issues 1–2, 1996, pp. 1–54.

[27]. W. C. Wang, E. T. Kang, K. G. Neoh, Electroless plating of copper on polyimide films modified by plasmagraft copolymerization with 4-vinylpyridine, *Applied Surface Science*, Vol. 199, 2002, pp. 52–66.

[28]. C. M. Chan, T. M. Koa, H. Hiraoka, Polymer surface modification by plasmas and photons, *Surface Science Reports*, Vol. 24, Issues 1–2, 1996, pp. 1–54.

[29]. H. Yasuda, Plasma Polymerization, *Academic Press*, New York, 1985.

[30]. M. R Wertheimer, H. R. Thomas, M. J. J. E. Perri, Klemberg Sapieha, L. Martinu, Plasmas and polymers: From laboratory to large scale commercialization, *Pure and Applied Chemistry,* Vol. 68, Issue 5, 1996, pp. 1047-1053.

[31]. S. Kurosawa, B. G. Choi, J. W. Park, H. Aizawa, K. B. Shim, K. Yamamoto, Synthesis and characterization of plasma-polymerized hexamethyldisiloxane films, *Thin Solid Films*, Vol. 506, 2006, pp. 176-179.

[32]. H. Yasuda, M. O. Bumgarner, J. J. Hillman, Polymerization of organic compounds in an electrodeless glow discharge. IV. Hydrocarbons in a closed system, *Journal of Applied Polymer Science,* Vol. 19, Issue 2, 1975, pp. 531–543.

[33]. H. Yasuda, Q. S. Yu, Creation of polymerizable species in plasma polymerization, *Plasma Chemistry &Plasma Processing,* 2004, Vol. 24, Issue 2, pp. 325-351.

[34]. A. Bogaerts, E. Neyts, R. Gijbels, J. van der Mullen, Gas discharge plasmas and their applications, *Spectrochimica Acta Part B*, Vol. 57, 2002, pp. 609–658.

[35]. P. Favia, L. C. Lopez, E. Sardella, R. Gristina, M. Nardulli, R. d'Agostino, Low temperature plasmaprocesses for biomedical applications and membrane processing, *Desalination*, Vol. 199, 2006, pp. 268–270.

[36]. A. I. Drachev, A. B. Gil'man, E. S. Obolonkova, A. A. Kuznetsov, Semiconductive polymer layers obtained by plasma polymerization of 1-amino-9, 10-anthraquinone in dc discharge, *Synthetic Metals*, Vol. 142, Issues 1–3, 2004, pp. 35–40.

[37]. L. G. Paterno, S. Manolache, F. Denes, Synthesis of polyaniline-type thin layer structures under low-pressure RF-plasma conditions, *Synthetic Metals*, Vol. 130, 2002, pp. 85–97.

[38]. G. A. Zaharias, H. H. Shi, S. F. Bent, Characterization of polyconjugated thin films synthesized by hot-wire chemical vapor deposition of aniline, *Thin Solid Films*, Vol. 501, Issues 1–2, 2006, pp. 341–345.

[39]. C. J. Mathai, S. Saravanan, M. R. Anantharaman, S. Venkitachalam, S. Jayalekshmi, Characterization of low dielectric constant polyaniline thin film synthesized by AC plasma polymerization technique, *J. Phys. D: Appl. Phys*. Vol. 35, 2002, pp. 240-249.

[40]. C. J. Mathai, S. Saravanan, S. Jayalekshmi, S. Venkitachalam, M. R. Anantharaman, Conduction mechanism in plasma polymerized aniline thin films, *Materials Letters*, Vol. 57, 2003, pp. 2253 – 2257.

[41]. R. Gupta, Nidhi, T. Sharma, S. Aggarwal, S. Kumar, Effect of thermal annealing on optical properties ofCR-39 polymeric track detector, *Indian J. Phys*. Vol. 83, Issue 7, 2009, pp. 921-926.

[42]. N. V. Bhat, N. V. Joshi, Structure and properties of plasma-polymerized thin films of polyaniline, *Plasma Chemistry and Plasma Processing,* Vol. 14, Issue 2, 1994, pp. 151-161.

[43]. M. G. Olayo, J. Morales, G. J. Cruz, R. Olayo, E. Ordoñez, S. R. Barocio, On the influence of electron energy on iodine-doped polyaniline formation by plasma polymerization, *Journal of Polymer Science Part B: Polymer Physics,* Vol. 39, Issue 1, 2001, pp. 175–183.

[44]. M. Vásquez, G. J. Cruz, M. G. Olayo, T. Timoshina, J. Morales, R. Olayo, Chlorine dopants in plasma synthesized heteroaromatic polymers, *Polymer*, Vol. 47, Issue 23, 2006, pp. 7864–7870.

[45]. J. Jensen, G. Possnert, Y. Zhang, Temperature effect on low-k dielectric thin films studied by ERDA, *Journal of Physics: Conference Series*, Vol. 100, 2008, pp. 12041-12055.

[46]. J. Wang, K. G. Neoh, L. Zhao, E. T. Kang, Plasma Polymerization of Aniline on Different Surface Functionalized Substrates, *Journal of Colloid and Interface Science*, Vol. 251, 2002, pp. 214–224.

[47]. W. Chen, L. Dai, H. Jiang, T. J. Bunning, Controlled Surface Engineering and Device Fabrication of Optoelectronic Polymers and Carbon Nanotubes by Plasma Processes, *Plasma Processes and Polymers,* Vol. 2, Issue 4, 2005, pp. 279–292.

[48]. M. S. Silverstein, I. Visoly Fisher, Plasma polymerized thiophene: molecular structure and electrical properties, *Polymer*, Vol. 43, 2002, pp. 11-20.

[49]. K. J. Kim, N. E. Lee, M. C. Kim, J. H. Boo, Chemical interaction, adhesion and diffusion properties at the interface of Cu and plasma-treated thiophene-based plasma polymer (ThioPP) films, *Thin Solid Films*, Vol. 398 – 399, 2001, pp. 657–662.

[50]. A. Uygun, L. Oksuz, A. Yavuz, A. Guleç, S. Sen, Characteristics of nanocomposite films deposited by atmospheric pressure uniform RF glow plasma, *Current Applied Physics*, Vol. 11, Issue 2, 2011, pp. 250–254.

[51]. M. C. Kim, S. H. Cho, S. B. Lee, Y. Kim, J. H. Boo, RF-plasma polymerization and characterization of polyaniline, *Thin Solid Films*, Vol. 592, 2004, pp. 447-448.

[52]. M. Vasquez, G. J. Cruz, M. G. Olayo, T. Timoshina, J. Morales, R. Olayo, Chlorine dopants in plasma synthesized heteroaromatic polymers, *Polymer*, Vol. 47, 2006, pp. 7864–7870.

[53]. L. M. H. Groenewoud, A. E. Weinbeck, G. H. M. Engbers, J. Feijen, Effect of dopants on the transparency and stability of the conductivity of plasma polymerized thiophene layers, *Synthetic Metals*, Vol. 33, Issue3, 2002, pp. 373-380.

[54]. L. M. H. Groenewoud, G. H. M. Engbers, J. Feijen, Langmuir, Plasma Polymerization of Thiophene Derivatives, *Mendeley*, Vol. 19, 2003, pp. 1368.

[55]. A. Uygun, L. Oksuz, A. G. Yavuz, A. Guleç, S. Sen, Characteristics of nanocomposite films deposited by atmospheric pressure uniform RF glow plasma, *Current Applied Physics*, Vol. 11, Issue 2, 2011, pp. 250–254.

[56]. S. J. Cho, I. S. Bae, J. H. Boo, S. Lee, D. Jung, Optical and Mechanical Properties of Toluene-TEOS Hybrid Plasma-Polymer Thin Films Deposited by Using PECVD, *Journal of the Korean Physical Society,* Vol. 55, Issue 5, 2009, pp. 1780-1784.

[57]. O. S. Nakagawa, S. Ashok, C. W. Sheen, J. Märtensson1, D. L. Allara, GaAs Interfaces with Octadecyl Thiol Self-Assembled Monolayer: Structural and Electrical Properties, *Jpn. J. Appl. Phys.,* Vol. 30, 1991, pp. 3759-3562.

[58]. N. V. Bhat, D. S. Wavhal, Preparation and characterization of plasma-polymerized thiophene films, *Journal of Applied Polymer Science*, Vol. 70, Issue 1, 1998, pp. 203–209.

[59]. N. V. Bhat, D. S. Wavhal, Characterization of plasma-polymerized thiophene onto cellulose acetatemembrane and its application to pervaporation, *Separation Science and Technology*, Vol. 35, Issue 2, 2000, pp. 227-242.

[60]. D. H. Shina, S. D. Leea, K. P. Leeb, S. Y. Parkb, D. H. Choib, N. Kimb, Characteristics of heterojunction consisting of plasma polymerized thiophene and n-type silicon, *Synthetic Metals*, Vol. 71, Issues 1–3, 1995, pp. 2263–2264.

[61]. S. M. Park, K. Ebihara, T. Ikegami, B. J. Lee, K. B. Lim, P. K. Shin, Enhanced performance of the OLED with plasma treated ITO and plasma polymerized thiophene buffer layer, *Current Applied Physics*, Vol. 7, Issue 5, 2007, pp. 474–479.

[62]. J. L. Delattre, R. d'Agostino, F. Fracassi, Plasma-polymerized thiophene films for enhanced rubber–steel bonding, Applied Surface Science, Vol. 252, Issue 10, 2006, pp. 3912–3919.

[63]. Y. Zhou, Y. Yuan, L. Cao, J. Zhang, H. Pang, J. Lian, X. Zhou, Improved stability of OLEDs with Mild oxygen plasma treated PEDOT: PSS, *Journal of Luminescence*, Vol. 122–123, 2007, pp. 602–604.

[64]. J. Morales, M. G. Olayo, G. J. Cruz, M. M. Castillo-Ortega, R. Olayo, Electronic conductivity of pyrroleand aniline thin films polymerized by plasma, *Journal of Polymer Science, Part B: Polymer Physics*, Vol. 38, 2000, pp. 3247-3255.

[65]. Z. S. Tong, M. Z. Wu, T. S. Pu, Z. Y. Zhang, J. Zhang, R. P. Jin, D. Z. Zhu, F. Y. Zhu, D. X. Cao, J. Q. Cao, The conductive behavior and structural characteristics in the I^+-beam-implanted layer of plasma-polymerized pyrrole film, *J. Appl. Polym. Sci,.* Vol. 69, 1998, pp. 1743-1751.

[66]. L. Martin, J. Esteve, S. Borros, Proceedings of Symposium D on Thin Film and Nano-Structured Materials for Photovoltaics, in *Proc. of the E-MRS 2003 Spring Conference*, Vol. 74, 2004, pp. 451-459.

[67]. G. J. Cruz, J. Morales and R. Olayo, Films obtained by plasma polymerization of pyrrole, *Thin Solid Films,* Vol. 342, 1999, pp. 119-127.

[68]. J. M. Corona, J. A. López-Barrera, A. A. Ortega, G. J. Cruz, M. G. Olayo, M. O. López, M. V. Ortega, H. Vazquez, R. Olayo, Luminescent polydibenzothiophene thin film obtained by glow discharge method, *Journal of Applied Polymer Science*, Vol. 123, Issue 2, 2012, pp. 1120–1124.

[69]. B. Paosawatyanyong, K. Tapaneeyakorn, W. Bhanthumnavin, AC plasma polymerization of pyrrole, Surface and Coatings Technology, Vol. 204, Issues 18–19, 2010, pp. 3069–3072.

[70]. Hosono, K., Matsubara, I., Murayama, N., Shin, W., Izu, N. The sensitivity of 4-ethylbenzenesulfonic acid-doped plasma polymerized polypyrrole films to volatile organic compounds. *Thin Solid Films*, Vol. 484, 2005, pp. 396-399.

[71]. R. K. John, D. S. Kumar, Structural, electrical and optical studies of plasma polymerized and iodine dopedpoly pyrrole, *Journal of Applied Polymer Science*, Vol. 83, 2002, pp. 1856-1869.

[72]. D. Shi, Z. Yu, S. X. Wang, W. J. van Ooij, L. M. Wang, J. G. Zhao, Multi-Layer Coating of Ultrathin Polymer Films on Nanoparticles of Alumina by a Plasma Treatment, *Mat. Res. Soc. Syrmp. Proc.,* Vol. 635, 2001, pp. 1-6.

[73]. Morales, M. G. Olayo, G. J. Cruz, R. Olayo, Plasma polymerization of random polyaniline–polypyrrole– iodine copolymers, *Journal of Applied Polymer Science*, Vol. 85, Issue 2, 2002, pp. 263–270.

[74]. T. Hirotsu, Z. Hou, A. Partridge, Plasma Copolymerization of Pyrrole with Si-compounds, *Plasmas and Polymers*, Vol. 4, 1999, pp. 1-15.

[75]. F. Nastase, I. Stamatin, C. Nastase, D. Mihaiescu, A. Moldovan, Advanced Functional Nanomaterials: from Nanoscale Objects to Nanostructered Inorganic and Hybrid Materials, *Progress in Solid State, Chemistry*, Vol. 34, 2006, pp 191-199.

[76]. J. Morales, M. G. Olayo, G. J. Cruz, R. Olayo, Synthesis by plasma and characterization of bilayeraniline–pyrrole thin films doped with iodine, *Journal of Polymer Science Part B: Polymer, Physics,* Vol. 40, Issue 17, 2002, pp. 1850–1856.

[77]. O. Goossens, E. Dekempeneer, D. Vangeneugden, R. Van de Leest, C. Leys, Application of atmospheric pressure dielectric barrier discharges in deposition, cleaning and activation, *Surface & Coatings Technology*, Vol. 142, 2001, pp. 474-481.

[78]. Y. Sawada, S. Ogawa, M. Kogoma, Synthesis of plasma-polymerized tetraethoxysilane and hexamethyldisiloxane films prepared by atmospheric pressure glow discharge, *Journal of Physics D: Applied Physics,* Vol. 28, 1995, pp. 1661.

[79]. H. R. Lee, D. J. Kim, K. H. Lee, Anti-reflective coating for the deep coloring of PET fabrics using an atmospheric pressure plasma technique, *Surf. Coat Tech.*, Vol. 142, 2001, pp. 468-473.

[80]. G. R. Prasad, S. Daniels, D. C. Cameron, B. P. McNamara, E. Tully, R. O'Kennedy, PECVD of Biocompatible Coatings on 316L Stainless Steel. Surface And Coatings Technology, Vol. 200, Issue 1-4, 2005, pp. 1031-1035.

[81]. F. Massines, N. Gherardi, A. Fornelli, S. Martin, Atmospheric pressure plasma deposition of thin films by Townsend dielectric barrier discharge, *Surface and Coatings Technology*, Vol. 200, Issues 5, 2005, pp. 1855–1861.

[82]. C. P. Klages, M. Eichler, R. Thyen, Atmospheric pressure PA-CVD of silicon and carbon-based coatings using dielectric barrier discharges, New Diamond and Frontier Carbon Technology, Vol. 13, 2003, pp. 175-189.

[83]. L. J. Ward, W. C. E. Schofield, J. P. S. Badyal, A. J. Goodwin, P. J. Merlin, Atmospheric Pressure Plasma Deposition of Structurally Well-Defined Polyacrylic Acid Films, *Chemistry of Materials*, Vol. 15, 2003, pp. 1466-1479.

[84]. P. Koidl, C. Wild, B. Dischler, J. Wagner, M. Ramsteiner, Plasma deposition, properties and structure of amorphous hydrogenated carbon films, *Mater. Sci. Forum*, Vol. 41, 1990, pp. 52-53.

[85]. J. Robertson, Amorphous carbon, *Advances in Physics*, Vol. 35, 1986, pp. 317–374.

[86]. P. Marcus, I. Platzner, A mass-spectrometric investigation of polymerization processes in used by microwave irradiation in methane, methane – argon and methane -krypton mixtures during the formation of pyrolytic carbon, *International Journal of Mass Spectrometry and Ion Physics,* Vol. 32, Issue 1, 1979, pp. 77–86.

[87]. A. R. Westwood, Glow discharge polymerization–II. The structure of glow discharge polymers, *European Polymer Journal*, Vol. 7, Issue 4, 1971, pp. 377–385.

[88]. F. F. Shi, Recent advances in polymer thin films prepared by plasma polymerization Synthesis, structural characterization, properties and applications, *Surface and Coatings Technology*, Vol. 82, Issues 1–2, 1996, pp. 1–15.

[89]. S. D. Phadke, Dielectric properties of plasma-polymerized ferrocene films, *Thin Solid Films,* Vol. 48, Issue 3, 1978, pp. 319–324.

[90]. Yasuda and C. E. Lamaze, Polymerization of styrene in an electroless glow discharge, *J. Appl. Poly. Sci.*, Vol. 15, 1971, pp. 2277-2291.

[91]. N. Paul, M. G. Krishna Pillai, Dielectric studies of poly-p-toluidine films produced by the electrodeless glow discharge method, *Thin Solid Films*, Vol. 76, Issue 3, 1981, pp. 201–205.

[92]. L. Ledernez, F. Olcaytug, H. Yasuda, G. Urban, Ultraviolet light in glow discharges, *J. Appl. Phys*. Vol. 104, Issue 10, 2008, pp. 103303-103317.

[93]. H. Yasuda, Magnetron–AF Plasma Polymerization, *Plasma Processes and Polymers*, Vol. 5, Issue 3, 2008, pp. 215–227.

[94]. A. E. D Heylen, The relationship between electron—molecule collision cross sections, experimental Townsend primary and secondary ionization coefficients and constants, electric strength and molecular structure of gaseous hydrocarbons, *Proceedings of the Royal Society,* Vol. 456, Issue 2004, 2000, pp. 3005-3040.

[95]. M. J. Druyvesteyn and F. M. Penning, The mechanism of electrical discharges in gases of low pressure, *Reviews of Modern Physics*, Vol. 12, Issue 2, 1940, pp. 87–174.

[96]. A. A. Kruithof, Townsend's ionization coefficients for neon, argon, krypton and xenon, *Physica*, Vol. 7, Issue 6, 1940, pp. 519–540.

[97]. F. Paschen, Ueber die zum funkenuebergang in luft, wasserstoff und kohlensaeure bei verschiedenendrucken erforderliche potentialdifferenz, *Annalen der Physica*, 273, Issue 5, 1889, pp. 69–96.

[98]. Mikki Larner Stephen L. Kaplan, The Challenge of Plasma Processing – Its Diversity, in *Proceedings of the ASM Materials and Processes for Medical Devices Conference*, August 25-27, 2004, St. Paul, Minnesota.

[99]. Z. Jiang, Y. Meng, Y. Shi, Synthesis of Proton-Exchange Membranes by a Plasma Polymerization Technique, *Jpn. J. Appl. Phys.,* Vol. 47, 2008, pp. 6891-6895.

[100]. J. W. Coburn, Surface processing with partially ionized plasmas, *IEEE Trans. Plasma Sci.,* Vol. 19, 1991, pp. 1048-1062.

[101]. C. Petcu, B. Mitu, G. Dinescu, characterization of a Tubular Plasma Reactor with external Annular Electrodes, *Romanian Reports in Physics*, Vol. 57, Issue 3, 2005, pp. 390 -395.

[102]. E. E. Johnston, B. D. Ratner, Surface characterization of plasma deposited organic thin films, *Journal of Electron Spectroscopy and Related Phenomena*, Vol. 81, Issue 3, 1996, pp. 303–317.

[103]. T. J. Donahue, R. Reif, Silicon epitaxy at 650-800 °C using low pressure chemical vapor deposition with and without plasma enhancement, *J. Appl. Phys*., Vol. 57, 1985, pp. 2757-2765.

[104]. R. Reif, W. Kern, J. L. Vossen, W. Kern, Plasma-enhanced chemical vapor deposition, in Thin Film Processes II, Eds. by J. Vossen et al., *Academic. Press,* San Diego, 1991, pp. 525-564.

[105]. G. Lucovsky, D. V. Tsu, R. A. Rudder, R. J. Markunas, J. L. Vossen, W. Kern, Formation of inorganic films by remote plasma-enhanced chemical-vapor deposition, in Thin Film Processes II, *Academic. Press*, San Diego, 1991, pp. 565-619.

[106]. J. F. Evans and G. W. Prohaska, Preparation of thin polymer films of predictable chemical functionality using plasma chemistry, *Thin Solid Films*, Vol. 118, 1984. pp. 171-180.

[107]. Z. Jiang, Y. Meng, Y. Shi, Synthesis of Proton-Exchange Membranes by a Plasma Polymerization Technique, *Jpn. J. Appl. Phys*. Vol. 47, 2008, pp. 6891-6895.

[108]. Q. Yu1, C. E. Moffitt, D. M. Wieliczka, H. Yasuda, DC cathodic polymerization of trimethylsilane in aclosed reactor system, *J. Vac. Sci. Technol. A*, Vol. 19, 2001, pp. 2163.

[109]. M. Matsushita, A. Kashem, S. Morita, Co-operation process of plasma CVD and sputtering, using methane, SF_6 and Ar mixture gas, and gold plate discharge electrode, *Thin Solid Films*, Vol. 407, Issues 1–2, 2002, pp. 50–53.

[110]. S. Majumder, A. H. Bhuiyan, Direct optical transition in plasma polymerized vinylene carbonate thin Films, *Indian Journal of Physics*, Vol. 85, Issue 8, 2011, pp. 1287-1297.

[111]. H. Biederman, P. Hlídek, J. Pešička, D. Slavínská, V. Stundžia, Deposition of composite metal/C:H films-the basic properties of Ag/C:H, *Vacuum*, Vol. 47, Issue 11, 1996, pp. 1385–1389.

[112]. D. B. Chrisey, J. M. Fitz-Gerald, R. A. McGill, R. C. Y. Auyeung and H. D. Wu, S. Lakeou, V. Nguyen, R. Chung, M. Duignan, Direct writing of electronic and sensor materials using a laser transfer technique, A. Pique(a), *J. Mater. Res.*, Vol. 15, Issue 9, 2000, pp. 1-8.

[113]. A. Schutze, J. Y. Jeong, S. E. Babayan, J. Park, G. S. Selwyn and R. F. Hicks, The atmospheric-pressure plasma jet: A review and comparison to other plasma sources, *IEEE Trans Plasma Sci.,* Vol. 26, 1998, pp. 1685-1694.

[114]. J. Salge, Plasma-assisted deposition at atmospheric pressure, *Surface and Coatings Technology*, Vol. 80, 1996, pp. 1-7.

[115]. A. P. Napartovich, Overview of atmospheric pressure discharges producing non thermal plasma, *Plasma and Polymers*, Vol. 6, 2001, pp. 1-14.

[116]. S. E. Alexandrov, M. L. Hitchman, Chemical Vapor Deposition Enhanced by Atmospheric Pressure Non-thermal Non-equilibrium Plasmas, *Chemical Vapor Deposition*, Vol. 11, Issue 11-12, 2005, pp. 457–468.

[117]. J. Engemann, D. Korzec, Assessment of discharges for large area atmospheric pressure plasma enhanced chemical vapor deposition (AP PE-CVD), *Thin Solid Films*, Vol. 442, 2003, pp. 36–39.

[118]. J. Chen, J. H. Davidson, Electron Density and Energy Distributions in the Positive DC Corona: Interpretation for Corona-Enhanced Chemical Reactions, *Plasma Chemistry and Plasma Processing,* Vol. 22, 2002, Issue 2, pp. 199-224.

[119]. K. Yan, E. J. M. van Heesch, A. J. M. Pemen, and P. A. H. J. Huijbrechts, From chemical kinetics to streamer corona reactor and voltage pulse generator, *Plasma Chem. Plasma Process*., Vol. 21, 2001, pp.107–137.

[120]. R. H. Stark, K. H. Schoenbach, Direct current glow discharges in atmospheric air, *Appl. Phys. Lett*., Vol. 74, 1999, pp. 3770–3772.

[121]. O. Goossens, T. Callebaut, Y. Akishev, A. Napartovich, N. Trushkin, C. Leys, The DC glow discharge at atmospheric pressure, *IEEE Transactions on Plasma Science*, Vol. 30, Issue 1, 2002, pp. 176-177.

[122]. A. Pfuch, R. Cihar, Deposition of SiO_x thin films by microwave induced plasma CVD at atmospheric pressure, *Surface and Coatings Technology*, Vol. 183, Issue 2-3, 2004, pp. 134-140.

[123]. S. D. Anghel, Generation of a Low-Power Capacitively Coupled Plasma at Atmospheric Pressure, *IEEE Transactions on Plasma Science,* Vol. 30, Issue 2, 2002, pp. 660-664.

[124]. Y. Mori, K. Yoshii, K. Yasutake, H. Kakiuchi, H. Ohmi, K. Wada, High rate growth of epitaxial silicon at low temperatures (530–690 °C) by atmospheric pressure plasma chemical vapor deposition, *Thin Solid Films*, Vol. 444, 2003, pp. 138–145.

[125]. L. O'Neill, L. A. O'Hare, S. R. Leadley, A. J. Goodwin, Atmospheric Pressure Plasma Liquid Deposition- A Novel Route to Barrier Coatings, *Chemical Vapor Deposition*, Vol. 11, Issue 11-12, 2005, pp. 477–479.

[126]. J. Park, I. Henins, H. W. Herrmann, G. S. Selwyn, J. Y. Jeong, R. F. Hicks, D. Shim, C. S. Chang, An atmospheric pressure plasma source, *Appl. Phys. Lett.*, Vol. 76, 2000, pp. 288-292.

[127]. Y. Sawadat, S. Ogawat, M. Kogomaf, Synthesis of plasma-polymerized tetraethoxysilane andhexamethyldisiloxane films prepared by atmospheric pressure glow discharge, *J. Phys. D. Appl. Phys.,* Vol. 28, 1995, pp. 1661-1669.

[128]. X. Zhu, F. Arefi-Khonsari, C. Petit-Etienne, M. Tatoulian, Open air deposition of SiO_2 films by an atmospheric pressure line-shaped plasma, *Plasma Process Polym.,* Vol. 2, 2005, pp. 407-413.

[129]. H. Kakiuchi, Y. Nakahama, H. Ohmi, K. Yasutake, K. Yoshii and Y. Mori, Investigation of deposition characteristics and properties of high-rate deposited silicon nitride films prepared by atmospheric pressure plasma chemical vapor deposition, *Thin Solid Films,* Vol. 479, 2005, pp. 17-23.

[130]. A. M Wróbel, M Kryszewski, Effect of plasma treatment on surface structure and properties of polyesterfabric, *Polymer*, Vol. 19, Issue 8, 1978, pp. 908–912.

[131]. D. K. Owens, Mechanism of corona–induced self-adhesion of polyethylene film, *Journal of Applied Polymer Science,* Vol. 19, Issue 1, 1975, pp. 265–271.

[132]. G. Franz, Low Pressure Plasmas and Microstructuring Technology, *Springer*, 2009.

[133]. S. J. Moss, Polymer degradation in reactive-gas plasmas, *Polymer Degradation and Stability,* Vol. 17, Issue 3, 1987, pp. 205–222.

[134]. H. Abe, Y. Sonobe, Tatsuya Enomoto, Etching Characteristics of Silicon and its Compounds by Gas plasma, *Jpn. J. Appl. Phys*., Vol. 12, 1973, pp. 154-155.

[135]. L. W. Shacklette, R. R. Chance, D. M. Ivory, G. G. Miller, R. H. Baughman, Electrical and optical properties of highly conducting charge-transfer complexes of poly(p-phenylene), *Synthetic Metals*, Vol. 1, Issue 3, 1980, pp. 307–320.

[136]. F. Hellegouarc'h, F. Arefi-Khonsari, R. Planade, J. Amouroux, PECVD prepared SnO_2 thin films for ethanol sensors, *Sensors and Actuators B,* Vol. 73, 2001, pp. 27–34.

[137]. A. Chaturvedi, V. N. Mishra, R. Dwivedi, S. K. Srivastava, Selectivity and sensitivity studies on plasmatreated thick film tin oxide gas sensors, *Microelectronics Journal,* Vol. 31, 2000, pp. 283–290.

[138]. S. Liqin, Y. Hasegawa, T. Katsube, K. Onoue, K. Nakamura, Highly sensitive NO_2 gas sensor fabricated with RF induction plasma deposition method, *Sensors and Actuators,* Vol. 99, 2004, pp. 361–366.

[139]. F. Kraus, S. Cruz, J. Muller, Plasma polymerized silicon organic thin films from (HMDSM) for capacitive humanity sensors, *Sensor and Actuators, B.,* Vol. 88, 2003, pp. 300-311.

[140]. R. V. Dabhade, Dhananjay S. Bodas, S. A. Gangal, Plasma-treated polymer as humidity sensing material-a feasibility study, *Sensors and Actuators, B.,* Vol. 98, 2004, pp. 37–40.

Chapter 5
MEMS Non-Silicon Fabrication Technologies

Poonam Goel

5.1. Introduction

Recently, MEMS technology has emerged as a revolutionary technology to realize sensors, communication systems, biosensors, etc. Because Silicon micromachining is a mature technology, MEMS are preferred to be fabricated on Silicon. Silicon technology is most suitable for VLSI because of its high accuracy in achieving small dimensions, but we don't need small features with high precision in MEMS technology.

Since years, research is being carried out to develop MEMS non-Silicon fabrication technologies such as printed circuit board (PCB), low temperature cofired ceramics (LTCC), liquid crystal polymers (LCP), polymer core conductor and poly-dimethyl siloxane (PDMS). There are many groups working on MEMS non-Silicon fabrication technologies. Pagel et al [1] reported microfluidic microsystems using multilayer stacks, Goel et al [2] and Ramadoss et al [3] has demonstrated monolithic phased antenna array, Ghodsian et al reported RF MEMS switch [4], Cetiner et al explored tunable antenna [5] on printed circuit board using PCB micromachining processes. Golonka group at Wroclaw University of Technology has explored MEMS and MEMS sensors [6], Gongora-Rubio et al manufactured meso-scale EMS (electro-mechanical system) thermal flow sensor [7] on LTCC substrate. Reconfigurable phased antenna array [8, 9], RF modules [10-13] and MEMS sensors [14, 15] has been reported using liquid crystal polymer (LCP) technology. In polymer core conductor technology, polymer core is coated with metal which is more favourable for RF MEMS applications. It is due to skin effect in which most of the current is confined to outermost portions of the conductor. Passive RF modules such as spiral and solenoid inductor structures, transformer, co-axial core inductor [16, 17] have been reported using polymer core conductor technology. PDMS is being considered for microfluidic MEMS which are most suitable for bio applications [18], microfuel cell array [19]. Many fabrication processes has been

Poonam Goel
School of Electrical and Electronics Engineering, Nanyang Technological University
Singapore, 639798

developed for these technologies depending on the substrate material characteristics; but still there is need to develop improved & cost effective fabrication processes.

This paper provides an insight on substrate materials and their characteristics, and micro-fabrication processes for each of non-Silicon fabrication technologies of PCB, LTCC, LCP, polymer core conductor and PDMS technology. A brief overview of their applications in MEMS and RF MEMS is also provided.

5.2. Printed Circuit Board (PCB) Technology

Printed circuit board (PCB) was originated in 1904 by Frank Sprague, but it was commercially used 50 years later. PCB is made of thin dielectric base material coated with conductor on both sides. Base dielectric provides mechanical support to electrical circuits as well as components and coated conductor is used to make electrical contacts. Copper is preferred conductor in PCB technology due to its high electrical conductivity [20, 21]. There are different resins being used as PCB dielectric. Some of the PCB base materials with their dielectric properties are listed in Table 5.1.

Table 5.1. PCB base materials.

PCB base materials	Glass transition temperature (Tg)°C	Dielectric constant ε_r	Loss tangent $\tan\delta$	Dielectric strength (V/mil)
Alumina	1700	4.5	-	220 [22]
*FR4 epoxy	125	4.1	0.02	1100 [20]
Multifunctional epoxy	145	4.1	0.022	1050 [20]
Tetra-functional epoxy	150	4.1	0.022	1050 [20]
*BT/Epoxy	185	4.1	0.013	1350 [20]
Cyanate Ester	245	3.8	0.005	800 [20]
Polyimide	285	4.1	0.015	1200 [20]
Teflon	NA	2.2	0.0002	450 [20]
Flourene based resin	230	3.2	0.002	5000 [23]

*FR4: Diglycidyl ether of tetrabromobisphenol A
*BT: Bismaleimide trazine

Initially, PCB substrates were used only for electronic circuitry. Recently, fabrication of MEMS using PCB technology is reported due to low cost fabrication equipments and ease of micromachining technology. Advantages of using PCB technology over Silicon technology for MEMS are listed below:

1. Manufacturing cost of PCB substrates is low compared to conventional semiconductor substrates such as Silicon or Gallium Arsenide;
2. PCB technology does not need high class clean room and sophisticated fabrication equipments like conventional semiconductor technology;
3. PCB substrate is compatible with organic substrate which makes it suitable for

MCML (Multi Chip Module Laminate) technology [15];
4. Packaging is required to protect the MEMS device against environmental harsh conditions. Packaging of MEMS fabricated with Silicon micromachining is still a critical issue. Mostly, MEMS packaging cost is high or comparable to the device fabrication cost. Packaging related issues and package cost can be avoided with PCB technology because in this, substrate itself acts as a package;
5. PCB substrates of different thicknesses and wide range of dielectric constants with low loss tangents are commercially available. These properties make it suitable for RF MEMS components;
6. PCB fabrication utilizes multilayer technology which makes it highly economic process. All layers are patterned individually and aligned using alignment marks. An epoxy of certain viscosity [1, 24, 25] is used to bond all layers under certain temperature and pressure [24, 25] which is a low cost process compared to the Silicon wafer bonding which needs very sophisticated and expensive equipments.

Disadvantages of PCB Technology are the following:
1. Surface roughness of PCB copper clad is high which limits the fabrication to attain minimum feature size;
2. PCB substrate cannot be used for devices which need to be exposed to high temperatures.

5.2.1. Fabrication Processes for PCB Technology

Silicon technology fabrication processes cannot be used for PCB technology because PCB substrate cannot sustain at high temperature. PCB fabrication processes are low temperature (< 250 °C) processes. Fabrication processes reported for PCB technology are discussed:

5.2.1.1. Thin Film Deposition

Deposition is one of the high temperature processes of Silicon micro-fabrication. A low temperature deposition process of high density inductively coupled plasma chemical vapor deposition (HDICP CVD), which is developed specifically for PCB substrates.

High Density Inductively Coupled Plasma Chemical Vapor Deposition (HDICP CVD) [26]: Chemical vapor deposition (CVD), Plasma Enhanced chemical vapor deposition (PECVD) and other conventional deposition techniques are high temperature processes and cannot be used for PCB substrate. A low temperature (90 °C -170 °C) deposition process of high density inductively coupled plasma chemical vapor deposition (HDICP CVD) technique is reported specific for PCB substrates.

HDICP CVD utilizes high density plasma of 10^{10}-10^{11} ions/cm^3 place of plasma density used in PECVD of 10^9 ions/cm^3. HDICP CVD method uses a commercially available inductively coupled plasma (ICP) reactor (Bethel Materials Research, Irvine, CA) to inductively couple RF power, with designed antenna array and a set of magnets with Faraday shield copper tapes wrapped around them to increase the plasma density and to adjust the plasma profile. HDICP CVD is being used to deposit SiN_x on PCB substrate. It uses SiH_4+N_2 (in place of NH_3), less hydrogen reactants which deposits high dielectric properties SiN_x. Comparison between HDICP CVD and PECVD processes and deposited film characteristics are listed in Table 5.2 [26].

Table 5.2. Comparison between Silicon nitride film deposited by PECVD and HDICP CVD [26]. Reproduced from H.-P. Chang, J. Qian, B. A. Cetiner, F. D. Flaviis, M. Bachman, Design and process considerations for fabricating RF MEMS switches on printed circuit boards. *Journal of Microelectromechanical Systems,* 14, no. 6, 2005, pp. 1311-1322, © IEEE.

Process parameters	**PECVD**		**HDICP CVD**	
Process gases	SiH4:NH3:N2	SiH4:NH3:N2	SiH4:N2	SiH4:N2
Process temperature	250 °C	300-400 °C	< 190 °C	< 100 °C
Wet etch rate in BHF (A°/min)	600-650	200-300	50	200-300
Concentration	>25@%	2025@%	<15@%	<20@%
Film stress (MPa)	100-600 (compressive)	100-800 (compressive)	150-600 (compressive)	100-800 (compressive)
Surface roughness (nm)	2.5	3.1	2.0	1.6
Dielectric strength (MV/cm)	~8	~8	>9	>9

5.2.1.2. Planarization

PCB uses thick metal clad layers which make the surface rough. Couples of surface planarization techniques are mentioned below to improve its surface profile for successive processes. PCB substrates are commercially available with Copper clad of 18 μm to 105 μm thickness deposited on 100 μm to 1500 μm thick base [24] where large thickness of copper clad causes rough surface. In some cases, because of this copper clad surface roughness, reliability and working performance of the devices is significantly impaired. To reduce the surface roughness of copper clad, techniques adopted are called as planarization techniques. Some of these are:

a. Polishing and planarization
b. Compressive molding planarization (COMP)

5.2.1.2.1. Polishing and Planarization Process

This is a two steps process which includes planarization and polishing.

Planarization: Conductor traces are patterned on the PCB copper clad Fig. 5.1 (a). As shown in Fig. 5.1 (b), 18-20 μm thick polyimide (depending on the copper clad thickness) is spin coated on patterned copper of 17.5 μm thickness and then placed on hot plate by raising temperature from room temperature to 220 °C with rate of 6.9 °C/min and kept for 1hr at 220 °C to cure the polyamide layer. Polyamide film surface profile is same as that of lower copper surface. This is followed by next step of polishing [4].

Polishing: Polyimide is polished in two steps. First, polyimide surface is polished with Silicon carbide fine powder sheet to remove the polyimide up to Copper surface Fig. 5.1(c). Although it may cause scratches and cuts on the surface, but this can be reduced in second step. Second step is to polish the surface with diamond paste of 6um particles on nylon cloth. This technique will in turn reduce the root mean square surface roughness up to 30 nm or even less [4].

5.2.1.2.2. Compressive Molding Planarization (COMP)

Compressive molding planarization (COMP) is utilized if device has membrane structures such as RF-Switch, varactors, filters, resonators, phase shifters etc. Copper clad is pattered Fig. 5.2 (a). In this technique, thick layer of photo-resist (PR) is spin coated on patterned copper as shown in Fig. 5.2 (b) and a smooth surface like glass is kept on the top of the spun coated photoresist Fig. 5.2 (b). Second step is to apply high temperature (100 °C) and pressure (40 Psi). High temperature is to melt and to re-flow the photoresist under smooth surface to planarize the PR Fig. 5.2 (c). This planarized PR acts as sacrificial for membrane. Later on PR is etched using acetone. Only drawback of COMP is low reliability of device because voids formation (Fig. 5.2 (d)) may take place in the PR during this process which may cause air bubble entrapment when high temperature is applied [4, 27]. Surface roughness can be reduced to less than 50 nm and 10 % uniformity in film thickness can be obtained over 2.5×2.5 cm^2 by controlled COMP process [26].

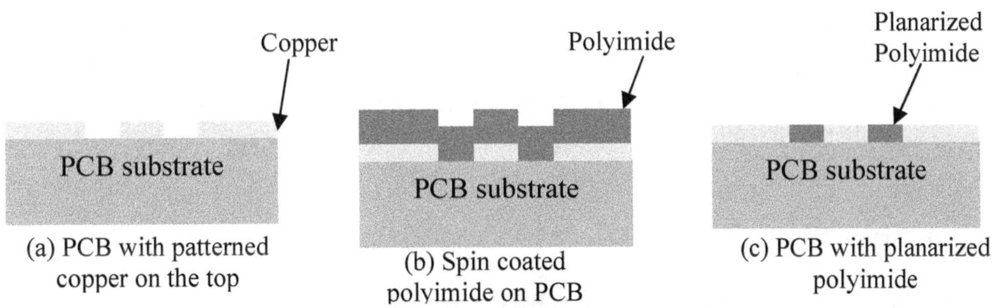

Fig. 5.1. Polishing and planarization process [4]. Reproduced from B. Ghodsian, C. Jung, B. A. Cetiner, F. D. Flaviis, Development of RF-MEMS switch on PCB substrates with polyimide planarization. *IEEE Sensors Journal,* 5, No. 5, 2005, pp. 950-955 © IEEE

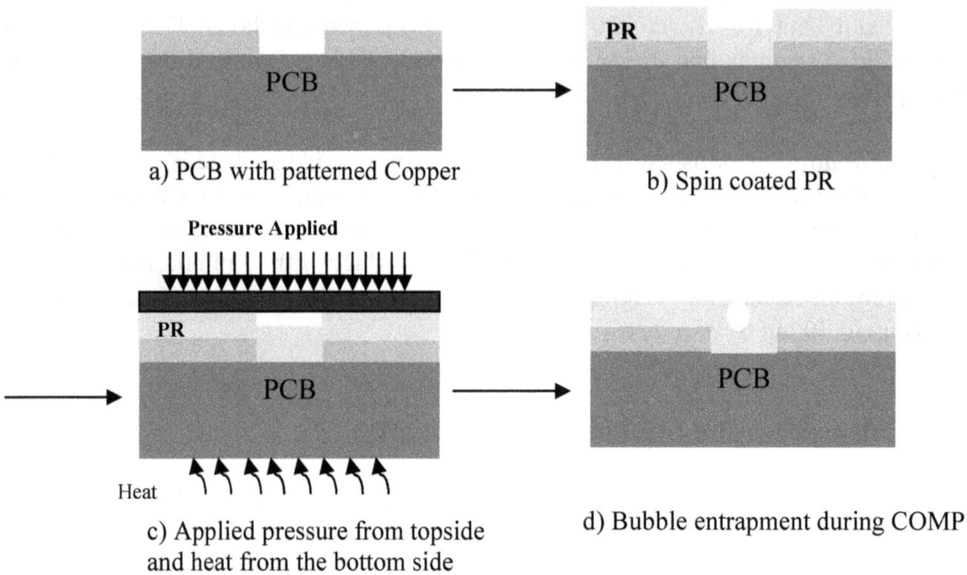

Fig. 5.2. COMP process steps [4]. Reproduced from B. Ghodsian, C. Jung, B. A. Cetiner, F. D. Flaviis, Development of RF-MEMS switch on PCB substrates with polyimide planarization, *IEEE Sensors Journal,* 5, No. 5, 2005, pp. 950-955 © IEEE.

5.2.1.3. Micro-vias in PCB

Mechanical drilling can be used to drill through holes of diameter > 200 µm in PCBs. Smaller size holes < 50 µm is prelim requirement for high package density. Laser is reliable method to drill micro-vias in PCB with high resolution depending on the laser wavelength, beam energy density, composition and thickness of the substrate. Some of the lasers types used for drilling are radio-frequency excited carbon dioxide laser, transversely excited atmospheric carbon oxide laser, 3rd harmonic Nd:YAG laser, Krypton fluoride (KrF) excimer laser [22]. Lasers are utilized depending on the substrate materials as listed in Table 5.3.

5.2.1.4. Other Processes

In PCB lithography, ink-jet printer printout is used as mask (no sophisticated or costly mask production equipments are required to make mask) [28]. Lithography and etching processes can be performed similar to Silicon fabrication, but PCB lithography and etching do not need high class clean room. Dry film photoresist technique is used for lithography which does not need spinner; and also there is no substrate size constraint.

Table 5.3. Comparison in drilling techniques [22].

	Radio-frequency excited carbon dioxide laser	Transversely Excited atmospheric carbon oxide laser	3rd harmonic Nd:YAG laser	KrF excimer laser
Wavelength	CO_2 laser emits light in far infra-red region 10.6	Emits Same wavelength as CO_2 laser	operated ultra violet spectrum 355 nm	Operates at short wavelength 248 nm
Laser Pulse width and output power	Pulse width of range 30-100 μs with average output power of 200 W at 10 kHz	Short pulses of 1 μs with high peak power of 10^6~10^7 W at 100 Hz	Few nanoseconds and high output peak power	20 ns High peak power of several megawatts
Drilling speed	Drilling is done sequentially	Depending on the optics demagnification, processing area can be varied from 2*2 mm^2 to 5*5 mm^2	Microvias are drilled sequentially	Microvias exposed within laser beam area drilled simultaneously
Compatibility of the PCB material	This is used to drill holes in FR4 with narrower range than and of lower quality but with low taper walls than TEA CO_2.	This technique is good for FR4 with E- glass fiber reinforcement material to drill high quality vias with consistent taper angle and smooth walls with optimal energy density range 12-18 J/cm^2	This technique can be used to drill holes with smooth walls of high aspect ratio 20:1. minimum achievable diameter is 50 μm	It is used to drill holes in polyamides for high resolution vias with laser energy density of 3J/cm^2. It can provide vias of size <50 μm

5.2.2. Applications of PCB Technology

The PCB technology is compatible for multilayer fabrication which includes stack of PCB substrates together to construct the device. The multilayer PCBs are used to fabricate fluidic channels, temperature sensors, capacitance bubble detector, valves, pressure sensors, micropump. These sensors/devices can be integrated with electronics circuitry on the same PCB substrate [1, 24, 25]. Multilayer stack of PCB substrates has been demonstrated to develop a micropump. This micropump is made of four double sides 70 μm copper coated PCB with total thickness 800 μm aligned as sown in Fig. 5.3 [24]. Patterned 8 μm thin kapton foils and four doubled sided copper plated PCB are properly aligned and fixed using adhesives. It is worth to note that PCB technology has found applications in optics [29, 30] also. The conductivity sensor, temperature sensor, and pressure sensor are combined on the PCB substrate to make salinity sensing device [31]; which is used to monitor and analyze the marine water, fresh water, industrial water etc.

The complete RF system cannot be fabricated on conventional substrates i.e. Silicon or Gallium arsenide because radiating elements (antenna) do not work efficiently on high dielectric constant substrates. RF MEMS devices fabricated on conventional substrates are mounted on PCB using wire bonding to construct RF system. Wire bonding may cause impedance mismatching and increase in signal losses at the interface which need to be compensated by adding matching network. RF MEMS devices fabricated on conventional substrates need to be packaged individually which further add in device cost [32]. This can be avoided by integrating RF MEMS devices and other components of the RF system on the same PCB substrate. There is full ESA (electronically steerable antenna) system consisting of power divider, RF MEMS phase shifter and antenna on single PCB substrate [33]. A monolithic phased antenna array is shown in Fig. 5.4 [2]. A monolithic electronically steerable antenna (ESA) is reported by Sundaram et al [34] also. An antennas array with 50 % impedance bandwidth from 4 GHz to 6 GHz with CPW feed utilizes RF MEMS switches with antenna on PCB substrates [5, 35]. The PCB MEMS phase shifter with very low insertion loss of 0.56 dB at 9 GHz [3] and RF MEMS switches with insertion loss 0.4-0.45 dB with isolation greater than 20 dB for frequency range 1-30G Hz [36] are mentioned in literature. Except that there are switches fabricated by depositing Silicon nitride using HDCIP CVD deposition process at low temperature compatible to PCB to get high breakdown voltage and COMP planarization technique to get smooth surface for further layers [26, 37, 38]. PCB based low cost coplanar patch antenna with varactors of tunable frequency from 5.545 GHz at 0 V to 5.185 GHz at 116 V having return loss of the CPA is better than 40 dB over the tuning range of 360 MHz [39].

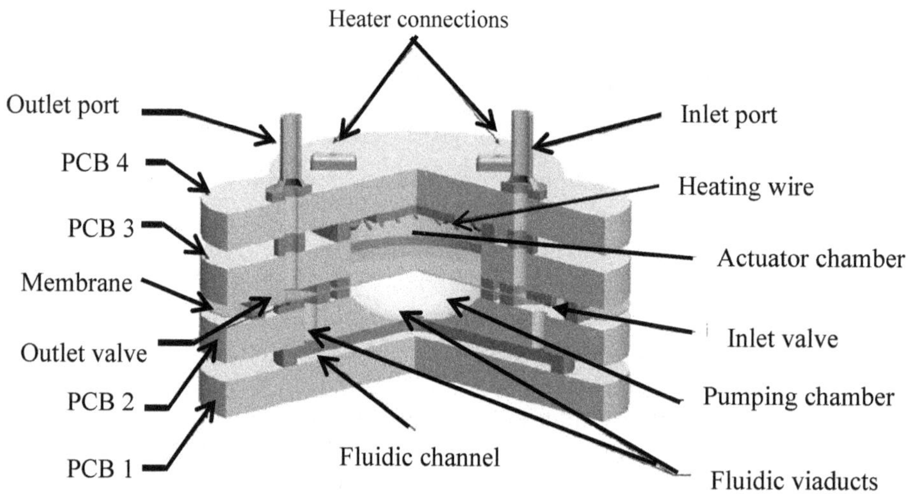

Fig. 5.3. Cross-sectional view of the micropump using multilayer technology [24]. Reproduced from A. Wego, S. Richter, L. Pagel, Fluidic MEMS based on printed circuit board technology, *Journal of Micromechanics and Microengineering,* IOP publisher, 11, 2001, pp. 528–531, doi:10.1088/0960-1317/11/5/313 © IOP Publisher.

5.3. Low Temperature Co-fired Ceramics (LTCC)

Multilayer ceramic substrate was originated in late 1950's by RCA Corporation using current process technologies. These ceramics were demonstrated by firing at high temperature of 1600 °C, there called as HTCC (high temperature co-fired ceramic).

During early 1990s many Japanese and American electronics developed low temperature co-fired ceramic (LTCC) multilayer boards. Fujitsu and IBM were first to use LTCC in commercial applications [40].

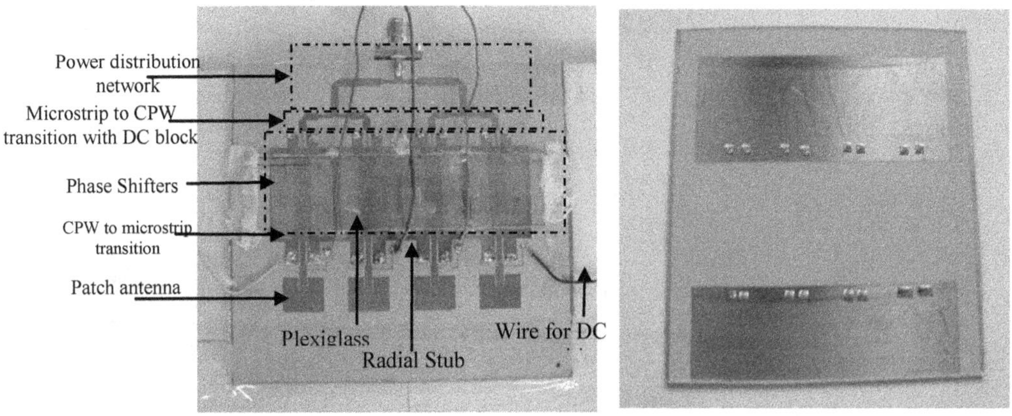

(a) (b)

Fig. 5.4. PCB based System (a) top side, (b) backside of phased antenna array [2]. Reproduced from P. Goel, K. J. Vinoy, A low-cost phased array antenna integrated with phase shifters cofabricated on the laminate, *Progress in Electromagnetics Research B,* 30, 2011, pp. 255-277, *© Progress in Electromagnetics Research. Reproduced courtesy of the Electromagnetic Academy.*

LTCC is made of Green sheet which in turn is made by casting slurry of ceramic powder (45 % Al_2O_3, 40 % glass) and organic components (15 % solvent, plasticizer and binder) [41] into tape using doctor blade. This is a flexible material. Vias for conduction between layers and patterns are screen printed on green sheets using conductive paste. Screen printed layers are aligned under heat and pressure to laminate. A multilayer substrate is obtained by firing conductor and ceramic together and driving off organic binder [40]. A typical LTCC multilayer process is illustrated in Fig. 5.5.

LTCC can be used to fabricate both MEMS and ICs, to implement monolithic devices. Some of the advantages of LTCC usage are listed below:

1. LTCC board can be used for substrate as well as package;
2. It is a multilayer technique which helps in miniaturizing and making highly dense interconnections in the system. Now days, up to 80 layers can be arranged with proper alignment. LTCC's multilayer nature helps in fabricating 3D structures [7]. 3D fabrication helps in reducing device size which further decreases interconnection length and consequently parasitics could be reduced [42]. Secondly LTCC substrates are patterned separately for each layer. Therefore unlike any conventional fabrication technology, manufacturing defect in any layer can be rectified by repeating the process for that particular layer [43];

3. Thick and thin film components, active and passive components can be fabricated all together and on both sides of LTCC [44].
4. Important features of LTCC are: wider range of electrical properties, mechanical and thermal properties comparable to Silicon or pyrex substrates [7].
5. Earlier LTCC was specifically used for RF devices because of their high density interconnection feasibility, low dielectric constant and low loss materials. But because of their flexibility in fabrication to make 3D structures made it useful for various MEMS.

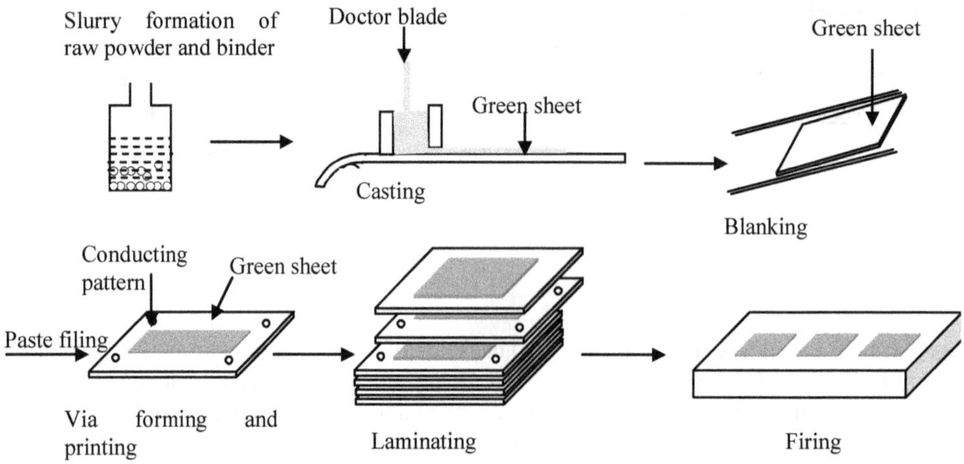

Fig. 5.5. LTCC manufacturing process [40]. Reproduced from Y. Imanaka, *Multilayer low temperature cofired ceramic (LTCC) technology,* Fujitsu Laboratories Ltd. Japan Springer (2005) © Springer.

On the other hand, disadvantages of LTCC technology are:
1. Low mechanical strength, low thermal conductivity, high tolerance (±0.3 % along x, y and ±0.5 % along z direction) [7];
2. Features size of < 1 mm cannot be attained (generally 1 mm to 100 mm) [7, 42].

5.3.1. Fabrication Processes for LTCC Technology

LTCC is made using dielectric tapes, connecting vias, external and internal conductors and passive components (resistors, capacitors, inductors). Passive components (resistors, capacitors and inductors) can be made using screen printing. Vias can be fabricated using mechanical punching, laser micromachining, numerically controlled milling or jet vapor etching. Casting and embossing are to create cavities and channels. All layers are stacked, registered, laminated and then cofired to make 3D structures [7, 44].

5.3.1.1. Screen Printing

Screen printing is done to pattern conductor traces on LTCC substrate. As shown in Fig. 5.6 [45], mask and screen is mesh of wires kept in aligned fashion to pattern metal. Screen is a mesh structure made of stainless steel wires. Diameter of wires, number of strands in the mesh and viscosity of the emulsion are optimized according to the requirement. RF MEMS applications need very narrow width conductor tracks which are possible using small diameter screen to get large aperture ration with thin emulsion [45]. Conductor paste made by pressing metallic powder, solvent and glass binder on the LTCC substrate using comb or squeeze through printing screen [40, 45].

5.3.1.2. Micromachining

CO_2 laser and excimer laser, or a diamond cutter can be used for micromachining small structures [7].

5.3.1.3. Via Punching

Vias are made in green sheet for interconnection between layers. Vias can be punched using numerically controlled (NC) drilling, milling or laser micromachining techniques. Recently Nd-YAG laser has been used for via punching and also fine holes can be made depending on the power and focus of the beam [40]. Via holes are made using either any of these methods or mechanically with die. Vias of diameter 80 μm can be made and more than 1,11,000 vias can be made in 100 mm^2 area [46].

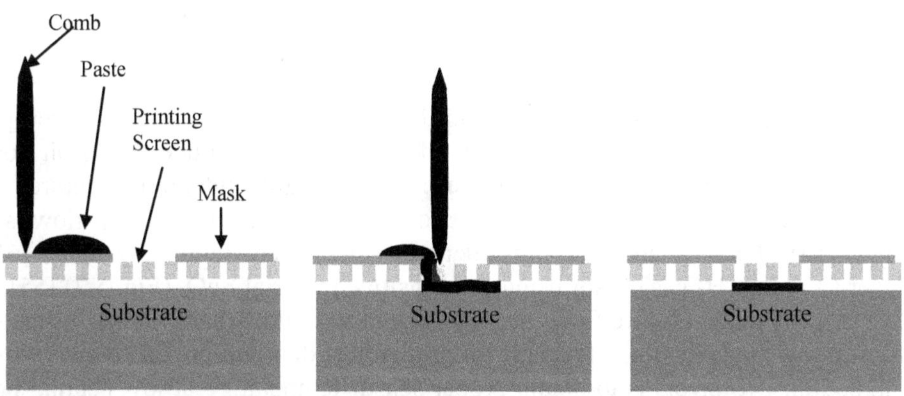

Fig. 5.6. Screen printing process flow to get pattered conductor tracks [45]. Reproduced from A. Lamminen, Design of millimeter-wave antennas on low temperature co-fired ceramic substrates, *Master's Thesis* Espoo, 2006, © Lammimen.

5.3.1.4. Lamination

All the patterned layers are stacked in aligned manner under heat and pressure to laminate the stack. Lamination is done at temperature range 60 °C-110 °C under pressure of 2200-3600 psi uniformly. Lamination process should be homogenous because in-homogeneity causes shrinkage of the substrates. Lamination temperature and pressure vary from system to system [7, 40, 46].

5.3.1.5. Co-firing

Cofiring is done to make single system by fixing LTCC substrates altogether. Laminated stack of LTCC is placed in furnace, first at 500 °C to evaporate binder material and then at 900 °C for conductor and ceramic sintering together [40].

5.3.2. Applications of LTCC Technology

Although the major application of LTCC technology is in RF/millimeter wave devices, but versatility of LTCC material properties makes it useful for many other applications including especially MEMS, and bio-MEMS. Initially this technology was used for fabrication of MEMS devices whereas microelectronic components were fabricated on conventional semiconductor substrate but later it was found compatible for microelectronics technology too. Some of the LTCC applications are listed as: MEMS based thermistor flow sensors [7], almost same TCE of silicon and LTCC made it possible to transfer probes from silicon to LTCC to fabricate probe card using LTCC multilayer technology [47, 48]. Bio-compatibility of LTCC substrates found applications in making lab-on chip sort of MEMS in which major issue is of fabricating microchannels, which is possible using laser micromachining, milling etc techniques [6, 41, 43, 49, 50].

Apart from that LTCC found applications in the field of chemical sensors, micropump [51], solid micro-thrusters for micro-spacecraft for accurate altitude and orbit control [52], base material for optical fibers because of it's highly integrable features [53], optics integrated with fluidic micro-channels to see the impact of optics on flow is also discussed [54], heating and cooling microsystems, micro-fluidic systems [6]. A multilayer CO and CH_4 gas sensor using gas sensitive material SnO_2 (Fig. 5.7) [55]. Gas sensitive layer is made of pure SnO_2 or SnO_2 doped with Pd, where Pd acts as catalyst. Smaller window made in all the layers is for proper heat distribution, and bigger window made in bottom two layers is to obtain proper sensor temperature at low heating power [55].

In the beginning, major application of LTCC was in RF devices due to low loss substrate and dense interconnection properties. Macroscale mechanical switches cause low loss but bulky, whereas semiconductor electronics switches are light but lossy. LTCC overcomes both the challenges because of its integration compatibility. 9-layers LTCC substrate with 32 switches integrated to create 4x4 switch matrix with excellent RF performance up to 7 GHz is demonstrated [56]. Secondly, LTCC is reported to integrate antenna, transceiver, amplifier, and antenna switch, all RF modules as single system, without limiting its performance of power, efficiency, and directivity unlike Silicon technology [57]. LTCC could be used as substrate as well as packaging material for RF-devices, as reported [58]. Beauty of this technology is to make 3D structures (Fig. 5.8 [59]) which takes 75 % lesser space than planar inductors. It is clear that all the sides of the spiral inductor are fabricated on different metallic layers [59]. Zhao et al [60] demonstrated PHEMT and LDMOS power amplifier utilizing LTCC technology.

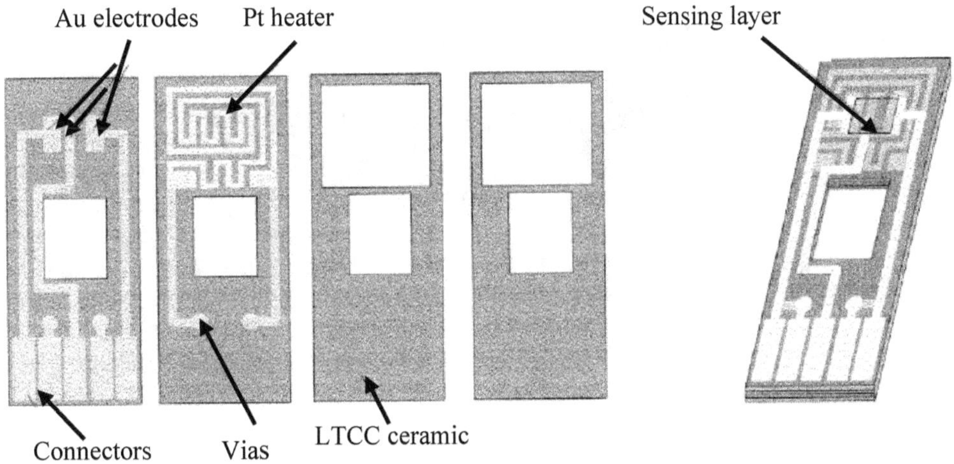

Fig. 5.7. Multilayer SnO_2 gas sensor using stack of LTCC substrates [55]. Reproduced from H. Teterycz , J. Kita, R. Bauer, L. J. Golonka, B.W. Licznerski, K. Nitsch, K. Wiśniewski, New design of an SnO_2 gas sensor on low temperature cofiring ceramics, *Sensors and Actuators B, Elsevier*, 47, pp. 100-103 (1998) © *Elsevier.*

5.4. Liquid Crystal Polymer (LCP) Technology

LCP is a thermoplastic material and came into picture in 1970's. These were commercially known only in 1980s when its polymer resins were available. It consists of densely packed polymer chains of aromatic polyesters in aligned manner. Molecular structure of liquid crystal polymer is made of two aromatic rings A and A′ which can be either benzene ring or derivative of benzene. Side chain group R and R' can be alkyl (C_nH_{2n+1}), alkoxy ($C_nH_{2n+1}O$), acyloxyl, alkylcarbonate, alkoxycarbonyl, etc. Linkage Xs are made of stilbene (-CH=CH-), ester (-COO-), azoxy (-N ≡ N-), schiff base (-CH=N-), acetylene (-C≡C-), and diacetylene (—C≡C-C≡C-) [61].

LCP layers can be bonded using thermocompression bonding (Fig. 5.9). Polymer such as Kapton is used in MEMS fabrication but LCP is being considered better due to cost which is around 50-80 % lower than Kapton. Secondly LCP polymers can be directly bonded to other substrate by thermal lamination, whereas Kapton need adhesive to make bond with other substrate. Comparison in properties of LCP and Kapton is listed in Table 5.4.

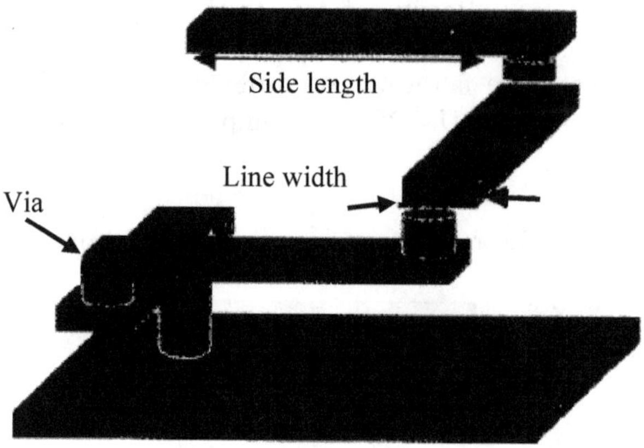

Fig. 5.8. Schematic diagram of 3D inductor [59]. Reproduced from R. J. Pratap, S. Sarkar, S. Pinel, J. Laskar, Gary S, Modeling and optimization of multilayer LTCC inductors for RF/Wireless applications using neural network and genetic algorithms, *IEEE Electronic Components and Technology Conference*, 2004, © IEEE.

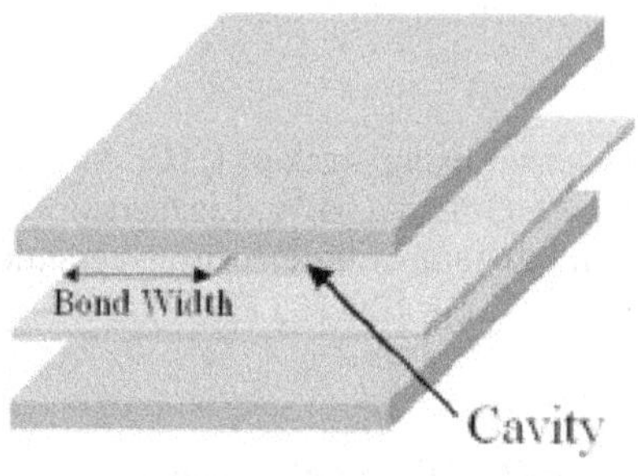

Fig. 5.9. Two high melting points and one (centre) low melting point LCP for thermo-compression bonding [12]. Reproduced from, N. Kingsley, S. K. Bhattacharya, J. Papapolymerou, Moisture lifetime testing of RF MEMS switches packaged in liquid crystal polymer, *IEEE Transactions on Components and Packaging Technologies,* 31, No. 2, 2008, pp. 345-350.

Some of the important properties of LCP are mentioned below:

1. LCP polymers are with loss factor 0.004 and relative permittivity of ~3, which is approximately constant for 0.5 to 40 GHz frequency range. Therefore LCP are suitable for wide band RF applications [14, 62];
2. Negligible moisture effects even at elevated temperatures which is good for RF MEMS applications [14, 62];

3. Excellent chemical compatibility as it is not attacked by commonly used solutions such as: HF, buffered HF, photoresist, developers, acetone and alcohol [62];
4. Low substrate cost as its manufacturing is easy [62];
5. It is near hermetic material and low thermal coefficient of expansion which can be controlled during fabrication [62, 63];
6. The unique feature of polymers technology over Silicon technology is high mechanical fracture limit and biocompatibility [14];
7. LCPs properties: high thermal stability, low flammability, low moisture absorption and its thermoplastic in nature which is recyclable. All these properties of LCP make it suitable material for packaging [14]

Table 5.4. Comparison between LCP and Kapton properties [14]. Reproduced from X. Wang, J. Engel, C. Liu, Liquid crystal polymer (LCP) for MEMS: processes and applications, *J. Micromech. Microeng,* IOP publisher, 13, 2003, pp. 628-633, doi:10.1088/0960-1317/13/5/314 © IOP publisher.

Parameters	**LCP (Vectra A-50)**	**Kapton (HN200)**
Melting temperature (°C)	280	>400
Dielectric constant	2.8	3.5
Loss factor, tan δ	0.004	0.002
Moisture absorption (%)	<0.02	2.8
Coefficient of thermal (ppm/°C)	0–30	20
Expansion controllable	Controllable	
Tensile strength	30 kpsi	34 kpsi
Tensile modulus	1.3 Mpsi	370 kpsi
Specific gravity	1.4 kgm^{-3}	1.42 kgm^{-3}

LCP is a thin and flexible material which needs specific fabrication processes to fabricate MEMS. Fabrication processes developed specifically for LCP technology are discussed here.

5.4.1. Fabrication Processes of LCP Technology

LCP films are thin and flexible which makes difficult to conduct processes of spin coating, lithography and deposition on it directly. Secondly LCP are made of polymers which needs special kind of etching processes. Fabrication processes for LCP technology are described below.

5.4.1.1. Spin Coating/lithography and Metallization

LCPs are thin and flexible which may cause warpage during spin coating of photoresist. Therefore LCP films are bonded to planar surfaces such as Silicon wafer to avoid warpage during spinning. LCP films are bonded to Silicon wafer with dissolvable adhesives to remove Silicon wafer on completion of process. There may be warpage issues during high temperature processes which can be avoided by performing process in intervals which would help in limiting the heat [14].

5.4.1.2. Etching

A suspended flap supported by two flexural cantilevers is reported in [14] as shown in Fig. 5.10 using RIE in oxygen plasma. Surface roughness increased from 190 nm to 1.138 μm before and after RIE process respectively. Etch rate of vectra A-950 LCP in oxygen plasma is of 0.22~0.27 μm/min at power of 350 W under chamber pressure of 500 mTorr. Laser micromachining is commonly used for etching the LCP films. Minimum feature size of 25 μm can be attained. Conventional MEMS fabrication techniques of lithography, metallization, and plasma etching are used with proper modification to obtain lower feature size.

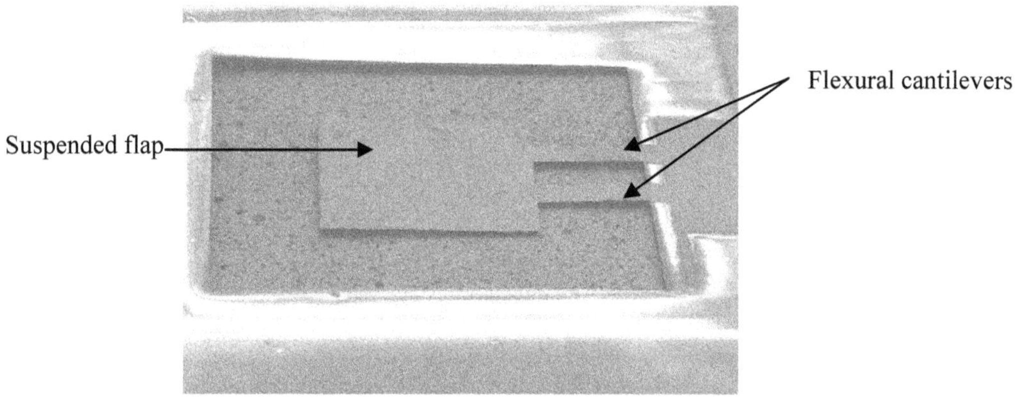

Fig. 5.10. The SEM image of a suspended flap supported by two flexural cantilevers [14]. Reproduced from X. Wang, J. Engel, C. Liu, Liquid crystal polymer (LCP) for MEMS: processes and applications, *J. Micromech. Microeng.,* IOP publisher, 13, 2003, pp. 628-633, doi:10.1088/0960-1317/13/5/314 © IOP Publisher.

5.4.1.3. Drilling Holes

Excimer laser is used to etch the LCP film in controlled manner to make cavity or steps. Laser micromachining is used for LCP films etching. CO_2 engraving laser of 10 μm wavelength is used to make through holes in LCP films because of its high power and fast cutting speed [63]. LCP and metals can be etched selectively by using laser pulses of different fluency [64].

LCP polymers are patterned with these processes to realize MEMS. MEMS applications using LCP technology are discussed next.

5.4.2. MEMS Applications of LCP Technology

The MEMS capacitive sensor is demonstrated using multilayer LCP lamination technique because LCP polymers layers can be strongly bonded using thermo-compression bonding at 150 psi and 280 °C (280 °C is melting temperature of one of the multilayer structure). Reported sensitivity of fabricated novel capacitive pressure sensor is 4.8 mV/kPa [65]. The LCP polymers can be used for environmental sensing system using PCBMEMS based technology to measure the conduction, temperature, depth of ocean under harsh conditions such as salinity, temperature, and pressure of ocean [66]. LCP found applications to fabricate micro-fluidic channels, planar neuron probes, and micro-tactile sensors. It can be bonded to conventional substrate such as glass, silicon with strong force using lamination bonding to make sealed micro-fluidic channels [62]. LCP polymers can be used to fabricate flow sensor, tactile sensor because of its flexibility [14]. LCP technology is compatible for multilayer stack arrangements to reduce the space without compromising the system performance. Schematic of phased array antenna system fabricated by stacking LCP films using multilayer technology is shown in Fig. 5.11 [8]. In this, LCP has been used as package which is similar to system in package technology.

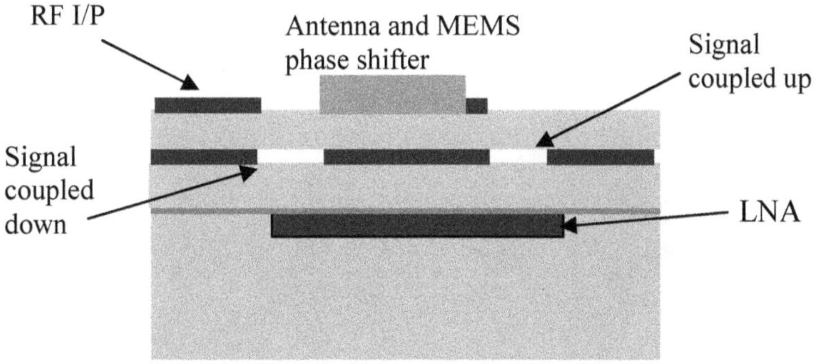

Fig. 5.11. Multilayer system-on-package (SOP) phased array antenna utilizing LCP multilayer technology [8]. Reproduced from N. Kingsley, G. E. Ponchak, J. Papapolymerou, Reconfigurable RF MEMS phased array antenna integrated within a liquid crystal polymer (LCP) system-on-package, *IEEE Transactions on Antennas and Propagation*, 56, No. 1, 2008, pp. 108-118. © IEEE.

Dielectric properties of LCP make it useful for RF MEMS applications. LCP can be used to realize system-in-packaged (SIP) RF devices. Most important application of LCP polymer based SIP are RF MEMS because of low dielectric constant and low loss tangent of LCP materials which are comparable to FR4 and LTCC [67, 68] as listed below in Table 5.5.

Its measured dielectric constant variation over frequency range of 2 GHz-110 GHz is 2.36 to 2.37, which makes it suitable for wide band applications [69]. MEMS phase shifter in LCP package is already demonstrated with reliable performance [11], an integrated RF-module fabricated on multilayer LCP substrates with top and bottom PCB substrate to provide mechanical strength to the system is also reported [10], low cost and light weight 4X8 antenna array with 2-bit phase shifter using MEMS switches with multiple layers of LCP substrate with loss less than 0.5 dB per bit [9]. RF MEMS switch fabricated using LCP thin films have applications in phase shifter, filters etc. [63]. Delay line microstrip 2-bit phase shifter fabricated on LCP using MEMS switches showed performance of average return loss 22.5 dB and average insertion loss 0.98 dB per bit [13]. RF MEMS switch fabricated on LCP is shown in Fig. 5.12 [12].

Table 5.5. Comparison in substrate materials [68]. Reproduced from D. C. Thompson, O. Tantot, H. Jallageas, G. E. Ponchak, M. M. Tentzeris, J. Papapolymerou, Characterization of liquid crystal polymer (LCP) material and transmission lines on LCP substrates from 30 to 110 GHz, *IEEE Transactions on Microwave Theory and Techniques,* 52, No 4, 2004, pp. 1343-1352, © IEEE.

Substrate Material	Dielectric Constant	Loss tangent	Frequency (GHz)
FR4 PCB	4	0.025	<10
LTCC	5.7-9.1	0.0012-0.0063	<12
LCP	2.9-3.2	0.002-0.0045	<105

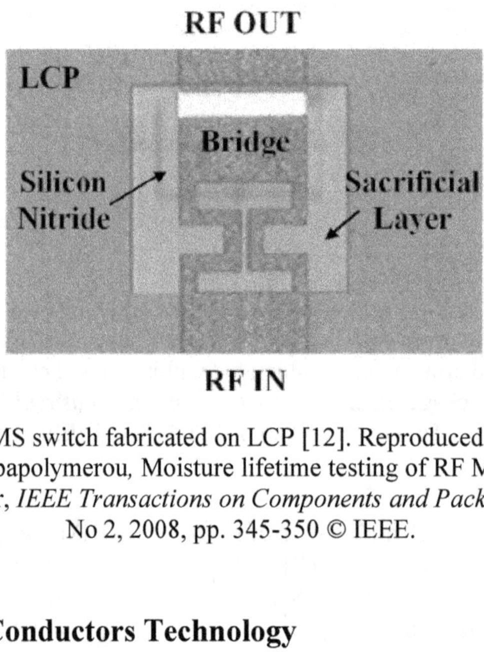

Fig. 5.12. RF MEMS switch fabricated on LCP [12]. Reproduced from N. Kingsley, S. K. Bhattacharya, J. Papapolymerou, Moisture lifetime testing of RF MEMS switches packaged in liquid crystal polymer, *IEEE Transactions on Components and Packaging Technologies,* 31, No 2, 2008, pp. 345-350 © IEEE.

5.5. Polymer Core Conductors Technology

As is clear from the name that polymer is being used as core with metal coating. It is difficult to process thick metallic layer compared to polymers which is made convenient with this technology. Major use of this technology is in RF-devices. This is because, at high frequency, signal attenuates while moving from conductor surface to inside along depth/thickness. Therefore a hollow conductor of thickness five times the skin depth, is same as solid conductor at high frequencies [16]. Until now, SU-8 is being considered as the most promising material for core realization due to its versatile nature. SU-8 polymer was developed in 1980 by IBM. In the beginning it was used in microelectronics as high resolution negative photoresist. Later this material, because of its ultra thick spin coating on substrate as photoresist and structural material, found applications in MEMS. Basic structure of SU-8 is shown in Fig. 5.13.

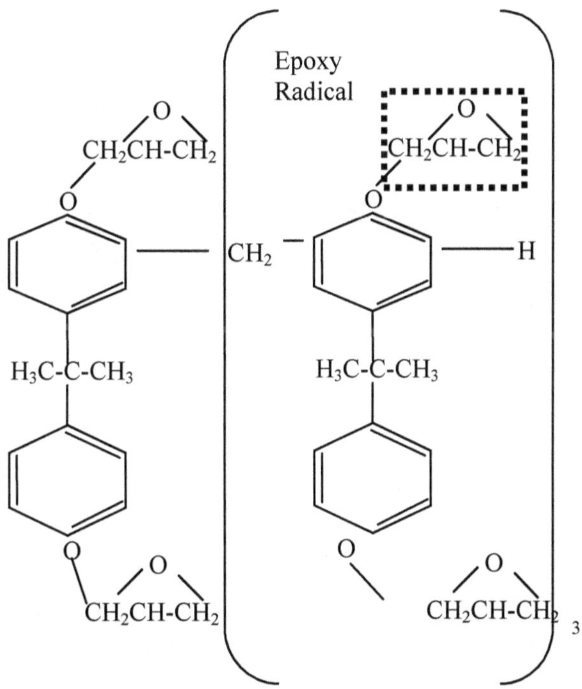

Fig. 5.13. SU-8 molecular model [70]. T. Namazu, S. Inoue, K. Takio, T. Fujita, R Maenaka, K. Koterazawa, Visco-elastic properties of micron-thick SU-8 polymers measured by two different types of uniaxial tensile tests, *IEEE,* 2005, pp. 447-450. © IEEE.

Major features of polymer core conductor are:

1 Polymer core conductors can be used for high aspect ratio metallic structures [16]. This process can be used to make thick (up to thousands of microns) structures;
2 In applications, were substrate to metal gap is required to reduce substrate effects on device which are useful especially at high frequency [71];
3 Double expose single develop process of SU-8 made this technology very easy and cheap to process [16].

5.5.1. Fabrication Processes of Polymer Core Conductors Technology

A typical process to realize polymer core conductors is discussed here. A thick layer of SU-8 is spun coated Fig. 5.14 (a). SU-8 is patterned using double exposure and single development to get bridge structure Fig. 5.14 (b, c). After double exposure, SU-8 polymer links are only at the bridge structure and remaining SU-8 gets etched away during development Fig. 5.14 (d) [16]. Thin metal layer can be electroplated on the substrate as well as below and top of SU-8 bridge to make polymer core conductor.

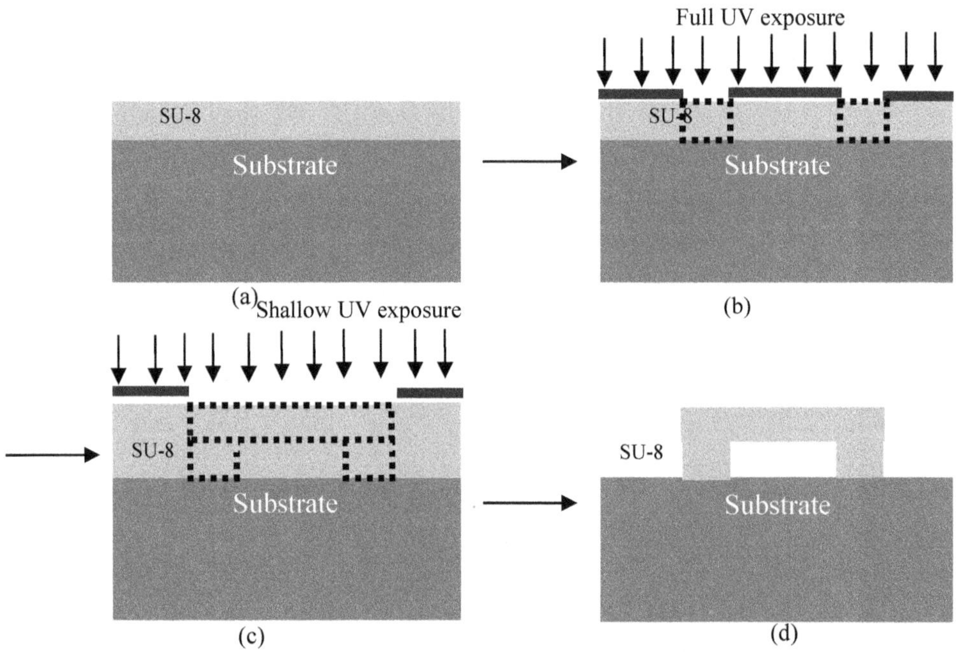

Fig. 5.14. Double exposure single development technique [16]. Reproduced from Yong-Kyu Yoon, Jin-Woo Park, Mark G. Allen, Polymer-core conductor approaches for RF MEMS, *Journal of Microelectromechanical Systems*, 14, No. 5, 2005. © IEEE.

5.5.2. MEMS Applications of Polymer Core Conductor Technology

3D structures of polymer core conductor found applications in making microelectrode arrays (MEAs) with microfluidic ports for profusion channels to study complex networks morphologies and tissue slices [72]. Polymer core conductor properties made it useful for transmission over ACSR (aluminum conductor steel reinforced) even at macro scale [73].

The performance of a device operating at millimeter wavelength is highly influenced by substrate properties. Substrate effects can be reduced using polymer core conductors. High Q inductors and antennae are already reported in literature. A 50 μm thick spiral inductor with Q factor 95 for 1.1 nH and solenoid inductor of 500 μm height with Q factor 71 are reported [74, 75]. Air core inductor fabricated using double exposure single develop fabrication process is shown in Fig. 5.15. Patch antenna is the most popular antenna where performance is degraded due to substrate effects but 3D antennae formed with this technology can overcome these drawbacks. As monopole driven YAGI-UDA antenna with wide bandwidth 12 %, air lifted patch antenna having BW 7 %, magnetically lifted monopole antenna with BW 20.7 %, CPW fed quarter wavelength monopole antenna for W-band (75 GHz-110 GHz) [71], integrated filters in air cavity on silicon wafer [76].

Fig. 5.15. SEM photograph of air core inductors double expose single develop fabrication process [16]. Reproduced from Yong-Kyu Yoon, Jin-Woo Park, Mark G. Allen, Polymer-core conductor approaches for RF MEMS, *Journal of Microelectromechanical Systems,* 14, no. 5, 2005. © IEEE.

5.6. Polydimethylsiloxane (PDMS) Technology

The Polydimethylsiloxane (PDMS) is an elastomer polymer synthesized by mixing liquid monomer and cross-linker. Polymerization reaction takes place on heating and forms solid elastomer [77]. Chemical formula of PDMS is as shown in Fig. 5.16.

Elastomer polymers can be stretched 2-10 times of the original dimension like a rubber and its elasticity is controlled by precursors i.e. monomer and cross-linker. PDMS is one of the most widely used polymers in MEMS technology [77, 79]. Patterning of PDMS using mold is simple and fast compared to etching and bonding in semiconductor technology [80]. In some applications, flexible substrate is better to use because of light weight, low cost, and more flexibility over conventional semiconductor substrate [81]. These can develop feature size of the order of nm by casting. It is one time process to make master mould of desired structure [82].

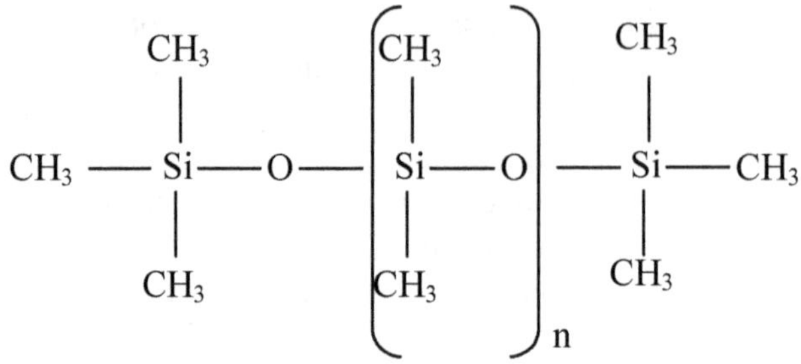

Fig. 5.16. Polydimethylsiloxane (PDMS) polymer structure [78]. Reproduced from P. L. Baron, J. Casanovas, J. P. Guelfucci, R. L. S. Hoi, Photoconductivity induced by VUV photons in Polydimethylsiloxane and Polymethylphenylsiloxane oils, *IEEE Transactions on Electrical Insulation,* 23, No 4, 1988, pp. 563-570. © IEEE.

5.6.1. Fabrication Process of PDMS Technology

PDMS structures are fabricated using mould where SU-8 [83] and Silicon [84, 85] are most commonly used master moulds to realize structures. PDMS fabrication process using silicon mould is shown in Fig. 5.17 which is carried out in six steps. Initially, Silicon wafer is spin coated with photo resist (Fig. 5.17 (a)). Photoresist is patterned using lithography (Fig. 5.17(b)) and developed using developer to attain pattern on photoresist (Fig. 5.17(c)). Silicon etching is done using DRIE from the exposed parts of the wafer (Fig. 5.17(d)). PDMS is poured on the clean master mould and cured in oven (Fig. 5.17(e)). Cured PDMS is peeled from the silicon wafer and PDMS cantilever array is ready (Fig. 5.17(f)).

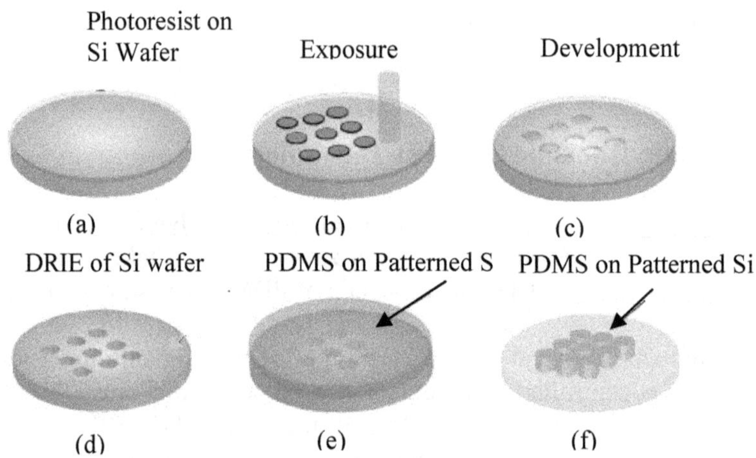

Fig. 5.17. Fabrication steps for making the microcantilever array on PDMS (a) Spin photoresist on Si wafer (b) define patterns on photoresist by exposure (c) develop exposed wafer (d) perform DRIE to pattern silicon (e) Pour PDMS on cleaned wafer, and cure in oven (f) Peel cured PDMS from silicon wafer to obtain device [85]. Reproduced from K. A. Addae-Mensah, S. Retterer, S. R. Opalenik, D. Thomas, N. V. Lavrik, J. P. Wikswo, Cryogenic etching of silicon: an alternative method for fabrication of vertical microcantilever master molds, *Journal of Microelectromechanical Systems*, Vol. 19, No. 1, 2010, pp. 64-74 © IEEE.

5.6.2. MEMS Applications of PDMS Technology

The 3D micro channels fabricated in PDMS are reported in [80], which is made of many thin PDMS layers where each layer is patterned using master mould made of silicon. An array of temperature sensor on PDMS [81] is shown in Fig. 5.18. Silicon is used as the base as well as mould. Base is required during spin coating whereas mould is used for patterning the PDMS.

Fig. 5.18. Array of 8 × 8 temperature sensor of size 2.5 mm × 5.5 mm on PDMS [81]. Reproduced from S. Y. Xiao, L. F. Che, X. X. Li, Y. L. Wang, A novel fabrication process of MEMS devices on polyimide flexible substrates, *Microelectronic Engineering,* Elsevier, 85, pp. 452-457 (2008) © Elsevier.

5.7. Conclusion

Presently, microsystem technology is one of the most promising technologies. MEMS technology created the feasibility of making monolithic systems which was not feasible with IC technology. Due to its benefits of compactness, low DC power consumption, and to realize monolithic integrated system, this is one of the fastest growing technologies. Despite many advantages of MEMS technology, this technology is not much used in field due to its issues of high fabrication cost and integration issues. In conventional systems, MEMS component are realized on semiconductor substrate using micromachining techniques and packaged individually to integrate in system. These issues can be overcome with non-silicon fabrication technologies where MEMS as well as other components can be fabricated on the same board. This paper provides an idea about various non-silicon MEMS fabrication technologies with an overview of their applications in MEMS/RF MEMS.

Recent years witnessed significant research activities and remarkable advances in development of microsystems. In this regard, need was felt to review the evolution of non-silicon microfabrication technologies for microsystems. This paper will help the researchers to select fabrication technology depending on application, available fabrication facility, and cost.

Acknowledgment

The author would like to thank Prof. K. J. Vinoy, Microwave Laboratory, Department of Electrical Communication Engineering, Indian Institute of Science, Bangalore, for his guidance and support throughout this work. I would like to thank Mukesh, working in Samsung, Bangalore, India for his help in finalizing the manuscript.

References

[1]. L. Pagel, S. Gabmann, Microfluidic systems in PCB technology, in *Proceedings of the 31st IEEE Annual Conference (IECON' 2005),* 6-10 November 2005, pp. 2368-2371.
[2]. P. Goel, K. J. Vinoy, A low-cost phased array antenna integrated with phase shifters cofabricated on the laminate, *Progress in Electromagnetics Research B,* 30, 2011, pp. 255-277.
[3]. R. Ramadoss, A. Sundaram, L. M. Feldner, RF MEMS phase shifters based on PCB MEMS technology, *Electronics Letters,* 41, No. 11, 2005.
[4]. B. Ghodsian, C. Jung, B. A. Cetiner, F. D. Flaviis, Development of RF-MEMS switch on PCB substrates with polyimide planarization, *IEEE Sensors Journal,* No. 5, 2005, pp. 950-955.
[5]. B. A. Cetiner,J. Y. Qian, H. P. Chang, M. Bachman, G. P. Li, F. De Flaviis, Monolithic integration of RF MEMS switches with a diversity antenna on PCB substrate, *IEEE Transactions on Microwave Theory and Technique,* 51, No. 1, 2003, pp. 332-335.
[6]. L. J. Golonka, Technology and applications of low temperature cofired ceramic (LTCC) based sensors and microsystems, *Bulletin of The Polish Academy of Sciences Technical Sciences,* 54, No. 2, 2006, pp. 221-231.
[7]. M. Gongora-Rubio, L. M. Sola-Laguna, P. J. Moffett, J. J. Santiago-Avilés, The utilization of low temperature co-fired ceramics (LTCC-ML) technology for meso-scale EMS a simple thermistor based flow sensor, *Sensors and Actuators A: Physical,* Vol. 73, 1999, pp. 215-221.
[8]. N. Kingsley ,G. E. Ponchak, J. Papapolymerou, Reconfigurable RF MEMS phased array antenna integrated within a liquid crystal polymer (LCP) system-on-package, *IEEE Transactions on Antennas and Propagation,* 56, No. 1, 2008, pp. 108-118.
[9]. D. J. Chung, D. E. Anagnostou, G. Ponchak, M. Tentzeris, J. Papapolymerou, Integration of a 4x8 antenna array with a reconfigurable 2-bit phase shifter using RF MEMS switches on multilayer organic substrates, in *Proceedings of the IEEE APS/URSI International Symposium,* Honolulu, HI, USA, 10-15 June 2007, pp. 93-96.
[10].M. Swaminathan, A. Bavisi, W. Yun, V. Sundaram, V. Govind, P. Monajemi, Design and fabrication of integrated RF modules in liquid crystalline polymer (LCP) substrates, in *Proceedings of the 31st Annual Conference of IEEE (IECON' 2005),* 6-10 November 2005, pp. 2346-2351.
[11].M. J. Chen, A.-V. Pham, N. Vers, C. Kapusta, J. Iannotti, W. Kornrumpf, J. Maciel, Multilayer organic multichip module implementing hybrid microelectromechanical systems, *IEEE Transactions on Microwave Theory and Techniques,* 56, No. 4, 2008, pp. 952-958.
[12].N. Kingsley, S. K. Bhattacharya, J. Papapolymerou, Moisture lifetime testing of RF MEMS switches packaged in liquid crystal polymer, *IEEE Transactions on Components and Packaging Technologies,* 31, No. 2, 2008, pp. 345-350.
[13].N. Kingsley, G. Wang, J. Papapolymerou, 14 GHz microstrip MEMS phase shifters on flexible organic substrate, *Microwave Conference European,* Vol. 1, 2005.

[14]. X. Wang, J. Engel, C. Liu, Liquid crystal polymer (LCP) for MEMS: processes and applications, *J. Micromech. Microeng,* 13, 2003, pp. 628-633.

[15]. J. N. Palasagaram, R. Ramadoss, MEMS-capacitive pressure sensor fabricated using printed-circuit-processing techniques, *IEEE Sensors Journal*, No. 6, 2006, pp. 1374-1375.

[16]. Y.-K. Yoon, J.-W. Park, M. G. Allen, Polymer-core conductor approaches for RF MEMS, *Journal of Microelectromechanical Systems,* 14, No. 5, 2005, pp. 886-894.

[17]. J.-M. Yook, Y.-S. Kwon, Spiral inductors using polymer-core conductors, *IEEE Microwave and Wireless Components Letters,* 17, No. 7, 2007, pp. 495-497.

[18]. N. C. Cady, S. Stelick, M. V. Kunnavakkam, C. A. Batt, Real-time PCR detection of *Listeria monocytogenes* using an integrated microfluidics platform, *Sensors and Actuators B,* 107, 2005, pp. 332-341.

[19]. Z. Xiao, C. Feng, P. C. H. Chan, I-M. Hsing, Monolithically integrated planar microfuel cell arrays, *Sensors and Actuators B*, 132, 2008, pp. 576-586.

[20]. C. F. Coombs, Jr. Combs' Printed Circuits Handbook, 5th edition, *McGraw-Hill*, New York, 2001.

[21]. R. S. Khandpur, Printed Circuit Boards Design Fabrication and Assembly, *McGraw-Hill*, 2006.

[22]. E. K. W. Gan, Z. H. Y, Lim G. C, Laser drilling of micro-vias in PCB substrates, in *Proceedings of the IEEE Electronics Packaging Technology Conference*, 2000, pp. 321-326.

[23]. T. Shimoto, K. Matsui, K. Kikuchi, Y. Shimada, K. Utsumi, New high-density multilayer technology on PCB, *IEEE Transactions on Advanced Packaging,* 22, 2, 1999, pp. 116-122.

[24]. A. Wego, S. Richter, L. Pagel, Fluidic microsystems based on printed circuit board technology, *Journal of Micromechanics and Microengineering,* 11, 2001, pp. 528-531.

[25]. T. Merkel, M. Graeber, L. Pagel, A new technology for fluidic microsystems based on PCB technology, *Sensors and Actuators A Physical,* 77, 1999, pp. 98-105.

[26]. H-P Chang, J Qian, B A Cetiner, F D Flaviis, M Bachman, Design and process considerations for fabricating RF MEMS switches on printed circuit boards, *Journal of Microelectromechanical Systems,* 14, No. 6, 2005, pp. 1311-1322.

[27]. B. Ghodsian, C. Jung, B. Cetiner, F. De Flaviis, Polyimide planarization for RF-MEMS switch on PCB, *URSI Meeting at the University of Colorado Boulder,* USA, 2004, pp 1-1.

[28]. J. Branson, J. Naber, G. Edelen, A simplistic printed circuit board fabrication process for course projects, *IEEE Transactions on Education,* 43, No. 3, 2000, pp. 257-261.

[29]. S. Lehniacher, A. Neyer Integration of polymer optical waveguides into printed circuit boards, *Electronics Letters,* 36, No. 72, 2000, pp. 1052-1053.

[30]. D. Fries, S. Ivanov, H. Broadbent, R. Willoughby, E. Sheehan, Micro ion-optical systems technology [MIST] for mass spectrometry using PCBMEMS, *IEEE,* 2007, pp. 3278-3281.

[31]. H. A. Broadbent, S. Z. Ivanov, D. P. Fries, PCB-MEMS environmental sensors in the field, *in Proceedings of the IEEE International Symposium on Industrial Electronics (ISIE' 2007),* 4-7 June 2007, pp. 3582-3586.

[32]. H.-P. Chang, J. Qian, B. A. Cetiner, F. D. Flaviis, M. Bachman, Design and process considerations for fabricating RF MEMS switches on printed circuit boards, *Journal of Microelectromechanical Systems,* 14, No. 6, 2005, pp. 1311-1322.

[33]. A. Sundaram, R. Ramadoss, L. M. Feldner, Electronically steerable antenna array using PCB-based MEMS phase shifters, in *Proceedings of the* 31st *IEEE Annual Conference (IECON' 2005)*, pp. 2341-2345.

[34]. A. Sundaram, M. Maddela, R. Ramadoss, MEMS-based electronically steerable antenna array fabricated using PCB technology, *Journal of Microelectromechanical Systems,* Vol. 17, No. 2, April 2008, pp. 356-362.

[35]. B. A. Cetiner, L. Jofre', C. H. Chang, J. *Y.* Qian, M. Bachman, G .P. Li, F. D. Flaviis, Integrated MEMS antenna system for wireless communications, *IEEE MTT-S Digest*, 2002, pp. 1333-1336.

[36]. R. Ramadoss, Simone Lee, Y. C. Lee, V. M. Bright, K. C. Gupta, RF-MEMS capacitive switches fabricated using printed circuit processing techniques, *IEEE Journal of Microelectromechanical Systems,* 15, No. 6, 2006, pp. 1595-1604.

[37]. H.-P. Chang, Qian, B. A. Cetiner, F. D. Flaviis, M. Bachman, G. P. Li, RF MEMS switches fabricated on microwave-laminate printed circuit boards, *IEEE Electron Device Letters,* 24, No 4, 2003, pp. 227-229.

[38]. C. H. Chang, J. Y. Qian, B. A. Cetiner, Q. Xu, M. Bachman, H. K. Kim, Y. Ra, F. D. Flaviis, G. P. Li, RF MEMS capacitive switches fabricated with HDICP CVD SIN, *IEEE MTT-S Digest,* 2002.

[39]. M Maddela, R Ramadoss, R Lempkowski, A MEMS-based tunable coplanar patch antenna fabricated using PCB processing techniques, *J. Micromech. Microeng.,*17, 2007, pp. 812-819.

[40]. Y. Imanaka, Multilayer Low Temperature Cofired Ceramic (LTCC) Technology, Fujitsu Laboratories Ltd. Japan, *Springer*, 2005.

[41]. N. I.-García, C. S. M.-Cisneros, F. Valdés, J. Alonso, I.-García, Green-tape ceramics: new technological approach for integrating electronics and fluidics in Microsystems, *Trends in Analytical Chemistry,* 27, No. 1, 2008, pp. 24-33.

[42]. R. Chanchani, D. T. Bethke, D. B. Webb, C. Sandoval, G. Wouters, Integrated substrate technology, in *Proceedings of the IEEE Electronic Components and Technology Conference,* 2004, pp. 1232-1236.

[43]. D. J. Sadler, R. Changrani, P. Roberts, C.-F. Chou, F. Zenhausern, Thermal management of BioMEMS: Temperature control for ceramic-based PCR and DNA detection devices, *IEEE Transactions on Components and Packaging Technologies,* 26, No. 2, 2003, pp. 309-316.

[44]. L. J. Golonka, New application of LTCC technology, *IEEE* 28th *Int. Spring Seminar on Electronics Technology*, 2005, pp. 148-152.

[45]. A. Lamminen, Design of millimeter-wave antennas on low temperature co-fired ceramic substrates, Master's Thesis, Espoo, 2006.

[46]. Y. Shimada, K. Utsumi, M. Suzuki, H. Takamizawa, M. Nitta, T. Watari, Low firing temperature multilayer glass-ceramic substrate, *IEEE Transactions on Components Hybrids and Manufacturing Technology,* 6, No. 4, 1983, pp. 382-388.

[47]. S.-H. Choe, S. Tanaka, M. Esashi, MEMS-based probe array for wafer level LSI testing transferred on to low CTE LTCC substrate by Au/Sn eutectic bonding, in *Proceedings of the 14th International Conference on Solid State Sensors Actuators and Microsystems, IEEE Transducers and Eurosensors,* Lyon France, 2007.

[48]. S.-H. Choe, S. Tanaka, M. Esashi, A matched expansion MEMS probe card with low CTE LTCC substrate in *Proceedings of the IEEE International Test Conference,* 2007, pp. 1-6.

[49]. G. Schlottig, L. Rebenklau, J. Uhlemann, C. Nytsch-Geusen, K.-J. Wolter, Modeling LTCC-based microchannels using a network approach, in *Proceedings of the IEEE Electronics System Integration Technology Conference,* Dresden Germany, 2006, pp. 1374-1377.

[50]. D. J. Sadler, R. Changrani, P. Roberts, C.-F. Chou, F. Zenhausem, Thermal Management of BioMEMS, in *Proceedings of the IEEE Intersociety Conference on Thermal Phenomena,* 2002, pp. 1025-1032.

[51]. T. Thelemann, H. Thust, M. Hintz, Using LTCC for Microsystems, *Microelectronics International,* 2002, pp. 19-23.

[52]. K. Zhang, S. Kiang Chou, S. S. Ang, A low temperature co-fired ceramic solid propellant microthruster for micropropulsion applications, in *Proceedings of the IEEE The 13th International Conference on Solid-State Sensors Actuators and Microsystems* , Seoul Korea, 2005, pp. 672-675.

[53]. D. Kara, S. Patela, L. Golonka, Preparation of optical fiber to the LTCC technological process, in *Proceedings of the IEEE International Students and Young Scientists Workshop on Photonics and Microsystems*, 2004, pp. 21-23.

[54]. K. Malecha, L. J. Golonka, CFD simulations of LTCC based microsystems, *IEEE ISSE*, Germany, 2006, pp. 156-160.

[55]. H. Teterycz , J. Kita, R. Bauer, L. J. Golonka, B.W. Licznerski, K. Nitsch, K. Wiśniewski, New design of an SnO_2 gas sensor on low temperature cofiring ceramics, *Sensors and Actuators B*, 47, 1998, pp. 100-103.

[56]. B. Yassini, S. Choi, A. Zybura, M. Yu, R. E. Mihailovich, Jeffrey. F. DeNatale, A novel MEMS LTCC switch matrix, *IEEE MTT-S Digest*, 2004, pp. 721-724.

[57]. A. C. W. Lu, K. M. Chual, L. H. Guo, Emerging manufacturing technologies for RFIC antenna and RF-MEMS integration, in *Proceedings of the RFIT - IEEE International Workshop on Radio-Frequency Integration Technology*, Singapore, 2005, pp. 142-146.

[58]. K.-I. Kim, J.-M. Kim, J.-M. Kim, G.-C. Hwang, C.-W. Baek, Y.-K. Kim, Packaging for RF MEMS devices using LTCC substrate and BCB adhesive layer, *J. Micromech. Microeng.*, 16, 2006, pp. 150-156.

[59]. R. J. Pratap, S. Sarkar, S. Pinel, J. Laskar, Gary S, Modeling and optimization of multilayer LTCC inductors for RF/wireless applications using neural network and genetic algorithms, in *Proceedings of the IEEE Electronic Components and Technology Conference*, 2004, pp. 248-254.

[60]. L. Zhao, A. Pavio, B. Stengel B. Thompson, A 6 Watt LDMOS broadband high efficiency distributed power amplifier fabricated using LTCC technology, *IEEE MTTS Digest*, 2000, pp. 897-900.

[61]. Iam-Choon Khoo, Liquid Crystals, 2nd edition, *John Wiley & Sons*, 2007.

[62]. X. Wang, L.-H. Lu, C. Liu, Micromachining techniques for liquid crystal polymer, in *Proceeding of the 14th IEEE International Conference on Micro Electro Mechanical Systems (MEMS' 2001)*, 2001, pp. 126-130.

[63]. D. Thompson, N. Kingsley, G. Wang, J. Papapolymerou, M. M. Tentzeris, RF characteristics of thin film liquid crystal polymer (LCP) packages for RF MEMS and MMIC integration, *IEEE Transactions on Microwave Theory and Techniques*, 52, No. 4, 2004, pp. 857-860.

[64]. E. E. Guzzo, J. S. Preston Laser ablation as a processing technique for metallic and polymer layered structures, *IEEE Transactions on Semiconductor Manufacturing*, 7, No. 1, 1994, pp. 73-78.

[65]. J. N. Palasagaram, R. Ramadoss, MEMS capacitive pressure sensor array fabricated using printed circuit processing techniques, in *Proceedings of the 31st IEEE Annual Conference (IECON' 2005), 6-10 November* 2005, pp. 2357-2362.

[66]. D. Fries, H. Broadbent, G. Steimle, S. Ivanov, A. Cardenas-Valencia, J. Fu, M. Janowiak, T. Weller, S. Natarajan, L. Guerra, PCBMEMS for environmental sensing systems, in *Proceedings of the 31st IEEE Annual Conference (IECON' 2005), 6-10 November* 2005, pp. 2352-2356.

[67]. G. Zou, H. Grönqvist, J. P. Starski, J. Liu, Characterization of liquid crystal polymer for high frequency system-in-a-package applications, *IEEE Transactions on Advanced Packaging*, Vol. 25, No. 4, 2002, pp. 503-508.

[68]. D. C. Thompson, O. Tantot, H. Jallageas, G. E. Ponchak, M. M. Tentzeris, J. Papapolymerou, Characterization of liquid crystal polymer (LCP) material and transmission lines on LCP substrates from 30 to 110 GHz, *IEEE Transactions on Microwave Theory and Techniques*, 52, No. 4, 2004, pp. 1343-1352.

[69]. D. Thompson, P. Kirby, J. Papapolymerou, M. M. Tentzeris, W-band characterization of finite ground coplanar transmission lines on liquid crystal polymer (LCP) substrates, *IEEE Electronic Components and Technology Conference*, 2003, pp. 1652-1655.

[70]. T. Namazu, S. Inoue, K. Takio, T. Fujita, R Maenaka, K. Koterazawa, Visco-elastic properties of micron-thick SU-8 polymers measured by two different types of uniaxial tensile tests, in *Proceedings of the 18th IEEE International Conference on Micro Electro Mechanical Systems, (MEMS' 2005)*, 30 January-3 February 2005, pp. 447-450.

[71]. Y.-K. Yoon, B. Pan, J. Papapolymerou, M. Tentzeris, M. G. Allen, Surface-micromachined millimeter-wave antennas, in *Proceedings of the 13th IEEE International Conference on Solid-State Sensors Actuators and Microsystem*, Seoul Korea, 2005, pp. 1986-1989.

[72]. S. Rajaraman, S.-O. Choi1, R. H. Shafer, J. D. Ross, J. Vukasinovic, Y. Choi, S. P. DeWeerth, A. Glezer, M. G Allen, Microfabrication technologies for a coupled three-dimensional microelectrode microfluidic array, *J. Micromech. Microeng.*, 17, 2007, pp. 163-171.

[73]. A. Alawar, E. J. Bosze, Steven, A composite core conductor for low sag at high temperatures, *IEEE Transactions on Power Delivery,* 20, No. 3, 2005, pp. 2193-2199.

[74]. Y.-H. Joung, M. G. Allen, Chip-to-board micromachining for interconnect layer passive components, *IEEE Transactions on Components and Packaging Technologies,* 30, No. 1, 2007, pp. 15-23.

[75]. J.-M. Yook, K.-M. Kim, S.-K. Yeo, J.-I. Yu, Y.-S. Kwon, RF transmission lines and spiral inductors using polymer-core conductor, in *Proceedings of the 37th European Microwave Conference,* 2007, pp. 1145-1148.

[76]. B. Pan, Y. Li, M. M. Tentzeris, J. Papapolymerou, Surface micromachining polymer-core-conductor approach for high-performance millimeter-wave air-cavity filters integration, *IEEE Transactions on Microwave Theory and Techniques,* 56, 4, 2008, pp. 959-970.

[77]. S. A. Wilson, R. P. J. Jourdain, Q. Zhang, R. A. Dorey et al, New materials for micro-scale sensors and actuators an engineering review, *Materials Science and Engineering,* R 56, 2007, pp. 1-129.

[78]. P. L. Baron, J. Casanovas, J. P. Guelfucci, R. L. S. Hoi, Photoconductivity induced by VUV photons in Polydimethylsiloxane and Polymethylphenylsiloxane oils, *IEEE Transactions on Electrical Insulation,* 23, No. 4, 1988, pp. 563-570.

[79]. C.-F. Lin, G. Lee, C. Wang, H. Lee, W. Liao, T. Chou, Microfluidic pH-sensing chips integrated with pneumatic fluid-control devices, *Biosensors and Bioelectronics,* 21, 2006, pp. 1468-1475.

[80]. B.-H. Jo, L. M. V. Lerberghe, K. M. Motsegood, D. J. Beebe, Three-dimensional micro-channel fabrication in Polydimethylsiloxane (PDMS) elastomer, *IEEE Journal of Microelectromechanical Systems,* 9, No. 1, 2000, pp. 76-81.

[81]. S. Y. Xiao, L. F. Che, X. X. Li, Y. L. Wang, A novel fabrication process of MEMS devices on polyimide flexible substrates, *Microelectronic Engineering,* 85, 2008, pp. 452-457.

[82]. K. Huikko, P. Östman, K. Grigoras, S. Tuomikoski, V.-M. Tiainen, A. Soininen, K. Puolanne, A. Manz, S. Franssila, R. Kostiainen, T. Kotiaho, Poly(dimethylsiloxane) electrospray devices fabricated with diamond-like carbon–poly(dimethylsiloxane) coated SU-8 masters, *The Royal Society of Chemistry,* 3, 2003, pp. 67-72.

[83]. B. Matthews, J. W. Judy, Design and fabrication of a micromachined planar patch-clamp substrate with integrated microfluidics for single-cell measurements, *Journal of Microelectromechanical Systems,* Vol. 15, No. 1, 2006, pp. 214-222.

[84]. S.-H. Yoon, V. Reyes-Ortiz, K.-H. Kim, Y. H. Seo, M. R. K. Mofrad, Analysis of circular PDMS microballoons with ultralarge deflection for MEMS design, *Journal of Microelectromechanical Systems*, Vol. 19, No. 4, August 2010, pp. 854-864.

[85]. K. A. Addae-Mensah, S Retterer, S R. Opalenik, D Thomas, N V. Lavrik, J. P. Wikswo, Cryogenic etching of Silicon: an alternative method for fabrication of vertical microcantilever master molds, *Journal of Microelectromechanical Systems*, Vol. 19, No. 1, February 2010, pp. 64-74.

Chapter 6
MEMS Applications in Medical Industries: Review

Kochuthomman Joseph Mampilly, Arjun Ashok, Sudha Ramasamy and Prabhu Ramanathan

6.1. Introduction

MEMS (Micro Electro Mechanical Systems) are a part of semiconductor electronics-VLSI (very large scale integrated) technologies that was developed in the 1980s. MEMS have wide spread applications from mechanical, electrical to microsurgery, diagnostics. It has mechanical parts, which are movable whereas VLSI has no mechanical parts. MEMS have a complex fabrication processes which are used in manufacturing of very high precision devices. In VLSI devices are mainly circuits of commercial purpose manufactured in large scale and have applications in day-to-day life. MEMS devices are made in clean room facilities available in many places across the world. The principle of MEMS is based on silicon as the main component used in manufacturing and they are used for making Biosensors. These silicones wafer are bio compactable with the human body. And some of the MEMS devices used in medical field are discussed in this paper.

6.2. MEMS in Surgical Fields

Today MEMS is widely used in the field of surgery and it has given the surgery as risk less and successful process. Due to MEMS technology miniaturization, today the cost of surgerical process is very much reduced. In the early days the surgery was done with cutting and sewing of the body tissues. This process of surgery was called open surgery which is painful and takes a lot of time to recover. In present, these methods are replaced by minimally invasive surgery procedures. It helps the patient to recover quickly and pain is reduced. So cost of operation and short hospital stays helps to reduce the total cost for surgery. But main disadvantage is surgeon has very limited view on the surgical area. Here comes the importance of MEMS technology, it helps to give better view and control over surgical procedures. Since MEMS devices are very small and have better

Kochuthomman Joseph Mampilly
School of Electrical Engineering, VIT University, Vellore - 632014, India

performance than any other devices, it helps the surgeon to be precise and accurate during minimally invasive surgerical procedures [1].

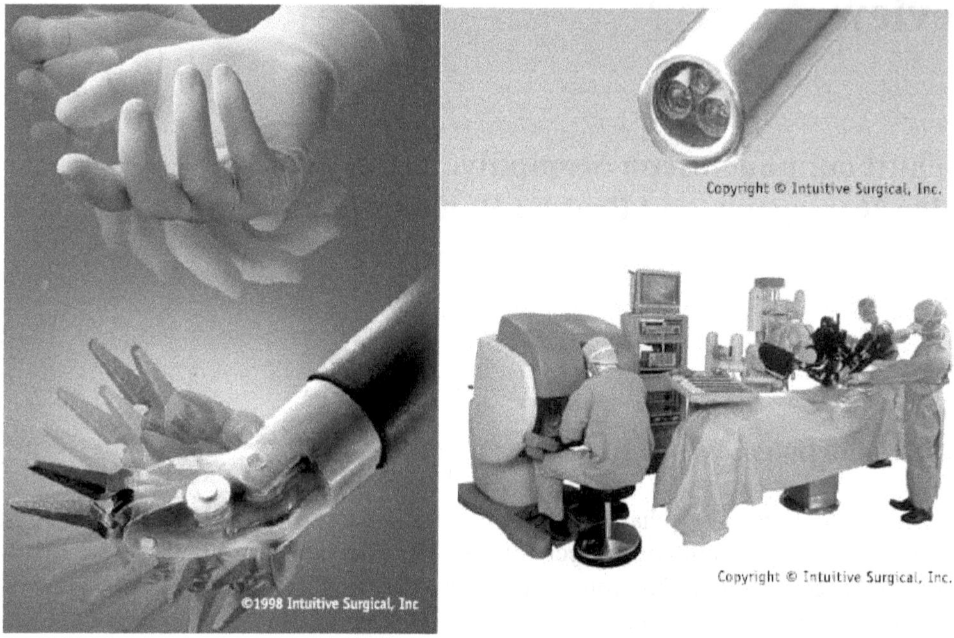

Fig. 6.1. MEMS based Intuitive Surgical da Vinci Robot [2].

6.2.1. MEMS Based Micro-Machined Cutting Tools

The making of cutting tool which is miniature helped to make incision precise and accurate even for small one. It also helps to reduce the bleeding caused by these incisions. Previously we use diamond and stainless steel blades for the incision but it is replaced by using silicon as scalpels. The main advantage is that we can make it shaper to atomic level of their crystalline structure. And also we can integrate sensors and their electronic circuit parts to silicon scalpels to increase precision and accuracy in the surgery (See Fig. 6.2) [2].

6.2.2. MEMS Based Eye-Surgery (Minimally Invasive Type)

MEMS based ultrasonic cutting tool is commonly used in cataract removing process in modern world now. The cutter is connected with a piezoelectric material and when the tip which is in contact with piezoelectric material resonates at ultrasonic frequencies it cut very easily through larger tissues like cataract affected Len of the patient. Today piezoelectric sensor is connected with cataract removing hand pieces to ease the process of cataract removal and thus gives a general feed back to surgeon about thickness of tissues in different area.

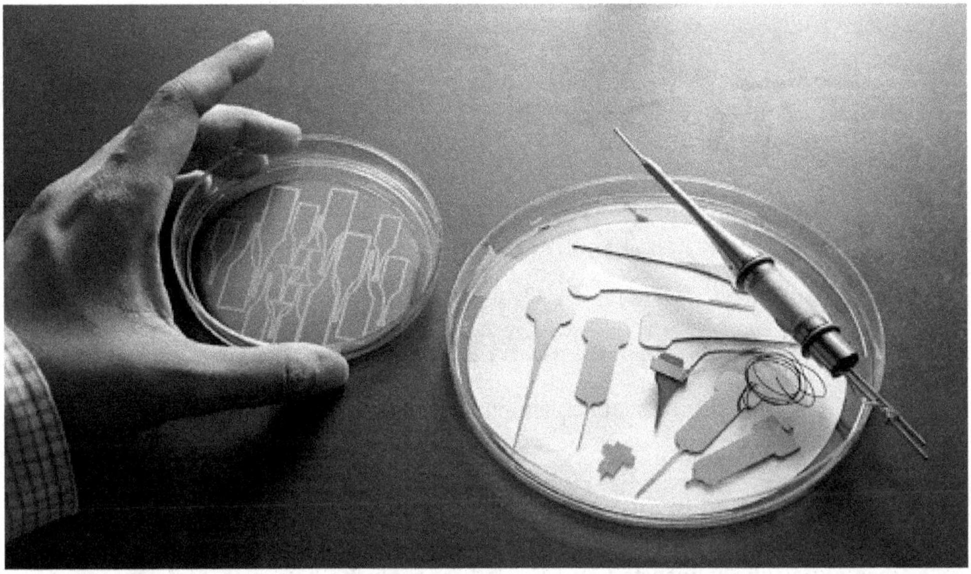

Fig. 6.2. MEMS based micro machine cutting tools [3].

Other important part of MEMS in eye surgery is the retinal implant. Retina is the part of eye which converts light into neural impulse and transmit those into brain. It contains many photoreceptive cells which are in surface of the retina. These cells are work with photo transduction. In some case retina of eye will damaged and optic nerves remain undamaged, here we use retinal implant. Retinal implant copies the function of retina by exciting photo restive cell. Recently MIT Researchers have created implants which consist of electrodes that can stimulate correct ganglion cells by making current pulses which is controlled by a chip attached to the white part of the eye (Fig. 6.3) [4, 5]. The microprocessor in the video camera fixed with *spectacles* make light wave to radio signals.

6.2.3. MEMS Based Catheters

Today MEMS based catheters are very much in small and have multifunctional properties and also displays high performance. Pressure sensors designed using MEMS are used in catheters and for ultrasonic intravascular imaging, MEMS based transducer are used. MEMS based sensors placed in catheters for diagnostics purposes like to measure the pressures, oxygen level, blood flow rate, temperature and chemical compositions. Conventional type catheters cannot move actively and often causes the patient more pain and damage. Today MEMS based catheters are made up of shape memory alloy and gel type actuators which reduces pain and increase the comfort for the patients [1].

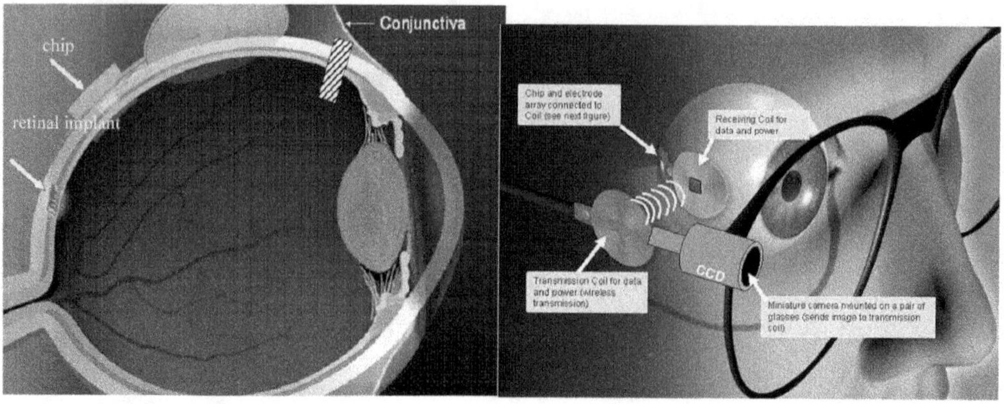

Fig. 6.3. Retinal implant on the eye during surgery [5].

6.2.4. MEMS Based Endoscopy

Here a fiber optic tube is inserted and passed through the patient's gastrointestinal tract and is able to view the condition of each parts and perform procedures like biopsies to rectify the problems. The important part of endoscope is its end. It consists of an objective Len piece and with camera on one end and on other end is having light source. In conventional type MEMS used in mirror technology. Mirrors are integrated to endoscope and it reduced the size of the end of endoscope inserted into gastrointestinal tract. The new model of endoscope is similar to that of a capsule. This type contains one optical fiber for illumination and six for receiving the light back. The four LED light are used. The three LED (red, green, blue) are used to color output and one (IR) to give thermal imaging of the area. The fiber inside endoscope spins and bounces according to electric impulses given through a wireless radio wave transmitter. The video camera integrated to endoscope records the images and transmits the data to image processor which convert data to 2-D pictures. The capsules are made up of bio-compatible materials which don't digest in stomach [4, 6, 7].

6.2.5. MEMS Based Tactile Sensing

For a minimally invasive procedure, determining the force used on tissue is very important. So the tactile sensing is used for this purpose. It gives the feedback to the surgeon about firmness and elasticity of the tissue and this achieved through force feedback devices. This type sensor contain a sensing layer made up of piezoelectric, capacitive, piezoresistive, mechanical and optical elements and also have an electronic layer with a protective layer and supportive layer. A tactile sensor can measure the magnitude and the direction of force applied, softness of the tissue and slippage of the device. Tactile sensor is integrated with many minimally invasive surgery tools using MEMS technology. Piezoelectric films (sensory layer) like PVDF produce electrical output when the beam bends inward due to tissues and electrical output produce is directly proportional to the bending stress [8].

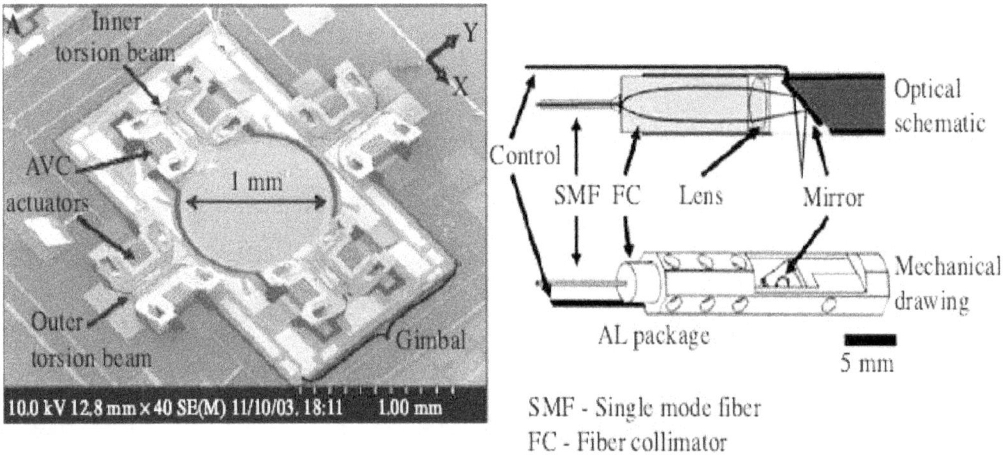

Fig. 6.4. Mirror technology used in endoscope [6].

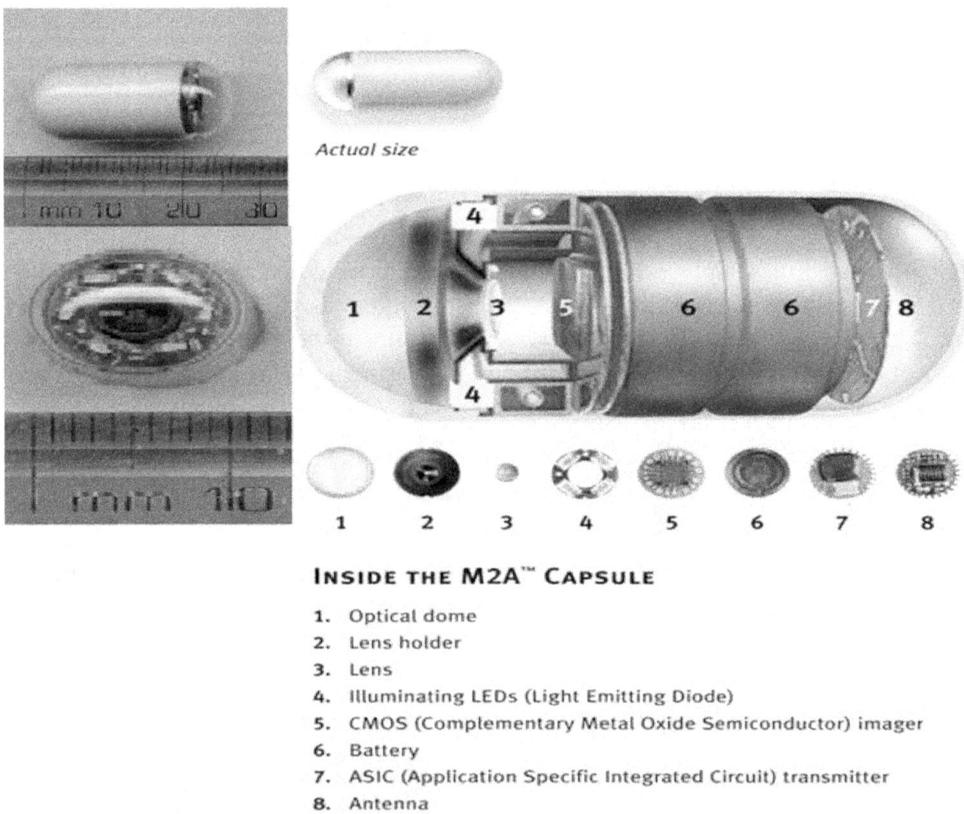

Fig. 6.5. Capsule type endoscope [8, 9].

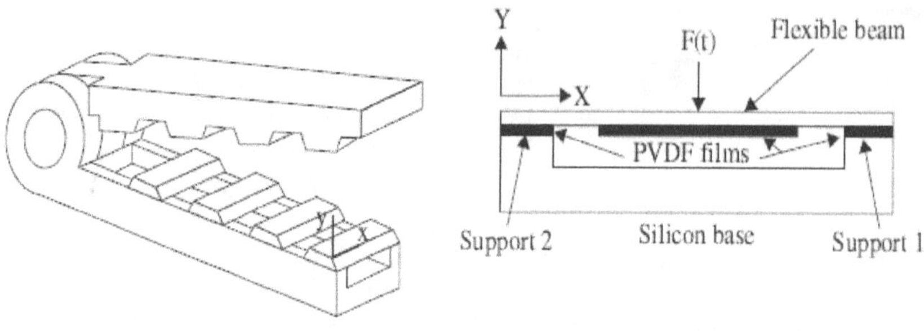

Fig. 6.6. An example of MEMS based tactile sensor [10].

6.3. Drug Delivery Systems Based on MEMS

The drug's effect and its efficiency depend on the method of delivery of the drugs to the body. MEMS based delivery system increases parameters like controlling, targeting and gives us a precise delivery of the drugs with automated feedback. So MEMS based drug delivery system emerging high potential platform for delivery system of high precision. Now new ways to improve timed delivery are microencapsulation, implants and transdermal patches. Any delivery system based on MEMS consists of micro pumps, micro needles, valves and micro reservoirs components.

6.3.1. MEMS Based Micro Reservoirs

We know that drug supply or depot is needed for drug delivery systems. The silicon micro particles are used for drug reservoir in oral type and the surface of these micro devices is designed to attach to specific tissues or cells in the digestive system to deliver therapeutic agents. If it is injected type, the surface is made from starch or gelatin which is used deliver drug for cancer treatment. The supply of anodic voltage in the presence of chloride ion induces the electrochemical dissolution of the cell tissue causing tissue to be weak. Since the tissue is weaken, it is easy to dissolve drugs within the reservoir and thus it diffuse into the surrounding cells .the MEMS based reservoir is digitalized so that release of drug can be chosen in complex profile. One of the problems raised was sealing of components like electronics, reservoir, etc. This was rectified by Santini et al [11] by using cold compression technique.

6.3.2. MEMS Based Micro Needles

Micro needles are very important and considered to primarily developed technique for transdermal drug delivery. Recently HP (Hewlett-Packard) Co. has integrated with their inkjet technology in printer with transdermal patch to reduce pain because of low penetration (below 0.75 mm) and increases the absorption of bigger molecules which cannot pass through the skin directly. This patch has about 400 micro reservoir and 150 micro needles [12].

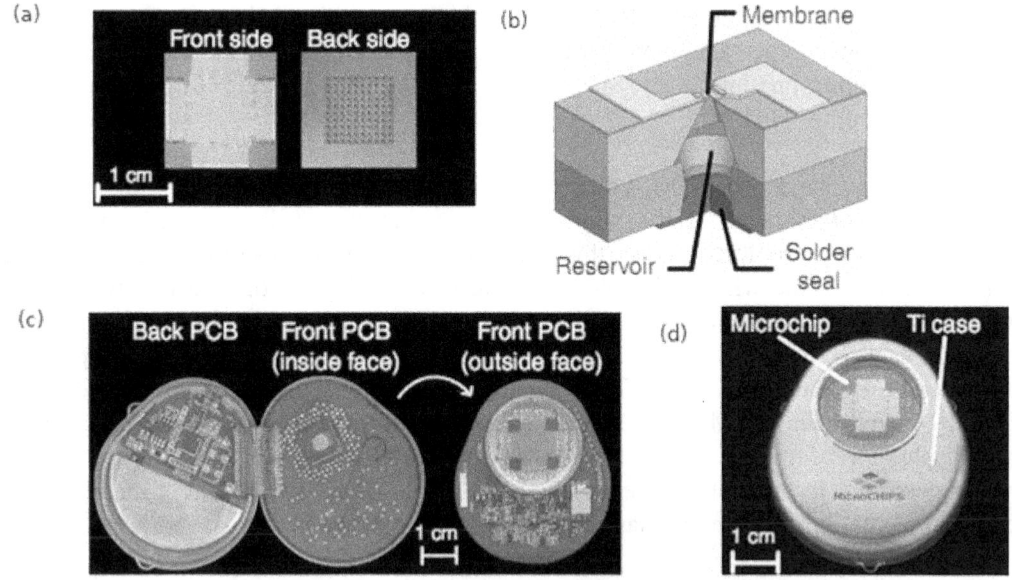

Fig. 6.7. MEMS based microchips [11].

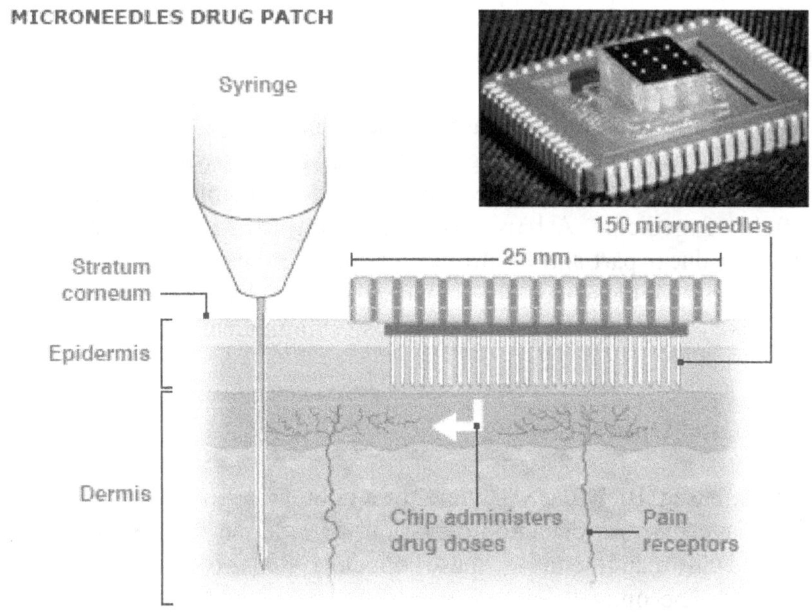

Fig. 6.8. HP micro needles [13].

6.3.3. MEMS Based Micro Pumps and Values

The controlling fluids flow and precise measurement of the fluids are very importance in drug delivery systems. For these purpose we use MEMS based micro pumps. Here major designs are based on size, pressure and biocompatibility. Micro pumps are of two types-mechanical type and non mechanical type. The electrostatic, piezoelectric, bimetallic, thermo-pneumatic and shape memory alloy (SMA) are examples of MEMS based micro pumps [14].

In electrostatic type when the voltage is applied across pump diaphragm and the electrode, the diaphragm of the pump will go either up or down based on applied voltage. But in piezoelectric type, this applied voltage will cause deformation on the surface of membrane where piezoelectric materials are deposited. And this will act as push mechanism to throw fluid out of the micro pump. In case of thermo pneumatic type of micro pumps, the chamber under the diaphragm will get expanded and compressed using a pair of cooler and heater. The change in volume will cause momentum for fluid flow by membrane. In SMA type, the shape of deformation is used as force to actuate the diaphragm of micro pump. Here SMA can retain their shape after a cycle of heating/cooling. In bimetallic type, the deformation cause by two different metals is the cause for actuating force in the diaphragm of the micro pump [15].

MEMS based micro valves are used for switching flow control, sealing of liquids, gas and vacuum. The mechanism of actuation in micro valves is electric, magnetic, thermal, piezoelectric are as described in the case of the MEMS based micro pumps.

6.4. MEMS in Diagnostics

It is very important to detect the diseases in early manner and give proper treatment to the patient in medical field. Biological testing for knowing disease and certain substances are made quicker, flexible and sensitive by combination of MEMS to the diagnostic field of medicine. A biosensor made using MEMS technology consists of a sensing part, transducer part and electronics circuit to process the data and generate the signals. MEMS based biosensors had made the high end platform for commercial applications like drug discovery, genomics, biomedical research and medical diagnostics.

6.4.1. MEMS Based Optical Sensing

Optical sensing based BioMEMS devices use light as a source of excitation. It has photosensitive layer used in detection of various samples in the field of medicine. The techniques used in manufacturing these optically sensitive devices are based on photolithography. The optical device consists of layer which is exposed using photo-mask by UV light. The active regions are denoted by the antibody bindings. And the grating produces a diffraction pattern with laser as its source [16]. The example for these types of BioMEMS sensors is gene chip made by affymetrix.

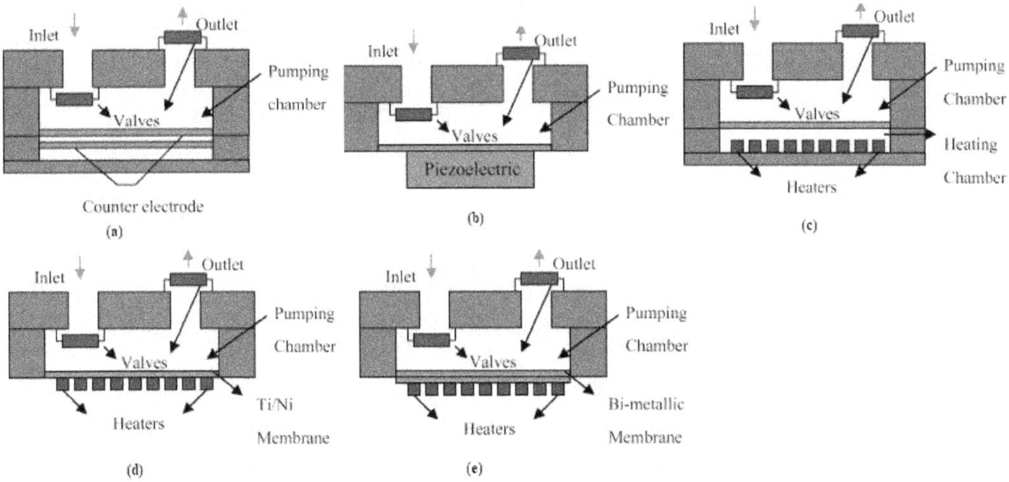

Fig. 6.9. Actuating mechanism in micropumps [15].

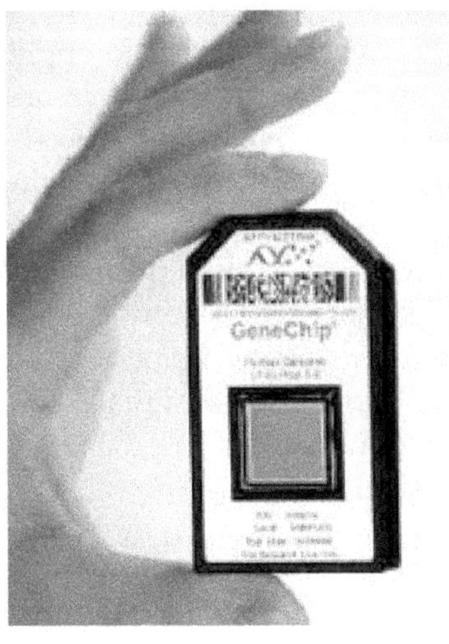

Fig. 6.10. Gene chip made by affymetrix [17].

6.4.2. MEMS based Micro-cantilever Beam Sensor

Micro-cantilever works on the principle of mechanical deflection of the beam. The cantilever beam is coated with bio-molecules like antibodies. The materials used in making the beam are mostly gold. There are two modes of operation. One is static mode and another one is dynamic mode. In static mode the deflection of the cantilever beam signifies the presence of analyte molecules. In the dynamic mode, the principle of detection of analyte molecules depends on the change in resonance frequency of cantilever beam [18].

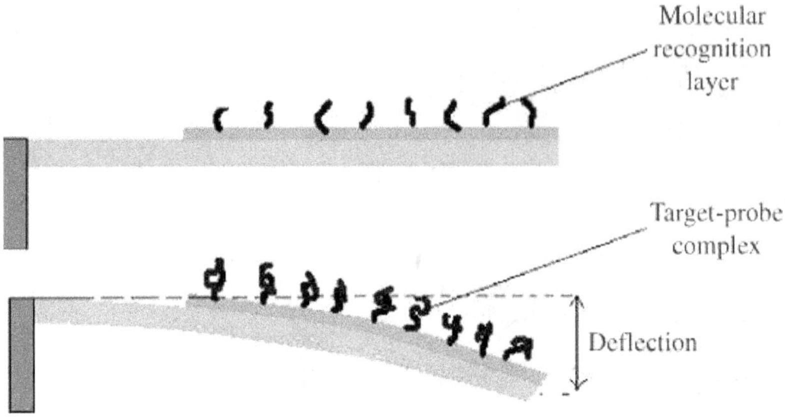

Fig. 6.11. Working of MEMS based Micro-cantilever beam sensor [18].

6.5. Conclusions

MEMS discussed in this paper have very high end applications in everyday life. And they play a major role in shaping the role of MEMS in medical field related application in the future as well. There are lots of challenges to faced in developing multifunctional devices in surgery. Also the integration of nanotechnology into MEMS will play an important role in miniaturization of these devices. This can in turn result in smart surgical devices that can communicate with surgeons. All these advancements can be achieved with proper funding, collaboration between surgeons and MEMS engineers. These MEMS device are of fourth generation type and they posses fusion of the sensor element with analog amplification, analog-to-digital Converter and digital intelligence for linearization and compensation on the same chip. As well as Memory for storing data related to analysis of biological systems.

References

[1]. Keith J. Rebello, Applications of MEMS in surgery, *Proceedings of the IEEE*, Vol. 92, No. 1, January 2004.

[2]. http://www.mims.com/Taiwan/pub/topic/JPOG/2011-08/Robotics in Minimally Invasive Gynecologic Surgery the Final Frontier.

[3]. A. Lal, Silicon-based ultrasonic surgical actuators, in *Proceedings of the 20th Annual International Conference of the IEEE Engineering in 'Medicine and Biology Society'*, Vol. 20, Hong Kong, China, 29 October 1998, pp. 2785–2790.

[4]. Teena James, Manu Sebastian Mannoor, Dentcho V. Ivanov, BioMEMS –advancing the frontiers of medicine, *Sensors,* Vol. 8, 2008, pp. 6077-6107.

[5]. The Retinal Implant Project, Accessed on 3 November 2011, http://www.rle.mit.edu/media/pr147/13.pdf

[6]. A. D. Aguirre, P. R. Hertz, Y. Chen, J. G. Fujimoto, W. Piyawattanametha, L. Fan, and M. C. Wu, Two-axis MEMS scanning catheter for ultrahigh resolution three-dimensional and en face imaging, *Optical Express*, Vol. 15, 2007, pp. 2445–2453.

[7]. D. Panescu, Emerging technologies. An imaging pill for gastrointestinal endoscopy, *IEEE Engineering Medical Biology Magazine*, Vol. 24, No. 4, July–Aug, 2005.
[8]. M2A capsule, http://petrifiedunrest.net/wordpress/?attachment_id=50
[9]. M2A capsule, http://wapedia.mobi/en/Given_Imaging
[10]. S. Sokhanvar, M. Packirisamy, and J. Dargahi, A multifunctional PVDF-based tactile sensor for minimally invasive surgery, *Smart Mater Structure*, Vol. 16, 2007, pp. 989–998.
[11]. J. H. Prescott, S. Lipka, S. Baldwin, N. F. Sheppard Jr., J. M. Maloney, J. Coppeta, B. Yomtov, M. A. Staples, and J. T. Santini Jr. Chronic, Programmed polypeptide delivery from an implanted, multireservoir microchip device, *Nat Biotechnology*, Vol. 24, 2006, pp. 437–438.
[12]. S. Kaushik, A. H. Hord, D. D. Denson, D. V. McAllister, S. Smitra, M. G. Allen, and M. R. Prausnitz, Lack of pain associated with microfabricated microneedles, *Anesth Analg*, 2001, pp. 502–504D.
[13]. http://newsimg.bbc.co.uk/media/images/44125000/gif/_44125474_drug_microchip _hp_416.gif
[14]. R. Zengerle, J. Ulrich, S. Kluge M. Richter, A. Richter, A bidirectional silicon micropump, *Sensors and Actuators A: Physical*, 50, No. 1-2, Aug. 1995, pp. 81-86.
[15]. N. C. Tsai and C. Y. Sue, Review of MEMS-based drug delivery and dosing systems, *Sensors and Actuators B: Chemical*, Vol. 134, No. 2, 15 March 2007, pp. 555–564.
[16]. S. P. Mohanty and E. Kougianos, Biosensors: a tutorial review, *IEEE Potentials*, Vol. 25, No. 2, March–April 2006, pp. 35–40.
[17]. GeneChip, http://www.affymetrix.com/about_affymetrix/outreach/lesson_plan/downloads/ student_manual_activities/activity1/activity1_applicationofgenechips_lessonplan.pdf
[18]. J. Fritz, M. K. Baller, H. P. Lang, T. Strunz, E. Meyer, H. J. Guntherodt, E. Delamarche, C. Gerber, and J. K. Gimzewski, *Langmuir*, Vol. 16, 2001, p. 96-94.

Chapter 7
MEMS Switches for RF Applications

Ashish Kumar Sharma, Navneet Gupta

7.1. Introduction

MEMS (Micro Electro Mechanical System) technology has been achieved a remarkable attention of researchers since past two decades due to its immense radio frequency (RF) applications. MEMS devices are small integrated system that includes electrical and mechanical components. MEMS technology include switches, voltage tunable capacitors, high quality integrated inductors, film bulk acoustic resonators (FBAR), MEMS mechanical resonators and filters. Researchers have also developed various other devices like microsensors, transducers, cantilevers based on MEMS technology. RF MEMS devices are manufactured using bulk micromachining and surface micromachining. MEMS and micromachining have been identified as two of the most significant enabling technologies in developing miniaturized, low-cost and low-energy RF components and RF systems [1-3]. RF MEMS devices have experienced immense use for telecommunication application due to their high performance compared to other microelectronic switches. Also the reduction in dimension of electromechanical systems offers advantages such as soft spring, high resonance frequency and low thermal mass, and leads to dramatic decrease in power consumption. Therefore RF MEMS devices provide the tremendous performance in terms of low power consumption, higher isolation and low insertion losses to their solid state counterparts (P-I-N diodes and FET's). MEMS devices applications include portable wireless system, accelerometer, micro actuators, reconfigurable antennas and switching networks. The MEMS technology enables the ability to reduce the number of passive components on RF system by combining many switching element into single circuit. RF MEMS devices utilized micromachining technology, which provides a higher performance to several RF applications. Therefore these advantages produce a significant performance impact on RF system [4].

Ashish Kumar Sharma
Department of Electrical and Electronics Engineering
Birla Institute of Technology and Science Pilani, Rajasthan, India, 333 031

Among MEMS devices RF MEMS switches are the fundamental component, which can be used to develop other RF MEMS devices. RF MEMS switches are the important device in the area of RF and microwave. First RF MEMS switch was developed by Larsen in 1990's for radio frequency application. These RF MEMS switches achieve an open or short circuit on a transmission line by mechanical movement of the switch beam in vertical or lateral directions. These switches are the ideal component for electrical signal switching which can be used in various RF and microwave devices for different communication applications. These switches have been designed to present nearly 50 Ω impedance across a broad range of frequency when it is closed and nearly an open circuit when there is no connection. This property makes RF MEMS switch an attractive choice for microwave application [5-8].

The forces required for the mechanical movement of the switch beam to achieve make and break contacts between the transmission line and RF MEMS switch, can be obtained using actuation mechanism. The actuation mechanism in RF MEMS switch can be electrostatic or electromagnetic. Electrostatic actuation mechanism is the most prevalent technique in use today due to its very low power consumption, small electrode size, relatively short switching time, while electromagnetic actuation mechanism requires higher current requirement and shows more switching time as compared to electrostatic actuation mechanism [9-11]. Petersen K. E. demonstrated the first electrostatic controlled RF MEMS and achieved the switching of electrical signals with high of-off impedance ratio at low frequencies.

The important switch design parameters used to evaluate the performance of the RF MEMS switches are:

(a) RF power handling; RF power handling is a measure of how efficiently a switch passes the RF signal at the time of signal transmission. This can be defined as the output power level follows the input power level with linear ratio.

(b) Insertion loss; The insertion loss of RF MEMS switch refers to the RF signal power dissipated in the switch at the time of signal transmission. Insertion loss occurs at low microwave frequencies as well as at high microwave frequencies. This occurs at low frequencies due to the resistive loss between the finite resistances of transmission line and switch beam contact area. At higher frequencies insertion loss occurs because of skin depth effect. For design aspect the insertion loss should be very less for efficient signal transmission.

(c) Isolation; The isolation of RF MEMS switch can be defined as the RF signal power isolation between the input and output terminals of the switch in its signal blocking state. A large value (in decibels) of isolation indicates very small coupling between input and output terminals. In RF MEMS switches isolation occurs due to capacitive coupling between the moving switch beam and the stationary transmission line as a result of leakage currents. One of the design objectives of communication system is to achieve a very high isolation at low and high microwave frequencies.

(d) Return loss; The return loss of the RF MEMS switch refers to the RF signal power reflected back by the device, at the input terminal of the switch in its signal transmission state. Return loss occurs due to the mismatching of the total characteristic impedance between the switch and transmission line.

(e) Actuation voltage; Actuation voltage can be defined as the minimum voltage required to pull down the switch beam of the RF MEMS switch. One of the design objectives of state-of-the-art microelectromechanical switching systems is to achieve the low actuation voltage, depending on the switch design and application.

(f) Resonant frequency; The resonant frequency of the switch can be defined in terms of the effective spring constants and resonating mass of the mechanical switch model. At this particular frequency, the stored potential energy and the kinetic energy of the switch tend to resonate. The natural frequency of a simple mechanical system consisting of a weight suspended by a spring can be formulated as $\frac{1}{2\pi}\sqrt{\frac{k}{m}}$, where k is the spring constant of the spring and m is the resonating mass of mechanical system.

The resonant frequency can also be defined for an electrical system. In electrical system the resonant frequency can be achieved by using a resonant circuit, which consists of an inductor (L) and a capacitor (C) in series configuration. This LC circuit stores an oscillating electrical energy at the circuit's resonant frequency due to the collapsing magnetic field created by the inductor and capacitor. The fundamental resonant frequency of the electrical circuit can be formulated as $\frac{1}{2\pi}\sqrt{\frac{1}{LC}}$, where L is the inductance and C is the capacitance of the LC electrical circuit.

(g) Switching speed; The switching speed can be defined as the time for toggling from one state of the switch to another. Switching speed can also be called as switching rate. RF MEMS switches possess low switching speed as compared to the semiconductor switching devices.

(h) Quality factor; The quality factor (Q-factor) is one of the most important parameter of RF MEMS switch and can be explained as the ratio of the energy stored to the energy dissipated per cycle at the resonant frequency. Higher Q-factor indicates a lower rate of energy loss relative to the stored energy of the resonant circuit.

The RF MEMS switch performs RF signal switching by physically blocking or opening the transmission path in a microwave devices to achieve transmit/receive operation. So these RF MEMS switches have a broad range of application in RF and microwave based modern communication systems. Brown E. R. [5] categorized the RF and microwave devices based on the location of RF MEMS component in the system design.

(i) RF extrinsic: In this the RF MEMS component is situated outside the RF circuit and actuates or controls other component in the RF circuit.

(ii) RF intrinsic: In this the RF MEMS component is situated inside the RF circuit having actuation and RF circuit function.

(iii) RF reactive: In this the MEMS component is located inside the RF circuit where it performs an RF function.

RF MEMS switches have proved their ability to sense and transmit the electromagnetic energy for several communication related application, which are difficult to achieve with FET's and p-i-n diodes. These switches can replace the many RF and microwave components used in satellite and mobile communication systems having the ability of reduced size, weight and power consumption.

Recently many efforts have been carried out to make antennas reconfigurable by using RF MEMS switches to improve the performance and portability of wireless system. By implementing these RF MEMS switches with antenna system design, it is possible to achieve the electronic switchable antenna. Transmission of electromagnetic energy can be accomplished by using microwave and millimeter wave antennas, these antennas can be monolithically fabricated with other electrical and mechanical component to develop a reconfigurable antenna by changing the physical structure of the antenna with RF MEMS switches. Anagnorto D. et.al [12] developed multiple frequency fractal reconfigurable antennas for high frequency application to achieve the additional degree of resonance frequencies. The resonance behavior of the developed fractal reconfigurable antenna segment can be changed by controlling the current on each conductive part of the antenna by using RF MEMS switches.

Despite of rapid advancement in RF MEMS switches, still many issues need to resolve in terms of switch lifetime, packaging, power handling capability and switching speed. Out of these switch lifetime and packaging are the most important issues for their survival in recent industry developments.

The objective of this chapter is to review the development of RF MEMS switches in the past two decades. The organization of this chapter is as follows: Section 7.2 describes about the classification of the RF MEMS switches. Section 7.3 includes the characteristics and performances of RF MEMS switches and finally Section 7.4 describes the conclusion of this review chapter.

7.2. Classification of RF MEMS Switches

The RF MEMS switches are classified on the basis of contact and on the basis of circuit configuration. Based on contact type there are two types of switches: metal to metal contact and capacitive contact switches. On the basis of circuit configuration, these switches are classified as series and shunt RF MEMS switches. RF MEMS switches are often used in series circuit configuration for low frequency application and in shunt circuit configuration for high frequency application.

7.2.1. Metal to Metal Contact RF MEMS Switches

The schematic view of metal to metal contact switch is shown in Fig. 7.1. In this switch metal to metal direct contact is used to achieve an ohmic contact between the switch membrane and the lower electrode. These switches are capable to maintain excellent electrical isolation in off state and minimal insertion loss in on state. Metal to metal contact switches shows excellent electrical contact while maintaining the minimum contact resistance and low parasitic capacitive coupling to provide a large dynamic range of on-off state impedances. Therefore these switches are suitable for low frequency RF application from DC to 60 GHz radio frequency range [13, 14]. However the main disadvantage of these switches is the contact lifetime which affects the switch performance in terms of insertion loss and isolation. Metal to Metal contact switch in series configuration performs a make and break operation in transmission line typically with metal–metal contacts.

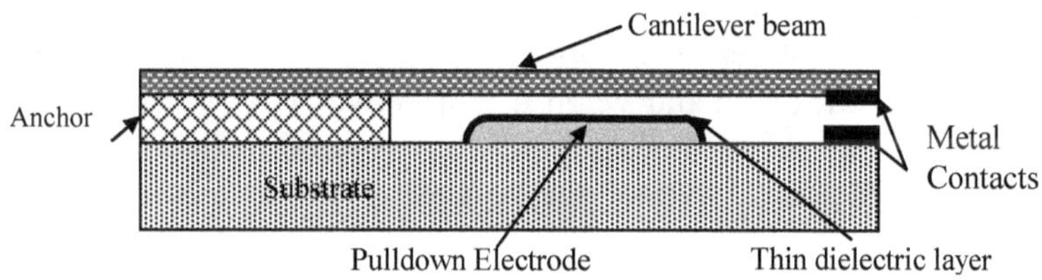

Fig. 7.1. Sketch of a typical Metal to Metal contact RF MEMS switch.

Larson L. E. et.al [15] demonstrated a cantilever based metal to metal contact RF MEMS switch, which is connected in series configuration with the transmission line. Santos H .J. D. L. et. al [16] discussed cantilever based RF MEMS switches is series configuration with transmission line.

7.2.2. Capacitive Coupling RF MEMS Switches

The schematic view of capacitive coupled RF MEMS switch is illustrated in Fig. 7.2. A capacitive coupled RF MEMS switch uses a metal membrane which exhibits high impedance due to an air gap between the membrane and bottom plate in unactuated state. When a DC voltage is applied between the membrane and bottom plates, the top membrane gets capacitive coupled with the bottom plate which causes the low impedance and high capacitance between the switch contacts. This low impedance path allows the RF signal to the ground and it stops the signal transmission through the MEMS switch. The capacitance ratio between down state to the up state capacitance is the key parameter and a high value of the capacitance ratio is always desirable for high frequency application. Capacitive coupled RF MEMS switch uses a thin dielectric layer on the bottom plates to avoid the stiction problem between switch membrane and bottom plates of the switch [17, 18].

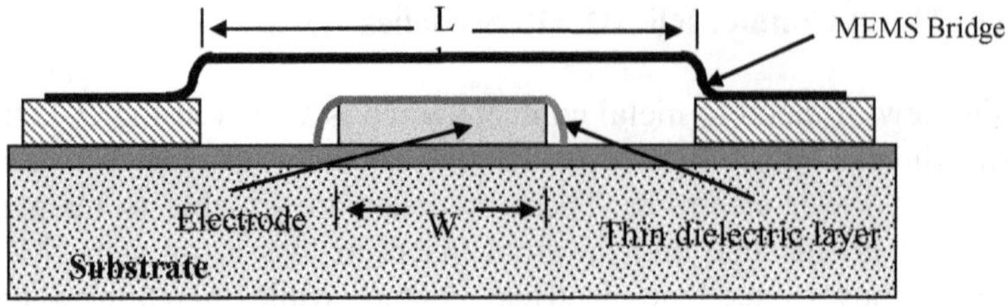

Fig. 7.2. Sketch of a typical RF-MEMS capacitive coupled switch.

These capacitive coupled RF MEMS switches have been integrated in shunt circuit configuration with a coplanar-waveguide (CPW) transmission line. In a CPW configuration, the anchors of the MEMS switch are connected to the CPW ground planes. When a voltage is applied between a switch beam and the pull-down electrode, an electrostatic force is induced on the beam. The switch operates as a digitally tunable capacitor with on and off states of the switch. The beam over the electrode acts as a parallel plate capacitor. When the switch beam is in up-state, transmission line experiences a small capacitance and when it is in down state the switch membrane get touched with the transmission line therefore transmission line experiences a high capacitance with low impedance path to the ground.

Goldsmith C. et.al [17-19] developed a capacitive coupled RF MEMS switch and this switch was used in shunt circuit configuration with coplanar waveguide (CPW) transmission line for 35 GHz frequency RF application. Tan G. L. et.al [20] also demonstrated a fixed-fixed metal to metal contact RF MEMS switch in shunt configuration with transmission line by using two pull down electrodes and a central dc-contact area.

7.3. Characteristics and Performances of RF MEMS Switches

A large number of RF MEMS switch design parameters has been discussed in the section 7.1. These design parameters exists with conflicting requirements and trade off's in various switch design structures. This section shows the research efforts that have done in past two decades to mature this technology. These research efforts have been done to improve the performance of the RF MEMS switch in terms of actuation voltage, insertion loss, isolation, switch lifetime, RF signal power handling, temperature sensitivity and switching speed.

7.3.1. Actuation Voltage

The first MEMS switch was developed by the Petersen K. E. [21] in 1979 where cantilever based switch design was used on the silicon substrate to achieve the switching of electrical signals at actuation voltage of ~70 V. Since then a large number of papers have been published in various journals and conferences with various structural designs of RF MEMS switch to reduce the actuation voltage. Yao J. J. et. al [13] developed a surface micromachined miniature RF MEMS switch using a cantilever beam structure of SiO_2 and achieved an actuation voltage of 28 V. Pacheco S. et.al [22] reported two different designs of RF MEMS capacitive switches by using serpentine and cantilever spring structure with low spring constant to achieve an low actuation voltage of 14 V to 16 V for DC to 40 GHz frequency range application.

Pacheco S. P. et.al [23] proposed a modified design of low loss RF MEMS switch design incorporating serpentine folded spring structure as illustrated in Fig. 7.3, to achieve a low actuation voltage of 9 V and a minimum off state isolation. This proposed switch was designed with a folded suspension varying from 1 to 5 mender suspension attached with a electroplate nickel membrane. The structure was attached with the actuation plates situated at center conductor of the finite ground coplanar wave guide (FGCPW), the switch membrane clamps down thereby a high capacitance at center conductor provide a short path to the RF signal. The proposed switch design concluded that by increasing the number of mender structure a reduced actuation voltage can be achieved. Actuation voltage and spring constant for various hinge structures has been tabulated in Table 7.1.

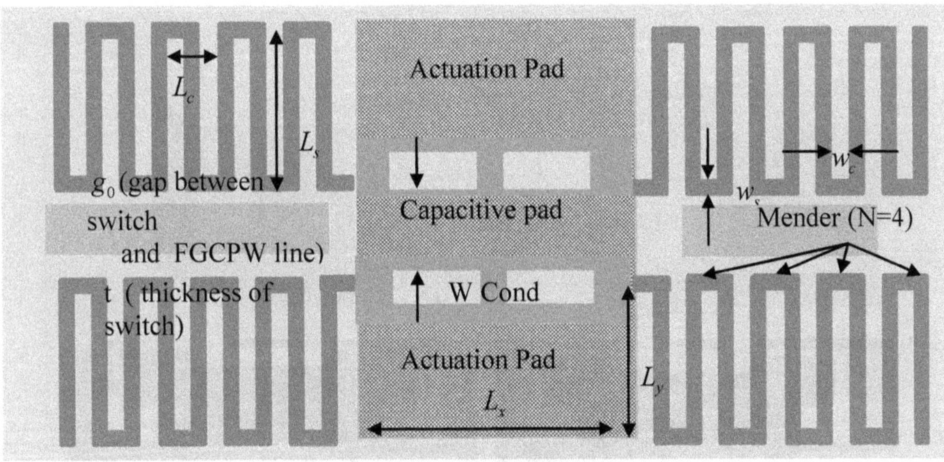

Fig. 7.3. A schematic view of RF MEMS switch with serpentine folded suspensions containing four meanders. Pacheco S. P. et. al [23].

Balaraman D. et. al [24] also proposed a design of RF MEMS switch with various hinge (spring) structures with high resistive silicon substrate to achieve a low actuation voltage. This switch was developed by using sputtered copper based switch membrane with four various hinge geometries. Lee S.-D. et .al [25] developed a RF MEMS switch with a freely moving contact structure to achieve low actuation voltage. The schematic view of this switch is illustrated in Fig. 7.4. This proposed switch design used a freely moving structure which opens and closes the switch through electrostatic actuation. This switch design was capable to reduce the sufficient amount of actuation voltage because actuation energy was not used in elastic deformation of a suspension beam.

Table 7.1. Actuation voltages and spring constant for various hinge structures of capacitive shunt switch RF MEMS switches.

Hinge structures	**Actuation Voltage (V)**	**Spring constant (N/m)**	**References**
Cantilever	5.97	0.654	[22]
Serpentine spring (Each suspension contains 2 menders)	4.95	0.478	[22]
4 suspensions bridge (Each suspension contains 5 menders)	9	----	[23]
4 suspension bridge (Each suspension contains 1 mender)	16.7	14.8	[24]
2 suspension bridge (Each suspension contains 1 mender)	5.7	1.75	[24]
Solid cantilever bridge	17.8	16.7	[24]
2 Mender cantilever bridge	11.5	7.4	[24]

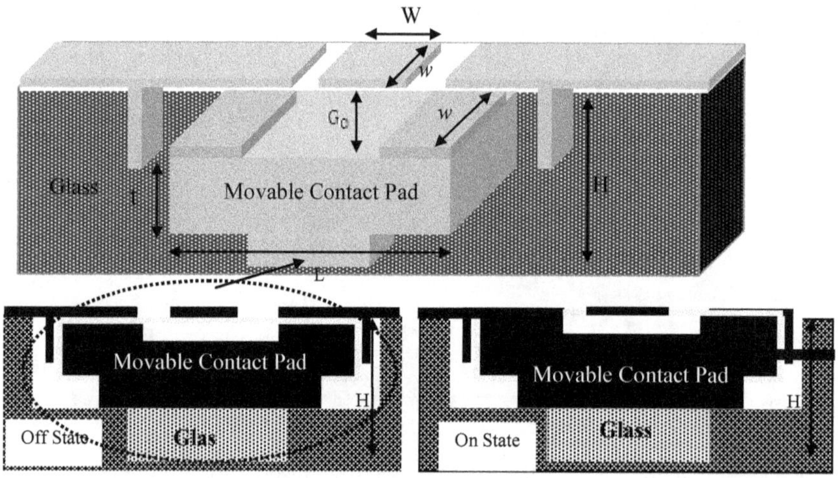

Fig. 7.4. A schematic view of RF MEMS switch with movable contact pad. Lee S.-D. et.al [25].

Actuation voltages and spring constant for various hinge structures of capacitive shunt switch RF MEMS switches are listed in Table 7.1.

To further achieve a lower actuation voltage, Polcawich R.G. et.al [26] presented a RF MEMS series switch design using piezoelectric actuation mechanism. This actuation mechanism retains large restoring force and excellent RF performance. In this RF

MEMS switch design a piezoelectric actuation mechanism was proposed as compared to the electrostatic actuation mechanism because piezoelectric actuation mechanism required extremely low currents and voltages for operation along with the ability to close the large vertical gaps. Lee H.-C. et.al [27] also developed a shunt type ohmic RF MEMS switch by using piezoelectric actuation mechanism and reported that these switches can be operated at a low operating voltage of 5 V with very low power consumption. This proposed switch was developed with single and four cantilevers based piezoelectric actuators. Kim J. et.al [28] proposed a DC contact type RF MEMS switch design to achieve a low actuation voltage. This proposed switch was developed by using a thick and stiff silicon membrane with a seesaw mode operation design to mitigate the effect of the bending of the membrane due to an internal stress gradient. This switch was designed with uniform small gap between the electrode and the switch membrane. Actuation voltages for various actuation mechanism are listed in Table 7.2.

Table 7.2. Actuation voltages for different actuation mechanism.

Actuation Mechanism	Actuation voltage (V)	References
Electrostatic	9 to 70	[13 to 33]
Piezoelectric	2 to12	[26, 56]
Electromagnetic	2	[57]
Electro-thermal	2.5-3.5	[58]

Kundu A. et.al [29] presented a new design of RF MEMS which required a very low actuation voltage with improved switching time. The actuation voltage was reduced by a sufficient amount by using a concept of moving transmission line of coplanar waveguide (CPW). This proposed switch used two movable plates where the first moving plate was switch membrane and the second moving plate was CPW transmission line. This CPW transmission line acted as a movable bottom electrode to initiate the actuation process. This proposed switch was able to reduce the actuation voltage and switching time around 20 % for optimized design parameter for the two movable plates switch.

7.3.2. RF Characterization (Isolation, Insertion loss and Return Loss)

The RF performance at low frequencies should be characterized by the 'on' resistance in signal passing state and 'off' resistance in signal blocking state. At high frequencies RF performance can be described by its insertion and return loss in signal passing state and isolation in signal blocking state. Till now various papers have been published to achieve a high RF performance at low as well as at high frequencies. Here we will describe the research efforts that have been done to achieve high RF performance in RF MEMS switches.

Goldsmith C. et.al [17, 18] developed a low-cost, low loss RF MEMS capacitive shunt switch using electrostatic actuation mechanism. This proposed switch design reported that insertion loss can be minimized by reducing the length of the transmission lines. Because the narrow input/output interconnects between the transmission line and switch contact area produces more insertion loss during signal passing state. High isolation can be achieved by removing the imperfection in the deposited switch membrane because these imperfections provide the improper contact between the bottom electrodes and switch membrane. Goldsmith C. et.al [19] modified the RF MEMS capacitive shunt for microwave and millimeter wave frequencies with significant improvement in terms of low insertion loss, high isolation. This proposed switch design used an aluminum based coplanar waveguide (CPW) and switch membrane in order to achieve low insertion loss and high isolation. Muldavin J. B. et. al [30] demonstrated a tuned cross switch incorporating four membrane shunt MEMS switches. The proposed tuned cross switch proved the advantage of the tuning approach to achieve high isolation in down-state and excellent return loss in up-state in comparison with individual RF MEMS switch. This proposed tuned cross switch design was suitable for low-loss high-isolation communication application at 28 GHz frequency. Rizk J. B. et. al [31] reported a significant improvement in isolation and insertion loss of X-Band RF MEMS switches by using a π-circuit network with two MEMS shunt switches.

Muldavin J. B. et.al [32] developed a RF MEMS shunt switches to achieve a very high isolation for X-band frequency range from 7 GHz to 12 GHz. This proposed switch design achieved a very high isolation by increasing the series inductance of the switch. This series inductance was maximized by adding a short section of transmission line between the MEMS bridge and ground plane thereby the resonant frequency was pushed down to the X band frequency range. The proposed switch design resulted in a lower insertion loss and much higher isolation of better than 30 dB at X-band frequencies. It was proposed that the insertion loss and isolation may be improved by using thicker metal underlying between the metal membrane and transmission line.

Tan G.-L. et.al [33] reported the effect of bias lines on the insertion loss. This proposed switch design used a resistive bias line which was attached to each of the two pull down electrode. This schematic view of this developed switch design is illustrated in Fig. 7.5. The proposed switch design reported that resistive bias line with low value can result with improved insertion loss in dc contact MEMS shunt switches.

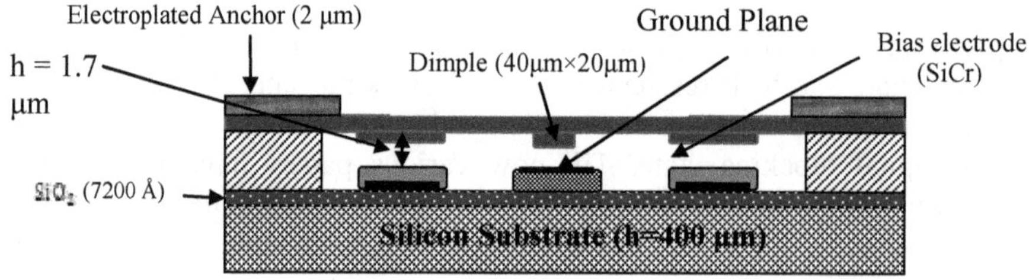

Fig. 7.5. A schematic view of RF MEMS DC-contact shunt switch. Tan G.-L. et.al [33].

Rizk J. B. et.al [34] reported a significant improvement in isolation and insertion loss of W-Band RF MEMS switches by using T-match and π-circuit network with two MEMS shunt switches. Ghodsian B. et.al [35] developed a DC-contact MEMS series switch by using cantilever beam with fork tip design. This cantilever beam with fork tip design reduced the upstate capacitance by minimizing the electrode overlap area between the tip of cantilever beam and the underlying signal line. Thereby this proposed metal to metal contact RF MEMS switch has shown the high isolation at high frequencies because of up-state capacitance.

Jang Y.-H. et. al [36] proposed a RF MEMS contact switch design to achieve the high isolation using two directional movements, namely vertical and lateral movement of the contact part by using comb and parallel plate actuators. Yamane D. et.al [37] demonstrated a bidirectional electrostatic actuation mechanism for a dual SPDT (single-pole double-through) switch. This proposed switch was designed with silicon on insulator (SOI) wafer to minimize the dielectric loss because suspension beam over the coplanar or micro strip line exhibited the electric field of travelling wave which prevents the condensed matter of substrate and increases the dielectric loss. The insertion loss increases due to the dielectric loss. In order to reduce the insertion loss this switch was designed by adapting the layer-separation technique. Pacheco S. P. et. al [23] suggested that this isolation can also be reduced by decreasing the thickness of the dielectric between the switch and the conductor in such a way that dielectric break down should not happen.

7.3.3. Switch Lifetime

Another important parameter of RF MEMS switch is the switch lifetime. The lifetime of the switch can be measured by inspecting the degradation of the mechanical structure (switch membrane) of the switch. Switch lifetime of capacitive coupled and metal to metal contact switches has been intensively studied in the last few years. In metal to metal contact switches, the mechanism that limits lifetime is the degradation of the metal contacts with repeated actuations. In capacitive membrane switches, the limiting mechanism is the dielectric charging within the switch dielectric layer. This switch lifetime can be divided into cold switching and hot switching life time. Cold switching lifetime defined as zero signals passes through the switch contact area and hot switching lifetime refers to that of a switch when a specific signals level passes through the switch contact area [38, 39].

Goldsmith C. et.al [40] demonstrated a capacitive RF MEMS switch and characterized the switch's lifetime as a function of actuation voltage. It was reported that the applied electric field in RF MEMS switch is responsible for dielectric charging which results in sticking (failure) of the switch membrane. The relation between electric field and actuation voltage is given as follows [40].

$$J \approx V e^{+2a\sqrt{V}/T - q\varphi_B/KT}, \tag{7.1}$$

where J is the current density, V is the applied voltage, T is the temperature in Kelvin, K is the Boltzman's constant, φ_B is the barrier height and a is a constant composed of electron charge, insulator dynamic permittivity, and film thickness. Equation 7.1 represented that switch lifetime (electric field) has exponential relationship with the applied actuation voltage. Czarnecki P. et.al [41] reported that the charging of the substrate can also affect the lifetime of capacitive shunt RF MEMS switch. They had used two substrates (silicon and glass) for switch design and concluded that the substrate can also trap charges which degrade the switch lifetime. Wong W. S. H. et. al [42] proposed a novel actuation voltage technique to achieve the sufficient reduction in dielectric charging in RF MEMS capacitive switches, leading to a longer switch lifetime. In this work actuation voltage technique is used to analyze the dielectric charging generated by different actuation-voltage. It was concluded that a suitable actuation voltage can reduce the charge buildup and thereafter extends the lifetime of the switch. Rottenberg X. et. al [43] proposed an analytical model of RF MEMS capacitive switch with distributed dielectric charging, which was demonstrated with a nonuniform air-gap distribution and a nonuniform volume distribution of charges in the dielectric layer. This switch design provided a deep insight into the irreversible stiction problem in electrostatic actuation mechanism.

7.3.4. RF Signal Power Handling

RF power handling refers the power at which the MEMS device fails to operate properly. RF signal power can be applied to the switch at a broad range of frequencies with different power levels to achieve high performance for microwave frequencies. RF signal power handling capability of RF MEMS switches are mainly limited by the two factors [44].

(i) **RF latching:** - RF latching can be defined as the applied RF power provides the enough force on the membrane to hold the switch down when DC bias is removed.
(ii) **RF self-actuation:** - RF self-actuation is referred as a situation in which the high RF power actually creates enough potential to pull the membrane down in unactuated state without using a DC bias voltage across the switch.

Lot of work has been done in past towards the RF signal power handling capabilities of RF MEMS switches. To enhance the power handling capability of metal to metal contact type RF MEMS switch, Jensen B. D. et. al [45] reported an electro-thermal model, to predict the current density in the metal to metal contact area of the switch using electromagnetic modeling. With the help of this model the accurate power handling capability can be calculated without sacrificing the switch performance. For high power application, Peroulis D. et. al [46] proposed a new RF MEMS switch design by using DC and RF pads with a top electrode as shown in Fig. 7.6, which considerably increases the RF power range with improved power handling capability. This design also investigated the self actuation and stiction problem of switch membrane in high RF power handling situation. Grenier K. et. al [47] proposed a specific RF MEMS switch with two bridge level topology to encounter the failure mechanism related to the self actuation and electromigration for high RF power handling. This topology was developed with a mobile bridge structure, which was attached with two non movable top and bottom electrodes to achieve the capacitive contact between the switch membrane and transmission line. High RF power self actuation and electromigration phenomenon were encountered by using pull up electrodes and large bridge dimensions respectively. Various RF MEMS switch designs and studies has been investigated in order to improve the power handling capability up to 6 W of RF power [48- 50].

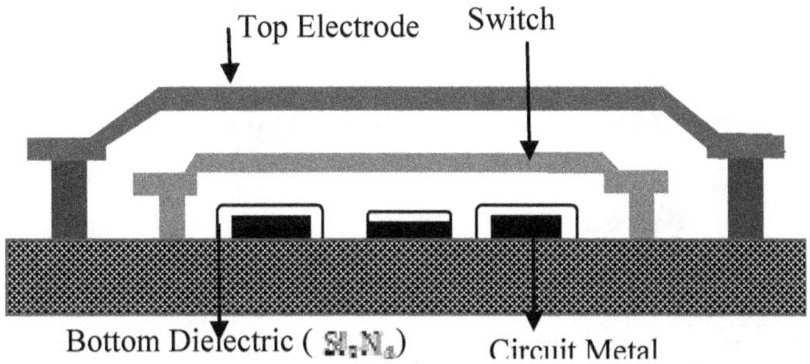

Fig. 7.6. A schematic illustration showing MEMS shunt switch with top electrode. Peroulis D. et.al [46].

7.3.5. Temperature Sensitivity

Past few years has seen an increase in the development of temperature sensitive RF MEMS switch designs to attain a reliable performance in fast temperature changing environment. Palego C. et. al [51] proposed RF MEMS capacitive shunt switches with molybdenum membrane because molybdenum switch membrane exhibited a significantly reduced sensitivity to ambient temperature as compared to aluminum switch membrane. Mahameed R. et. al [52] presented a RF MEMS switch which utilizes the lateral thermal buckle-beam actuator design in order to reduce the switch sensitivity to thermal stress. This switch demonstrated very low temperature sensitivity of ~ -50 mV/°C at 25 °C to 125 °C. Mahameed R. et. al [53] proposed a design by using separate and interdigitated RF and actuation electrodes which prevents dielectric charging under high actuation voltages. This design has shown the low temperature sensitivity with actuation voltage. Various switch design with different switch membrane structures has been developed in order to develop temperature sensitive RF MEMS switch for wider temperature range applications [54, 55].

7.3.6. Switching Speed

Switching speed is an important performance parameter of RF MEMS switches for transmit/ receive switching application. Even though RF MEMS switches exhibit excellent RF characteristics such as low insertion losses and high isolation but these switches have a very slow switching speed. This switching speed depends on mass of the switch membrane and switch construction. This switching speed can be increased by using a switch membrane with lower mass. Generally an RF MEMS switch operates at a speed of 2 to 50 ms for several million switching operations. The low mass of membrane based MEMS capacitive shunt switches make it suitable for relatively fast switching operation in comparison with cantilever-style MEMS switches [11]. Yao J. J. et. al [13] demonstrated a RF MEMS switch on semi-insulating GaAs substrate and achieved a low switching speed. Goldsmith C. et.al [18] developed a switch whose switching speed is comparatively low with respect to semiconductor switches, however these switches are most suitable for beam steering application in phased array antennas. Goldsmith C. et.al [19] modified the RF MEMS capacitive shunt switch design and achieved increased switching speed by using thin aluminum switch membrane. A comparison table of characteristics and performances for various developed RF MEMS switches are listed in Table 7.3.

Table 7.3. Characteristics and performances of different RF MEMS switches.

Switch Design	Contact Type	Substrate used	Actuation voltage (V)	Insertion loss (dB)	Isolation (dB)	Switching speed	Frequency range	Ref.
Cantilever based series switch	Metal to metal	GaAs	~28	0.1 at 4 GHz	-50 at 4 GHz	30 μs	DC to RF frequency	[13]
Cantilever based series switch	Metal to metal	Silicon	2	1 up to 40 GHz	20 up to 60 GHz	40-60 ms	---	[26]
Fixed-fixed beam shunt switch	Capacitive	Silicon	30-50	<0.25 at 35 GHz	35 at 35 GHz	<5 μs	Up to 40 GHz	[19]
Fixed-fixed beam shunt switch	Capacitive	--	14-16	<0.2 at 20 GHz	-30 at 40 GHz	--	--	[22]
Fixed-fixed beam shunt switch	Capacitive	Silicon	15-20	0.6 at 22-38 GHz	50	--	DC to 40 GHz	[30]
Fixed-fixed beam shunt switch	Capacitive	Silicon	--	<0.2	35 at 10 GHz	--	7-12 GHz	[32]
Fixed-fixed beam shunt switch	Metal to metal	Silicon	65	-0.1 to -0.15 up to 30 GHz	-34 at 0.1 to 2 GHz -20 at 18 GHz	--	0.1 to 20 GHz	[33]
Fixed-fixed beam shunt switch	Capacitive	Quartz	--	-0.2 to -0.5 at W band	≤ -30 at W band	--	75 to 110 GHz	[34]
Fixed-fixed beam shunt switch	Capacitive	Silicon	25	0.29 at 24 GHz	30.1 at 24 GHz	8 ms	24 GHz	[59]

7.4. Conclusion

The recent progress in RF MEMS switch was detailed in this chapter. This chapter discussed the different type of MEMS switch such as metal to metal contact and capacitive coupled MEMS switches for series and shunt circuit configuration with coplanar waveguide (CPW) transmission line. Important switch parameters like actuation voltage, switch lifetime, temperature sensitivity and switching speed were highlighted, as they make significant contribution to the performance of the switch. RF performance parameters like insertion loss, isolation, RF power handling and return loss have observed for different switch designs. The main intention of this chapter is to provide a good knowledge to the newcomers in this field and empower an overall device picture, current status, and vision of their ultimate performance capabilities. This RF MEMS switch field has really been shaped within the past two decades. The future of this miniature MEMS device should be bright and unlimited.

References

[1]. G. M. Rebeiz, RF MEMS: Theory, Design, and Technology, 3rd ed., *John Wiley & Sons,* New Jersey, USA, 2003.

[2]. R. Yamase, T. Maeda, I. Khmyrova, E. Shestakova, E. Polushkin, A. Kovalchuk, S. Shapoval, Study of fringing effects in multi-cantilever HEMT-based resonant MEMS, *International Journal of Applied Electromagnetics and Mechanics*, Vol. 38, Issue 2-3, 2012, pp. 93-100.

[3]. W. Wang, Y. Zhao, Q. Lin, G. Yuan, A three-axial micro-force sensor based on MEMS technology, *International Journal of Applied Electromagnetics and Mechanics*, Vol. 33, Issue 3-4, 2012, pp. 991-999.

[4]. V. T. Srikar, S. M. Spearing, Material selection for microfabricated electrostatic actuators, *Sensors & Actuators*, Vol. 102, 2003, pp. 279–285.

[5]. E. R. Brown, RF-MEMS switches for reconfigurable Integrated circuits, *IEEE Transaction on Microwave Theory and Techniques*, Vol. 46, No. 11, 1998, pp. 1868-1880.

[6]. C. T.-C. Nguyen, RF MEMS in Wireless Architectures, *Association for Computing Machinery,* 2005, pp. 416-420.

[7]. V. Puyal, D. Dragomirescu, C. Villeneuve, J. Ruan, P. Pons, R. Plana, Frequency scalable model for MEMS capacitive shunt switches at millimeter-wave frequencies, *IEEE Transaction on Microwave Theory and Techniques,* Vol. 57, No. 11, 2009, pp. 2824-2833.

[8]. J. Y. Park, G. H. Kim, K. W. Chung, J. U. Bu, Monolithically integrated micromachined RF MEMS capacitive switches, *Sensors & Actuators,* Vol. 89, 2001, pp. 88-94.

[9]. P. Ekkels, X. Rottenberg, R. Puers, H. A. C. Tilmans, Evaluation of platinum as a structural thin film material for RF-MEMS devices, *Journal of Micromechanics and Microengineering,* Vol. 19, 2009, pp. 065010-065018.

[10]. K. V. Caekenberghe, RF MEMS on the radar, *IEEE Microwave Magazine*, Vol. 10, 2000, pp. 99-116.

[11]. V. K. Varadan, K. J. Vinoy, K. A. Jose, RF MEMS and their applications, *John Wiley & Sons.,* UK, 2003.

[12]. D. Anagnorto, M. Khodier, J. C. Lyke, C. G. Christodoulou, Fractal antenna with RF MEMS switches For multiple frequency applications, *IEEE Antenna and Propagation Society International Symposium,* San Antonio, Texas, 16-21 June 2002, pp. 22-25.

[13]. J. J. Yao, M. F. Chang, A surface micromachined miniature switch for telecommunications applications with signal frequencies from dc up to 4 GHz, in *Proceedings of the 8th International Conference on Solid-state Sensors and Actuators, and Eurosensors IX,* Stockholm, Sweden, June 1995, pp. 384-387.

[14]. J. J. Yao, Topical review RF MEMS from a device perspective, *Journal of Micromechanics and Microengineering,* Vol. 10, 2000, pp. 9-38.

[15]. L. E. Larson, R. H. Hackett, R. F. Lohr, Microactuators for GaAs-based microwave integrated circuits, in *Proceedings of the 6th International Conference on Solid-State Sensors and Actuators*, 1991, pp. 743–746.

[16]. H. J. D. L. Santos, Y.-H. Kao, A. L. Caigoy, E. D. Ditmars, Microwave and mechanical considerations in the design of MEM switches for aerospace applications, in *Proceedings of the IEEE Aerospace Conference,* Vol. 3, 1997, pp. 235–254.
[17]. C. Goldsmith, T.-H. Lin, B. Powers, W.-R. Wu, B. Norvell, Micromechanical membrane switches for microwave applications, *IEEE Microwave Theory and Techniques Symposium,* Vol. 1, 1995, pp. 91-94.
[18]. C. Goldsmith, J. Randall, S. Eshelman, T.-H. Lin, D. Denniston, S. Clhen, B. Norvell, Characteristics of micromachined switches at microwave Frequencies, *IEEE Microwave Theory and Techniques Symposium,* Vol. 2, 1996, pp. 1141-1144.
[19]. C. L, Goldsmith, Z. Yao, S. Eshelman, D. Denniston, Performance of low-loss RF MEMS capacitive switches, *IEEE Microwave Guided Wave Letters*, Vol. 8, 1998, pp. 269-271.
[20]. G. L. Tan and G. M. Rebeiz, DC-26 GHz MEMS series-shunt absorptive switches, *IEEE Microwave Guided Wave Letters,* Vol. 1, 2001, pp. 325-328.
[21]. K. E. Petersen, Micromechanical membrane switches on silicon, *IBM Journal of Research and Development*, Vol. 23, 1979, pp. 376–385.
[22]. S. Pacheco, C.-T. Nguyen, L. P. B. Katehi, Micromechanical electrostatic K-band switches, *IEEE Microwave Theory and Techniques Symposium,* Vol. 3, 1998, pp. 1569-1572.
[23]. S. Pacheco, C.-T. Nguyen, L. P. B. Katehi, Design of low actuation voltage RF MEMS switch, *IEEE Microwave Theory and Techniques Symposium,* Vol. 1, 2000, pp. 165-168.
[24]. D. Balaraman, S. K. Bhattacharya, F. Ayazi, J. Papapolymerou, Low-Cost low actuation voltage copper RF MEMS Switches, *IEEE Microwave Theory and Techniques Symposium,* Vol. 2, 2002, pp. 1225-1228.
[25]. S. D. Lee, B. C. Jun, S. D. Kim, H. C. Park, J. K. Rhee, K. Mizuno, An RF-MEMS switch with low-actuation voltage and high reliability, *Journal of Microelectromechanical Systems,* Vol. 15, 2006, pp. 1605 - 1611.
[26]. R. G. Polcawich, J. S. Pulskamp, D. Judy, P. Ranade, S. Trolier-McKinstry, M. Dubey, Surface micromachined microelectromechanical ohmic series switch using thin-film piezoelectric actuators, *IEEE Transaction on Microwave Theory and Techniques,* Vol. 55, 2007, pp. 2642-2654.
[27]. H. C. Lee, J. C. Park, Y. H. Park, Development of shunt type ohmic RF MEMS switches actuated by piezoelectric cantilever, *Sensors and Actuators*, Vol. 136, 2007, pp. 282–290.
[28]. J. Kim, S. Kwon, H. Jeong, Y. Hong, S. Lee, I Song, B. Ju, A stiff and flat membrane operated dc contact type RF MEMS switch with low actuation voltage, *Sensors and Actuators,* Vol. 153, 2009, pp. 114–119.
[29]. A. Kundu, S. Sethi, N. C. Mondal, B. Gupta, S. K. Lahiri, H. Saha, Analysis and optimization of two movable plates RF MEMS switch for simultaneous improvement in actuation voltage and switching time, *J Microelectro*, Vol. 41, 2010, pp. 257-265.
[30]. J. B. Muldavin, G. M. Rebeiz, 30 GHz tuned MEMS switches, *IEEE Microwave Theory and Techniques Symposium,* 1999, pp. 1511-1514.
[31]. J. B. Rizk, J. B. Muldavin, G. L. Tan, G. M. Rebeiz, Design of X-Band MEMS Microstrip Shunt Switches, in Proceedings of the 30th European *Microwave Conference,* 2000, pp. 1-4.
[32]. J. B. Muldavin, G. M. Rebeiz, High-isolation inductively-tuned x-band MEMS shunt switches, *IEEE Microwave Theory and Techniques Symposium,* Vol. 1, 2000, pp. 169-172.
[33]. G. L. Tan, G. M. Rebeiz, A DC-contact MEMS shunt switch, *IEEE Microwave Guided Wave Letters*, Vol. 12, 2002, pp. 212-214.
[34]. J. B. Rizk, G. M. Rebeiz, W-band CPW RF MEMS circuits on quartz substrates, *IEEE Transaction on Microwave Theory and Techniques,* Vol. 51, 2003, pp. 1857-1862.
[35]. B. Ghodsian, P. Bogdanoff, D. Hyman, Wideband DC-contact MEMS series switch, *Journal of Micro & Nano Letters*, Vol. 3, 2008, pp. 66-69.

[36]. Y. H. Jang, Y. S. Lee, Y. K. Kim, J. M. Kim, High isolation RF MEMS contact switch in V-band and W-band using two dimensional motions, *Electronics Letters,* Vol. 46, 2010, pp. 153-155.

[37]. D. Yamane, W. Sun, H. Seita, S. Kawasaki, H. Fujita, H. Toshiyoshi, Ku-band dual-SPDT RF-MEMS switch by double-side SOI bulk micromachining, *Journal of Microelectromechanical Systems*, Vol. 20, 2011, pp. 1211 - 1221.

[38]. X. Yuan, J. C. M. Hwang, D. Forehand, C. L. Goldsmith, Modeling and characterization of dielectric-charging effects in RF MEMS capacitive switches, *IEEE Microwave Theory and Techniques Symposium,* 2005, pp. 753-756.

[39]. J. R. Reid, R. T. Webster, Measurements of charging in capacitive microelectromechanical switches, *Electronics Letters,* Vol. 38, 2002, pp. 1544-1545.

[40]. C. Goldsmith, J. Ehmke, A. Malczewski, B. Pillans, S. Eshelman, Z. Yao, J. Brank, M. Eberly, Lifetime characterization of capacitive RF MEMS switches, *IEEE Microwave Theory and Techniques Symposium,* Vol. 1, 2001, pp. 227-230.

[41]. P. Czarnecki, X. Rottenberg, P. Soussan, P. Ekkels, P. Muller, P. Nolmans, W. De Raedt, H. A. C. Tilmans, R. Puers, L. Marchand, I. De Wolf, Influence of the substrate on the lifetime of capacitive RF MEMS switches, in Proceedings of the *21st IEEE International Conference on Micro Electro Mechanical Systems* 2008, pp. 172 – 175.

[42]. W. S. H. Wong, C. H. Lai, Longer MEMS switch lifetime using novel Dual-pulse actuation voltage, *IEEE Transaction on Devices and Material Reliability*, Vol. 9, 2009, pp. 569-575.

[43]. X. Rottenberg, I. D. Wolf, B. K. J. C. Nauwelaers, W. De Raedt, H. A. C. Tilmans, Analytical model of the DC actuation of electrostatic MEMS devices with distributed dielectric charging and nonplanar electrodes, *Journal of MEMS*, Vol. 16, 2007, pp. 1243-1253.

[44]. B. Pillans, J. Kleber, C. Goldsmitht, M. Eberly. RF power handling of capacitive RF MEMS devices, *IEEE Microwave Theory and Techniques Symposium,* Vol. 1, 2002, pp. 329-332.

[45]. B. D. Jensen, Z. Wan, L. Chow, K. Saitou, K. Kurabayashi, J. L. Volakis, Integrated electrothermal modeling of RF MEMS switches for Improved power handling capability, *IEEE Topical Conference on Wireless Communication Technology,* 2003, pp. 10-11.

[46]. D. Peroulis, S. P. Pacheco, L. P. B. Katehi. RF MEMS switches with enhanced Power-handling capabilities, *IEEE Microwave Theory and Techniques Symposium,* Vol. 52, 2004, pp. 59-68.

[47]. K. Grenied, D. Dubuc, E. Ducnrouge, V. Conedera, D. Bourrier, E. Ongareau, P. Derderian, R. Plana, High power handling RF MEMS design and technology, in Proceedings of the *18th IEEE International Conference on Micro Electro Mechanical Systems (MEMS'2005),* pp. 155 – 158.

[48]. L. Dussopt, G. M. Rebeiz, Intermodulation distortion and power handling in RF MEMS switches, varactors, and tunable filters, *IEEE Transaction on Microwave and Theory Techniques*, Vol. 51, 2003, pp. 1247-1256.

[49]. R. Mahameed, G. M. Rebeiz. Power handling of temperature-stable thin-RF MEMS capacitive switches, in *Proceedings of the 40th European Microwave Conference (EuMC),* 2010, pp. 97-100.

[50]. A. Stehle, C. Siegel, V. Ziegler, B. Schonlinner, U. Prechtel, H. Seidel, U. Schmid. High-power handling capability of low complexity RF-MEMS switch in Ku-band, *Electronics letters*, Vol. 43, 2007, pp. 1367 - 1368.

[51]. C. Palego, J. Deng, Z. Peng, S. Halder, J. C. M. Hwang, D. I. Forehand, D. Scarbrough, C. L. Goldsmith, I. Johnston, S. K. Sampath, A. Datta, Robustness of RF MEMS capacitive switches with molybdenum membranes, *IEEE Transaction on Microwave and Theory Techniques,* Vol. 57, 2009, pp. 3262-3269.

[52]. R. Mahameed, G. M. Rebeiz, A high-power temperature-stable electrostatic RF MEMS capacitive switch based on a thermal buckle-beam design, *Journal of Microelectromechanical Systems*, Vol. 19, 2010, pp. 816 - 826.
[53]. R. Mahameed, G. M. Rebeiz, Electrostatic RF MEMS tunable capacitors with analog tunability and low temperature sensitivity, *IEEE Microwave Theory and Techniques Symposium,* 2010, pp. 1254–1257.
[54]. J. Muldavin, C. O. Bozler, S. Rabe, P. W. Wyatt, C. L. Keast, Wafer-scale packaged RF microelectromechanical switches, *IEEE Transaction on Microwave and Theory Techniques,* Vol. 56, 2008, pp. 522-529.
[55]. R. Mahameed, G. M. Rebeiz, RF MEMS Capacitive Switches for Wide Temperature Range Applications Using a Standard Thin-Film Process, *IEEE Transaction on Microwave and Theory Techniques,* Vol. 59, 2011, pp. 1746-1752.
[56]. M. Hoffmann, H. Kuppers, T. Schneller, U. Bottger, U. Schnakenberg, W. Mokwa, R. Waser, Theoretical calculations and performance results of a PZT thin film actuator, *IEEE Transactions on Ultrasonics, Ferroelectrics, and Frequency Control*, Vol. 50, 2003, pp. 1440-1446.
[57]. H. A. C. Tilmans, E. Fullin, H. Ziad, M. D. K. J. Van de Peer, J. Kesters, E. Van Geffen, J. Bergqvist, M. Pantus, E. Beyne, K. Baert, F. Naso, A fully-packaged electromagnetic microrelay, *IEEE International Conference on Micro Electro Mechanical Systems*, 1999, pp. 25-30.
[58]. Y. Wang, Z. Li, D. T. McCormick, C. N. Tien, A low-voltage lateral MEMS switch with high RF performance, *Journal of MEMS*, Vol. 13, 2004, pp. 902-911.
[59]. Park J., Shim E. S., Choi W., Kim Y., Kwon Y., Cho D-i, A non-contact-type Rf-MEMS switch for 24-GHz radar applications, *Journal of Microelectromechanical Systems*, Vol. 18, 2009, pp. 163 - 173.

Chapter 8
Advances in Amperometric Acetylcholinesterase Biosensors

Xia Sun, Chen Zhai, Xiangyou Wang

8.1. Introduction

Organophosphorous (OP) and carbamate pesticides (see structures in Fig. 8.1) [1] have been used widely in agriculture to protect crops and seeds before and after harvesting. They also have contributed significant health and economic benefits to society. At the same time, these compounds inhibit acetylcholinasterase (AChE) that hydrolyses the neurotransmitter acetylcholine (ACh), often causing severe impairment of nerve functions of human or even death [2-4].

At the present, the identification and quantification of pesticides are generally based on classical standardized chromatographic analysis techniques, such as gas chromatography (GC), high-pressure liquid chromatography (HPLC), capillary electrophoresis (CE) and mass spectrometry (MS). These analytical techniques have been described and reviewed extensively in the literature. They are very sensitive and reliable. However, they have some disadvantages such as complex, time consuming, require costly, bulky instrumentation and so on. Therefore, the development of rapid determination and reliable quantification detection methods has become increasingly important for human health and environment protection.

Electrochemical biosensors have emerged the past few years as the most promising alternative to detect pesticides due to the high sensitivity inherent to the electrochemical detection and the possibility of portability and miniaturization [5-7]. Among them, amperometric AChE biosensors as a combination of enzymatic reactions with the electrochemical methods have shown satisfactory results for pesticides analysis, where the enzyme activity was employed as indicator of quantitative measurement of insecticides [8]. When AChE is immobilized on the working electrode surface, its interactions with the substrate of acetylthiocholine (ATCl) produce the electroactive product of thiocholine [3].The reaction equation is shown as follows [9-10].

Xia Sun
School of Agriculture and Food Engineering, Shandong University of Technology,
No. 12, Zhangzhou Road, Zibo 255049, Shandong Province, P. R. China

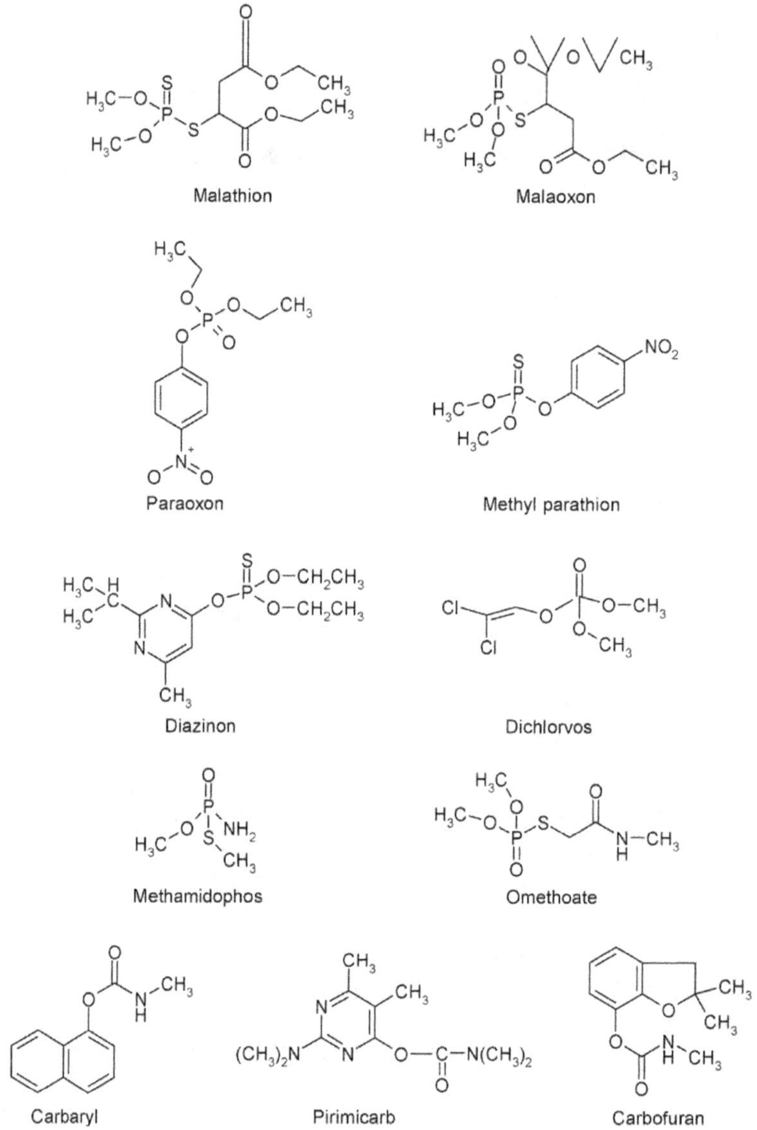

Fig. 8.1. Structures of the main pesticides used as targets in AChE biosensors.

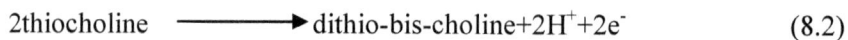

$$\text{acetylcholine} + H_2O \xrightarrow{\text{AChE}} \text{thiocholine} + \text{acetic acid} \qquad (8.1)$$

$$\text{2thiocholine} \longrightarrow \text{dithio-bis-choline} + 2H^{+} + 2e^{-} \qquad (8.2)$$

Furthermore, the anodic oxidation current is inversely proportional to the concentration of organophosphorous pesticides (OPs) in samples, and the exposed time as well. The procedure of the preparation of AChE biosensor and pesticides detection is shown in Fig. 8.2 [11-12].

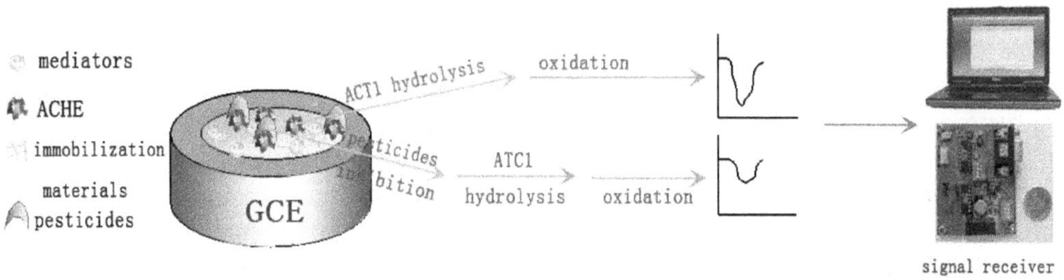

Fig. 8.2. The procedure of the preparation and detection of amperometric AChE biosensor.

Amperometric AChE biosensors are considered as sensitive and accurate methods in the pesticides determination due to their liner relationship between the current and concentration of the pesticides at relatively low substrate concentration. Many groups have reported these satisfied results about amperometric AChE biosensors for pesticides detection [2, 5, 8, 13-23]. Moreover, these biosensors had been the subject of several reviews [24-25] before five years ago. However, there is a time gap between current status in the field and the most recent reviews. Thus, in this review, we specifically provide an overview of the research carried out during the last 5 years relative to AChE inhibition-based biosensors for food and environment safety. We will review the key steps to construct an AChE biosensor including enzyme resource, the immobilization protocols used for formation of a bio-recognition interface, the electrode modification. We also will discuss the trends and challenges associated with designing a reliable AChE sensor for practical applications in detail.

8.2. AChE Source

It is well known that Drosophila melanogaster and the Electric Eel are commercially available and are the most widely used for biosensor fabrication. AChE has different substrate specificity and susceptibility to inhibitors from various extraction sources. Some research groups have reported the influences of different AChE sources on the enzyme activity and specificity to OP and carbamate pesticides [26-27]. Hence, in order to improve the performance characteristics of the AChE biosensors, the design and production of the appropriate enzymes is crucial. It is very difficult to find a natural enzyme with all the features required to construct the ideal biosensor. Nevertheless, protein engineering is helping researchers to manipulate enzymes in order to produce tailor-designed bio-recognition molecules for their integration into biosensing platforms.

Initial biochemical studies revealed that Drosophila melanogaster acetylcholinesterase (Dm. AChE) is 8-fold more sensitive than Electric Eel enzyme toward Ops, but the sensitivity has been increased to 12-fold by introducing a mutation [28]. Latter some research groups also have verified that the use of genetically modified AChE in biosensors has significantly increased their sensitivity to inhibition by OP pesticides [20-21, 29-30]. The design of mutation enzyme is to increase the affinity for the target analyte favouring the accessibility of the active site. Since the active site of Drosophila melanogaster (Dm) AChE, the most active wild form, is buried 20Å inside the protein and the entrance to the active site is very narrow, mutations have been mainly addressed to alter some residues of that region [31]. The replacement of glutamic acid 69 (Glu69), located at the rim of the active site gorge, by amino acids with bulky side chains, such as tryptophan (Trp) or tyrosine (Tyr), has been demonstrated to increase significantly the inhibition constant, k_i, for dichlorvos [32]. This effect has been suggested to be due to interactions between those chains and the insecticide, which favour its way towards the buried active site. Double mutations, e.g. replacing Tyr71 by aspartic acid (Asp) additionally to the Glu69 substitution, have led to even more sensitive enzymes.

Since AChE can be inhibited by several compounds (not only organophosphate and carbamate insecticides, but also some toxins). Bucur et al. [33] developed biosensors for carbaryl, carbofuran and pirimicarb. Although they identified the best mutant for each pesticide, the high sensitivity provided by all biosensors towards all analytes did not allow the discrimination between them. However, to this day, the combination of engineered AChEs with different sensitivity levels for different analytes can not obtain satisfied selective biosensors.

8.3. AChE Immobilization

The immobilization of AChE is a key step in optimizing the analytical performance of an AChE biosensor in terms of response, sensitivity, stability, and reusability. The immobilization strategies most generally employed are physical or chemical methods. In general, they mostly fall into following methodologies:

(1) Physical adsorption: Physical adsorption generally consists of simple deposition of AChE onto the surface of working electrode and attachment of AChE through weak bonds such as Van der Waals forces and electrostatic interactions between the AChE and the transducer. It is the easiest and the least denaturing method, and it is generally non-destructive for enzyme activity because this technique does not involve any functionalization of the electrode materials or covalent links. However, this mode of immobilization are random orientation and weak attachment, therefore, the biosensors with adsorbed enzyme suffer from a poor operational and storage stability. Sotiropoulou's group immobilized AChE into the nano-structured conductive carbon with this technique. Using this biosensor, the monitoring of the organophosphorous pesticide dichlorvos at 1×10^{-12} M [34].

(2) Covalent coupling: Covalent coupling method immobilization AChE is the most widely used procedure. AChE can be covalently linked to the surfaces of a transducer through formation of a stable covalent bond between functional groups of AChE and the transducer. By comparing the performances of covalent with non-covalent immobilization method for the immobilization of AChE onto screen-printed carbon electrodes, it was concluded that when AChE was immobilized with non-covalent, the sensors detection limit was found to be about 10^{-10} M for the organophosphate pesticide

dichlorvos, lower than covalent immobilization method [7]. The procedure provides increased stability of the AChE but decreases the activity of AChE and is generally poorly reproducible. Recently, this approach has also been utilized to immobilize acetylcholinesterase with glutaraldehyde [8, 10, 35, 36].

(3) Physical entrapment: Physical immobilization methods such as entrapment in sol-gel matrices and photo-polymerized monomers have also been used for AChE biosensors. The sol-gel process is a synthetic inorganic procedure, well known in material chemistry. Some research groups have reported the protocols using sol-gel to immobilize AChE. Shi et al. reported an AChE biosensor by entrapping AChE in Al_2O_3 sol-gel matrix in a screen-printed 3-electrode plastic chip. The biosensor can be found the highest current signal at an operating potential of 0.25 V, and the detection limit for dichlorvos is achieved at 10 nM in the simulated seawater for 15 min inhibiting time [37]. Zejli et al. developed an AChE biosensor by entrapping AChE in Al_2O_3 sol-gel matrix on the sonogel-carbon electrode. The biosensor showed optimal activity at an operating potential of 210 mV, and the detection limit achieved for chlorpyriphos-ethyl-oxon was 2.5×10^{-10} M at a 10 min incubation time [38]. Yin et al. reported that using gold nanoparticles (AuNPs) and silk fibroin (SF) colloid immobilized AChE, and the detection limits were found to be 2×10^{-11} M for methyl paraoxon [39]. Zinc oxide sol-gel has also been used as a matrix for immobilization of AChE. The immobilized enzyme retained its enzymatic activity up to three months when stored in phosphate buffered saline (pH 7.4) at 4^0 C [40]. Some research groups also obtained satisfied results that using the silica sol-gel incorporating gold nanoparticles (AuNPs) composite immobilized AChE onto GCE surface [11, 13].

Another physical entrapment method that could be applied for AChE immobilization involves encapsulation of enzymes during the electrodeposition of a conducting polymer such as polypyrrole (PPy) or polyaniline (PANI) in the presence of enzyme [41]. The method enables control of polymer/enzyme layer thickness, is extremely simple and rapid but requires a significant amount of enzyme. Polypyrrole and polyaniline copolymer doped with multi-walled carbon nanotubes was used to immobilize AChE. The synthesized PANI-PPy-MWCNTs copolymer presented a porous and homogeneous morphology which provided and ideal size to entrap enzyme molecules [5]. AChE was immobilized onto a three-dimensional porous network based on Au nanoparticles-polypyrrole nanowires, it exhibited a strong synergetic effect on improving the sensing properties of OP pesticides [3]. Valdés-Ramírez et al. reported that AChE was immobilized by entrapment in a photocrosslinkable PVA-SbQ polymer on a screen printed graphite electrode [30]. The principle of entrapment of enzymes in the photo-sensitive polymer was shown in Fig. 8.3 [19].

However, these methods suffer from several limitations such as leaking of biocomponent and possible diffusion barriers which restrict the performance of biosensor.

(4) Self-assembled monolayer (SAM): SAM may be generated by the spontaneous chemisorption of molecules onto a gold surface, thus, SAM as a platform can link

biomolecules either using direct chemical linkages or by encapsulation with the help of polymeric supports. It is built of long-chained n-alkylthiols with derivatized organic functional groups, which are easily linked to the gold film via the thiol groups [42]. This technique was used to immobilize analyte derivatives onto the surface of gold-coated sensors.

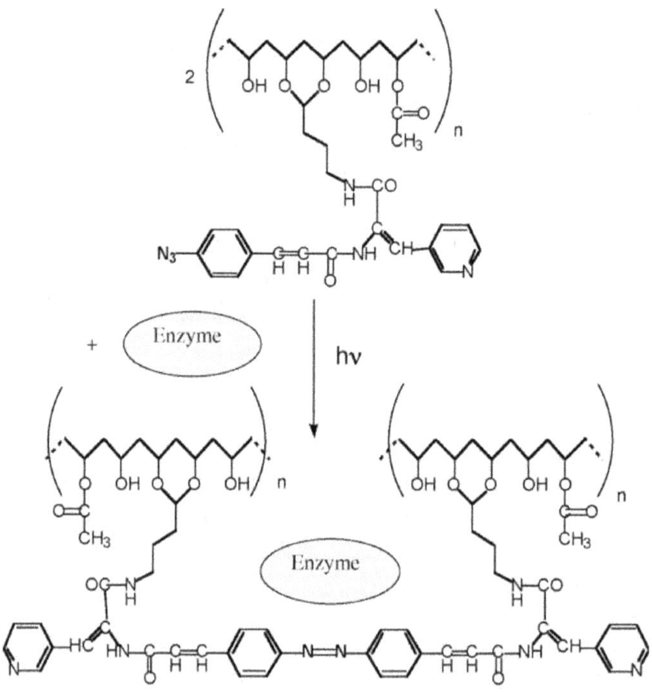

Fig. 8.3. The entrapment of enzymes in photo-sensitive polymer.

For example, Viswanathan et al. reported the method of the self-assembled monolayers (SAMs) of single walled carbon nanotubes (SWCNT) wrapped by thiol terminated single strand oligonucleotide (ssDNA) on gold to prepare nano-size polyaniline matrix for acetylcholinesterase (AChE) enzyme immobilization. The pesticides were determined through inhibition of enzyme reaction. The biosensor has been applied for the determination of methyl parathion and chlorpyrifos in spiked river water samples with the detection limit of 1×10^{-12} M [43]. Du et al. reported an immobilization method of AChE on cysteamine self assembled monolayers modified gold electrode for carbaryl detection. The detection linear ranges were from 1 to 50 ng/mL and from 50 to 500 ng/mL, with a detection limit of 0.6 ng/mL [15]. Another kind of AChE/dendrimers polyamidoamine (PAMAM)-Au/Carbon nanotubes (CNTs) layer- by-layer self-assembled technique is employed in the detection of carbofuran in samples with the range of 4.8×10^{-9} M to 0.9×10^{-7} M and a detection limit of 4.0×10^{-9} M [44]. Similar strategies are also reported for the preparation of AChE biosensor [17, 45-47]. In summary, this method possesses easy formations of ordered, pinhole free and stable monolayers, ensures the orientation and spatial control of the enzymes in the immobilization process and the absence of diffusion barriers, and provides suitable microenvironment for biomolecule immobilization, but the reproducible layering of biomolecules remains a major limitation.

The one-step protocol, based on the incubation of the gold electrode into an enzyme solution, not only simplified the construction of the biosensor but also provided more reproducible results. Du et al. reported one-step electrochemically deposited interface of chitosan-gold nanoparticles and one-step synthesis of multiwalled carbon nanotubes-gold nanocomposites for fabricating amperometric AChE biosensor, respectively. Gong et al also reported one-step fabrication of three-dimensional porous calcium carbonate-chitosan composite film as the immobilization matrix of acetylcholinesterase. These one-step protocols all obtained satisfying results [6, 18].

(5) Oriented immobilization: The aforementioned immobilized methods such as adsorption, covalent binding, entrapment within a polymeric matrix and crossing-linking may involve multiple-point attachment, loss of biomolecule activity, biomolecule leaking, random biomolecule orientation and large diffusion barriers, which render biosensors poorly sensitive, unstable and with long response times. New trends focus on the development of protocols for the oriented immobilization of enzymes through specific functional groups located at their surface. In this way, active sites may be faced towards the target analytes present in the sample, and substrates and products may freely diffuse in the biological layer [1]. The procedure of oriented immobilization enzymes for pesticides detection was shown in Fig. 8.4.

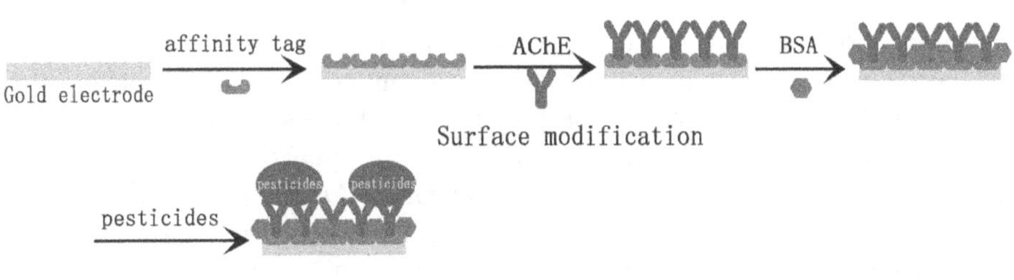

Fig. 8.4. Schematic representation of oriented immobilization enzymes.

Istamboulie et al. oriented immobilization AChE by metal chelate-functionalised magnetic microbeads. This method is based on the ability of some metal ion (Ni2+, Cu2+, Zn2+) to bind strongly but reversibly to peptides and proteins containing histidine (His) or cysteine (Cys) residues. In affinity chromatography, chelating agent nitrilotriacetic acid (NTA) and iminodiacetic acid (IDA) are generally used to bind these metal ions on solid supports to perform His-taggedenzyme purification. In this work, pre-activated magnetic beads carrying Ni-IDA complexes have been used to immobilize a genetically modified AChE (B394) having engineered a hexa (histidine) tail. This method provides a basis for controlled and oriented immobilization of the enzyme on different supports, opening the way for new approaches to enzyme immobilization. The general principle of this procedure is presented in Fig. 8.5 [19].

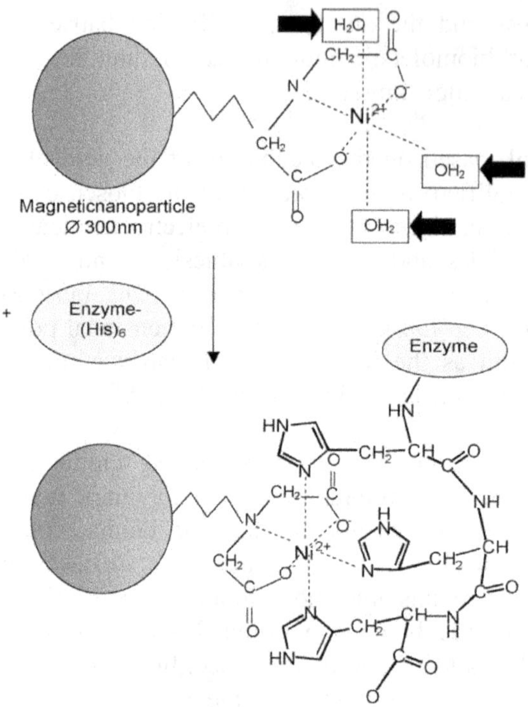

Fig. 8.5. Schematic representation of enzyme immobilisation on magnetic microbeads via Ni–His affinity. Black arrows show histidine-exchangeable water molecules.

(6) Immobilization on membrane materials: Enzyme was usually immobilized on electrode surfaces directly, which has many disadvantages, such as weak protection of the immobilized enzyme, elaborate pretreatment of the electrode surface before the actual immobilization, etc. Therefore, in the recent years, the membrane support materials become important alternative immobilization AChE strategies. Because it can prolong the enzyme storage life, possess the convenience of storing only the membranes, not whole electrodes and it can be easily replaced when enzyme's activity is lost, and have not the necessity of constantly cleaning the electrode surface after each immobilization procedure. Moreover, there are multiple options for analyte detection (one electrode-multiple membranes-multiple enzymes). Recently, many research groups reported that enzyme was immobilized on different membrane materials, indicating the higher activity of AChE, higher sensitivity of sensor and simplifier operation [10, 48-52].

8.4. Electrode Modification

It is well known that the principle of amperometric AChE biosensor is to detect the concentration of the pesticides by measuring the change of the oxidation current of thiocholine. However, thiocholine oxidation occurs at relatively high potential that cause high background current and interference from other electroactive compounds [53].

Thus, appropriate selection of the electrode modified material and immobilization chemistry is essential for a reliable biosensor [24]. However, in most of the cases, native materials do not meet these requirements and therefore additional modifications are necessary in order to allow further attachment of the enzyme. To overcome this problem, mediators, such as tetracyanoquinodimethane (TCNQ) [33], Prussian blue [54, 10], cobalt-phthalocyanine (CoPC) [55, 30] and some nanoparticles including carbon nanotubes (CNT) [13, 18, 52], gold nanoparticles [2, 11], CdTe quantum dots [15] have been successfully used to decrease the applied potential versus Ag/AgCl or saturated calomel electrode (SCE), hindering the oxidation of other compounds thus reducing the interferences. These electronic mediators are generally modified on the surface of the working electrode.

Recently, some new trends about electrode modification focus on the new modification materials and composite modification materials applications. Istamboulie et al. demonstrated poly (3,4-ethylenedioxythiophene) (PEDOT), a new conducting polymers materials for AChE biosensor applications. The new material as mediator is suitable for thiocholine oxidation, allowing the measurement of AChE activity at 100 mV vs. Ag/AgCl [22]. In addition, the nanocomposites are also gradually used for constructing AChE biosensor, for example, it is well known that carbon nanotubes (CNT) are an important class of materials with unique and interesting properties, like high mechanical strength, high aspect ratio, high electrical and thermal conductivity, metallic and structural characteristics. At the same time, Au nanoparticles have also been reported to provide immobilization sites for the enzymes retaining their biocatalytic activity. They also permit direct and fast electron transfer between redox species and the electrode. These properties make Au nanoparticles an attractive material for sensing biological reactions. At present, some research groups reported the synthetical effect of Au nanoparticles and CNTs for OP pesticides sensing [4, 18]. Others composite modification materials were also reported such as CdTe quantum dots/gold nanoparticles by Du et al. [16] and multi-walled carbon nanotube/cobalt phthalocyanine by Moraes et al. [56], respectively.

Various immobilization protocols and electrode modification materials (mediators) used for constructing AChE biosensors were show in Table 8.1 in detail.

8.5. New Trends and Challenges

8.5.1. Miniaturization

New analytical approaches are oriented to the development of portable systems with high accuracy, low-cost, short-time response, and that can provide qualitative information about the composition of a sample with minimum preparation. With this aim, Steinberg and Lowe [58] described the development of a potentiostat with low-power consuming and high accuracy. A portable electro-chemical instrument which can be used in neurotoxic agents in water have been designed by Nagatani et al. [59] and Hildebrandt et al. [12, 60], respectively. Fig. 8.6a was the picture of the miniaturized electronic plate that functions as a potentiostat designed by Hildebrandt group.

8.5.2. High Throughput and Quantification Detection

AChE-based biosensors have a major drawback: they give a sum parameter of AChE-inhibition without any qualitative or quantitative information about the individual analytes. One approach to solve this problem involves the application of multi-sensor arrays that are combined with the data processing of artificial neural networks. Using micro or nano-sensor arrays will likely become a new trend. Thus, the ability to construct arrays of enzymes will likely allow current multianalyte detection of several compounds to be expanded to accommodate the analysis of perhaps hundreds or thousands of separate compounds [61]. One of the challenges that must be met for this type of system would be the development of parallel computational methods to convert electronic responses for each analyte into meaningful concentration data. In this respect progress has been reported for artificial neural network implementation in single low-cost chip for the detection of insecticides by modeling of enzymatic sensors response [20, 62]. The design of single chip to selectively quantify mixtures of the pesticides chlorpyrifos oxon and chlorfenvinfos by an artificial neural network implementation was shown in Fig. 8.6 b, and the dedicated system was shown in Fig. 8.6 c [62].

8.5.3. Integration of Detection System

To improve sample throughput or allow online monitoring of the inhibition processes, flow-injection analysis (FIA) systems have proved to be a popular approach and capable of performing continuous analysis. Recently, the use of micro-fluidic devices as minute FIA systems has attracted increasing attention due to their significant reduction of reagent consumption and low operating costs as well as high throughput capability. He et al. described the development of a silica based, immobilized enzyme micro-reactor, with an integrated micro-fluidic electrochemical detector for the rapid evaluation of enzyme kinetics [63]. The micro-fluidic device was shown in Fig. 8.7.

8.5.4. Real Samples Detections

The majority of AChE biosensors reported to date have been designed for detection of pesticides in water samples. Application to other matrices such as food samples (fruits and vegetates) has been restricted due to the problems related to the use of these devices in the presence of organic solvents extracts. Several works reported that enzyme biosensors can function in a mixed aqueous-organic phase in low amounts of organic solvents [44]. However, these devices suffered from a low sensitivity and have not been used for the quantification of pesticides in solid sample.

Simple sample pretreatment methods for AChE biosensor need further research, because it hinders the real application of biosensor.

Table 8.1. Amperometric AChE biosensors for pesticides detection.

Inhibitors	Ache Source	Immobilization matrix	Mediator/ Electrode	Lod	Samples	Incubation time (min)	Reference
Paraoxon, dichlorvos, chlorpyriphos-ethyl-oxon	Electric eel	Al_2O_3 sol-gel	Sonogel-carbon electrode	7.5×10^{-9}M 5×10^{-10}M 2.5×10^{-10}M	-	10	[38]
dichlorvos	Electric eel	Al_2O_3 sol-gel	-/SPE	10nM	river water	15	[37]
methyl parathion Chlorpyrifos	Electric eel	Polyaniline matrix	SWCNT/Au	1×10^{-11}M 1×10^{-12}M	river water	15	[43]
methyl paraoxon Carbofuran Phoxim	Electric eel	AuNPs-SF	AuNPs/pt	2×10^{-11}M 1×10^{-10}M 2×10^{-10}M	-	15	[39]
dimethoate	Electric eel	MWCNTs-β cyclodextrins	MWCNTs/GCE	2nM	garlic	8	[17]
malathion	Electric eel	MWCNTs-Au-CHIT	MWCNTs-Au/GCE	0.6ng/ml	garlic	8	[18]
monocrotophos	Electric eel	AuNPs-SiSG sol-gel	AuNPs/GCE	0.6ng/ml	-	10	[2]
malathion	Electric eel	PANI-PPy-MWCNTs copolymer	MWCNTs/GCE	1.0ng/ml	-	15	[5]
paraoxon	Electric eel	ZnO sol-gel	ZnO	0.035ppm	-	10-30	[40]
Aldicarb Carbaryl	Electric eel	-	PB/SPE	24ppb	water	30	[54]
Chlorpyriphoso-oxon	Drosophila B394	Magnetic microbeads	Magnetic microbeads/ SPE	25ppb 1.3×10^{-11}M	-	-	[19]
Coumaphos	Electric eel	PB	PB/SPE	8ng/ml	honey	20	[57]
Chlorpyrifosoxon	Electric eel	Entrapment in a AWP	PEDOT:PSS/ SPE	4×10^{-9}M	-	10	[22]
Paraoxon ethyl Sarin	Electric eel	Cross-linking with glutaraldehyde	Au-Pt NPs/GCE	150nM 40nM	-	25	[36]
Aldicarb		AChE droped on modified electrode	Au-MWNTs/ GCE	40μM		30	[4]
Carbaryl	Electric eel	CdTe-Cys SAMs	CdTe-Cys SAMs/Au	0.6ng/ml	garlic	10	[14]
Carbofuran	Electric eel	PAMAM-Au/ MWNTs	PAMAM-Au/ MWNTs/GCE	4×10^{-9}M	onion, lattuce, cabbage	6	[44]
Dichlorvos Omethoate Trichlorfon Phoxim	Dm	CHIT membrane	PB/GCE	2.5ng/l 15ng/l 5ng/l 10ng/l	-	10	[10]
Dichlorvos	Dm mutant B394	Entrapment in a PVA-SbQ polymer	CoPC/SPE	7×10^{-11}M	apple	10	[30]

LOD: limit of detection, SWNT:single wall carbon nanotube, gold nanoparticles: AuNPs, SF: silk fibroin, MWNTs:multiwalled carbon nanotubes, chitosan:CHIT, silica sol-gel:SiSG, polyaniline: PANI, polypyrrole:PPy, SPE: screen printed electrode, PB:Prussian-blue, PEDOT: poly(3,4-ethylenedioxythiophene), PSS: poly(styrenesulfonate), AWP:polyvinylalcohol-based photopolymer, GCE: glassy carbon electrode, NPs: nanoparticles, SAMs:self-assembled monolayers, Cys:cysteamine, PAMAM:dendrimers polyamidoamine, Dm:Drosophila melanogaster, CoPC:Cobalt phthalocyanine.

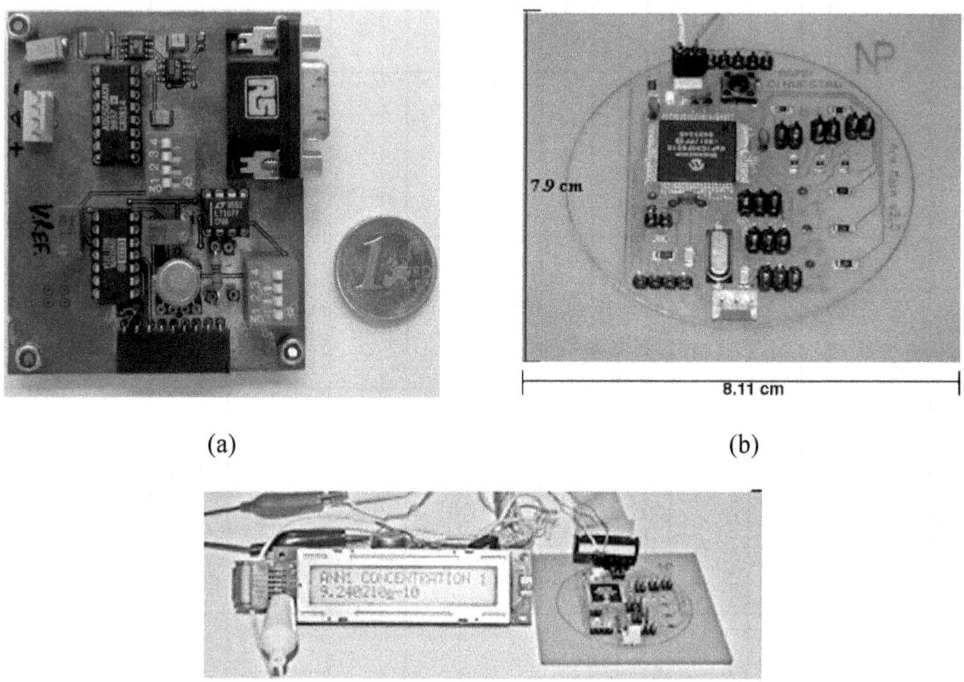

(a) (b)

(c)

Fig. 8.6. Picture of the miniaturized electronic plate that functions as a potentiostat.

8.5.5. Combined Sensor for OP and Carbamates Detection

Further new trends will likely focus on combined sensor for OP and carbamates detection. For example, multifunctional biosensor will be applied for detecting different analytes simultaneously at food samples. Teller et al. developed a dual piezoelectric/amperometric sensor for the detection of two unrelated analytes in one experiment, which can simultaneously completed the detection of cocaine with a dynamic range from 10^{-9} to 10^{-7} M and the organophosphate chlorpyrifos-oxon range from 10^{-6} to 10^{-8}M [64]. In addition, AChE-choline oxidase based bienzymatic or multienzyme sensor will be paid more attention for OP and carbamates detection due to high sensitivity and reliable [29, 36, 57]. Moreover, Wei et al. developed photoelectron-synergistic catalysis for detecting organophosphorous pesticides exhibits another trends of AChE biosensor. Using composite nanostructured PbO2/TiO2/Ti modification material decrease applied potential. This strategy was found to catalyze the oxidative reaction of thiocholine effectively, make the AChE/PbO2/TiO2/Ti biosensor detect the substrate at 0.30 V, hundreds millvolts lower than others reported [65].

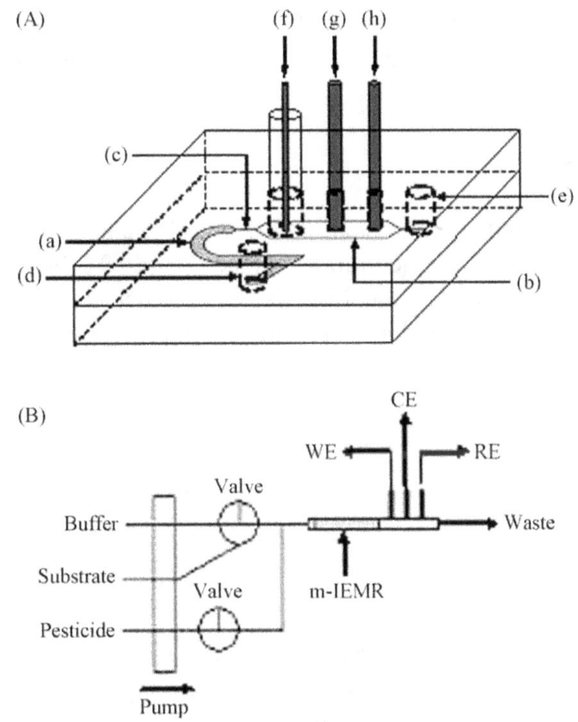

Fig. 8.7. Schematic of the micro-fluidic device: (a) monolith channel, 600 μm wide, 50 μm deep and 20 mm long; (b) electrochemical detection channel, 1.5 mm wide, 50 μm deep and 20 mm long; (c) link channel, 100 μm wide, 50 μm deep and 5 mm long; (d) inlet, 1.5 mm diameter, to which syringe a pump was linked through a plastic tube (0.5 mm diameter); (e) outlet, 1.5 mm diameter; (f) working electrode (WE, pt disc, 0.5 mm diameter), (g) Pt wire (1 mm diameter) counter electrode (CE) and (h) Ag/AgCl (1 mm diameter) reference electrode (RE).

8.6. Conclusions

In summary, amperometric acetylcholinesterase biosensors are strong candidates for screening pesticide residues and they become more and more relevant in environmental and food analysis. Compared to traditional chromatography and other methods, the strengths of AChE biosensor can be described as follows:

- They are very selective and sensitive;
- They can be carried out for use in the field;
- They can work with complete automation and give the results after a short period of time.

However, these AChE biosensors still face many challenges hindering their real applications:

- The data are analyzed on the electrochemical analysis instrument, and they are not miniaturization and portable;
- They only give a sum parameter of AChE-inhibition without quantitative information and high throughput detection;
- They do not combine with FIA system to build integration instrument;
- They have not specific and suitable real samples pretreatment.

Acknowledgments

This work was supported by the National Natural Science Foundation of China (No.30972055, 31101286), Agricultural Science and Technology Achievements Transformation Fund Projects of the Ministry of Science and Technology of China (No.2011GB2C60020) and Shandong Provincial Natural Science Foundation, China (No.Q2008D03).

References

[1]. M. Campàs, B. Prieto-Simón, J. L. Marty, A review of the use of genetically engineered enzymes in electrochemical biosensors, *Seminars in Cell & Developmental Biology*, Vol. 20, Issue 1, 2009, pp. 3-9.

[2]. D. Du, S. Chen, J. Cai, A. D. Zhang, Immobilization of acetyl-cholinesterase on gold nanoparticles embedded in sol-gel film for amperometric detection of organophosphorous insecticide, *Biosensors and Bioelectronics*, Vol. 23, Issue 1, 2007, pp. 130-134.

[3]. J. Gong, L. Wang, L. Zhang, Electrochemical biosensing of methyl parathion pesticide based on acetylcholinesterase immobilized onto Au-polypyrrole interlaced network-like nanocomposite, *Biosensors and Bioelectronics*, Vol. 24, Issue 7, 2009, pp. 2285-2288.

[4]. N. Jha, S. Ramaprabhu, Development of Au nanoparticles dispersed carbon nanotube-based biosensor for the detection of paraoxon, *Nanoscale*, Vol. 2, 2010, pp. 806-810.

[5]. D. Du, X. X. Ye, J. Cai, J. Liua, A. D. Zhang, Acetylcholinesterase biosensor design based on carbon nanotube-encapsulated polypyrrole and polyaniline copolymer for amperometric detection of organophosphates, *Biosensors and Bioelectronics*, Vol. 25, Issue 11, 2010, pp. 2503-2508.

[6]. J. Gong, T. Liu, D. Song, X. Zhang, L. Zhang, One-step fabrication of three- dimensional porous calcium carbonate-chitosan composite film as the immobilization matrix of acetylcholinesterase and its biosensing on pesticide, *Electrochemistry Communications*, Vol. 11, Issue 10, 2009, pp. 1873-1876.

[7]. A. Vakurov, C. E. Simpson, C. L. Daly, T. D. Gibson, P. A. Millner, Acetylcholinesterase-based biosensor electrodes for organophosphate pesticide detection: I. Modification of carbon surface for immobilization of acetylcholinesterase, *Biosensors and Bioelectronics*, Vol. 20, Issue 6, 2004, pp. 1118-1125.

[8]. D. Du, X. Huang, J. Cai, A. D. Zhang, Comparison of pesticide sensitivity by electrochemical test based on acetylcholinesterase biosensor, *Biosensors and Bioelectronics*, Vol. 23, Issue 2, 2007, pp. 285-289.

[9] H. Z. Wu, Y. C. Lee, T. K. Lin, H. C Shih, F. L. Chang, H. P. P. Lin, Development of an amperometric micro-biodetector for pesticide monitoring and detection, *Journal of the Taiwan Institute of Chemical Engineers*, Vol. 40, Issue 2, 2009, pp. 113-122.

[10]. X. Sun, X. Y. Wang, Acetylcholinesterase biosensor based on Prussian blue-modified electrode for detecting organophosphorous pesticides, *Biosensors and Bioelectronics*, Vol. 25, Issue 12, 2010, pp. 2611-2614.

[11]. D. Du, S. Chen, J. Cai, A. D. Zhang, Electrochemical pesticide sensitivity test using acetylcholinesterase biosensor based on colloidal gold nanoparticle modified sol-gel interface, *Talanta*, Vol. 74, Issue 4, 2008, pp. 766-772.

[12]. A. Hildebrandt, J. Ribas, R. Bragós, J. L. Marty, M. Tresànchez, S. Lacorte, Development of a portable biosensor for screening neurotoxic agents in water samples, *Talanta*, Vol. 75, Issue 5, 2008, pp. 1208-1213.

[13]. D. Du, S. Chen, J. Cai, D. Song, Comparison of drug sensitivity using acetylcholinesterase biosensor based on nanoparticles-chitosan sol-gel composite, *Journal of Electroanalytical Chemistry*, Vol. 611, Issue 1-2, 2007, pp. 60-66.

[14]. D. Du, J. W. Ding, Y. Tao, X. Chen, Application of chemisorption/desorption process of thiocholine for pesticide detection based on acetylcholinesterase biosensor, *Sensors and Actuators B*, Vol. 134, Issue 2, 2008, pp. 908-912.

[15]. D. Du, W. Chen, J. Cai, J. Zhang, F. Qu, H. Li, Development of acetylcholinesterase biosensor based on CdTe quantum dots modified cysteamine self-assembled monolayers, *Journal of Electroanalytical Chemistry*, Vol. 623, Issue 1, 2008, pp. 81-85.

[16]. D. Du, S. Chen, D. Song, H. Li, X. Chen, Development of acetylcholinesterase biosensor based on CdTe quantum dots/gold nanoparticles modified chitosan microspheres interface, *Biosensors and Bioelectronics*, Vol. 24, Issue 3, 2008, pp. 475-479.

[17]. D. Du, M. H. Wang, J. Cai, A. D. Zhang, Sensitive acetylcholinesterase biosensor based on assembly of β-cyclodextrins onto multiwall carbon nanotubes for detection of organophosphates pesticide, *Sensors and Actuators B: Chemical*, Vol. 146, Issue 1, 2010, pp. 337-341.

[18]. D. Du, M. H. Wang, J. Cai, Y. H. Qin, A. D. Zhang, One-step synthesis of multiwalled carbon nanotubes-gold nanocomposites for fabricating amperometric acetylcholinesterase biosensor, *Sensors and Actuators B*, Vol. 143, Issue 2, 2010, pp. 524-529.

[19]. G. Istamboulie, S. Andreescu, J. L. Marty, T. Noguer, Highly sensitive detection of organophosphorous insecticides using magnetic microbeads and genetically engineered acetylcholinesterase, *Biosensors and Bioelectronics*, Vol. 23, Issue 4, 2007, pp. 506-512.

[20]. G. Istamboulie, M. Cortina-Puig, J. L. Marty, T. Noguer, The use of Artificial Neural Networks for the selective detection of two organophosphate insecticides: Chlorpyrifos and chlorfenvinfos, *Talanta*, Vol. 79, Issue 2, 2009, pp. 507-511.

[21]. G. Istamboulie, D. Fournier, J. L. Marty, T. Noguer, Phosphotriesterase: A complementary tool for the selective detection of two organophosphate insecticides: Chlorpyrifos and chlorfenvinfos, *Talanta*, Vol. 77, Issue 5, 2009, pp. 1627-1631.

[22]. G. Istamboulie, T. Sikora, E. Jubete, E. Ochoteco, J. L. Marty, T. Noguer, Screen-printed poly (3, 4-ethylenedioxythiophene) (PEDOT): A new electrochemical mediator for acetylcholinesterase-based biosensors, *Talanta*, Vol. 82, Issue 3, 2010, pp. 957-961.

[23]. V. Dounin, A. J. Veloso, H. Schulze, T. T. Bachmann, K. Kerman, Disposable electrochemical printed gold chips for the analysis of acetylcholinesterase inhibition, *Analytica Chimica Acta*, Vol. 669, Issue 1-2, 2010, pp. 63-67.

[24]. S. Andreescu, J. L. Marty, Twenty years research in cholinesterase biosensors: From basic research to practical applications, *Biomolecular Engineering*, Vol. 23, Issue 1, 2006, pp. 1-15.

[25]. A. Amine, H. Mohammadi, I. Bourais, G. Palleschi, Enzyme inhibition-based biosensors for food safety and environmental monitoring, *Biosensors and Bioelectronics*, Vol. 21, Issue 8, 2006, pp. 1405-1423.
[26]. X. Sun, X. Y. Wang, X. Y. Wang, Z. Liu, A Comparative Study of Sensitivity of Acetylcholinesterase in Detection of Organophosphorous Pesticide Residues, *International Journal of Food Engineering*, Vol. 4, Issue 3, 2008, article 7.
[27]. Y. H. Zheng, T. C. Hua, D. W. Sun, J. J. Xiao, F. Xu, F. F. Wang, Detection of dichlorvos residue by flow injection calorimetric biosensor based on immobilized chicken liver esterase, *Journal of Food Engineering*, Vol. 74, Issue 1, 2006, pp. 24-29.
[28]. F. Villatte, V. Marcel, S. Estrada-Mondaca, D. Fournier, Engineering sensitive acetylcholinesterase for detection of organophosphate and carbamate insecticides, *Biosensors and Bioelectronics*, Vol. 13, Issue 2, 1998, pp. 157-164.
[29]. M. Waibel, H. Schulze, N. Huber, T. T. Bachmann, Screen-printed bienzymatic sensor based on sol-gel immobilized Nippostrongylus brasiliensis acetylcholinesterase and a cytochrome P450 BM-3 (CYP102-A1) mutant, *Biosensors and Bioelectronics*, Vol. 21, Issue 7, 2006, pp. 1132-1140.
[30]. G. Valdés-Ramírez, D. Fournier, M. T. Ramírez-Silva, J. L. Marty, Sensitive amperometric biosensor for dichlorovos quantification: Application to detection of residues on apple skin, *Talanta*, Vol. 74, Issue 4, 2008, pp. 741-746.
[31]. Y. Boublik, P. Saint-Aguet, A. Lougarre, M. Arnaud, F. Vilatte, S. Estrada-Mondaca, Acetylcholinesterase engineering for detection of insecticide residues, *Protein Engineering*, Vol. 15, 2002, pp. 43-50.
[32]. S. Sotiropoulou, D. Fournier, N. A. Chaniotakis, Genetically engineered acetylcholinesterase-based biosensor for attomolar detection of dichlorvos, *Biosens Bioelectron*, Vol. 20, Issue 23, 2005, pp. 47-52.
[33]. B. Bucur, D. Fournier, A. Danet, J. L. Marty, Biosensors based on highly sensitive acetylcholinesterases for enhanced carbamate insecticides detection, *Analytica Chimica Acta*, Vol. 562, Issue 1, 2006, pp. 115-121.
[34]. S. Sotiropoulou, N. A. Chaniotakis, Tuning the sol-gel microenvironment for acetylcholinesterase encapsulation, *Biomaterials*, Vol. 26, Issue 33, 2005, pp. 6771-6779.
[35]. K. Gabrovska, T. Nedelcheva, T. Godjevargova, O. Stoilova, N. Manolova, I. Rashkov, Immobilization of acetylcholinesterase on new modified acrylonitrile copolymer membranes, *Journal of Molecular Catalysis B: Enzymatic*, Vol. 55, Issue 3-4, 2008, pp. 169-176.
[36]. S. Upadhyay, G. R. Rao, M. K. Sharma, B. K. Bhattacharya, V. K. Rao, R. Vijayaraghavan, Immobilization of acetylcholineesterase-choline oxidase on a gold-platinum bimetallic nanoparticles modified glassy carbon electrode for the sensitive detection of organophosphate pesticides, carbamates and nerve agents, *Biosensors and Bioelectronics*, Vol. 25, Issue 4, 2009, pp. 832-838.
[37]. M. Shi, J. Xu, S. Zhang, B. H. Liu, J. Kong, A mediator-free screen-printed amperometric biosensor for screening of organophosphorous pesticides with flow-injection analysis (FIA) system, *Talanta*, Vol. 68, Issue 4, 2006, pp. 1089-1095.
[38]. H. Zejli, J. L. H. Cisneros, I. N. Rodriguez, B. Liu, K. R. Temsamani, J. L. Marty, Alumina sol-gel/sonogel-carbon electrode based on acetylcholinesterase for detection of organophosphorous pesticides, *Talanta*, Vol. 77, Issue 1, 2008, pp. 217-221.
[39]. H. Yin, S. Ai, J. Xu, W. Shi, L. Zhu, Amperometric biosensor based on immobilized acetylcholinesterase on gold nanoparticles and silk fibroin modified platinum electrode for detection of methyl paraoxon, carbofuran and phoxim, *Journal of Electroanalytical Chemistry*, Vol. 637, Issue 1-2, 2009, pp. 21-27.

[40]. R. Sinha, M. Ganesana, S. Andreescu, L. Stanciu, AChE biosensor based on zinc oxide sol-gel for the detection of pesticides, *Analytica Chimica Acta*, Vol. 661, Issue 2, 2010, pp. 195-199.

[41]. S. Cosnier, A. Lepellec, Poly(pyrrole-biotin): a new polymer for biomolecule grafting on electrode surfaces, *Electrochimica Acta*, Vol. 44, Issue 11, 1999, pp. 1833-1836.

[42]. T. Wink, S. J. Van Zuilen, A. Bult, W. P. Van Bennkom, Self-assembled monolayers for biosensors, *Analyst*, Vol. 122, Issue 4, 1997, pp. 43R-50R.

[43]. S. Viswanathan, H. Radecka, J. Radecki, Electrochemical biosensor for pesticides based on acetylcholinesterase immobilized on polyaniline deposited on vertically assembled carbon nanotubes wrapped with ssDNA, *Biosensors and Bioelectronics*, Vol. 24, Issue 9, 2009, pp. 2772-2777.

[44]. Y. Qu, Q. Sun, F. Xiao, G. Shi, L. Jin, Layer-by-Layer self-assembled acetylcholinesterase/PAMAM-Au on CNTs modified electrode for sensing pesticides, *Bioelectrochemistry*, Vol. 77, Issue 2, 2010, pp. 139-144.

[45]. S. Zhang, L. G. Shan, Z. R. Tian, Y. Zheng, L. Y. Shi, D. S. Zhang, Study of enzyme biosensor based on carbon nanotubes modified electrode for detection of pesticides residue, *Chinese Chemical Letters*, Vol. 19, Issue 5, 2008, pp. 592-594.

[46]. W. Xue, T. Cui, Thin-film transistor based acetylcholine sensor using self- assembled carbon nanotubes and SiO2 nanoparticles, *Sensors and Actuators B: Chemical*, Vol. 134, Issue 2, 2008, pp. 981-987.

[47]. Z Chen., F. Xi, S. Yang, Q. Wu, X. Lin, Development of a bienzyme system based on sugar–lectin biospecific interactions for amperometric determination of phenols and aromatic amines, *Sensors and Actuators B: Chemical*, Vol. 130, Issue 2, 2008, pp. 900-907.

[48]. X. Sun, X. Y. Wang, Z. Liu, Study on Immobilization Methods of Acetylcholinesterase, *International Journal of Food Engineering*, Vol. 4, Issue 8, 2008, Article 4.

[49]. I. Marinov, Y. Ivanov, K. Gabrovska, T. Godjevargova, Amperometric acetylthiocholine sensor based on acetylcholinesterase immobilized on nanostructured polymer membrane containing gold nanoparticles, *Journal of Molecular Catalysis B: Enzymatic*, Vol. 62, Issue 1, 2010, pp. 66-74.

[50]. Y. Ivanov, I. Marinov, K. Gabrovska, N. Dimcheva, T. Godjevargova, Amperometric biosensor based on a site-specific immobilization of acetylcholinesterase via affinity bonds on a nanostructured polymer membrane with integrated multiwall carbon nanotubes, *Journal of Molecular Catalysis B: Enzymatic*, Vol. 63, Issue 3-4, 2010, pp. 141-148.

[51].T. Shimomura, T. Itoh, T. Sumiya, F. Mizukami, M. Ono, Amperometric biosensor based on enzymes immobilized in hybrid mesoporous membranes for the determination of acetylcholine, *Enzyme and Microbial Technology*, Vol. 45, Issue 6-7, 2009, pp. 443-448.

[52]. X. Sun, X. Y. Wang, W. P. Zhao, Multiwall Carbon Nanotube-based Acetylcholinesterase Biosensor, *Sensor Letters*, Vol. 8, Issue 2, 2010, pp. 247-252.

[53]. N. A. Pchelintsev, A. Vakurov, P. A. Millner, Simultaneous deposition of Prussian Blue and creation of an electrostatic surface for rapid biosensor construction, *Sensors and Actuators B: Chemical*, Vol. 138, Issue 2, 2009, pp. 461-466.

[54]. F. Arduini, F. Ricci, C. S. Tuta, D. Moscone, A. Amine, G. Palleschi, Detection of carbamic and organophosphorous pesticides in water samples using a cholinesterase biosensor based on Prussian Blue-modified screen-printed electrode, Analytica Chimica Acta, Vol. 580, Issue 2, 2006, pp. 155-162.

[55]. S. Laschi, D. Ogończyk, I. Palchetti, M. Mascini, Evaluation of pesticide-induced acetylcholinesterase inhibition by means of disposable carbon-modified electrochemical biosensors, *Enzyme and Microbial Technology*, Vol. 40, Issue 3, 2007, pp. 485-489.

[56]. F. C. Moraes, L. H. Mascaro, S. A. S. Machado, C. M. A. Brett, Direct electrochemical determination of carbaryl using a multi-walled carbon nanotube/cobalt phthalocyanine modified electrode, *Talanta*, Vol. 79, Issue 5, 2009, pp. 1406-1411.

[57]. M. D. Carlo, A. Pepe, M. Sergi, M. Mascini, A. Tarentini, D. Compagnone, Detection of coumaphos in honey using a screening method based on an electrochemical acetylcholinesterase bioassay, *Talanta*, Vol. 81, Issue 1-2, 2010, pp. 76-81.

[58]. M. D. Steinberg, C. R. Lowe, A micropower amperometric potentiostat, *Sensors and Actuators B: Chemical*, Vol. 97, Issue 2-3, 2004, pp. 284-289.

[59]. N. Nagatani, A. Takeuchi, M. A. Hossain, T. Yuhi, T. Endo, K. Kerman, Y. Takamura, E. Tamiya, Rapid and sensitive visual detection of residual pesticides in food using acetylcholinesterase-based disposable membrane chips, *Food Control*, Vol. 18, Issue 8, 2007, pp. 914-920.

[60] A. Hildebrandt, R. Bragós, S. Lacorte, J. L. Marty, Performance of a portable biosensor for the analysis of organophosphorous and carbamate insecticides in water and food, *Sensors and Actuators B: Chemical*, Vol. 133, Issue 1, 2008, pp. 195-201.

[61] K. R. Rogers, Recent advances in biosensor techniques for environmental monitoring, *Analytica Chimica Acta*, Vol. 568, Issue 1-2, 2006, pp. 222-231.

[62]. G. A. Alonso, G. Istamboulie, A. Ramírez-García, T. Noguer, J. L. Marty. R. Muñoz, Artificial neural network implementation in single low-cost chip for the detection of insecticides by modeling of screen-printed enzymatic sensors response, *Computers and Electronics in Agriculture*, Vol. 74, Issue 2, 2010, pp. 223-229.

[63]. P. He, J. Davies, G. Greenway, S. J. Haswell, Measurement of acetylcholinesterase inhibition using bienzymes immobilized monolith micro-reactor with integrated electrochemical detection, *Analytica Chimica Acta*, Vol. 659, Issue 1-2, 2010, pp. 9-14.

[64]. C. Teller, J. Halámek, J. Žeravík, W. F. M. Stöcklein, F. W. Scheller, Development of a bifunctional sensor using haptenized acetylcholinesterase and application for the detection of cocaine and organophosphates, *Biosensors and Bioelectronics*, Vol. 24, Issue 1, 2008, pp. 111-117.

[65]. Y. Wei, Y. Li, Y. Qu, F. Xiao, G. Shi, L. Jin, A novel biosensor based on photoelectro-synergistic catalysis for flow-injection analysis system/ amperometric detection of organophosphorous pesticides, *Analytica Chimica Acta*, Vol. 643, Issue 1-2, 2009, pp. 13-18.

Chapter 9
Quartz Crystal Microbalance DNA Based Biosensors for Diagnosis and Detection: A Review

Thongchai Kaewphinit, Somchai Santiwatanakul, and Kosum Chansiri

9.1. Introduction

Biosensor is a new nanotechnology, which started in 1962, when Clark and Lyons published their results for the first successful application of their biosensor, called "enzyme sensor," to measure the amount of glucose in human blood. It has become an indispensable modern-day analytical tool that offers high performance sensitivity and selectively. Also, it is superior to any other diagnostic device to produce either discrete or continuous digital electronic signals which are proportional to a single analyte or a related group of analytes [1].

First of all, it is important to define the word "biosensor". A biosensor is an analytical measuring tool comprised of a biological element of known molecular properties tightly coupled to a physical transducer responsible for converting the biological signal into quantifiable information, as in Fig. 9.1. Therefore, the biosensor selectivity is induced by the immobilization, in the sensitive area of the detector, of the biological component (enzyme, DNA receptor, antibody, antigen, microorganism, cell, etc.) specific to the target analyte. The molecular recognition then corresponds to the association of the biological element and its target molecule (analyte) through an association such as: enzyme-substrate, antibody-antigen, receptor-hormone, and complementary DNA sequencing, etc. These associations maximize the capacity of the bimolecular to recognize a unique substance among various substances [2].

Thongchai Kaewphinit
Innovative Learning Center, Srinakharinwirot University, Sukhumvit 23,
Bangkok 10110, Thailand

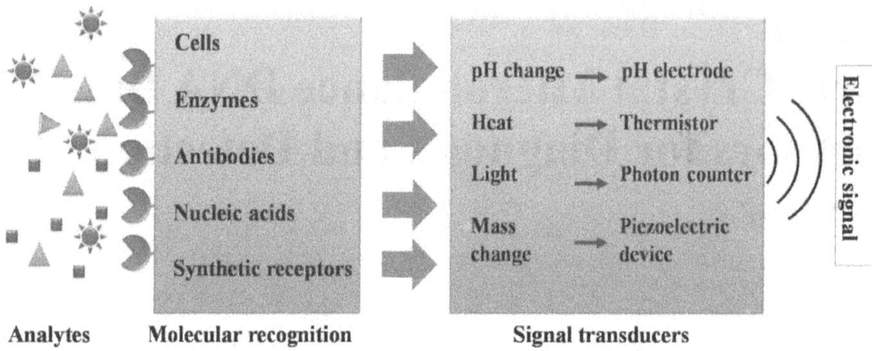

Fig. 9.1. Principle of biosensing operation.

The combinations of recognition-transducer systems are numerous and this explains many definitions and nomenclatures of these types of sensors. The main methods of transduction that are the most current and well developed, from both a fundamental and experimental point of view, are: electrochemical [3], optical [4], and piezoelectric [5, 6]. Besides this classification of biosensors, which is based on the energetic type of transduction method, it is also interesting to cite their classification based on the chemical nature of the alliance between the biological element and the target molecule. The resulting chemical transduction may be governed by:

- The transformation of the target analyte by one (or several) biomolecular recognition processes involving the biological component fixed on the sensor's sensitive area. This is the case with biosensors containing enzymes, cells or biological tissues.
- A specific interaction between biological components fixed in the sensor's sensitive area and the target analyte, resulting in a particular bioaffinity between them and their binding. This is the case with antibody-antigen, complementary DNA sequences, etc [1].

9.2. Quartz Crystal Microbalance Biosensor

9.2.1. Background

Piezoelectric (PZ) biosensor or Quartz Crystal Microbalance is a mass sensitive capability of the measuring and label-free detection of the molecular level. An applied AC-potential causes the quartz crystal to vibrate at a resonance frequency. As molecules interacting on the crystal bind to the surface, the vibration frequency changes are recorded on a frequency counter. Therefore, the quartz crystal can be used to detect mass change due to its piezoelectric property.

9.2.2. Piezoelectricity

The term "piezoelectric" is derived from the Greek word *piezen* meaning "to press" [7]. The first described piezoelectricity occurred in the year 1880 when Pierre and Jacques

Curie used the term to describe the generation of electrical charges on opposing surfaces of a solid material upon deformation (torsion, pressure, bending, etc.) along an appropriate direction. Conversely, mechanical deformation of the material induced by an external electric field is called the *converse piezoelectric effect* [8]. These findings were the discovery of the piezoelectric effect. Piezoelectricity did not receive lot of interest in the beginning and a more detailed study of piezoelectricity was not started until 1917, when it was shown that quartz crystals could be used as transducers and receivers of ultrasound in water. In 1919, several devices of everyday interest based on the piezoelectricity of Rochelle salt were described i.e. loudspeakers, microphones, and sound pick-ups. In 1921, the first quartz crystal controlled oscillator was described. These first quartz crystal controlled oscillators were based on X-cut crystals, which have the drawback of being very temperature sensitive. Therefore, the X-cut crystals are nowadays used in applications where the large temperature coefficient is of little importance, such as transducers in space sonar.

The dominance of the quartz crystal for all kinds of frequency control applications started in 1934, when the AT-cut quartz crystal was introduced. The advantage with the AT-cut quartz crystal is that it has nearly zero frequency drift with temperature around room temperature. From the very beginning of using quartz crystal resonators as frequency control elements, it was common to increase the frequency of the resonator by drawing pencil marks on the electrodes, or decreasing the frequency by rubbing off some electrode material with an eraser. The understanding of this mass induced frequency shift was only known on a qualitative basis. However, in 1959, Sauerbrey published a paper that showed that the frequency shift of a quartz crystal resonator was directly proportional to the added mass [9]. A prerequisite for the occurrence of piezoelectricity in crystals is an inversion center. Twenty-one point groups fulfill this requirement but only 20 classes have a nonzero piezoelectric. Although a large number of crystals exhibit piezoelectricity, only quartz provides the unique combination of mechanical, electrical, chemical, and thermal properties, which has led to its commercial significance [8].

The Curies employed what they called a *quartz electric balance*, which was in turn the first application of a piezoelectric device as a chemical sensor. Nowadays, the so-termed *quartz crystal microbalance* technique is well established in non-biological applications. The core component of the device is a thin quartz disc, which is sandwiched between two evaporated metal electrodes, and is commonly referred to as *thickness shear mode resonator* (TSM resonator) or *bulk acoustic wave sensor*. As this quartz crystal is piezoelectric in nature, an oscillating potential difference between the surface electrodes leads to corresponding shear displacements of the quartz disk. This mechanical oscillation responds very sensitively to any changes that occur at the crystal surfaces. It was Sauerbrey who first established in 1959 that the resonance frequency of such a quartz resonator alters linearly when a foreign mass is deposited on the quartz surface in the air or in a vacuum. From these resonance frequency readings, it was possible to detect mass deposition on the quartz surface in the sub-nanogram regime, and accordingly, the device was named a *microbalance* [8]. The system functions as a sensitive QCM, and with the improved sensitivity of recent devices resulted in the alternative name of "nanobalance". In this way, the amount of molecules bound on the sensitive area of electrodes can be easily quantitatively measured as a decrease of the resonant frequency [10].

At present, quartz crystal microbalance is currently used commercially for frequency control in communications equipment, selective filters in electronic equipment, the measurement of the temperature and the dew point of gases and in very accurate clocks. The first demonstrated quartz crystals, sometimes called bulk acoustic wave (BAW) devices or thickness shear mode devices, could be used as sorption detectors (mass-sensitive device) by coating the crystals with liquid GC stationary phases; subsequently, they have been used in a wide range of chemical sensor applications.

The properties of a QCM depend on the plane, which can be cut in different angles, and which gives different quartz crystal types their specific properties. The AT-cut is normally employed, which has an orientation of approximately 35° to the z-axis or optical axis [11]. It is known to have a particularly low temperature coefficient. The sensor, a thin wafer about one tenth of a millimeter, is cut from the quartz, polished and then electrodes deposited on the opposite faces of the quartz. The electrodes are deposited on the quartz wafer by evaporation or sputtering, with Aluminum, Nickel, Silver, Titanium often being used. However, for sensor application and higher tolerance devices, gold and silver are preferred (Fig. 9.2.). The frequency of oscillation of the crystal is typically 10 to 30 MHz [12, 13].

Hence, one can describe the QCM to be an ultra-sensitive mass sensor. The heart of the QCM in Fig. 9.2 is the piezoelectric AT-cut quartz crystal sandwiched between a pair of electrodes. When the electrodes are connected to an oscillator and an AC voltage is applied over the electrodes, the quartz crystal starts to oscillate at its resonance frequency due to the piezoelectric effect. This oscillation is generally very stable due to the high quality of the oscillation (high Q factor) [8].

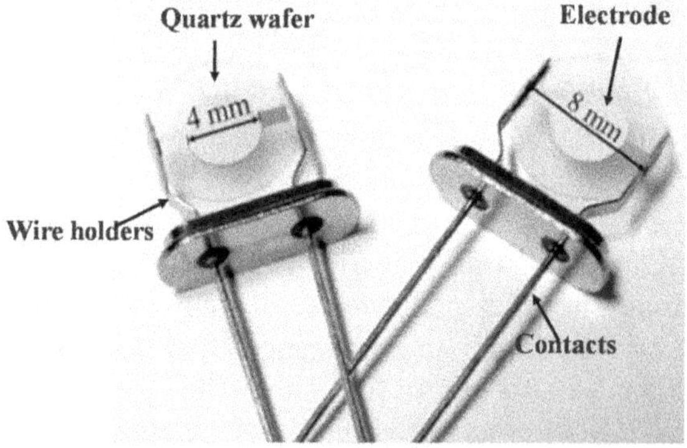

Fig. 9.2. Quartz crystal microbalance structures.

9.2.3. Relationship between Added Mass and Frequency Shift

The resonant frequency of the quartz crystal depends on the physical dimensions of the quartz plate and the thickness of the electrode deposited. For use as a piezoelectric detector, only AT-or BT-cut quartz plates are useful. AT and BT cut refer to the orientation of the plate with respect to the crystal structure (35°15' and -49°00', respectively). The AT-cut crystal is more stable than most other piezoelectric cuts and has a temperature coefficient of about 1 ppm/°C over a temperature range of 10-50 °C [14].

The thickness of the quartz plate θ is related to the fundamental frequency by

$$\theta = \frac{N}{f}, \tag{9.1}$$

where N is the frequency constant. For an AT-cut quartz crystal N is 1.67×10^5 cm• Hz.

The thickness of the quartz plate θ is also related to its mass by the equation

$$\theta = \frac{m}{A\rho_q}, \tag{9.2}$$

where ρ_q is the density of quartz (2.65 g/cm^3), m is the mass (g) of oscillating quartz, and A is the area (cm^3) of the quartz plate undergoing oscillation (between the two electrodes on opposite faces).

Combining Equations (9.1) and (9.2) yields

$$f = \frac{\rho_q NA}{m}, \tag{9.3}$$

where f is the fundamental oscillation frequency (MHz). If an added mass Δm is attached uniformly to one of the electrodes of area A, then

$$\Delta F + f = \frac{\rho_q NA}{m + \Delta m} \tag{9.4}$$

Rearrangement of Equation (9.4) provides

$$\Delta F = -\frac{f^2}{\rho_q NA}\frac{\Delta m}{1 + \Delta m / m} \tag{9.5}$$

Substituting numerical values for N and ρ_q in Equation (9.5) yields

$$\Delta F = -2.26x10^6 f^2 \frac{\Delta m}{A(m+\Delta m)} \tag{9.6}$$

When Δm is very small compared to m, interfacial mass changes Δm are related in a simple manner to changes in oscillation frequency changes ΔF :

$$\Delta F = -2.26x10^6 f^2 \frac{\Delta m}{A} \tag{9.7}$$

Equation (9.7) describes the effect of vacuum deposition of metals on the quartz crystal. The piezoelectric crystal is used as a sensing device for the measurement of the thickness of thin films. Sauerbrey demonstrated that the shift in the resonant frequency of the crystal was proportional to the mass deposited to within ±2 % [14] or the mass change on quartz crystal in gas phase [15, 16] as shown below.

$$\Delta F = \frac{-2 f_0^2 \Delta m}{A\sqrt{\mu_q \rho_q}}, \tag{9.8}$$

where ΔF = measured frequency shift (Hz);
f_0^2 = the fundamental resonant frequency of the crystal (Hz);
Δm = mass change (g);
A = area of electrode surface (cm^2);
μ_q = shear modules of quartz crystal = 2.947×10^{11} g/$cm^2 \times s^2$
ρ_q = density of quartz crystal = 2.648 g/cm^3

It is obvious from Equation (9.7) that when ΔF is very small in relation to f (0.17 % or less) ΔF is a linear function of added mass Δm. The Sauerbrey equation is appropriate for many situations; however, its applicability is only good for thin and rigid layers, that is, the mass added should not experience any shear deformation during oscillation. This equation predicts that a commercially available 9 MHz crystal has a sensitivity of about 400 Hz/μg or a 15 MHz crystal has a sensitivity of 2600 Hz/μg [17].

The Sauerbrey equation assumes a uniform distribution of mass on the entire electrode portion of an AT-cut quartz crystal. Mass sensitivity decreases monotonically with the radius, in a Gaussian manner becoming negligible at and beyond the electrode boundary [18, 19]. Another assumption of this equation is that the mass added or lost at the crystal surface does not experience any deformation during the oscillation. This is true for thin and rigid layers. For thicker, less rigid layers, a more complex theory is necessary [20]. It had previously been thought that a piezoelectric quartz crystal could not oscillate in solutions and the crystal had been used only in the vapor phase. Improvements in electrical circuits and contact of only one surface of the crystal with the solution allow it to oscillate stably in solution [21].

There are situations where the Sauerbrey equation does not hold, for example, when the added mass is a) Not rigidly deposited on the electrode surface; b) Slips on the surface or c) Not deposited evenly on the electrode [8].

Another crucial quantity used to describe mechanical and electrical resonators is the quality factor Q that provides information about the energy dissipation in relation to the energy that is stored in the oscillation per cycle Q is defined as:

$$Q = \omega \frac{L}{R} = \frac{1}{\omega RC}$$

The inverse of the quality factor is a measure of energy dissipation and is thus called the *dissipation factor* D =1/Q.

Piezoelectric quartz crystal biosensors in gas phase detections are advantageous in that the frequency signal can in some cases be interpreted through the use of the Sauerbrey equation relation, although, the sensitivity method may result in errors due to humidity, hydration, and solvent retention. The measurement error is due to the crystal ceasing to oscillate when submerged in solution. To sense liquid phase analytes, either the sample is converted into a gas or a tedious dipping procedure is used [22]. These problems have been addressed by developing oscillator circuits that allow for crystal immersion in a liquid [23] or through the use of specially designed quartz crystal flow through and/or flow cell reaction where solution contacts only one side of crystal surface as shown in Fig. 9.3. In 1980, the new quartz crystal microbalance in liquid solution phase was developed where one crystal face was exposed to a flowing organic solution [24], but the theoretical foundation for using the piezoelectric effect was first pioneered by Kanazawa and Gordon [25] in 1985. They described the frequency change response in liquid phases in terms of the physical parameters of the quartz crystal and analyte in liquid by considering the coupling of quartz crystal shear wave to a dampened shear wave propagating into the liquid [26, 27].

The frequency shift (ΔF) related to changes in viscosity and density of the aqueous solution with the frequency shift of a quartz crystal that contacted with the Newtonian liquid (density (ρ_L), viscosity (η_L)) is expressed by the Kanazawa's equation [15, 28] as shown below.

$$\Delta F = f_0^{3/2} \left(\frac{\rho_L \eta_L}{\pi \mu_q \rho_q} \right)^{1/2},$$

where ΔF = measured frequency shift (Hz);

$f_0^{3/2}$ = resonant frequency of the unloaded quartz crystal (Hz);

ρ_L = density of liquid in contact with the quartz crystal;

η_L = viscosity of liquid in contact with the quartz crystal;

μ_q = shear modulus of quartz crystal = 2.947×10^{11} g/cm^2×s^2

ρ_q = density of quartz crystal = 2.648 g/cm^3

Liquid phase setup measurement for affinity studies consisted of the piezoelectric biosensor, driving electronics and a flow-through system. The quartz crystal was sandwiched between two soft rubber o-rings because reasonable force should be applied to prevent damage of thin and fragile quartz crystals shown in Fig. 9.3.

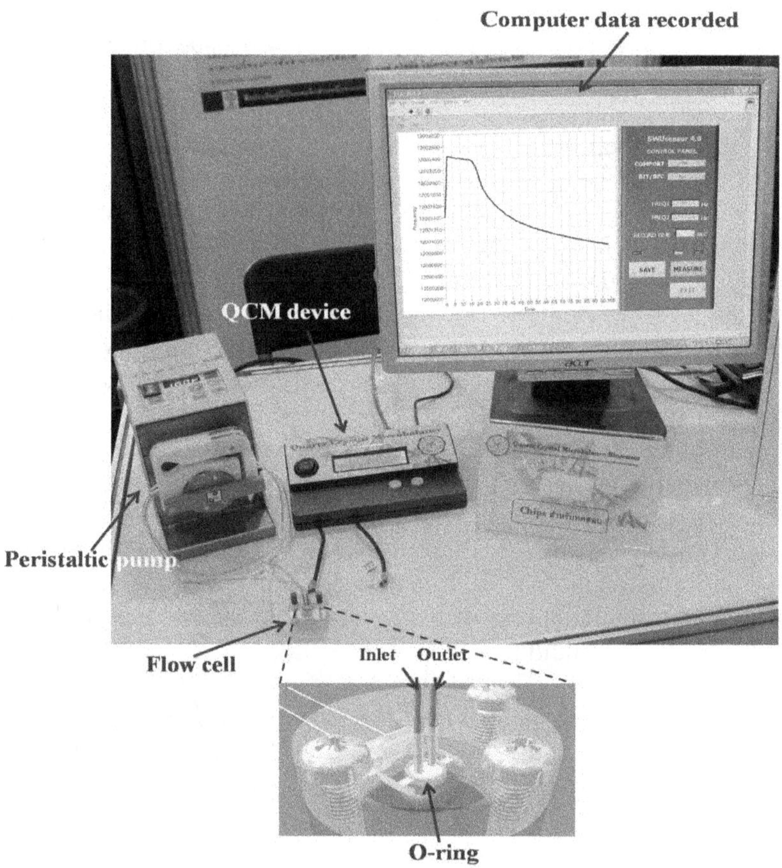

Fig. 9.3. QCM DNA based biosensor system in real time component 4 parts. (1) data recorded by a computer with the aid of LabVIEW interface software; (2) QCM device as oscillation counting device was used for measuring the frequency shift of the quartz crystal after the addition of immobilization material; (3) flow cell detection and (4) peristaltic pump.

The modified quartz crystals were fixed in a flow-through cell between two soft rubber o-rings in which only one side of the crystal was in contact with the flowing solution. The electrodes of the crystal were connected to a gate oscillating circuit based on the integrated TTL oscillator driver 74LS320. Consequently, the oscillating circuit driving the quartz crystal should provide enough energy to the crystal for smooth oscillation in the presence of liquid. The integrated form of this oscillator will provide much higher energy to the quartz crystal resulting in improved performance under variable conditions. Better stability was obtained by using carefully designed lever oscillators consisting of individual transistors. The oscillator circuit was chosen from several conditions in the laboratory by focusing on high frequency stability and IC suppliers in Thailand.

The output frequency was measured by AVR microcontroller to computer. The system used LabVIEW software (for Windows) interface for data storage. The delay interval between two data points was 10 seconds, whereas achieved resolution of the frequency from the counter was 1 Hz. The system has the advantage of being an inexpensive sensor with digital output that can be operated online and in real time, as well cheap, easy to use, and rapid for detection. This biosensor can avoid sensitivity to errors due to hydration, humidity, time consumption, solvent retention, and expect to be analytical routine procedure.

9.3. DNA Analysis

Since the completion of the human genome mapping, interest has shifted to the study of the molecular functions of genes and how they might lead to a diseased state. Consequently, DNA biosensors have rapidly been applied to fields such as gene identification, genetic expression analysis, DNA sequencing, and clinical diagnostics. The detection and analysis of point mutations, and single nucleotide polymorphisms (SNP), is one of the main applications of DNA biosensors. However, in order to detect such mutations, an optical fluorescent label or redox label must attach to target DNAs directly or indirectly, which requires the extra label binding processes and the use of hazardous materials and expensive lab equipment. Also, it takes a relatively long time to analyze the results, and it is difficult to detect quantitatively the absolute amount of hybridization with those labeling, because one has to normalize the data [29].

Molecular detection has shown a great potential for rapid identification of diseases and for food and environmental monitoring. After the recent successes in sequencing the human genome, the detection of specific DNA sequences in biological samples has been playing a fundamental role in genetic diagnostics and in the detection of pathogens in cells. The large demand for low-cost genetic assays has lead to the development of portable and easy-to-use biosensors. These systems should be able to perform the analysis in a very short time and with a very limited amount of specimen. Micro-fabricated structures, based on micro and nano-technology, can satisfy these requirements and also allow a high degree of parallelism and sensitivity.

Numerous genetic diseases are known, which are caused by single base mutations or deletions in the human genome. Methods capable of identifying those aberrations from small sample volumes with high specificity and sensitivity have great potential in the diagnosis and therapy of genetic diseases. The public is also interested in fast methods for the control of food products regarding genetically engineered ingredients. In fundamental research, the sequencing of the genomes of many organisms currently produces a wealth of DNA material that has to be deciphered. For these reasons, much effort is currently made in order to improve existing surface-based DNA detection systems and to develop novel detection schemes, which are compatible with microarray formats. The improvement of available detection schemes requires a detailed knowledge of the DNA hybridization reaction. Surface coupling chemistries and the control of the lateral density of DNA probes are crucial for the efficiency of those methods. Analytical approaches that have been proven efficient for this objective include methods that utilize either fluorescence for end-point measurements, evanescent field sensors for the analysis of reaction kinetics, and piezoelectric and electrochemical biosensors.

9.3.1. Structure and Stability of DNA

The monomers of DNA, which are referred to nucleotides (nt), consist of three subunits: a deoxyribose sugar, a base and a phosphate group. Linking the 3' and 5' OH of the sugar units via phosphodiester bonds creates a DNA strand. Therefore, the ends of a DNA strand are designated as 3' and 5'-terminus. The C1 atom of the ribose is attached to one of the four naturally occurring bases, the purines, adenine and guanine, or the pyrimidines, cytosine and thymine. In single-stranded (ss) DNA, the distance between two successive phosphates is about 0.7 nm. The factors ruling DNA stability are manifold. In a DNA hybridization reaction, two complementary single strands of DNA become oriented in an anti-parallel manner to form double-stranded (ds) DNA via Watson Crick base pairing like depicted in Fig. 9.4 [30].

Base pairing is not solely driven by H-bonding between the complementary GC and AT base pairs, but also by nearest neighbor dependent base stacking interactions and the hydrophobic effect. DNA profits entropically from burying the hydrophobic bases in the inner part of the duplex by contacting the aqueous environment primarily via the charged phosphate-sugar backbone. A DNA duplex can adopt several conformations; the B-form is by far the most common one. The wound double helix is not perfectly symmetrical: it forms a 0.6 nm wide minor and a 1.2 nm wide major groove.

The helix has a pitch height of 10.4 base pairs (bp) with a spacing of 0.34 nm between two stacked bases. The various factors stabilizing B-DNA also govern its local rigidity. The persistence length of ds-DNA is 45 nm, while ss-DNA is more flexible with a persistence length of ~1 nm at 150 mM monovalent salt. Having a line charge density of 2 electrons per base pair, electrostatic forces contribute substantially to stability and flexibility of the DNA chain.

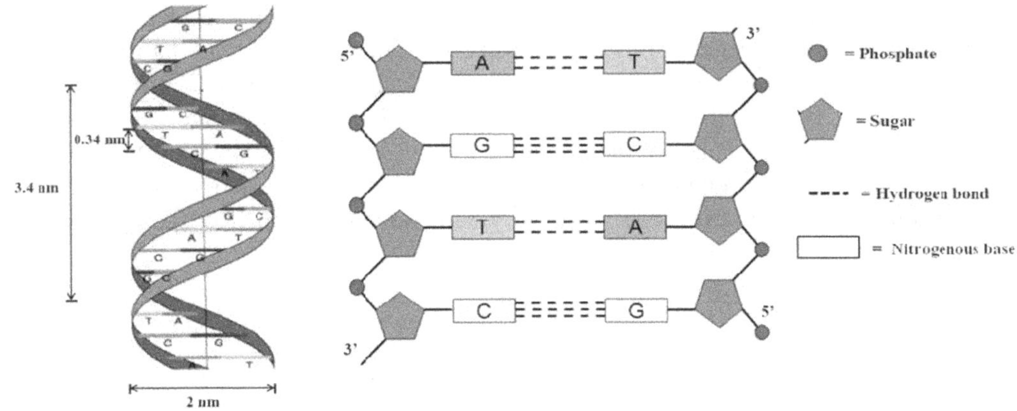

Fig. 9.4. Structure of DNA.

Since DNA stability is ruled by many factors, it is difficult to predict the hybridization behavior of a certain sequence only on the basis of the calculated gain in Gibbs free energy. This is particularly true for hybridization reactions on metal surfaces, because the planar electric double layer formed at the metal/water interface superimposes with the electrical field of the immobilized DNA probes in a complex manner. Especially at high surface densities of DNA probes, steric hindrances and electrostatic repulsions come into play, which are not accounted for by conventional computer routines calculating ΔG. However, the general features important for DNA stability are summarized:

- Content of GC base pairs: A high content stabilizes DNA, since GC base pairs are thermodynamically more stable than AT base pairs. In genomic DNA, the GC and AT content is identical, but the effect can be dramatic for short strands like applied in biosensors;
- Length of the DNA sequence: The melting temperature increases with length of the sequence;
- Nearest neighbor and end effects: The extent of base stacking interactions does not only depend on the number of neighbors, but also on their identity. The following trend has been reported for the influence of the two directly neighbored bases: GC>CG>GG>GA~GT~CA>CT>AA>AT>TA frayed ends contribute an additional term to the stability [31];
- Presence of mismatched bases: As rule of the thumb, the melting temperature of ds-DNA is reduced by 1 % per introduced base mismatch for DNA strands < 100 bp. The destabilizing effect decreases if the overall length of DNA increases;
- Hairpin formation and self-complementarity: If DNA probes form stable hairpin structures, opening of the secondary structure consumes energy. Hairpin stability depends on the identity and number of bases in the loop region as well as on the structure of the stem region. The minimum number of bases required to form a loop region is 4-5. The stability of the loop bases follows the trend: T-loop>C-loop>G-loop>A-loop;

- Buffer conditions: Destabilization of the DNA structure can be intentionally achieved by exposure to H-bond disrupting agents like urea or formamide or chaotropic salts that interrupt base stacking interactions. The pH and ionic strength are crucial to hybridization, because duplex formation requires shielding of the charged phosphate backbone. In order to enhance the potential for single base mismatch discrimination, DNA hybridization can be carried out under destabilizing, so called "low stringency" conditions that enable exclusively the fully complementary target DNA to hybridize [32].

9.3.2. DNA-based Biosensors

DNA sensors are usually in the form of electrodes, chips, and crystals; hence, hybridization on a sensory surface is a solid-phase reaction. Solution hybridization is more rapid than hybridization on solid-supports, but unless the assay is a homogeneous assay, a separation step is required before final detection. Solid phase or filter hybridization is the longest established and the most often used.

The kinetics of nucleic acid hybridization at a solid substrate-solution interface is still not well understood due to the lack of suitable methods for the continuous measurement of the hybridization process. The process is also difficult to predict from theoretical considerations, partly because the exact concentration of the immobilized nucleic acid and its availability for hybridization are unknown. Furthermore, the actual mechanism of strand association in solid-support hybridization remains obscure [33].

The kinetics and mechanism of nucleic acid single strand hybridization are widely studied only in cases where both strands are in free solution. The hybridization reaction between DNA probe and target oligonucleotide can be expressed by the following equation:

$$[probe] + [target] \longrightarrow [hybrid] \tag{9.9}$$

The association constant (K_a) is defined as:

$$K_a = \frac{[hybrid]}{[target][probe]} \tag{9.10}$$

$$= \frac{[V]}{[target][V_{max} - V]} \tag{9.11}$$

The initial reaction rate of the hybridization reaction between the immobilization and detected DNA is represented as V and V_{max} which indicates the maximum initial reaction rate of the hybridization reaction. Because the amount of immobilization is very small, the hybridized DNA was also very small. Thus the equilibrium concentration can be considered as the initial concentration $[t \arg et]_0$. So the above equation can be written as:

$$\frac{[t \arg et]_0}{V} = \frac{[tagrget]_0}{V_{\max}} + \frac{1}{V_{\max} K_a} \tag{9.12}$$

Reciprocal plots between $[t \arg et]_0 / V$ and $[t \arg et]$ give a simple straight line. The association constant and the maximum initial reaction rate are calculated from the slope and the linear correlation respectively [34, 35].

9.3.3. DNA Biosensors based on QCM Detection

Piezoelectric materials are used in a variety of configurations as microgavimetric detectors and their general theory and use are well reviewed. They offer an attractive, near-universal mode of transducing the biomolecular recognition event, but only if the changes in detector mass that accompany analyst binding are sufficiently large. Resolution of mass changes to $<1\times10^{-9}$ g•cm^{-2} is possible in liquid media, and at least for high molecular mass substances, this provides for a viable transduction strategy. DNA biosensors using piezoelectric crystals are developed more extensively compared to other types of transducers.

Piezoelectric transducers also offer the advantages of a solid-state construction, chemical inertness, durability, and, ultimately, the possibility of low cost mass production. Attention to data is mainly on AT-cut quartz crystals with the piezoelectric material that can function in a "microbalance" mode. In order to carry out a measurement, an external voltage is used to deform the quartz crystal plate so that the relative motion between two parallel crystal surfaces, the crystal relaxation and oscillation at the resonant frequency, is maintained by means of an appropriate external circuit. The change in frequency (ΔF) resulting from any added mass (Δm) to the device can be described by the Sauerbrey equation. Thus, at least to a first approximation, the change in mass per unit area of the crystal is directly proportional to the change in frequency [33].

Immobilization is the important step in fabricating a biosensor device. For example, probe immobilization is a fundamental step in DNA based QCM biosensor development. Often, the detection limits and, in general, the analytical performance of the biosensor can be improved by optimizing the immobilization procedure on the quartz surface [36, 37]. Actually, the limitation of QCM devices is nonspecific adsorption of molecules present in real matrices, since QCM is a mass sensor and any molecule is able to bind or to be adsorbed on the surface, which is a potential interference. Moreover, a receptor, which is the DNA, must be attached to the solid support, retaining native conformation and binding activity. This attachment must be stable over the course of a binding assay and, in addition, sufficient binding sites must be presented to the solution phase to interact with the analyte. Many coupling strategies utilize a chemical linker layer between the sensor base (e.g., the gold layer) and the biological component to achieve these ends. Functionalized alkane thiols form stable self-assembled layers on planar surfaces [38] and act as ideal linkers [39]. Moreover, many published papers showed that immobilization techniques based on direct adsorption or on protein A coating resulted in appropriate sensor signals, but only crosslinker procedures using thiols or the interaction between avidin and biotinylated molecules provided a long sensor lifetime and an increased stability against degradation during the regeneration process [40].

The quartz crystal is fabricated by using a 12 MHz AT cut quartz crystal slab with a layer of a gold electrode on each side (0.125 cm^2 in area on each side). The gold electrode surface of quartz crystals was cleaned by using a Piranha solution (30 % H_2O_2: H_2SO_4 = 1:3) for 30 seconds. The crystals were then thoroughly washed with distilled water and used immediately afterward. There are different methods for immobilization between thiol or biotin modification of DNA probe on QCM electrodes by chemical bonding to form monolayer DNA sensing. Initially, the thiol-modified oligonucleotide probe on the gold electrode surface was immobilized by passing 4-(2-hydroxyethyl)-1-piperazine-ethanesulfonic acid (HEPES) buffer (0.05 M HEPES, 0.2 M NaCl, pH 7.5) through the flow system. The inlet/outlet pumps were controlled to deliver 50 μL of solution to the detection cell. After the baseline was stabilized, a thiol-modified oligonucleotide probe in buffer (1M KH_2PO_4, pH 3.8) was added into the cell through the reagent reservoir and the pump was stopped for 20 minutes. Then 1 mM of 6-mercapto- hexanol (MCH) was added for 30 minutes to help reduce nonspecific binding of the thiol-modified oligonucleotide to the gold surface. After the immobilization, the pump was restarted again. The detection cell was washed using the buffer for 5 minutes prior to injection of the DNA target (Fig. 9.5a) [6].

Another method for immobilizing the biotin-modified oligonucleotide probe on the gold electrode surface is by passing a HEPES buffer through the flow system. The inlet/outlet pumps were controlled to deliver 50 μL of solution to the detection cell. After the baseline was stabilized with a 1 mM 3-mercaptopropionic acid (MPA) aqueous solution for 1 hour, the cell was washed with buffer. To activate the monolayer, 1-ethyl-3 (3-dimethylaminopropil) carbodiimide ethanolic solution (EDC) was placed on the surface. Then *N*-hydroxysuccinimide aqueous solution (NHS) was immediately added and left to react on the surface of gold electrode to form an MPA monolayer for 30 minutes. The surface was then rinsed with the buffer. An aliquot of avidin HEPES buffer was placed on the electrode surface for at least 1 hour [5].

Then, the quartz crystal was exposed to ethanolamine for 30 minutes and rinsed with distilled water and the HEPES buffer. The DNA biotinylated probes were placed over the gold electrode surface for 20 minutes. Then, the probes were rinsed with immobilization buffer. In all experiments, at this stage, the quartz crystal was ready for the hybridization (Fig. 9.5b).

DNA sequences of even a few hundred base pairs have considerable molecular weights. It seems likely, therefore, that the mass change associated with DNA hybridization may be detectable by employing piezoelectric devices as in the study by Fawcett on immobilizing ssDNA onto quartz crystal and detecting the mass change after hybridization [41]. Nucleic acid strands are covalently attached to the polymer-modified surface of a piezoelectric crystal. When immobilized probe strands are melted and interact with complementary target strands in solution, the association of probe and target to form duplexes results in an increase in mass that is detectable as a decrease of several hundred hertz in the crystal's resonance frequency, relative to control crystals on which non-complementary strands are attached. Furthermore, the crystals used are inexpensive and the results are quantitative as well as qualitative. The solution is similar, but the frequency measurements are performed in the dry state. The limit of detection of the unoptimized technique was 2 ng of on-surface added nucleic acid [34].

Hybridization in solution is believed to be a two-step process involving nucleation and zippering up. Nucleation is the rate-limiting step, and a second order reaction equation can describe the process. The nature of the hybridization reaction on solid surfaces is assumed to approximate closely to that of hybridization in solution [42]. However, the rate of solid-phase hybridization is only about a tenth to a hundredth of that in solution. Although not systematically examined, it is suggested that efficient hybridization to DNA attached to solid supports can be impeded by several related phenomena. For example, the immobilized DNA molecules may link to solid surface at several points along the DNA chains; hence, some of the attached DNA may not be accessible for hybridization [39].

The hybridization with the oligonucleotides target was performed by using oligonucleotides target placed on the surface of a gold electrode, which had been previously immobilized with a DNA probe (thiol, thiol/MCH, biotin probe). After another interval of 30 minutes (pump stopping), the cell was washed with a buffer for 5 minutes. At the same time, the DNA target was driven by a pump to pass through and denatured (95 °C) for 5 min, and then rapidly passed through the cold down (50 °C) for 1 min prior to a rapid transfer to the detection cell. The flow rate was maintained with a slow flow at 30 μL/min of denaturing fragmentation of genomic DNA by restriction enzyme and the solution was transferred between reactors to deliver the samples to the detection cell. After the denatured fragmentation of genomic DNA was injected into the piezoelectric biosensor for 15 - 20 minutes, the cell was washed by the buffer again, prior to the recording of frequency shift after a stable frequency was reached. For the regeneration of the probe modified electrode, 1 mM HCl was added to the electrode surface for 30 sec, followed by a thorough washing using the buffer solution (Fig. 9.5).

DNA probe modification was linked to TTTTTT or polymer to decrease the steric hindrance upon protected attachment onto the surface of the quartz crystal and to assess the hybridization to the complementary DNA target. The frequency shift was decreased after the immobilization process. However, the steric hindrance could be increased in the hybridization efficiency by adding 6 dT to the 5' end of the probes. This confirmed the studies that reported the poly dT at 5' end of probes might reduce steric hindrance in three-dimensional space, and that the addition of the spacer during DNA hybridization caused the increase in molecular collisions [43, 44]. It is also proposed that during DNA target sequence hybridization, spacers have been shown to reduce steric interference, making the probe end closet to the surface of the device more accessible [44].

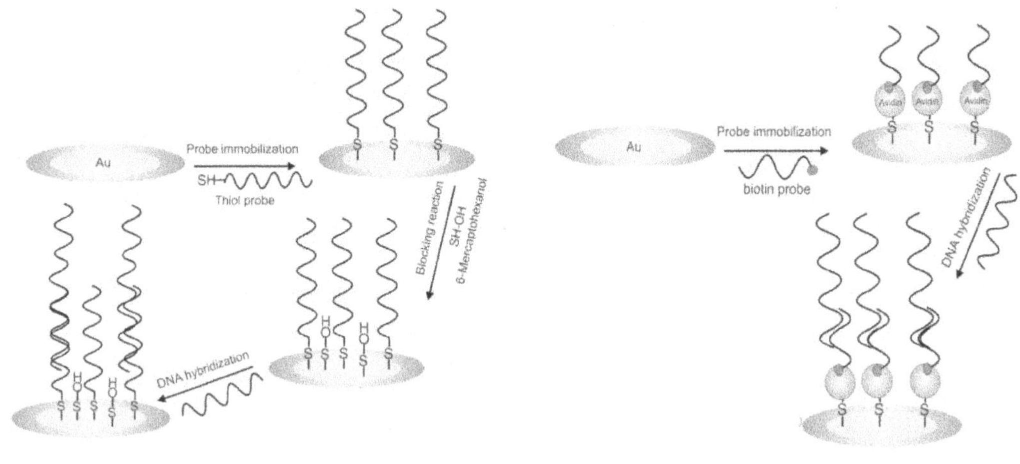

A. **B.**

Fig. 9.5. Schematic diagram of immobilization and hybridization method. (a) Thiol-modified oligonucleotide probe and Thiol-modified oligonucleotide probe/MCH immobilization, and (b) Biotin-modified oligonucleotide probe immobilization.

The shelf-life of both thiol-modified oligonucleide probe/MCH and biotin-modified oligonucleide probe after storage at 4 ^{o}C were 160 days and 10 days, respectively. However, the shelf-life of probe on DNA based QCM biosensor reported by Caruso and coworkers [45] and Tombelli and colleagues [46] was several weeks at 4 •C without losing its activity. Therefore, the thiol-modified oligonucleotide probe/MCH probe gave the most stable frequency shift after hybridization with the complementary DNA target.
Normally, thermal denaturation was used to denature the PCR amplified target DNA. However, in non-amplified genomic DNA, the blocking oligonucleotides were added in this denaturation step. This blocking could prevent the reannealing of ssDNA and increase the efficiency of DNA target hybridization. Moreover, the blocking oligonucleotides were used to develop a DNA target sequence hybridization of optimum concentration for target capture and to enhance the signal with protein, polymer, or gold nanoparticles (AuNPs). The frequency shift was decreased after the interaction process (Fig. 9.6).

However in non-amplified genomic DNA, the blocking oligonucleotides were added in this denaturation step. This blocking can prevent the reannealing of single-stranded DNA to increase the efficiency of DNA target hybridization (Fig. 9.7). Therefore, the blocking oligonucleotides were used for improving the DNA target hybridization. The thermal plus blocking oligonucleotides gave the frequency change by decreasing the resonant frequency higher than only thermal denaturation [47, 48, 5, 6].

The signal enhancement used 20 nm diameter of gold nanoparticles as optimal for the surface immobilization of quartz crystal and the consequent sensitivity improvement, because the hybridization had a highest maximum value when the average diameter of nanoparticles was 20 nm and then decreased with the increasing of particle size [49].

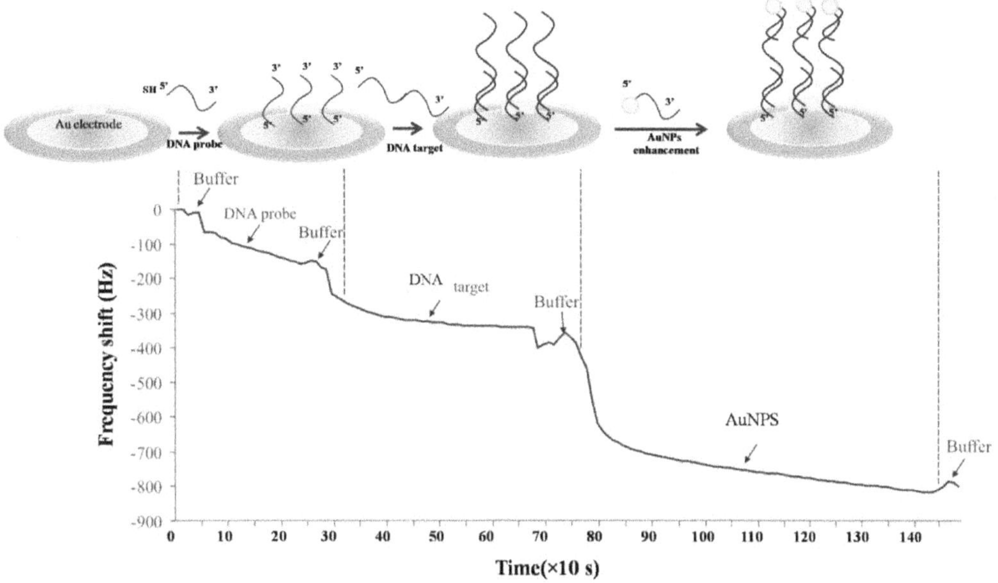

Fig. 9.6. The frequency shift was observed after probe immobilization, DNA target hybridization, and AuNPs as signal enhancement onto the gold surface, respectively.

Therefore, the frequency shift of the sensor was decreased, accompanied by an increase of nanoparticle size applied in the static solution [49 ,50]. Particles with an average diameter of 20 nm afforded the best hybridization rate, probably, as previously reported [49]. In addition, AuNPs modified blocking oligonuclotides deposited onto the piezoelectric sensor were dependant on the DNA hybridization efficiency between the specific probe and target sequences [51- 53].

The AuNPs could be attributed to the steric hindrance effect, where the larger nanoparticles could not move as independently as the smaller nanoparticles, and the larger nanoparticles connected to the DNA target hindered the approach of further nanoparticles [50].

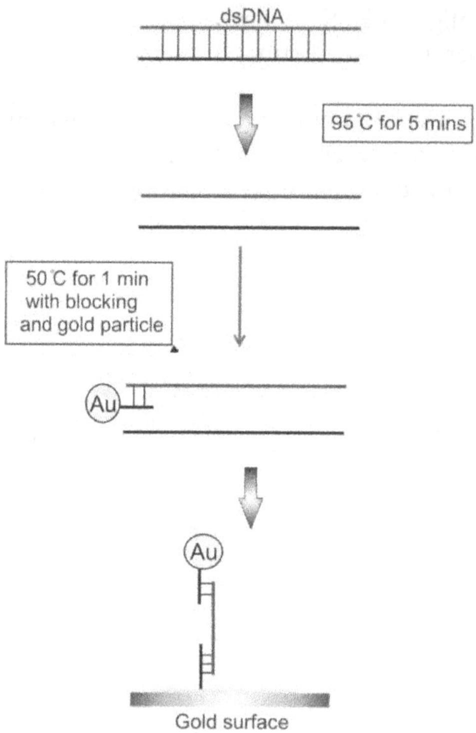

Fig. 9.7. Schematic diagram of denaturation method.

9.3.4. DNA Based QCM for Clinical Diagnosis

At present, the quartz crystal has been applied in wide areas. This device could show both qualitative and quantitative results. The major application of this device involves biological recognition, for example, of nucleic acids. Therefore, there are many type of QCM based biosensors categorized by type of biological elements. Moreover, a QCM biosensor could be developed with a DNA detection sensitivity based on AuNPs amplification, which could detect *E. coli* O157:H7 cells to limit of 2.0×10^3 CFU/mL [54], 2.67×10^2 CFU/mL [55], and detected *Staphylococcus epidermidis* to limit of 1.3×10^3 CFU/mL [56]. Chen and colleagues detected Dengue virus to a limit of 2 PFU/mL by AuNPs signal amplification [57], and Kaewphinit and coworker reported the data that QCM could detect serial dilution of *M. tuberculosis* DNA limited as 5 pg of genomic DNA [53]. The technique showed a significant specificity since no cross-hybridization with other mycobacteria was found. In testing in direct clinical sputum samples, the sensitivity of QCM was 100 %, and the specificity was 100 % by comparison with cultures they developed for diagnosing lymphatic filariasis, which showed the sensitivity of limited detection of genomic DNA as 0.05 μg/ml. The specificity of the sensor can be used to test the *Hha*I repetitive region gene and also to differentiate *B. malayi* from *B. pahagi* , which PCR-gel electrophoresis cannot detect [58]. Gill and co-works detected *M. tuberculosis* by using gold nanoparticle probes for colorimetric detection to the limit of 10 CFU/ml. Uludag and colleagues studied the hybridisation with neutrAvidin capture for signal enhancement for the detection of HSV viral nucleic acids at 5.2×10^{-11} M concentration [59]. Kaewphinit and coworker developed the DNA QCM biosensor for detecting non-amplified bacterial genomic DNA of *M. tuberculosis*. The method involved the immobilization of a specific probe by using a biotinylated and thioled modified DNA probe that was designed from IS*6110* gene-specific for *M. tuberculosis* [5, 6]. Minunni and coworker hybridized with non-amplified in highly repetitive genomic DNA; the genomic DNA was first enzymatically digested to obtain fragments that could be easily denatured and be recognized by the immobilized probe [60]. Mannelli and coworker analyzed and compared oligonucleotides of DNA amplified by polymerase chain reaction (PCR) and no PCR, but to allow detection of the target sequence of genomic DNA [61], the sample must be fragmented by enzymatic digestion as shown in Table 9.1.

Table 9.1. Quartz crystal microbalance DNA based biosensor for diagnosis.

DNA target diagnosis	DNA amplification	Detection limit	Signal enhancement	Reference
Escherichia coli O157:H7	PCR	2.0×10^{3} CFU/mL	AuNPs	[54]
Escherichia coli O157:H7	PCR	2.67×10^{2} CFU/mL	AuNPs	[55]
Staphylococcus epidermidis	PCR	1.3×10^{3} CFU/mL	AuNPs	[56]
Dengue virus	PCR	2 PFU/mL	AuNPs	[57]
Mycobacterium tuberculosis	PCR	5 pg of genomic DNA	AuNPs	[53]
	Non-amplified	0.5 µg/mL of genomic DNA	-	[5]
	Non-amplified	0.25 µ M of synthetic DNA	-	[6]
Brugia malayi	PCR	0.05 µg/mL of genomic DNA	-	[58]
Herpes simplex viral	PCR	5.2×10^{-11} M	neutrAvidin	[59]
Bostaurus	Non-amplified	5 µg/mL of genomic DNA	-	[61, 62]

The interaction between 3' end of thiol-modified oligonucleoyide capture nanoparticle on the surface was strongly affected by the particle size. The high accuracy was determined by counted particle size with TEM images in this study. Moreover, the particle size could be counted by scan electron microscope image (SEM) [57, 62 – 64], TEM image [50], UV spectroscopy [63], Atomic force microscope (AFM) [65] or gel electrophoresis [66].

The detection of genomic DNA target by using a quartz crystal normally has less probability to hybridize with the DNA probe due to the steric hindrance effect of the DNA secondary structure and the lower amount of specific DNA sequences than in the detection limit of general analytical technique. The alternative solution to solve this problem did restrict enzymatic digestion of the DNA samples without any previous amplification step to separate small DNA fragments [25, 47]. However, sensitivity of genomic detection may break the cells of the reagent.

The non amplified DNA combined with QCM assay required 4 hours for the preparation of the immobilized DNA probe and the hybridized DNA target on the quartz crystal. In another report, Wu and colleagues [67] studied the circulating-flow system of QCM biosensors for real-time detection of *E. coli* O157:H7 by using 1 day for preparation. The advantage of this biosensor is non-labeling detection, ease of use, and direct detection of non-amplified genomic DNA, which reduces the PCR step. PCR techniques require well-trained personnel to handle the process and this amplification process demands additional processing time, reagents and devices, which affects the cost of the assay.

9.4. Conclusions

DNA based biosensor technology provides rapid and sensitive detection, especially, the QCM biosensor by using an oligonucleotide hybridization detection method. This advantage of this biosensor is that the detection method is label-free from radioactive or fluorescent tags. The QCM biosensor is one of the candidate devices of biosensor technology for the detection of DNA hybridization. DNA based QCM biosensor is the measurement of frequency change between the frequency of the oligonucleotides probe immobilized on the quartz crystal and the frequency after the hybridization of the DNA target. There are many reports about the development of quartz crystal specific DNA-based biosensor for detecting many pathogens in clinical samples.

Finally, this QCM biosensor could be beneficial for further routine clinical development as a portable device that is sensitive, specific, cheap, easy to use, and rapid for detection.

Acknowledgments

I would also like to acknowledge Srinakhariwirot University.

References

[1]. F. Bedioui. Biosensors, *Analusis*, Vol. 27, 1999, pp. 555-559.

[2]. W. Joseph, R. Gustavo, C. Xiaohua, D. Narasaiah S. Haruki, L. Denbai. et al., Sequence-specific electrochemical biosensing of *M. tuberculosis* DNA, *Anal Chim Acta*, Vol. 337, 1997, pp. 41-48.

[3]. K. Masaaki, K. Takashi, S. Masato, K. Sakiko, O. Miyuki, I. Shinichiro, et al., Electrochemical DNA quantification based on aggregation induced by Hoechst 33258, *Electrochem Commun*, Vol. 6, 2004, pp. 337-343.

[4]. L. Jun, T. Weihong, W. Kemin, X. Dan, Y. Xiaohai, H. Xiaoxiao, et al., Ultrasensitive optical DNA biosensor based on surface immobilization of molecular beacon by a bridge structure, *Analyt Scienc*, Vol. 17, 2001, pp. 1149-1153.

[5]. T. Kaewphinit, S. Santiwatanakul, C. Promptmas, K. Chansiri, Development of piezoelectric DNA-based biosensor for direct detection of *Mycobacterium Tuberculosis* in Clinical Specimens, *Sensors & Transducers*, Vol. 113, Issue 2, February 2010, pp. 115-126.

[6]. T. Kaewphinit, S. Santiwatanakul, C. Promptmas, K. Chansiri, Detection of non-amplified *Mycobacterium Tuberculosis* genomic DNA using piezoelectric DNA-based biosensor, *Sensors*, Vol. 10, 2010, pp. 1847-1858.

[7]. S. Tombelli, M. Minunni, M. Mascini, Piezoelectric biosensors: strategies for coupling nucleic acids to piezoelectric devices, *Methods*, Vol. 37, 2005, pp. 48-56.

[8]. A. Janshoff, C. Steinem, Quartz crystal microbalance for bioanalytical application, *Sensors Update*, Vol. 9, Issue 1, 2001, pp. 313-354.

[9]. G. Sauerbrey, The use of quartz oscillators for weighing thin layers and for microweighing, *J Phys*, Vol. 155, 1959, pp. 206-222.

[10]. S. Petr, Piezoelectric quartz crystal sensors applied for bioanalytical assays and characterization of affinity interactions, *J Braz Chem Soc*, Vol. 14, Issue 4, 2003, pp. 491-502.

[11]. R. L. Bunde, E. J. Jarvi, J. J. Rosentreter, Piezoelectric quartz crystal biosensor, *Talanta,* Vol. 46, 1998, pp. 1223-1236.

[12]. A. Sharma, K. R. Rogers, Biosensors, *Meas Sci Technol*, Vol. 5, 1994, pp. 461-472.

[13]. C. K. O'Sullivan, G. G. Guilbault, Commercial quartz crystal microbalances-theory and applications, *Biosens Bioelectron*, Vol. 14, 1999, pp. 663-670.

[14]. G. C. Guilbault, J. M. Jordan, Analytical uses of piezoelectric crystal, *CRC Crit Rev Anal Chem*, Vol. 19, 1890, pp. 1.

[15]. M. Kaspar, H. Stadler, T. Wei, C. H. Ziegler, Thickness shear mode resonators (mass-sensitive devices) in bioanalysis, *Fre-senius J Anal Chem*, Vol. 366, 2000, pp. 602-610.

[16]. Y. C. Liu, C. M. Wang, K. P. Hsiung, Comparison of different protein immobilization methods on quartz crystal microbalance surface in flow injection immunoassay, *Analytical Biochemistry*, Vol. 299, 2001, pp. 130-135.

[17]. L. J. Blum, P. R. Coulet, Biosensor: principles and application, *M. Dekker*, New York, 1991, pp. 108-137.

[18]. G. C. Guilbault, J. H. T. Luong, Gas phase biosensor, *J Biotechnol*, Vol. 9, 1988, pp. 1-9.

[19]. M. T. Gomes, A. C. Duarte, J. P. Oliveira, Comparison of two methods for coating piezoelectric crystals, *Anal Chim Acta*, Vol. 300, 1995, pp. 329-334.

[20]. A. C. Hillier, M. D. Ward, Sensitivity mapping of the quartz crystal microbalance in liquid media, *Anal Chem*, Vol. 64, 1992, pp. 2539-2554.

[21]. M. R. Deakin, D. A. Buttry, Electrochemical applications of the quartz crystal microbalance, *Anal Chem*, Vol. 61, 1989, pp. 1147A-1154A.

[22]. J. F. Alder, J. J. McCallum, Piezoelectric crystals for mass and chemical measurements. A review, *Analyst*, Vol. 108, 1983, pp. 1169-1189.

[23]. B. Christopher, Development of quartz crystal oscillators for under-liquid sensing, *Sens Actuat A: Physic,* Vol. 29, 1991, pp. 59-69.

[24]. P. L. Konash, G. J. Bastiaans, Piezoelectric crystals as detectors in liquid chromatography, *Anal Chem,* Vol. 52, 1980, pp. 1929-1931.

[25]. K. K. Kanazawa, J. G. Gordon, Frequency of a quartz microbalance in contact with liquid, *Anal Chem*, Vol. 57, 1985, pp. 1770-1771.

[26]. K. K. Kanasawa, J. G. Gordon, A liquid phase piezoelectric detector, *Anal Chem*, Vol. 57, 1985, pp. 1771-1775.
[27]. K. K. Kanazawa, J. G. Gordon, The oscillation frequency of a quartz resonator in contact with a liquid, *Anal Chim Acta*, 175, 1985, pp. 99-105.
[28]. A. Itohand, M. Ichihashi, A frequency of the quartz crystal microbalance (QCM) that is not affected by the viscosity of a liquid, *Meas Sci Technol*, Vol. 19, 2008, pp. 1-9.
[29]. P. In'acio, J. N. Marat-Mendes, C. J. Dias, Development of a biosensor based on a piezoelectric film, *Taylor & Francis*, Vol. 293, 2003, pp. 351-356.
[30]. J. D. Watson, F. H. C. Crick, A structure for deoxyribose nucleic acid, *Nature*, Vol. 171, 1953, pp. 737-738.
[31]. A. Ulman, Formation and Structure of Self-Assembled Monolayers, *Chem Rev*, Vol. 96, 1996, pp. 1533-1554.
[32]. V. Vorgelegt, S. Gudrun, H. Aus, Real-time monitoring of DNA hybridization and replication using optical and acoustic biosensors, January 2004, Retrieved November 30, 2006, from URL: http//www.mpip-mainz.mpg.de/knoll/publications/thesis/stengel 2004.pdf
[33]. Z. Junhui, C. Hong, Y. Ruifu, DNA based biosensors, *Biotechnol Advances*, Vol. 5, 1997, pp. 43-58.
[34]. L. Jun, T. Weihong, W. Kemin, X. Dan, Y. Xiaohai, H. Xiaoxiao, et al., Ultrasensitive optical DNA biosensor based on surface immobilization of Molecular Beacon by a bridge structure, *Analyt Scienc*, Vol. 17, 2001, pp. 1149-1153.
[35]. T. Kaewphinit, S. Santiwatanakul, K. Chansiri, Quartz crystal microbalance DNA based biosensor for diagnosis: A review, *Sensors & Transducers,* Vol. 143, Issue 8, August 2012, pp. 44-59.
[36]. S. Tombelli, M. Mascini, APF. Turner, Improved procedures for immobilisation of oligonucleotides on gold-coated piezoelectric quartz crystals, *Biosens Bioelectron*, Vol. 17, 2002, pp. 929-936.
[37]. X. C. Zhou, L. Q. Huang, S. F. Y. Li, Microgravimetric DNA sensor based on quartz crystal microbalance: comparison of oligonucleotide immobilisation methods and the application in genetic diagnosis, *Biosen Bioelectron*, Vol. 16, 2001, pp. 85-95.
[38]. E. Uttenthaler, C. Kosslinger, S. Drost, Characterization of immobilization methods for African swine fever virus protein and antibodies with a piezoelectric immunosensor, *Biosens Bioelectron*, Vol. 13, 1998, pp. 1279-1286.
[39]. T. Bunemann, Immobilization of denatured DNA to macroporous supports: II. Steric and kinetic parameters of heterogeneous hybridization reactions, *Nucleic Acids Res*, Vol. 10, 1982, pp. 7181-7196.
[40]. N. F. Campbell, J. A. Evans, N. C. Fawcett, Detection of poly (U) hybridization using azido modified poly (A) coated piezoelectric crystals, *Biochem Biophys Res Commun*, Vol. 196, 1993, pp. 858-863.
[41]. N. C. Fawcett, J. A. Evans, L. C. Chen, K. A. Drozda, N. Flowers, A quartz crystal detector for DNA, *Anal Lett*, Vol. 21, 1988, pp. 1099-1110.
[42]. P. Vadgama, P. W. Grump, Biosensors: present trends, *Analyst*, Vol. 117, 1992, pp. 1657-1670.
[43]. Z. Mo, H. Wang, Y. Liang, F. Liu, Y. Xue, Highly reproducible hybridization assay of zeptomole DNA based on adsorption of nanoparticle bioconjugate, *Analyst*, Vol. 130, 2005, pp. 1589-1594.
[44]. M. S. Shchepinov, S. C. Case-Green, E. M. Southern, Steric factors influencing hybridisation of nucleic acids to oligonucleotide arrays, *Nucleic Acids Res*, Vol. 25, 1997, pp. 1155-1161.
[45]. F. Caruso, E. Rodda, D. N. Furlong, Quartz crystal microbalance study of DNA immobilization and hybridization for nucleic acid sensor development, *Anal Chem*, Vol. 69, 1997, pp. 2043-2049.

[46]. S. Tombelli, M. Mascini, A. P. F. Turner, Improved procedures for immobilization of oligonucleotides on gold coated piezoelectric quartz crystals, *Biosens Bioelectron*, Vol. 17, 2002, pp. 929-936.
[47]. M. Minunni, I. Mannelli, M. M. Spiriti, S. Tombelli, M. Mascini, Detection of highly repeated sequences in non-amplified genomic DNA by bulk acoustic wave (BAW) affinity biosensor, *Anal Chim Acta*, Vol. 526, 2004, pp. 19-25.
[48]. M. Minnuni, S. Tombelli, J. Fonti, M. M. Spiriti, M. Mascini, P. Bogani, et al., Detection of fragmented genomic DNA by PCR-free piezoelectric sensing using a denaturation approach, *J Am Chem Soc*, Vol. 127, 2005, pp. 7966-7967.
[49]. T. Liu, J. Tang, M. Han, L. Jiang, Surface modification of nanogold particles in DNA detection with quartz crystal microbalance, *Chin Sci Bull*, Vol. 48, 2003, pp. 873-875.
[50]. L. B, Nie, Y, Yang, S, Li, N. Y, He, Enhanced DNA detection based on the amplification of gold nanoparticles using quartz crystal microbalance, *Nanotechnology*, Vol. 18, 2007, pp. 305501.
[51]. H. Q. Zhao, L. Lin, J. R. Li, J. A. Tang, M. X. Duan, L. Jiang, DNA biosensor with high sensitivity amplified by gold nanoparticles, *J Nanopart Res,* Vol. 3, 2001, pp. 321-323.
[52]. T. Liu, J. Tang, M. Han, L. Jiang, A novel microgravimetric DNA sensor with high sensitivity, *Biochem Biophys Res Commun*, Vol. 304, 2003, pp. 98-100.
[53]. T. Kaewphinit, S. Santiwatanakul, K. Chansiri, Gold nanoparticle amplification combined with quartz crystal microbalance DNA based biosensor for detection of *Mycobacterium tuberculosis*, *Sensors & Transducers*, Vol. 146, Issue 11, 2012, pp. 156-163.
[54]. L. J. Wang, Q. S. Wei, C. S. Wu, Z. Y. Hu, J. Ji, P. Wang, The *Escherichia coli* O157:H7 DNA detection on a gold nanoparticle-enhanced piezoelectric biosensor, *Chin Sci Bull*, Vol. 53, 2008, pp. 1175-24.
[55]. X. Mao, L. Yang, XL. Su, Y. Li, A nanoparticle amplification based quartz crystal microbalance DNA sensor for detection of Escherichia coli O157:H7, *Biosens Bioelectron*, Vol. 21, 2006, pp. 1178-1185.
[56]. H. Xia, F. Wang, Q. Huang, J. Huang, M. Chen, J. Wang, Detection of *Staphylococcus epidermidis* by a quartz crystal microbalance nucleic acid biosensor array using Au nanoparticle signal amplification, *Sensors,* Vol. 8, 2008, pp. 6453-6470.
[57]. S. H. Chen, Y. C. Chuang, Y. C. Lu, H. C. Lin, Y. L. Yang, C. S. Lin, A method of layer-by-layer gold nanoparticle hybridization in a quartz crystal microbalance DNA sensing system used to detect dengue virus, *Nanotechnology*, Vol. 20, 2009, pp. 215501
[58]. T. Kaewphinit, S. Santiwatanakul, S. Areekit, K. Chansiri, Quartz crystal microbalance DNA based biosensor for the detection of *Brugia malayi*, *Sensors & Transducers*, Vol. 144, Issue 9, September 2012, pp. 153-160.
[59]. Y. Uludag, X. Li, H. Coleman, S. Efstathiou, M. A. Cooper, Direct acoustic profiling of DNA hybridisation using HSV type 1 viral sequences, *Analyst*, Vol. 33, 2008, pp. 52-57.
[60]. M. Minunni, I. Mannelli, M, M. Spiriti, S. Tombelli, M. Mascini, Detection of highly repeated sequences in non-amplified genomic DNA by bulk acoustic wave (BAW) affinity biosensor, *Anal Chim Acta*, Vol. 526, Issue 1, 2004, pp. 19-25.
[61]. I. Mannelli, M. Minunni, S. Tombelli, R. Wang, M. M. Spiriti, M. Mascini, Direct immobilisation of DNA probes for the development of affinity biosensors, *Bioelectrochem*, Vol. 66, 2005, pp. 129-138.
[62]. Sf. Liu, Yf. Li, J. Li, L. Jiang, Enhancement of DNA immobilization and hybridization on gold electrode modified by nanogold aggregates, *Biosens Bioelectron*, Vol. 21, 2005, pp. 789-795.
[63]. T. Takahagi, G. Tsutsui, S. Huang, H. Sakaue, S. Shingubara, Scanning electron microscope observation of heterogeneous three-dimensional nanoparticle arrays using DNA, *Jpn J Appl Phys*, Vol. 40, 2001, pp. L521-L523.

[64]. B. Nie, M. R. Shortreed, L. M. Smith, Quantitative detection of individual cleaved DNA molecules on surfaces using gold nanoparticles and scanning electron microscope imaging, *Anal Chem*, Vol. 78, 2006, pp. 1528-1534.

[65] Z. Liang, J. Zhang, L. Wang, S. Song, C. Fan, G. Li, Acentrifugation-based method for preparation of gold nanoparticles and its Application in biodetection, *Int J Mol Sci*, Vol. 8, 2007, pp. 526-532.

[66]. X. Xu, K. K. Caswell, E. Tucker, S. Kabisatpathy, K. L. Brodhacker, W. A. Scrivens, Size and shape separation of gold nanoparticles with preparative gel electrophoresis, *J Chromatography A*, Vol. 1167, 2007, pp. 35-41.

[67]. V. C. Wu, S. H. Chen, C. S. Lin, Real-time detection of Escherichia coli O157:H7 sequences using a circulating-flow system of quartz crystal microbalance, *Biosens Bioelectron*, Vol. 22, 2007, pp. 2967-75.

Chapter 10
Recent Advance in Antibody or Hapten Immobilization Protocols of Electrochemical Immunosensor for Detection of Pesticide Residues

Xia Sun, Ying Zhu, Xiangyou Wang, Yaoyao Cao, Lu Qiao

10.1. Introduction

Pesticides derived from synthetic chemicals are essential inputs in increasing agricultural production by preventing control pest and crop losses before and after harvesting. One-third reduction in crop yield would be happened if pesticides are not used against pest [1, 2].

Depending upon the species of pest, these chemicals have been divided into groups (e.g., herbicides, insecticides, fungicides, rodenticides, and nematocides). However, their indiscriminate use, apart from being an operational hazard, is posing a serious threat to human health [3]. By transformation through the food chain, their bio-accumulation in animal and human body and eventually show their adverse effects, like: cancer, hormone disruption, birth defect and neurological effects [4]. Therefore, there is a growing need to introduce and develop new, sensitive, reproducible and rapid methods for monitoring of pesticide residues in agricultural products at trace levels.

Numerous analysis methods such as gas chromatography [5], high-performance liquid chromatography [6], capillary electrophoresis [7], flow injection immunoanalysis [8-10] and fluorimetry [11] have been developed for detection of pesticides residues. These analytical methods are standardized techniques. However, these methods have some drawbacks such as complexity, poor selectivity, high cost, slow response, poor stability and time-consuming [12]. Moreover, they can only be performed by highly trained technicians and are not convenient for on-site or in-field detection, which limit their

Xia Sun
School of Agriculture and Food Engineering, Shandong University of Technology,
No. 12, Zhangzhou Road, Zibo 255049, Shandong Province, P.R. China

application for real-time detection. For these reasons, the development of rapid and efficient monitoring methods becomes more and more important.

In this respect, biosensors are potentially useful as suitable complementary tools for the real-time detection of pesticides residues and have been an active research area for some years [13]. The biosensors supply important advantages such as simplicity, sensitivity, selectivity, easy use, high efficiency, possibility of portability and miniaturization, fast response times (few seconds). Among the biosensor technology available for the detection of pesticides, enzyme-based biosensors for pesticide determination have caused public interest due to their reliability, fast response, high sensitivity and selectivity. In recent years, enzyme-linked immunosorbent assays (ELISA) have grown rapidly as powerful and flexible tools for pesticide measurement [14-19]. However, false positive may easily appear and this method also need some improvements (e.g. for continuous detection).

Many biosensors which are used for pesticide detection are based on the inhibition reaction or catalytic activity of several enzymes in the presence of pesticides [20-22]. Enzyme-based biosensors (e.g. acetylcholinesterase biosensor) for pesticide determination have been widely reported in the literature [23-27]. Since a number of pesticides have a similar mode of action affecting the activity of the same enzyme, most of enzyme-based biosensors are used for screening purposes and are unspecific for individual pesticides. They can only detect total pesticide content and do not provide specific information about a particular pesticide [28].

Immunosensors have been used to detect or quantify the specific pesticide based on the binding interactions between immobilized biomolecules (Ab or hapten) on the transducer surface with the analyte of interest (hapten or Ab), resulting in a detectable signal. The sensor system takes advantage of the high selectivity provided by the molecular recognition characteristics of an Ab, which binds reversibly with a specific hapten. In solution phase, Ab molecules interact specifically and reversibly with a hapten to form an immune complex (Ab–hapten) according to the following equilibrium equation:

$$Ab + hapten \underset{Kd}{\overset{Ka}{\rightleftarrows}} Ab - hapten,$$

where K_a and K_d are the rate constants for association and dissociation, respectively [29]. They appear to be appropriate for identification of a single pesticide or, in some cases, small groups of similar pesticides in environmental monitoring as they are rapid, specific, sensitive and cost-effective analytical devices [30]. Currently, many electrochemical, optical and piezoelectric immunosensors have been developed for pesticides detection [28]. Among them, electrochemical immunosensors have received increasing attention due to their lower cost, high sensitivity, simple instrumentation, and easy signal amplification [31]. Excellent reviews that focused on electrochemical immunosensors for detection of different pesticide molecules have been reported [32, 33].

However, there is a time gap between current status in the field and the most recent reviews. Thus, in this review, we specifically provide an overview of the research carried out during the last 5 years relative to electrocheimical immunosensor for pesticide residues detection. We will review several types of electrochemical immunosensors developed for their applications in pesticide analysis, various immobilization protocols used for formation of a biorecognition interface. We also will discuss the trends and challenges associated with designing a reliable immunosensor for practical applications in detail.

10.2. Electrochemical Immunosensors

Formation of Ab–Ag complex in electrochemical transducers alters the change in ion concentration or electron density on the electrode surface, which, in turn, is measured by electrodes. Electrochemical transducers, classified as amperometric, potentiometric, conductometric, capacitive and impedimetric measure changes in current, potential (voltage), conductance, capacitance and impedance respectively [34-36]. Depending on if labels are used or not, immunosensors are divided into two categories: labeled type and label-free type.

Electrochemical immunosensors could be competitive and revolutionize analysis, because of their simplicity, rapidity and cheap technology. For pesticide detection, most of them use impedance or amperometry in label or label-free format and label-free format is a tendency. Some examples of electrochemical immunosensors for the detection of pesticides residues are presented in Table 10.1.

10.3. Immobilization Protocols

In terms of the development of electrochemical immunosensor, the Ab/hapten immobilization onto a transducer or a support matrice is a key step in optimizing the analytical performance, such as response, reproducibility, stability, selectivity and regeneration. A good immobilization method should meet the following requirements: (1) be simple and fast; (2) produce immobilized reagents that are stable and do not leach from the substrate; and (3) maintains its biological integrity flexibility, and proper active site orientation toward the bulk solution. Thereby, Ab/hapten immobilization has been a critical issue in immunosensor technology [57-61].

Ab/hapten immobilization consists of physical adsorption and chemical binding, which depends on the driving force [62]. In general, they mostly fall into following methodologies.

Table 10.1. Some examples of electrochemical immunosensors for the detection of pesticides residues.

Pesticide	Detector	Label	Electrode Modification	Detection limit	Sample	Assay time	Ref.
Carbofuran	Amperometric	free	carbofuran/BSA/Ab/GNPs/ TU/GNPs/GCE	0.11 ng/mL	cabbage	15 min	[37]
Carbofuran	Amperometric	free	carbofuran/BSA/Ab/SiSG/GCE	0.33 ng/mL	cabbage lettuce	20 min	[38]
Carbofuran	Amperometric	free	carbofuran/HRP/Ab/GNPs/ L-cysteine/Au electrode	40 ng/mL	-	12 min	[39]
Carbofuran	Amperometric	free	carbofuran/BSA/Ab/PA/DpAu/Au electrode	0.192 ng/mL	Chinese chive, celery, cabbage	15 min	[31]
Carbofuran	Amperometric	free	carbofuran/BSA/Ab/ {DpAu/DMDPSE}$_2$/Au electrode	0.06 ng/mL	Lettuce, cabbage, pepper, tomatoe, chive, strawberry	40 min	[40]
2,4-D	Potentiometric	HRP	2,4-D-HRP/Ab/GA/Graphite electrode	40 ng/mL	water, serum	12 min	[41]
2,4-DB	Impedimetric	free	2,4-DB/BSA/Ab/GA/L-Cys/Au	10^{-7} g/L	soybean	-	[42]
Simazine	Potentiometric	HRP	simazine-HRP/ glycine/Ab/PA/ GA/ISFET	1.25 ng/mL	-	50 min	[43]
Diuron	Impedimetric	free	diuron/Ab/GNPs/SPE	5.46 ng/mL	water	-	[44]
Diuron	Amperometric	free	Ab/DCPU-BSA/PB-GNP/LC-LAGE	1 ppt	-	-	[33]
Paraoxon	Amperometric	free	paraoxon/Ab/Nafion-GNPs/GCE	12 ng/mL	aqueous samples	20 min	[45]
Paraoxon	Amperometric	free	Ab/Paraoxon/EDC/NHS/FDMA/SWNTs/GCE	2 ppb	PBS (pH 7)	-	[46]
Picloram	Amperometric	HRP	HRP-G, anti-RIgG/picloram/ Ab/BSA-picloram/GNPs/ GCE	0.5 ng/mL	peach	-	[47]
Atrazine	Conducti-metric	GNPs	Ab_2/Ab_1/atrazine/ GPTS/ N-acetylcysteamine/IDµE	0.1 ng/mL	buffers	-	[48]
Atrazine	Amperometric	HRP	atrazine/atrazine–BSA/Ab/ immobilon membrane/H_2O_2 electrode	5.0×10^{-11} M	buffalo milk, vegetal samples	15 min	[49]
Atrazine	Amperometric	HRP	atrazine/atrazine-HRP scAb/PANI/ PVSA/SPE	0.1 ng/mL	-	-	[50]
Atrazine	Impedimetric	free	atrazine/bio-Fab/ neutravidin/ Gold/MHDA+biotinyl-PE	20 ng/mL	PBS (pH 7)	-	[51]

Table 10.1. Some examples of electrochemical immunosensors for the detection of pesticides residues (Continued).

Pesticide	Detector	Label	Electrode Modification	Detection limit	Sample	Assay time	Ref.
Atrazine	Impedimetric	free	atrazine/bio-Fab K47/ BSA/PPy/ neutravidin/Au electrode	0.1 ng/mL	PBS (pH 7)	-	[52]
Atrazine	Amperometric	HRP	atrazine-HRP/Ab/ ProtA-GEB	6 μg/mL	orange juices	-	[51]
Atrazine	Impedimetric	free	Atrazine/BSA/ histidine-Ab/poly NTA -Cu^{2+}/Au electrode	10 pg/mL	PBS (pH 7)	-	[53]
Atrazine	Impedimetric	free	Ab_{11}/antigen2d-BSA/3-(glycidoxypropy)trimethoxysilane/N-acetylcysteamine/IDμE	0.19 μg/mL	red wine	-	[32]
Atrazine	Impedimetric	free	Atrazine/BSA/ histidine-Ab/poly NTA -Cu^{2+}/Au electrode	10 pg/mL	PBS (pH 7)	-	[54]
Atrazine	Amperometric	free	ATZ/α-ATZ/poly (JUG-HATZ)/GCE	0.2 ng/L	-	-	[55]
Chlorpyrifos-methyl	Amperometric	HRP	HRP-Ab/BSA-Ag/Pt/ SiO_2/SPCEs	22.6 ng/L	grape, soil	40 min	[56]

10.3.1. Physical Adsorption

Physical adsorption is generally based on interactions such as van der Waals forces, electrostatic interactions and hydrophobic interactions between the Ab/hapten and the transducer. Physical adsorption is simple and easy, but nonspecific attractive forces easily causes Ab/hapten desorption [62]. In addition, the immobilized Ab/hapten can be susceptible to the reduction of biological activity by an inappropriate orientation caused by physical adsorption [63]. This leads to the limitation that the sensing elements have decreasing response with time and, thus, short life-times.

Gobi et al., created a functional sensing surface of the immunosensor by immobilizing an ovalbumin conjugate of 2,4-D (2,4-D-OVA) by simple physical adsorption on a thin-film gold chip. It has been established that the Au surface of the sensor chip was completely covered by 2,4-D-OVA up to a monomolecular layer and that the 2,4-D-OVA immobilized sensor chip was highly resistive to non-specific binding of proteins [64].

10.3.2. Covalent Coupling

More specific and stronger attachment of Ab/hapten can be obtained by covalent modification through formation of a stable covalent bond between functional groups of Ab/hapten and the transducer. Covalent modification requires a bifunctional cross-linker, which has one functional group that reacts with a base support, and another group that interacts with an active group of Ab/hapten [65, 66, 60]. The procedure provides increased stability of the Ab. However, the immobilization by covalent coupling may results in the random orientation of Ab/hapten, decreases the activity of Ab/hapten and is generally poorly reproducible due to the chemical modification of critical residues and random protein orientations [67]. In addition, blocking steps are usually necessary to limit the nonspecific binding.

An example of where this approach has been exploited is that the Ab immobilization was carried out by using carboxylic groups activated with EDC/NHS as a cross-linker to connect the NH2- group of the antibody with the surface of carboxylized transducer (Fig. 10.1).

By using 2,4-D immobilized through its carboxylic group covalently to the silanized surface of the gold working electrode, Kalib et al., developed a disposable immunochemical biosensor for the herbicide 2,4-dichlorophenoxyacetic acid (2,4-D), with a detection limit close to 0.1 μg/L of free 2,4-D. For covanlent immobilizations, the 2,4-D molecule was activated by isobutyl chloroformate and then it was linked to the free amino group, which was obtained: (1) directly from the APTS moledule, (2) using gultaraldehyde/hexametyhlenediamine spacer, (3) using glutaraldehyde/albumin spacer [68].

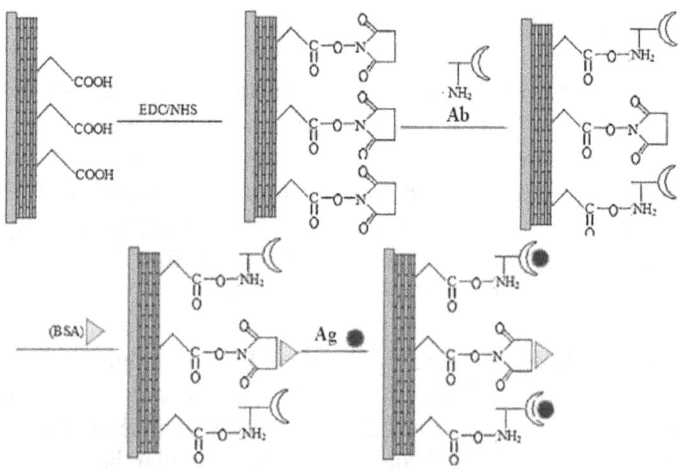

Fig. 10.1. Covalent immobilization of antibody onto carboxylized transducer activated with EDC/NHS as a cross-linker.

GNPs: colloidal gold nanoparticles; TU: thiourea; GCE: glassy carbon electrode; SiSG: silica sol-gel; HRP: Horseradish peroxidase; DpAu: deposited gold nanocrystals; PA: staphylo-coccal protein A; DMDPSE: 4,4'-thiobisbenzenethiol; GA: glutaraldehyde; 2,4-DB: 2,4-Dichlorophenoxybutyric acid; ISFET: ion-selective field effect transistor; SPE: screen-printed electrodes; PB: Prussian blue; IDμE: interdigitated microelectrodes;

PANI: polymer polyaniline; PVSA: poly(vinylsulphonic acid); biotinyl-PE: phospholipid dipalmitoyl-sn-glycero-3-phospho-ethanolamine-N-(biotinyl); ProtA-GEB: Protien A-graphite-epoxy biocomposite; Poly(JUG-HATZ): poly[N-(6-(4-hydroxy-6-isopropylamino - 1, 3, 5 – triazin - 2 –ylamino) hexyl) 5-hydroxy-1, 4-naphthoquinone-3-propionamide] and tween is usually used as the appropriate material to block nonspecific binding sites; SPCEs: screen printed carbon electrode.

Another example is that Ramon-Azcon et al., developed a novel impedimetric immunosensor based on an array of interdigitated μ-electrodes (IDμE) and immunoreagents specifically developed to detect atrazin. In this study, an atrazine-haptenized protein was covalently immobilized on the surface of the IDμE area (interdigits space) previously activated with (3-glycidoxypropyl)trimethoxysilane. With this configuration, the immunosensor detects atrazine with a limit of 0.19 μ/gL in red wine, far below the Maximum Residue Level (MRL) established by EC for residues of this herbicide in wine (Fig. 10.2) [32].

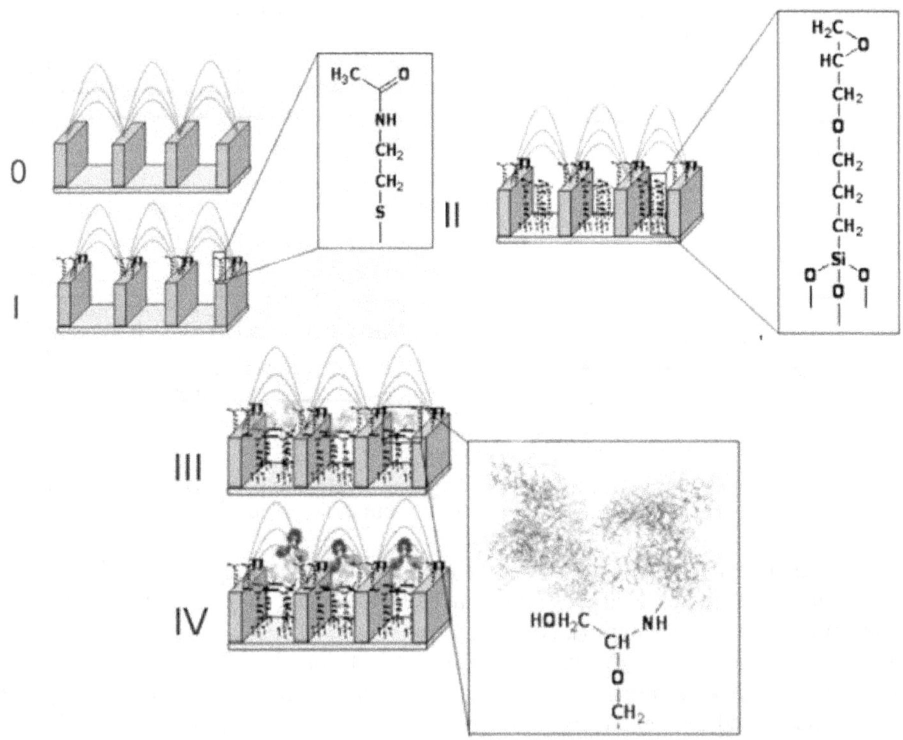

Fig. 10.2. Scheme showing steps used to prepare the immunosensor surfaces and antibody binding: (a) IDμE, (b) step I: N-acetylcysteamine, gold protection, (c) step II: functionalization of Pyrex substrate with (3-glycidoxypropyl)trimethoxy-silane, (d) step III: coating antigen 2d-BSA, covalent immobilization, and (e) step IV: antibody Ab11.

Valera et al., has designed and developed a novel conductimetric immunosensor for atrazine detection using covalent immobilization of the competitor antigen which was performed on the interdigitated μ-electrodes surface via the side chain amino groups of lysines or arginines with the epoxy groups on the device surface. The immunosensor developed detects atrazine with limits of detection in the order of 0.1–1 μg/mL (Fig. 10.3) [48].

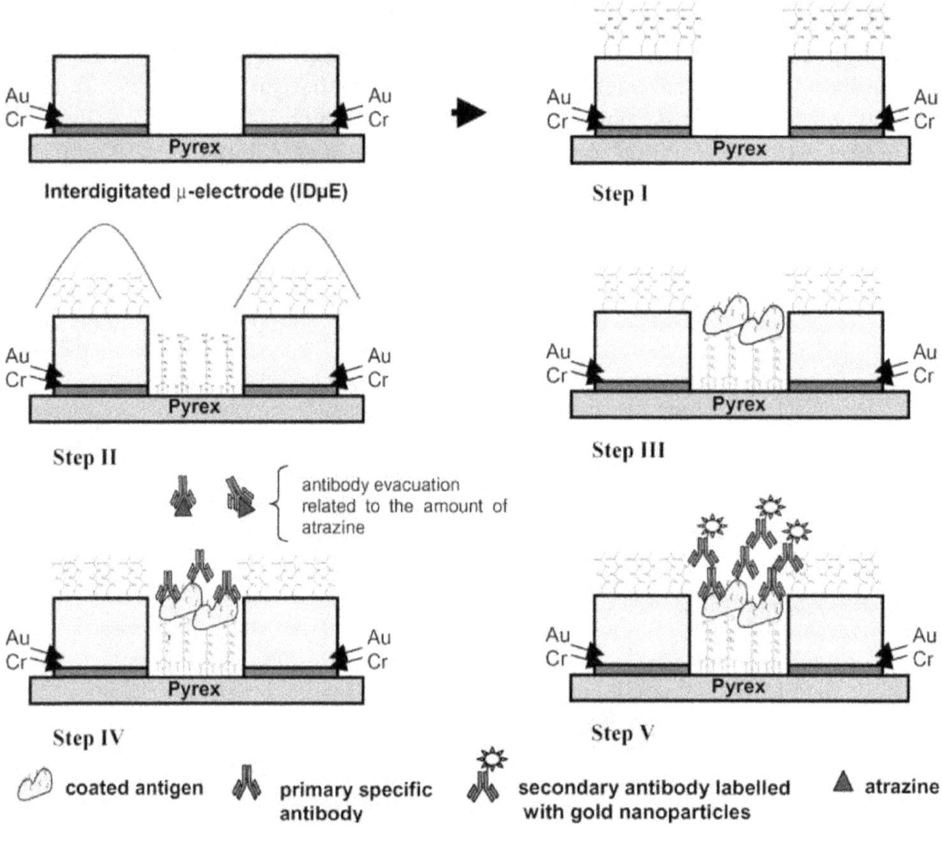

Fig. 10.3. Schematic diagram of the complete assay system performed on the IDμEs: Step I, protection of interdigitated μ-electrodes with N-acetylcysteamine; Step II, immunosensor surface functionalization with GPTS; Step III, covalent immobilization of the antigen on the IDμE; Step IV, specific primary antibody (Ab_1) capture in the competition step; Step V, secondary labelled with gold antibody (Ab_2) capture. In the Step IV, an amount of the specific antibody (Ab_1) is bounded on the coated antigen layer, whereas other amount is evacuated of the IDμEs, this amount is related to the atrazine concentration. In the Step V, an amount of the secondary antibody (Ab_2) is bounded on the specific antibodies.

In addition, Sun et al., has proposed an amperometric immunosensor for the detection of chlorpyrifos with a detection limit of 0.046 ng/mL using the multi-walled carbon nanotubes-thionine-chitosan nanocomposite film and GA, which can combine the (-NH_2) between thionine, chitosan and anti-chlorpyrifos (Fig. 10.4) [69].

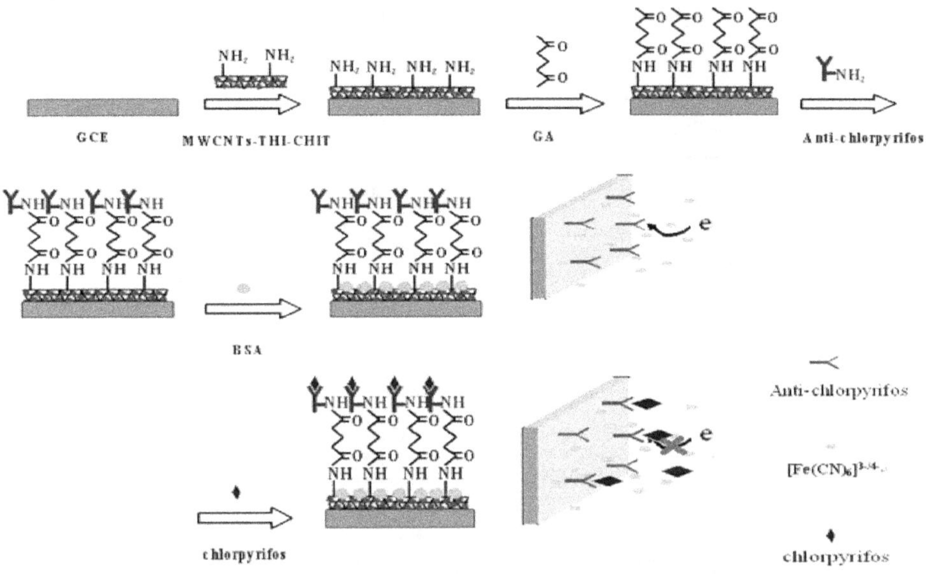

Fig. 10.4. Schematic illustration of the immunosensor fabrication process.

Using this technique, Valera et al., have developed a simple and low-cost method for the fabrication of mechanically flexible interdigitated μ-electrodes (FIDμEs) for the development of a conductimetric immunosensor for atrazine detection recently [70].

Recently, Li et al., have reported a label-free impedimetric immunosensor for sensitive detection of 2,4-dichlorophenoxybutyric acid (2,4-DB) in soybean. In this experiment, synthetic 2,4-DB antibodies were immobilized on the electrode by the crosslinking of L-Cysteine/glutaraldehyde, and 2,4-DB were measured by the increase of electron-transfer resistance when the immune reaction occurred (Fig. 10.5). Under optimal conditions, the change of resistance is in a linear relationship with the logarithm of the concentration with the detection limit of 1.0×10^{-7} g/L (0.1 ppb). This method bears such merits as simplicity in operation, high sensitivity, wide linear range, specificity, reproducibility and good stability. The actual soybean samples were analyzed with the recovery of 82.8 %-102.3 % [42].

Recently, Liu et al., developed a label-free immunosensor based on SWNTs modified GCE for the direct detection of paraoxon. Based on aryldiazonium salt chemistry, forest of SWNTs can be vertically aligned on mixed monolayers of aryldiazonium salt modified GC electrodes by C-C bonding, which provides an interface showing efficient electron transfer between biomolecules. PEG molecules were introduced to the interface to resist non-specific protein adsorption. Ferrocenedimethylamine (FDMA) was subsequently attached to the ends of SWNTs through the amide bonding followed by the attachment of epitope i.e. paraoxon hapten to which a paraoxon antibody would bind (Fig. 10.6). This immunosensor shows good selectivity and high specificity to paraoxon, and is functional for the detection of paraoxon in both laboratory and field by a displacement assay [46].

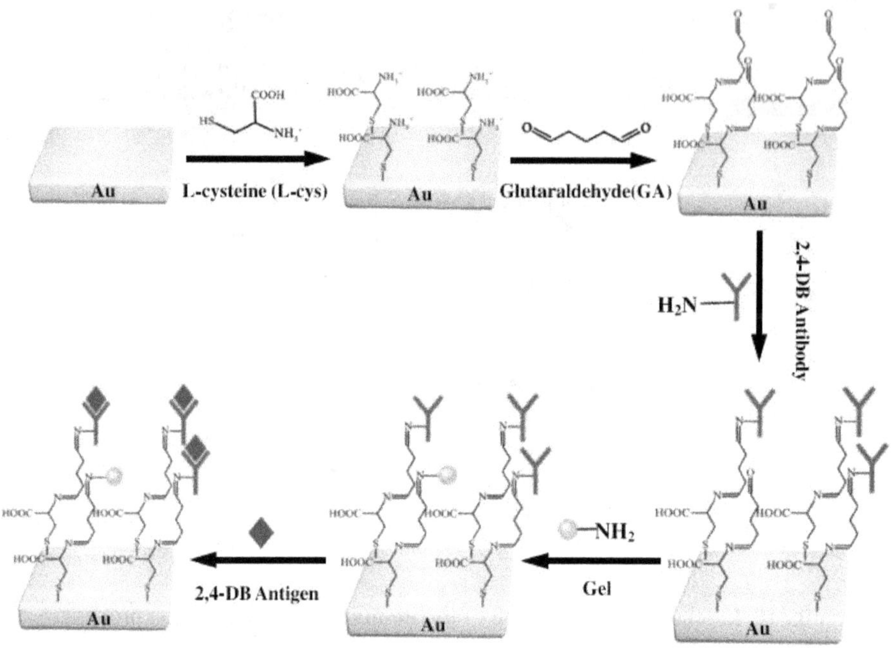

Fig. 10.5. The schematic illustration of preparation of novel immunosensor and the interaction between antibody and antigen.

10.3.3. Entrapment

In an encapsulation method, the reagent is physically trapped within a porous matrix. It is simple and compatible with various reagents. The reagents trapped in the matrix usually do not leach out or leach out very slowly when an appropriate entrapment procedure is used. It appears that the encapsulation method avoids the disadvantages and combines the advantages of the first two methods. Organic polymeric matrices have been widely used for entrapment of sensing agents [71].

10.3.3.1. Sol-gel Entrapment

Recent development in the area of electrochemical immunosensors with sol-gel encapsulation of Ab/hapten as an immobilization matrix is very encouraging and offers potential advantages. These advantages include the ability of sol-gel (1) to form at low temperatures and under chemical, mechanical stability and offers negligible swelling, (2) open to a wide variety of chemical modifications based on the inclusion of various polymer additives, redox modifiers and organically modified silanes, resulting in electrically conducting materials and (3) to exhibit tunable pore size and pore distribution, which allows small molecules and ions to diffuse into the matrix while larger biomolecules remain trapped in the pores, simplicity of preparation without any kinds of modifications [57].

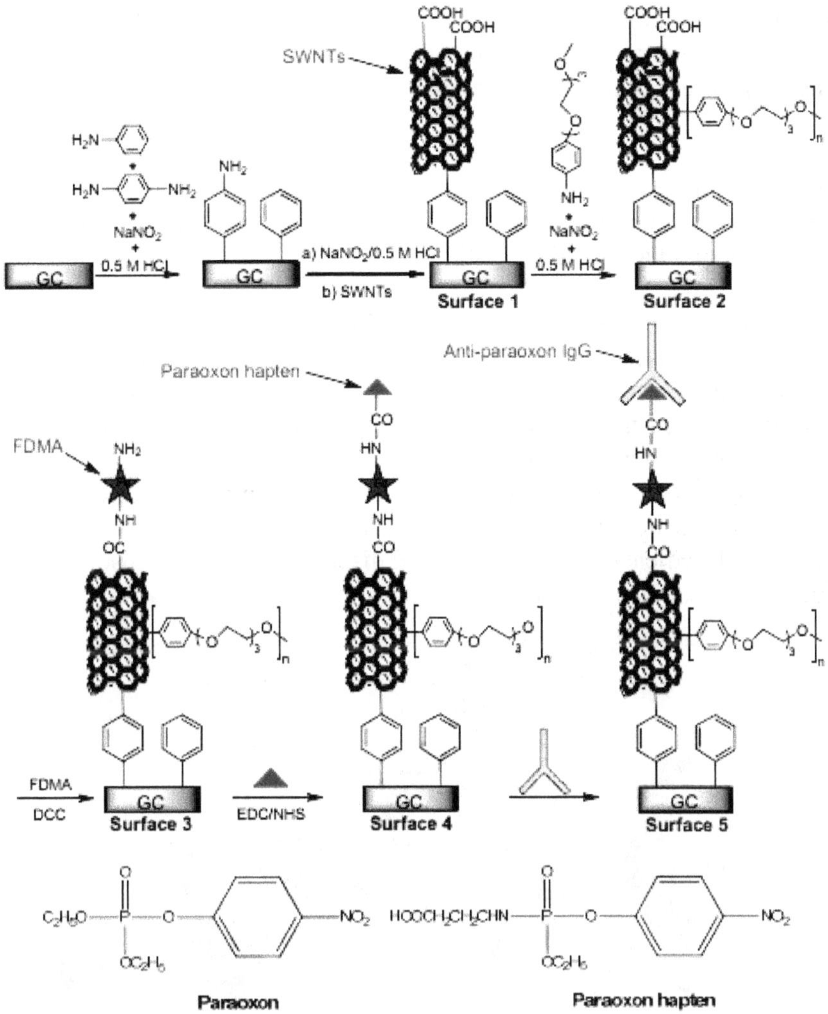

Fig. 10.6. Schematic of SWNTs modified sensing interface for the detection of paraoxon.

Although it has many advantages over other methods, the sol-gel method has some disadvantages: low response (as long as several minutes) in aqueous media and slightly change biological activities due to reduced degree of freedom in the pores and/or interactions with the inner surface of the pores [57].

Turniansky et al., report the successful entrapment of an anti-atrazine antibody in a SiO_2 sol-gel matrix, retaining its ability to bind antigen from aqueous solutions based methods for monitoring pesticide residues and other organo-synthetic environmental contaminants. Under appropriate sol-gel-forming conditions, high amounts of atrazine were bound to the sol-gels, ranging between 60 % and 91 % of the amount applied to the column. The combination of the properties of the sol-gel matrix (e.g., stability, inertness, high porosity, high surface area and optical clarity), together with the selectivity and sensitivity of the antibodies, enable extension of this feasibility study to development of a novel group of immunosensors which could be used for purification, concentration and monitoring of a variety of residues from different sources [72].

Sun et al., developed a novel label-free impedance immunosensor for the direct detection of carbofuran using silica sol–gel (SiSG) as immobilizing agent. Sol–gel technology provides a unique means to prepare a three-dimensional network suited for the encapsulation of a variety of Ab [38].

Recently, Wei et al., reported a disposable amperometric immunosensor for chlorpyrifos-methyl based on immunogen/ platinum doped silica sol–gel film modified screen-printed carbon electrode (Fig. 10.7). The immobilisation of BSA-Ag on the nanocomposite retained its immunoactivities, which allowed the immobilized BSA-Ag to effectively capture unbound Ab-HRP in the detection solution. A linear response to CM concentration was exhibited, ranging from 0.4 to 20 ng/mL. Detection of CM with the presented method in soil or grape samples treated with CM matched the reference values well, which indicated that the proposed disposable immunosensor hold promising applications in environmental and food monitoring [56].

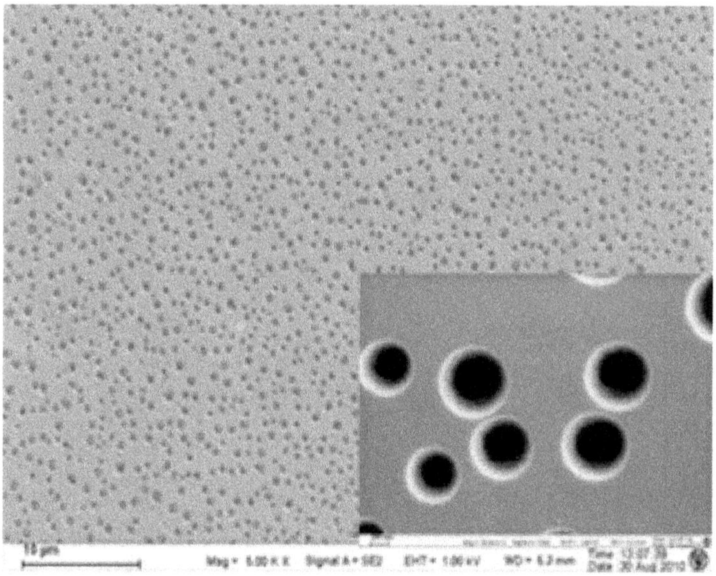

Fig. 10.7. SEM images of Pt colloid doped silica sol-gel.

10.3.3.2. Electrically Conducting Polymers Entrapment

The electrically conducting polymers (CP) are known to possess numerous features, which allow them to act as excellent materials for immobilization of biomolecules and rapid electron transfer for the fabrication of efficient biosensors [73].

Recently, CP, such as polyaniline(PANI) etc, has captured attention of scientific community due to its applications including those in biosensors because of a number of useful features such as 1) Direct and easy deposition on the sensor electrode; 2) Control of thickness; 3) Redox conductivity and polyelectrolyte characteristics; 4) High surface area; 5) Chemical specificities; 6) Long term environmental stability and 7) Tuneable properties. Fig. 10.8 is the 3D and 2D structure of PANI [74].

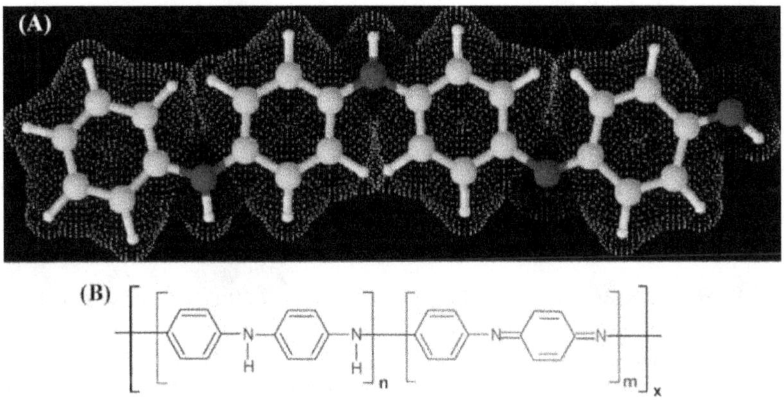

Fig. 10.8. (A) 3D and (B) 2D structure polyaniline.

A new electropolymerizable monomer, [N-(6-(4-hydroxy-6-isopropyl-amino-1,3,5-triazin-2-ylamino) hexyl)5-hydroxy-1,4-naphthoquinone-3-propionamide], has been designed for use in a label-free electrochemical immunosensor when polymerized on an electrode and coupled with a monoclonal anti-atrazine antibody for the detection of atrazine (Fig. 10.9). This monomer contains three functional groups: hydroxyl group for electropolymerization, quinone group for its transduction capability, and hydroxyatrazine as bio-receptor element. This constitutes a direct, label-free and signal-on electro- chemical immunosensor with a very low detection limit of 0.2 ng/L, one of the lowest reported for such immunosensors [55].

Ionescu et al., reported a label-free impedimetric immunosensor for the determination of atrazine, based on a poly(pyrrole-nitrilotriacetic acid) (poly NTA) film and combined with an impedimetric detection of atrazine without reagent and label. The poly NTA film constituted a convenient tool for the easy anchoring of histidine-labelled antibody directed against atrazine, allowing the detection of extremely low atrazine concentration namely 10 pg/mL [54].

Grennan et al., described the development of an electrochemical immunosensor for the analysis of atrazine using recombinant single-chain antibody (scAb) fragments. The sensors are based on carbon paste screen-printed electrodes incorporating the conducting polymer polyaniline (PANI)/poly(vinylsulphonic acid) (PVSA), which enables direct mediatorless coupling to take place between the redox centres of antigen-labelled horseradish peroxidase (HRP) and the electrode surface (Fig. 10.10) [50].

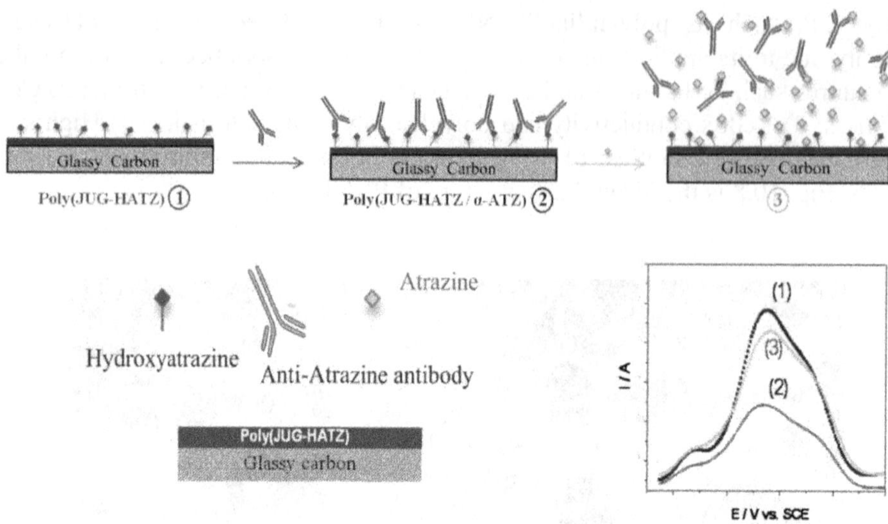

Fig. 10.9. Strategy for the electrochemical detection of atrazine based on the change in electroactivity of polymer film, poly(JUG-HATZ). SWV recorded with (1) poly(JUG-HATZ)-modified electrode; (2) after complexation with α-ATZ, poly(JUG-HATZ/α-ATZ)-modified electrode; (3) after addition of ATZ in solution.

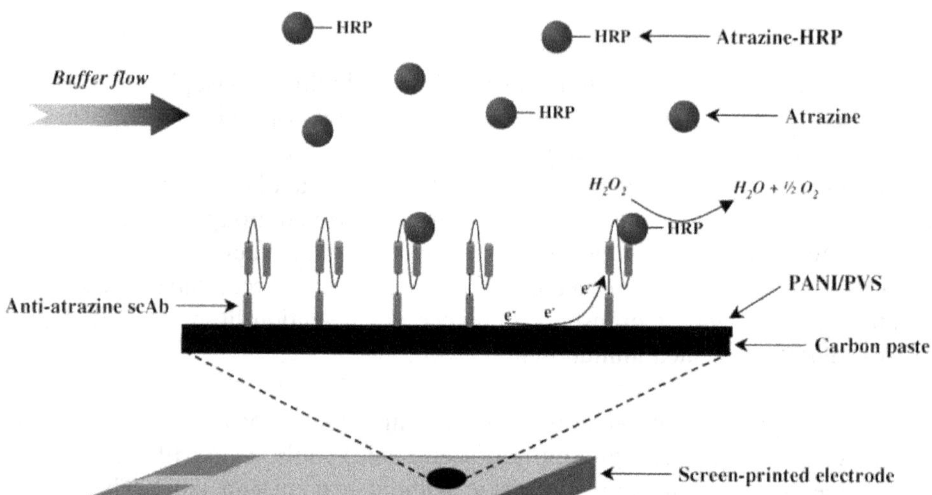

Fig. 10.10. Schematic diagram of the electrochemical real-time sensing process for atrazine detection.

10.3.4. Oriented Immobilization

Antibodies immobilized by these methods such as physical adsorption or covalent coupling, however, often suffer from reduced hapten binding ability due to a combination of denaturation, random orientation, and chemical modification of the antibodies [75]. Antibody-binding proteins (protein A, G, A/G, and L) have been widely used to overcome the aforementioned drawbacks [76, 77]. These proteins specifically bind the Fc region of an Ab and, thus, properly orient the bound Ab for optimal hapten binding [78]. Moreover, because the antibody-binding proteins capture antibodies without any chemical modifications, bound antibodies fully retain their function. Improved surface orientation of these engineered antibody-binding proteins enhanced the subsequent Ab/hapten immobilization. Despite the evident advantages of using antibody-binding proteins for Ab immobilization, these proteins have limitations; for example, they are susceptible to denaturation and are difficult to use in site-specific modifications [79]. This technology has been widely used in electrochemical immunosensors through the antibody's oriented immobilization for pesticides detection.

An immunological reaction for the detection of atrazine performed on the Protien A (2 %)-graphite-epoxy biocomposite (ProtA-GEB) biosensors is based on the antibody bonding through Fc fragment to Protein A and a direct competitive assay using atrazine-HRP tracer as the enzymatic label (Fig. 10.11). The electrochemical detection is thus achieved through a suitable substrate and a mediator for the enzyme HRP. The detection limit for atrazine in orange juices was found to be 6μg/mL [51].

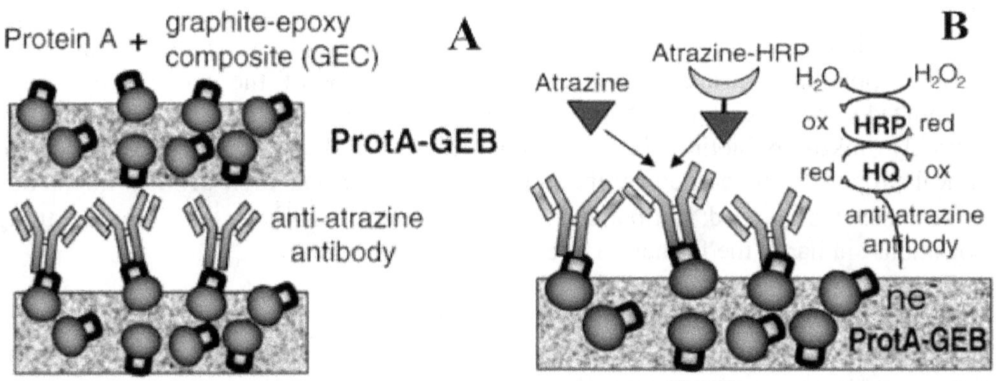

Fig. 10.11. The immobilization of anti-atrazine antibodies on the surface of the electrochemical transducer for the detection of atrazine in orange juice with ProtA-GEB-based electrochemical immunosensors (A) and the competitive immunological reaction (B).

Recently, Sun et al., introduced a strategy for preparing a new label-free amperometric immunosensor, which successfully immobilized the anti-carbofuran antibody on the PA/DpAu modified electrode surface for the detection of carbofuran. Due to PA's specially binding ability of the Fc fragment of the antibody molecules, the application of PA improves the capacity of antibody, thus enhance the detection sensitivity. With this strategy, a detection limit of 0.1924 ng/mL was achieved for carbofuran (Fig. 10.12) [31].

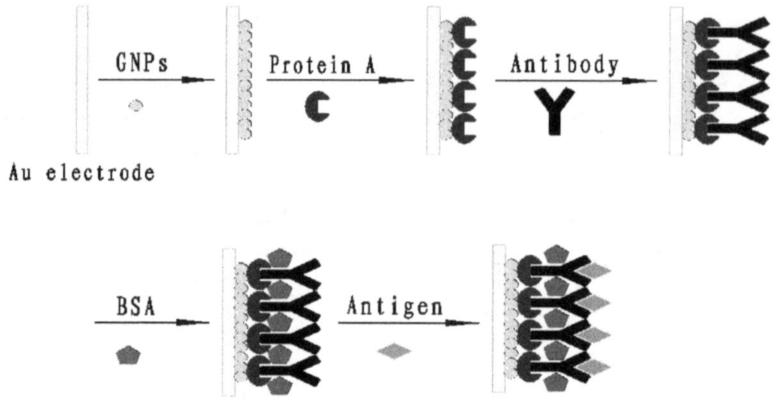

Fig. 10.12. Fabrication process of the stepwise Amperometric immunosensor based on a protein A/deposited gold nanocrystals modified electrode for carbofuran detection.

10.3.5. Avidin–biotin Affinity Reaction

One of the most valuable strategies for t he effective immobilization of biomaterial on different substrate s is based on the avidin–biotin affinity reaction [80].This interaction is highly resistant to a wide range of chemical (detergents, protein denaturants), pH range variations and high temperatures [81]. In addition, the avidin–biotin based immobilization method maintains the biological activity of the biomolecule being immobilized more successfully than other commonly used methods [82]. Ab/hapten can be readily linked to biotin without serious effects on their biological, chemical or physical properties. In particular, the extremely specific and high affinity interaction between the biotinylated antibodies and avidin (Ka ~10^{15} M^{-1}) leads to strong associations similar to the formation of a covalent bonding.

This technique was used to immobilize anti-atrazine antibodies on the surface of avidin-graphite-epoxy biocomposite based (Av-GEB-based) electrochemical transducer for the detection of atrazine in orange juice (Fig. 10.13) [83].

Another example of this technique is to attach the biotinylated anti-triazine Fab fragment to the polypyrrole (PPy)/ neutravidin modified electrode throughout the well-studied biotin–neutravidin interaction for the detection of atrazine. The immunosensor was very sensitive to atrazine antigen in the range of 0.1–200 ng/ml and the detection limit attained 0.1 ng/ml (Fig. 10.14) [52].

10.3.6. Self-assembled Monolayer (SAM)

Self-assembled monolayers (SAMs) have aroused much interest due to their potential applications in biosensors, biomolecular electronics and nanotechnology. This has been largely attributed to their inherent ordered arrangement and controllable properties.

SAMs can be formed by chemisorption of organic molecules containing groups like thiols, disulphides, amines, acids or silanes, on desired surfaces to fabricate immunosensors [84].

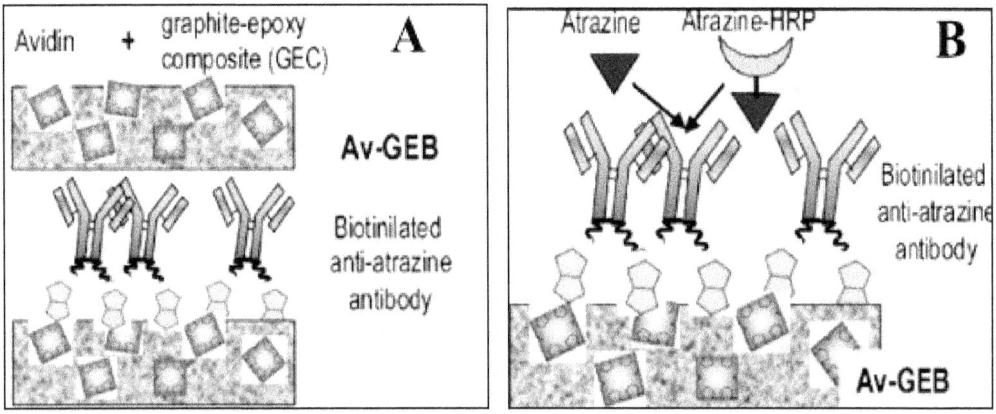

Fig. 10.13. The immobilization of anti-atrazine antibodies on the surface of the electrochemical transducer for the detection of atrazine in orange juice with Av-GEB-based electrochemical immunosensors (A) and the competitive immunological reaction (B).

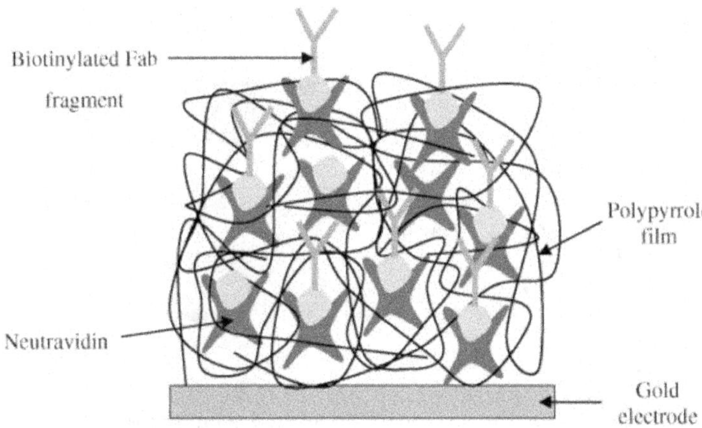

Fig. 10.14. Schematic representation of the immunosensor architecture based on the immobilization of biotinylated Fab fragment through the interaction biotin–neutravidin within the electro-generated polypyrrole for the detection of atrazine.

The stability, uniform surface structure and relative ease of varying thickness of a SAM make it suitable for development of biosensors. And the immobilization of biomolecules on a SAM requires very small amount and desired analytes can be easily detected via various transduction modes. The use of an appropriate SAM helps in oriented and controlled immobilization of biomolecules [85-87]. SAMs can be used to prevent protein denaturation at an electrode surface and for enhancing stability of biomolecules [88, 89].

A novel label-free amperometric immunosensor for the detection of carbofuran residues was developed based on immobilization anti-carbofuran antibody on deposited gold nanocrystals (DpAu)/4,4'-thiobisbenzenethiol (DMDPSE) multilayers ({DpAu/DMDPSE}$_n$) through Au-S bond by layer-by-layer self-assembly technology. Compared with a separate layer of DpAu/DMDPSE, the presence of the multiple membranes not only promoted electron-transfer reactions, but also increased the surface area to capture a large amount of antibodies, thus increased detection sensitivity with a detection limit of 0.06 ng/mL (Fig. 10.15) [40].

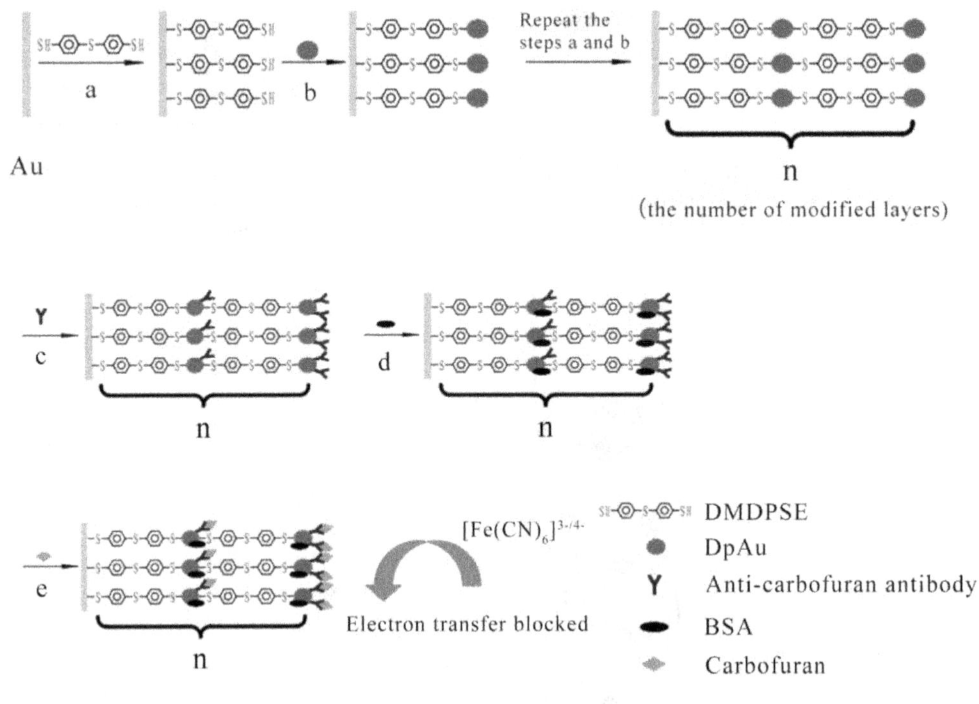

Fig. 10.15. Fabrication process of the stepwise the stepwise immunosensor based on deposited gold nanocrystals/ 4,4'-thiobisbenzenethiol for determination of carbofuran.

Recently, describes the development of an electrochemical immunosensor for the analysis of atrazine associated to biotinylated-Fab fragment K47 antibody. The sensors are based on mixed self-assembled monolayer consisting of 1,2 dipalmitoyl-sn-glycero-3-phosphoethanolamine-N-(biotinyl) (biotinyl-PE) and 16-mercapto-hexadecanoic acid (MHDA). The tethered neutravidin was used the biotin sites present in the mixed monolayer, with those associated to the biotinyl-Fab fragment K47 antibody (Fig. 10.16) [51].

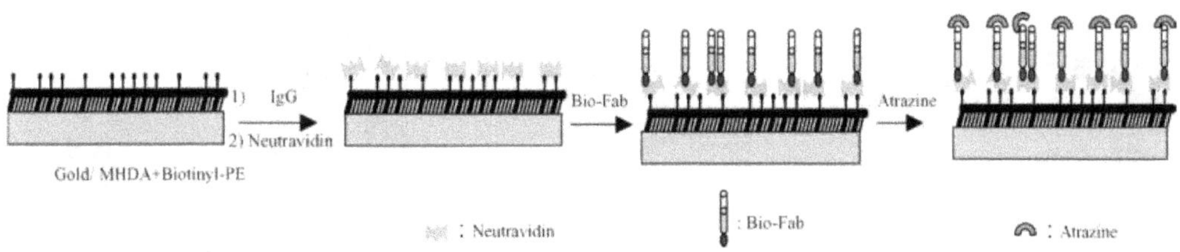

Fig. 10.16. Schematic showing the assembly of a mixed SAM based immunosensor.

10.3.7. Nanoparticles

In addition to these conventional methods, new materials such as nanoparticles have been employed in immobilizing Ab/ hapten when constructing immunosensors. Gold nanoparticles (GNPs) have been widely used for immobilization of biomolecules due to their large specific surface area, high surface free energy and biocompatibility. GNPs can adsorb biomolecules and play an important role in the immobilization of biomolecules for biosensor construction [90]. So far, GNPs have been widely applied in the biosensors for detection of pesticide residues [91-93]. Biological interactions, such as biotin/streptavidin interactions can be used to easily immobilize the Ab on the surface of nanoparticales. Combining the catalytic and protein-adsorptive characteristics of gold nanoparticles, Hu et al., prepared a label-free electrochemical immunosensor with paraoxon antibodies loaded on the gold nanoparticles to monitor the concentration of paraoxon in aqueous samples with a detection limit of 12 μg/L. TEM experiment of colloidal gold indicated an average diameter to be 10±0.5 nm (Fig. 10.17) [45].

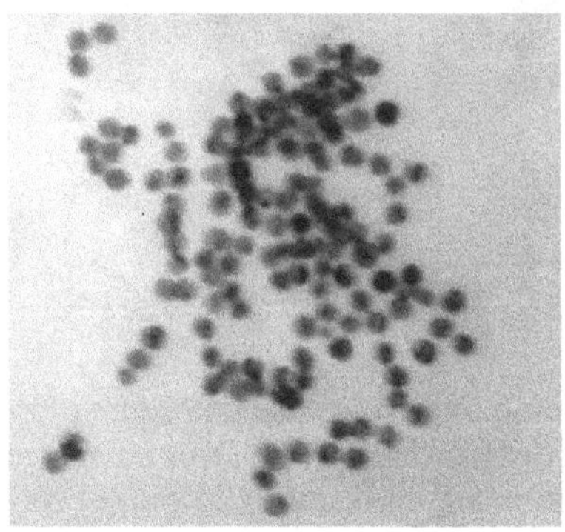

Fig. 10.17. TEM of gold nanoparticles.

Recently, a type of ordered three-dimensional (3D) gold (Au) nanoclusters obtained by two-step electrodeposition using the spatial obstruction/direction of the polycarbonate membrane was reported. The electrodeposited Au nanoclusters built direct electrical contact and immobilization interface with protein molecules without post-modification and positioning (Fig. 10.18) [47].

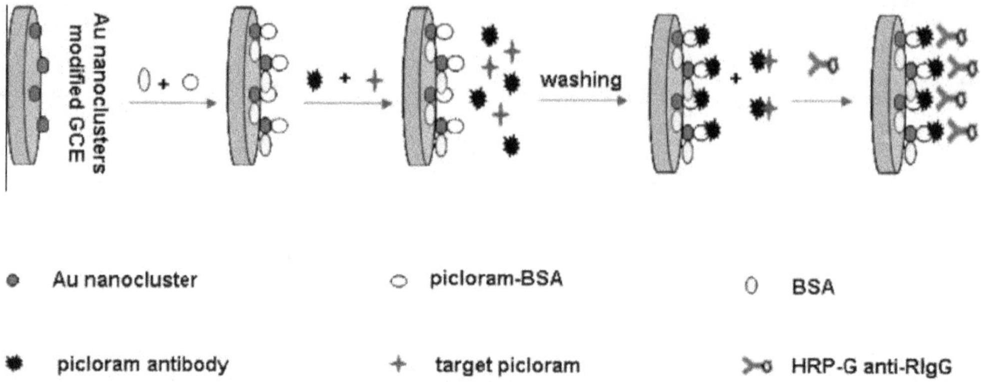

Fig. 10.18. Schematic diagram of the immunosensor based on 3D gold (Au) nanoclusters and competitive immunoreaction.

Recently, Sun et al., developed a novel immunosensor for direct determination of carbofuran concentration by immobilizing anti-carbofuran antibody on the gold nanoparticles(GNPs)/Thiourea (TU)/GNPs composite film with the detection limit 0.11 ng/mL. The presence of GNPs can enhance electron transfer between Ab and electrode surface and provide a favorable microenvironment for immunoreaction. In addition, GNPs on the composite film had a profound influence on enhancing the conductivity and biocompatibility (Fig. 10.19) [37].

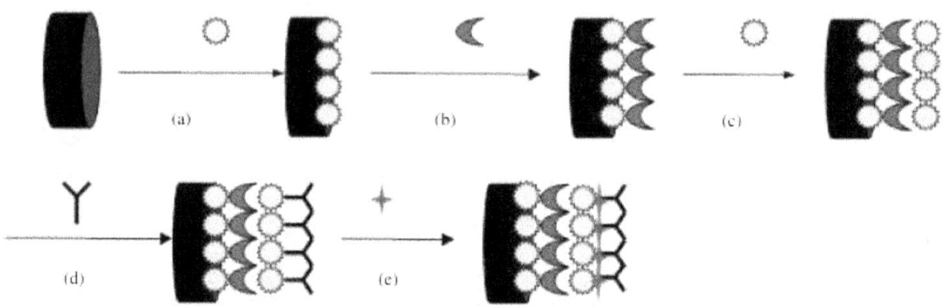

Fig. 10.19. Schematic illustration of the stepwise procedure of the immunosensor preparation: (a) electrodeposited GNPs; (b) modified TU; (c) electrodeposited the second layer of GNPs; (d) adsorption o f anti-carbofuran and (e) BSA blockin.

In addition, Bhalla et al., reported a label-free detection of phenylurea herbicides by impedance spectroscopy based on immobilization class specific anti-diuron antibodies on gold nanoparticles (~20 nm). Gold nanoparticles, used as signal enhancers cum immobilization matrix, were electrodeposited on carbon screen-printed electrodes (SPE) and functionalized with specific anti-diuron antibodies for the development of bio-interface (Fig. 10.20) [44].

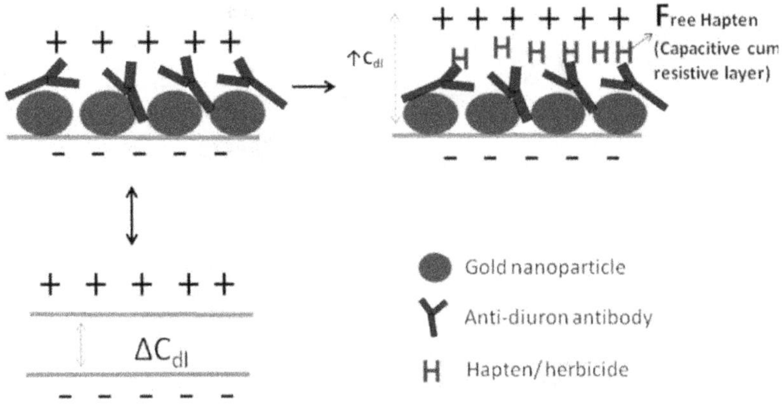

Fig. 10.20. Schematic illustration of the stepwise procedure of the immunosensor preparation.

Another example based on nanoparticles technology is reported for sensitive atrazine determination based on magnetic beads. The immuno-method is a competitive solid-phase immunoassay where the anti-atrazine antibody is immobilized on the magnetic beads surface and fixed at the reaction cell bottom using a simple magnet, which generates a magnetic field. The performance of magnetic beads-based immunoassay for atrazine determination was evaluated demonstrating that the magnetic beads-based immunoassay is one of the most sensitive methods for atrazine determination (Fig. 10.21) [94].

In addition, chitosan (CHIT) containing large groups of $-NH_2$ and $-OH$ has been widely used as dispersion because of its good biocompatibility and film-forming ability. Multiwall carbon nanotubes (MWCNTs) have received increasing interest due to their great chemical stability, large aspect ratio, excellent electrical conductivity, and extremely high mechanical strength and stiffness and demonstrated to be an excellent material for the development of electrochemical sensors. Recently, a simple, stable, sensitive, and low-cost amperometric immunosensor based on $\{MWCNTs\text{-}COOH\text{-}CHIT\}_2$/GNPs/GCE. The poly (diallyldimethylammonium chloride) (PDDA), positively-charged, was used to assemble the second layer of MWCNTs-CHIT (negatively charged) was reported (Fig. 10.22) [95].

10.4. New Trends and Challenges

10.4.1. Miniaturization

New analytical approaches are oriented to the development of portable systems with high accuracy, low-cost, short-time response, and that can provide qualitative information about the composition of a sample with minimum preparation. Future advances in immobilization will likely focus on directing biorecognition elements to addressable locations on micro or nano-sensor arrays.

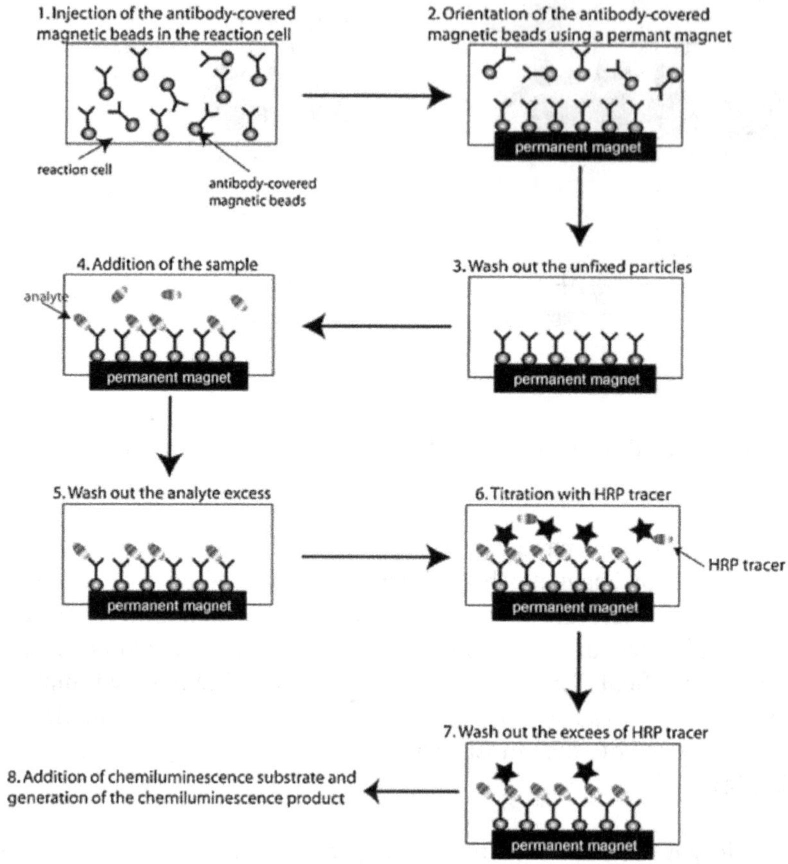

Fig. 10.21. Principle steps for performing the magnetic beads-based immunoassay.

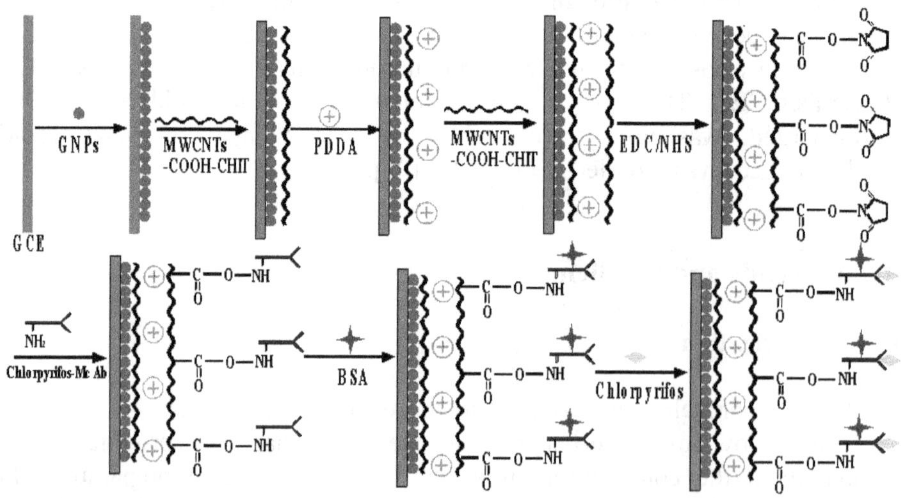

Fig. 10.22. Schematic illustration of the stepwise immunosensor fabrication process.

A microelectrode, its dimensions are in the micrometer range, which become a trend to replace common electrode due to its miniaturization, faster response, greater sensitivity and increased response per unit electrode surface area (greater current density, increasing the signal-to-noise ratio). Ramon-Azcon et al., have reported an array of interdigitated μ-electrodes (IDμE) for atrazine detection [32].

10.4.2. High Throughput of Detection Samples

The ability to construct arrays of microelectrode will likely allow current multianalyte detection of several compounds to be expanded to accommodate the analysis of perhaps hundreds or thousands of separate compounds. The combination of microelectrode and microfluidic devices as analytical systems will become a trend to realized high throughput due to their significant reduction of reagent consumption and low operating costs as well as high throughput capability.

10.4.3. Integration of Detection System

One of the challenges that must be met for this type of system would be the development of parallel computational methods to convert electronic responses for each analyte into meaningful concentration data. Recently, silicabased monoliths, coupled with micro-fluidic devices, have been used as an attractive alternative to packed columns for the analysis of proteins, peptides and nucleic acids with special features of low diffusion resistance during mass transfer, controllable porosity and low back pressure compared to packed columns.

10.4.4. Real Samples Detections

Despite the promise of immunosensors, they do have certain limitations. For example, few immunosensors are commercially available at the present time and are yet to be established as research or routine tools, due to a lack of validated protocols for a wide range of sample matrices.

10.4.5. Using Aptamer to Replace Antibody

Immunosensor, itself still has several problems, such as biomolecule deactivation or leaking and high diffusion resistance of the substrate/biocomponent, which are also key factors in the development of immunosensors that can be successfully applied to pesticide detection. Aptamers are short, single-stranded, functional DNA or RNA molecules selected from random-sequencenucleic acid combinatorial libraries by Systematic Evolution of Ligandsby Exponential Enrichment (SELEX).The aptamers are more chemically stable, smaller in size, cheaper and can bind nearly any target with high afnity and specicity compared to antibody [96].

Acknowledgements

This work was supported by the National Natural Science Foundation of China (No. 30972055, 31101286), Agricultural Science and Technology Achievements Transformation Fund Projects of the Ministry of Science and Technology of China (No. 2011GB2C60020) and Shandong Provincial Natural Science Foundation, China (No.Q2008D03).

References

[1]. M. I. Pinto, G. Sontag, R. J. Bernardino, J. P. Noronha, Pesticides in water and the performance of the liquid-phase microextraction based techniques. A review, *Microchemical Journal*, Vol. 96, Issue 2, 2010, pp. 225-237.

[2]. J. L. Tadeo, C. Sanchez-Brunete, B. Albero, L. Gonzalez, Analysis of pesticide residues in juice and beverages, *Critical Reviews in Analytical Chemistry*, Vol. 34, Issue 2, 2004, pp. 165-175.

[3]. S. B. Valdez, D. E. I. Garcia, M. S. Wiener, Impact of pesticides use on human health in Mexico: a review, *Reviews on Environmental Health*, Vol. 15, Issue 4, 2000, pp. 399-412.

[4]. C. M. Kin, T. G. Huat, Headspace solid-phase microextraction for the evaluation of pesticide residue contents in cucumber and strawberry after washing treatment, *Food Chemistry*, Vol. 123, Issue 3, 2010, pp. 760-764.

[5]. E. P. Syrago-Styliani, G. Evagelos, T. Anthony, A. S. Panayotis, Gas chromato-graphic-tandem mass spectrometric method for the quantitation of carbofuran, carbaryl and their main metabolites in applicators' urine, *Journal of Chromatography A*, Vol. 1108, Issue 1, 2006, pp. 99-110.

[6]. O. Eva, S. Manuel, Monitoring some phenoxyl-type N-methylcarbamate pesticide residues in fruit juices using high-performance liquid hromatography with peroxyoxalate-chemiluminescence detection, *Journal of Chromatography A*, Vol. 1007, Issue 1-2, 2003, pp. 197–201.

[7]. T. B. Jiang, H. B. Halsall, W. R. Heineman, T. Giersch, B. Hock, Capillary enzyme immunoassay with electrochemical detection for the determination of atrazine in water, *Journal of Agricultural and Food Chemistry*, Vol. 43, Issue 4, 1995, pp. 1098-1104.

[8]. J. Ramón-Azcón, E. Valera, Á. Rodríguez, A. Barranco, B. Alfaro, F. Sanchez-Baeza, M.-P. Marco, An impedimetric immunosensor based on interdigitated microelectrodes (IDμE) for the determination of atrazine residues in food samples, *Biosensors and Bioelectronics*, Vol. 23, Issue 9, 2008, pp. 1367–1373.

[9]. J. Gascón, A. Oubiña, B. Ballesteros, D. Barceló, F. Camps, M.-P. Marco, M. A. González-Martínez, S. Morais, R. Puchades, A. Maquieira, Development of a highly sensitive enzyme-linked immunosorbent assay for atrazine Performance evaluation by flow injection immunoassay, *Analytica Chimica Acta*, Vol. 347, Issue 1-2, 1997, pp. 149-162.

[10]. Y. H. Zheng, T. C. Hua, D. W. Sun, J. J. Xiao, F. Xu, F. F. Wang, Detection of dichlorvos residue by flow injection calorimetric biosensor based on immobilized chicken liver esterase, *Journal of Food Engineering*, Vol. 74, Issue 1, 2006, pp. 24-29.

[11]. N. L. Pacioni, A. V. Veglia, Determination of carbaryl and carbofuran in fruits and tap water by cyclodextrin enhanced fluorimetric method, *Analytica Chimica Acta*, Vol. 488, Issue 2, 2003, pp. 193–202.

[12]. Z. M. Yu, G. H. Zhao, M. C. Liu, Y. Z. Lei, M. F. Li, Fabrication of a novel atrazine biosensor and its subpart-per-trillion levels sensitive performance, *Environmental Science & Technology*, Vol. 44, Issue 20, 2010, pp. 7878-7883.

[13]. J. L. Marty, D. Garcia, R. Rouillon, Biosensor: potential in pesticide detection, *Trends in Analytical Chemistry*, Vol. 14, Issue 7, 1995, pp. 329-333.

[14]. S. Bhand, I. Surugiu, A. Dzgoev, K. Ramanathan, P. V. Sundaram, B. Danielsson, Immuno-arrays for multianalyte analysis of chlorotriazines, *Talanta*, Vol. 65, Issue 2, 2005, pp. 331-336.

[15]. M. A. Kumar, R. S. Chouhan, M. S. Thakur, B. E. Amita Rani, B. Mattiasson, N. G. Karanth, Automated flow enzyme-linked immunosorbent assay (ELISA) system for analysis of methyl parathion, *Analytica Chimica Acta*, Vol. 560, Issue 1-2, 2006, pp. 30-34.

[16]. A. P. Deng, H. Yang, A multichannel electrochemical detector coupled with an ELISA microtiter plate for the immunoassay of 2, 4-dichlorophenoxyacetic acid, *Sensors and Actuators B: Chemical*, Vol. 124, Issue 1, 2007, pp. 202-208.

[17]. G. L. Qian, L. M. Wang, Y. R. Wu, Q. Zhang, Q. Sun, Y. Liu, F. Q. Liu, A monoclonal antibody-based sensitive enzyme-linked immunosorbent assay (ELISA) for the analysis of the organophosphorous pesticides chlorpyrifos-methyl in real samples, *Food Chemistry*, Vol. 117, Issue 2, 2009, pp. 364-370.

[18]. Y. H. Liu, C. M. Wang, W. J. Gui, J. C. Bi, M. J. Jin, G. N. Zhu, Development of a sensitive competitive indirect ELISA for parathion residue in agricultural and environmental samples, *Ecotoxicology and Environmental Safety*, Vol. 72, Issue 6, 2009, pp. 1673-1679.

[19]. H. Barchanska, E. Jodo, R. G. Price, I. Baranowska, R. Abuknesha, Monitoring of atrazine in milk using a rapid tube-based ELISA and validation with HPLC, *Chemosphere*, Vol. 87, Issue 11, 2012, pp. 1330-1334.

[20]. T. M. Anh, S. V. Dzyadevych, M. C. Van, N. J. Renault, C. N. Duc, J. M. Chovelon, Conductometric tyrosinase biosensor for the detection of diuron, atrazine and its main metabolites, *Talanta*, Vol. 63, Issue 2, 2004, pp. 365-370.

[21]. R. P. Deo, J. Wang, I. Block, A. Mulchandani, K. A. Joshi, M. Trojanowicz, F. Scholz, W. Chen, Y. Lin, Determination of organophosphate pesticides at a carbon nanotube/ organophosphorous hydrolase electrochemical biosensor, *Analytica Chimic Acta*, Vol. 530, Issue 2, 2005, pp. 185-189.

[22]. S. Laschi, D. Ogonczyk, I. Palchetti, M. Mascini, Evaluation of pesticide-induced acetylcholinesterase inhibition by means of disposable carbon-modified electrochemical biosensors, *Enzyme and Microbial Technology*, Vol. 40, Issue 3, 2007, pp. 485-489.

[23]. S. P. Zhang, L. G. Shan, Z. R. Tian, Y. Zheng, L. Y. Shi, D. S. Zhang, Study of enzyme biosensor based on carbon nanotubes modified electrode for detection of pesticides residue, *Chinese Chemical Letters*, Vol. 19, Issue 5, 2008, pp. 592-594.

[24]. T. Liu, H. C. Su, X. j. Qu, P. Ju, L. Cui, S. Y. Ai, Acetylcholinesterase biosensor based on 3-carboxyphenylboronic acid/reduced graphene oxide–gold nanocomposites modified electrode for amperometric detection of organophosphorous and carbamate pesticides, *Sensors and Actuators B: Chemical,* Vol. 160, Issue 1, 2011, pp. 1255-1261.

[25]. N. Chauhan, C. S. Pundir, An amperometric biosensor based on acetylcholinesterase immobilized onto iron oxide nanoparticles/multi-walled carbon nanotubes modified gold electrode for measurement of organophosphorous insecticides, *Analytica Chimica Acta*, Vol. 701, Issue 1, 2011, pp. 66-74.

[26]. P. Raghu, T. Madhusudana Reddy, B. E. Kumara Swamy, B. N. Chandrashekar, K. Reddaiah, M. Sreedhar, Development of AChE biosensor for the determination of methyl parathion and monocrotophos in water and fruit samples: A cyclic voltammetric study, *Journal of Electroanalytical Chemistry*, Vol. 665, 2012, pp. 76-82.

[27]. N. Chauhan, C. S. Pundir, An amperometric acetylcholinesterase sensor based on Fe_3O_4 nanoparticle/multi- walled carbon nanotube-modified ITO-coated glass plate for the detection of pesticides, *Electrochimica Acta*, Vol. 67, 2012, pp. 79-86.

[28]. X. S. Jiang, D. Y. Li, X. Xu, Y. B. Ying, Y. B. Li, Z. Z. Ye, J. P. Wang, Immunosensors for detection of pesticide residues, *Biosensors and Bioelectronics*, Vol. 23, Issue 11, 2008, pp. 1577-1587.

[29]. C. Raman Suri, R. Boro, Y. Nangia, S. Gandhi, P. Sharma, N. Wangoo, K. Rajesh, G. S. Shekhawat, Immunoanalytical techniques for analyzing pesticides in the environment, *Trends in Analytical Chemistry*, Vol. 28, Issue 1, 2009, pp. 29-39.

[30]. E. Mallat, C. Barzen, R. Abuknesha, G. Gauglitz, D. Barcelo, Part per trillion level determination of isoproturon in certified and estuarine water samples with a direct optical immunosensor, *Analytica Chimica Acta*, Vol. 426, Issue 2, 2001, pp. 209-216.

[31]. X. Sun, Y. Zhu, X. Y. Wang, Amperometric immunosensor based on a protein A/ deposited gold nanocrystals modified electrode for carbofuran detection, *Sensors*, Vol. 11, Issue 12, 2011, pp. 11679-11691.

[32]. J. Ramon-Azcon, E. Valera, A. Rodrıguez, A. Barranco, B. Alfaro, F. Sanchez-Baeza, M.-P. Marco, An impedimetric immunosensor based on interdigitated microelectrodes (IDμE) for the determination of atrazine residues in food samples, *Biosensors and Bioelectronics*, Vol. 23, Issue 9, 2008, pp. 1367-1373.

[33]. P. Sharma, K. Sablok, V. Bhalla, C. R. Suri, A novel disposable electrochemical immunosensor for phenyl urea herbicide diuron, *Biosensors and Bioelectronics*, Vol. 26, Issue 10, 2011, pp. 4209-4212.

[34]. M. F. Yulaev, R. A. Sitdikov, N. M. Dmitrieva, E. V. Yazynina, A. V. Zherdev, B. B. Dzantiev, Development of a potentiometric immunosensor for herbicide simazine and its application for food testing, *Sensors and Actuators B*, Vol. 75, Issue 1, 2001, pp. 129-135.

[35]. K. Grennan, G. Strachan, A. J. Porter, A. J. Killard, M. R. Smyth, Atrazine analysis using an amperometric immunosensor based on single-chain antibody fragments and regeneration-free multi-calibrant measurement, *Analytica Chimica Acta*, Vol. 500, Issue 1, 2003, pp. 287-298.

[36]. E. Valera, J. Ramon-Azcon, A. Barranco, B. Alfaro, F. Sanchez-Baeza, M.-P. Marco, A. Rodríguez, Determination of atrazine residues in red wine samples. A conductimetric solution, *Food Chemistry*, Vol. 122, Issue 3, 2010, pp. 888-894.

[37]. X. Sun, Q. Q. Li, X. Y. Wang, Amperometric immunosensor based on gold nanoparticles and saturated thiourea for carbofuran detection, *IEEE Sensors Journal*, Vol. 12, Issue 6, 2012, pp. 2071-2076.

[38]. X. Sun, S. Y. Du, X. Y. Wang, W. P. Zhao, Q. Q. Li, A label-free electrochemical immunosensor for carbofuran detection based on a sol-gel entrapped antibody, *Sensors*, Vol. 11, Issue 10, 2011, pp. 9520-9531.

[39]. X. Sun, X. Y. Wang, S. Y. Du, Label-free amperometric immunosensor for the detection of carbofuran pesticide, *Sensor Letters*, Vol. 9, Issue 3, 2011, pp. 958-963.

[40]. X. Sun, Y. Zhu, X. Y. Wang, Amperometric immunosensor based on deposited gold nanocrystals/4,4'- thiobisbenzenethiol for determination of carbofuran, *Food Control*, Vol. 28, Issue 1, 2012, pp. 184-191.

[41]. B. B. Dtantiev, A. V. Zherdev, Electrochemical immunosensors for determination of the pesticides 2,4-dichlorophenoxyatietic and 2,4,5-trichlorophenoxyacetic acids, *Biosensors and Bioelectronics*, Vol. 11, Issue l-2, 1996, pp. 179-185.

[42]. L. Zhang, M. R. Wang, C. Y. Wang, X. Y. Hu, G. X. Wang, Label-free impedimetric immunosensor for sensitive detection of 2,4-dichlorophenoxybutyric acid (2,4-DB) in soybean, *Talanta,* Vol. 101, 2012, pp. 226-232.

[43]. N. F. Starodub, B. B. Dzantiev, V. M. Starodub, A. V. Zherdev, Immunosensor for the determination of the herbicide simazine based on an ion-selective field-effect transistor, *Analytica Chimica Acta*, Vol. 424, Issue l, 2000, pp. 37-43.

[44]. Vijayender Bhalla, Priyanka Sharma, Satish K. Pandey, C. Raman Suri, Impedimetric label-free immunodetection of phenylurea class of herbicides, *Sensors and Actuators B,* Vol. 171–172, August–September 2012, pp. 1231–1237.
[45]. S. Q. Hu, J. W. Xie, Q. H. Xu, K. T. Rong, G. L. Shen, R. Q. Yu, A label-free electrochemical immunosensor based on gold nanoparticles for detection of paraoxon, *Talanta*, Vol. 61, Issue 6, 2003, pp. 769-777.
[46] Guozhen Liu, Dandan Song, FengJuan Chen, Towards the fabrication of a label-free amperometric immunosensor using SWNTs for direct detection of paraoxon, *Talanta*, Vol. 104, 30 January 2013, pp.103–108.
[47]. L. J. Chen, G. M. Zeng, Y. Zhang, L. Tang, D. L. Huang, C. Liu, Y. Pang, J. Luo, Trace detection of picloram using an electrochemical immunosensor based on three- dimensional gold nanoclusters, *Analytical Biochemistry*, Vol. 407, Issue 2, 2010, pp. 172-179.
[48]. E. Valera, J. Ramon-Azcon, F. J. Sanchez, M. P. Marco, A. Rodriguez, Conductimetric immunosensor for atrazine detection based on antibodies labelled with gold nano- particles, *Sensors and Actuators B*, Vol. 134, Issue 1, 2008, pp. 95-103.
[49]. L. Campanella, S. Eremin, D. Lelo, E. Martini, M. Tomassetti, Reliable new immunosensor for atrazine pesticide analysis, *Sensors and Actuators B*, Vol. 156, Issue 1, 2011, pp. 50-62.
[50]. K. Grennan, G. Strachan, A. J. Porter, A. J. Killard, M. R. Smyth, Atrazine analysis using an amperometric immunosensor based on single-chain antibody fragments and regeneration-free multi-calibrant measurement, *Analytica Chimica Acta*, Vol. 500, Issue 1-2, 2003, pp. 287-298.
[51]. S. Hleli, C. Martelet, A. Abdelghani, N. Burais, N. Jaffrezic-Renault, Atrazine analysis using an impedimetric immunosensor based on mixed biotinylated self-assembled monolayer, *Sensors and Actuators B*, Vol. 113, Issue 2, 2006, pp. 711-717.
[52]. C. Esseghaier, S. Helali, H. B. Fredj, A. Tlili, A. Abdelghani, Polypyrrole–neutravidin layer for impedimetric biosensor, *Sensors and Actuators B*, Vol. 131, Issue 2, 2008, pp. 584-589.
[53]. E. Zacco, R. Galve, M. P. Marco, S. Alegret, M. I. Pividori, Electrochemical biosensing of pesticide residues based on affinity biocomposite platforms, *Biosensors and Bioelectronics*, Vol. 22, Issue 8, 2007, pp. 1707-1715.
[54]. R. E. Ionescu, C. Gondran, L. Bouffier, N. Jaffrezic-Renault, C. Martelet, S. Cosnier, Label-free impedimetric immunosensor for sensitive detection of atrazine, *Electrochimica Acta*, Vol. 55, Issue 21, 2010, pp. 6228-6232.
[55]. H. V. Tran, R. Yougnia, S. Reisberg, B. Piro, N. Serradji, T. D. Nguyen, L. D. Tran, C. Z. Dong, M. C. Pham, A label-free electrochemical immunosensor for direct, signal-on and sensitive pesticide detection, *Biosensors and Bioelectronics*, Vol. 31, Issue 1, 2012, pp. 62-68.
[56]. Wei Weia, Xiaomin Zong, Xuan Wang, Lihong Yin, Yuepu Pu, Songqin Liu, A disposable amperometric immunosensor for chlorpyrifos-methyl based on immunogen/platinum doped silica sol-gel film modified screen-printed carbon electrode, *Food Chemistry*, 135, 2012, pp. 888–892.
[57]. J. Lin, C. W. Brown, Sol-gel glass as a matrix for chemical and biochemical sensing, *Trends in Analytical Chemistry*, Vol. 16, Issue 4, 1997, pp. 200-211.
[58]. J. M. Nam, C. S. Thaxton, C. A. Mirkin, Nanoparticle-based biobar codes for the ultrasensitive detection of protein, *Science*, Vol. 301, No 5641, 2003, pp. 1884-1886.
[59]. J. Wang, A. Ibanez, M. P. Chatrathi, Microchip-based amperometric immunoassays using redox tracers, *Electrophoresis*, Vol. 23, Issue 21, 2002, pp. 3744-3749.
[60]. J. S. Yuk, S. J. Yi, H. G. Lee, H. J. Lee, Y. M. Kim, K. S. Ha, Characterization of surface plasmon resonance wavelength by changes protein concentration on protein chips, *Sensors Actuators B*, Vol. 94, Issue 2, 2003, pp. 161-164.

[61]. X. F. Yuan, D. Fabregat, K. Yoshimoto, Y. Nagasaki, Development of a high- performance immunolatex based on "soft landing" antibody immobilization mechanism, *Colloids and Surfaces B: Biointerfaces*, 99, 1 November 2012, pp.45-52.
[62]. J. M. Peula, R. Hidalgo-Alvarez, F. J. de Las Nieves, Covalent binding of proteins to acetal-functionalized latexes. I. Physics and chemical adsorption and electrokinetic characterization, *Journal of Colloid and Interface Science*, Vol. 201, Issue 2, 1998, pp. 132-138.
[63]. J. A. Schwarz, C. Contescu, K. Putyera, Dekker Encyclopedia of Nanoscience and Nanotechnology, *CRC Press,* Vol. 3, 2004, pp. 2423-2433.
[64]. K. V. Gobi, S. J. Kim, H. Tanaka, Y. Shoyama, N. Miura, Novel surface plasmon resonance (SPR) immunosensor based on monomolecular layer of physically-adsorbed ovalbumin conjugate for detection of 2, 4-dichlorophenoxyacetic acid and atomic force microscopy study, *Sensors and Actuators B*, Vol. 123, Issue 1, 2007, pp. 583-593.
[65]. H. Zhu, M. Snyder, Protein chip technology, *Current Opinion in Chemical Biology*, Vol. 7, Issue 1, 2003, pp. 55-63.
[66]. J. S. Yuk, S.-H. Jung, J. W. Jung, J.-A. Han, Y.-M. Kim, K.-S. Ha, *Proteomics*, Vol. 4, Issue 1, 2004, pp. 3468-3476.
[67]. S. H. Jung, H. Y. Son, J. S. Yuk, J. W. Jung, K. H. Kim, C. H. Lee, H. Hwang, K. S. Ha, Oriented immobilization of antibodies by a self-assembled monolayer of 2-(biotinamido) ethanethiol for immunoarray preparation, *Colloids and Surface B: Biointerfaces*, Vol. 47, Issue 1, 2006, pp. 107-111.
[68]. T. Kalib, P. Sklidal, A disposable amperometric immunosensor for 2, 4-dichloro-phenoxyacetic acid, *Analytica Chimica Acta*, Vol. 304, Issue 3, 1995, pp. 361-368.
[69].X. Sun, Y. Y. Cao, Z. L. Gong, X. Y. Wang, Y. Zhang, and J. M. Gao, An Amperometric Immunosensor Based on Multi-Walled Carbon Nanotubes-Thionine- Chitosan Nanocomposite Film for Chlorpyrifos Detection, *Sensors*, Vol. 12, Issue 12, 2012, pp. 17247-17261.
[70]. E. Valera, D. Muniz, A. Rodriguez, Fabrication of flexible interdigitated μ-electrodes (FIDμEs) for the development of a conductimetric immunosensor for atrazine detection based on antibodies labelled with gold nanoparticles, *Microelectronic Engineering*, Vol. 87, Issue 2, 2010, pp. 167-173.
[71]. K. Eguchi, T. Hashiguchi, K. Sumiyoshi, H. Arai, Optical detection of nitrogen monoxide by metal porphine dispersed in an amorphous silica matrix, *Sensors and Actuators B*, Vol. 1, Issue 1-6, 1990, pp. 154-157.
[72]. A. Turniansky, D. Avnir, A. Bronshtein, N. Aharonson, M. Altstein, Sol-gel entrapment of monoclonal anti-atrazine antibodies, *Journal of Sol-Gel Science and Technology*, Vol. 7, 1996, pp. 135-143.
[73]. M. Gerard, A. Chaubey, B. D. Malhotra, Application of conducting polymers to biosensors, *Biosensors and Bioelectronics*, Vol. 17, Issue 5, 2002, pp. 345-359.
[74]. C. Dhand, M. Das, M. Datta, B. D. Malhotra, Recent advances in polyaniline based biosensors, *Biosensors and Bioelectronics*, Vol. 26, Issue 6, 2011, pp. 2811-2821.
[75]. M. Nisnevitch, M. A. Firer, The solid phase in affinity chromatography: strategies for antibody attachment, *Journal of Biochemical and Biophysical Methods*, Vol. 49, Issue 1-3, 2001, pp. 467-480.
[76]. R. Danczyk, B. Krieder, A. North, T. Webster, H. Hogen Esch, A. Rundell, Comparison of antibody functionality using different immobilization methods, *Biotechnology and Bioengineering*, Vol. 84, Issue 2, 2003, pp. 215-223.
[77]. K. Bonroy, F. Frederix, G. Reekmans, E. Dewolf, R. De Palma, G. Borghs, P. Declerck, B. Goddeeris, Comparison of random and oriented immobilisation of antibody fragments on mixed self-assembled monolayers, *Journal of Immunological Methods*, Vol. 312, Issue 1-2, 2006, pp. 167-181.

[78]. B. Guss, M. Eliasson, A. Olsson, M. Uhlen, A. K. Frej, H. Jornvall, J. I. Flock, M. Lindberg, Structure of the IgG-binding regions of streptococcal protein G, *EMBO Journal*, Vol. 5, Issue 7, 1986, pp. 1567-1575.

[79]. Y. Jung, H. J. Kang, J. M. Lee, S. O. Jung, W. S. Yun, S. J. Chung, B. H. Chung, Controlled antibody immobilization onto immunoanalytical platforms by synthetic peptide, *Analytical Biochemistry*, Vol. 374, Issue 1, 2008, pp. 99-105.

[80]. P. B. Luppa, L. J. Sokoll, D. W. Chan, Immunosensors-principles and applications to clinical chemistry, *Clinica Chimica Acta*, Vol. 314, Issue 1-2, 2001, pp. 1-26.

[81]. M. L. Jones, G. P. Kurzban, Noncooperativity of biotin binding to tetrameric streptavidin, *Biochemistry*, Vol. 34, Issue 37, 1995, pp. 11750-11756.

[82]. D. Hernández-Santos, M. Díaz-González, M. B. González-García, A. Costa-García, *Analytical Chemistry*, Vol. 76, Issue 37, 2004, pp. 6887.

[83]. O. Lazcka, F. J. Del Campob, F. X. Munoz, Pathogen detection: A perspective of traditional methods and biosensors, *Biosensors and Bioelectronics*, Vol. 22, Issue 7, 2007, pp. 1205-1217.

[84]. S. K. Arya, P. R. Solanki, M. Datta, B. D. Malhotra, Recent advances in self-assembled monolayers based biomolecular electronic devices, *Biosensors and Bioelectronics*, Vol. 24, Issue 9, 2009, pp. 2810-2817.

[85]. S. H. Jung, H. Y. Son, J. S. Yuk, J. W. Jung, K. H. Kim, C. H. Lee, H. Hwang, K. S. Ha, Oriented immobilization of antibodies by a self-assembled monolayer of 2-(biotinamido) ethanethiol for immunoarray preparation, *Colloids and Surfaces B: Biointerfaces*, Vol. 47, Issue 1, 2006, pp. 107-111.

[86]. S. J. Kim, K. V. Gobi, H. Tanaka, Y. Shoyama, N. Miura, A simple and versatile self-assembled monolayer based surface plasmon resonance immunosensor for highly sensitive detection of 2,4-D from natural water resources, *Sensors and Actuators B*, Vol. 130, No. 1, 2008, pp. 281-289.

[87]. F. Cheng, L. J. Gamble, D. G. Castner, XPS, TOF-SIMS, NEXAFS, and SPR characterization of nitrilotriacetic acid-terminated self-assembled monolayers for controllable immobilization of proteins, *Analytical Chemistry*, Vol. 80, Issue 7, 2008, pp. 2564-2573.

[88]. A. K. M. Kafi, D. Y. Lee, S. H. Park, Y. S. Kwon, Development of peroxide biosensor made of thiolated- viologen and hemoglobin modified gold electrode, *Microchemical Journal*, Vol. 85, Issue 2, 2007, pp. 308-313.

[89]. V. R. Sarath Babu, M. A. Kumar, N. G. Karanth, M. S. Thakur, Stabilization of immobilized glucose oxidase against thermal inactivation by silanization for biosensor applications, *Biosensors and Bioelectronics*, Vol. 19, Issue 10, 2004, pp. 1337-1341.

[90]. B. Hong, K. A. Kang, Biocompatible, nanogold-particle fluorescence enhancer for fluorophore mediated, optical immunosensor, *Biosensors and Bioelectronics*, Vol. 21, Issue 7, 2006, pp. 1333-1338.

[91]. P. Sharma, K. Sablok, V. C. Bhalla, A novel disposable electrochemical immunosensor for phenyl urea herbicide diuron, *Biosensors and Bioelectronics*, Vol. 26, Issue 10, 2011, pp. 4209-4212.

[92]. D. Du, M. H. Wang, J. Cai, Y. H. Qin, A. D. Zhang, One-step synthesis of multiwalled carbonnanotubes-gold nanocomposites for fabricating amperometric acetylcholinesterase biosensor, *Sensors and Actuators B: Chemical*, Vol. 143, Issue 2, 2010, pp. 524-529.

[93]. O. Shulga, J. R. Kirchhoff, An acetylcholines terase enzyme electrode stabilized by an electrodeposited gold nanoparticle layer, *Electrochemistry Communications*, Vol. 9, Issue 5, 2007, pp. 935-940.

[94]. M. Tudorache, A. Tencaliec, C. Bala, Magnetic beads-based immunoassay as a sensitive alternative for atrazine analysis, *Talanta*, Vol. 77, Issue 2, 2008, pp. 839-843.

[95]. X. Sun, L. Qiao, X. Y. Wang, A novel amperometric immunosensor based on {MWCNTs-COOH-CHIT}2/ GNPs for detection of chlorpyrifos, *Sensors and Transducers*, Vol. 146, Issue 11, 2012, pp. 96-109.

[96]. A. D. Girolamo, M. McKeague, J. D. Miller, M. C. De Rosa, A. Visconti, Determination of ochratoxin A in wheat after clean-up through a DNA aptamer-based solid phase extraction column, *Food Chemistry*, Vol. 127, Issue 3, 2011, pp. 1378-1384.

Chapter 11
Review on Interaction between Electromagnetic Field and Biological Tissues

Zulkarnay Zakaria, Ruzairi Abdul Rahim,
Pick Yern Lee, Muhammad Saiful Badri Mansor,
Azian Azamimi Abdullah, Sazali Yaacob,
Siti Zarina Mohd. Muji

11.1. Introduction

There are a lot of studies on the electromagnetic effect on biological tissues over the years, hence exposure to the electromagnetic field too frequently is harmful to our health. Thus computation of electromagnetic field distribution into human body is important in today imaging modalities. Different types of biological tissue have different electromagnetic properties values such as electrical permittivity and electrical conductivity included the tumour cells and malignant cells. The studies to measure the conductivity and permittivity values of different types of biological tissues can be classified into two general categories: in-vivo and in-vitro. Ex-vivo tissue measurements are typically less accurate, since the dielectric properties of tissue change rapidly and considerably after removal from the body [1]. The electrical conductivity and permittivity of biological materials will vary characteristically depending on the frequency applied. There are three main dielectric spectrums which are α, β and γ dispersions. These three kind of dispersion occurred in different frequency regions. Generally, α-dispersion occurred at low frequency region, β-dispersion at medium frequency region and γ-dispersion at high frequency region [2-4].

11.2. Mathematical Description of EM Field

In mathematical expression, EM field can be represented by Maxwell equations, where through these equations electric and magnetic fields arising from varying distributions of

Zulkarnay Zakaria
Biomedical Electronic Engineering Department, School of Mechatronic Engineering,
Universiti Malaysia Perlis, 02600 Arau, Perlis, Malaysia

electric charges and currents can be described. This is because all interaction between electromagnetic field and biological tissues follow the Maxwell's rules [5] which are:

$$\nabla \times E = -j\omega B \tag{11.1}$$

$$\nabla \times H = (\sigma + j\omega\varepsilon)E \tag{11.2}$$

$$\nabla . D = \rho \tag{11.3}$$

$$\nabla . B = 0 \tag{11.4}$$

Equation (11.1) is Faraday's law while equation (11.2) is Ampere's law and equation (11.3) is Coulomb's law. Faraday's law explains that time varying magnetic field density, B [Wbm^{-2}] with frequency ω [$rads^{-1}$] will induce electric field E [Vm^{-1}], while Ampere's law relates the magnetic field, H [Am^{-1}] to the induced conductance current density, J_c [Am^{-2}] and electric displacement field, D [Cm^{-2}]. However Coulomb's Law relates free electric charge density, ρ [Cm^{-3}] with D. Equation (11.4) represent that there is no magnetic field exist in that area. σ and ε are conductivity and permittivity value respectively.

Barber and Brown [6] have simplified the equations by assuming that biological tissues are non-magnetic and isotropic media. Hence for field frequencies under ~100 kHz and at current densities of less than ~100 Am^{-2} the tissues can also be assumed linear. Through these assumptions the conditions will be slightly different where:

$$B = \mu H \tag{11.5}$$

$$D = \varepsilon E \tag{11.6}$$

$$J = \sigma E \tag{11.7}$$

11.3. EM Field in General

11.3.1. Near Field and Far Field Region

'Far field' and 'near field' are the terms that normally used to describe the field or area around any electromagnetic-radiation sources as in Fig. 11.1. The measurement of Specific Absorption Rate (SAR) in exposed biological subjects at radio frequencies is a challenging task, both under near- and far-field exposure conditions. In the far-field case, internal fields are highly dependent on the size, shape, orientation (with respect to polarization), and composition (complex permittivity) of the object. EM waves in far fields are characterized by a single polarization type (horizontal, vertical, circular, or elliptical) whereas for the near field region all four polarization types can be present [7].

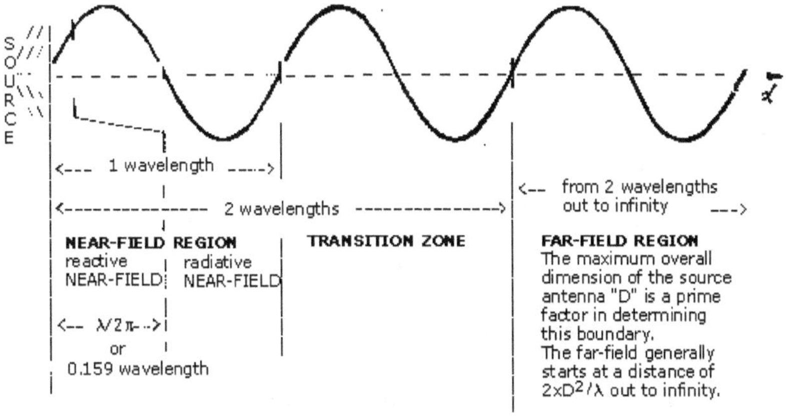

Fig. 11.1. Antenna Field region for typical antennas [5].

IEEE Standard C95.3-2002 defined near field region as a region in the field of an antenna, located near antenna, in which the electric and magnetic fields do not have a substantially plane-wave character, but vary considerably from point to point. However for far field region which is also known as free space region is defined as a region the angular field distribution is essentially independent of the distance from the antenna [8].

11.3.2 Mechanism of EM Field Interaction with Biological Tissues

There are two condition of mechanism in the EM interaction; with thermal effect and non thermal effects. Different condition will involve different types of mechanism of interactions.

11.3.2.1. Thermal Effects

Thermal effects are indirect interaction with biological tissues due to the high RF radiation where through this interaction EM field generates heat and causing temperature rising due to absorption of EM energy within the biological tissues. Specific absorption rate (SAR) is used for measuring biological thermal response. It is established by the International Electrotechnical Commission (IEC) to not exceed 8 Watts per kilogram (W/kg) of tissue for any 5-minute period or 4 W/kg for a whole body averaged over 15 minutes. SAR is defined as the time derivative of the incremental energy (dW) absorbed by an incremental mass contained in a volume of a given mass density (ρdV) and is expressed in units of Watts per Kilogram (W/kg).

$$SAR = \frac{d}{dt}\left(\frac{dW}{\rho \cdot dV}\right) = \frac{\sigma |E|^2}{\rho}, \qquad (11.8)$$

where σ is the conductivity, |E| is the RMS amplitude of electric field and ρ is the mass density of the object [3, 8].

11.3.2.2. Non Thermal Effects

Non-thermal effects are direct interactions of EMF with biological cells and happened due to low radio frequency radiation. IEEE standard C95.3-2005 defined non-thermal effect as *"any effect of EM energy absorption not associated with or dependent on the production of heat or a measurable rise in temperature"* [8]. Although the energy associated with RF radiations is not large enough to cause ionization of atom and molecules, non-thermal biological effects can still exist within these energy levels. One of the most detectable non-thermal effects is the ion-flux (ion-efflux), which describes the movement of calcium ions under the influence of external oscillating electric fields [9].

11.4. Properties of Biological Tissue

11.4.1. Biological Tissues in General

Cells are the structural and functional units of all living organisms. Some organisms, such as bacteria, are unicellular, consisting of a single cell and some such as human are multi cellular. In the body, there are many systems forms by several organs, while organs are made by tissues and tissues are consist of millions of cells and an extracellular matrix. The extra cellular matrix is a complex network of macromolecules, water molecules, ions, and other small molecules, and it helps to hold cells and different tissues together. The extracellular of plants tissue is rigid while in animals it has more forms many functions [2]. The examples of plant tissue and animal tissue are as in Fig. 11.2.

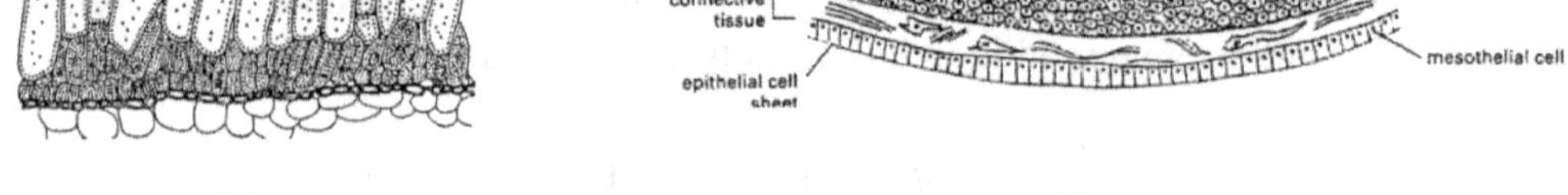

(a) (b)

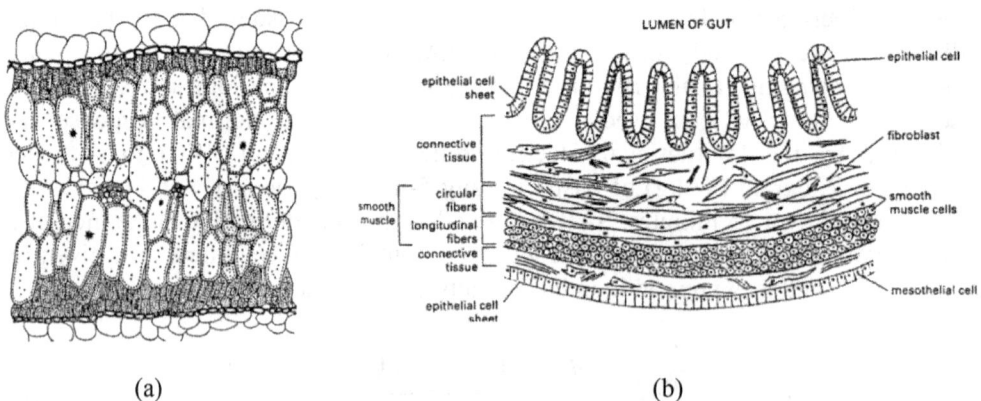

Fig. 11.2. Cross section of a) plant tissue, and b) animal tissue [7].

11.4.2. Types of Biological Tissues and Cell Membrane

Tissues are formed by a group of cells a collection and similar cells that group together to perform a specialized function. Generally, there are four primary types of tissues in human or animal body which are connective tissues, epithelial tissues, muscular tissues and nervous tissues [10] as in Fig. 11.3.

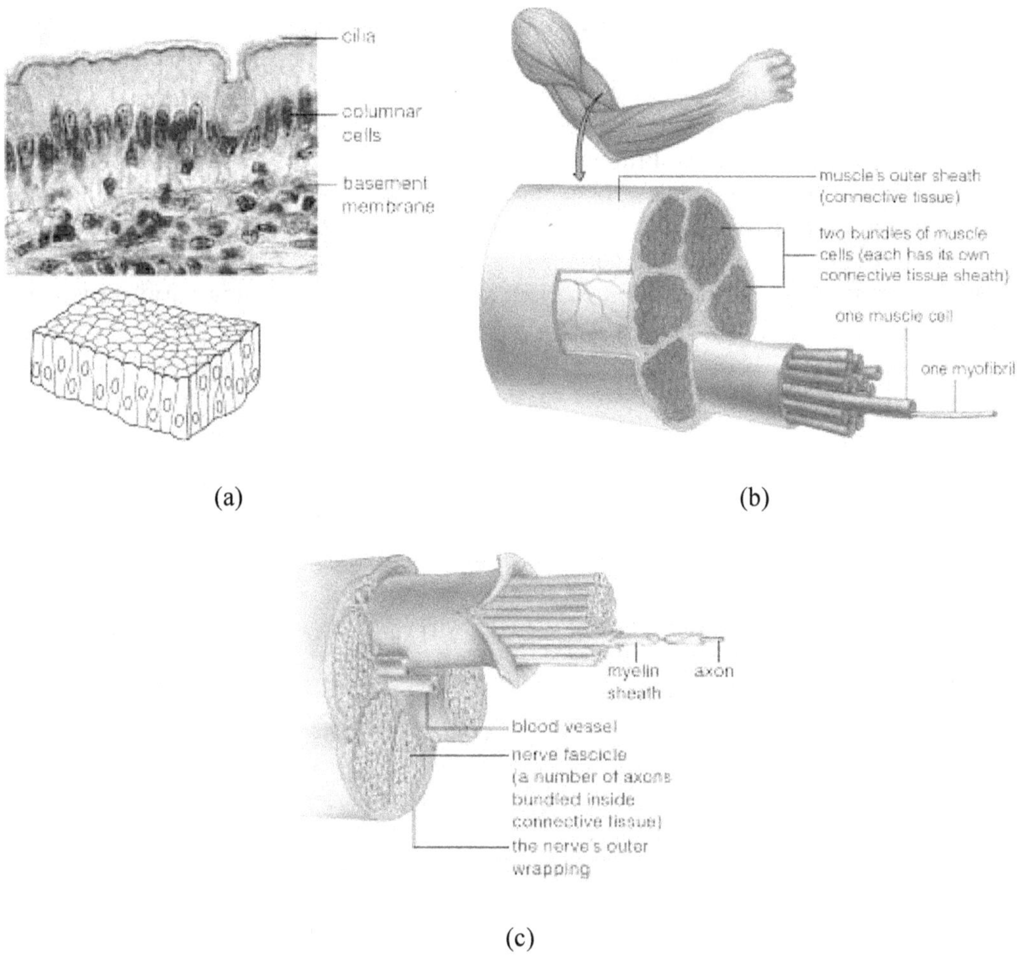

Fig. 11.3. Four primary types of tissues; a) epithelial tissue; b) muscle tissue, and connective tissue; c) nervous tissue [10].

Epithelia tissues act as membranes lining organs and it also help to protect and separated the body organ. Fibrous strands of the protein collagen can be found in most of the connective tissues. They help to strengthen connective tissue. Blood, ligaments, cartilage bone and fat tissue are the example of connective tissues. Muscle tissue contains the specialized proteins actin and myosin that slide past one another and allow movement. There are two types of nervous cells which are neurons and glial cells. Nerve tissue has the ability to generate and conduct electrical signals in the body. These electrical messages are managed by nerve tissue in the brain and transmitted down the spinal cord to the body.

11.4.3. Cell Membrane and Electrical Equivalent

Cell membrane is among the very important part of a cell. Plasma membrane act as a semi-porous barrier and it surround all living cells, prokaryotic and eukaryotic as in Fig. 11.4. The plasma membrane consists of a combination of phospholipids and protein, where the phospholipids are formed a thin, flexible sheet and arranged in a bilayer with the polar hydrophilic phosphate head facing outwards and their non polar hydrophobic fatty acid tails facing inwards. Proteins are randomly floated in the phospholipid sheet. The plasma membrane regulates some molecules enters and leaves the cell through passive transport such as diffusion or osmosis [11]. The plasma membrane protein acts as a selective transport channel of certain substances across the phospholipid bilayer. It also functioned as receptors, exhibit enzymatic activity and catalysing various reaction related to the plasma membrane.

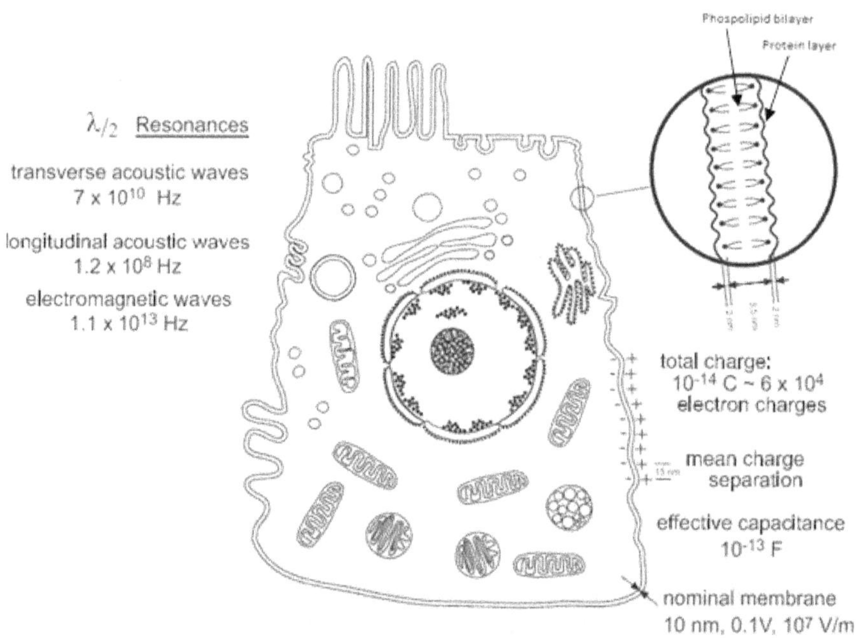

Fig. 11.4. Structure of a cell membrane from electrical perspective [11].

Related to electrical properties, lipid bilayer of a cell membrane acts like an insulator separating two conducting media and high capacitance in properties. This geometry constitutes an electric capacitor where the two conducting plates are the ionic media and the membrane is the dielectric [12] as shown in Fig. 11.5 a.

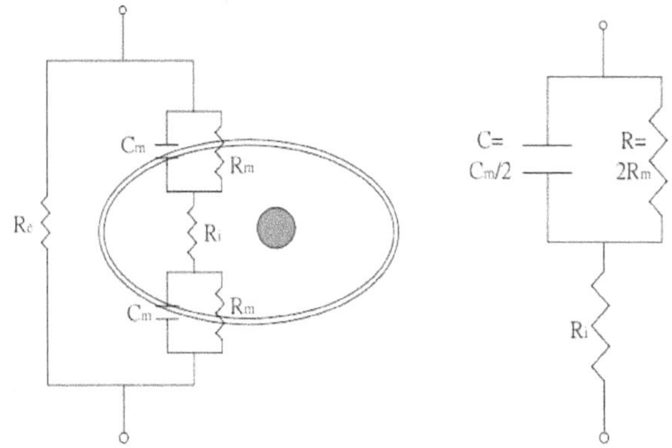

(a) (b)

Fig. 11.5. Electrical model of a cell membrane; a) General model, and b) Simplified model [12].

The conduction of electric current through such a structure is highly frequency dependent [13]. Since electrical impedance of human tissue contains both resistance and capacitance, it is complex and can be described by a serial representation

$$Z=R+jX, \tag{11.9}$$

where Z is the impedance, R is the resistance and X is the reactance. In parallel representation

$$Y=G+j\omega C, \tag{11.10}$$

where $Y=1/Z$ is the admittance; G is the conductance; C is the capacitance; and ω is the angular frequency. In simpler model, tissue can be represented as the circuit in Fig. 11.5 b where R_e is the resistance of extra cellular, R_i is the resistance of intra cellular and C_m is the membrane capacitance of the cell [14].

11.5. Electrical Properties of Biological Tissues

The electrical interactions of polar molecules and ions determined the electrical properties of biological tissues. Intracellular and extracellular of living organism spaces containing anionic and cationic species which produces conductive paths for current flow and thus having neutral molecular dipoles known as dielectric. The properties included dielectric permittivity, ε, which is related to the dielectric behaviour of the material; and conductivity, σ, which interacts with the electric field applied to the tissue. Dielectric permittivity and conductivity show three principal behaviours known as dispersions which it is depending on frequency [15]. Fig. 11.6 below has shown the typical frequency dependence of the complex permittivity of biological tissues.

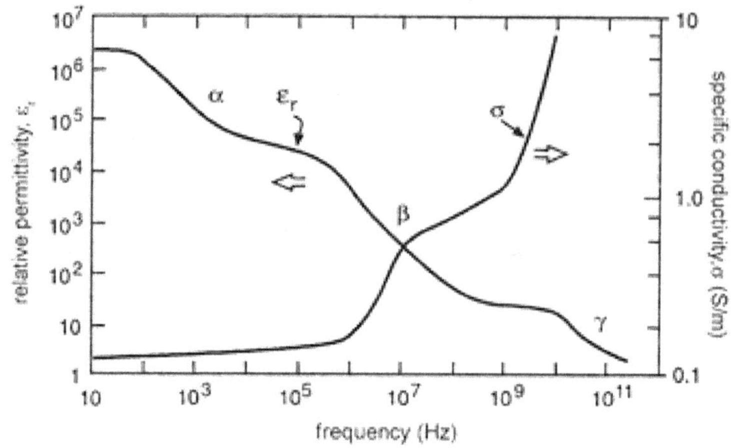

Fig. 11.6. Typical frequency dependence of the complex permittivity of a heterogeneous material such as biological tissues [15].

The impedance of living tissue varies, depending on its dielectric permittivity and conductivity. The value of the current and the attenuation of the electromagnetic field inside the tissue are therefore strongly depending on these two parameters. For biological tissues, both the dielectric permittivity and the conductivity are strongly non-linear functions of frequency. Moreover, if the frequency of an externally applied electromagnetic field changes, the interaction between the field and the tissue also changes. In particular, at low frequencies, electromagnetic fields interact at the cellular or multicellular level; as frequency progressively increases, bio-electromagnetic interactions occur with the cellular membrane and intracellular organelle, followed by molecular interaction and, finally, at microwave frequencies the field interacts only with water molecules [9]. Electrical properties are varying with different types of cell tissues as shown in Table 11.1 below.

11.6. Main Features of the Dielectric Spectrum of a Biological Tissue

The main features of the dielectric spectrum of tissues consist of three main relaxation regions. At low, medium and high frequencies consist of alpha-α, Beta-β and gamma-γ regions as in Fig. 11.6. A model for the dielectric spectrum of tissues is expressed in Debye expression [4]. Each of these relaxation regions is the manifestation of a polarization mechanism characterized by a single time constant, which to a first order approximations, the Debye expression is expressed for the complex relative permittivity as a function of angular frequency. The relative permittivity of a tissue may reach values up to 10^6 and 10^7 at frequencies below 100 Hz. It decreases at high frequencies in three main steps known as α, β and γ dispersions.

Table 11.1. Electrical properties (relative permittivity and conductivity) of cell tissues at frequency 4.33 MHz [3].

Electrical Properties of Tissues at 433 MHz

Tissue	ε_r	σ_{eff} (S/m)	Tissue	ε_r	σ_{eff} (S/m)
Air (vacuum)	1	0	Lens cortex	52.75	0.6742
Aorta	49.15	0.7395	Lens nucleus	38.76	0.38
Bladder	17.67	0.3128	Liver	50.34	0.68
Blood	57.3	1.72	Lung deflated	52.83	0.7147
Bone (cancellous)	21.08	0.02275	Lung inflated	21.58	0.3561
Bone (cortical)	13.77	0.1032	Muscle	64.21	0.9695
Bone (marrow)	5.137	0.03575	2/3 muscle	42.81	0.6463
Breast fat	5.62	0.04953	Nerve	35.7	0.500
Cartilage	43.64	0.65	Ovary	51.55	1.033
Cerebellum	52.9	0.91	Skin (dry)	42.48	0.5495
Cerebrospinal fluid	68.97	2.32	Skin (wet)	51.31	0.72
Cervix	44.17	1.020	Small intestine	74.1	2.053
Colon	60.88	0.96	Spleen	60.62	1.041
Cornea	54.4	1.070	Stomach	74.55	1.120
Dura	51.03	0.8	Tendon	50.53	0.7554
Eye tissues	57.69	1.010	Testes	65.2	1.137
Fat	5.028	0.04502	Thyroid	60.02	0.8183
Gall bladder	60.06	1.035	Tongue	58.79	0.8993
Gall bladder bile	76.55	1.613	Trachea	42.93	0.673
Gray matter	54.27	0.8775	Uterus	64.73	1.117
Heart	60.74	0.9866	Vitreous humour	66.16	0.3931
Kidney	57.3	1.152	White matter	39.84	0.5339

Source: Gabriel, C., *Compilation of the dielectric properties of body tissues at RF and microwave frequencies,* Final technical report, Occupational and Environmental Health Directorate Radiofrequency Radiation Division, Brooks Air Force Base, TX, 1996.

α- dispersion dominates below 100 kHz and is characterized by a very large permittivity variation which is assigned to counter ion diffusion effects and the charging of the cell membranes. The low frequency α-dispersion is associated with ionic diffusion processes at the site of the cellular membrane. There is an enough cycle time to allow charging and discharging of the cell membrane at very low frequencies, this causes a large tissue capacitance and a high permittivity. However β-dispersion occurred in the radiofrequency (RF) range of 100 kHz to 100 MHz [3, 15]. Between intra and extra cellular media, polarization of cellular membrane will act as a barrier to the flow of ions. The polarization of protein and other organic macromolecules also causes β-dispersion to occur [17]. The cell's capacitive reactance decreases as the frequency increases and this allowing a current flow through the intracellular medium. Charging of the cell membrane is reduced due the increasing frequency resulting in the further decrease of permittivity and increase of conductivity altogether [18]. γ-dispersion is dominated at microwave frequencies above 1 GHz due to the polarization of water molecules [15].

11.7. Electrical Properties of Normal and Tumour Cells

A lot of investigation on the electromagnetic properties of both normal and abnormal cell tissues has been carried out. The database of dielectric properties such as relative permittivity and conductivity of all the body tissues are used in electromagnetic studies. The dispersion effects of the dielectric parameters with increasing frequency are explained by various relaxation mechanisms occurring on the cellular level at different frequencies. Calculations of dielectric properties of biological tissues have been carried out from the previous study. It was performed by two different cell patch model and the calculation are performed at 16 different frequencies in experiment carried out by. The average permittivity and conductivity of the whole cell patch were calculated using the complex current density, thus calculating the complex impedance of an equivalent series circuit of a capacitance and a resistance. Generally, tumours have higher water content than normal cells because of cellular necrosis but also irregular and fenestrated vascularisation. Experimentally tumour has about half the electrical resistivity of normal tissue below 20 kHz, but similar resistivity above 500 kHz [4].

11.8. A Case Study: Normal and Tumour Cells of Liver

Experiment of dielectric properties of in vivo and ex vivo normal, malignant and cirrhotic human liver tissues has been done from 0.5 to 20 GHz of frequency range. The analysis includes a comparison of the narrowband (915 MHz and 2.45 GHz) and wideband (0.5 to 20 GHz) dielectric properties of normal and diseased human liver tissues as shown in Table 11.2.

The results indicate that statistically significant differences exist in the dielectric properties of ex-vivo normal and malignant liver tissue, as well as in vivo and ex-vivo normal liver tissues at 915 MHz and 2.45 GHz. The analysis also shown that at 915 MHz and 2.45 GHz, the dielectric properties of ex-vivo malignant liver tissue are 19 to 30 % higher than normal tissue. The differences in the dielectric properties of in vivo malignant and normal liver tissue are not statistically significant (with the exception of effective conductivity at 915 MHz, where malignant tissue properties are 16 % higher than normal). Fig. 11.7 has shown the data of permittivity and conductivity of liver normal cell obtained from the experiment in the range of frequency 10 kHz to 20 kHz.

11.9. EM Field Applications in Biological Tissue Imaging

There are several modalities which apply EM field in their imaging process. Among them are Magnetic Resonance Imaging (MRI) and Magnetic Induction Tomography (MIT).

Table 11.2. Relative permittivity and effective conductivity for in-vivo (above) and ex-vivo (below) liver normal, malignant and cirrhotic cells at frequency 915 MHz and 2.45 GHz [1].

	Normal (n = 11)		Malignant (n = 14)		Cirrhotic (n = 3)	
	ε_r	σ (S m^{-1})	ε_r	σ (S m^{-1})	ε_r	σ (S m^{-1})
915 MHz	59.94*	1.16*,†	64.09	1.34*,†	61.77	1.38
	±3.05	±0.14	±3.78	±0.13	2.58	±0.15
2.45 GHz	57.55*	1.95*	62.44	2.18	61.26	2.21
	±3.92	±0.18	±3.18	±0.13	±2.70	±0.17

n: number of data samples.
* $p < 0.05$ in a comparison of the *in vivo* and *ex vivo* tissue properties at the same frequency.
† $p < 0.05$ in a comparison of the normal and malignant tissue properties at the same frequency.

Table 4. Average and standard deviation values for the relative permittivity and effective conductivity for *ex vivo* samples at 915 MHz and 2.45 GHz.

	Normal (n = 20)		Malignant (n = 24)		Cirrhotic (n = 3)	
	ε_r	σ (S m^{-1})	ε_r	σ (S m^{-1})	ε_r	σ (S m^{-1})
915 MHz	48.11*,†	0.81*,†	57.09†	1.05*,†	51.60	0.94
	±7.67	±0.15	±3.00	±0.07	±2.69	±0.07
2.45 GHz	45.79*,†	1.68*,†	54.88†	1.99†	50.16	1.83
	±7.53	±0.27	±3.10	±0.11	±2.36	± 0.11

n: number of data samples.
* $p < 0.05$ in a comparison of the *in vivo* and *ex vivo* tissue properties at the same frequency.
† $p < 0.05$ in a comparison of the normal and malignant tissue properties at the same frequency.

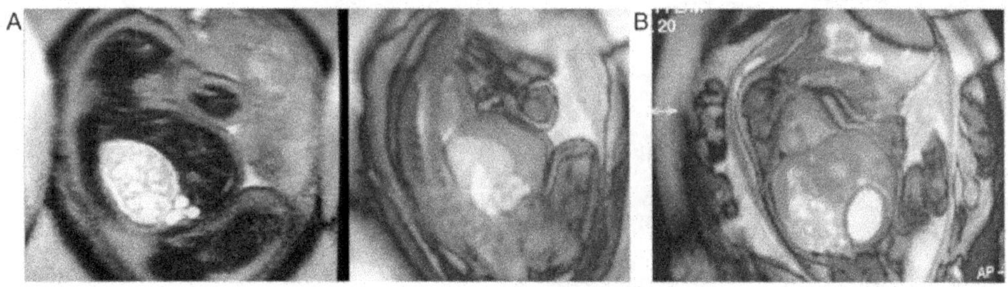

Fig. 11.7. Images of neonatal using MRI [20].

11.9.1. Magnetic Resonance Imaging (MRI)

MRI is known as the latest technology in the modern imaging with precise and accurate images. Magnetic resonance imaging (MRI) can be categorized as in the same group as Nuclear Magnetic Resonance Imaging (NMRI) and Magnetic Resonance Tomography (MRT). MRI applies the concept of atoms alignment in the body through the use of very high magnitude of magnetic field which may not suitable for certain types of patients such as those with pacemaker. Compare to other imaging modalities, MRI has the capability of providing detailed internal structures of the body parts through its good contrast between different tissues of the body as in Fig. 11.7. Current applications of MRI are the scanning of tumors or cancers in brain, lung, heart and other body parts including neonatal [20].

11.9.2. Magnetic Induction Tomography (MIT)

MIT has been introduced somewhere in 90's with the applications is more towards biological tissue imaging and has been the alternative to those who are not allowed for MRI. MIT is a low resolution imaging modality which interested in the three dimensional (3D) reconstruction of the electrical conductivity in the objects from alternating magnetic fields [21]. MIT consist of several components which are sensors (excitation coils, detection coils, and screen), interface electronics and host computer [22] as in Fig. 11.8. Object of interest is located within the region of interest (ROI). MIT the interaction concept of an oscillating magnetic field known as primary field, B_0 generated from excitation coils with the conductive medium (object under investigation), which is accompanied by the excitation eddy currents within the medium itself. These eddy currents then generate it own field known as secondary field and will be measured by the receiver coils. Since biological tissue is very low in conductivity, thus high frequency signal within the range of β - dispersion region is required. The measured data then undergo signal processing parts and finally through the use of image reconstruction algorithm the image of the object is reconstructed.

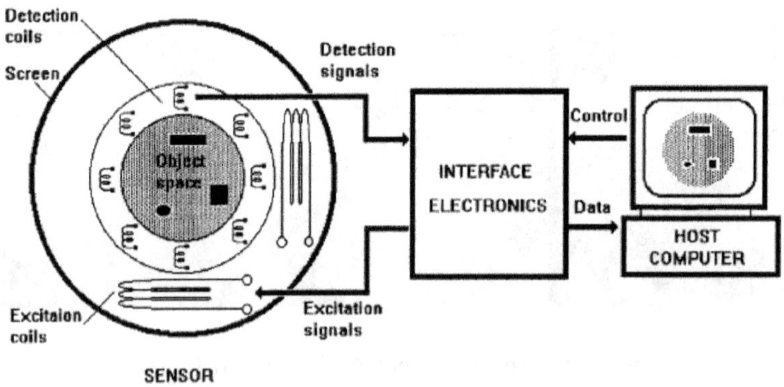

Fig. 11.8. Block diagram of a basic Electrical Inductance Tomography or MIT system [22].

11.10. Conclusion

The study on electromagnetic properties of biological tissues such as relative permittivity and conductivity are important in order to investigate on interaction between electromagnetic field and biological tissues and finally the application in biological tissue imaging. However, there are also some complications in dielectric measurements of tissues such as homogeneity of tissues, physiological factors and changes of tissues, anisotropy of tissues and electrode polarization. By comparing dielectric properties of tumour and healthy surrounding tissues, it is noticed that a tumour has much higher specific conductivity. It is obvious that the electrical conductivity and permittivity of biological materials will vary characteristically depending on the frequency applied.

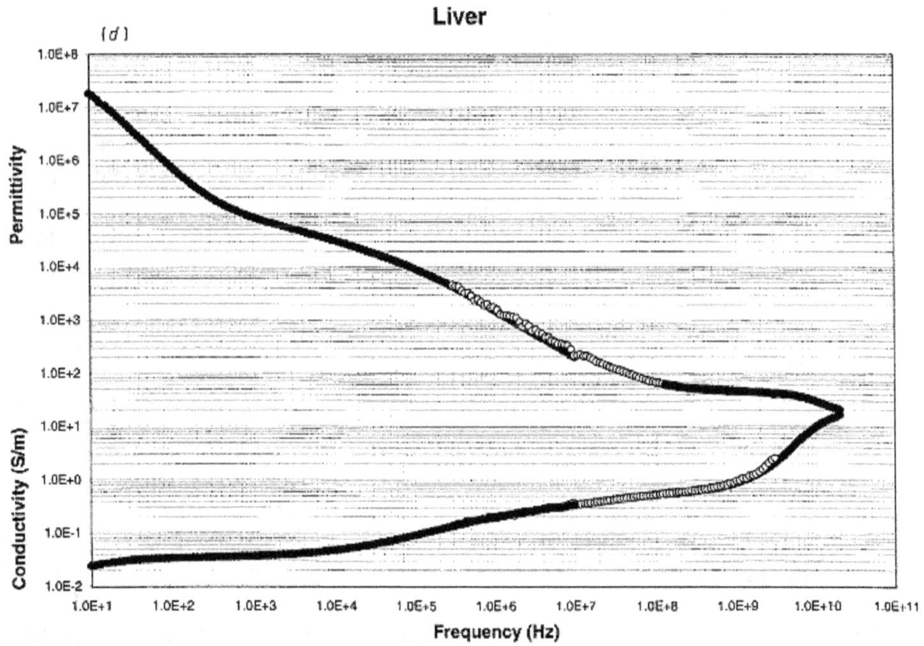

Fig. 11.9. The permittivity and conductivity of liver cell tissues from measurements on three experiments conducted with overlapping frequency coverage [19].

Acknowledgment

This work was supported in part by the Malaysian Government under FRGS Grant 9003-00248 and Science Fund Grant 06-01-06-SF0889.

References

[1]. A. P. Rourke, M. Lazebnik, J. M. Bertram, M. C. Converse, S. Hagness, J. G. Webster, D. M. Mahvi, Dielectric properties of human normal, malignant and cirrhotic liver tissue: in vivo and ex vivo measurements from 0.5 to 20 GHz using a precision open-ended coaxial probe, *Phys. Med. Biol.*, Vol. 52, 2007, pp. 4707–4719.

[2]. W. Kuang, S. O. Nelson, Low-Frequency Dielectric Properties of Biological Tissues: A Review with Some New Insights, *American Society of Agricultural Engineers,* Vol. 41, 1998, pp. 173-184.
[3]. L. Nicolas, N. Burais, F. Buret, O. Fabreque, L. Krähenbühl, A. Nicolas, C. Poignard, Interactions between Electromagnetic Fields and Biological Tissues: Questions, Some Answers and Future Trends, *ICS Newsletter,* July 2003, Vol. 10, Issue 2.
[4]. C. Gabriel, S. Gabriel, E. Corthout, The dielectric properties of biological tissues: I. Literature survey, *Phys. Med. Biol.,* Vol. 41, 1996, pp. 2231–2249.
[5]. A. J. Peyton, Z. Z. Yu, G. Lyon, S. Al-Zeibak, J. Ferreira, J. Velez, F. Linhares, A. R. Borges, H. L. Xiong, N. H. Saunders, and M. S. Beck, An overview of electromagnetic inductance tomography: Description of three different systems, *Measurement Science and Technology*, Vol. 7, 1996, pp. 261-271.
[6]. D. C. Barber, B. H. Brown, Applied potential tomography, *J. Phys. Eng: Sci. Instrum,* Vol. 17, 1984, pp. 723-732.
[7]. OSHA Cincinnati Laboratory, Ohio (http://www.icnirp.de/documents/emfgdl.pdf).
[8]. IEEE Recommended Practice for Measurements and Computations of Radio Frequency Electromagnetic Fields With Respect to Human Exposure to Such Fields, 100 kHz–300 GHz, *United State: IEEE-SA Standards Board*, 2008.
[9]. Z. Wang, Electromagnetic Field Interaction with Biological Tissues and Cells, Ph. D. Dissertation, Dept. School of Electronic Engineering and Computer Science, Queen Mary, *University of London*, 2009.
[10]. C. Starr, C. A. Evers, L. Starr, Biological: Concepts and Applications, Eight Editions, *Brooks Cole*, 2010.
[11]. R. H. W. Funk, T. Monsees, N. Ozkucur, Electromagnetic effects - From cell biology to medicine, *Progress in Histochemistry and Cytochemistry,* Vol. 43, 2009, pp. 177-264.
[12]. D. Haemmerich, D. J. Schutt, RF Ablation at Low Frequencies for Targeted Tumour Heating: In Vitro and Computational Modeling Results, *IEEE Transactions on Biomedical Engineering*, Vol. 58, Issue 2, 2011, pp. 404-410.
[13]. D. A. Dean, T. Ramanathan, D. Machado, R. Sundararajan, Electrical Impedance Spectroscopy Study of Biological Tissues, *Journal of Electrostatics*, Vol. 66, 2008, pp. 165-177.
[14]. Y. Zou, Z. Guo, A review of electrical impedance techniques for breast cancer detection, *Medical Engineering & Physics*, Vol. 25, 2003, pp. 79-90.
[15]. D. Miklavcic, N. Pavselj, F. X. Hart, Electric Properties of Tissues, Wiley Encyclopedia of Biomedical Engineering, *Wiley & Sons,* New York, 2006.
[16]. M. A. Golombeck, C. H. Riedel, O. Dossel, Calculation of the Dielectric Properties of Biological Tissue Using Simple Models of Cell Patches, *Biomed. Tech. (Berl).,* Vol. 47, Issue 1, 2002, pp. 253-256.
[17]. C. Gabriel, A. Peyman, E. H. Grant, Electrical conductivity of tissue at frequencies below 1 MHz, *Phys. Med. Biol.,* Vol. 54, 2009, pp. 4863–4878.
[18]. R. Pethig, D. B. Kell, The passive electrical properties of biological systems: their significance in physiology, biophysics and biotechnology, *Phys. Med. Biol.,* Vol. 32, Issue 8, 1987, pp. 933-970.
[19]. D. Haemmerich, D. J. Schutt, A. W. Wright, J. G. Webster, D. M. Mahvi, Electrical conductivity measurement of excised human metastatic liver tumours before and after thermal ablation, *Physiol Meas.,* Vol. 30, Issue 5, 2009, pp. 459–466.
[20]. Hosny, I., Elghawabi, H. S. Ultrafast MRI of the fetus: an increasingly important tool in prenatal diagnosis of congenital anomalies, *Magnetic Resonance Imaging,* 28, 2010, pp. 1431-1439.

[21]. Scharfetter, H., Merwa, R., Pilz, K. A new type of gradiometer for the receiving circuit of magnetic induction tomography (MIT), *Physiological Measurement*, 26, 2, 2005, pp. 307- 318.

[22]. Peyton, A. J., Yu, Z. Z., Lyon, G., Al-Zeibak, S., Ferreira, J., Velez, J., Linhares, F., Borges, A. R., Xiong, H. L., Saunders, N. H., Beck, M. S. An overview of electromagnetic inductance tomography: description of three different systems, *Measurement Science and Technology,* 7, 3, 1996, pp. 261-271.

Chapter 12
Application of Biotoxin Determination Using Advanced Miniaturized Sensing Platform

Zhaowei Zhang, Yuxin Liu, Peiwu Li,
Li Yu, Xiaofeng Hu, Hu, Jing Li

12.1. Introduction

In the field of food safety, environmental monitoring, and more importantly, biosecurity, the emerging sensing technologies provide guarantee the reliable information about the targets of interest, such as their activity and contents, owing to their rapidness, reliability, portability. In this respect, one of the most attractive targets of interest is the biotoxins. Biotoxins are the poison produced by and derived from plants, animals and microorganism, herein mainly covering the biological or chemical agents that are inanimate and not capable of reproducing themselves. These biologically produced toxins are produced from various organisms (fungus, bacteria, higher animal, and plant, etc.) and have highly complex (the venom of the cone snail contains dozens of small proteins, each targeting a specific nerve channel or receptor). Naturally, biotoxins bear two primary functions, namely, predation (by spider, snake, scorpion, jellyfish, and wasp) and defense (by bee, ant, termite, honeybee, wasp, and poison dart frog). Therefore, it can seriously affect the health and life of human and other animals. For example, hemotoxins, produced by snake, are biotoxins that cause hemolysis, disrupt blood clotting, and/or cause organ degeneration and generalized tissue damage, therefore usually resulting in painful injury and death [1, 2]. Necrosis can injure cell that induce the premature death of cells in living tissue, in which fiddle back spider, bitis arietans and the "flesh eating" bacteria can produce necrosis [3, 4]. Mycotoxins are toxic secondary metabolite produced by fungi, commonly known as molds [5]. They can readily enter the food chain by infecting food and feed, thus threatening health and life of human and livestock, such as the notorious turkey "X" disease that was caused by aflatoxins.

Zhaowei Zhang
Oil Crops Research Institute of the Chinese Academy of Agricultural Sciences,
Wuhan 430062, China

Tremendous health cases and economy losses caused by biotoxins have aroused great efforts on the detection and monitoring of biotoxins. Among the diversified detective platforms, there is an increased interest for developing rapid and reliable sensing platforms for their detection. In recent years, miniaturized sensing platform promise an excellent future towards the powerful analytical tools. These sensing platforms can serve as converters that record a physical quantity and convert it into a signal prior to record by an electronic readout instrument. For example, an antibody electrochemical sensor converts the measured antigen-antibody signals into electrochemical signals. For reliability, most sensing platforms require calibration against known standards. An ideal sensing platform has several requirements: (i) high specifically sensitivity to the targets of interest; (ii) insensitivity to any other property likely to be encountered in its application; and (iii) no interference the measured property.

Typical minimized sensing platforms usually combine a biological component with a physicochemical detector. The sensitive biological or chemical element act as the recognition of the biotoxins using antibody-antigen interaction, Van der Waals' force, molecular recognition and so on. The transducer or the detector element (works in a physicochemical way; optical, piezoelectric, electrochemical, etc.) that transforms the signal resulting from the interaction of the analyte with the biological element into another signal that can be more easily measured and quantified. Sensing readout device with the associated electronics or signal processors provided results in a user-friendly way.

In this chapter, we focus on the application of biotoxin determination by miniaturized sensing platform, such as microarray, microfluidic chip, test strip, SPR chip and other format. A selected literatures survey is divided into the following section: material for microsensing platform; sensing formats; application of microsensing platform for biotoxin determination.

12.2. Material for Microsensing Platform

12.2.1. Metal Nanomaterials

Metal nanomaterials (NMs) with interesting size-dependent electrical, optical, magnetic, and chemical properties, as a kind of modish materials, have been intensively pursued [6]. At present, size, shape, architecture, composition, hybrid and microstructure of metal NMs are key parameters, which enhance their functions and broaden their potential applications. In principle, the physical and chemical properties of metal NMs can be finely tuned by controlling these parameters. The flexibility and scope of change are highly sensitive to some specific parameters. For instance, Au nanoparticles (NPs) have size-dependent surface plasmon resonance (SPR) property and exhibit visible SPR absorption whereas gold NRs, gold nanocage and hollow gold nanospheres own strong near-infrared (NIR) absorption [7]. These novel gold nanostructures with NIR absorption are important for photothermal therapy and bioimaging in the NIR region because blood and soft tissue in the NIR region are relatively transparent in this region, so that collateral damage to surrounding healthy tissue is minimized; Ag nanostructures with proper size, complex sharp structure or more edges and corners have higher surface-enhanced Raman scattering (SERS) activity than spherical Ag NPs [8]; certain noble nanoclusters (NCs) (Au, Ag and Pt in particular), consisting of several to a hundred atoms and possessing sizes comparable to the Fermi wavelength of electrons, can exhibit molecule-like properties and strong size-dependent fluorescent emission [9]. Thus, the control of these pivotal parameters provides the good opportunity for enhancing their application potentials in the fields of catalysis, electronics, photography, photonics, sensing, imaging, medicine and information storage [10]. To date, diverse methods have been developed to synthesize metal NMs in a variety of shapes such as rod, wire, polyhedron, dendrite, dimer, belt, star, and cage, etc. The research on metal NMs has been flourishing in the last decades and many research papers as well as some review papers have been dedicated to this topic [11-13].

On the other hand, nanoanalytical sensing system is a rising interdisciplinary field. It combines inherent characteristics of analytical techniques (e.g., high sensitivity, rapid detection and low cost) with unique electronic, optical, magnetic, mechanical, and catalytic properties of NMs, thus becoming one of one most exciting topic [14]. Particularly, with the gradual appearance of novel or enhanced properties of metal NMs, different analytical techniques or strategies have been developed to construct high-sensitivity sensors for detecting biotoxins. Three notable techniques include electrochemical, fluorescent and colorimetric sensing ones, respectively. (1) Metal NMs-based electroanalytical technique exhibits enormous potentials for constructing enhanced sensing platforms. This is because that metal NMs can effectively catalyze the redox processes of targets due to their high conductivity, large surface area and good surface chemistry property, thus allowing an improved analytical performance (lower detection limit and shorter deposition time) of voltammetric techniques in comparison to conventional electrodes [15]. (2) Metal NMs-based NCs, as a new class of fluorophore, can be used as environmentally friendly and biocompatible fluorescence probes for detection of low-concentration analytes, owing to its low toxicity, inexpensive and good biocompatibility [16]. (3) metal NMs-based colorimetric methods (the change of SPR signals from Au or Ag NPs) are also extremely attractive because that they can be easily read out with the naked eye, and allow onsite, real-time qualitative or semi-quantitative detection without complicated analytical instruments [17]. Therefore, metal NMs-based sensing devices bear advantages for detecting different targets, which have important significance in the aspects of environmental pollution, serious diseases, human health and food safety.

12.2.1.1. Metal NMs for Electrochemical Sensing

The development of nanoscience provides enormous opportunities for analytical chemists. Metal NMs are one of the most extensively used materials in electroanalytical detection and have good potentials to construct electrochemical sensing platforms with high sensitivity and selectivity to detect target molecules based on different analytical strategies. For example, Au NPs and inorganic NMs could be used as enhanced electrode materials for electrochemical sensing applications [14]. In order to further improve the sensitivity of electrochemical detection, better electrode materials for electroanalytical applications are requested. The combination of different nanoscaled inorganic NMs with good conductivity may open a new avenue for utilizing nobel metal-based hybrid NMs as enhanced elements for constructing electrochemical sensing platform. This is because composite nanomaterials offer larger electrochemically active surface areas for the adsorption of target molecules and effectively accelerate the electron transfer between electrode and detection molecules, which could lead to a more rapid and sensitive current response [18].

12.2.1.2. Metal NMs for Colorimetric Sensing

Colorimetric sensors are attractive due to their simplicity, high sensitivity, low cost, easily read out with the naked eye or concisely performed with UV/vis spectrometry instead of the complex instruments. Metal NPs are emerging as an interesting colorimetric reporters because that their extremely high visible-region extinction coefficients (10^8 —10^{10} M^{-1} cm^{-1}) are often several orders of magnitude higher than those of organic dyes. Generally, this colorimetric sensing strategy relies on the different colors from dispersed gold NPs solution (red) and the aggregated gold NPs solution (purple to blue) [19]. In order to improve the sensitivity and selectivity of colorimetric sensors, engineering gold NPs with functional molecules having high recognition ability is also necessary. Nowadays, significant research interest has been directed toward gold NPs-based colorimetric assays for DNA, enzyme activity, small molecules, metal ions, carbohydrates, and proteins, etc. using their unique SPR as sensing elements [20-22]. In all the Au NPs-based colorimetric sensors, DNA-functionalized Au NPs as sensing elements have received more interest because DNA has many unique functions, which could be easily tuned by the binding of target molecules [23-25]. The Mirkin's group pioneered the use of Au NPs—thiolated—DNA conjugates [26], which led to a series of novel colorimetric sensors for the ultrasensitive detection of polynucleotide [20], small molecule [27], enzyme activity [28], DNA [29], protein [30], and metal ion [31, 32].

12.2.1.3. Metal NMs for Fluorescent Sensing

The fluorescent sensors are also attractive mainly due to their intrinsic advantages over other optical methods such as high sensitivity, easy operation, and multiplicity of measurable parameters. Particularly, recent advances in new properties of metal NMs open rich opportunities for chemical and biological fluorescent sensors design. At present, metal NMs-based fluorescent sensors can be generally categorized into the following four schemes. (1) Molecular-scale noble metal NCs have been exported a new class of fluorescence reporters for detecting different targets mainly based on fluorescence quenching mechanism of metal NCs induced by target. (2) Second frequently adopted scheme was based on the ultra-efficient quenching ability of metal NPs to the fluorescence of nearby fluorophores through non-radioactive energy/electron transfer processes. Heavy metals [33], drug residue [34] have been detected in this manner. (3) The third interesting scheme was the use of the inner filter effect (IFE) of metal NPs, where metal NPs act as an absorber to modulate the emission of the fluorophore. One important feature for this sensing strategy is the absence of link between the absorber and the fluorophore. Typical example is from Shang and Dong [35], who demonstrated that Au NPs could function as a powerful absorber in the IFE-based fluorescent assays for the detection of CN^- and H_2O_2. (4) Metal-enhanced fluorescence (MEF) is a unique sensing method for an increased sensitivity [36, 37]. A recent contribution in this aspect was given by Shtoyko et al. [36], who found that electrochemically deposited Ag fractal-like nanostructures on glass slide could act as an excellent MEF substrate for immunoassays.

12.2.2. Carbon Nanotubes

Over the last two decades, carbon nanotubes (CNTs), including single wall CNTs (SWCNTs) and multiwall CNTs (MWCNTs), have formed part of extensive and multidisciplinary research due to their superior properties and wide range of applications over other NMs. In fact, they have even been designated in several occasions as the most researched materials in the last decade. CNTs has unique properties such as having a high tensile strength, thermal conductivity, stability and resilience, specific electrical properties and their ability to establish different types of interactions with organic and inorganic analytes.

The most common CNTs synthesizing methods include CVD [38], arc discharge method [39] and laser vaporization or laser ablation method [40], which allow to synthesize moderately high quantities of CNTs with a relatively precise number of layers. Even though, the fact that large amounts of these materials are also nowadays internationally required, has encouraged research in this filed in which alternative and effortlessly methods have been suggested. These include the use of arc-jet plasma, floating catalyst or simplified carbon-arc, among others. In general, their chemical synthesis is a noteworthy challenge, since their dimensions (diameter and length), alignment (zig-zag, armchair or chiral) and the number of walls can be controlled. The purification and characterization steps are of particular importance, since it is frequent to find them together with impurities of amorphous carbon, carbon nanoparticles, fullerenes or metal catalyst particles that may affect their properties. In this sense, purified CNTs of variable dimensions are nowadays commercially available.

As to their chemistry, it has been widely recognized of a high number of procedures for the modification of their side walls in order to clearly enlarge their potential. It should be indicated that CNTs applications are often limited by their insolubility in most solvents due to strong intertube van der Waals interactions. Therefore, solubilization approaches like their functionalization are, in many cases, necessary [41]. In these cases, it is necessary to modify them by covalent or non-covalent functionalization [42]. In a covalent method, normally involves a previous oxidation step (with HNO_3 or H_2SO_4 at high temperature) in which –COOH, –OH and –C=O– groups are introduced in their structure [43]. After that, different substitution reactions are carried out, constructing diversified chemical structures. Nevertheless, one challenging problem is to unequivocally verify the successfully covalent decoration, for which various techniques are requested (i.e. TGA, infrared, Raman or X-ray photoelectron spectroscopy). Apart from the previously indicated characteristics, it is worth mentioning the existence of the so-called "point of zero charge" or "isoelectric point", at which modified CNTs' surface has zero net charge. When pH value is higher than this point, the surface is negatively charged and thus cationic species can be adsorbed by electrostatic interactions. On the contrary, when pH value decreases enough, protons compete with cations for the same sites on CNTs and the neutralization of CNTs provides a decrease in the adsorption. This phenomenon can be used, for example, to accumulate metals at a specific pH and to later elute those using acidic solutions. It should also be noted that the holes in CNTs structure by oxidation can fine-tune their properties. Functionalized CNTs are easier to disperse than its pristine form, thus facilitating their characterization and purification.

CNTs are easy to form non-covalent aggregates with surfactants or supramolecular complexes with polymers, where Van der Waals forces, π-π stacking interactions, hydrogen bonds and electrostatic forces, among others, are responsible for solubilization of modified CNTs. With this method, conjugating CNTs with molecules like carbohydrates, proteins, enzymes or DNA can be used in biosensors or as drug carriers [41]. To date, it has also been widely investigated the possibility of storing guest molecules (i.e. fullerenes or metallofullerenes, metals, etc.) in their inner cavity, especially for SWCNTs.

12.2.3. Quantum Dots

Quantum dots (QDs) are colloidal nanocrystalline semiconductors possessing broad continuous absorption spectra from ultraviolet to visible light, depending on the particle size. QDs have unique properties due to quantum confinement effects. The broad excitation and narrow size-tunable emission spectra (usually 20–40 nm full width at half maximum intensity), negligible photobleaching, as well as high photochemical stability are some extraordinary properties. Fluorescence is a powerful tool in biotoxins studies, owning to its diversified fluorescent probes available. The optical and spectroscopic unprecedented features of QDs make them convincing alternatives to traditional fluorophores, especially for multiplex bioanalysis.

QDs have bandgap energies that vary as a function of size (the higher size of the QD, the lower the energy of the bandgap). The bandgap energy corresponds to the minimum energy to move an electron from the valence to the conduction band creating a hole behind. The electron–hole pair, an exciton, may recombine immediately to produce heat or light with proper energy. It is more likely that trap states within the material trap either the electron or the hole [44]. Therefore, the trap states are one critical influencing factor of QDs surface on photoluminescence, which can be caused by structural defects, such as atomic vacancies, local lattice mismatches, dangling bonds, or adsorbates at the surface [45]. This increases the photostability of the core and the quantum yield, so that core-shell structured QDs are more favorable for fluorescence-based applications [46].

Several synthetic methods have been reported since the first synthesis of monodisperse CdE (E=S, Se, Te) nanocrystals and extensive reviews on the synthesis of colloidal nanocrystals can be found [47, 48]. Usually, QDs are synthesized using a coordinating solvent (e.g., trioctylphosphine oxide/trioctylphosphine, TOPO/TOP) that makes them soluble and stable in organic solvents by preventing their agglomeration. For biotoxins sensing, these hydrophobic QDs can be transferred to aqueous solutions using various routine methods, such as ligand exchange, surface silanization, embedding in a polymer shell or incorporation in micelles. Therefore, the knowledge of the surface chemistry of QDs is needed to understand their optical properties and to manipulate them to achieve a desired application.

Owing to the above mentioned fascinating optoelectronic properties, important applications of these NMs are vast. Together with the growing modification methods towards QDs surface by conjugation with appropriate functional molecules, sensing applications in food safety, environmental monitoring and biosecurity is in expansion [49, 50]. QDs have already been tagged to multiple different biomolecules, proving their potential to provide sufficient information on multiplexed-biotoxin events essential for toxin monitoring [51-54]. Thus, QD technology holds a great potential for multi-biotoxin sensing.

12.2.4. Graphene

Graphene is a 2D material, composed of layers of carbon atoms forming six-membered rings [55], which is the mother of all graphitic forms including zero-dimensional fullerenes, 1D CNTs, and 3D graphite [56]. Compare with the CNTs, graphene does not contain metallic impurities, which dominate the electrochemistry of CNTs and lead to misleading conclusions. In addition, the production of graphene uses graphite, which is cheap and accessible. Besides, Graphene exhibits fascinating properties such as electrical, mechanical, anomalous quantum Hall effect, thermal, and optical. Its unique form factor and exceptional physical properties have the potential to enable an entirely new generation of technologies beyond the limits of conventional materials. To obtain high-quality graphene films, several methods have been developed, such as mechanical cleavage of graphite crystals, chemical exfoliation of graphite through its intercalation compounds, chemical vapor deposition (CVD) on different substrates, and some other chemical synthetic routes. Among them, chemical methods are effective for producing graphene sheets from various precursors on a large scale and at low costs, which enables technical applications in a variety of fields [57].

12.3. Sensing Formats

12.3.1. Electrochemical Sensing

Electrochemistry is employed to monitor electrochemical reactions, usually in a solution, at the interface of an electron conductor (an electrode made of metal or semiconductor) and an ionic conductor (an electrolyte). It involves the electron transfer between the electrode and the electrolyte or species in solution. In an electrochemical reaction, a chemical reaction is driven by an external applied voltage (in electrolysis), or else, a voltage is created by a chemical reaction (in a battery). A first observation of an electrical phenomenon was in the 16th century, when scientists fabricated the first electric generator. In the late of the 18th century, electrochemistry was revealed, using a bridge between chemical reactions and electricity by the Italian physician and anatomist Luigi Galvani in 1791. In the 19th century, many scientists put extensive efforts on various theories or laws towards electrochemistry. In 1902, The Electrochemical Society (ECS) was established. Arne Tiselius developed the first sophisticated electrophoretic apparatus in 1937 prior to his awards from Nobel Prize for his work in protein electrophoresis in 1948. One year later, the International Society of Electrochemistry (ISE) was established.

During electrochemical process, electrode reactions and/or the charge transport can modulate chemically and serve as the basis of the sensing process. Electrochemical sensors are devices that generate a chemical or bio-logical change and convert that into a measurable signal. The sensor contains a recognition element. It enables the specific response towards one analyte or a group of analytes. The transducer, one key component of an electrochemical sensor, produces an electrical signal proportional to the concentration of analytes.

After the introduction of electrochemical sensors for oxygen monitoring in 1950s, many scientists have explored novel electrochemical sensors for clinical, environmental, industrial, and agricultural applications. Currently, electrochemical sensors have been used extensively as a key sensing element.

A typical electrochemical sensor has three electrodes, including a sensing electrode (or working electrode), a counter electrode, and a reference electrode. Electrochemical format included three main categories: current, potential and impedance. Some chemical or biological reactions generate a measurable current (amperometry), or a measurable charge accumulation or potential (potentiometry), while others alter the conductive properties in the medium between electrodes (conductometry). Thus, electrochemical sensors can be categorized into conductivity/capacitance, potentiometric, and voltammetric sensors. The electrochemical impedance spectroscopy by monitoring both resistance and reactance in an electrochemical sensor is a promising sensing platform.

12.3.1.1. Voltammetric Sensing

Voltammetric techniques are conducted by applying a potential to a working (or indicator) electrode versus a reference electrode and measuring the current. The current comes from the electrolysis. The process is an electrochemical reduction or oxidation on the working electrode. The electrolysis current to the electrode is limited by the mass transport rate of molecules. The mass transfer can be conducted through (i) an ionic migration from an electric potential gradient; (ii) a diffusion from a chemical potential difference or concentration gradient, and (iii) a bulk transfer by natural or forced convection. The mass transfer rate of targets towards the electrode surface affects the current potential characteristics.

In voltammetry method, the current response is recorded as a peak or a plateau that is proportional to the analyte concentration. Voltammetric sensing records the concentration of biotoxins on the current-potential characteristics. In an amperometric sensing platform, a fixed potential is applied to the electrochemical cell, followed by the obtaining of a corresponding current for biotoxin quantification.

Voltammetric methods include linear sweep voltammetry, cyclic voltammetry, hydrodynamic voltammetry, differential pulse voltammetry, square-wave voltammetry and stripping voltammetry. Voltammetric sensors can be operated in a linear or cyclic sweep mode. Linear sweep voltammetry involves an increase in the imposed potential linearly at a constant scanning rate from an initial potential to a defined upper potential limit. This is the so-called potential window. The current-potential curve shows a potential peak at the point of redox reaction. The peak height of the current can be used for the quantification of biotoxins. Cyclic voltammetry is similar to the linear sweep voltammetry except that the electrode potential returns to its initial value at a fixed scanning rate. The cyclic sweep generates the current peaks, corresponding to the redox reactions. Under these circumstances, the peak current value is in accordance to redox reaction. However, the voltammogram was highly complicated for adsorption (nonfaradaic processes) and charge processes (faradaic processes). The potential scanning rate, diffusivity of the reactant, and operating temperature are essential sensing parameters.

A voltammetric sensor is can be used very effectively to carry out qualitative and quantitative analyses of chemical and biochemical species. The fundamental of this sensing technique has been well established, especially in a complex, practical environment. For example, electrochemical signal changed with enzymatic suppression can be measured by cyclic voltammetric measurement. On the other hand, caffeine competitively inhibits enzymatic activity of alkaline phosphatase (ALP). Akyilmaz et al. [58] immobilized alkaline phosphatise (ALP) on a gold screen printed electrode via glutaraldehyde cross-linking, Based on inhibition theory, caffeine in soft drinks was determined with a LOD of 0.08 μM. Yeh et al. [59] fabricated an amperometric sensor by modifying indium tin oxide (ITO) electrode with conducting polymer and molecularly imprinted polymer (MIP). On this electroactive film that catalyzes morphine oxidation and lowered the oxidization potential, while MIP served as recognition element. It was found to be a sensitivity of 91.86 μA/cm^2 per mM and a LOD of 0.2mM. Carter et al. [60] reported an immunoelectrochemical biosensor for the detection of STX and PbTx. Antibody was conjugated with glucose oxidase (GOD), while STX and PbTx were immobilized covalently on membrane via BSA. H_2O_2 can be easily detected amperometrically as antibody brought GOD adjacent to the electrode.

12.3.1.2. Conductivity/Capacitance Electrochemical Sensing

Conductivity in an electrochemical cell can be served as an electrochemical sensing method. In this manner, the change of conductivity results from a given solute concentration, in which this solute is main platform of sensing. Conductometric detection is based on the changes in electrical conductivity of the solution, or a medium such as nanowires, when the composition of the solution/medium changes during the chemical reaction. SEB can be detected by flow-injection capacitive biosensor with an ultralow LOD of 0.3 pg mL^{-1} with a wide linear detecting range from 2.8 pg mL^{-1} to 2.8 ng mL^{-1}. The detection mechanism was based on the electrical double layer. The biosensors can determinate SEB in low level with simplicity, high sensitivity, and wide capability. In addition, the capacitive biosensor can run continuously for hours until it encounters the target analyte and does not consume antibody reagents [61].

12.3.1.3. Electrochemical Impedance Spectroscopy

Electrochemical impedance spectroscopy (EIS) is a powerful characterization technique of electrochemical systems. In this way, a small amplitude sinusoidal excitation signal to the system and its response (current or voltage or another signal of interest) are recorded. The changes of electrical impedance from a medium or a reaction solution can be recorded. The reactions results in an increase in both conductance and capacitance, causing impedance decrease. Therefore, impedance, conductance, capacitance, and resistance are different methods of monitoring the test system and are all interrelated.

Dinckaya et al. fabricated an impedimetric biosensor based on a DNA probe and gold NPs for detection of aflatoxin M_1. The biosensor provided a linear response to aflatoxin M_1 over the concentration range of 1-14 ng/mL with a standard deviation of ±0.36 ng/mL [62]. Detection of aptamer-protein interactions using a label-free electrochemical detection system based on impedance spectroscopic transduction has been reported [63].

12.3.1.4. Potentiometric Sensing

A potentiometric sensor is on the basis of the potential of an electrochemical cell by drawing negligible current. When a redox reaction takes place at an electrode surface, a potential may occur at the electrode-electrolyte interface. This potential can be used to quantify the activity (on concentration) of the targets. This kind of sensors can be turned into biosensors after coated with an enzyme. The reduction reaction at the cathode surface is defined as a half-cell reaction.

Based on whether the electrode, potentiometric sensors can be classified into inert one or active one. An inert electrode provides the surface for the electron transfer and/or a catalytic reaction. However, an active electrode can incorporate chemical or biocatalysts and act as either an ion donor or acceptor in the half-cell reaction. Generally, there are three types of active electrodes: the metal/metal ion, the metal/insoluble salt or oxide, and metal/metal chelate electrodes.

Potentiometric sensors can be used to detect biotoxin. For instance, caffeine can be detected by potentiometric pH probe. In one proposal, caffeine inhibited specifically phosphodiesterase 3',5'-cyclic nucleotide, hence reducing the H_3O^+ production. In this way, the caffeine concentration was correlated to the concentration of H_3O^+, which could be ultimately converted into potential signal. The potential response changes were proportional to the concentration of caffeine in the range 0-4 mg mL^{-1} [64].

In the laboratory, the lower detection limits of this sensing platform make it a promising alternative analytical technique. The advantage of distinguishing oxidation states is highly important. The electrochemical approach allow a rapid responding time. However, many electrochemical sensors require further development. There is a long way to reach the end of the wide consumer use in point of care test (POCT).

12.3.2. Optical Sensing

In an optical sensing method, it is based on the photons, rather than the electrons. Compared with electrochemical sensor, optical sensors require no reference electrodes. Optical sensors have been extensively studied for portable devices due to their sensitivity and portability.

Optical methods can be generally categorized as spectroscopic methods (SPR & SERS) and fluorescence methods (quenching & turning-on, FRET). Spectroscopic methods using infrared or visible light bear advantages of sensitivity, cost-effectivity, portability, and the ability to detect a wide range of biotoxins. Among the optical methods, fluorescence and SPR have achieved a most success [65].

12.3.2.1. SPR Sensing

Surface plasmon polaritons are surface electromagnetic waves that propagate in a direction parallel to the metal/dielectric (or metal/vacuum) interface. Since the wave is on the boundary of the metal and the external medium (air or water for example), these oscillations are highly sensitive to any change of this boundary. SPR is the collective oscillation of valence electrons in a solid stimulated by incident light. The resonance condition is established when the frequency of light photons matches the natural frequency of surface electrons oscillating against the restoring force of positive nuclei. In the case of *p*-polarized light (polarization occurs parallel to the plane of incidence), the light passes through specific glass to increase the wave number (and the momentum), and the resonance can be obtained at a specific wavelength and angle. However, s-polarized light (polarization occurs perpendicular to the plane of incidence) cannot excite electronic surface plasmons.

SPR is extensively used to measure adsorption of material onto planar metal surfaces or metal nanoparticles (gold and silver), where the reflection minimum angle (absorption maximum) can be measured. The mechanism lies in the changes caused by adsorbed molecules in the local index of refraction.

The first SPR immunoassay was proposed in 1983. The human IgG was adsorbed onto a 600-angstrom silver film for sensing anti-human IgG in water solution. Dorokhin et al. [66] reported an elaborate SPRi microarray platform for the detection of deoxynivalenol (DON) and zearalenone (ZEA) via a competitive inhibition immunoassay format. A continuous flow microspotter device was used in the fabrication. It should be noted that an SPR immunoassay is one label-free sensing format.

12.3.2.2. Surface Enhanced Raman Spectroscopy (SERS)

Raman scattering occurs during photonic inelastic collision. In this scattering process, photons may obtain energy from molecules, or transfer it to molecules. A change in the photon energy can produce a change in the frequency. It is relative to the excitation photons by the energy of characteristic molecular vibrations. Therefore, a Raman spectrum comprising several different "Raman lines" generated by scattering from different molecular vibrations provides a vibrational "fingerprint" of a molecule.

Raman scattering is a very weak effect. Fluorescence spectroscopy exploits effective cross sections between 10^{-17} cm^2 and 10^{-16} cm^2, whiled typical Raman cross section are between 10^{-30}–10^{-25} cm^2 per molecule with the larger values. It merely occurs under favorable resonance-Raman conditions when the excitation light matches the related electronic transition energy in the molecule.

The results in SERS come from specific spectral signatures (i.e., Raman signals) of certain organic and inorganic molecules. When strongly adsorbed on metallic substrates, these molecules can be affected by an intense local electromagnetic field generated from resonating electron gas cloud (surface plasmons) that surrounds the metal surface. As a result, the Raman fingerprints of these molecules are enhanced significantly. Excitation of surface plasmons due to intense localized electromagnetic fields is primarily responsible for the enhancement of Raman scattering effect. The charge transfer between metal substrate and adsorbed molecules contributes to signal enhancement to a certain extent.

SERS comes into being a promising method in bioanalytics. It exhibits rich molecular structural information using a vibrational spectroscopy with ultrasensitive detection limits, allowing both molecules detection and identification of their structures, even in a single-molecule level. For instance, Wu et al. acquired the high-quality SERS spectra of aflatoxins using silver nanorod (AgNR) array substrates fabricated by oblique angle deposition method [67].

Compared to conventional fluorescence methods, single biomolecules can be detected using the spectral signature of a SERS tag or their intrinsic surface enhanced Raman spectrum. SERS spectra are comprised of different vibrational modes. It provides sufficient structural information content on the target molecule. It is a powerful tool to detect a known molecule and monitor its distribution in a label-free manner.

12.3.2.3. Fluorescent Sensing

Fluorescent sensing provides high sensitivity because that the emission signal is measured on an ultra low background. Fluorescence detection is highly tunable, as the reliance on excited-state charge-transfer for detection permits wide latitude for the chemical composition of the indicator and analyte. Zhang's group [68] developed an immune-fluorescence based assay for determination of Naja Kaouthia Venom. Micro-scale polystyrene beads were used as substrate for antivenom antibody, forming the "capture-bead". Thereafter, the venom bonded to the capture-bead to form the complex through the antibody-antigen interaction.

Tremendous researches have focused on fluorescence quenching, in which the fluorescence signal is diminished in the presence of a quencher. In such format, analyte reduces the light intensity emitted from a fluorescent indicator. The origin emission intensity is quite high, while the intensity in the presence of analyte depends on ratio of analyte or quencher. Static quenching depends on the binding of analyte to indicator, as characterized by the association constant, whereas collisional quenching is a function of the intrinsic fluorescence lifetime and the rate of analyte-indicator collision. Both static and collisional quenching is derived from photoinduced electron transfer between the excited-sate of the indicator, and the ground state of the analyte. Zeineldin et al. using the fluorescence quenching technique to detect melittin. When melittin interacted with and lyse the lipid bilayer, superquenching occurred, which has positive relation with the concentration of melittin [69].

Amid the quest to improve sensitivity of the fluorescent sensors, the traditional focus has been on increasing quenching efficiency by increasing binding affinity for analyte. The other side to fluorescence sensing is fluorescence enhancement, where in the presence of the analyte, a fluorescent signal is turned-on. Turn-on fluorescence is less common than quenching, but we feel that it is the future for detection strategies. There are two reasons. First, the appearance of a bright signal on a completely dark background is qualitatively easier to detect than the dimming of an already bright signal. Second, turn-on signals result from a stoichiometric binding event, rather than from a collisional encounter. In turn-on fluorescence, a non-fluorescent precursor is converted to a fluorescent indicator only in the presence of analytes. The challenge is to identify chemistry by which analytes can selectively react with the appropriate precursors. It is worth noting that comparing fluorescence at several wavelengths, as by converting one fluorophore into a second fluorophore upon analyte binding (ratiometric sensing), would minimize noise from "bright" backgrounds. Nowadays, the turn-on fluorescent sensor has been applied widely in detecting heavy metal ions and cyanide. A simple design of "turn-on" fluorescent sensor for mercury was demonstrated based on structure-switching DNA with a low detection limit of 3.2 nM and high selectivity [70].

Some limitations with fluorescence detection methods include photodegradation and photobleaching of the fluorescent materials, slow response times due to analyte transport, as well as a potential for non-specific responses. Nevertheless, fluorescence detection is intrinsically cost-effective and portable, requiring little more than the equivalent of a digital camera and a blacklight.

12.3.2.4. Fluorescence Resonance Energy Transfer (FRET)

Fluorescence resonance energy transfer is a mechanism describing energy transfer between two chromophores. A donor chromophore, initially in its electronic excited state, may transfer energy to an acceptor chromophore through nonradiative dipole-dipole coupling. Measurements of FRET efficiency can be used to determine if two fluorophores are within a certain distance of each other. Such measurements are used as a research tool in fields including biology and chemistry.

Recently, FRET has been developed and various applications have been explored. FRET can be used to measure distances between domains a single protein and therefore to provide information about protein conformation. FRET can also detect interaction between proteins. Applied in vivo in living cells, FRET has been used to detect the location and interactions of genes and cellular structures including intergrins and membrane proteins. FRET can also be used to obtain information about metabolic or signaling pathways. Ma and Cheng developed cholera toxin sensor by using fluorescence resonance energy transfer (FRET) mechanism. The sensor offered a fast, robust, and sensitive, label-free measurement of bacterial toxin with simple operative procedures [71]. Another FRET-based sensor which can detect six o BoNT serotypes BoNT/A, B, D, E, F, and G serotypes in real time was reported by Ruge et al. femtomolar to picomolar detection limit was achieved [72].

12.3.2.5. Chemiluminescence

Chemiluminescence (CL) as a detection method for microsensing systems has the advantage of high sensitivity, low detection limits, and simple instrumentation compared with other spectrophotometric techniques, due to the exclusion of an external light source. Sauceda-Friebe et al [73]. developed a regenerable microarray for screening OTA in green coffee extract on a fully-automated flow-through device with a CL readout. The method reduced assay time of 12 min. The LOQ (7 μg/kg) is comparable to the EU MRL of 10.0 μg/kg. An ELISA-based sensor has been employed for detection of Staphylococcal enterotoxin B (SEB) by chemiluminescence analysis. The Detection limit was 3.3 pg mL^{-1} and the linear detection range of 6.0 to 564 pg mL^{-1} [74]. Morphine can be determined via chemiluminescence detection. A morphine sensor with ultralow detection limit of 5×10^{-8} M was developed based on resultant chemiluminescence (CL) emission from the reaction of morphine and acidic potassium permanganate ($KMnO_4$) in the presence of tetraphosphoric acid [75].

However, the drawback of this detection technique is the limited number of chemiluminescence reagents. Furthermore, the chemiluminescence reagent needs to be mixed with the separated analytes before detection.

12.3.3. Electronic Sensing

The field-effect transistor (FET) is a transistor that uses an electric field to control the shape and the conductivity of a channel of one type of charge carrier in a semiconductor material. FETs are unipolar transistors. All FETs have source, drain, and gate terminals. Most FETs have a fourth terminal called the body, base, bulk, or substrate. FETs are majority-charge-carrier devices. The device consists of an active channel through which majority charge carriers, electrons or holes, flow from the source to the drain. Source and drain terminal conductors are connected to the semiconductor through ohmic contacts. The conductivity of the channel is a function of potential applied across the gate and source terminals.

There are many kinds of FET. Chemical field-effect transistor is a type of a field-effect transistor acting as a chemical sensor. For such sensor, the charge on the gate electrode is applied by a chemical process. It may be used to detect atoms, molecules, and ions in liquids and gases. ISFET, an ion-sensitive field-effect transistor, is the best known subtype of ChemFET devices. It is used to detect ions in electrolytes. An alternative electronic transduction method of biorecognition events involves the application of ion-selective field-effect transistors (ISFETs). The control of the gate potential of FET devices as a result of biorecognition processes that occur on the gate surface became a common principle to develop biosensor devices.

Lim et al. have detected deoxynivalenol (DON) using a metal-oxide-semiconductor field-effect-transistor (MOSFET)-based biosensor. The MOSFET-based biosensor is fabricated by a standard complementary metal-oxide-semiconductor (CMOS) process. MOSFET-based biosensor can detect deoxynivalenol at concentrations as low as 0.1mg/mL [76]. Chan et al. propose a new type of photosensitive biosensor with a CMOS compatible Si photovoltaic integrated circuit, for the high-sensitive detection of small mycotoxin molecules requiring competitive assay approach. In this work, a photodiode is connected to the gate of a field effect transistor (FET) so that the open circuit voltage (VOC) of the illuminated photodiode is transferred into the drain/source current of the FET. The sensing scheme employs competitive binding of toxin molecules (within the sample solution) and toxin-BSA conjugates (immobilized on the photodiode surface) with Au-nanoparticle-labeled antibodies, followed by silver enhancement to generate opaque structures on the photodiode surface. As the amount of toxin increases, less opaque becomes the photodiode surface resulting in a higher VOC. By monitoring the channel current of the FET whose gate is driven by the VOC, quantitative detection of Aflatoxin B1 has been achieved in the range of 0-15 ppb [77].

A nanoelectronic sensing based on one-dimensional (1D) semiconducting NMs is an emerging sensing modality that offers high sensitivity, high temporal resolution, simple label-free detection scheme, and suitability for development of lab-on-a-chip devices. Silicon nanowire (SiNW) is perhaps the mostly explored 1D material for nanoelectronic sensing with successes. However, a major limitation of SiNW sensors is that their detection relies essentially on the induced field-effect. Therefore, they are only suitable to the detection charged analytes or electrogenic events. Nowadays, many new NMs have been successfully synthesized. For example, two-dimensional graphene has been added as a new building block for nanoelectronic sensors, taking advantages of its extraordinary electrical properties. It provides vast new possibilities. For instance, many proteins possess aromatic-ring-containing amino acids on the surface. Therefore, they can firmly bind to graphene via π-π interaction and therefore may be detected through the doping effect. Ohno et al. used a pristine graphene device to detect bovine serum albumin (BSA) with a LOD as low as 0.3 nM [78].

12.3.4. Piezoelectric Sensing

Over the past two decades, several advances have been made in micromachined sensors and actuators. As the field of microelectromechanical systems (MEMS) has advanced, a clear need for the integration of materials other than silicon and its compounds into micromachined transducers has emerged. Piezoelectric materials are high energy density materials. It has scaled very favorably upon miniaturization and led to an ever-growing interest in piezoelectric films for MEMS applications [79]. Piezoelectric materials are commonly used in sensor and actuator technologies due to their unique ability to couple electrical and mechanical displacements, i.e., to change electrical polarization in response to an applied mechanical stress or mechanically strain in response to an applied electric field [80]. Transducers using piezoelectric materials can be configured as either actuators or sensors. For actuators, piezoelectric effect is used when the design of the device is optimized for generating strain or stress. For sensors, it is used when the design of the device is optimized for the generation of an electric signal in response to mechanical input [79]. Typically, piezoelectric sensors are configured as direct mechanical transducers or as resonators. The observed resonance frequency and amplitude are determined by the physical dimension properties and materials comprising the device and most importantly by mechanical and interfacial inputs to the device. Piezoelectric immunosensor can be employed as an alternative method for bacterial toxin quantification. Lin and Tsai fabricated a piezoelectric (PZ) crystal immunosensor for the detection of staphylococcal enterotoxin B (SEB). The Au electrode was coated with polyethyleneimine (PEI) to attain high immunochemical activity [81].

A quartz crystal microbalance (QCM) measures a mass per unit area by measuring the change in frequency of a quartz crystal resonator. The resonance is disturbed by the addition or removal of a small mass due to oxide growth/decay or film deposition at the surface of the acoustic resonator. Microgravimetric analyses on piezoelectric quartz crystals have been used as an alternative electronic transduction of biological interactions. The QCM can be used under vacuum, in gas phase and in liquid environments. It is useful for monitoring the rate of deposition in thin film deposition systems under vacuum. In liquid, it is highly effective at determining the affinity of molecules (proteins, in particular) to surfaces functionalized with recognition sites. Larger entities such as viruses or polymers are investigated, as well. QCM has also been used to investigate interactions between bio-molecules. Frequency measurements are easily made to high precision; hence, it is easy to measure mass densities down to a level of below 1 $\mu g/cm^2$.

Several articles reported QCM based sensor for biotoxin detection. Percival et al. casted highly specific noncovalently MIP on to the surface of a gold-coated QCM electrode which was used as a thin permeable film. Frequency shift quantified by piezoelectric microgravimetry with the QCM indicated selective rebinding of the target analyte, monoterpene L-menthol onto the surface. The sensor had LOD of 200 ppb with a response range of 0-1.0 ppm and was able to distinguish between the D- and L-enantiomers of menthol owing to the enantioselectivity of the imprinted sites [82]. Tang et al. developed a piezoelectric immunosensor for the detection of okadaic acid (OA), a kind of DSP toxin. QCM detector was used as the transducer and Low LOD of 3.6ng mL^{-1} was achieved using a non-competitive format with the antibody immobilized in a hydrogel [83].

12.4. Application of Microsensing Platform for Biotoxin Determination

12.4.1. Phytotoxin

By board definition, phytotoxin are toxic metabolites/by-products produced by a plant or substance that is toxic to the plant. Herein, we only discuss phytotoxin that are produced by plants and that have significantly relevance to clinical application, environmental monitoring and food safety. Recent development of microsensor for detection of phytotoxin (e.g. caffeine and terpenes) paved an avenue towards simple, fast, cost-effective, portable assay for toxin sensing and quantification.

12.4.1.1. Caffeine

Caffeine (1,3,7-trimethylxanthine), an alkaloid found naturally in foods such as cacao beans, kola nuts , coffee and tea leaves, present in beverage (including cola and sports drink) and many pain killers and antimigraine pharmaceuticals. It is a powerful stimulant of the central nervous system and also stimulates the cardiac muscle and excessive doses of caffeine can cause anxiety, jitteriness, and upset stomach [84]. Other than traditional sensing formats [85, 86], recent advancement in microsensor hold the potential to provide a accurate, rapid high recovery rate, limit of detection (LOD), easy to use and portable detection assay, solving the inherited problem of current technology mentioned above.

Antibody, acting as specific recognition element, has been widely used in development of immunosensors. A highly sensitive immunoassay based on anti-caffeine monoclonal antibody achieved low LOD of 1 ng L^{-1} without any sample pre-treatment [87]. It can be used a tool to detect caffeine in beverages, shampoo, and caffeine tablets. However, lacking of stability of antibody made it unsuitable for long-term sensing application. Molecularly imprinted polymers (MIPs), known as "synthetic" antibodies, can solve this problem, since it is more stable, easy to modify and relatively cheaper in some cases. MIP possesses multiple recognition sites within polymer matrix which are complementary to the analytes (e.g. caffeine) with respect to the positioning of functional groups. Alizadeh and co-workers developed a MIP based voltammetric sensor with ultralow detection limit of 1.5×10^{-8} mol L^{-1} and broad linear range of 6×10^{-8} to 2.5×10^{-5} mol L^{-1} [88]. The selective binding mechanism of MIP-caffeine endowed high specificity against various interferences (> 500 fold for ions and >100 fold for Aniline, benzoic acid, pyridoxine, thiamine, ascorbic acid and oxalic acid). The caffeine was oxidized to 4,5-diol analogue of uric acid after selective extraction on MIP, the electrochemical signal of which was transuded via differential pulse voltammetry.

Caffeine can also be detected and quantified via its suppressive effect on certain enzymatic activities. For instance, caffeine can specifically inhibit phosphodiesterase 3',5'-cyclic nucleotide and hence reduce the H_3O^+ production, which can be detected by potentiometric pH probe [64]. In this way, the concentration of caffeine can be correlated to the concentration of H_3O^+, which can be ultimately converted into potential signal. The potential response changes of the sensor were proportional to the concentration of caffeine in the range 0-4 mg mL^{-1}. Similarly, caffeine competitively inhibits enzymatic activity of alkaline phosphatase (ALP). Based on the inhibition theory, Akyilmaz et al. fabricated a biosensor for detection of caffeine with LOD of 0.08 μM. [58] Alkaline phosphatise (ALP) was chemically immobilized on a gold screen printed electrode via a cross-linking agent, glutaraldehyde. Electrochemical signal perturbed by enzymatic suppression can be measured by cyclic voltammetric measurement. Caffeine in the soft drinks was determined sensitively using the biosensor.

NMs have played important roles in biosensing. The emerging of 1D and 2D NMs, such as CNTs, silicon nanowire, graphene, etc., becomes a strong force in development of biosensor area. A sensitive electrochemical biosensor for the determination of caffeine based on multi-walled carbon nanotube and a Nafion-modified glassy carbon electrode (MWNT-Nafion/GCE) was developed by Zhang et al. [89] The sensor was successfully used for the determination of caffeine in real samples of "Sanlietong" tablets (caffeine containing medicine) and Cola, exhibiting good recovery rate and selectivity. Yang et al. proved the synergic effect of MWCNT/Nafion sensor, which contribute to a low LOD of 2.3×10^{-7} mol L^{-1} [90]. The cation exchanger of Nafion has selective cation exchange enriched ability due to the electrostatic interaction. In addition, MWNTs have larger specific surface area, excellent adsorptive ability and catalytic ability towards caffeine oxidation. Sanghavi and co-workers developed an ultrasensitive sensor (a LOD of 8.83×10^{-8} M) by using adsorptive stripping differential pulse voltammetry (AdSDPV) [91]. An *in situ* surfactant-modified MWNT was used to construct the electrode to facilitate oxidation reaction. A novel sensitive molecularly imprinted electrochemical sensor was described for the detection of caffeine, utilizing both nanomaterial and molecularly imprinting technique [92]. The carbon electrode was modified with MWCNTs/vinyltrimethoxysilane (VTMS) recovered by a molecularly imprinted siloxane (MIS) film prepared by sol–gel process. Graphene and its derivatives can also be used as sensing materials for caffeine sensors [93, 94]. The Sensors based on graphene and graphene oxide (GO) achieved low detection limit of 1.2×10^{-7} M and 2.0×10^{-7} M respectively.

12.4.1.2. Morphine

Morphine, a phytotoxin extracted from *Pupaver somnifemm* (opium poppies), can be used as the treatment of choice for moderate-to-severe cancer pain [95]. A large overdose can cause asphyxia and death by respiratory depression and hence monitoring morphine concentration in whole blood, plasma or serum is of great importance in biomedical and forensic toxicology. In addition, accurate means to determine morphine are required for samples such as raw plant materials, industrial process streams and pharmaceutical formulations. Raidoimmunoassay (RIA) Gas chromatography (GC), high-performance liquid chromatography (HPLC) with various detection systems (e.g. UV detection or MS detection) are the common analytic techniques [96-100]. However, these techniques often have some disadvantages such as sample pre-treatment step requirement, time-consuming nature and cost of operation and incompatibility with portable device. Therefore, the development of sensitive, rapid and selective microsensors for determination of morphine is needed.

Morphine can be determined via chemiluminescence detection. A morphine sensor with ultralow detection limit of 5×10^{-8} M was developed based on resultant chemiluminescence (CL) emission from the reaction of morphine and acidic potassium permanganate ($KMnO_4$) in the presence of tetraphosphoric acid [101]. Flow injection analysis (FIA) was used to ensure continuous flow and automated process. It was suggested that the manganese(III)-alkaloid complex was reduced to an electronically excited manganese(II) complex which subsequently emitted a photon. Methyltriphenylphosphonium permanganate was reported as an alternative to potassium permanganate [102]. FIA can also be incorporated with continuous cyclic voltammetry for morphine determination. Norouzi and co-workers reported morphine monitoring system employing fast Fourier transformation with continuous cyclic voltammetry (FFTCV) and achieved a detection limit and a quantitation limit of 95.5 and 285 $pgmL^{-1}$, respectively [103]. The biosensor required no prior extraction step or oxygen removal process.

Yeh et al. fabricated an amperometric sensor by modifying indium tin oxide (ITO) electrode with conducting polymer, poly(3,4-ethylenedioxythiophene), PEDOT and MIP. [59] PEDOT, acted as electroactive film that catalyzes morphine oxidation and lowered the oxidization potential, while MIP served as recognition element. Sensitivity of 91.86 $\mu A/cm^2$ per mM and LOD of 0.2 mM were achieved. Moreover, the biosensor can discriminate between morphine and its analogs (e.g. codeine). The MIP–PEDOT morphine sensor can be further incorporated into a microfluidic chip, which ensured continuous morphine measurement and automatic detection format [104]. The advantages of microfluidic analytical and detection systems also include high sensitivity, low sample consumption, portability, low power consumption, low cost and compatibility with advanced integrated circuits (IC) technology. Poly(dimethylsiloxane) (PDMS) microchip with electrochemical (EC) detection was employed as morphine determination method and used for monitoring the amount of morphine in human urine with satisfied result [105]. Compared with the conventional methods, such PDMS based EC chip has lower instrument cost, less reagent consumption and shorter analysis time.

Electrodes (e.g. glassy carbon electrode and ITO) can be modified by various modifiers (e.g. nanoparticles and electrocatalyst). Li et al. developed a morphine sensor based on ordered mesoporous carbon (OMC) modified glassy carbon electrode (GCE) [106]. The mesoporous carbon can significantly decrease the overpotential (ca. 82 mV) and increase in the peak current (80 times), indicating that OMC/GCE remarkably enhanced electrocatalytic activity towards the oxidation of morphine. As the result, high sensitivity of 1.74 μA/μM and low detection limit of 10 nM were achieved. Incorporated with a medium-exchange procedure, the OMC/GCE was successfully applied to the detection and quantification of morphine in urine samples with a low detection limit of 50 nM and satisfied recovery of 96.4 %. CNT was used as sensing material owing to its large surface area and prominent electrical properties (Fig. 12.1).

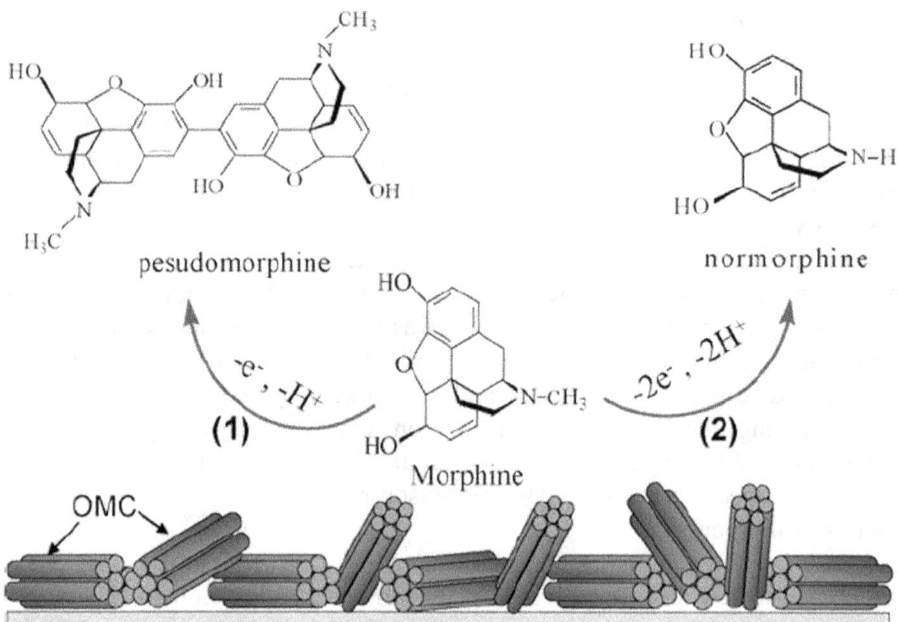

Fig. 12.1. Electrochemical oxidation of morphine at OMC/GCE.
Reprinted from Ref. [106] © 2009 Elsevier B.V.

A novel modified carbon paste electrode with vinylferrocene/multiwall carbon nanotubes was fabricated, of which relative standard deviation for multiple successive assays or using different electrodes were below 3 % [107]. In addition, the electrochemical sensor can be used for real biological and pharmaceutical samples. Babaei et al. fabricated a multi-walled carbon nanotubes/chitosan polymer composite modified glassy carbon electrode [108]. Wide linear dynamic range, low detection limit and good repeatability were achieved, attributing to high water stability and high mechanical strength of the MWCNT/chitosan nano biocomposite electrode and the antifouling effect of chitosan. 4-hydroxy-2-(triphenylphosphonio) phenolate (HTP) and MWCNT composite can be used to enhance the performance of GCE. A HTP-MWCNT-CPE based sensor achieved low detection limit of 66 nM, partially owing to the fact that HTP caused to decrease the overpotential, 300 mV. In addition, differential pulse voltammetry (DPV), which had lower charging current contribution to the background current and a much higher current sensitivity than cyclic voltammetry was used to estimate concentration of morphine [109]. Graphene, a 2D nanomaterial electrochemically similar to CNT, was used to decorate GCE. Navaee et al. fabricated graphene nanosheets (GNSs) modified glassy carbon (GC) electrode for detection of morphine [110]. The modified electrode showed excellent electrocatalytic activity toward oxidation of morphine, in wide pH range. In addition, the GNC modified electrode can be used for simultaneous detection of three major narcotic components, heroin, noscapine and morphine at micromolar concentration without any separation or pretreatment steps.

Zhao et al. reported a morphine sensor based on ITO electrode modified with directly electrodeposited gold nanoparticles [111]. AuNPs assembled on the ITO electrode surface had high electrocatalytic activity for the morphine oxidation and effectively reduced the morphine oxidation overpotential on the ITO electrode. As the result, board dynamic range from 8.0×10^{-7} to 1.6×10^{-5} M and low LOD of 2.1×10^{-7} M were achieved. Prussian blue (PB), a promising artificial peroxidase for morphine, can also be used as modifier for ITO electrode. [112] PB is an excellent mediator which facilitates electron transfer and considerably lowers the overpotential. Most importantly, the PB-ITO based sensor can readily discriminate MO analogs lacking the phenolic –OH group, such as codeine, ensuring high selectivity.

In addition to electrochemical sensing, a photonic mechanism can also be used for detection of morphine. Bonanno and DeLouise developed chip-based label free porous silicon (PSi) photonic sensor [113]. As shown in Fig. 12.2, the morphine sensor was appealing for clinical and POC applications since the straightforward optical detection was label free and does not require any secondary label amplification in currently available enzyme immunoassays. They further tested with clinic sample and achieved 96 % concordance with liquid chromatography-mass spectrometry/tandem mass spectrometry (LC-MS/MS) [114]. Strip-based immunochromatographic assay can be used for visual detection of morphine. Using specific egg yolk antibodies (IgY), a strip-based immunochromatographic assay achieved a detection range of 1–1000 ng/mL and LOD of 2.5 ng/mL [115]. Such high throughput strip-based sensor is compatible of commercialization due to its low-cost and simplicity.

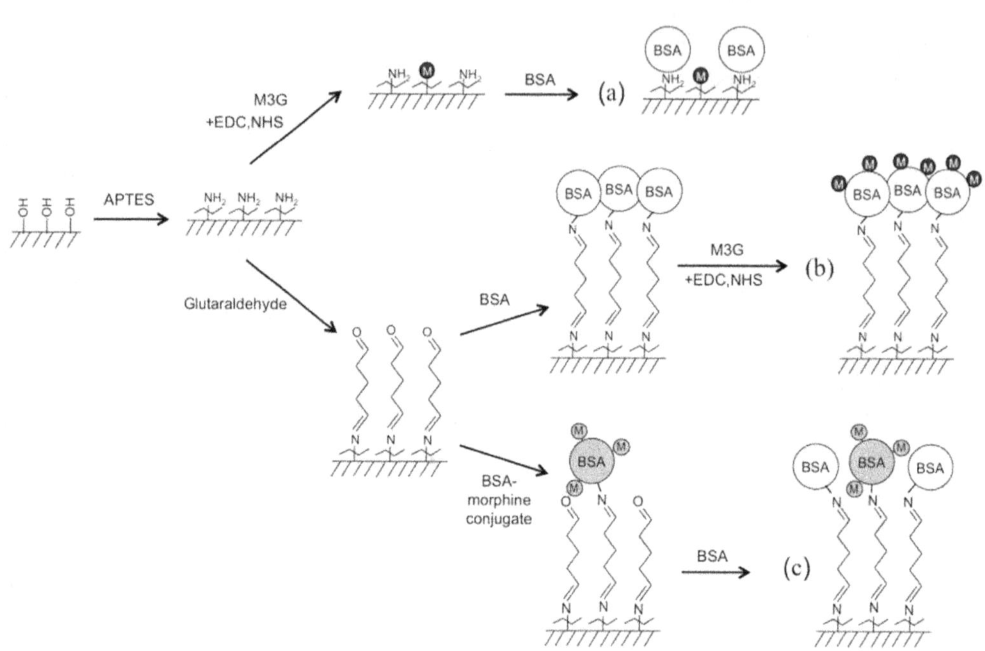

Fig. 12.2. Surface Chemistry to Attach Opiate Analogue to PSi Substrate for Opiate Competitive Inhibition Assay. Reprinted from Ref. [113] © 2010 American Chemical Society.

12.4.1.3. Terpenes

Terpenes, the primary constituents of the essential oils of many types of plants and flowers, are synthesized from acetyl CoA or basic intermediates of glycolysis. It is a large and diverse class of organic compounds, including monoterpenes, sesquiterpenes, diterpenes, triterpenes, glycosides et al. Many of terpenes are defensive agents against insects secreted by plants. Some are toxic and even lethal to human. For instance, Cyanogenic glycosides are extremely poisonous since it can influence NA^+/K^+ activated ATPases in human heart and cause the heart rate deviating from normal.

Well-established analytic methods for determination of terpene includes: Chromatographic techniques (e.g. GC and HPLC), multi-component identification techniques (e.g. MS and NMR spectrometry) and hyphenated techniques (e.g. GC-MS and GC-UV). They are comprehensively discussed in previous reviews [116, 117]. Headspace solid-phase microextraction (HS-SPME) can be used to achieve fast and sensitive detection. A sensor for the detection of β-caryophyllene was developed by coupling HS-SPME with gas chromatography with flame ionization detection (GC-FID) [118]. LOD and LOQ of the sensor were 0.03 and 0.14 μg mL^{-1}, respectively. UV ion mobility spectrometry was used to detect terpenes as well. Analyte molecules are ionized under the exposure of UV and accelerated in an external electrical field in the direction normally against drift gas flow. Therefore, the properties of the analytes, such as charge, shape and molecular weight can be reflected from the mobility of the ions. With the help of pre-separation by multi-capillary column (MCC), the sensor achieved a

LOD of 5 μg/m^3 (about 1 ppbv) for selected terpenes and the results were independent on relative humidity of the analyte [119].

Pre-concentration or pre-separation can be viewed as one of the effective steps in biosensing. The application of capillary electrophoresis for the separation of various active components in plants, including terpenes, has become increasingly widespread because of its short analysis time, minimal sample volume requirement and high separation efficiency. Usually, capillary electrophoresis (CE) is coupled with electrochemical detection. Chen et al. developed a CE-electrochemical sensor for detection of glycosides in *Paeonia suffruticosa* (Moutan Cortex). LOD of 1.1 μM was achieved. The use of NaOH aqueous solution as the background electrolyte can minimize the fouling of the inner wall of the capillary and the surface of the working electrode during the CE analysis so that the high repeatability was achieved [120]. Capillary electrophoresis can also be incorporated with fluorometric detection. Kubesová and co-workers has developed a CE-fluorometric detection based sensor for detection of farnesol, an acyclic sesquiterpenoids [121]. Terpenes were dynamically labeled by the non-ionogenic tenside poly(ethylene glycol) pyrenebutanoat and ultralow LOD of 10^{-17} mol L^{-1} was achieved under excitation wavelength 335 nm and the emission wavelength 463 nm. Qi et al. reported a triterpenes sensor which was based on nonaqueous capillary electrophoresis separation [122]. Three bioactive triterpenes: oleanolic acid (OA), ursolic acid (UA) and 2α,3β,24-trihydroxy-urs-12-en-28-oic acid (TA) of Chinese herb, Rabdosia japonica, were simultaneously separated by nonaqueous capillary electrophoresis (NACE) in a fused-silica capillary tube. The use of organic solvent in NACE offered several advantages including reduced electrophoretic currents, and improved mass spectrum compatibility, solubility and stability of hydrophobic compounds.

The QCM is well suited as transducer elements for chemical sensors, being portable, rapid, and sensitive. Recognition elements are usually added to the acoustic wave device for selectively binding the analyte to the device surface. Once target analytes binds to recognition molecule, the added weight result in decrease in their resonant frequency. Several articles reported QCM based sensor for terpene detection. Percival et al. casted highly specific noncovalently imprinted polymer (MIP) on to the surface of a gold-coated QCM electrode which was used as a thin permeable film [82]. Frequency shift quantified by piezoelectric microgravimetry with the QCM indicated selective rebinding of the target analyte, monoterpene L-menthol onto the surface. The sensor had LOD of 200 ppb with a response range of 0-1.0 ppm and was able to distinguish between the *D*- and *L*-enantiomers of menthol owing to the enantioselectivity of the imprinted sites.

Iqbal et al. reported a MIP based QCM-arrays for sensing terpenes in fresh and dried Lamiaceae family species, i.e., rosemary (Rosmarinus Officinalis L.), basil (Ocimum Basilicum) and sage (Salvia Officinalis) [123]. The sensor had linear detection range of 20–250 ppm and capable of discriminating molecules of similar molar masses and even isomers such as *α*-pinene and *β*-pinene. Employing similar mechanism, Lieberzeit et al. fabricated a QCM sensor array for monitoring terpene emissions from odoriferous plants [124]. The sensor comprised of six sensing channels each coated with a MIP selectively interacting with alpha-pinene, thymol, estragol, linalool, and camphor, respectively. The sensor can detect and determinate various terpenes in a concentration range below 70 ppm and accurately monitor the terpenes emanated from fresh leaves of herbaceous plants in real time.

12.4.2. Animal Toxin

Animal toxins are roughly divided into venoms and poisons. The purpose of the venoms is offensive, and often used in the quest for food, while that of poisons is defensive. Human beings come in contact with these toxins, usually by accident or as a result of the animal defending itself. In an emergent medical setting, rapid and accurate identification of animal toxin is of great importance. In addition, Animal venoms and poisons are now recognized as major sources of bioactive molecules that may be tomorrow's new drug leads. Therefore, detection assays for monitoring their concentration in drugs and in human body sensitively and selectively are required.

12.4.2.1. Apitoxin

Apitoxin, often known as honey bee venom, is colorless acidic liquid produced in the abdomen of worker bees. Apitoxin usually do not exhibit severe toxicity to human body. However, bee stings can cause allergic reactions in 1% of human population. Therefore, methods for the separation of the venom components, their characterization, and their determination are of great interest for both diagnostic and therapeutic purposes. The principle active component of apitoxin, melittin has inhibition to many essential physiological process and proteins such as Na^{+}-K^{+}-ATPase, H^{+}-K^{+}-ATPase and Ca^{2+}/calmodulin-dependent protein kinase II. Other components include apamin, phospholipase a2, adolapin, histamine, hyaluronidase et al.

Chromatographic techniques were extensively reported as a method for identification of a variety of bee venom peptides including melittin, apamine, adolapin and mast cell degranulating peptide (MCDP) [125, 126]. Meanwhile, in contrast with chromatographic techniques, high-performance capillary electrophoresis requires less analytical reagent (~nanoliters), time and considerably lower cost. Kokot et al. developed a high-performance capillary electrophoresis (CE) method for determination of four major constituents (apamine, MCDP, phospholipase A2, and melittin) of honeybee venom [127]. The LOD and LOQ were comparable to HPLC (at μM level). Principal component analysis (PCA) was applied to enhance the accuracy and standardize the bee venom sample classification. According to strong correlations between apamine, MCDP, phospholipase A2, and melittin revealed by PCA, the strain of bee is the only criteria for bee venom sample classification. Similarly, Pacakova and Stulik fabriacated a CE based sensor for detection of phospholipase A2 and melittin [128]. UV detector at 190 nm was incorporated into the system and linear response ranges were approximately 2 concentration decades with correlation coefficients of 0.9994 for phospholipase A2 and 0.9997 for melittin. In addition, the limits of detection and quantitation were 4.5 and

15 μg/mL, respectively, for phospholipase A2 and 1.6 and 6 μg/mL, respectively, for melittin. High recovery rate of 98.8 % for phospholipase A2 and 101.7 % for melittin were achieved.

The concentration of bee venom peptides can also be determined by fluorescence spectrometry. In the work of Banks et al., melittin, apamin, and peptide 401 (MCD peptide), were first readily separated on columns of Heparin Sepharose CL-6B using a linear salt gradient [129]. Subsequently, the fluorescence detection (emission:354 nm) was performed. The linear detection range for melittin was found to be from 3×10^{-7} to 6×10^{-5} g mL^{-1}.

Many optical based sensors utilized the fact that melittin usually forms pores in zwitterionic lipid, whereas it exhibited a detergent-like action in disrupting a lipid bilayer composed of anionic lipids. For instance, a lipid bilayer were "insert" in between microsphere-supported cationic conjugated polyelectrolyte poly(pphenyleneethynylene) (PPE) and an electron-transfer quencher, 9,10-anthraquinone-2,6-disulfonic acid (AQS) [69]. Consequently, the presence of lipid bilayer resulted in attenuation of superquenching by AQS. However, when melittin interacted with and lyse the lipid bilayer, superquenching occurred, which has positive relation with the concentration of melittin. Employing the same lysis mechanism of melittin, Rozner et al. developed colorimetric assay based on supramolecular assemblies of lipid–polydiacetylene (PDA) vesicles [130]. The lipid–PDA aggregates undergo visible and quantifiable blue-to-red transitions following interfacial interactions and perturbation by melittin. The degree of lipid–bilayer perturbation by membrane positively related with the concentration of melittin. The LOD at level of μM was achieved. Heip et al. reported a label-free melittin sensor based on localized SPR (LSPR) and electrochemistry measurements. Core-shell structure nanoparticle was used to make the analysis in SPR detection more tuneable. The silica nanoparticles were used as the "core" and thin gold films "shell" were coated on the "core". As the result, the excitation mode of the plasmon absorption spectra of the core-shell nanoparticle structure can be controlled by adjusting the core-to-shell ratio. Similarly, hybrid bilayer membrane (HBM) was used as the recognition element to assess membrane-disturbing properties of melittin, which reflected its own concentration [131]. (Fig. 12.3) The employment of SPR ensured a label free, real time detection format and low LOD of 10 ng/mL was achieved. LSPR microfluidic format was also developed to lower consumption of reagents and to gain the ability of multiplexing. Gheorghiu and co-workers reported a SPR sensor by modifying L1 sensor chip with lipid, which provided an accessible platform for SPR exploration of peptide–membrane interaction [132]. The sensing platform can quantitatively describe the interaction process based on a mathematical model encompassing the distinct stages involved in peptide–lipid interaction: association, insertion of melittin into lipid matrix, pore formation and destabilization of lipid membrane via related rate constants and threshold concentrations. Relatively low LOD of 0.45 μM was achieved. In the work of Shahal et al., thin films of poly(dimethylsiloxane) (PDMS) was spin coated on thiol-modified gold layer of SPR devices, serving as the substrate of homogeneous zwitterionic supported lipid bilayers (SLBs) [133]. DMS is a good substrate for the formation of stable, homogeneous SLBs due to its hydrophibicity. One limitation of biosensors based on lytic peptide-membrane interaction is that they usually have low specificity against other enzymes which are capable of digest lipids.

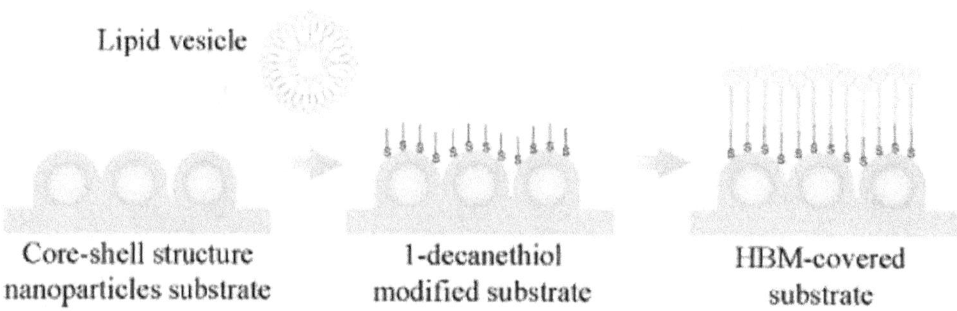

Fig. 12.3. Fabrication of the membrane-based sensor using the core-shell structure nanoparticle substrate. Reprinted from Ref. [131] © 2008 American Chemical Society.

12.4.2.2. Spider Toxin

Spider toxins are a family of proteins produced by spiders which function as neurotoxins. The venom of spiders is a complex mixture of neuroactive proteins and other chemicals. The venoms of approximately 200 species of spiders represent a risk to humans. Therefore, it is of great importance to develop sensitive and selective detection methods for determination of spider toxin.

Escoubas et al. developed a detection method for crude venoms of tarantula species in the genus *Brachypelma* (a kind of spider) by using high-performance liquid chromatography and matrix-assisted laser desorption/ionization time-of-flight mass spectrometry (MALDI-TOFMS) [134]. This method allowed rapid mass fingerprinting of large numbers of samples in a reproducible and non-destructive manner, and offered a powerful systematic tool in combination with morphological methods for the classification of tarantula species. Escoubas's group further improved the multidimensional biochemical analysis by incorporating HPLC, capillary zone electrophoresis and matrix assisted laser desorption/ionization time-of-flight mass spectrometry for the identification of tarantula venom samples [135]. This accurate assay was successfully applied to detect trace level of toxin from 5 live spiders with high reproducibility. His group further developed a combined cDNA and mass spectrometric approach for detection of Australian funnel-web spider venoms [136]. The approach was capable of sensing minor molecular species, revealing the comprehensive venom landscapes of funnel-web spider.

Enzyme linked immunosorbent assays (ELISA) were also used for spider toxin detection. Châvez-olortegui et al. reported a ELISA for the detection of venom antigens in experimental and clinical envenoming by *Loxosceles intermedia* spiders [137]. Immunoaffinity chromatography was used to set up a wich-type ELISA with hyperimmune horse anti-Loxosceles intermedia IgGs. The immunosensor was capable of discrimination of circulating antigens in mice that were experimentally inoculated with L. intermedia venom from those inoculated with L. gaucho, L. laeta, and Phoneutria nigriventer spider venoms, Tityus serrulatus scorpion venom and Bothrops jararaca, Crotalus durissus terricus, Lachesis muta and Micrurus frontalis snake venoms. The same group developed another ELISA based sensor for Phoneutria nigriventer spider venom employing similar mechanism [138]. Measurable absorbance signals were detectable at low dose of 0.8 ng.

Although many analytic assays (e.g. ELLISA and MS) have been explored so far, there is lacking of effective and rapid determination method for identification and quantitation of spider toxin. Current methods are either too time consuming or too expensive for clinic application. Future development of spider toxin sensor may utilize the perturbation of electrochemical action of the toxin on neuron activities. Therefore, the chemical signal can be converted to electrochemical signal, which can be recorded by electrophysiological approach [139].

12.4.2.3. Snake Toxin

Snake bites can be lethal, but snake venoms also contain bioactive constitutes with biomedical and biotechnological value. Thus, determination of the snake venom has a number of potential benefits for basic research, clinical diagnosis, and development of new research tools and drugs of potential clinical use. Snake toxin has diverse mechanism to disrupt human physiology, ranging from disrupting the nervous system, to cardiotoxins, to myotoxins, to increasing or decreasing coagulative functions.

Snake toxin is complex mixture of a large number of distinct proteins. Fractionation of the crude venom by reverse-phase HPLC is usually required. The characterization of various components of snake toxin can be done by N-terminal sequencing, sodium dodecyl sulfate polyacrylamide gel electrophoresis (SDS-PAGE), mass spectrometric determination of the molecular masses and the cysteine (SH and S–S) content, SDS-PAGE, MALDI-TOF mass fingerprinting, and collision-induced dissociation tandem mass spectrometry [140, 141].

On the other hand, immune based assay, such as radioimmunoassay, enzyme-linked immunosorbent assay (ELISA) and fluorescence immunoassay, can sever as a simpler alternative for snake toxin detection. Affinity-purified venom-specific antibodies were demonstrated to be the ideal for snake venom detection. Various immune based assay methods were discussed in a comprehensive review [142]. Selvanayagam et al. reported a species-specific avidin-biotin microtitre ELISA for detection four medically important snakes (Bungarus caeruleus, Naja naja, E. carinatus and Daboia russelli) in autopsy specimens of human victims of snake bite [143]. The assay successfully detected venom concentrations as low as 100 pg/mL of tissue homogenate. Dong et al. reported an ELISA test kit for four common snake venoms and detection limit of 0.2–1.6 ng/mL (depending species of the snake) was achieved [144]. Three-step affinity purification protocol was set up for preparation of species-specific antivenom antibodies, including affinity chromatography of IgG from hyper-immunized rabbit sera with protein A columns, immuno-affinity chromatography of monovalent antivenom antibodies with respective homologous venom columns, and immuno-absorption of cross-species reacting antibody molecules with heterologous venom columns. Snake Venom Detection Kit (SVDK) (CSL Australia, Parkville, Australia), a commercial tool for identification of snake venom, were constructed by using rapid, freeze-dried, sandwich enzyme immunoassay [145, 146]. It was reported that no false-positive reactions was occurred in 50 dog and 25 cat urine samples and that the specificity was 100 %.

Several optical based sensors have been reported for detection of snake toxin. Zhang's group developed an immune-fluorescence based assay for determination of Naja kaouthia Venom [68]. Micro-scale polystyrene beads were used as substrate for antivenom antibody, forming the "capture-bead". Thereafter, the venom bonded to the capture-bead to form the complex through the antibody-antigen interaction. Lastly, the Qdot conjugated second antibody was added, the emitting fluorescence of which can be directly observed under UV-microscope. Low LOD of 5–10 ng/mL was achieved. Biardi et al. reported a fluorometric method for the quantitative analysis of snake venom metalloproteases (a biotoxin that is responsible for the hemorrhagic effects) [147]. The sensor was sensitive enough to detect differences among venoms using < 2 ng of whole venom protein. Dong et al developed a optical silicon chip based immunosensor for snake venom of *Trimeresurus albolabris, Calloselasma rhodostoma, Naja kaouthia and Ophiophagus Hannah* in blood, plasma, urine, wound exudates, blister fluid or tissue homogenates [148]. The specific binding of snake toxin on an optical silicon chip resulted changes in thickness of molecular thin film (Fig. 12.4). The reflection of light through the thin film resulted in destructive interference of a particular wavelength of the light from gold to purple-blue depending on the thickness of the thin film, which represented the amount of snake venom. The sensor was two times more sensitive than ELISA (0.4 and 0.8 ng/mL, respectively) in detection of N. kaouthia venom. Hartono and co-workers reported a liquid crystal (LC)-based sensor for real-time and label-free identification of phospholipase-like toxins, Beta-bungarotoxin, a Bungarus snake venom [149]. The LCs was coupled to the presence of phospholipids. Therefore, the enzymatic events acting on the phospholipids can be amplified.

12.4.3. Marine Toxin

Phycotoxins, non-proteinaceous compounds synthesized by marine microalgae, are toxic chemicals that enter into the food chain as components of the phytoplankton. Shellfish and some fish spices ingest these toxins and act as vectors, transmitting them to humans. The toxin accumulation in vector's body cause minimal poisonous effect, but when consumed by human, it could cause diarrheic, paralytic, amnesic syndromes and even death. Therefore, monitoring and controlling marine toxins is a serious concern for public health, environmental protection, fishing industry and tourist economy. Several analytic techniques and biosensors has been reported for detection of phycotoxins, including MS based analysis, immunoassays, immunosensors, SPR based sensor, enzyme inhibition-based biosensors, MIP-based sensors, sodium channel-based biosensors et al. [150] In the previous issue (Sensors 2012), most marine toxin detection methods have been comprehensively reviewed. Herein, only biosensors with prominent performance will be discussed.

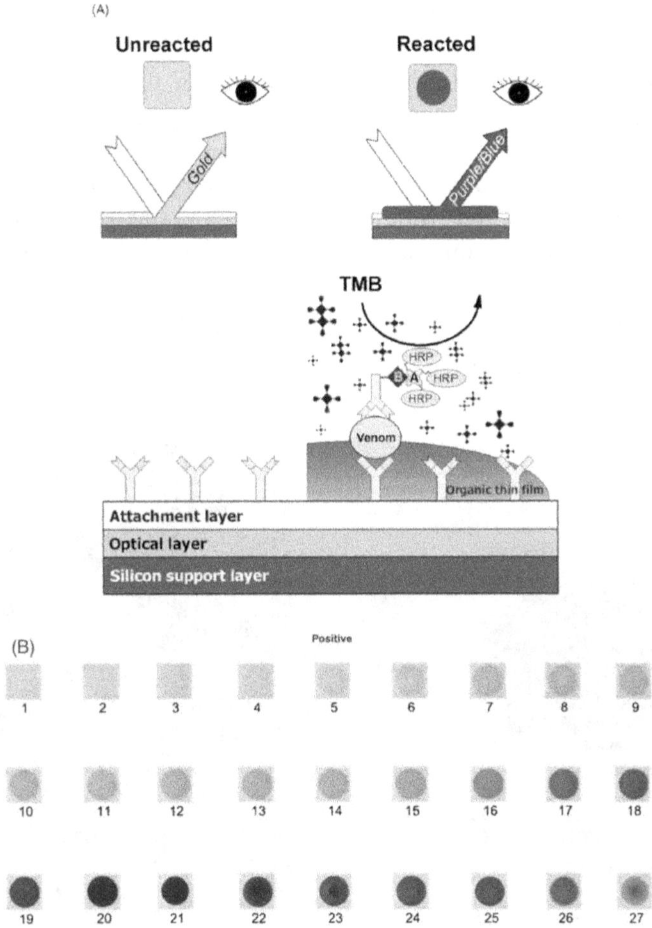

Fig. 12.4. Principles of optical immunoassay for snake venom detection. (A) The assay was done based on the detection of physical change in the thickness of molecular thin films resulting from specific binding events on the top of an optical silicon surface. The silicon wafer comprises three layers: (i) silicon supporting layer, (ii) an optical layer which reflects the light (iii) and the attachment layer on the top which enhances the binding of capture antibody. Formation of immunocomplex with HRP enzyme attached via avidin-biotin interaction and the precipitation of TMB substrate produce a biological thin layer on the optical reflecting surface. Presence of the thin film changes the reflected light from gold colour into purple or blue depend on the thickness of the optical layer or the amount of immunocomplex formed. (B) Signal strength chart, which scores the full reflection colour spectrum of this optical surface from 1 to 27 according to the thickness of the layer formed on top of it. Reprinted from Ref. [148] © 2003 Elsevier B.V.

12.4.3.1. Shellfish Poisoning Toxin

Shellfish poisoning toxin can be further divided into 5 categories based on poisoning syndromes: diarrheic shellfish poisoning (DSP), paralytic shellfish poisoning (PSP), amnesic shellfish poisoning (ASP), neurologic shellfish poisoning (NSP), azaspiracid shellfish poisoning (AZP). Currently, the most commonly used assay is in vivo mouse bioassay which has low reproducibility and involves ethical issues. The development sensitive and rapid biosensors could potentially solve this problem.

Tang et al. developed a piezoelectric immunosensor for the detection of okadaic acid (OA), a kind of DSP toxin [83]. A QCM detector was used as the transducer and Low LOD of 3.6 ng mL^{-1} was achieved using a non-competitive format with the antibody immobilized in a hydrogel. Similar to antibody, MIP also has specific recognition property [151] (Fig. 12.4). In the work of Lotierzo et al., a MIP-modified SPR chip was fabricated for detection of domoic acid (DA), a kind of ASP. Competitive assay with free DA and HRP-DA conjugate was used to attain LOD (5 ng mL^{-1}). In addition, the sensor can be put in continuous use over a period of 2 months.

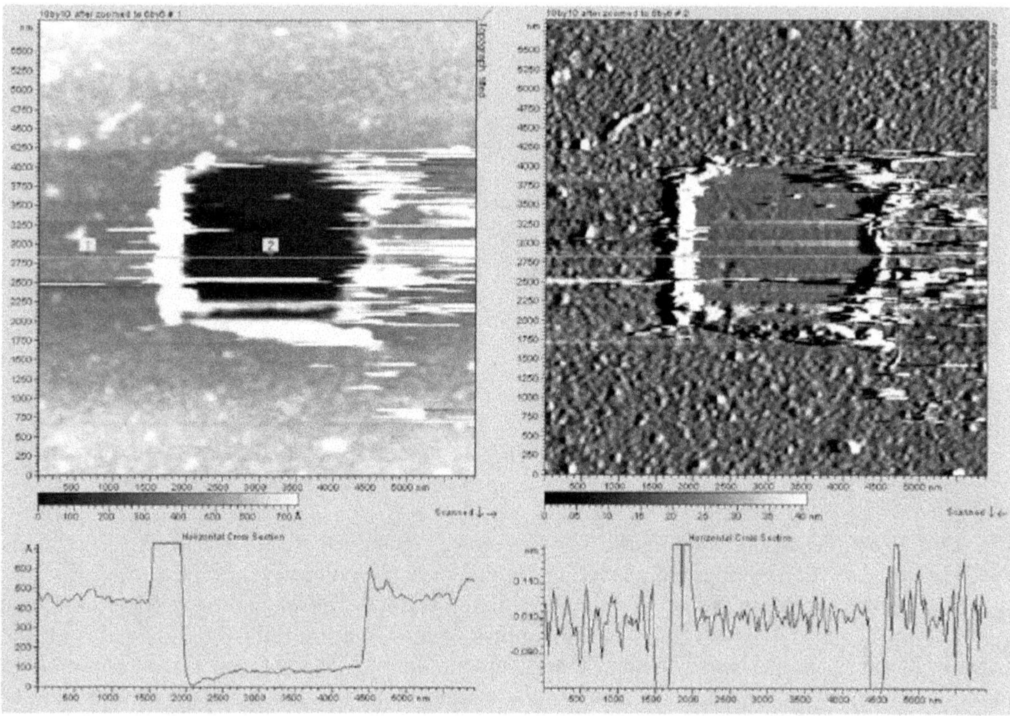

Fig. 12.5. 6 μm × 6μm AFM image of a 2 μm × 2 μm scraped area of a MIP grafted chip and the cross-section analysis (below). Reprinted from Ref. [151] © 2004 Elsevier B.V.

In addition to common biosensors, several groups has developed marine toxin based on their specific effect on natural activities/process at tissue, cell, molecular level. For instance, PSP toxins (e.g. STX) can block as sodium channel of neuron cell. Kulagina et al. constructed a neuronal network biosensor for detection of PSP toxins by employing this natural process [152]. Mammalian neurons from spinal cord tissue of embryonic mice were grown on a 64-site microelectrode array. The presence of PSP toxin can be detected by the decrease of mean spike rate of spinal cord neuronal networks. The detection limits for STX is 12 pg mL^{-1} in buffer and 28 pg mL^{-1} in diluted seawater, which was 30 000 times below the in vivo mouse bioassay detection limit. OA can bind to the receptorial site of phosphatases 2A (PP2A), blocking their activity. Hamada-Sato and co-workers reported a biosensor which combined the PP2A inhibition with the phosphate ion consumption by pyruvate oxidase (PyOx) into a flow injection analysis (FIA) system.[153] Ultralow LOD of 0.1 ng mL^{-1} was achieved, which was 50 times more sensitive compared with ELISA.

Electrochemical biosensor can be easily miniaturized, which makes it candidate of point of detection. Carter et al. reported an immunoelectrochemical biosensor for the detection of STX and PbTx [60]. Antibody was conjugated with glucose oxidase (GOD), while target analytes (antigen), STX and PbTx were immobilized covalently on membrane via BSA. H_2O_2 can be easily detected amperometrically as antibody brought GOD adjacent to the electrode. Alternatively, alkaline phosphatase (ALP) as label and paminophenyl phosphate as enzyme substrate achieved better LOD: 1.5, 1.0, and 2.0 $ngmL^{-1}$ for OA, PbTx-3 and DA respectively [154-156].

12.4.3.2. Fish Poisoning Toxin

There are mainly 2 types of fish poisoning toxin, i.e. puffer fish poisoning (PFP) toxin and ciguatera fish poisoning (CFP) toxin. Fish poisoning toxin are resulted from various sources, such as bacterial spoilage, dinoflagellates (microscopic sea plants) and certain organs of pufferfish. Less application of sensors for detection of fish poisoning toxin were developed compared with that of shellfish poisoning toxin.

Tetrodotoxin, commonly known puffer fish poisoning (PFP) toxin, is one of the most dangerous sea-born toxins, which threaten the food safety and human health preservation. Anti-tetrodotoxin specific antibody can be used as recognition element to construct an immunobiosensor for detection of tetrodotoxin. Amperometric detection was performed on a screen-printed electrode via an indirect competition format where the amount of current generated by p-aminophenol, the product of the enzymatic activity of the alkaline phosphatase label of the specific antibody, was inhibited by the presence of tetrodotoxin [155]. Ultralow LOD of 0.016 ppb was achieved. Similarly, Neagu et al. reported direct competitive immunoenzymatic sensor by measuring the p-aminophenol product of alkaline phoshatase activity [157]. The dynamic detection range was reported as 2–50 ng mL^{-1} and the LOD was 1 ng mL^{-1}. By employing the protection effect of tetrodotoxin to mouse neuroblastoma cells in the presence of ouabain and veratridine, Gallacher and co-worker developed a hemolysis based assay with board detection range of 3.2–160 ng/mL [158]. The quantity of surviving cells, which was positively related to tetrodotoxin concentration, were determined by their uptake of the vital dye Neutral red which was quantified by a microtire plate reader.

Ciguatera fish poisoning (CFP) are commonly caused by ingestion of some species of tropical and subtropical reef fish. CFP toxin can be detected and quantified via various detection format and mechanisms, including, LC-MS [159], ELISA [160], RIA [161] and fluorimetric assays based on changes in membrane potential [162].

12.4.4. Microbial Toxin

Microbial toxins are toxins produced by micro-organisms, i.e. bacteria and fungi. Microbial biosensors are analytic devices with microbial recognition element and transducer which convert signal into readable form. They are mainly applied for food safety, and pharmaceutics military defence and environmental monitoring. In this section, sensors for mycotoxin (produced by fungi) and bacterial toxin (produced by bacteria) are discussed respectively.

12.4.4.1. Mycotoxin

Mycotoxins are toxic secondary metabolites produced by filamentous fungi. Mycotoxin contamination in the food chain has caused serious health issues in humans and animals. Thus, a rapid on-site and lab-independent detection method for mycotoxins is desirable.

Electrochemical immunoassay-based microarrays have been successfully used for mycotoxin analysis because of their sensitivity, selectivity, versatility, and simplicity. Microelectrodes have greater sensitivity because of enhanced mass-transport on microscale electrodes. The most widely used method for mycotoxin recognition on a microarray depends on the antibody molecule, which can offer the high specificity and sensitivity required for low-level mycotoxin detection. A microelectrode immunosensor-based microarray for aflatoxin M1 (AFM1), a kind of mytotxin, has been investigated [163]. (Fig. 12.6) The microelectrode arrays consisted of 35 microsquare electrodes with dimensions of 20 μm × 20 μm and an edge-to-edge spacing of 200 μm. The on-chip reference and counter electrodes were fabricated via standard photolithographic methods. Cyclic voltammetry was used to determine the characteristics of the microelectrode arrays and the behavior of the on-chip electrodes. To analyze for AFM1 directly in milk samples, antibodies against AFM1 were immobilized by cross-linking with 1,4-phenylene diisothiocyanate on the surface of the microarray, which was pre-functionalized with silanization reagent. Without matrix interference, a cELISA assay format was conducted on the microarray electrode surface using 3,3,5',5'-tetramethylbenzidine dihydrochloride/H_2O_2 with horseradish peroxidase as the enzyme label in an electrochemical detection mode. The LOD for AFM1 in milk was 8 ng/L with a dynamic detection range of 10–100 ng/L, which was lower than the current EU legislative maximum residue limit (MRL) of 50 ng/L. Compared with other recent reports on the analysis of AFM1 using electrochemical detection (e.g., an electrochemical immunosensor based on magnetic NPs coated with antibody and screen-printed carbon electrodes (LOD: 50 ng/L) reported by Paniel et al. [164] and an impedimetric biosensor based on a DNA probe and gold NPs (LOD: 39 ng/L) reported by Dinckaya et al. [62], this method demonstrated a lower LOD and comparable dynamic detection range. This highlighted the advantages of using microarray technology.

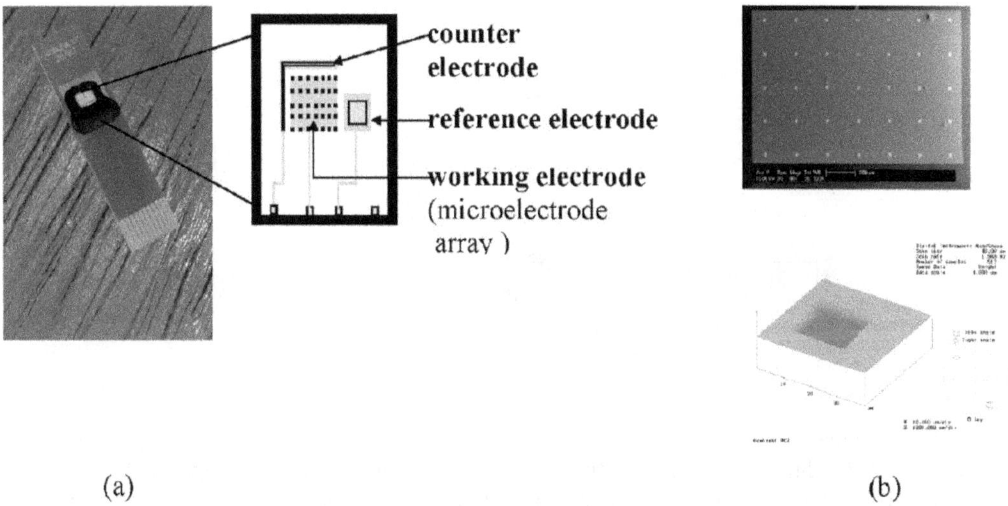

Fig. 12.6. (a) The three-electrode chips were fabricated with one working electrode area (35 electrodes in the array), a counter electrode, and a reference electrode area. (b) The whole working microelectrode of the untreated surface at 80 × magnification using a sFEG. Atomic force microscopy image of a single element of the array for the untreated working microelectrode (image 40 μm × 40 μm). Reprinted from Ref. [163] © 2009 American Chemical Society.

Radi et al. [165] used a screen-printed gold electrode for the sensitive detection of ochratoxin A (OTA), one of the most abundant food-contaminating mycotoxins produced by *Aspergillus ochraceus*. After modification via electrochemical and chemical reactions, amine-modified microelectrodes can be covalently bound with antibodies against OTA. A competitive immunoassay was further demonstrated via a competition between OTA and a horseradish peroxidase-labeled OTA (OTA-HRP) for the immobilized antibodies. The activity of the bound OTA-HRP was electrochemically analyzed by chronoamperometry using 3,3',5,5'-tetramethylbenzidine as the substrate. The achieved LOD of 12 ng/mL and a dynamic range up to 60 ng/mL for OTA analysis was comparable to the EU MRL of 10 ppb. Mak et al. [166] advanced multiplex mycotoxin analysis via the integration of the typical sandwich immunoassay into a modified magnetic nanotag (MNT) detection platform (Fig. 12.7). MNT was adapted to detect target molecules that are smaller in size (< 300 Da) and insoluble. Real-time detection was realized via the addition of MNTs onto the spin-valve sensor surface that had been previously immobilized with capture antibodies against mycotoxins and secondary antibodies. Capture antibodies were immobilized on different parts of the giant magneto-resistive (GMR) sensor surface. After the addition of the MNT into the sandwich immunoassay system, the signal related to the streptavidin-avidin binding kinetics was recorded in real time. The results demonstrated high specificity, high capability of multiplexed analysis, and reduced LODs of 50 pg/mL for AFB1, 0.05 ng/mL for ZEA, and 333 pg/mL for HT-2. These LODs met the EU MRLs (AFB1: 5–12 μg/kg; ZEA: 20–350 μg/kg).

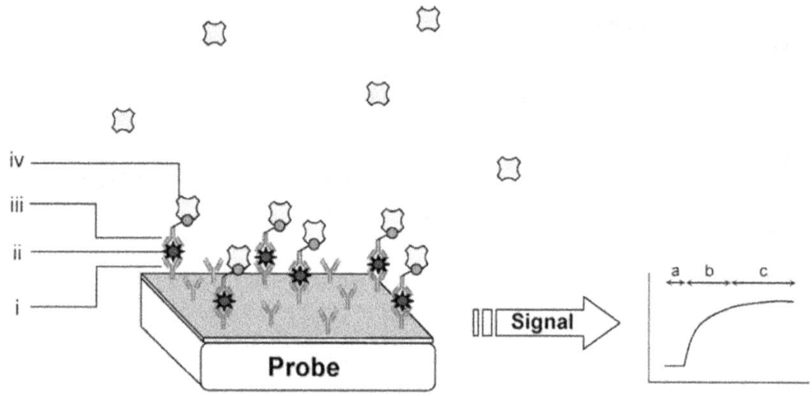

Fig. 12.7. Schematic illustration of the magnetic nanotag-based immunoassay for mycotoxins and typical MNT binding curve. (i) Positive probes are immobilized with capture antibodies of interest, (ii) analytes are mixed into a single pool for incubation, (iii) finally biotinylated detection antibodies are added, and (iv) the binding of streptavidincoated MNTs with detection antibodies and the detection of magnetic signal in real time. (a) Signal baseline after the rinsing and the subsequent removal PBS solution; (b) binding kinetics upon the addition of MNTs, defined as t = 0; and (c) signal saturation achieved when the MNTs have saturated the available binding sites. Reprinted from Ref. [166] © 2009 Published by Elsevier B.V.

More recently, SPR has been explored as a rapid label-free screening method for the detection of food contaminants, in particular mycotoxins. SPRi allows multiplex screening of tens of different biointeractions using a microarray of sensing spots. The maximum number of mycotoxins detected using an SPR microarray is limited to 4 [167]. Dorokhin et al. [66] reported an elaborate SPRi-based microarray platform used for the detection of deoxynivalenol (DON) and zearalenone (ZEA) via a competitive inhibition immunoassay format. A continuous flow microspotter device was used in the fabrication. LODs of 84 and 68 μg/kg for DON and 64 and 40 μg/kg for ZEA were found for maize and wheat samples, respectively, suggesting that this microarray method could satisfy EU MRLs (ZEA: 350 μg/kg for maize and 100 μg/kg for wheat; DON: 1,750 μg/kg for both maize and wheat). The results using a single microarray chip were in good agreement with LC-MS/MS data. This method easily met the EU regulatory limits for mycotoxin contaminants in food and feed samples. Furthermore, the results suggested that the SPRi microarray chip platform holds promise for the development of a rapid multiplex screening method for up to 40 different mycotoxins.

A regenerable, reusable microarray is of growing interest because of its improved reusability and environmentally-friendly characteristics. Sauceda-Friebe et al. [73] developed a regenerable microarray for screening OTA in green coffee extract on a fully-automated flow-through device with a CL readout. After synthesis of a water-soluble peptide-OTA conjugate and peptide-biotin conjugate for covalent immobilization on a glass support by contact spotting, the microarray was used in an indirect competitive immunoassay format with flow-through reagent addition and CL detection. On-line mixing and sequential pumping of solutions over the microarray surface modified with the analytes of interest were carried out. Over 20 assay-regeneration cycles of the microarray surface were completed by repeated covalent conjugate immobilizations. The results indicated a reduced assay time of 12 min and a LOQ of OTA in green coffee extract of 0.3 μg/L, which corresponds to 7 μg/kg. The LOQ is comparable to the EU MRL of 10.0 μg/kg. Before starting a new assay cycle, a rebinding procedure was performed with a regeneration solution having acidic pH and high ionic strength. The solution was pumped through the flow cell to remove the streptavidin-HRP. There was a continuous reduction in signal intensity over several regeneration cycles because of the imperfect disruption of the biotin-streptavidin complex, leading only to its partial removal from the chip surface.

Microfluidic chip based immunosensor technology is one of the most promising methods for sensitive and rapid mycotoxin assays. There are two major microfluidic immunosensor types for mycotoxin detection, i.e. flow-through (CE chip) and lateral flow (LFTS) technologies. The development of CE chip based miniaturized biosensors has attracted much interest because of their potential for the simultaneous assay of numerous samples, rapid analysis, ultrahigh throughput, and low sample and regent consumption [168]. Hervas et al. [169] proposed a microfluidic immunosensor integrated with an electrokinetic magnetic bead based electrochemical immunoassay on a microfluidic chip for ZEA detection in infant foods. Using this simple microchip layout, total integration of whole immunoassay was realized. In a double-T microfluidic immunosensor, a competitive ELISA was developed using protein G-modified magnetic beads. Results showed a low LOD of 0.4 μg/L, an extremely low systematic error of 2 %, and an excellent recovery of 103 % for solid sample and 101 % for liquid samples. An electrochemical microfluidic immunosensor incorporated in a microfluidic cell was established by Arevalo and co-workers [170] for the quantitation of CIT in rice samples on a gold surface, electrodeposited on a glassy carbon electrode, illustrating a wide linear range (from 0.5 to 50 ng/mL), low LOD and LOQ (0.1 and 0.5 ng/mL, respectively), and reduced analysis time (within 2 min). An electrochemical microfluidic immunosensor was investigated by Hervas and co-workers [171] to determine ZEA in baby foods, showing a reliable LOD of less than 1 μg/kg. To avoid the typical time-consuming and laborious four-parameter logistic curve fit, a sequential, fast, and simplified calibration was performed, which reduced a total analytical time of about 200 s. A regenerable CE microfluidic immunosensor is a current trend in food safety, such as detecting antibiotics in raw milk [172] and leptin [173]. Knopp's group [73] developed a regenerable glass microfluidic immunosensor to realize an ic-ELISA with chemiluminescence detection. The results showed a reduced analysis time of 12 min for

OTA, and a LOQ of 7 μg/kg in green coffee extract, with detection over 20 successful assay-regeneration cycles.

LFTS-based microfluidic immunosensors, also called test strips, immunochromatographic strips, immunocapillary tests, and sol particle immunoassays, are a simplified microfluidic device to detect the presence or absence of an analyte, such as mytotoxin. Anfossi et al. [174] reported a LFTS to analyze AFB1 with a LOD of 1 mg/kg and a wide dynamic range of 2–40 mg/kg, showing excellent interday precision. To achieve high selectivity and sensitivity of LFTS for AFB1 detection, a novel monoclonal antibody based LFTS was developed by Zhang and co-workers [175], showing a LOD of 1 ng/mL with the naked eye. Wang and co-workers [88] reported an on-site direct competitive ELISA with a colloid gold LFTS for detecting AFM1 with a LOD of 1.0 ng/mL in milk and milk products. Polyclonal antibodies have also been used in LTFS for mycotoxin detection. Sun et al. [176] reported a LFTS based on a colloidal gold labeled polyclonal antibody to detect AFB, with a LOD of 2.5 ppb by naked eyes, and of 0.05–0.1 ppb by a photometric strip reader. To amplify the signal, Saha et al. [177] developed a super catalyzed reporter deposition method, also called tyramide signal amplification, which was used for the rapid detection of AFB1, with a LOD of 12.5 μg/kg.

Other mycotoxin detection systems based on LFTSs have been developed for various sample extracts [178]. For FB1, within 10min, the LFTS results showed a LOD of 199 μg/kg, as reported by Molinelli et al. [179], and a visual LOD of 1.0 μg/kg noted by Wang and co-workers [180] and 5 μg/kg by Yu's group [181]. Kolosova et al. [182] developed LFTSs to detect deoxynivalenol (DON) with wide ranges of 250–500 and 1000–2000 μg/kg by visual readout methods, respectively. Xu et al. [183] also reported a sensitive LFTS for DON detection with a LOD of 50 μg/kg in a 10-min assay. LFTSs have been used for ZEA assays, with a LOD of 2.5 ng/mL and 30 μg/kg for a standard solution and spiked sample, respectively, as reported by Shim and co-workers [184]. Molinelli et al. [185] reported a rapid assay for T-2 toxin by LFTS, resulting in a cut-off level of around 100 μg/kg within 4 min for naturally contaminated wheat and oats. OTA has been detected with a LOD of 10 μg/kg [186], 5 μg/kg [187], 1.9 μg/kg [188], and 1.5 μg/kg [189], respectively. The results of these LFTS methods have shown high consistency with those from HPLC-MS methods.

The multiplexed detection of mycotoxins in food and feed samples with LFTS is one of the most attractive approaches for practical application, providing a simple and low-cost strategy.

Kolosova et al. [190] reported the multiplexed detection of DON and ZEA, with cut-off levels of 1500 and 100 μg/kg for DON and ZEA, respectively. Li's group [191] developed an ultrasensitive LFTS for the simultaneous detection of total AFTs (AFB1, B2, G1, G2). The visual LODs for AFB1, B2, G1, and G2 in peanut matrix were 0.03, 0.06, 0.12, and 0.25 ng/mL, respectively. The results of AFTs detection in peanuts were in excellent agreement with those by HPLC. LFTSs were compared with direct competitive ELISA for AFB1, where the LOD of LFTS showed better performance in sample throughput and simplicity [192].

Other particles have been used to enhance the sensitivity of LFTS. Knopp's group [193] used magnetic nanogold microspheres grafted with nano-Fe_2O_3 particles as labels for AFB2 detection in food, with a threefold lower LOD of 0.9 ng/mL (Fig. 12.8). Colloidal Au coated with Ag strongly improved detection sensitivity with comparable reproducibility and stability as well [194].

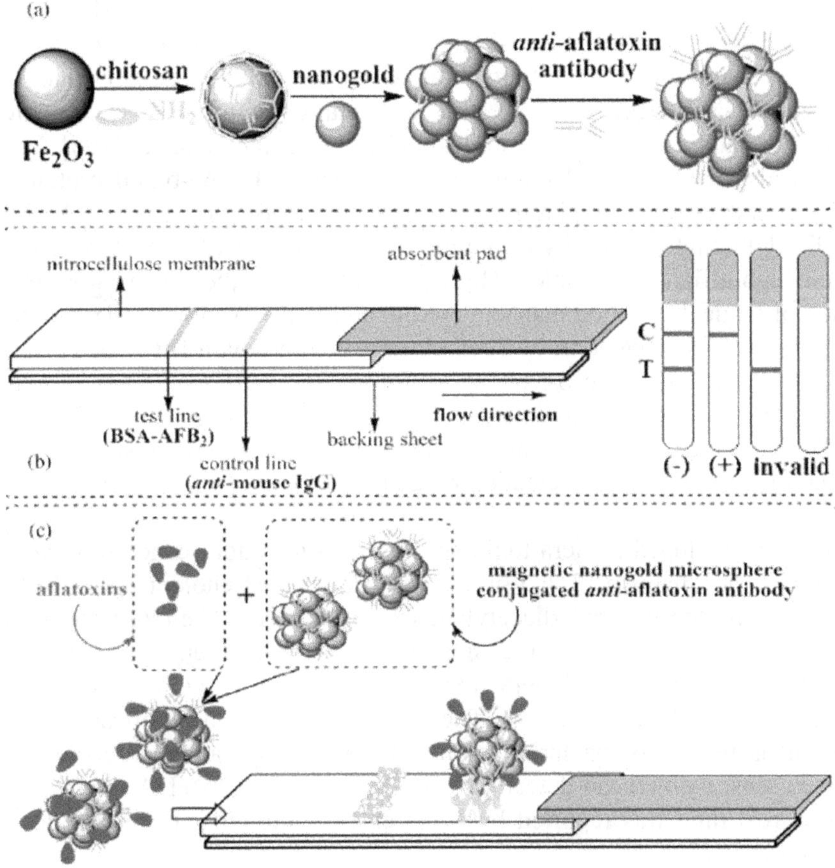

Fig. 12.8. (a) The fabrication process of the synthesized MnGMs, (b) the schematic illustration of the immunodipstick (left) and readout (right), and (c) principle of the detection method. Reprinted from Ref. [193] © 2009 Elsevier B.V.

12.4.4.2. Bacterial Toxin

Bacterial generated toxin can be classified as either exotoxins or endotoxins. Exotoxins are constantly and actively secreted while endotoxins remain inside bacterial outer membrane, until the bacterium is killed. Several bacterial toxins (e.g. Botulinum neurotoxin) can cause severe inflammation and deadly food poisoning disease. In this part, sensors for detection of common bacterial toxin (mainly Botulinum neurotoxin, Tetanus toxin, Cholera toxin and Staphylococcal toxins) are reviewed.

Optical sensor has been widely employed in bacterial toxin detection. Weingart et al. reported an array-based fluorescence sensor for determination of Botulinum Neurotoxin type A (BoNT/A), Staphylococcal enterotoxin B (SEB) [195]. SEB and BoNT/A can be recognized by biotin labeled antibodies, which can be subsequently visualized by Cy5 (fluorescent tracer) labelled streptavidin. The toxins were detectable at levels as low as 0.5–1 $ng{\cdot}mL^{-1}$ in buffer or in raw milk. Monosaccharide, such as N-acetyl galactosamine (GalNAc) and N-acetylneuraminic acid (Neu5Ac) derivatives, can be used as recognition element for sensing of cholera toxin and tetanus toxin. Low LOD (100 ng/mL) and high specificity (against Salmonella typhimurium, Listeria monocytogenes, Escherichia coli and staphylococcal enterotoxin B) were achieved due to the selective binding between the toxins and GalNAc and Neu5Ac [196]. In the work of Zhang et al., enzyme-linked immunosorbent assay (ELISA)-based protein antibody microarray was developed for simultaneous detection of BoNT serotypes A, B, C, D, E, and F [197]. Ultrahigh sensitivity arranging from 1.3 fM to 14.7 fM can be ascribed to engineered high-affinity antibodies. High performance of the sensor was observed in application of clinical and food matrices (serum and milk). Another ELISA-based sensor has been employed for detection of Staphylococcal enterotoxin B (SEB) [74]. After the ELISA procedure, 3-(4-hydroxyphenyl propionate) (PHPPA) was reacted with Hydrogen peroxide-urea, with catalysis by HRP-conjugated anti-SEB, to produce PHPPA Dimer, which was determined by chemiluminescence analysis. Detection limit was 3.3 pg mL^{-1} and the linear detection range of 6.0 to 564 pg mL^{-1} was achieved.

Ma and Cheng developed cholera toxin sensor by using fluorescence resonance energy transfer (FRET) mechanism [71]. Without the presence of cholera toxin, BO-GM dye spontaneously inserted into polydiacetylene (PDA) vesicles. The close proximity of the dye molecules to the conjugated chains of PDA enabled energy transfer, causing fluorescence quenching. The cholera toxin can bind with BO-GM1 and consequently prohibit it from membrane insertion. The sensor offered a fast, robust, and sensitive, label-free measurement of bacterial toxin with simple operative procedures. Another FRET-based sensor which can detect six o BoNT serotypes BoNT/A, B, D, E, F, and G serotypes in real time was reported by Ruge et al. femtomolar to picomolar detection limit was achieved [72].

Microfluidcs and lab-on-a-chip (LOC) technology has been widely explored for point of care diagnosis, food analysis, and biohazard, and microbial toxin analysis. The potential for miniaturization and other prominent properties makes microfluidcs and LOC technology desirable for being incorporated into biosensors. Frisk et al. developed an immunofluorescence sensor with parallel sensing ability realized via arrayed microchannels [198]. The microscale approach required minimal sample (7 μL) and no pre-processing (i.e. dilution, centrifugation, filtering) were needed. LOD of 16.6 pg (for BoNT/B) in whole milk was achieved. Hana et al reported a plastic ELISA-on-a-chip (EOC) device for measurement of botulinum neurotoxin A (BoNT/A) [199]. Based on the dose–response curve, the detection limit of BoNT/A was 2.0 $ngmL^{-1}$, which is 5 times lower than commercial-version detection kit employing colloidal gold tracer. The EOC system offers a low cost in manufacturing, high robustness, and a confirmation feature of the signals by bare eyes without instrumentation. Ozanich et al. developed a multiplexed flow cytometric assay for botulinum neurotoxin detection using an automated fluidic microbead-trapping flow cell [200]. It was demonstrated that the sandwich assays performed on the fluidic system gave 2-4 times higher fluorescence intensity signals compared with that performed manually.

The performance of fluorescence based sensor can be further improved with the help of other commonly used material or tools, such as nanomaterial, QDs and magnetic beads.

An eight channel Lab-on-a-Chip (LOC) device with optical detection of Staphylococcal Enterotoxin B (SEB) was developed by Yang et al. [201]. Because of its large specific surface area, CNT was used for antibody immobilization, which increased the sensitivity of detection six fold. Unctionalized fluorescent core–shell silica nanoparticles were employed in sensing of staphylococcal enterotoxin B [202]. QD was also used as reporter for botulinum neurotoxin in fluorescence sandwich immunoassay, because it had high extinction coefficients over a wide wavelength range, size dependent optical emission and relatively high quantum yields in aqueous media (as high as 25–30 %) [203]. Sensitive magnetic and fluorescent multiplex beads array was used to detect BoNT/A, and BoNT/B with excellent sensitivities between 2 ng/L and 546 ng/L from a minimal sample volume of 50 μL [204]. Nitrocellulose membrane strip based device was reported for botulinum toxin (BT) measurement and low LOD of 15 $pgmL^{-1}$ was achieved [205]. BT was first bound to the trisialoganglioside (GT1b) on the liposomes (as carrier) and then captured by the antibodies in the analytical zone during capillary migration on the test strip.

Liu et al. ultrasensitive SPR sensor for detection of botulinum toxin [206]. The detection system coupled a polymerization initiator to a biospecific interaction and induced inline atom transfer radical polymerization (ATRP) for amplifying SPR response. The board dynamic detection range (8.23×10^{-15} to 3.61×10^{-12} mol/cm2) and ultralow LOD (6.27×10^{-15} mol cm^{-2}) can be attributed to the two consecutive ATRP steps, resulting large amount of polymer brushes of poly- (hydroxyl-ethyl methacrylate) (PHEMA) (Fig. 12.9). SPR based sensors have also been used to detect other bacterial toxin, such as BoNT/B, staphylococcal enterotoxin B, tetanus toxin [207-209].

In addition to optical sensor, bacterial toxin can be determined by electrical and electrochemical means. Bacterial toxin staphylococcal enterotoxin B (SEB) can be detected at concentration as low as 5 pg/mL via electrochemical sensor chip [210]. Super avidin-biotin system (SABS) was used to amplify electrochemical signal intensity by using high-affinity streptavidin (SA)-biotin binding to create layers of poly-HRP, resulting in the creation of an enzyme scaffold. Lillehoj et al developed a sensitive electrochemical sensor with self-pumping lab-on-a-chip component optimized fluidic network and a robust hydrophilic PDMS coating, ensuring autonomous delivery of liquid samples and rapid detection.[211] The sensor can sense down to 1 pg of BoNT type A (in a 1 mL sample) within 15 min in an autonomous manner.

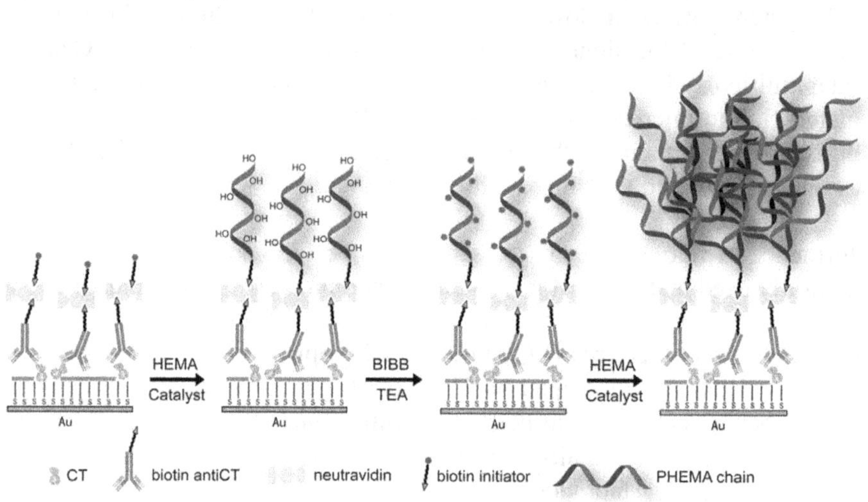

Fig. 12.9. Cartoon Representation of the Biotinylated Initiator Coupled Surface and Consecutive Two Steps of in Situ Surface ATRP Reactions for SPR Signal Amplification. Reprinted from Ref. [206] © 2010 American Chemical Society.

Piezoelectric immunosensor can be employed as an alternative method for bacterial toxin quantification. Lin and Tsai fabricated a piezoelectric (PZ) crystal immunosensor for the detection of staphylococcal enterotoxin B (SEB) [81]. The Au electrode was coated with polyethyleneimine (PEI) to attain high immunochemical activity. Recently, Salmaina et al. fabricated a piezoelectric immunosensor using QCM with dissipation (QCM-D) as transduction method [212]. Staphylococcal enterotoxin A (SEA) can be detected at concentration as low as 20 ng mL^{-1}. The sensing layer including the anti-SEA antibody was constructed by chemisorption of a self-assembled monolayer of cysteamine on the gold electrodes placed over the quartz crystal sensor followed by covalent linking of binding protein. Mutharasan's group developed piezoelectric-excited millimeter-sized cantilever (PEMC) sensors which can detect Staphylococcus aureus enterotoxin B (SEB) at picogram levels [213, 214]. The sensors were successfully applied for sensing SEB in apple juice, and milk. Limit of detection in spiked milk and apple juice samples is 10 and 100 fg, respectively.

Yang et al devised a novel electrical percolation-based sensor for detection of SEB [215]. Similar to the mechanism of semiconductor, single-walled carbon nanotubes (SWNTs)–antibody complex formed a network functioning as a "Biological Semiconductor" (BSC). Once SEB binds with SWNT–antibody complex, connected ends of the SWNT are disrupted, resulting in increased tunneling distance and subsequent resistance. Secondary antibody can be used to amplify the electrical signal. SEB can also be detected by flow-injection capacitive biosensor with ultralow LOD of 0.3 pg mL^{-1} and broad linear detection range from 2.8 pg mL^{-1} to 2.8 ng mL^{-1} [61]. The detection mechanism was based on the theory of the electrical double layer. The biosensor can measure low levels of SEB with simplicity, high sensitivity, and multiple use capability. In addition, the capacitive biosensor can run continuously for hours until it encounters the target analyte and does not consume antibody reagents.

12.5. Conclusion

Miniaturized sensing platform make a highly promising future to provide rapid and specific sensing for food quality, environmental monitoring and biosecurity. Sensing biotoxins at the required legislative limit requires highly sensitive and specific micro devices that allow rapid, reliable, and point-of-care determination. The minimaized format allows a portable device because that a large number of biotoxin targets could benefit from on-site testing for risk assessment and management. Thereafter, an urgent request emerges for simple and sensitive sensing methods that can identify multiple mycotoxins which occur at ultra low concentrations from different food and feed matrices. Miniaturized sensing platform provides a promising future in this respect.

It should be noted that, however, micro sensing platforms require to be further developed to face new challenges such as multiplex analysis of several biotoxins using microfludic chip and mircroarray. On the other hand, analysis software and high throughput that can handle directly the diversified samples can make these sensors of great potential in this application area. Moreover, novel emerging nanomaterials in the development of sensors for biotoxins will make these devices highly sensitive and more applicable for lab-on-a-chip. This lab-on-a-chip, as investigated in this chapter, exhibits its significant potential in practical applications at a rapid pace.

The existing and emerging miniaturized sensing platforms will pave a way to eliminate these biotoxins from entering the food chain and keep life from toxic affect. For these reasons, these rapid platforms are in urgent to be developed further to bring the technology from lab to commercial products.

Acknowledgments

The authors gratefully acknowledge the financial supports from National Natural Science Foundation of China (31171702, 31101299), Key Project of the Ministry of Agriculture (2011-G5), Project of National Science & Technology Pillar Plan (2012BAB19B09, 2012BAK08B03).

Reference

[1]. Y. Bourne, P. Taylor, P. Marchot, Acetylcholinesterase inhibition by fasciculin - crystal-structure of the complex, *Cell*, Vol. 83, 1995, pp. 503-512.

[2]. B. G. Fry, N. Vidal, J. A. Norman, F. J. Vonk, H. Scheib, S. F. R. Ramjan, S. Kuruppu, K. Fung, S. B. Hedges, M. K. Richardson, W. C. Hodgson, V. Ignjatovic, R.

Summerhayes, E. Kochva, Early evolution of the venom system in lizards and snakes, *Nature*, Vol. 439, 2006, pp. 584-588.

[3]. L. X. Zheng, G. Fisher, R. E. Miller, J. Peschon, D. H. Lynch, M. J. Lenardo, Induction of apoptosis in mature T-cells by tumor-necrosis-factor, *Nature*, Vol. 377, 1995, pp. 348-351.

[4]. R. A. Black, C. T. Rauch, C. J. Kozlosky, J. J. Peschon, J. L. Slack, M. F. Wolfson, B. J. Castner, K. L. Stocking, P. Reddy, S. Srinivasan, N. Nelson, N. Boiani, K. A. Schooley, M. Gerhart, R. Davis, J. N. Fitzner, R. S. Johnson, R. J. Paxton, C. J. March, D. P. Cerretti, A metalloproteinase disintegrin that releases tumour-necrosis factor-alpha from cells, *Nature*, Vol. 385, 1997, pp. 729-733.

[5]. T. J. Ward, J. P. Bielawski, H. C. Kistler, E. Sullivan, K. O'Donnell, Ancestral polymorphism and adaptive evolution in the trichothecene mycotoxin gene cluster of phytopathogenic Fusarium, in *Proceedings of the National Academy of Sciences of the United States of America*, Vol. 99, 2002, pp. 9278-9283.

[6]. M. C. Daniel, D. Astruc, Gold nanoparticles: Assembly, supramolecular chemistry, quantum-size-related properties, and applications toward biology, catalysis, and nanotechnology, *Chemical Reviews*, Vol. 104, 2004, pp. 293-346.

[7]. S. E. Skrabalak, J. Y. Chen, Y. G. Sun, X. M. Lu, L. Au, C. M. Cobley, Y. N. Xia, Gold Nanocages: Synthesis, Properties, and Applications, *Accounts of Chemical Research*, Vol. 41, 2008, pp. 1587-1595.

[8]. M. J. Mulvihill, X. Y. Ling, J. Henzie, P. D. Yang, Anisotropic Etching of Silver Nanoparticles for Plasmonic Structures Capable of Single-Particle SERS, *Journal of the American Chemical Society*, Vol. 132, 2010, pp. 268-274.

[9]. H. X. Xu, K. S. Suslick, Water-Soluble Fluorescent Silver Nanoclusters, *Advanced Materials*, Vol. 22, 2010, pp. 1078-1082.

[10]. V. Mazumder, Y. Lee, S. H. Sun, Recent Development of Active Nanoparticle Catalysts for Fuel Cell Reactions, *Advanced Functional Materials*, Vol. 20, 2010, pp. 1224-1231.

[11]. A. R. Tao, J. X. Huang, P. D. Yang, Langmuir-Blodgettry of Nanocrystals and Nanowires, *Accounts of Chemical Research*, Vol. 41, 2008, pp. 1662-1673.

[12]. R. W. Murray, Nanoelectrochemistry: Metal nanoparticles, nanoelectrodes, and nanopores, *Chemical Reviews*, Vol. 108, 2008, pp. 2688-2720.

[13]. Z. M. Peng, H. Yang, Designer platinum nanoparticles: Control of shape, composition in alloy, nanostructure and electrocatalytic property, *Nano Today*, Vol. 4, 2009, pp. 143-164.

[14]. S. J. Guo, E. K. Wang, Synthesis and electrochemical applications of gold nanoparticles, *Analytica Chimica Acta*, Vol. 598, 2007, pp. 181-192.

[15]. S. J. Guo, J. Li, W. Ren, D. Wen, S. J. Dong, E. K. Wang, Carbon Nanotube/Silica Coaxial Nanocable as a Three-Dimensional Support for Loading Diverse Ultra-High-Density Metal Nanostructures: Facile Preparation and Use as Enhanced Materials for Electrochemical Devices and SERS, *Chemistry of Materials*, Vol. 21, 2009, pp. 2247-2257.

[16]. R. J. Zhou, M. M. Shi, X. Q. Chen, M. Wang, H. Z. Chen, Atomically Monodispersed and Fluorescent Sub-Nanometer Gold Clusters Created by Biomolecule-Assisted Etching of Nanometer-Sized Gold Particles and Rods, *Chem.-Eur. J.*, Vol. 15, 2009, pp. 4944-4951.

[17]. G. Q. Wang, Y. Q. Wang, L. X. Chen, J. Choo, Nanomaterial-assisted aptamers for optical sensing, *Biosensors & Bioelectronics*, Vol. 25, 2010, pp. 1859-1868.

[18]. S. J. Guo, D. Wen, Y. M. Zhai, S. J. Dong, E. K. Wang, Platinum Nanoparticle Ensemble-on-Graphene Hybrid Nanosheet: One-Pot, Rapid Synthesis, and Used as New Electrode Material for Electrochemical Sensing, *Acs Nano*, Vol. 4, 2010, pp. 3959-3968.

[19]. C. A. Mirkin, R. L. Letsinger, R. C. Mucic, J. J. Storhoff, A DNA-based method for rationally assembling nanoparticles into macroscopic materials, *Nature*, Vol. 382, 1996, pp. 607-609.

[20]. R. Elghanian, J. J. Storhoff, R. C. Mucic, R. L. Letsinger, C. A. Mirkin, Selective colorimetric detection of polynucleotides based on the distance-dependent optical properties of gold nanoparticles, *Science*, Vol. 277, 1997, pp. 1078-1081.
[21]. H. Wei, B. L. Li, J. Li, S. J. Dong, E. K. Wang, DNAzyme-based colorimetric sensing of lead (Pb(2+)) using unmodified gold nanoparticle probes, *Nanotechnology*, Vol. 19, 2008, pp. 095501.
[22]. D. Li, A. Wieckowska, I. Willner, Optical analysis of Hg(2+) ions by oligonucleotide-gold-nanoparticle hybrids and DNA-based machines, *Angew. Chem.-Int. Edit.,* Vol. 47, 2008, pp. 3927-3931.
[23]. Z. Z. Lv, H. Wei, B. L. Li, E. K. Wang, Colorimetric recognition of the coralyne-poly(dA) interaction using unmodified gold nanoparticle probes, and further detection of coralyne based upon this recognition system, *Analyst*, Vol. 134, 2009, pp. 1647-1651.
[24]. J. Zhang, L. H. Wang, D. Pan, S. P. Song, F. Y. C. Boey, H. Zhang, C. H. Fan, Visual cocaine detection with gold nanoparticles and rationally engineered aptamer structures, *Small*, Vol. 4, 2008, pp. 1196-1200.
[25]. F. Xia, X. L. Zuo, R. Q. Yang, Y. Xiao, D. Kang, A. Vallee-Belisle, X. Gong, J. D. Yuen, B. B. Y. Hsu, A. J. Heeger, K. W. Plaxco, Colorimetric detection of DNA, small molecules, proteins, and ions using unmodified gold nanoparticles and conjugated polyelectrolytes, in *Proceedings of the National Academy of Sciences of the United States of America*, Vol. 107, 2010, pp. 10837-10841.
[26]. N. L. Rosi, C. A. Mirkin, Nanostructures in biodiagnostics, *Chemical Reviews*, Vol. 105, 2005, pp. 1547-1562.
[27]. M. S. Han, A. K. R. Lytton-Jean, B. K. Oh, J. Heo, C. A. Mirkin, Colorimetric screening of DNA-binding molecules with gold nanoparticle probes, *Angew. Chem.-Int. Edit.,* Vol. 45, 2006, pp. 1807-1810.
[28]. Y. Choi, N. H. Ho, C. H. Tung, Sensing phosphatase activity by using gold nanoparticles, *Angew. Chem.-Int. Edit.,* Vol. 46, 2007, pp. 707-709.
[29]. M. S. Han, A. K. R. Lytton-Jean, C. A. Mirkin, A gold nanoparticle based approach for screening triplex DNA binders, *Journal of the American Chemical Society*, Vol. 128, 2006, pp. 4954-4955.
[30]. J. W. Liu, Z. H. Cao, Y. Lu, Functional Nucleic Acid Sensors, *Chemical Reviews*, Vol. 109, 2009, pp. 1948-1998.
[31]. J. W. Liu, Y. Lu, Optimization of a Pb2+-directed gold nanoparticle/DNAzyme assembly and its application as a colorimetric biosensor for Pb2+, *Chemistry of Materials*, Vol. 16, 2004, pp. 3231-3238.
[32]. S. J. He, D. Li, C. F. Zhu, S. P. Song, L. H. Wang, Y. T. Long, C. H. Fan, Design of a gold nanoprobe for rapid and portable mercury detection with the naked eye, *Chemical Communications*, Vol., 2008, pp. 4885-4887.
[33]. L. Shang, J. Y. Yin, J. Li, L. H. Jin, S. J. Dong, Gold nanoparticle-based near-infrared fluorescent detection of biological thiols in human plasma, *Biosensors & Bioelectronics*, Vol. 25, 2009, pp. 269-274.
[34]. J. Zhang, L. H. Wang, H. Zhang, F. Boey, S. P. Song, C. H. Fan, Aptamer-Based Multicolor Fluorescent Gold Nanoprobes for Multiplex Detection in Homogeneous Solution, *Small*, Vol. 6, 2010, pp. 201-204.
[35]. L. Shang, S. J. Dong, Design of Fluorescent Assays for Cyanide and Hydrogen Peroxide Based on the Inner Filter Effect of Metal Nanoparticles, *Analytical Chemistry*, Vol. 81, 2009, pp. 1465-1470.
[36]. T. Shtoyko, E. G. Matveeva, I. F. Chang, Z. Gryczynski, E. Goldys, I. Gryczynski, Enhanced fluorescent immunoassays on silver fractal-like structures, *Analytical Chemistry*, Vol. 80, 2008, pp. 1962-1966.

[37]. L. Touahir, E. Galopin, R. Boukherroub, A. C. Gouget-Laemmel, J. N. Chazalviel, F. Ozanam, S. Szunerits, Localized surface plasmon-enhanced fluorescence spectroscopy for highly-sensitive real-time detection of DNA hybridization, *Biosensors & Bioelectronics*, Vol. 25, 2010, pp. 2579-2585.
[38]. Y. Kobayashi, H. Nakashima, D. Takagi, Y. Homma, CVD growth of single-walled carbon nanotubes using size-controlled nanoparticle catalyst, *Thin Solid Films*, Vol. 464, 2004, pp. 286-289.
[39]. M. Keidar, Factors affecting synthesis of single wall carbon nanotubes in arc discharge, *Journal of Physics D-Applied Physics*, Vol. 40, 2007, pp. 2388-2393.
[40]. M. Kusaba, Y. Tsunawaki, Production of single-wall carbon nanotubes by a XeCl excimer laser ablation, *Thin Solid Films*, Vol. 506, 2006, pp. 255-258.
[41]. L. J. Meng, C. L. Fu, Q. H. Lu, Advanced technology for functionalization of carbon nanotubes, *Progress in Natural Science*, Vol. 19, 2009, pp. 801-810.
[42]. C. Y. Hu, Y. J. Xu, S. W. Duo, R. F. Zhang, M. S. Li, Non-Covalent Functionalization of Carbon Nanotubes with Surfactants and Polymers, *Journal of the Chinese Chemical Society*, Vol. 56, 2009, pp. 234-239.
[43]. K. A. Wepasnick, B. A. Smith, J. L. Bitter, D. H. Fairbrother, Chemical and structural characterization of carbon nanotube surfaces, *Analytical and Bioanalytical Chemistry*, Vol. 396, 2010, pp. 1003-1014.
[44]. B. O. Dabbousi, J. Rodriguez Viejo, F. V. Mikulec, J. R. Heine, H. Mattoussi, R. Ober, K. F. Jensen, M. G. Bawendi, (CdSe)ZnS core-shell quantum dots: Synthesis and characterization of a size series of highly luminescent nanocrystallites, *Journal of Physical Chemistry B*, Vol. 101, 1997, pp. 9463-9475.
[45]. C. J. Murphy, Optical sensing with quantum dots, *Analytical Chemistry*, Vol. 74, 2002, pp. 520A-526A.
[46]. J. K. Jaiswal, H. Mattoussi, J. M. Mauro, S. M. Simon, Long-term multiple color imaging of live cells using quantum dot bioconjugates, *Nature Biotechnology*, Vol. 21, 2003, pp. 47-51.
[47]. I. L. Medintz, H. T. Uyeda, E. R. Goldman, H. Mattoussi, Quantum dot bioconjugates for imaging, labelling and sensing, *Nature Materials*, Vol. 4, 2005, pp. 435-446.
[48]. U. Resch-Genger, M. Grabolle, S. Cavaliere-Jaricot, R. Nitschke, T. Nann, Quantum dots versus organic dyes as fluorescent labels, *Nature Methods*, Vol. 5, 2008, pp. 763-775.
[49]. T. Jamieson, R. Bakhshi, D. Petrova, R. Pocock, M. Imani, A. M. Seifalian, Biological applications of quantum dots, *Biomaterials*, Vol. 28, 2007, pp. 4717-4732.
[50]. J. B. Delehanty, H. Mattoussi, I. L. Medintz, Delivering quantum dots into cells: strategies, progress and remaining issues, *Analytical and Bioanalytical Chemistry*, Vol. 393, 2009, pp. 1091-1105.
[51]. S. M. Nie, Y. Xing, G. J. Kim, J. W. Simons, Nanotechnology applications in cancer, in: *Annual Review of Biomedical Engineering*, 2007, pp. 257-288.
[52]. A. M. Smith, H. W. Duan, A. M. Mohs, S. M. Nie, Bioconjugated quantum dots for in vivo molecular and cellular imaging, *Advanced Drug Delivery Reviews*, Vol. 60, 2008, pp. 1226-1240.
[53]. R. C. Somers, M. G. Bawendi, D. G. Nocera, CdSe nanocrystal based chem-/bio-sensors, *Chemical Society Reviews*, Vol. 36, 2007, pp. 579-591.
[54]. R. Gill, M. Zayats, I. Willner, Semiconductor quantum dots for bioanalysis, *Angew. Chem.-Int. Edit.*, Vol. 47, 2008, pp. 7602-7625.
[55]. L. J. Zhi, K. Mullen, A bottom-up approach from molecular nanographenes to unconventional carbon materials, *Journal of Materials Chemistry*, Vol. 18, 2008, pp. 1472-1484.
[56]. C. Berger, Z. M. Song, T. B. Li, X. B. Li, A. Y. Ogbazghi, R. Feng, Z. T. Dai, A. N. Marchenkov, E. H. Conrad, P. N. First, W. A. de Heer, Ultrathin epitaxial graphite:

2D electron gas properties and a route toward graphene-based nanoelectronics, *Journal of Physical Chemistry B*, Vol. 108, 2004, pp. 19912-19916.

[57]. P. Avouris, Graphene: Electronic and Photonic Properties and Devices, *Nano Letters*, Vol. 10, 2010, pp. 4285-4294.

[58]. E. Akyilmaz, M. Turemis, An inhibition type alkaline phosphatase biosensor for amperometric determination of caffeine, *Electrochimica Acta*, Vol. 55, 2010, pp. 5195-5199.

[59]. W. M. Yeh, K. C. Ho, Amperometric morphine sensing using a molecularly imprinted polymer-modified electrode, *Analytica Chimica Acta*, Vol. 542, 2005, pp. 76-82.

[60]. R. M. Carter, M. A. Poli, M. Pesavento, D. E. T. Sibley, G. J. Lubrano, G. G. Guilbault, Immunoelectrochemical Biosensors for Detection of Saxitoxin and Brevetoxin, *ImmunoMethods*, Vol. 3, 1993, pp. 128-133.

[61]. M. Labib, M. Hedstrom, M. Amin, B. Mattiasson, A capacitive biosensor for detection of staphylococcal enterotoxin B, *Analytical and Bioanalytical Chemistry*, Vol. 393, 2009, pp. 1539-1544.

[62]. E. Dinckaya, O. Kinik, M. K. Sezginturk, C. Altug, A. Akkoca, Development of an impedimetric aflatoxin M1 biosensor based on a DNA probe and gold nanoparticles, *Biosensors & Bioelectronics*, Vol. 26, 2011, pp. 3806-3811.

[63]. M. C. Rodriguez, A. N. Kawde, J. Wang, Aptamer biosensor for label-free impedance spectroscopy detection of proteins based on recognition-induced switching of the surface charge, *Chemical Communications*, Vol., 2005, pp. 4267-4269.

[64]. A. Pizzariello, J. Svorc, M. Stred'ansky, S. Miertus, A biosensing method for detection of caffeine in coffee, *Journal of the Science of Food and Agriculture*, Vol. 79, 1999, pp. 1136-1140.

[65]. M. S. Steiner, A. Duerkop, O. S. Wolfbeis, Optical methods for sensing glucose, *Chemical Society Reviews*, Vol. 40, 2011, pp. 4805-4839.

[66]. D. Dorokhin, W. Haasnoot, M. C. R. Franssen, H. Zuilhof, M. W. F. Nielen, Imaging surface plasmon resonance for multiplex microassay sensing of mycotoxins, *Analytical and Bioanalytical Chemistry*, Vol. 400, 2011, pp. 3005-3011.

[67]. X. M. Wu, S. M. Gao, J. S. Wang, H. Y. Wang, Y. W. Huang, Y. P. Zhao, The surface-enhanced Raman spectra of aflatoxins: spectral analysis, density functional theory calculation, detection and differentiation, *Analyst*, Vol. 137, 2012, pp. 4226-4234.

[68]. R. Gao, Y. Zhang, P. Gopalakrishnakone, Single-bead-based immunofluorescence assay for snake venom detection, *Biotechnology Progress*, Vol. 24, 2008, pp. 245-249.

[69]. R. Zeineldin, M. E. Piyasena, L. A. Sklar, D. Whitten, G. P. Lopez, Detection of membrane biointeractions based on fluorescence superquenching, *Langmuir*, Vol. 24, 2008, pp. 4125-4131.

[70]. Z. D. Wang, J. H. Lee, Y. Lu, Highly sensitive, turn-on, fluorescent sensor for Hg(2+) in aqueous solution based on structure-switching DNA, *Chemical Communications*, 2008, pp. 6005-6007.

[71]. G. Y. Ma, Q. Cheng, Manipulating FRET with polymeric vesicles: Development of a, mix-and-detect, type fluorescence sensor for bacterial toxin, *Langmuir*, Vol. 22, 2006, pp. 6743-6745.

[72]. D. R. Ruge, F. M. Dunning, T. M. Piazza, B. E. Molles, M. Adler, F. N. Zeytin, W. C. Tucker, Detection of six serotypes of botulinum neurotoxin using fluorogenic reporters, *Analytical Biochemistry*, Vol. 411, 2011, pp. 200-209.

[73]. J. C. Sauceda-Friebe, X. Y. Z. Karsunke, S. Vazac, S. Biselli, R. Niessner, D. Knopp, Regenerable immuno-biochip for screening ochratoxin A in green coffee extract using an automated microarray chip reader with chemiluminescence detection, *Analytica Chimica Acta*, Vol. 689, 2011, pp. 234-242.

[74]. P. Xue, Y. M. Li, Z. J. Zhang, A. H. Fu, F. Liu, X. M. Zhang, Y. J. Sun, L. L. Chen, B. Q. Jin, K. Yang, Novel chemiluminescent assay for staphylococcal enterotoxin B, *Microchimica Acta*, Vol. 174, 2011, pp. 167-174.

[75]. N. W. Barnett, S. W. Lewis, D. J. Tucker, Determination of morphine in process streams by sequential injection analysis with chemiluminescence detection, *Fresenius Journal of Analytical Chemistry*, Vol. 355, 1996, pp. 591-595.

[76]. M. M. Ngundi, S. A. Qadri, E. V. Wallace, M. H. Moore, M. E. Lassman, L. C. Shriver-Lake, F. S. Ligler, C. R. Taitt, Detection of deoxynivalenol in foods and indoor air using an array biosensor, *Environmental Science & Technology*, Vol. 40, 2006, pp. 2352-2356.

[77]. C. W. Park, J. H. Yang, C. S. Ah, C. G. Ahn, Y. H. Choi, K. H. Chung, W. J. Kim, G. Y. Sung, Detection of small toxin molecules by Si photovoltaic integrated circuit, *Eurosensors Xxv*, Vol. 25, 2011, pp. 912-915.

[78]. Y. Ohno, K. Maehashi, Y. Yamashiro, K. Matsumoto, Electrolyte-Gated Graphene Field-Effect Transistors for Detecting pH Protein Adsorption, *Nano Letters*, Vol. 9, 2009, pp. 3318-3322.

[79]. S. Tadigadapa, K. Mateti, Piezoelectric MEMS sensors: state-of-the-art and perspectives, *Measurement Science & Technology*, Vol. 20, 2009, pp. 250-254.

[80]. E. Aksel, J. L. Jones, Advances in Lead-Free Piezoelectric Materials for Sensors and Actuators, *Sensors*, Vol. 10, 2010, pp. 1935-1954.

[81]. H. C. Lin, W. C. Tsai, Piezoelectric crystal immunosensor for the detection of staphylococcal enterotoxin B, *Biosensors & Bioelectronics*, Vol. 18, 2003, pp. 1479-1483.

[82]. C. J. Percival, S. Stanley, M. Galle, A. Braithwaite, M. I. Newton, G. McHale, W. Hayes, Molecular-imprinted, polymer-coated quartz crystal microbalances for the detection of terpenes, *Analytical Chemistry*, Vol. 73, 2001, pp. 4225-4228.

[83]. A. X. J. Tang, M. Pravda, G. G. Guilbault, S. Piletsky, A. P. F. Turner, Immunosensor for okadaic acid using quartz crystal microbalance, *Analytica Chimica Acta*, Vol. 471, 2002, pp. 33-40.

[84]. L. M. Juliano, R. R. Griffiths, A critical review of caffeine withdrawal: empirical validation of symptoms and signs, incidence, severity, and associated features, *Psychopharmacology*, Vol. 176, 2004, pp. 1-29.

[85]. K. Shrivas, H. F. Wu, Rapid determination of caffeine in one drop of beverages and foods using drop-to-drop solvent microextraction with gas chromatography/mass spectrometry, *Journal of Chromatography A*, Vol. 1170, 2007, pp. 9-14.

[86]. J. K. Zou, N. Li, Simple and environmental friendly procedure for the gas chromatographic-mass spectrometric determination of caffeine in beverages, *Journal of Chromatography A*, Vol. 1136, 2006, pp. 106-110.

[87]. J. J. Carvalho, M. G. Weller, U. Panne, R. J. Schneider, A highly sensitive caffeine immunoassay based on a monoclonal antibody, *Analytical and Bioanalytical Chemistry*, Vol. 396, 2010, pp. 2617-2628.

[88]. T. Alizadeh, M. R. Ganjali, M. Zare, P. Norouzi, Development of a voltammetric sensor based on a molecularly imprinted polymer (MIP) for caffeine measurement, *Electrochimica Acta*, Vol. 55, 2010, pp. 1568-1574.

[89]. J. Zhang, L. P. Wang, W. Guo, X. D. Peng, M. Li, Z. B. Yuan, Sensitive Differential Pulse Stripping Voltammetry of Caffeine in Medicines and Cola Using a Sensor Based on Multi-Walled Carbon Nanotubes and Nafion, *International Journal of Electrochemical Science*, Vol. 6, 2011, pp. 997-1006.

[90]. S. L. Yang, R. Yang, G. Li, L. B. Qu, J. J. Li, L. L. Yu, Nafion/multi-wall carbon nanotubes composite film coated glassy carbon electrode for sensitive determination of caffeine, *Journal of Electroanalytical Chemistry*, Vol. 639, 2010, pp. 77-82.

[91]. B. J. Sanghavi, A. K. Srivastava, Simultaneous voltammetric determination of acetaminophen, aspirin and caffeine using an in situ surfactant-modified multiwalled carbon nanotube paste electrode, *Electrochimica Acta*, Vol. 55, 2010, pp. 8638-8648.

[92]. W. D. R. Santos, M. Santhiago, I. V. P. Yoshida, L. T. Kubota, Electrochemical sensor based on imprinted sol-gel and nanomaterial for determination of caffeine, *Sensors and Actuators B-Chemical*, Vol. 166, 2012, pp. 739-745.

[93]. F. Y. Zhao, F. Wang, W. N. Zhao, J. Zhou, Y. Liu, L. N. Zou, B. X. Ye, Voltammetric sensor for caffeine based on a glassy carbon electrode modified with Nafion and graphene oxide, *Microchimica Acta*, Vol. 174, 2011, pp. 383-390.

[94]. J. Y. Sun, K. J. Huang, S. Y. Wei, Z. W. Wu, F. P. Ren, A graphene-based electrochemical sensor for sensitive determination of caffeine, *Colloids and Surfaces B: Biointerfaces*, Vol. 84, 2011, pp. 421-426.

[95]. P. A. Glare, T. D. Walsh, Clinical pharmacokinetics of morphine, *Ther Drug Monit*, Vol. 13, 1991, pp. 1-23.

[96]. W. E. Brewer, R. C. Galipo, K. W. Sellers, S. L. Morgan, Analysis of cocaine, benzoylecgonine, codeine, and morphine in hair by supercritical fluid extraction with carbon dioxide modified with methanol, *Analytical Chemistry*, Vol. 73, 2001, pp. 2371-2376.

[97]. M. Freiermuth, J. C. Plasse, Determination of morphine and codeine in plasma by HPLC following solid phase extraction, *Journal of Pharmaceutical and Biomedical Analysis*, Vol. 15, 1997, pp. 759-764.

[98]. A. B. Melent'ev, Gas chromatography-mass spectrometry determination of morphine and codeine in blood as their propionic esters, *Journal of Analytical Chemistry*, Vol. 59, 2004, pp. 566-570.

[99]. A. Dienes-Nagy, L. Rivier, C. Giroud, M. Augsburger, P. Mangin, Method for quantification of morphine and its 3-and 6-glucuronides, codeine, codeine glucuronide and 6-monoacetylmorphine in human blood by liquid chromatography-electrospray mass spectrometry for routine analysis in forensic toxicology, *Journal of Chromatography A*, Vol. 854, 1999, pp. 109-118.

[100]. U. Wieczorek, N. Nagakura, C. Sund, S. Jendrzejewski, M. H. Zenk, Radioimmunoassay determination of six opium alkaloids and its application to plant screening, *Phytochemistry*, Vol. 25, 1986, pp. 2639-2646.

[101]. N. W. Barnett, D. G. Rolfe, T. A. Bowser, T. W. Paton, Determination of morphine in process streams using flow-injection analysis with chemiluminescence detection, *Analytica Chimica Acta*, Vol. 282, 1993, pp. 551-557.

[102]. N. W. Barnett, C. E. Lenehan, S. W. Lewis, D. J. Tucker, K. M. Essery, Determination of morphine in water immiscible process streams using sequential injection analysis coupled with acidic permanganate chemiluminescence detection, *Analyst*, Vol. 123, 1998, pp. 601-605.

[103]. P. Norouzi, M. R. Ganjali, A. A. Moosavi-Movahedi, B. Larijani, Fast Fourier transformation with continuous cyclic voltammetry at an Au microelectrode for the determination of morphine in a flow injection system, *Talanta*, Vol. 73, 2007, pp. 54-61.

[104]. C. H. Weng, W. M. Yeh, K. C. Ho, G. B. Lee, A microfluidic system utilizing molecularly imprinted polymer films for amperometric detection of morphine, *Sensors and Actuators B-Chemical*, Vol. 121, 2007, pp. 576-582.

[105]. Q. L. Zhang, J. J. Xu, X. Y. Li, H. Z. Lian, H. Y. Chen, Determination of morphine and codeine in urine using poly(dimethylsiloxane) microchip electrophoresis with electrochemical detection, *Journal of Pharmaceutical and Biomedical Analysis*, Vol. 43, 2007, pp. 237-242.

[106]. F. Li, J. X. Song, C. S. Shan, D. M. Gao, X. Y. Xu, L. Niu, Electrochemical determination of morphine at ordered mesoporous carbon modified glassy carbon electrode, *Biosensors & Bioelectronics*, Vol. 25, 2010, pp. 1408-1413.

[107]. A. Mokhtari, H. Karimi-Maleh, A. A. Ensafi, H. Beitollahi, Application of modified multiwall carbon nanotubes paste electrode for simultaneous voltammetric determination of morphine and diclofenac in biological and pharmaceutical samples, *Sensors and Actuators B-Chemical*, Vol. 169, 2012, pp. 96-105.

[108]. A. Babaei, M. Babazadeh, Multi-walled carbon nanotubes/chitosan polymer composite modified glassy carbon electrode for sensitive simultaneous determination of levodopa and morphine, *Analytical Methods*, Vol. 3, 2011, pp. 2400-2405.

[109]. M. R. Shishehbore, H. R. Zare, D. Nematollahi, Electrocatalytic determination of morphine at the surface of a carbon paste electrode spiked with a hydroquinone derivative and carbon nanotubes, *Journal of Electroanalytical Chemistry*, Vol. 665, 2012, pp. 45-51.

[110]. A. Navaee, A. Salimi, H. Teymourian, Graphene nanosheets modified glassy carbon electrode for simultaneous detection of heroine, morphine and noscapine, *Biosensors & Bioelectronics*, Vol. 31, 2012, pp. 205-211.

[111]. Y. R. Zhao, Y. Wu, Y. Zhang, Z. G. Chen, X. Cao, J. W. Di, J. P. Yang, Electrocatalytic Behavior and Amperometric Detection of Morphine on ITO Electrode Modified with Directly Electrodeposited Gold Nanoparticles, *Electroanalysis*, Vol. 21, 2009, pp. 939-943.

[112]. K. C. Ho, C. Y. Chen, H. C. Hsu, L. C. Chen, S. C. Shiesh, X. Z. Lin, Amperometric detection of morphine at a Prussian blue-modified indium tin oxide electrode, *Biosensors & Bioelectronics*, Vol. 20, 2004, pp. 3-8.

[113]. L. M. Bonanno, L. A. DeLouise, Tunable Detection Sensitivity of Opiates in Urine via a Label-Free Porous Silicon Competitive Inhibition Immunosensor, *Analytical Chemistry*, Vol. 82, 2010, pp. 714-722.

[114]. L. M. Bonanno, T. C. Kwong, L. A. DeLouise, Label-Free Porous Silicon Immunosensor for Broad Detection of Opiates in a Blind Clinical Study and Results Comparison to Commercial Analytical Chemistry Techniques, *Analytical Chemistry*, Vol. 82, 2010, pp. 9711-9718.

[115]. S. Gandhi, N. Caplash, P. Sharma, C. R. Suri, Strip-based immunochromatographic assay using specific egg yolk antibodies for rapid detection of morphine in urine samples, *Biosensors & Bioelectronics*, Vol. 25, 2009, pp. 502-505.

[116]. I. Merfort, Review of the analytical techniques for sesquiterpenes and sesquiterpene lactones, *Journal of Chromatography A*, Vol. 967, 2002, pp. 115-130.

[117]. A. De Palma, R. Rossi, M. Carai, C. Cabras, G. Colombo, L. Arnoldi, N. Fuzzati, A. Riva, P. Morazzoni, P. L. Mauri, Pharmaceutical and Biomedical Analysis of Terpene Constituents in Salvia miltiorrhiza, *Current Pharmaceutical Analysis*, Vol. 4, 2008, pp. 249-257.

[118]. D. D. Dias, M. Colombo, R. G. Kelmann, T. P. De Souza, V. L. Bassani, H. F. Teixeira, V. F. Veiga, R. P. Limberger, L. S. Koester, Optimization of headspace solid-phase microextraction for analysis of beta-caryophyllene in a nanoemulsion dosage form prepared with copaiba (Copaifera multijuga Hayne) oil, *Analytica Chimica Acta*, Vol. 721, 2012, pp. 79-84.

[119]. W. Vautz, S. Sielemann, J. I. Baumbach, Determination of terpenes in humid ambient air using ultraviolet ion mobility spectrometry, *Analytica Chimica Acta*, Vol. 513, 2004, pp. 393-399.

[120]. G. Chen, L. Y. Zhang, Y. Z. Zhu, Determination of glycosides and sugars in Moutan Cortex by capillary electrophoresis with electrochemical detection, *Journal of Pharmaceutical and Biomedical Analysis*, Vol. 41, 2006, pp. 129-134.

[121]. A. Kubesova, M. Horka, F. Ruzicka, K. Slais, Z. Glatz, Separation of attogram terpenes by the capillary zone electrophoresis with fluorometric detection, *Journal of Chromatography A*, Vol. 1217, 2010, pp. 7288-7292.

[122]. S. D. Qi, L. Ding, K. Tian, X. G. Chen, Z. D. Hu, Novel and simple nonaqueous capillary electrophoresis separation and determination bioactive triterpenes in Chinese herbs, *Journal of Pharmaceutical and Biomedical Analysis*, Vol. 40, 2006, pp. 35-41.

[123]. N. Iqbal, G. Mustafa, A. Rehman, A. Biedermann, B. Najafi, P. A. Lieberzeit, F. L. Dickert, QCM-Arrays for Sensing Terpenes in Fresh and Dried Herbs via Bio-Mimetic MIP Layers, *Sensors*, Vol. 10, 2010, pp. 6361-6376.

[124]. P. A. Lieberzeit, A. Rehman, N. Iqbal, B. Najafi, F. L. Dickert, QCM sensor array for monitoring terpene emissions from odoriferous plants, *Monatshefte Fur Chemie*, Vol. 140, 2009, pp. 947-952.

[125]. K. Rader, A. Wildfeuer, F. Wintersberger, P. Bossinger, H. W. Mucke, Characterization of bee venom and its main components by high-performance liquid chromatography, *J Chromatogr*, Vol. 408, 1987, pp. 341-348.

[126]. Z. J. Kokot, J. Matysiak, Simultaneous Determination of Major Constituents of Honeybee Venom by LC-DAD, *Chromatographia*, Vol. 69, 2009, pp. 1401-1405.

[127]. Z. J. Kokot, J. Matysiak, B. Urbaniak, P. Derezinski, New CZE-DAD method for honeybee venom analysis and standardization of the product, *Analytical and Bioanalytical Chemistry*, Vol. 399, 2011, pp. 2487-2494.

[128]. V. Pacakova, K. Stulik, Validation of a method for determination of phospholipase A(2) and melittin in bee venom preparations by capillary electrophoresis, *Journal of Aoac International*, Vol. 83, 2000, pp. 549-554.

[129]. B. E. Banks, C. E. Dempsey, F. L. Pearce, C. A. Vernon, T. E. Wholley, New methods of isolating been venom peptides, *Anal Biochem*, Vol. 116, 1981, pp. 48-52.

[130]. S. Rozner, S. Kolusheva, Z. Cohen, W. Dowhan, J. Eichler, R. Jelinek, Detection and analysis of membrane interactions by a biomimetic colorimetric lipid/polydiacetylene assay, *Analytical Biochemistry*, Vol. 319, 2003, pp. 96-104.

[131]. H. M. Hiep, T. Endo, M. Saito, M. Chikae, D. K. Kim, S. Yamamura, Y. Takamura, E. Tamiya, Label-free detection of melittin binding to a membrane using electrochemical-localized surface plasmon resonance, *Analytical Chemistry*, Vol. 80, 2008, pp. 1859-1864.

[132]. M. Gheorghiu, A. Olaru, A. Tar, C. Polonschii, E. Gheorghiu, Sensing based on assessment of non-monotonous effect determined by target analyte: Case study on pore-forming compounds, *Biosensors & Bioelectronics*, Vol. 24, 2009, pp. 3517-3523.

[133]. T. Shahal, K. A. Melzak, C. R. Lowe, E. Gizeli, Poly(dimethylsiloxane)-coated sensor devices for the formation of supported lipid bilayers and the subsequent study of membrane interactions, *Langmuir*, Vol. 24, 2008, pp. 11268-11275.

[134]. P. Escoubas, M. L. Celerier, T. Nakajima, High-performance liquid chromatography matrix-assisted laser desorption/ionization time-of-flight mass spectrometry peptide fingerprinting of tarantula venoms in the genus Brachypelma: Chemotaxonomic and biochemical applications, *Rapid Communications in Mass Spectrometry*, Vol. 11, 1997, pp. 1891-1899.

[135]. P. Escoubas, B. J. Whiteley, C. P. Kristensen, M. L. Celerier, G. Corzo, T. Nakajima, Multidimensional peptide fingerprinting by high performance liquid chromatography, capillary zone electrophoresis and matrix-assisted laser desorption/ionization time-of-flight mass spectrometry for the identification of tarantula venom samples, *Rapid Communications in Mass Spectrometry*, Vol. 12, 1998, pp. 1075-1084.

[136]. P. Escoubas, B. Sollod, G. F. King, Venom landscapes: Mining the complexity of spider venoms via a combined cDNA and mass spectrometric approach, *Toxicon*, Vol. 47, 2006, pp. 650-663.

[137]. C. Chavez-Olortegui, V. C. Zanetti, A. P. Ferreira, J. C. Minozzo, O. C. Mangili, I. C. Gubert, ELISA for the detection of venom antigens in experimental and clinical envenoming by Loxosceles intermedia spiders, *Toxicon*, Vol. 36, 1998, pp. 563-569.
[138]. C. Chavez-Olortegui, K. Bohorquez, L. M. Alvarenga, E. Kalapothakis, D. Campolina, W. S. Maria, C. R. Diniz, Sandwich-ELISA detection of venom antigens in envenoming by Phoneutria nigriventer spider, *Toxicon*, Vol. 39, 2001, pp. 909-911.
[139]. J. A. Umbach, A. Grasso, S. D. Zurcher, H. I. Kornblum, A. Mastrogiacomo, C. B. Gundersen, Electrical and optical monitoring of alpha-latrotoxin action at Drosophila neuromuscular junctions, *Neuroscience*, Vol. 87, 1998, pp. 913-924.
[140]. C. L. Weldon, S. P. Mackessy, Biological and proteomic analysis of venom from the Puerto Rican Racer (Alsophis portoricensis: Dipsadidae), *Toxicon*, Vol. 55, 2010, pp. 558-569.
[141]. J. J. Calvete, P. Juarez, L. Sanz, Snake venomics. Strategy and applications, *Journal of Mass Spectrometry*, Vol. 42, 2007, pp. 1405-1414.
[142]. Z. E. Selvanayagam, P. Gopalakrishnakone, Tests for detection of snake venoms, toxins and venom antibodies: review on recent trends (1987-1997), *Toxicon*, Vol. 37, 1999, pp. 565-586.
[143]. Z. E. Selvanayagam, S. G. Gnanavendhan, K. A. Ganesh, P. V. S. Rao, ELISA for the detection of venoms from four medically important snakes of India, *Toxicon*, Vol. 37, 1999, pp. 757-770.
[144]. L. V. Dong, L. K. Quyen, K. H. Eng, P. Gopalakrishnakone, Immunogenicity of venoms from four common snakes in the South of Vietnam and development of ELISA kit for venom detection, *Journal of Immunological Methods*, Vol. 282, 2003, pp. 13-31.
[145]. J. Steuten, K. Winkel, T. Carroll, N. A. Williamson, V. Ignjatovic, K. Fung, A. W. Purcell, B. G. Fry, The molecular basis of cross-reactivity in the Australian snake venom detection kit (SVDK), *Toxicon*, Vol. 50, 2007, pp. 1041-1052.
[146]. R. K. C. Ong, K. Swindells, C. S. Mansfield, Prospective determination of the specificity of a commercial snake venom detection kit in urine samples from dogs and cats, *Australian Veterinary Journal*, Vol. 88, 2010, pp. 222-224.
[147]. J. E. Biardi, K. T. Nguyen, S. Lander, M. Whitley, K. P. Nambiar, A rapid and sensitive fluorometric method for the quantitative analysis of snake venom metalloproteases and their inhibitors, *Toxicon*, Vol. 57, 2011, pp. 342-347.
[148]. L. Van Dong, K. H. Eng, L. K. Quyen, P. Gopalakrishnakone, Optical immunoassay for snake venom detection, *Biosensors & Bioelectronics*, Vol. 19, 2004, pp. 1285-1294.
[149]. D. Hartono, S. L. Lai, K. L. Yang, L. Y. L. Yung, A liquid crystal-based sensor for real-time and label-free identification of phospholipase-like toxins and their inhibitors, *Biosensors & Bioelectronics*, Vol. 24, 2009, pp. 2289-2293.
[150]. M. Campas, B. Prieto-Simon, J. L. Marty, Biosensors to detect marine toxins: Assessing seafood safety, *Talanta*, Vol. 72, 2007, pp. 884-895.
[151]. M. Lotierzo, O. Y. F. Henry, S. Piletsky, I. Tothill, D. Cullen, M. Kania, B. Hock, A. P. F. Turner, Surface plasmon resonance sensor for domoic acid based on grafted imprinted polymer, *Biosensors & Bioelectronics*, Vol. 20, 2004, pp. 145-152.
[152]. N. V. Kulagina, C. M. Mikulski, S. Gray, W. Ma, G. J. Doucette, J. S. Ramsdell, J. J. Pancrazio, Detection of marine toxins, brevetoxin-3 and saxitoxin, in seawater using neuronal networks, *Environmental Science & Technology*, Vol. 40, 2006, pp. 578-583.
[153]. N. Hamada-Sato, N. Minamitani, Y. Inaba, Y. Nagashima, T. Kobayashi, C. Imada, E. Watanabe, Development of amperometric sensor system for measurement of diarrheic shellfish poisoning (DSP) toxin, okadaic acid (OA), *Sensors and Materials*, Vol. 16, 2004, pp. 99-107.

[154]. M. Kania, M. Kreuzer, E. Moore, M. Pravda, B. Hock, G. Guillbault, Development of polyclonal antibodies against domoic acid for their use in electrochemical biosensors, *Analytical Letters*, Vol. 36, 2003, pp. 1851-1863.
[155]. M. P. Kreuzer, M. Pravda, C. K. O'Sullivan, G. G. Guilbault, Novel electrochemical immunosensors for seafood toxin analysis, *Toxicon*, Vol. 40, 2002, pp. 1267-1274.
[156]. A. Tang, M. Kreuzer, M. Lehane, M. Pravda, G. G. Guilbault, Immunosensor for the determination of okadaic acid based on screen-printed electrode, *International Journal of Environmental Analytical Chemistry*, Vol. 83, 2003, pp. 663-670.
[157]. D. Neagu, L. Micheli, G. Palleschi, Study of a toxin-alkaline phosphatase conjugate for the development of an immunosensor for tetrodotoxin determination, *Analytical and Bioanalytical Chemistry*, Vol. 385, 2006, pp. 1068-1074.
[158]. S. Gallacher, T. H. Birkbeck, A tissue culture assay for direct detection of sodium channel blocking toxins in bacterial culture supernates, *FEMS Microbiol Lett*, Vol. 71, 1992, pp. 101-107.
[159]. R. J. Lewis, A. Jones, Characterization of ciguatoxins and ciguatoxin congeners present in ciguateric fish by gradient reverse-phase high-performance liquid chromatography mass spectrometry, *Toxicon*, Vol. 35, 1997, pp. 159-168.
[160]. H. Oguri, M. Hirama, T. Tsumuraya, I. Fujii, M. Maruyama, H. Uehara, Y. Nagumo, Synthesis-based approach toward direct sandwich immunoassay for ciguatoxin CTX3C, *Journal of the American Chemical Society*, Vol. 125, 2003, pp. 7608-7612.
[161]. Y. Hokama, A. H. Banner, D. B. Boylan, A radioimmunoassay for the detection of ciguatoxin, *Toxicon*, Vol. 15, 1977, pp. 317-325.
[162]. M. C. Louzao, M. R. Vieytes, T. Yasumoto, L. M. Botana, Detection of sodium channel activators by a rapid fluorimetric microplate assay, *Chemical Research in Toxicology*, Vol. 17, 2004, pp. 572-578.
[163]. C. O. Parker, Y. H. Lanyon, M. Manning, D. W. M. Arrigan, I. E. Tothill, Electrochemical Immunochip Sensor for Aflatoxin M-1 Detection, *Analytical Chemistry*, Vol. 81, 2009, pp. 5291-5298.
[164]. N. Paniel, A. Radoi, J. L. Marty, Development of an Electrochemical Biosensor for the Detection of Aflatoxin M-1 in Milk, *Sensors*, Vol. 10, 2010, pp. 9439-9448.
[165]. A. E. Radi, X. Munoz-Berbel, M. Cortina-Puig, J. L. Marty, An electrochemical immunosensor for ochratoxin A based on immobilization of antibodies on diazonium-functionalized gold electrode, *Electrochimica Acta*, Vol. 54, 2009, pp. 2180-2184.
[166]. A. C. Mak, S. J. Osterfeld, H. Yu, S. X. Wang, R. W. Davis, O. A. Jejelowo, N. Pourmand, Sensitive giant magnetoresistive-based immunoassay for multiplex mycotoxin detection, *Biosensors & Bioelectronics*, Vol. 25, 2010, pp. 1635-1639.
[167]. Y. Li, X. Liu, Z. Lin, Recent developments and applications of surface plasmon resonance biosensors for the detection of mycotoxins in foodstuffs, *Food Chemistry*, Vol. 132, 2012, pp. 1549-1554.
[168]. V. Dolnik, S. R. Liu, S. Jovanovich, Capillary electrophoresis on microchip, *Electrophoresis*, Vol. 21, 2000, pp. 41-54.
[169]. M. Hervas, M. A. Lopez, A. Escarpa, Integrated electrokinetic magnetic bead-based electrochemical immunoassay on microfluidic chips for reliable control of permitted levels of zearalenone in infant foods, *Analyst*, Vol. 136, 2011, pp. 2131-2138.
[170]. F. J. Arevalo, A. M. Granero, H. Fernandez, J. Raba, M. A. Zon, Citrinin (CIT) determination in rice samples using a micro fluidic electrochemical immunosensor, *Talanta*, Vol. 83, 2011, pp. 966-973.
[171]. M. Hervas, M. A. Lopez, A. Escarpa, Electrochemical microfluidic chips coupled to magnetic bead-based ELISA to control allowable levels of zearalenone in baby foods using simplified calibration, *Analyst*, Vol. 134, 2009, pp. 2405-2411.

[172]. K. Kloth, M. Rye-Johnsen, A. Didier, R. Dietrich, E. Martlbauer, R. Niessner, M. Seidel, A regenerable immunochip for the rapid determination of 13 different antibiotics in raw milk, *Analyst*, Vol. 134, 2009, pp. 1433-1439.
[173]. W. Chen, Y. Lei, C. M. Li, Regenerable Leptin Immunosensor Based on Protein G Immobilized Au-Pyrrole Propylic Acid-Polypyrrole Nanocomposite, *Electroanalysis*, Vol. 22, 2010, pp. 1078-1083.
[174]. L. Anfossi, G. D'Arco, M. Calderara, C. Baggiani, C. Giovannoli, G. Giraudi, Development of a quantitative lateral flow immunoassay for the detection of aflatoxins in maize, *Food Additives and Contaminants Part a-Chemistry Analysis Control Exposure & Risk Assessment*, Vol. 28, 2011, pp. 226-234.
[175]. D. H. Zhang, P. W. Li, Y. Yang, Q. Zhang, W. Zhang, Z. Xiao, X. X. Ding, A high selective immunochromatographic assay for rapid detection of aflatoxin B-1, *Talanta*, Vol. 85, 2011, pp. 736-742.
[176]. S. Xiulan, Z. Xiaolian, T. Jian, G. Xiaohong, Z. Jun, F. S. Chu, Development of an immunochromatographic assay for detection of aflatoxin B1 in foods, *Food Control*, Vol. 17, 2006, pp. 256-262.
[177]. D. Saha, D. Acharya, D. Roy, T. K. Dhar, Filtration-based tyramide amplification technique-a new simple approach for rapid detection of aflatoxin B-1, *Analytical and Bioanalytical Chemistry*, Vol. 387, 2007, pp. 1121-1130.
[178]. J. P. Meneely, F. Ricci, H. P. van Egmond, C. T. Elliott, Current methods of analysis for the determination of trichothecene mycotoxins in food, *Trac-Trends in Analytical Chemistry*, Vol. 30, 2011, pp. 192-203.
[179]. A. Molinelli, K. Grossalber, R. Krska, A rapid lateral flow test for the determination of total type B fumonisins in maize, *Analytical and Bioanalytical Chemistry*, Vol. 395, 2009, pp. 1309-1316.
[180]. S. Wang, Y. Quan, N. Lee, I. R. Kennedy, Rapid determination of fumonisin B-1 in food samples by enzyme-linked Immunosorbent assay and colloidal gold immunoassay, *Journal of Agricultural and Food Chemistry*, Vol. 54, 2006, pp. 2491-2495.
[181]. C. M. Shiu, J. J. Wang, F. Y. Yu, Sensitive enzyme-linked immunosorbent assay and rapid one-step immunochromatographic strip for fumonisin B1 in grain-based food and feed samples, *Journal of the Science of Food and Agriculture*, Vol. 90, 2010, pp. 1020-1026.
[182]. A. Y. Kolosova, L. Sibanda, F. Dumoulin, J. Lewis, E. Duveiller, C. Van Peteghem, S. De Saeger, Lateral-flow colloidal gold-based immunoassay for the rapid detection of deoxynivalenol with two indicator ranges, *Analytica Chimica Acta*, Vol. 616, 2008, pp. 235-244.
[183]. Y. Xu, Z. B. Huang, Q. H. He, S. Z. Deng, L. S. Li, Y. P. Li, Development of an immunochromatographic strip test for the rapid detection of deoxynivalenol in wheat and maize, *Food Chemistry*, Vol. 119, 2010, pp. 834-839.
[184]. W. B. Shim, K. Y. Kim, D. H. Chung, Development and Validation of a Gold Nanoparticle Immunochromatographic Assay (ICG) for the Detection of Zearalenone, *Journal of Agricultural and Food Chemistry*, Vol. 57, 2009, pp. 4035-4041.
[185]. A. Molinelli, K. Grossalber, M. Fuehrer, S. Baumgartner, M. Sulyok, R. Krska, Development of qualitative and semiquantitative immunoassay-based rapid strip tests for the detection of T-2 toxin in wheat and oat, *Journal of Agricultural and Food Chemistry*, Vol. 56, 2008, pp. 2589-2594.
[186]. W. H. Lai, D. Y. C. Fung, Y. Xu, R. R. Liu, Y. H. Xiong, Development of a colloidal gold strip for rapid detection of ochratoxin A with mimotope peptide, *Food Control*, Vol. 20, 2009, pp. 791-795.
[187]. B. H. Liu, Z. J. Tsao, J. J. Wang, F. Y. Yu, Development of a monoclonal antibody against ochratoxin A and its application in enzyme-linked immunosorbent assay and gold

nanoparticle immunochromatographic strip, *Analytical Chemistry*, Vol. 80, 2008, pp. 7029-7035.

[188]. L. B. Wang, W. Chen, W. W. Ma, L. Q. Liu, W. Ma, Y. A. Zhao, Y. Y. Zhu, L. G. Xu, H. Kuang, C. L. Xu, Fluorescent strip sensor for rapid determination of toxins, *Chemical Communications*, Vol. 47, 2011, pp. 1574-1576.

[189]. A. Laura, D. Gilda, B. Claudio, G. Cristina, G. Gianfranco, A lateral flow immunoassay for measuring ochratoxin A: Development of a single system for maize, wheat and durum wheat, *Food Control*, Vol. 22, 2011, pp. 1965-1970.

[190]. A. Y. Kolosova, S. De Saeger, L. Sibanda, R. Verheijen, C. Van Peteghem, Development of a colloidal gold-based lateral-flow immunoassay for the rapid simultaneous detection of zearalenone and deoxynivalenol, *Analytical and Bioanalytical Chemistry*, Vol. 389, 2007, pp. 2103-2107.

[191]. D. H. Zhang, P. W. Li, Q. Zhang, W. Zhang, Ultrasensitive nanogold probe-based immunochromatographic assay for simultaneous detection of total aflatoxins in peanuts, *Biosensors & Bioelectronics*, Vol. 26, 2011, pp. 2877-2882.

[192]. W. B. Shim, J. S. Kim, J. Y. Kim, J. G. Choi, J. H. Je, N. S. Kuzmina, S. A. Eremin, D. H. Chung, Determination of aflatoxin B1 in rice, barley, and feed by non-instrumental immunochromatographic strip-test and high sensitive ELISA, *Food Science and Biotechnology*, Vol. 17, 2008, pp. 623-630.

[193]. D. Tang, J. C. Sauceda, Z. Lin, S. Ott, E. Basova, I. Goryacheva, S. Biselli, J. Lin, R. Niessner, D. Knopp, Magnetic nanogold microspheres-based lateral-flow immunodipstick for rapid detection of aflatoxin B-2 in food, *Biosensors & Bioelectronics*, Vol. 25, 2009, pp. 514-518.

[194]. J. Y. Liao, H. Li, Lateral flow immunodipstick for visual detection of aflatoxin B-1 in food using immuno-nanoparticles composed of a silver core and a gold shell, *Microchimica Acta*, Vol. 171, 2010, pp. 289-295.

[195]. O. G. Weingart, H. Gao, F. Crevoisier, F. Heitger, M. A. Avondet, H. Sigrist, A Bioanalytical Platform for Simultaneous Detection and Quantification of Biological Toxins, *Sensors*, Vol. 12, 2012, pp. 2324-2339.

[196]. M. M. Ngundi, C. R. Taitt, S. A. McMurry, D. Kahne, F. S. Ligler, Detection of bacterial toxins with monosaccharide arrays, *Biosensors & Bioelectronics*, Vol. 21, 2006, pp. 1195-1201.

[197]. Y. F. Zhang, J. L. Lou, K. L. Jenko, J. D. Marks, S. M. Varnum, Simultaneous and sensitive detection of six serotypes of botulinum neurotoxin using enzyme-linked immunosorbent assay-based protein antibody microarrays, *Analytical Biochemistry*, Vol. 430, 2012, pp. 185-192.

[198]. M. L. Frisk, G. Y. Lin, E. A. Johnson, D. J. Beebe, Synaptotagmin II peptide-bead conjugate for botulinum toxin enrichment and detection in microchannels, *Biosensors & Bioelectronics*, Vol. 26, 2011, pp. 1929-1935.

[199]. S. M. Han, J. H. Cho, I. H. Cho, E. H. Paek, H. B. Oh, B. S. Kim, C. Ryu, K. Lee, Y. K. Kim, S. H. Paek, Plastic enzyme-linked immunosorbent assays (ELISA)-on-a-chip biosensor for botulinum neurotoxin A, *Analytica Chimica Acta*, Vol. 587, 2007, pp. 1-8.

[200]. R. M. Ozanich, C. J. Bruckner-Lea, M. G. Warner, K. Miller, K. C. Antolick, J. D. Marks, J. L. Lou, J. W. Grate, Rapid Multiplexed Flow Cytometric Assay for Botulinum Neurotoxin Detection Using an Automated Fluidic Microbead-Trapping Flow Cell for Enhanced Sensitivity, *Analytical Chemistry*, Vol. 81, 2009, pp. 5783-5793.

[201]. M. H. Yang, S. Sun, Y. Kostov, A. Rasooly, Lab-on-a-chip for carbon nanotubes based immunoassay detection of Staphylococcal Enterotoxin B (SEB), *Lab on a Chip*, Vol. 10, 2010, pp. 1011-1017.

[202]. X. M. Zhang, F. Liu, R. F. Yan, P. Xue, Y. M. Li, L. L. Chen, C. J. Song, C. Liu, B. Q. Jin, Z. J. Zhang, K. Yang, An ultrasensitive immunosensor array for determination of staphylococcal enterotoxin B, *Talanta*, Vol. 85, 2011, pp. 1070-1074.
[203]. M. G. Warner, J. W. Grate, A. Tyler, R. M. Ozanich, K. D. Miller, J. L. Lou, J. D. Marks, C. J. Bruckner-Lea, Quantum dot immunoassays in renewable surface column and 96-well plate formats for the fluorescence detection of botulinum neurotoxin using high-affinity antibodies, *Biosensors & Bioelectronics*, Vol. 25, 2009, pp. 179-184.
[204]. D. Pauly, S. Kirchner, B. Stoermann, T. Schreiber, S. Kaulfuss, R. Schade, R. Zbinden, M. A. E. Avondet, M. B. Dorner, B. G. Dorner, Simultaneous quantification of five bacterial and plant toxins from complex matrices using a multiplexed fluorescent magnetic suspension assay, *Analyst*, Vol. 134, 2009, pp. 2028-2039.
[205]. S. Ahn-Yoon, T. R. DeCory, R. A. Durst, Ganglioside-liposome immunoassay for the detection of botulinum toxin, *Analytical and Bioanalytical Chemistry*, Vol. 378, 2004, pp. 68-75.
[206]. Y. Liu, Y. Dong, J. Jauw, M. J. Linman, Q. Cheng, Highly Sensitive Detection of Protein Toxins by Surface Plasmon Resonance with Biotinylation-Based Inline Atom Transfer Radical Polymerization Amplification, *Analytical Chemistry*, Vol. 82, 2010, pp. 3679-3685.
[207]. G. Ferracci, S. Marconi, C. Mazuet, E. Jover, M. P. Blanchard, M. Seagar, M. Popoff, C. Leveque, A label-free biosensor assay for botulinum neurotoxin B in food and human serum, *Analytical Biochemistry*, Vol. 410, 2011, pp. 281-288.
[208]. X. Liu, D. Q. Song, Q. L. Zhang, Y. Tian, H. Q. Zhang, An optical surface plasmon resonance biosensor for determination of tetanus toxin, *Talanta*, Vol. 62, 2004, pp. 773-779.
[209]. W. C. Tsai, P. J. R. Pai, Surface plasmon resonance-based immunosensor with oriented immobilized antibody fragments on a mixed self-assembled monolayer for the determination of staphylococcal enterotoxin B, *Microchimica Acta*, Vol. 166, 2009, pp. 115-122.
[210]. J. Wojciechowski, D. Danley, J. Cooper, N. Yazvenko, C. R. Taitt, Multiplexed Electrochemical Detection of Yersinia Pestis and Staphylococcal Enterotoxin B using an Antibody Microarray, *Sensors*, Vol. 10, 2010, pp. 3351-3362.
[211]. P. B. Lillehoj, F. Wei, C. M. Ho, A self-pumping lab-on-a-chip for rapid detection of botulinum toxin, *Lab on a Chip*, Vol. 10, 2010, pp. 2265-2270.
[212]. M. Salmain, M. Ghasemi, S. Boujday, J. Spadavecchia, C. Techer, F. Val, V. Le Moigne, M. Gautier, R. Briandet, C. M. Pradier, Piezoelectric immunosensor for direct and rapid detection of staphylococcal enterotoxin A (SEA) at the ng level, *Biosensors & Bioelectronics*, Vol. 29, 2011, pp. 140-144.
[213]. D. Maraldo, R. Mutharasan, Detection and confirmation of staphylococcal enterotoxin B in apple juice and milk using piezoelectric-excited millimeter-sized cantilever sensors at 2. 5 fg/mL, *Analytical Chemistry*, Vol. 79, 2007, pp. 7636-7643.
[214]. G. A. Campbell, M. B. Medina, R. Mutharasan, Detection of Staphylococcus enterotoxin B at picogram levels using piezoelectric-excited millimeter-sized cantilever sensors, *Sensors and Actuators B-Chemical*, Vol. 126, 2007, pp. 354-360.
[215]. M. H. Yang, S. Sun, H. A. Bruck, Y. Kostov, A. Rasooly, Electrical percolation-based biosensor for real-time direct detection of staphylococcal enterotoxin B (SEB), *Biosensors & Bioelectronics*, Vol. 25, 2010, pp. 2573-2578.

Chapter 13
Ultralow Detection of Bio-markers Using Gold Nanoshells

Dhruvinkumar N. Patel, Xinghua Sun, Guandong Zhang, Martin G. O'Toole, Scott Cambron, Robert S. Keynton, Andre M. Gobin

13.1. Introduction

Although many types of biosensors have been around for decades, while some are simple and others are very complex, researchers are actively are trying to develop new methods of detections for improving the sensitivity and specificity of detection events. In recent years, researchers have demonstrated the successful employment of biosensors for various uses including, environmental and biomedical sensing applications. The over-riding principle concept for developing such sensors or biosensor is to produce a simple yet highly effective, sensitive and rapid, low cost, and easy to use system [3, 4]. Typical biosensors consist of a biological complex designed to detect and measure a specific substance and be able to differentiate between other non-selective substances. Specifically, biosensors are used as an analytical device with a biological sensing element in intimate communication with a signal transducer (typically through physical contact), which together generate a detectable signal of some sort from a collection of analytes presents in the system to produce a measurable signal [4-6]. "Sensors that implement immunological detection methods are amongst the most advanced type of biosensor technologies due to the ubiquity of traditional immunoassay techniques and because high detection specificity is readily obtained with receptor—ligand interaction chemistry [7]." Recent technological advancements have produced biosensors capable of detection of events over many orders of magnitude of analyte concentration using a wide variety of different devices designs, various biological interaction principles, and disparate materials [7].

Generally, analysis of biological samples requires a low threshold of detection, around micromolar concentrations. However, in many other applications such as detection of

[13] Dhruvinkumar N. Patel
Department of Bioengineering
University of Louisville, Louisville, KY 40292, USA

viruses, biomarkers, and bioterrorism agents, detection thresholds in the nanomolar to picomolar concentration range are required [7, 8]. Such analyses are typically done in laboratories that houses expensive equipment and trained researchers for measurements. The ideal biosensor would satisfy the following requirements:

- The device should be compact, self-contained, easy to fabricate, simple, and sensitive [9];
- Analysis of samples and the displayed results should be unaffected by whole blood, serum, and other biological samples introduced within the system [9, 10];
- The sensitivity of the system or the detection-limit should be appropriate to the analyte being used (micromolar to picomolar concentration range) [7, 8];
- The biosensor should be adaptable to many different analytes (viruses, immunochemistry, bioterrorism agents, clinical chemistry, enzymology, DNA probe, etc.), hold the potential for sorting of multiple substances or making simultaneous measurements of multiple and be cheap to manufacture [3, 4, 9, 11-13].

The development of nanotechnology had led to the application of nanoparticles in biosensors and bioassays [3]. Gold nanoparticles have attracted much attention due to their unique optical, physical, and chemical properties [14]. These properties offer excellent and versatile possibilities for chemical and biological sensing applications [3, 15-17]. Nanoparticles have been around for years and because of their unique properties; they have been used as sensitive tracers or markers for the electronic, optical, and micro-gravimetric transduction for many immunoassays [3, 18-20]. Colloidal gold and semiconductor quantum dot nanoparticles, in particular have attracted enormous attention in bioanalytical applications. The versatility of nanoparticles can be significantly improved by coupling them with biological recognition reactions and electrical processes [3, 7, 18-20]. This allows signal amplification with the use of bioconjugated nanoparticles through the formation of various nanoparticle–biomolecule components to provide an ultrasensitive and ultralow, optical and electrical, detection with exquisite sensitivity [3, 4]. The promise of such a system has resulted in an increase in interest of using different biomolecules to synthesize nanoparticle "architectures" that are tailored for specific applications *via* functionalizing the surfaces of nanoparticles for specific applications [20-24]. Thus, nanoparticle-based biosensors, smart particles, offer considerable advancement for DNA and protein based diagnostics [3, 15, 25]. According to Liu et. al., the novel usage of nanoparticles as biosensors is classified into two categories, specifically according to their functions: (1) nanoparticle-modified transducers for bioanalytical applications and (2) biomolecule–nanoparticle conjugates as labels for biosensing and bioassays [3, 15, 16, 25]. For the purpose of this chapter, we intend to focus on the usage of gold nanoshells as biosenors, specifically for detection of bio-marker and the potential to expand to various applications. Gold nanoshells as identification markers or labels can be designed for detection of antigens, analytes, biomarkers, or viral particles, in small sample volumes and concentrations with high sensitivity and specificity. These particles with the unique properties, discussed in Section 13.1.2, can be tailored for many applications, making them an ideal system for biosensor applications.

13.1.1. Radiation Exposure

Astronauts experience lengthy and repeated exposure to radiation such as heavy ions, solar particle events [26], and galactic cosmic rays (GCR) during extended missions outside of the earth's magnetosphere. These have been linked to the initiation and progression of carcinogenesis, cell apoptosis, and tissue degeneration by inducing DNA damage and resulting in various mutations [27, 28]. A significant fraction of radiation insults experienced by astronauts generate radiation-induced lesions including DNA double-strand breaks [27]. The cellular responses to such DNA strands breaks include production of a number of biomarkers that serve to mediate cell cycle arrest and DNA repair-related gene induction such as H2AX, p53 and p21CIP1/WAF1 have been identified as the principal biomarkers for these processes [29, 30]. The use of gold nanoshells as identification markers or labels can allow for detection of these biomarkers in small sample volumes at low concentrations, in a complex environment. By optimizing the gold nanoshells to selectively bind to these biomarkers expressed after such events, a sensitive, portable low weight platform for detection of radiation induced cellular damage can be developed, which would allow proactive decisions prior to return to earth for full diagnostics.

13.1.2. Gold Nanoshells as a Biosensor

Gold nanoshells are unique composite spherical nanoparticles consisting of a silica dielectric core covered by a thin gold shell [1, 14, 31]. Gold nanoshells have an advantage in biomedical applications due to the biocompatibility of gold and the ability to optically tune the resonant properties of these particles a specific wavelength [1, 31]. The properties of gold nanoshells (GNS) have had numerous applications for immunoassay [32], detection markers, drug delivery [33, 34], contrast enhancement, hyperthermia for tumor therapy [35], and combined with imaging and therapy [36]. GNS display low toxicity, thus gold-coated (with biomolecules) nanoshells have the same degree of biocompatibility as solid gold nanoparticles, which are used in a variety of applications today including theranostic applications [31, 37, 38]. In this chapter, we focus on a system capable of rapid detection on variety of analytes using GNS that can also be coupled with Lab-on-chip (LOC) devices [39] for a portable diagnostics tool. In addition to the following application in radiation damage detection, this system could be used in diagnosis of respiratory viruses, cancer, HIV, hepatitis B, C and E, Herpes simplex virus and exposure to bioterrorism agents in the modern era.

GNS are tunable spherical nanoparticles, composed of a dielectric core covered by a thin gold shell, with diameters ranging from 10 to 200 nm (Fig. 13.1B). GNS have optical, chemical and physical properties that make them an ideal candidate for enhancing cancer detection, cancer treatment, cellular imaging and medical biosensing [14]. Gold nanoshells specifically tend to have tunable plasmon resonance in the near infrared (nIR) region of the electromagnetic spectrum (Fig. 13.1A). The importance of the nIR region (650 nm – 900 nm) in medicine is due to the high transmission and low absorption of nIR light by native tissue components such as water and hemoglobin [2, 40]. The high nIR absorptivity of GNS combined with minimal native interference from biological tissue enables minimally invasive imaging and treatment [1, 35, 41, 42]. The geometry of the GNS (e.g. the overall nanoshell diameter and the ratio of the shell thickness to the radius of the dielectric core) determines the optical properties. The ratio of shell thickness to core diameter determines the peak plasmon field resonance maximum; while the overall size determines the absorbing vs. scattering properties of the particle [2, 14]. Furthermore, the inert gold surface provides several benefits, including affinity for thiol groups, which simplifies conjugation to antibodies or other biomolecules for both active tumor targeting and biosensing applications. The former property is important for chemical conjugation of antibodies, or other biomolecules for both active tumor targeting and bio sensing applications on the surface of the particle [14]. Not only does this system have the capability to be utilized in diagnosis of various viruses, but bioterrorism agents in the modern era [36]; thus, GNS can be fabricated with specific nIR resonance profiles and light scattering capabilities, with surfaces that can be decorated with molecules that can impart biocompatibility, chemical functionality, and the ability to have strong binding affinity for a variety of cell types via antibodies, ligands or bio-markers, and could be used to extend the use of GNS beyond therapeutic applications.

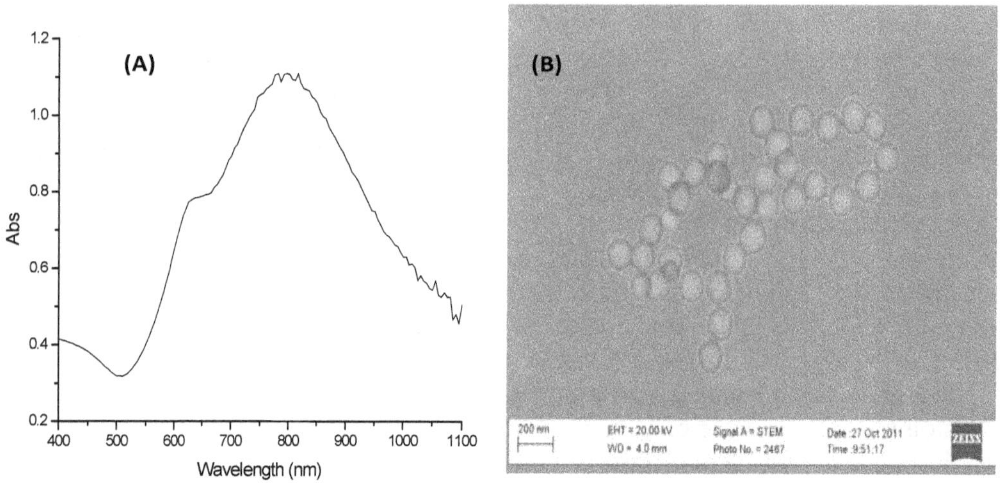

Fig. 13.1. Spectral profile (A) and STEM image (B) of the GNS that were fabricated with radius of ~74 nm and core of ~60 nm (Gold shell thickness = 14 nm).

13.1.3. Immunoassays

The fundamental concept of all immunosensors is the specificity of the biomolecule detection of antigens by antibodies used to form a stable recognition complex. The affinity ligand-based biosensor then couples the binding interaction with a transducer [13, 43, 44]. In contrast to traditional immunoassays, modern transducer technology enables the label-free detection and quantification of the immune complex.

Immunoassays play an important factor in clinical diagnostics, which are dependent on recognition of target antigens by specific antibodies [44, 45]. Immunosensors can be classified as according to whether they are competitive or noncompetitive systems, as well as by the detection principle applied such as radio-immunoassay, enzyme immunoassays, chemiluminescence immunoassay, and fluorescent immunoassay; Table 13.1 gives a brief overview of advantages and disadvantages of these immunoassays. In a competitive immunoassay, all the reactants are mixed together simultaneously, and labeled antigen competes with unlabeled antigen for the same binding site on an antibody. While, in a typical noncompetitive immunoassay it requires a bit more work; first, antibody is absorbed to a solid phase. Next, an unknown antigen is reacted with the antibody to be captured. Lastly, a multistep washing is done to remove unbound antigen and a second antibody with a label or a tag is added to the reaction [46]. Immunoassays have been used for numerous applications including, quantification of proteins and small molecules in medical diagnostics, proteomics, drug discovery, bio-marker detection, and various other biological fields [45, 47].

Table 13.1. Advantages and disadvantages of the different immunoassays presented [46].

Immunoassays	Advantages	Disadvantages
Radioimmunoassay	Extremely sensitive and precise technique for determining concentration of analytes in small sample size.	Health hazard involved with usage of radioactive substances.
Enzyme Immunoassay	No need for expensive instrumentation.	Some specimens may contain natural inhibitors.
Chemiluminescence Immunoassay	Excellent sensitivity, compared to enzyme immunoassays and radioimmunoassay, the reagents used are stable and relatively nontoxic.	May obtain false results if there is lack of precision injection of the hydrogen peroxide or quenching of the light emission.
Fluorescent Immunoassay	Has potential for high sensitivity and adaptability to various needs.	Autofluorescence produced by different organic substances normally present in serum.

To utilize the GNS as a biosensor, biomolecules will be added to the surface for specific detection of bio-markers. Two unique homogeneous immunoassays were developed in our lab using GNS tethered with Protein G and antibody; a solution-based detection assay and a surface bound sandwich enzyme-linked immunosorbent assay (ELISA). The liquid detection immunoassay is based on polyclonal antibody-Protein G-GNS aggregation upon interaction with an analyte present in solution. This allows particles to bind to multiple analyte molecules, which leads to the agglomeration of multiple GNS particles. The resulting close proximity of multiple nanoshells results in a shift of the plasmon resonance frequency due to the interaction of the plasmon fields of several individual nanoparticles. The strong absorption profiles of the GNS allows for a highly sensitive detection system. Biomarker detection via nanoshell aggregation has been demonstrated by Hirsch et. al. in 20 % whole blood with detection levels down to 0.8 ng/ml in 30 minutes [32]. By carefully designing the tethering mechanism to orient the antibody for maximum activity, we hypothesized that these levels of detection could be greatly increased. The system reported here, uses a modular conjugation assembly based on Protein G to create an antibody-Protein G-GNS nanoshell assembly. This particle system was exposed to samples of known analyte concentrations ranging from ng/mL to pg/mL levels. By monitoring the changes in the spectral profile over time, a detection level of 0.5 pg/ml within only 5-10 minutes was achieved.

The field of medical virology has seen numerous technological advancements in the past few decades for diagnosis of viral infections. ELISA was introduced in laboratories during the 1980s. The assay was designed to detect antibodies, specific immunoglobulins (IgG, IgM, IgA) to viruses or viral antigens/particles [36].The sandwich ELISA (Enzyme-Linked ImmunoSorbant Assay) follows more of the traditional concept used in clinical settings or various applications. These ELISAs are typically expensive, labor intensive, and often consume large quantities of reagents and patient specimen [46, 48]. A standardized sandwich-ELISA system using silver stain enhancement with targeted gold nanoparticles for cost effective diagnostics can be developed for detecting viral nanoparticles, instead gold nanoshells are used. The analyte detection system and versatile depending on the type of nanoparticle assembly used [36].

13.2. Methods and Materials

13.2.1. Gold Nanoshell Synthesis

Monodispersed silica nanoparticles were prepared by the Stöber method [49] (the process for synthesizing gold/silica nanoparticles is outlined in Fig. 13.2). Briefly, at 1 mL increments, 14 mL of Ammonium Hydroxide (NH_4OH, 30 %, Sigma) was mixed with 225 mL of absolute ethanol (C_2H_5OH, 99.5 %, Sigma). To achieve smaller size of these silica nanoparticles, lower volume of NH_4OH is used. Next, 7.5 mL of Tetraethyl Ortho-silicate (TEOS, 99.999 %, Sigma) solution was added into the mixture and reacted for minimum of 20 hours. To wash off any remaining NH_4OH, silica nanoparticles were centrifuged twice in 20-25 mL aliquots for 20 minutes at 1500 G and resuspended in absolute ethanol. The silica nanoparticles were then sized using a scanning electron microscope (SEM, SUPRA Carl Zeiss) and Malvern Zetasizer (ZS90). Only batches with polydispersity indices (PDI) of less than 10 % were used for all experiments. The average particle size was measured at ~120 nm with a PDI of ~7 %. Then, silica nanoparticles were then functionalized with amino groups by 3-Aminopropyl Trimethoxysilane (APTES, 99 %, Sigma), which provided amine groups on the surface of the silica nanoparticles to allow for adsorption of gold colloid on the surface. About 200 mg silica nanoparticles were dispersed in 45 mL 90 % ethanol solution which contain 4 % APTES in a 50 mL plastic tube. The tube was placed in a sonication bath at 75 °C for 4 hr, with sonication every 30 min. Then the sample was left on rotator at room temperature (RT) for 20 hrs. After amino-silanization, silica particles were boiled for 2 hours (additional absolute ethanol can be added to maintain a specific volume), then separated and washed with an ethanol solution via centrifugation at 1500 G for 20 minutes. Finally, resuspension of the pellet was done in neat alcohol.

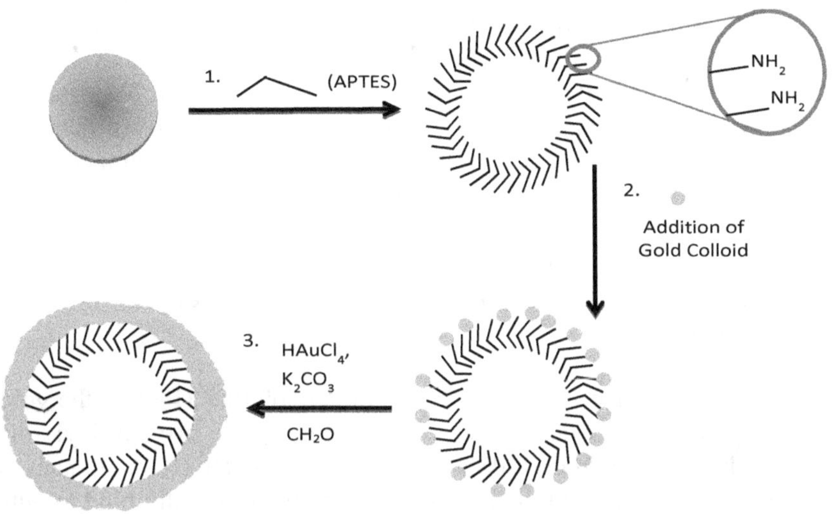

Fig. 13.2. Representation of gold nanoshell synthesis: (1) Monodispersed silica nanoparticles are synthesized using the Stöber method and surface is functionalized with amine groups using APTES, (2) gold colloids are synthesized using the Duff method and absorbed on to the surface, (3) reduction of gold colloids using HAuCl4 in the presence of formaldehyde to form a continuous shell.

Gold colloid was prepared to a size of 2-4 nm using Duff et al method [49], and aged 3 weeks at 4 °C. Briefly, 1wt % gold solution was prepared with 99 grams 18.2 MΩ/cm DI water and 1 gram Hydrogen Tetra-chloroaurate (III) Hydrate (chloroauric acid, HAuCl4, 99.99 %) purchased from Alfa Aesar. The 1wt % gold solution was stored in amber bottles at room temperature until use. Next, 400 µl of (hydroxymethyl) phosphonium chloride (THPC, 80 %) was mixed with 33 mL DI water as a stock solution. Mixing 180 mL DI water, 1.2 mL of 1M NaOH, 4 mL THPC solution, and 6.75 mL of 1 wt % gold solution produces the colloidal gold particles. The colloid sample was then concentrated 10X through rotary evaporation (rotovap, Buchi Rotovapor R215) at 10 mbar, and mixed with the amino-functionalized silica nanoparticles. This allows small gold colloid to attach to silica nanoparticles surface, to act as nucleation sites for the reduction step and to allow formation of a contiguous shell of gold.

The final step, growing the gold shell was accomplished by the reduction of $HAuCl_4$ in the presence of formaldehyde (HCOH, 37 %, Fisher). $HAuCl_4$ reduction solution was prepared by adding 3.0 mL of 1.0 wt % $HAuCl_4$ to 200 mL Potassium Carbonate (K_2CO_3, 99.7 %) solution (50 mg of K_2CO_3 in 200 mL DI water), and aged for two days before use. To synthesize gold nanoshells with varying shell thicknesses we varied the concentration of seed particles, while using the same amount of reduction solution. The spectral profile of each set of samples was examined for optimal conditions to produce desired nIR absorbing nanoshells in K_2CO_3 buffer. nIR absorption characteristics of the nanoshells were determined using a UV-Vis spectrophotometer (Cary 80 BIO UV Vis spectrophotometer). Samples with the appropriate nIR peak resonance (~ 810 nm) were scaled up linearly in K_2CO_3 buffer to provide nanoshells for the subsequent experiments. To wash off any remaining formaldehyde, GNS were centrifuged twice in 5-10 mL aliquots for 10 minutes at 650 G and resuspended in K_2CO_3 buffer. Nanoshell size, zeta, and morphology of the gold nanoshells were analyzed using a Malvern Zetasizer (ZS90) and TEM (FEI Tecnai F20). The GNS were store in 4° C at a maximum concentration of 1.36×10^{11} particles/mL or 25 optical density (OD).

13.2.2. Immobilization of PEG-ProG Conjugate and Antibodies to Nanoshell Surface

In order to better immobilize antibodies like immunoglobulins (Igs) on the gold surface nanoshells, the coshell particles were pre-coated with orthopyridyl disulfides–poly(ethylene) glycol—Protein G (OPSS-PEG-ProG) conjugate (Recombinant ProG was used here). Protein G is a protein that has the ability to specifically bind to antibody by Fc region, which is found on the surface of a variety of *staphylococci* and *streptococci*. From protein and gene sequence analysis, it was reported that ProG is similar to Protein A (ProA), but ProG binds human IgG of all four subclasses, especially to mouse or goat monoclonal IgG, and the affinity constant is usually higher than of ProA. Therefore, it is a great candidate that can be used as a cofactor to capture antibody on the surface of GNS to allow for maximum binding activity to its antigen. Recombinant ProG (For simplification, ProG is used there and in the following text) is a single chain protein that lacks serum-binding domains. As there are no free thiol groups for ProG to bind to gold surfaces, only physical adsorption of ProG on GNS leads to random and weak coating of ProG with potential loss of IgG binding activity. Therefore more oriented coupling of antibody can be achieved using the pre-mobilization of the genetically or chemically modified ProG on gold surface for further oriented antibody capture.

For the PEGylation of ProG, Orthopyridyl disulfides–poly(ethylene) glycol–succinimidyl valerate (OPSS-PEG-SVA, 2000 Da, LaysanBio) and ProG from *Streptococcus sp* (P4689, Sigma), were dissolved at 1 mg/mL in PBS and stored separately. Next, OPSS-PEG-SVA and ProG were mixed at a mole ratio of 20:1 and reacted for 3hr at room temperature with gently stirring then stored overnight at 4 °C. The sample was then purified using multiple NanoSep Centrifuge (8000×g for 10 min, OD010C33, VWR) and stored at -20 °C for later use. The purified PEG-ProG concentration was estimated using Bradford microassay (Sigma).The OPSS-PEG group couples with ProG when the SVA group is cleaved in an aqueous environment, which leaves an activated carboxylic terminus that is capable of reacting to a free primary amine group available on ProG. This will allow the formation of a peptide bond and covalently linking the OPSS-PEG and ProG to form OPSS-PEG-ProG (PEG-ProG), Fig. 13.3 (A).

OPSS-PEG-ProG (or PEG-ProG) conjugate were added onto GNS in K_2CO_3 buffer at a targeted mole ratio of 2000:1 (PEG-ProG:GNS), and reacted for 4 hrs at 4° C with agitation every ~30 min. Next, a solution of 1 mg/mL Goat Anti-Rabbit antibody (Anti-Rb, R2004, Sigma) at 1 mg/mL, was added to GNS-PEG-ProG conjugate at mole ratio of 5000:1 (Anti-Rb:GNS) and reacted under the same conditions as PEG-ProG conjugation on GNS. To backfill any remaining unspecific, "empty", sites on the particle, 10 μL of 10 μM PEG-SH (5000 Da, LaysanBio) was added after antibody incubation and reacted overnight. The addition of PEG molecules to the GNS also aids in the prevention of non-specific protein adsorption to the GNS surface. The reaction scheme is shown in Fig. 13.3 (B-D). Incubation of antibody to PEG-ProG prior to blocking allows for a higher concentration of antibody on the nanoshell surface, which maximizes the use of the antibody solution and allows for better analyte capturing. Prior to adding Anti-Rb or PEG-SH, the GNS sample was split into 5 mL aliquots and centrifuged at 650 G for 7 min. The supernatant was removed immediately and aliquots were resuspended in 5 mL K_2CO_3 buffer.

3A.

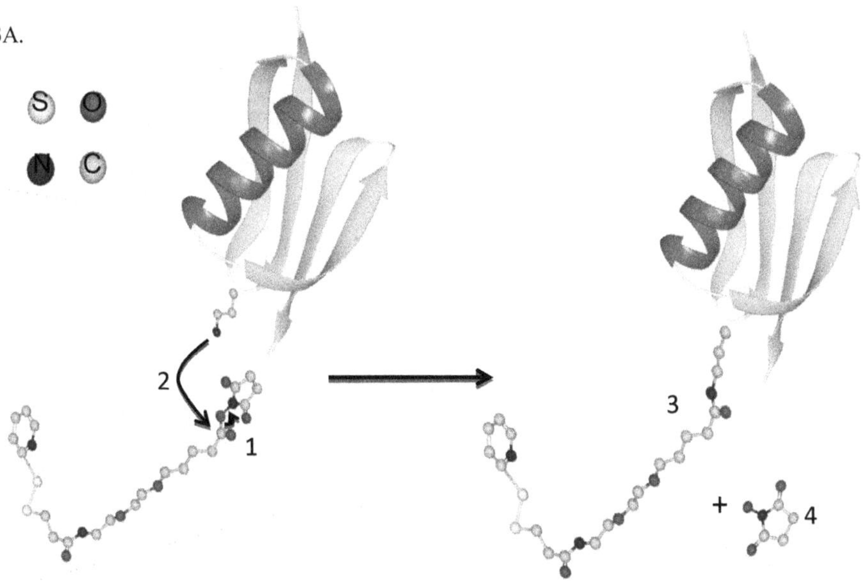

Fig. 13.3 (A). Step by step reaction of PEG-ProG conjucation. (1) Cleaving of the SVA group, (2) bond formation starting at activated carboxylic terminus, (3) formation of Peg-ProG, and (4) SVA group is free floating. ProG ribbon diagram represents one Ab binding domain of ProG.

13.2.3. ELISA on GNS to Determine the Number of Anti-Rb on Surface

An ELISA was carried out using a modification of the Lowery-Gobin quantification method [1]. Briefly, GNS-PEG-ProG-Anti-Rb was fabricated as mentioned above; next, the synthesized particles were washed and resuspended in PBS. For a negative control, GNS blocked with PEG-SH (5000 Da, LaysanBio) was used. Next, the samples are blocked with 3 % PBSA (3 % BSA in PBS), for preventing unspecific binding reaction, and washed afterwards via centrifugation (650 G for 10 min). Then, suspensions of nanoshells were incubated with Anti-Goat HRP (horseradish peroxidase labeled, A9452, Sigma) for 1 hr at room temperature; the final concentration of Anti-Goat HRP was 100 µg/ml. After the incubation of Anti-Goat IgG Ab-HRP, samples were washed by centrifugation twice (650 G for 10 min) to remove any unbound secondary antibody.

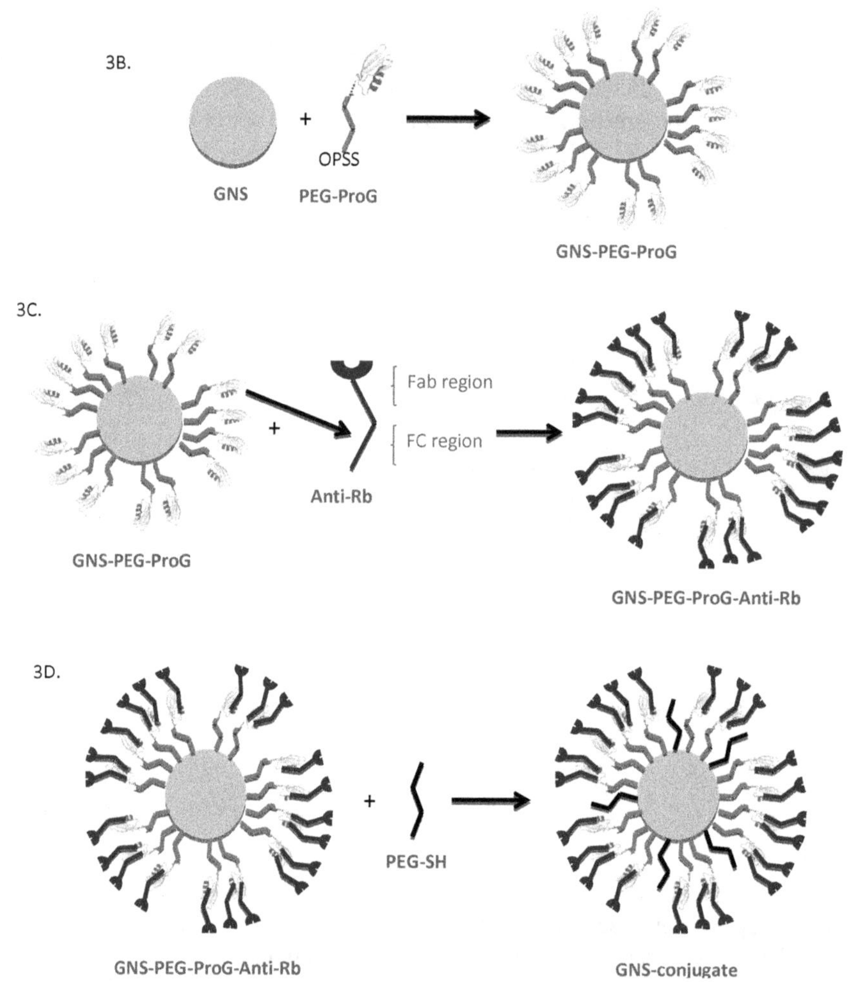

Fig. 13.3. (B, C, D) A step by step schematic of the GNS-PEG-ProG-Anti-Rb conjugation: (3B) GNS and PEG-ProG are mixedat a targeted mole ratio of 2000:1 (PEG-ProG:GNS). (3C) Anti-Rb was added to GNS-PEG-ProG conjugate at targeted mole ratio of 5000:1 (Anti-Rb:GNS). (32D) PEG is added to back fill “empty” spots on the surface of GNS.

Once the second centrifugation was completed, supernatant and GNS pellet were split for antibody quantification (GNS as positive control, supernatant as a negative control). Final GNS conjugates and supernatant, including PEG samples, were then reacted with

3,3′,5,5′-Tetramethylbenzidine (TMB) substrate with fresh hydrogen peroxide for maximum of 15 min; the reaction was stopped with 2M H_2SO_4 solution and read using BioTek ELx800 plate reader at 450 nm. Antibody standards were made using serial dilution at concentrations ranging from 2 ng/ml to 100 ng/ml. Prior to running the TMB assay, it was important to determine the reaction rate of the TMB development for both standards and unknown samples (for both GNS and PEG, positive and negative control) so that the best dilutions of samples and standard curve can be obtained. This was done using a diluted sample of the GNS and PEG only (including positive and negative control), and a high valued concentration.

13.2.4. Detection of Antigen/IgG Complex

GNS-PEG-ProG-Anti-Rb were fabricated as detailed above and blocked with PEG-SH to resist non-specific protein adsorption. The GNS-conjugates were washed and resuspended in PBS prior to use. Next, in separate 2 mL microcentrifuge tubes, IgG from rabbit serum (I5006 lyophilized powder, Sigma) was dissolved and diluted to 50, 5, 0.5, 0.05, and 0.005 ng/mL in PBS and stored at 4° C until needed. GNS-conjugates were aliquoted in a quartz cuvette and diluted to 1 OD using DI water, PBS, or Serum (FBS), and with 100 μL of the IgG, at different concentration as mentioned above, bringing the final concentration of IgG to 5, 0.5, 0.05, 0.005, and 0.0005 ng/mL (0.5 pg/mL). Preparing the solutions in this manner allows the concentration of GNS to remain at 1 OD. Samples of GNS and antigen were mixed and spectral scans of each sample were collected every 5 min for 1 hr (also referred to as kinetic scans). Once the particles are exposed to the IgG, they will start forming particles dimers or higher order aggregates, see Fig. 13.4. This antigen GNS interaction causes the nIR absorbance of the GNS to drop drastically.

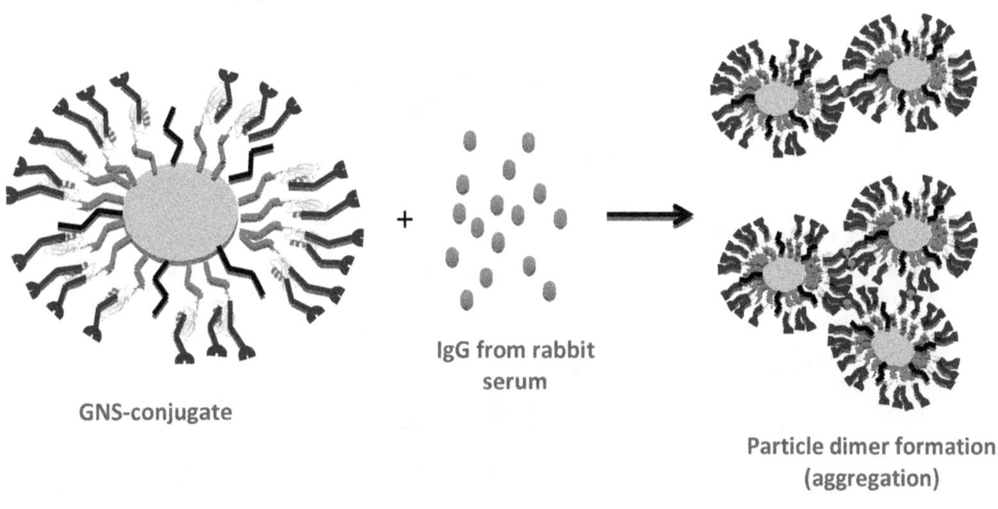

Fig. 13.4. Representation of particle aggregation when GNS-conjugate is exposed to various antigens.

13.2.5. Surface ELISA using GNS-Conjugate

Goat-Anti-Rabbit-IgG-Biotin (B8895, Sigma) was incubated on Streptavidin (SuperStreptavidin – SMS, Arrayit) coated glass slides for 1.5 hours with in a multi-well microarray. Next, the wells were blocked with SuperStreptavidin Blocking buffer (SBB, Arrayit) for 1 hour, followed by incubation of IgG from rabbit serum at various concentrations (in PBS or FBS) for 1 hr. The GNS-conjugate, GNS-PEG-ProG-Anti-Rb, was fabricated as mentioned above and blocked with PEG-SH, then washed and stored in PBS or FBS. Next, GNS-conjugate was incubated for 30 min on wells containing IgG from rabbit serum; a PEG-GNS without antibodies and blank controls were also used. The reaction scheme is shown in Fig. 13.5. After incubation, sample wells were then washed with PBS and a silver enhancer stain was performed. Slides were imaged and the grayscale intensity measurements were quantified using Nikon NIS viewer.

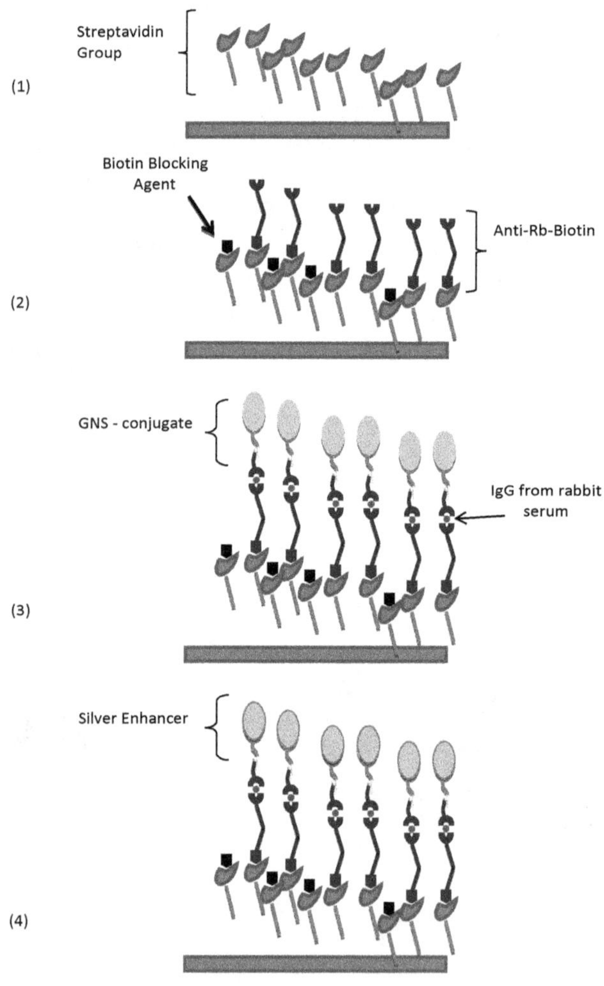

Fig. 13.5. Representation of the surface ELISA using Streptavidin, Biotin-Anti-Rb, GNS-Conjugate, and silver enhancer.

13.3. Results and Discussion

The goal of this research was synthesize GNS as biosensors to be used for detection of different biomarkers using the unique optical, chemical, and physical properties. These properties provide a great opportunity for advancement of biosensors in LOC platforms.

13.3.1. Assessment of GNS-PEG-ProG-Anti-Rb Conjugation

After each step of the GNS assembly, NIR peak, dynamic light scattering (DLS), and zeta potential (charge) measurements were taken to assess size and surface properties. From Table 13.2, we can see the size change from 195.2 nm to 241.2 nm, which is due to the addition of PEG-ProG, Anti-Rb, and PEG-SH, while the NIR spectral profile showing shows minimal change. The completeness of the gold shell on the silica core, with addition of APTES, colloids and finally shell formation (Fig. 13.6), can be correlated to the line width of the nIR spectra, Fig. 13.7. Incomplete shell formation results in a broadening of the nIR spectral features. To elucidate the changes occurring during synthesis and separation steps, a ratio of the absorbance of peaks, the absorbance values in the nIR region relative to the colloidal gold peak (~530 nm), is defined as a relative measure of the fraction of nIR particles and the overall purity of the process.

Table 13.2. Characteristic parameter shifts of particle size and zeta potential by the ProG-Ab conjugated GNS. Bare and conjugated GNS were suspended in K_2CO_3 and DI water, respectively.

GNS	Spectra (λ_{max})	Size (in K_2CO_3, nm)	Zeta (in K_2CO_3, mV)	Size (in DI, nm)	Zeta (in DI, mV)
Bare	814	195.2	-61.6	196.3	-32.7
ProG	816	204.4	-57.8	209.6	-33.2
ProG-Ab	823	239.6	-29.5	-	-
ProG-Ab-PEG	818	241.2	-27.2	239.2	2.09

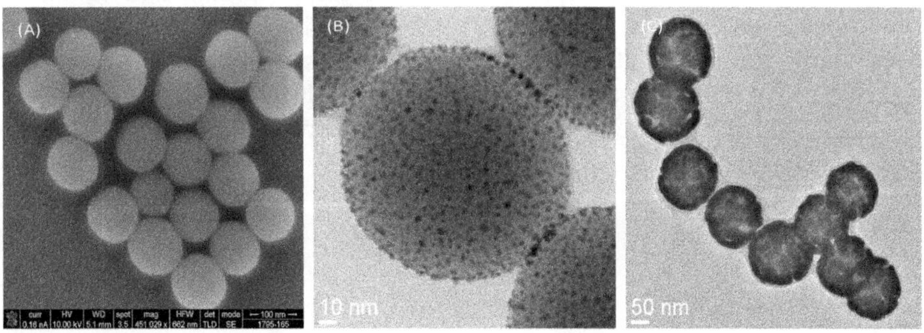

Fig. 13.6. (A) Silica cores with functionalized with APTES. (B) Silica cores with gold colloied attachedto the surface, and (C) after formaldehyde reduction and completion of gold shell.

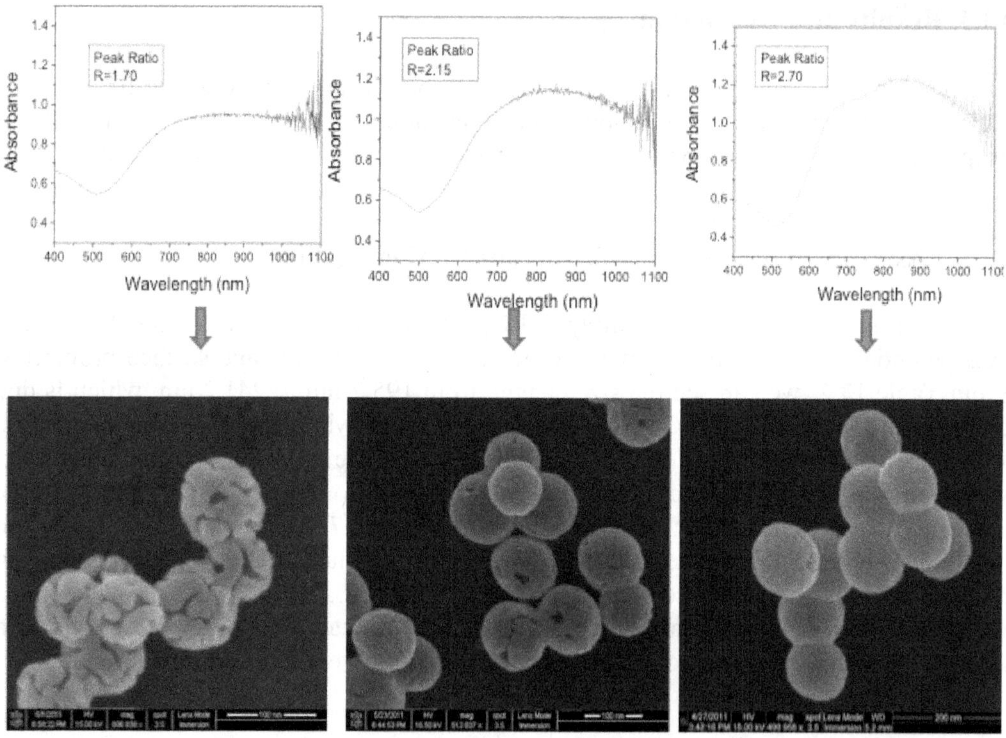

Fig. 13.7. Quality of GNS shell determined using spectral profile and TEM of different batches. Image above displays the quality of GNS from worst to best (left to right). Due to the "holes" or the "incompleteness" of the GNS shell the spectral profile will broaden out.

13.3.2. Quantification on GNS to Determine the Number of Active Antibodies on the Particle Surface Using an ELISA

Due to the ability of ProG to capture multiple antibodies per unit protein, there was a significantly higher amount of antibodies bound on the surface of fully assembled GNS compared to previously reported data and to PEG controls. Additionally it is hypothesized that the ProG allowed the antibodies that were captured to maintain a preferred orientation to allow more effective binding with the target antigen surfaces. Prior to running the ELISA, GNS-PEG-ProG-Anti-Rb samples were assembled using different molecular ratios of antibody to particles: 1000:1, 3000:1, and 5000:1. The 5000:1 ratio resulted in approximately 311 antibodies per GNS, from Fig. 13.8. The antibody density reported here is twice that obtained by methods reported previously ($p < 0.00001$) [1, 2].

13.3.3. Detection of Analyte Complex

The GNS conjugate, GNS-PEG-ProG-Anti-Rb, was fabricated as detailed above and blocked with PEG-SH, washed and stored in PBS or FBS. These initial studies are focused on the ability to detect analyte in the range of 5 ng/mL-0.5 pg/mL in saline solutions, using GNS-PEG-ProG-Anti-Rb, within 10 min of mixing.

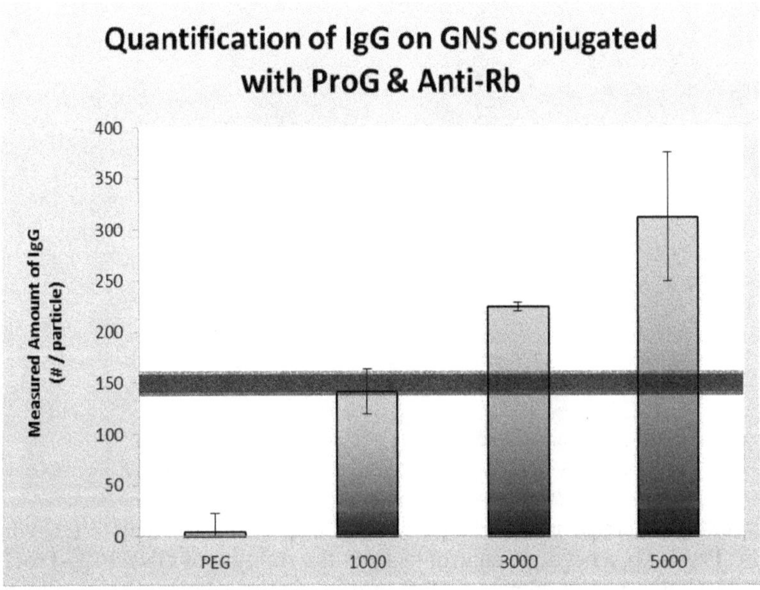

Fig. 13.8. ELISA-based Lowery-Gobin quantification method to determine the number of antibodies on GNS with different antibody to GNS ratios (1000, 3000, 5000:1), red bar represents the amount of antibodies on GNS reported previously [1, 2].

The analyte detection assay was performed using kinetic scans for five different analyte concentrations in PBS; Fig. 13.9 indicates the percent decrease in absorbance of GNS, from t=0 to 65 min, in presence of analytes, compared to analyte-free samples which shows. Little to no aggregation of the nanoshell assemblies was noted for the control groups. The aggregation or formation of multi-mer particle complexes was measured at 820 nm. For samples containing analyte, however, the assay demonstrates rapid aggregation kinetics at early times in a concentration dependent manner. STEM images were taken of GNS-PEG-ProG-Anti-Rb after exposure to 5 ng/mL of antigen and compared to the same sample without antigen present, Fig. 13.10. Samples exposed to antigen show a marked increase in aggregation compared to control groups. All five analytes concentrations were statistically different compared to control samples after 10 min of the kinetic scan, $p < 0.05$.

In order to detect the analytes in serum, the GNS conjugate GNS-PEG-ProG-Anti-Rb was fabricated as detailed above and blocked with PEG-SH. The GNS-conjugates were then washed and resuspended in 10 %, 30 %, or 50 % FBS prior to performing the assay. The analyte, IgG from rabbit serum, was dissolved and diluted to 50, 5, 0.5, 0.05, and 0.005 ng/mL in 10 %, 30 %, or 50 % FBS and stored at 4° C until needed.

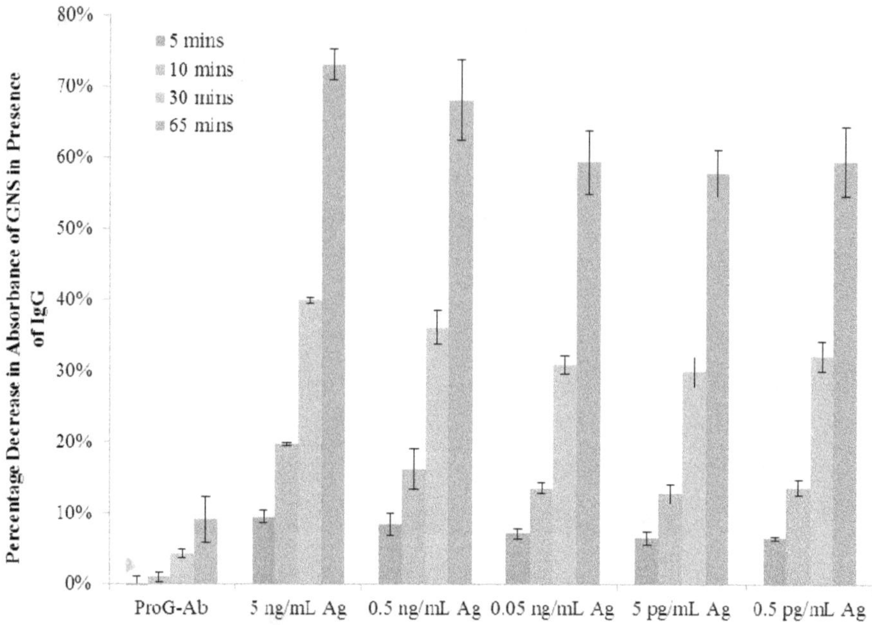

Fig. 13.9. Percentage decrease in absorbance of GNS in presence of Analyte/IgG complex. From left to right: ProG-Ab, a negative control to show the stability of GNS-PEG-ProG-Anti-Rb conjugate. ProG-Ab+5...0.0005 ng/mL (0.5 pg/mL), the sample incubated with analyte and kinetic change was observed over an hour.

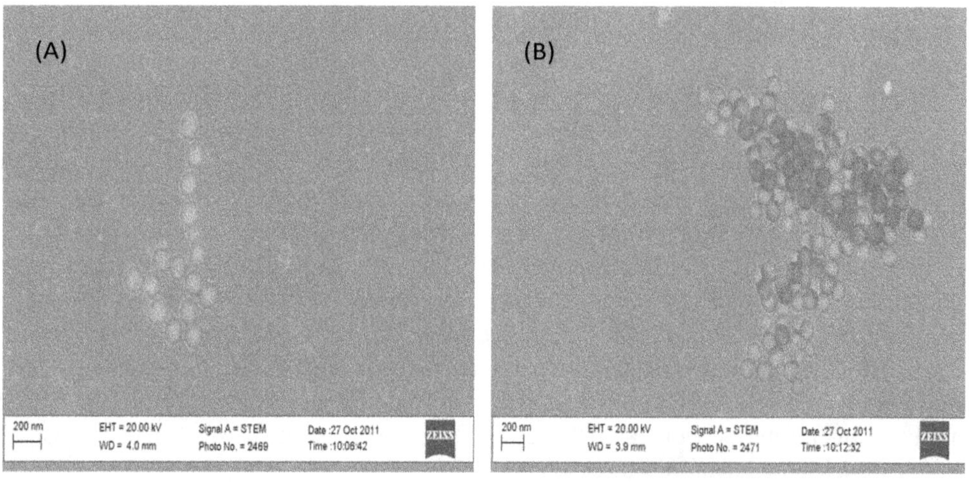

Fig. 13.10. STEM images taken of GNS-PEG-ProG-Anti-Rb with (A) ProG-Ab no antigen, and (B) with 5 ng/mL of antigen after the liquid detection assay.

GNS-conjugates were placed in a quartz cuvette and diluted to 1 OD using 10 %, 30 %, or 50 % FBS, and with 100 μL of the IgG, at different concentration as mentioned above, to bring the final concentration of IgG to 5, 0.5, 0.05, 0.005, and 0.0005 ng/mL

(0.5 pg/mL). GNS concentrations were maintained at 1 OD. Samples were mixed and spectral scans of each sample were collected every 5 min for 1 hr. No successful detection events were noted during the serum studies. It is hypothesized that proteins present in the serum adsorbed to the surface of the nanoshell assemblies and prevented detection events due to steric crowding around the antigen binding sites. This is especially likely as literature reports suggest a high affinity to protein G was found for serum albumins from rat, man, and mouse; while a medium affinity was found from rabbit, cow, hen, and horse [50]. In an attempt to prevent non-specific protein adsorption, the serum was mixed with the non-ionic surfactant Tween 20. Studies in the presence of Tween 20 showed results similar to the prior studies, with no significant signs of antigen binding events.

13.3.4. Assessment of Surface ELISA

In an attempt to overcome the lack of detection events with the GNS-PEG-ProG-Anti-Rb in serum, we adapted the GNS conjugate system to work in a surface-based sandwich ELISA assay. It is known that such assay systems can utilize gold nanoparticles as tags or labels to allow for detection of antigens or viral particles in small sample volumes in a very sensitive and specific environment. It was hoped that this system could be used as an advanced early detection system, enabling accurate readings using lab-on-chip or microfluidic device coupled with sandwich ELISA. Specifically, a standardized sandwich-ELISA system using silver stain enhancement with targeted gold nanoparticles for cost effective diagnostics can be developed for detecting viral nanoparticles. The analyte detection system is very quick, sensitive, and has more manipulability when certain nanoparticles are used [51].

As mentioned previously, Goat-Anti-Rabbit-IgG-Biotin was incubated on a Streptavidin coated glass slides for 1.5 hrs with multi-well microarray. Next, the wells were blocked with SuperStreptavidin Blocking buffer for 1 hr, followed by incubation of IgG from rabbit serum at various concentrations, in PBS or FBS for 1 hr. The GNS conjugate, GNS-PEG-ProG-Anti-Rb, was fabricated as mentioned above and blocked with PEG-SH, washed and stored in PBS or FBS. Next, GNS-conjugate was incubated for 30 min to wells with IgG from rabbit serum, a PEG-GNS and blank controls were also used. Sample wells were then washed and a silver stain was performed. The slides were then imaged and the grayscale intensity measurements were analyzed using Nikon NIS viewer. The image analysis results are shown in Fig. 13.11. By moving from a solution-based system to a surface based sandwich ELISA system analyte detection was possible in both PBS and in the presence of serum proteins down to 0.5 pg/mL analyte concentrations.

13.4. Conclusion and Future

ELISA based systems have been used in medical virology for detection of antibodies, immunoglobulins to viruses, viral antigens or nanoparticles. Traditional ELISA based detection systems have several limitations. First, antibodies used are generally laid on various types of surfaces in a random orientation, which makes the protein markers, viruses, viral antigens and nanoparticles inaccessible to antibody binding sites, reducing the effectiveness of the assay. Second, low ratios of antibodies to viral vectors may result in limited sensitivity. Third, a viral detector with many modifications can lead to inefficiency of its function and it is a time consuming process. By slight modification to standard ELISA systems with GNS, we have demonstrated a significant increase in detection sensitivity. The achievements were made by: (a) increased exposure of antibody to the binding regions due to the use of protein G during surface modifications, (b) amplification of signal due to GNS and Silver Stain, and (c) the small size of GNS functionalized with PEG-Ab conjugation targeting only specific viral particles/antigen. These factors combine to produce a highly sensitive detection system based on antibody-antigen/viral particle binding which can detect analytes at concentrations as low as 0.5 pg/mL. Future direction for the project includes incorporating the detection system into a microfluidic system with fiber-optics / LED (Fig. 13.12).

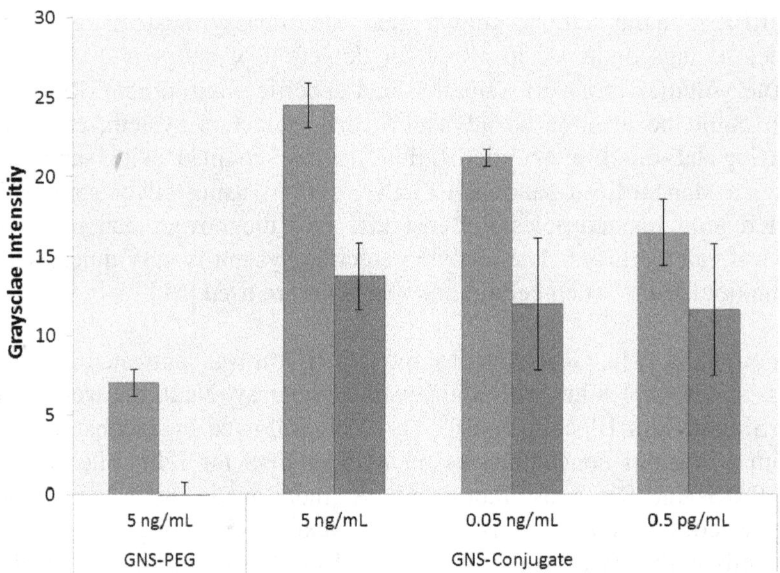

Fig. 13.11. Grayscale intensity results of GNS-PEG-ProG-Anti-Rb sandwich ELISA assay. Blue and Red bars represent samples incubated in PBS and FBS, respectively.

The initial concept of the LOC device for a solution-based detection assay includes five regions: (1) multiple inlets for insertion of GNS-conjugates and analytes, (2) droplet generation, allows the device to produce droplets to be introduced to (3) mixing region. Next, in order to combine the droplets, (4) merging region [52] is added to the device. Lastly, (5) imaging/optical analysis region to measure the spectral profile of the particles or the grayscale intensity after the device has been modified for Elisa steps.

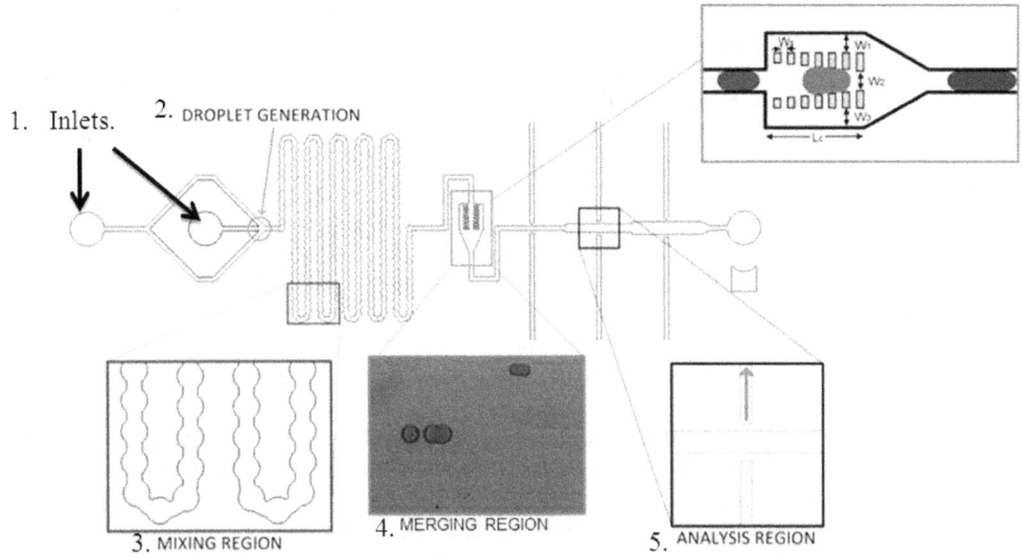

Fig. 13.12. Concept image of LOC device for a solution-based detection assay.

References

[1]. Lowery, A. R., et al., Immunonanoshells for targeted photothermal ablation of tumor cells, *Int J. Nanomedicine*, 1, 2, 2006, pp. 149-54.

[2]. Gobin, A. M., et al., Near-Infrared Resonant Nanoshells for Combined Optical Imaging and Photothermal Cancer Therapy, *Nano Letters*, 7, 7, 2007, pp. 1929-1934.

[3]. Guodong Liu, J. W., Yuehe Lin, and Joseph Wang, Nanoparticle-based biosensors and bioassays, in Electrochemical Sensors, Biosensors and Their Biomedical Applications, H. J. Xueji Zhang, Joseph Wang, (Eds.), *Elsevier*, 2008, pp. 441-457.

[4]. Vivek Babu Kandimalla, V. S. T., and Huangxian Ju, Biosensors based on immobilization of biomolecules in sol-gel matrices, in Electrochemical Sensors, Biosensors and Their Biomedical Applications, H. J. Xueji Zhang, Joseph Wang, (Eds.), *Elsevier*, 2008, pp. 503-529.

[5]. Kandimalla, V. B., V. S. Tripathi, and H. Ju, Immobilization of Biomolecules in Sol–Gels: Biological and Analytical Applications, *Critical Reviews in Analytical Chemistry*, 36, 2, 2006, pp. 73-106.

[6]. Wang, B., et al., Amperometric Glucose Biosensor Based on Sol−Gel Organic−Inorganic Hybrid Material, *Analytical Chemistry*, 70, 15, 1998, pp. 3170-3174.

[7]. James Gimzewski, J. R., Michael Teitell, Gordon Malan, Immunological Biosensor, in The Immunoassay Handbook, D. Wild (Ed.), *Elsevier,* 2005, pp. 265-280.

[8]. Liu, J. and Y. Lu, Preparation of aptamer-linked gold nanoparticle purple aggregates for colorimetric sensing of analytes, *Nat Protoc*, 2006. 1, 1, pp. 246-52.

[9]. R Shantilatha, Shailly Varma and Chanchal K. Mitra., Designing a Simple Biosensor, in Advances in Biosensors: Perspectives in Biosensors, Bansi Dhar Malhotra, Anthony Turner (Eds.), *JAI Press*, 2003, pp. 1-36.

[10]. Lowe, C. R., Overview of Biosensor and Bioarray Technologies, in Handbook of Biosensors and Biochips, C. R. L. Robert S. Marks, David C. Cullen, Howard H. Weetall, Isao Karube, (Eds.), *Wiley*, 2007, pp. 1-16.

[11]. Vo-Dinh, T. and B. Cullum, Biosensors and biochips: advances in biological and medical diagnostics, *J Anal Chem*, 366, 6-7, 2000, pp. 540-51.

[12]. Cruz, H. J., C. C. Rosa, and A. G. Oliva, Immunosensors for diagnostic applications, *Parasitol Res*, 88, 13, Suppl. 1, 2002, pp. S4-7.

[13]. Luppa, P. B., L. J. Sokoll, and D. W. Chan, Immunosensors--principles and applications to clinical chemistry, *Clin Chim Acta*, 314, 1-2, 2001, pp. 1-26.

[14]. Erickson, T. A., Tunnell, J. W., Gold Nanoshells in Biomedical Applications, in Nanotechnologies for the Life Sciences, *Wiley-VCH Verlag GmbH & Co. KGaA*, 2007.

[15]. Caruso, F., Nanoengineering of particle surfaces, *Advanced Materials*, 13, 1, 2001, p. 11.

[16]. Storhoff, J. J. and C. A. Mirkin, Programmed Materials Synthesis with DNA, *Chem Rev*, 99, 7, 1999, pp. 1849-1862.

[17]. Willner, I. and B. Willner, Functional nanoparticle architectures for sensoric, optoelectronic, and bioelectronic applications, *Pure and Applied Chemistry*, 74, 9, 2002, pp. 1773-1783.

[18]. Katz, E., I. Willner, and J. Wang, Electroanalytical and Bioelectroanalytical Systems Based on Metal and Semiconductor Nanoparticles, *Electroanalysis*, 16, 1-2, 2004, pp. 19-44.

[19]. Katz, E. and I. Willner, Integrated nanoparticle-biomolecule hybrid systems: synthesis, properties, and applications, *Angew Chem Int Ed Engl*, 43, 45, 2004, pp. 6042-108.

[20]. Rosi, N. L. and C. A. Mirkin, Nanostructures in biodiagnostics, *Chem Rev*, 105, 4, 2005, pp. 1547-62.

[21]. Lin, Y.-C., et al., Tailoring the surface potential of gold nanoparticles with self-assembled monolayers with mixed functional groups, *Journal of Colloid and Interface Science*, 340, 1, 2009, pp. 126-130.

[22]. Castellana, E. T. and D. H. Russell, Tailoring Nanoparticle Surface Chemistry to Enhance Laser Desorption Ionization of Peptides and Proteins, *Nano Letters*, 7, 10, 2007, pp. 3023-3025.

[23]. Wei, W., H. Quanguo, and C. Hong, Silane Bridged Surface Tailoring on Magnetite Nanoparticles, in *Proceedings of the 1st International Conference on Bioinformatics and Biomedical Engineering (ICBBE, 07)*, 6-8 July 2007, pp.76-79.

[24]. Karagoz, B., et al., Novel strategy for tailoring of SiO2 and TiO2 nanoparticle surfaces with poly(ε-caprolactone), *Colloid and Polymer Science*, 288, 5, 2010, pp. 535-542.

[25]. Niemeyer, C. M., Nanoparticles, Proteins, and Nucleic Acids: Biotechnology Meets Materials Science, *Angewandte Chemie International Edition*, 40, 22, 2001, pp. 4128-4158.

[26]. Ha, T. H. J., S. O., Lee, J. M., Lee, K. Y., Lee, Y., Park, J. S., Chung, B. H., Oriented immobilization of antibodies with GST-fused multiple F-c-specific B-domains on a gold surface, *Anal. Chem.*, 79, 2, 2007, pp. 546-556.

[27]. Redon, C. E., et al., gamma-H2AX as a biomarker of DNA damage induced by ionizing radiation in human peripheral blood lymphocytes and artificial skin, *Adv Space Res*, 43, 8, 2009, pp. 1171-1178.

[28]. Durante, M., Biomarkers of space radiation risk, *Radiat Res.*, 164, 4, 2005, pp. 467-73.

[29]. Marchetti, F., et al., Candidate protein biodosimeters of human exposure to ionizing radiation, *Int J Radiat Biol.* 2006, 82, 9, pp. 605-639.

[30]. Blakely, W. F., et al., Development and Validation of Radiation-Responsive Protein Bioassays for Biodosimetry Applications, in Proceedings of the *Radiation and Bioeffects and Countermeasures Meeting*, Bethesda, MD, USA, 21-23 June 2005, pp. 3-1-13-11.

[31]. Gobin, A. M., Biomedical Applications of Metal Nanoshells, in Methods in Bioengineering: Nanoscale Bioengineering and Nanomedicine, I. L. M. Kaushal Rege, Editor, *Artech House*, 2009, pp. 153-167.

[32]. Hirsch, L. R., et al., A whole blood immunoassay using gold nanoshells, *Anal Chem*, 75, 10, 2003, pp. 2377-81.

[33]. Sershen, S. R., et al., Temperature-sensitive polymer-nanoshell composites for photothermally modulated drug delivery, *J Biomed Mater Res*, 51, 3, 2000, pp. 293-8.
[34]. Bikram, M., et al., Temperature-sensitive hydrogels with SiO2–Au nanoshells for controlled drug delivery, *Journal of Controlled Release*, 123, 3, 2007, pp. 219-227.
[35]. Gobin, A. M., et al., Near-Infrared-Resonant Gold/Gold Sulfide Nanoparticles as a Photothermal Cancer Therapeutic Agent, *Small*, 6, 6, 2010, pp. 745-752.
[36]. Abraham, A. M., R. Kannangai, and G. Sridharan, Nanotechnology: a new frontier in virus detection in clinical practice, *Indian J Med Microbiol*, 26, 4, 2008, pp. 297-301.
[37]. Sun, X., et al., Targeted cancer therapy by immunoconjugated gold-gold sulfide nanoparticles using Protein G as a cofactor, *Ann Biomed Eng*, 40, 10, 2012, pp. 2131-9.
[38]. Zhang, G., et al., Gold/Chitosan Nanocomposites with Specific Near Infrared Absorption for Photothermal Therapy Applications, *Journal of Nanomaterials*, 2012, 2012, p. 9.
[39]. Guss, B., et al., Structure of the IgG-binding regions of streptococcal protein G, *EMBO J*, 5, 7, 1986, pp. 1567-75.
[40]. Day, E. S., et al., Antibody-conjugated gold-gold sulfide nanoparticles as multifunctional agents for imaging and therapy of breast cancer, *Int J Nanomedicine*, 5, 2010, p. 445-54.
[41]. Huang, X., et al., Gold nanoparticles: interesting optical properties and recent applications in cancer diagnostics and therapy, *Nanomedicine (Lond),* 2, 5, 2007, pp. 681-93.
[42]. Sun, X., et al., Targeted Cancer Therapy by Immunoconjugated Gold-Gold Sulfide Nanoparticles Using Protein G as a Cofactor, *Ann Biomed Eng.,* 40, 10, 2012, pp. 2131-2139.
[43]. Ekins, R. and F. Chu, Immunoassay and other ligand assays: present status and future trends, *J Int Fed Clin Chem*, 1997. 9, 3, pp. 100-9.
[44]. Wang, K. Y., et al., Multiplexed immunoassay: quantitation and profiling of serum biomarkers using magnetic nanoprobes and MALDI-TOF MS, *Anal Chem*, 80, 16, 2008, pp. 6159-67.
[45]. Weiming Zheng, L. H., Multiplexed Immunoassays, in Advances in Immunoassay Technology, T. K. C. Norman H. L. Chiu (Ed.), *InTech,* 2012, pp. 143-163.
[46]. Stevens, C. D., Labeled Immunoassays, in Clinical Immunology & Serology, A Laboratory Perspective, *F. A. Davis Company*, 2010, pp. 152-165.
[47]. Chen, J., C. Wang, and J. Irudayaraj, Ultrasensitive protein detection in blood serum using gold nanoparticle probes by single molecule spectroscopy, *J Biomed Opt*, 14, 4, 2009, pp. 040501.
[48]. Stober, W., Fink, A., Bohn, E., Controlled Growth of Monodisperse Silica Spheres in the Micron Size Range, *Journal of Colloid and Interface Science,* 26, 1968, pp. 62-69.
[49]. Duff, D. G. B., A., A New Hydrosol of Gold Clusters. 1. Formation and Particle Size Variation. *Langmuir,* 1993, 9, 2301.
[50]. Nygren, P. A., et al., Species-dependent binding of serum albumins to the streptococcal receptor protein G, *Eur J Biochem*, 193, 1, 1990, pp. 143-8.
[51]. Abraham, A. M., R. Kannangai, and G. Sridharan, Nanotechnology: a new frontier in virus detection in clinical practice, *Indian journal of medical microbiology*, 26, 4, 2008, pp. 297-301.
[523]. Niu, X., et al., Pillar-induced droplet merging in microfluidic circuits, *Lab on a Chip*, 8, 11, 2008, pp. 1837-1841.

Chapter 14
Anchoring Materials for Ultra-Sensitive Biosensors Modified with Au Nanoparticles and Enzymes

Solomon W. Leung, David Assan and James C. K. Lai

14.1. Introduction

We have previously developed some of the most sensitive biosensors/electrodes modified by Au nanoparticles and enzymes that are capable of detecting concentration levels below ppb [3]. These sensors were fabricated with composite layers of nanomaterials and enzymes anchored on conductive but non-reactive materials, such as glassy carbon and Pt. Performance of these sensors varies depending on the anchoring materials and as well as the composition of the biocomposite materials.

The important factors considered in modern era of chemical species detection are response time and concentration limit. Due to the advances of computer and nanotechnology, the fourth generation of sensor (with combinations of enzymes and composite materials) should have almost instantaneous response time and ultra-high detection sensitivity. The response time of early first and second generation biosensors were slow in response, measurement time could be in the order of minutes [1]. The more recent biosensors have much improved response time and detection limits, Table 14.1 in the following lists some of the Au nanoparticle-based biosensors (third and fourth generation) that are related to the research topic of this chapter, and their performance:

In this research project, we examined the detection limits and performance consistency of an ultra-high sensitive biosensor platform that was anchored by various conducting materials: Au, Ag, Pt and glassy carbon. These electrode sensors coated with various biocomposite materials were then challenged by various biological and environmental samples for their performance. The target species included lactate, ammonia, NO_2^-, and

Solomon W. Leung
Civil and Environmental Engineering Department, School of Engineering and Biomedical Research Institute, Idaho State University, Pocatello, ID 83209, USA

peroxide. The time duration for the testing of performance for the anchoring materials varied from one (freshly prepared) to 28 days.

Table 14.1. Gold nanoparticle-based electrochemical enzyme biosensors [2]

Enzyme(s) / Electrode	Immobilization mode	Detection	Life-Time	Analyte / Sample	Analytical Characte-ristics
GOx/gold	Covalent attachment of GOx to a nAu monolayer-modified Au E	Amperom. (E = 0.3 V vs. SCE)	$K^{app}{}_{m}$= 4.3 mM	Glucose	Linear range 2.0×10^{-5} - 5.7×10^{-3} M; LOD:8.2 µM
GOx/GCE	GOx and the redox mediator TTF coimmobilized by cross-linking with glutaraldehyde on gold-modified electrodes with either Cyst or MPA monolayers	Amperom. (E = 0.2 V)	Useful lifetime: 28 days 0.05M PBS, pH 7.4	Glucose	Linear range: 0.01–10 mM; LOD: 0.7×10^{-5} M
GOx/platinum	Layer-by-layer assembled chitosan/nAu/GOx multilayer films	Amperom. (E = 600 mV)	$K^{app}{}_{M}$= 10.5 mM, 0.1 M PBS, pH 6.8	Glucose /human serum	Linear range: 0.5–32 mM
HRP/gold	Au colloids associated with a cysteamine monolayer on gold electrode surface. HRP immob. on colloidal gold	Amperom. (E =–0.3 V)	$K^{app}{}_{M}$= 2.3 mM, R.S.D.= 3 % (n = 10), 0.1 M PBS pH 7	H_2O_2	Linear range: 0.0014–2.8 mM; LOD:0.58 µM
HRP/GCE	Covalent incorp. of CNTs and nAu onto poly(thionine) film deposited by electropolymerization on GCE. HRP immob. by incubation in 15 mgmL–1 HRP solution overnight at 4 °C	Amperom. (E = 0.4 V vs. SCE)	0.1 M PBS, pH 7.0	H_2O_2	Linear range: 5.0×10^{-6}– 7.0×10^{-3} M; LOD: 3.0×10^{-6} M
HRP/platinum	Polythionine (mediator) embedded in Nafion film on the surface of platinum electrode. HRP bound by gold nanoparticles and gelatin coated on the Nafion-modified electrode	Chronoamp	R.S.D.: 2.2 % (n=5)	H_2O_2	Linear range: 0.05–30.6 mM; LOD: 0.02 mM

Abbreviations: AChE: acetylcholinesterase, ADH: alcohol dehydrogenase, Amperom.: amperometry; APTMS: (3-aminopropyl)trimethoxysilane, ASV: anodic stripping voltammetry, AuE: gold electrode; chronoamp.: chronoamperometry, CPE: carbon paste electrode, CNT: carbon nanotubes, cov.: covalently, CPE: carbon paste electrode, CV: cyclic voltammetry, DPV: differential pulse voltammetry, EIS: electrochemical impedance spectroscopy, Electrochem.:

electrochemically, GC: glassy carbon, GCE: glassy carbon electrode, GCPE: glassy carbon paste electrode, GOx: glucose oxidase, Glu: glutaraldehyde, HRP: peroxidase, immob.: immobilization, incorp.: incorporation; ITO: indium tin oxide, KappM : apparent Michaelis–Menten constant, LOD: limit of detection, MP-11: microperoxidase 11, MPTS: (3-mercaptopropyl)trimethoxysilane, MWNT: multiwalled carbon nanotube, nAu: gold nanoparticles, p-ABSA: p-aminobenzene sulfonic acid, PBS, phosphate buffer solution; PDDA: poly(diallyldimethylammonium chloride), PVB: polivinil-butiral, PVS: poly(vinylsulfonic acid), SCE: saturated calomel electrode, TMB: tetramethyl benzidine, TTF: tetrathiafulvalene; Tyr: tyrosinase, XOD: xanthine oxidase.

14.1.1. Principle of Biosensors

An illustration of detection principle by means of an enzymatic reaction of biosensing is described in the following: glutamate and NAD^+ can be hydrolyzed to form α-ketoglutarate, NADH, and ammonium ion with the enzyme, glutamate dehydrogenase (GDH). The equilibrium constant is in favor of the formation of glutamate and thus the reverse reaction is faster kinetically:

$$\alpha\text{-Ketoglutarate} + NH_4^+ + NADH + H^+ \xleftrightarrow{GDH} \text{Glutamate} + NAD^+ + H_2O \quad (14.1)$$

Other enzymatic reactions with similar detecting principle included in this study were lactate dehydrogenase (LDH) for the detection of lactate, and hemoglobin for the detection of H_2O_2 and nitrite. The following are the governing redox reactions catalyzed by the respective enzymes.

LDH:

$$\text{Lactate} + NAD^+ \xleftrightarrow{LHD} \text{pyruvate} + NADH + 2H^+ + 2e^- \quad (14.2)$$

Hemoglobin (Hb):

$$H_2O_2 \xrightarrow{Hemoglobin} H_2O + O_2 \quad (14.3)$$

$$H_2O + NO_2^- \xrightarrow{Hemoglobin} NO_3^- + H_2 \quad (14.4)$$

A layer of biocomposite material comprised either one the above enzyme was deposited on the surface of the four anchoring electrode materials. A specified enzyme was used depending on the analyte to be detected and the reactions were described in equations (14.1) to (14.4). For Equation (14.3) and (14.4), the end products are likely to be different within a biological system than the testing solutions as shown in the equations, the reactions would be reversible and the electron(s) generated with the peroxide and nitrite would have reacted with other electron receptors to form metabolite(s). Equations (14.1) to (14.4) are important to biomedical systems, thus the measurements of any of the chemical species involved; in addition, NH_4^+, H_2O_2, NO_2^- are important species for the monitoring of environmental health. Hence, the biosensors that are being developed here can be valuable in many applications in both biomedical and environmental science.

14.2. Materials and Method

14.2.1. Electrodes

Au, Ag, Pt, and glassy carbon (GCE) electrode all had diameter of 0.2 cm. The platinum counter electrode had diameter of 0.1 cm and length of 5 cm. They all were purchased from Tianjin Aida Heng Sheng Co, Tianjin, China. These electrodes were coated with either GDH, LDH, or hemoglobin depending on the target chemicals to be detected [3].

14.2.2. Materials

All enzymes and biochemicals were purchased from Sigma-Aldrich Chemical Co, St. Louis, MO, USA. Cysteamine and α-ketoglutarate, $AuCl_3HCl{\cdot}4H_2O$ (Au %> 48 %) and Na_3citrate were purchased from Sigma. All the other chemicals were of analytical grade or highest grade available. All the experiments were carried out under deoxygenated condition in 0.1 M phosphate buffer solution at pH 7.0 or otherwise specified.

14.2.3. Nanoparticles and Electrode Preparations

Nanoparticles and electrodes were prepared according to methods reported previously [3, 4].

14.2.4. Detections

Cyclic voltammetry was conducted by using a Gamry 600 Potentiostat. Voltammetric potential was measured against a saturated chloride electrode (SCE).

14.3. Results and Discussions

14.3.1. GDH Coating on GCE, Au, and Pt Electrode for NH_4^+ Detection

Figs. 14.1 – 14.3 show the voltammetric responses of the modified GCE, Au, and Pt electrode with GDH biocomposite. As it is shown, the characteristic peak of α-ketoglutarate and NH_4^+ conversion to glutamate was detected at about 750 to 800 mV at pH 7.0. In comparison, the responses of Pt electrode (Fig. 14.3) were much preferred since they were "smoother" responses stepwise and larger in magnitude [5].

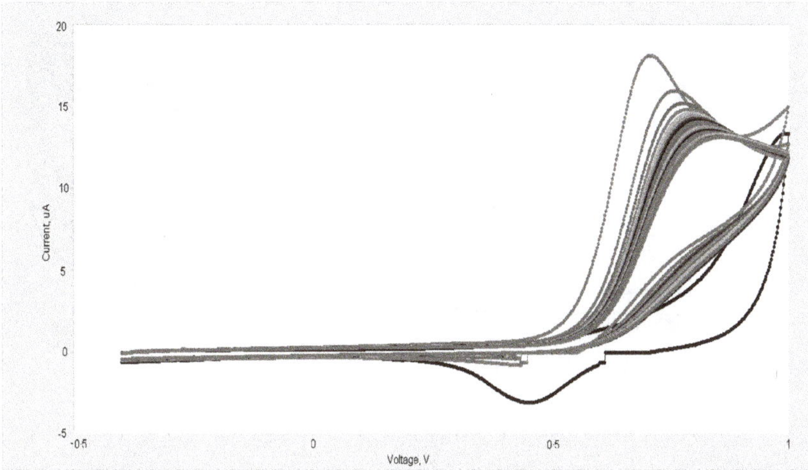

Fig. 14.1. Voltammetric responses of a GCE coated with GDH at pH 7.0 on Day 1. Responses were stepwise additions of α-ketoglutarate with NH_4^+ from 1.0×10^{-16} mol/L to 1.0×10^{-4} mol/L at 750 mV. The concentration of NADH was 3×10^{-3} mol/L.

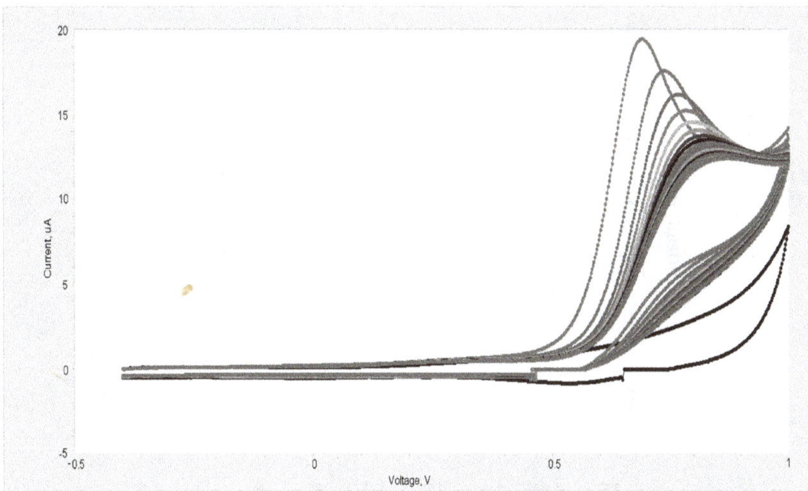

Fig. 14.2. Voltammetric responses of an Au electrode coated with GDH at pH 7.0 on Day 1. Responses were stepwise additions of α-ketoglutarate with NH_4^+ from 1.0×10^{-16} mol/L to 1.0×10^{-4} mol/L at 800 mV. The concentration of NADH was 3×10^{-3} mol/L.

Fig. 14.4 shows the accumulative current responses with the Au, Pt and GCE. As seen, current increased with concentration for the Pt electrode which indicated oxidation while current decreased with concentration for the Au and GCE which indicated reduction. Nevertheless, all three electrodes can be used for sensing purpose as long as their responses are consistent; however, Pt appeared to be the best anchoring material in this application of NH_4^+ detection. The concentration range used in this test was between 1×10^{-16} to 1×10^{-4} M.

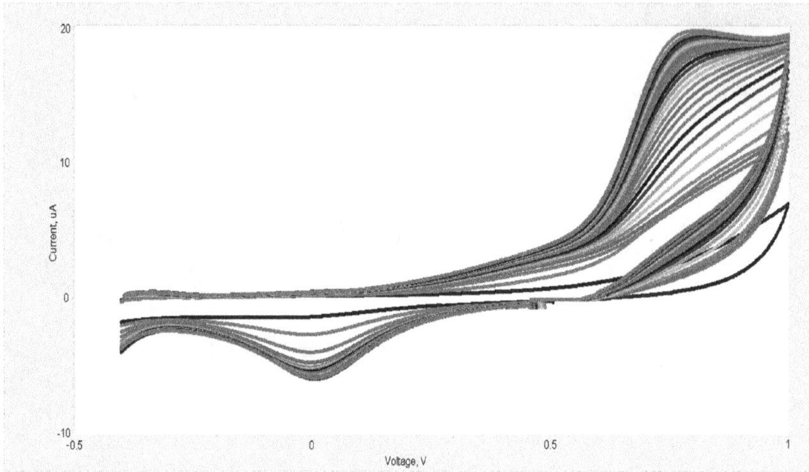

Fig. 14.3. Voltammetric responses of a Pt electrode coated with GDH at pH 7.0 on Day 1. Responses were stepwise additions of α-ketoglutarate with NH_4^+ from 1.0×10^{-17} mol/L to 1.0×10^{-4} mol/L at 800 mV. The concentration of NADH was $3\text{x}10^{-3}$ mol/L.

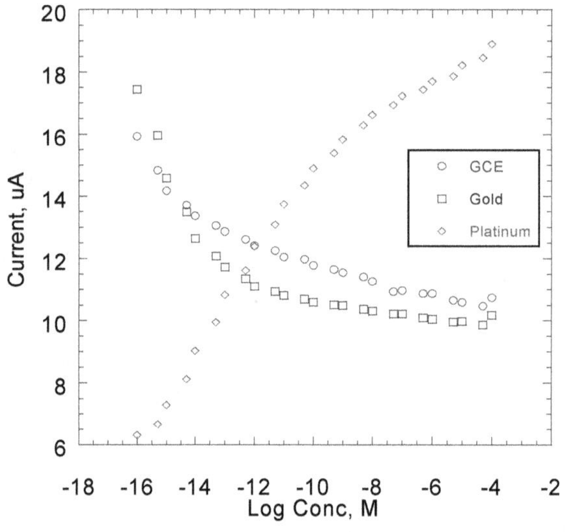

Fig. 14.4. Performance comparison by current measurements at peaks for NH_4^+ detection on Day 1 with Au, Pt and GCE at pH 7.0.

14.3.2. LDH Coating on Pt, Au, and GCE for Lactate Detection

Fig. 14.5 and Fig. 14.6 show the voltammetric responses of the modified Au and Pt electrode with LDH biocomposite. As it shown, the characteristic peak of lactate conversion to pyruvate was detected at 280 mV at pH 7.0. The characteristic peak was somewhat upstream from the peak at about 260 mV reported by some investigators, but was similar to other reports that found the peak at about 280 to 300 mV [6, 7]. The characteristic peak(s) of analyte is also influenced by pH of the solution.

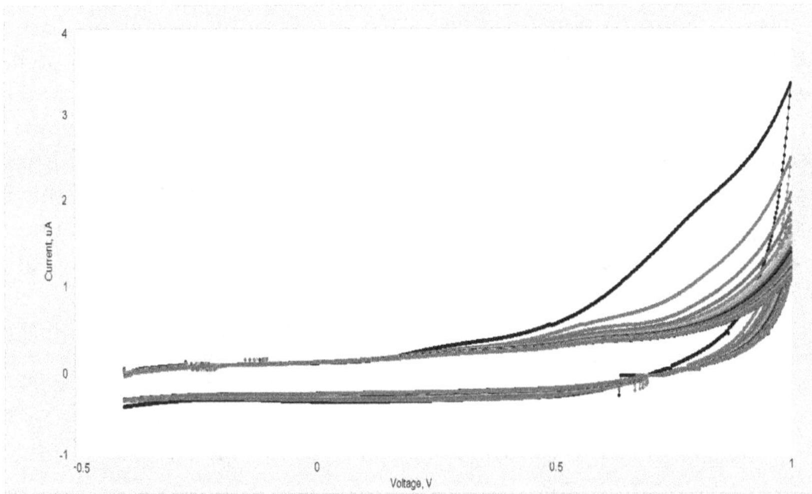

Fig. 14.5. Voltammetric responses of an Au electrode coated with LDH at pH 7.0 on Day 1. Responses were stepwise additions of lactate from 1.0×10^{-16} mol/L to 1.0×10^{-4} mol/L at 310 mV. The concentration of NAD^+ was 3×10^{-3} mol/L.

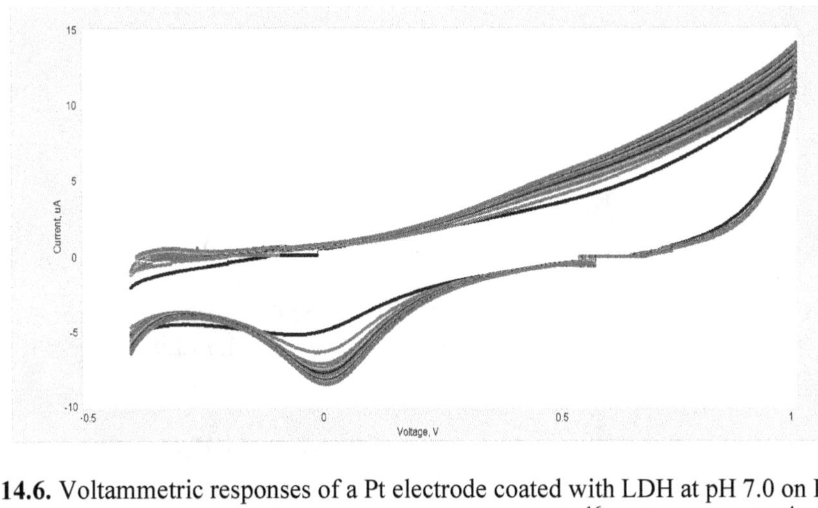

Fig. 14.6. Voltammetric responses of a Pt electrode coated with LDH at pH 7.0 on Day 1. Responses were stepwise additions of lactate from $1.0\text{x}10^{-16}$ mol/L to 1.0×10^{-4} mol/L at 310 mV. The concentration of NAD^+ was 3×10^{-3} mol/L.

Fig. 14.7 shows the comparison of performance for the GCE, Au, and Pt electrode. As seen in Equation (14.2), the reaction is a reversible catalytic reaction with which either oxidation or reduction can occur depending on the reacting condition (electrodes). Our results shown here indicate a relative weak reductive reaction for GCE and Au electrode, and a strong oxidative reaction for the Pt electrode in the tested concentration range. Equation (14.1), hence detection of lactate, proved to be difficult to control. Fig. 14.7. shows a weak reductive curve for the Au electrode; but we have observed that an oxidative curve can also be generated under the same testing procedures and condition, and the current signal detected with our biosensors is just as sensitive or better than reported in literature [6-8]. The reason that causes the oscillation of redox direction of Equation (14.2) with the Au electrode is still unclear at this point. Our results indicated that Pt electrode is the anchoring material of choice in lactate detection especially for very low detection concentrations that is below 10^{-7} M; the sensitivity of this ultra-high performance (Pt) for LDH detection is not found thus far in the literature. A slight drawback of the Pt electrode was that it tended to have higher signal perturbations at concentrations above 10^{-7} that were not seen for the other 2 electrodes.

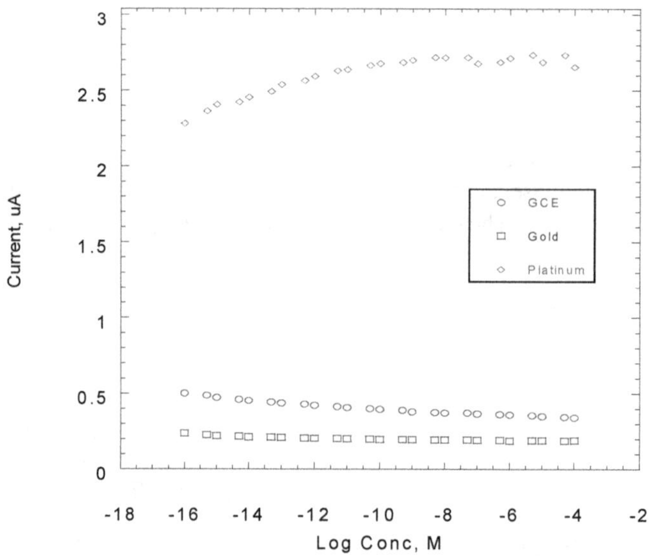

Fig. 14.7. Comparison of performance for the GCE, Au, and Pt electrode for lactate detection on Day 1 at 280 mV in concentrations from 1.0×10^{-16} mol/L to 1.0×10^{-4} mol/L.

14.3.3. Hemoglobin Coating on GCE, Pt, Au Electrode for H_2O_2 and Nitrite Detection

Hemoglobin is a biocatalyst that participates in many oxidation processes in biological systems. Similar redox reactions can be observed when oxidation and reduction agents are mixed in the presence of hemoglobin molecules. As shown in Fig. 14.8, a distinguished reductive peak (Equation (14.3)) was observed at about 1.1 V for a freshly prepared GCE at pH 7. A similar peak could also be detected by using a Pt electrode but the peak is an oxidative peak, and the current signals were diminishing with time for the two electrodes in the concentration range investigated in this study from 10^{-4} to 10^{-16} M.

However, the Au electrode revealed a very different behavior with time for the same biocomposite coating: its redox reaction shifted directions with time and eventually exhibited a reductive peak after 28 days (more discussion with the redox reaction oscillation of the Au electrode in later section).

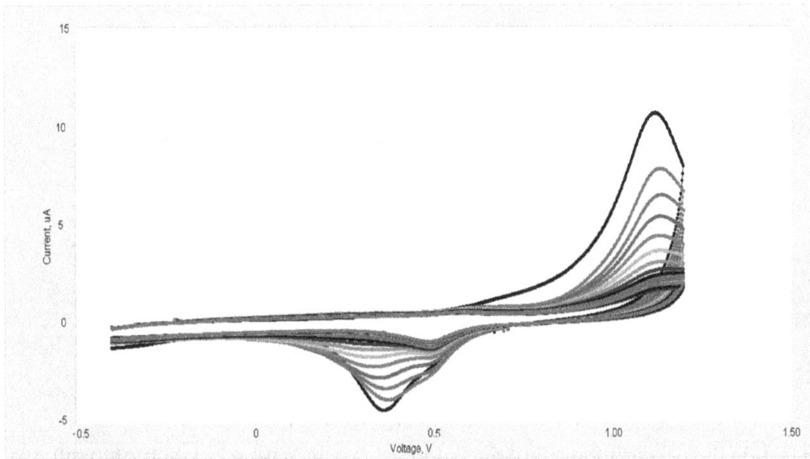

Fig. 14.8. Cyclic voltammogram of oxidation of H_2O_2 with a GCE coated with a layer of hemoglobin biocomposite at pH 7.0 with concentrations ranged from 10^{-4} to 10^{-16}M on Day 1. The characteristic redox peak of $H_2O_2$was detected at approximately 1.1 V.

Fig. 14.9 shows the cyclic voltammogram of a highly modified Pt electrode with hemoglobin biocomposite for the detection of NO_2^- on Day 28, which exhibited a characteristic oxidative peak at 800 mV. The characteristic peak may shift downstream/upstream with time which is also an indicator that there are intrinsic changes within the biocomposite which affect the activation potential of the electronic transfer for the oxidation of NO_2^-; for the Pt electrode, the peak shifted from 850 mV on Day 1 to 800 mV on Day 28. The lowering of activation potential implies lowering of potential requirement for the electron transfer of the oxidation of nitrite to nitrate, or a more efficient catalytic arrangement by the hemoglobin biocomposite.

Fig. 14.10 shows the performance comparison of GCE, Au, and Pt electrode for nitrite detection from concentrations of 10^{-4} to 10^{-16} M. As shown, Pt electrode provided the largest current signal for the concentration range thus is the best anchoring material for nitrite detection within these concentrations. In fact, other than at the extremely low concentrations, Pt electrode was the only electrode with moderate current increase with concentration until a "cutoff" concentration at about 10^{-6} M, then all three electrodes provides reasonable current responses to the testing upper limit in this paper of 1×10^{-4} M. In general, the three materials worked well as anchoring materials for the nitrite detecting sensor when concentrations were higher than 10^{-6} M, the ranking of materials was Pt > Au > GCE on Day 1; as the enzymatic biocomposite material aged, the ranking of materials herein remains the same, but the magnitude of current increased with time for Pt and Au, whilst the current decreased with time for GCE. Overall, Pt suited the best as anchoring material for the detection of nitrite with this biocomposite material.

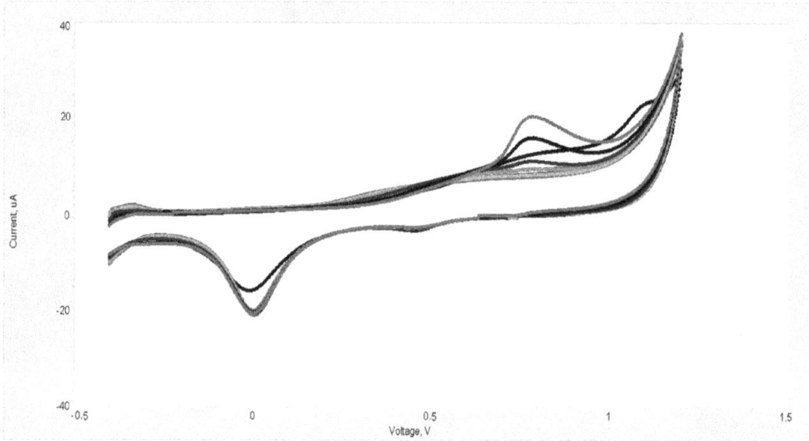

Fig. 14.9. Cyclic voltammogram of oxidation of nitrite with a Pt electrode with a layer of hemoglobin composite at pH 7.0 with concentrations ranged from 10^{-4} to 10^{-16} M on Day 28. The characteristic redox peak of nitrite was detected at approximately 0.8 V.

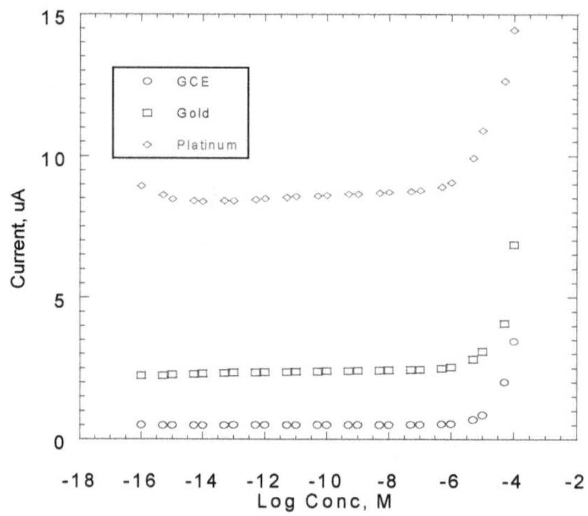

Fig. 14.10. Current outputs of oxidation of nitrite with the GCE, Au, and Pt electrode from 10^{-4} to 10^{-16} M on Day 1. Pt electrode provided the largest current output (signal) thus is the most ideal for nitrite detection.

14.3.4. Specificity of the Ultra-High Performing Electrode Sensor

Fig. 14.11 shows the cyclic voltammetry responses of the specificity test on GDH coated on a Pt electrode. In the specificity test, NO_2^-, and H_2O_2 were used to replace NH_4^+ in

Equation (14.1) and the concentration range used was between 10^{-16} and 10^{-6} M. Comparing with Fig. 14.3 (NH_4^+ detection), Fig. 14.11 shows that almost no separation (or difference in current signal) was observed at 800 mV due to concentration variations at which NO_2^-, or peroxide was added in the testing solutions, that implied the GDH coating has selectivity.

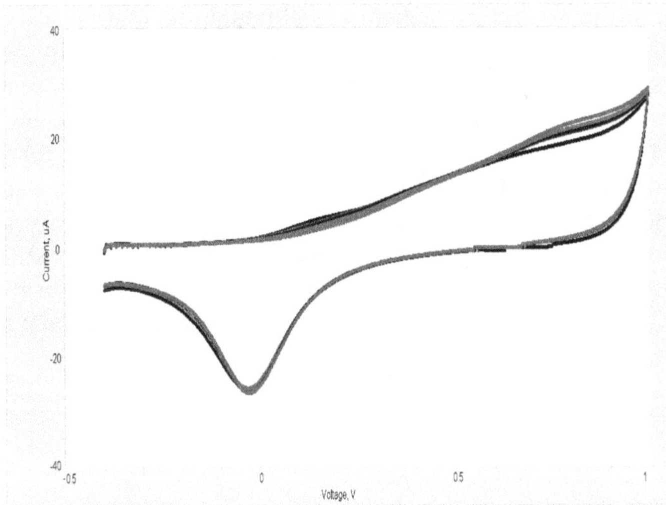

(a)

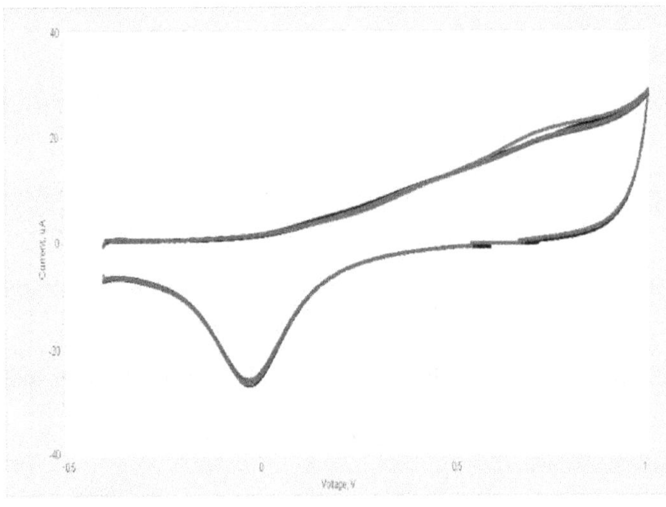

(b)

Fig. 14.11. Voltammetric responses of a Pt electrode coated with GDH biocomposite coating. The coating was designed for NH_4^+ detection as described in Equation (14.1), but the NH_4^+ was replaced by (a) NO_2^-, and (b) H_2O_2 as analyte tested.

14.3.5. Stability of Biosensor Electrodes with Time

In general, the sensitivity of these high performance biosensors (with the same platform) decreased with time after the sensors were prepared. The performance stability of these sensors was tested after 1, 5, and 28 days with the anchoring materials made of Pt, glassy carbon, and Au. As it is shown in Fig. 14.12, the performance of the sensor steadily declined with time, which is commonly expected for sensors using enzymatic reactions as part of the detecting mechanism. Although the matrix of the compositional materials stabilizes and decelerates the enzymatic decay substantially, the decay is still inevitable after a prolonged period in the buffer solution (pH 7). However, the duration of the decay, herein reversely proportionally to sensitivity of the sensor, varies greatly with the anchoring material and nature of the enzyme(s) used in the fabrication process. Fig. 14.13 shows the performance of a Pt electrode coated with hemoglobin for nitrite/nitrate detection on Day 1, Day 5, and Day 28. The sensitivity of sensor remained about the same with time and eventually increased slightly on Day 28, which implies that the hemoglobin molecular structure within the biocomposite layer remain intact and may even open the binding sites with NO_2^- slightly more after 28 days of aging, thus generating higher current or signaling.

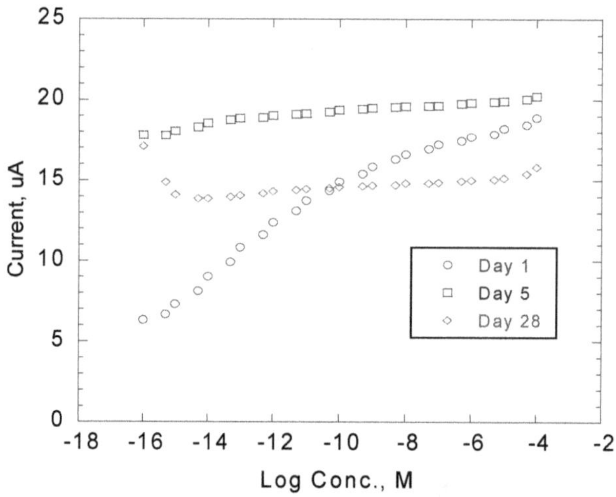

Fig. 14.12. Performance stability of Pt electrode coated for NH_4^+ detection on Day 1, Day 5, and Day 28 with cyclic voltammetry at 750 mV. Sensitive of the sensor steadily declined with time.

Fig. 14.14 shows the performance of an Au electrode coated with GDH biocomposite for NH_4^+ detection. Noted that the current consumption (oxidation) of the electrode on Day 28 is larger than on Day 1, while the current signal on the Day 5 was nearly flat or increased lightly (reduction) with time which signified the reverse reaction of the enzymatic reaction that was used for the NH_4^+ detection scheme (Equation (14.1)). The observation is rather interesting, definitely is not impossible in utilizing the enzymatic reactions for chemical species detection: GDH is a rather complex, large enzyme which exists as a multi-unit moiety; as a consequence, it is also a stable enzyme comparatively. It is believed that the enzyme opened up more binding sites with NH_4^+ on Day 28; its transition of configuration on Day 5 became more favorable for the reverse reaction (Equation (14.1)). The observation was consistent throughout our same repeated experiments and it only occurred with the Au electrode as anchoring material. Similar phenomena were also observed with the Au electrode when more modifications of the biocomposite layer were done to enhance the sensor performance [4, 9].

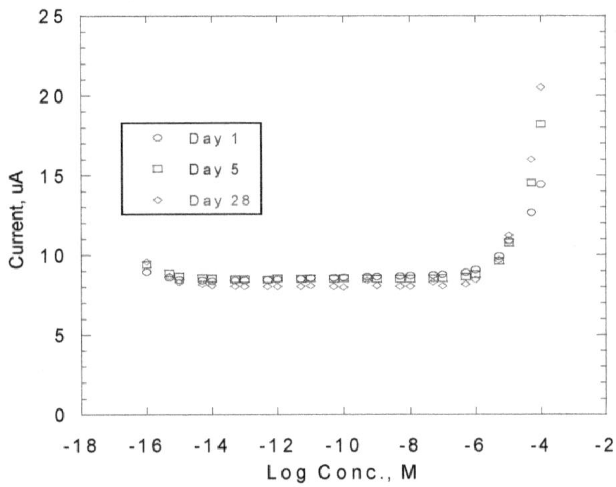

Fig. 14.13. Performance of a Pt electrode coated with hemoglobin for nitrite/nitrate detection on Day 1, Day 5, and Day 28. The sensitivity of sensor remained about the same with time.

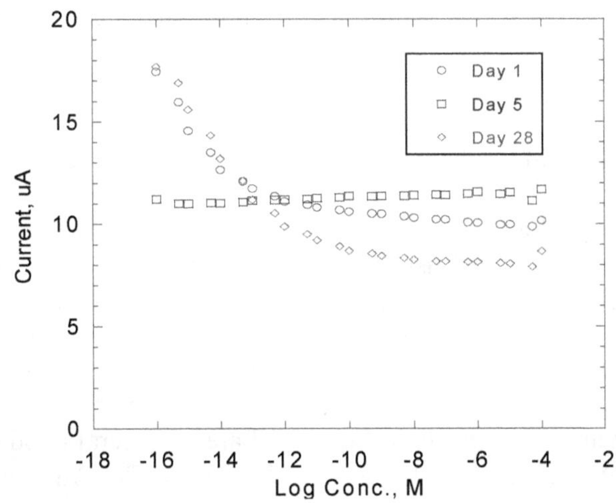

Fig. 14.14. Performance of an Au electrode coated with GDH for NH_4^+ detection on Day 1, Day 5, and Day 28.

Another evidence of the transition of enzymatic configuration within the biocomposite that can significantly affect the performance of the biosensor was observed with the Au anchoring electrode that is shown in Fig. 14.15. The signal outputs from Day 1 to Day 28 changed gradually: the sensor went through 2 cycles of redox reactions (Equation (14.2)) over the concentration range of 10^{-16} to 10^{-4} M for the Day 1 measurements, and slowly reduced to one cycle of redox reaction on Day 28. From the sensor detection point of view, performance of the sensor on the latter time was much preferred which provides an effective concentration range of 10^{-6} to 10^{-16} M (with only reductive reaction in the concentration range). The signals given by the Au electrode on Day 28 were similar to the signals given by a GCE on Day 1 measurements (Fig. 14.16). The sensor performance shown in Fig. 14.16 is more common than what is shown in Fig. 14.15.

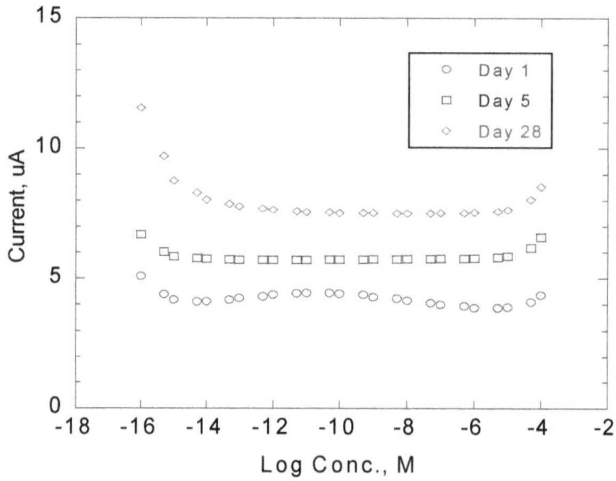

Fig. 14.15. Biosensor signal affected by transformation of enzymatic configuration. An Au electrode coated with hemoglobin for H_2O_2 detection.

14.3.6. Reproducibility of the Biosensor Electrodes after Repeated Uses

An important property of a biosensor is its' reproducibility that the sensor can produce predictable results after repeated usages. Fig. 14.17 is the reproducibility of our sensor platform anchored by an Au electrode coated with hemoglobin biocomposite for the measurement of NO_2^-. As shown in the figure, the slope of the response curves (current vs. time or concentration) remained nearly constant after approximately 300 measurement cycles, which indicates our sensor platform is reliable for high sensitive measurements. The results in Fig. 14.17 are in general good representations of the performance of the sensor platform anchored by Pt, Au, or GCE. However, performance of the sensor platform anchored by Ag electrode would not be as desirable in our study. The disadvantages of biosensors anchored by Ag electrode include excessive slope changes of response curve and high background interference, these could be the consequence of Ag's lower redox potential and it is subjected to oxidation easily. One advantage of using Ag electrode is the relative economic cost; Ag electrode is the least expensive among the four tested electrodes, which may be a factor in mass production for commercialization.

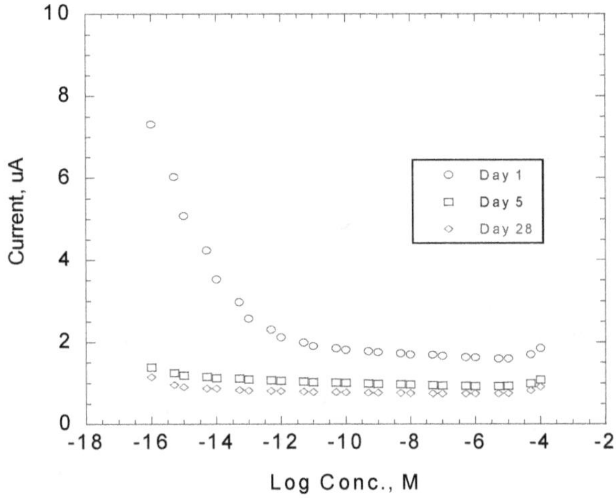

Fig. 14.16. GCE coated with hemoglobin for H_2O_2 detection at pH 7.0. The sensor lost essentially all its sensitivity for H_2O_2 detection on Day 5 and Day 28, but exhibited strong signals for the measurements on Day 1, in particular at very low H_2O_2 concentrations.

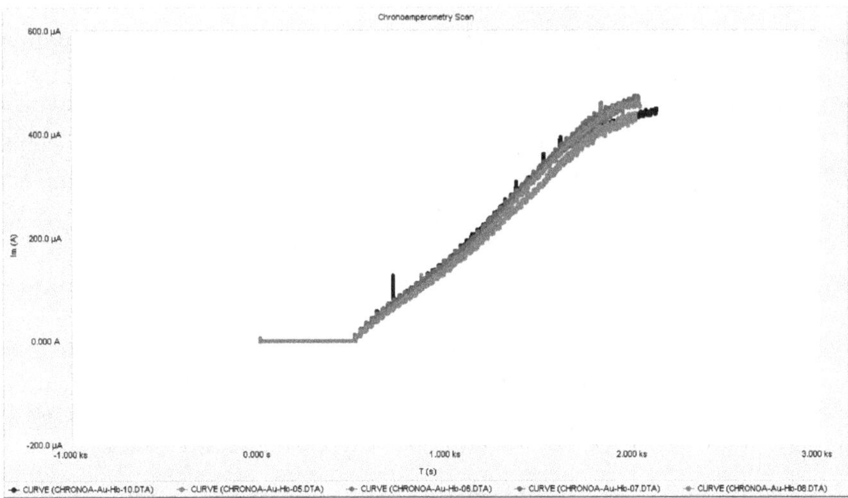

Fig. 14.17. Reproducibility of Biosensors: Cyclic voltammogram responses of Au electrode coated with Hb biocomposite and tested with analyte, NO_2^-. The responses represented current measurements with stepwise constant additions of NO_2^- at 30 seconds per addition for 50 steps. After the first 3 stepwise cycles (blue, red, and green line) for about 4500 seconds (150 steps), measurements of a fixed concentration of NO_2^- were performed for another 100 cycles, then a final 50 steps of constant addition of NO_2^- was performed (top blue line). The NO_2^- concentrations increased from 0 to 75 mM in the stepwise cycles. Total measurement cycles were approximately 300 for the electrode.

14.3.7. Identification of Analyte by Cyclic Voltammetry

There is always the issue of what is the correct potential (redox peaks) for analyte identification by using cyclic voltammetry. For example, the classic enzymatic reaction of LDH catalyzes the oxidation of lactate to pyruvate using NAD+ as a redox cofactor (Equation 14.2), most literature reports a characteristic oxidative peak of lactate at 260 to 280 mV, with a corresponding reductive peak at 130 to 150 mV for an overall redox potential of -130 mV [10]; however, the reductive peak of pyruvate to lactate is not always obvious in the reversible reaction, depending on how the redox system is set up. Moreover, some enzymes can be a catalyst for multiple chemical reactions, hemoglobin used in this study would be an example that can promote the redox reactions of H_2O_2 and NO_2-. Given the fact that cyclic voltammetry can be used to study enzymatic kinetics when the reactants are known; however, the exact location of characteristic redox peaks of analytes are not always the same, it is depending on multiple factors such as the materials and modifications of the electrodes. This is particularly valid for the identifications of bioanalytes such as proteins and DNAs in our studies. Thus, it is difficult to predict an analyte's characteristic redox peaks using cyclic voltammetry without previous experimentation, it is also important to ensure that the selected redox peaks for identification are unique for the detecting analyte especially when multiple analytes are present.

14.4. Conclusions

We successfully modified our ultra-high performance biosensor platform with Au, Ag, Pt, and glassy carbon as anchoring materials. These electrode sensors could detect target chemical (analyte) concentrations routinely at 1×10^{-16} M and lower depending on the enzyme couplings and anchoring materials, and the response time is instantaneous. Table 14.2 is the summary matrix of this research study that lists the anchoring materials, target analytes, coupling enzymes, and the lower measuring limits that were attempted with the locations of identification potentials of the analytes for cyclic voltammogram.

With these fourth generation of biosensors fabricated with the combinations of enzymes and composite layers that have instantaneous electron conducting property, we have examined the detecting concentration range from 1.0×10^{-16} to 1.0×10^{-4} M for four different types of anchoring electrodes with various enzyme/composite. As shown in our results, measurement responses (current) in the testing range were not all linear with concentrations; some electrode/enzyme/composite combinations worked more favorably in low concentrations (for example, Au/hemoglobin/composite for H_2O_2 detection) whilst others worked more favorably in high concentrations (for example, Pt/hemoglobin/composite for NO_2^- detection), and Pt/GDH/composite worked well in the entire testing range of the study. Thus, one should choose the right combination of anchoring electrode and biocomposite to have the optimal performance for a particular application.

Table 14.2. Performance summary of Au, GCE, and Pt electrode and the associated enzymes.

Element	Analyte	Enzyme	Detection limits, mol/L @ measuring potential (V)
Gold	NH_4^+	GDH	1.0×10^{-16} to 1.0×10^{-4} @ 0.75 V
Glassy Carbon			1.0×10^{-16} to 1.0×10^{-4} @ 0.8 V
Platinum			1.0×10^{-17} to 1.0×10^{-4} @ 0.8 V
Gold	Lactate	LDH	1.0×10^{-16} to 1.0×10^{-4} @ 0.31 V
Glassy Carbon			1.0×10^{-16} to 1.0×10^{-4} @ 0.31 V
Platinum			1.0×10^{-16} to 1.0×10^{-4} @ 0.31 V
Gold	H_2O_2	Hemoglobin	1.0×10^{-16} to 1.0×10^{-4} @ 1.1 V
Glassy Carbon			1.0×10^{-16} to 1.0×10^{-4} @ 1.1 V
Platinum			1.0×10^{-16} to 1.0×10^{-4} @ 1.1 V
Gold	NO_2^-	Hemoglobin	1.0×10^{-16} to 1.0×10^{-4} @ 0.8 V
Glassy Carbon			1.0×10^{-16} to 1.0×10^{-4} @ 0.8 V
Platinum			1.0×10^{-16} to 1.0×10^{-4} @ 0.8 V

Among the 4 metals used in the testing, Pt was thought to be the most ideal metal due to its inertness to chemical reactions and often provided the strongest signals. However, Pt appeared not to bind well with our linker polymers that was part of the biocomposite layer and thus did not generate characteristic signals as distinctively as other metals for at least the coupling enzyme, LDH; further research should be given to this direction. Despite of this observation, Pt electrode appeared to be functioning competitively with GDH and hemoglobin as part of the composite layer for the corresponding coupling reactions, thus the detections of target analytes. Glassy carbon worked competitively as compared to the metal electrodes, but is the most expensive among all electrodes tested. Ag electrode was easily oxidized and did not provide signals consistently in repeated testing and thus was not considered in the reported results, despite that it is the most inexpensive electrode.

Overall, it can be concluded that the anchoring materials have drastic effect to the performance of this ultra-high performance biosensor platform, their effect is mutually inclusive with the enzymes used and time (age) which affect the sensor's stability. These factors should be considered when diversified applications are to be developed.

Acknowledgement

This study was partially supported by a DOD USAMRMC Project Grant (Contract #W81XWH-07-2-0078).

References

[1]. B. Eggins, Chemical sensors and Biosensors, *John Wiley and Sons*, 2002.

[2]. J. Pingarron, P. Yanez-Sedeno, and A. Gonzalez-Cortes, Gold nanoparticle-based electrochemical biosensors. *Electrochimica Acta,* 53, 2008, pp. 5848–5866.

[3]. J. Lai, Y. Wang, W. Gao, H. Gu, S. Leung, Performance Comparisons of Nanoparticle Modified Sensor Electrodes for the Detection of Nitrite and Peroxide, in *Technical Proceedings of the 2009 Nanotechnology Conference and Expo,* Vol. 2, Chapter 4, Biosensors and Diagnostics, NSTI, 2009, pp. 233-235.
[4]. S. W. Leung, Y. Wang, J. C. K. Lai, Biomedical Applications of Modified Carbon Glassy Electrode Sensor with Nanoparticles and Dendrimers, *Sensors & Transducers*, Vol. 11, Special Issue, April 2011, pp. 74-82.
[5]. S. W. Leung, D. Assan, J. C. K. Lai, Evaluation of Anchoring Materials for Ultra-Sensitive Biosensors Modified with Au Nanoparticles and Enzymes, *Sensors & Transducers*, Vol. 15, Special Issue, October, 2012, pp. 59-70.
[6]. J. Cheng, D. Huang, J. Zhang, W. Yang, N. Wang, Y. Sun, K. Wnag, Z. Mo, S. Bi, Electrochemical Behavior of Lactate Dehydrogenase Immobilized on ''Silica Sol–Gel/Nanometre-sized Tridecameric Aluminium Polycation'' Modified Gold Electrode and Its Application, *Analyst*, Vol. 134, 2009, pp. 1392-1395.
[7]. M. M. Rahman, M. Shiddiky, M. A. Rahman, Y. Shim, A Lactate Biosensor Based on Lactate Dehydrogenase/Nictotinamide Adenine Dinucleotide (Oxidized Form) Immobilized on a Conducting Polymer/Multiwall Carbon Nanotube Composite Film, *Analytical Biochemistry*, Vol. 384, 1, 1, 2009, pp. 159–165.
[8]. J. Cheng, J. Di, J. Hong, K. Yao, Y. Sun, J. Zhuang, Q. Xu, H. Zheng, S. Bi, The Promotion Effect of Titania Nanoparticles on the Direct Electrochemistry of Lactate Dehydrogenase Sol-Gel Modified Gold Electrode, *Talanta*, Vol. 76, 5, 2008, pp. 1065-1069.
[9]. Y. Wang, L-glutamate Dehydrogenase Modified Biosensor Based with Gold Nanoparticles, M. S. Project Report, *Idaho State University*, Pocatello, Idaho, USA, 2009.
[10]. T. D. H. Bugg, Introduction to Emzyme and Coenzyme Chemistry, 3rd edition, *John Wiley and Sons*, 2012.

Chapter 15
Biomimetic Systems for Classification and Authentication of Beverages

P. C. Panchariya, Hepsiba K. Aanga, Santosh Kumar, P. Bhanu Prasad and A. L. Sharma

15.1. Introduction

The determination of quality of beverages whether alcoholic (wines, beer etc.) or non alcoholic (tea, coffee, soft drinks, juices etc.) has always been an ever evolving field of research. The main constituents of beverages are water, sugars, acids and ethanol (in alcohols). But the elements that distinguish one beverage from another, with regards to their quality, area of origin, year of manufacturing, etc. are a large number of constituent components that are present in very small proportions. Besides, the beverages can be considered as a living product because its chemical and organoleptic characteristics change over time. This makes the analysis of beverages a rather difficult task because, in addition to the determination of key components in small proportions, analysis must be performed in different stages of its manufacture, storage and consumption. Another reason for the interest in beverages analysis resides in the quality control and detection of possible spoilage and adulteration in case of expensive beverages. Currently the basic technique to assess beverages' quality is the use of human tasters-a panel of trained experts that carry out sensory analysis based on colour, taste and flavour. Human tasters are well known for their subjective evaluation. Moreover this evaluation methodology is often time-consuming and requires skilled personnel. Therefore, there is a need and increasing interest in beverage industry for the development of low-cost, easy-to-use analytical methodology and tools able for in situ and online monitoring of quality of beverages.

In the present scenario, the use of biomimetic systems is a good alternative to be used in the characterization of beverages and other foodstuffs [1]. Biomimetics or bionics is the application of biological methods and systems found in the nature to the study and design of engineering systems and modern technology. Based on the five senses – sight,

P.C. Panchariya
Digital Systems Group,
CSIR-Central Electronics Engineering Research Institute, Pilani-333031, India

hearing, touch, smell and taste – humans interact with the environment. Human effort to copy and learn from the nature resulted in the development of biomimetic systems such as Electronic Nose, Electronic touch, Electronic Vision and Electronic Tongue. In brief, Electronic noses or tongues are inspired in the mode mammalians recognize samples via olfaction and taste senses. In this approach, sensors do not have to be selective as in the case of traditional measurement techniques. Sensors that have low selectivity and consequently exhibit overlapping signals for different species or a group of related chemical species can be used to extract quantitative or qualitative information about the sample composition. The sensors are then integrated in an array and their response is analyzed by suitable pattern recognition techniques. The output of these systems usually indicates qualitative results rather than quantitative results. The operation of these biomimetic systems uses the concepts of the human tongue and nose, known as global selectivity, that is, the biological system does not identify a particular substance but brings to get all of the information into patterns that the brain decodes. Thus, a human being recognizes the taste of wine but does not understand that it is composed of different substances. The electronic sensor works in a similar way; it provides a global response, that is, a fingerprint of the sample, and the function of the brain is performed by a chemometric tool, such as principal component analysis (PCA) that decodes the information and discriminates standards for recognition [2].

In this chapter, the focus is on classification and authentication of two types of beverages; one alcoholic (wine) and the other non-alcoholic (tea) as case studies using electronic tongue and electronic nose respectively. The details of the systems are described in subsequent sections.

15.2. Electronic Nose

Faithful sensing of aroma or odor or volatile compounds by the human nose is of great importance and challenging in evaluating the quality of numerous items of everyday life specially foods. Therefore, over the years, researchers worldwide have put tremendous efforts to introduce instruments operating on a similar principle as the human nose. These systems would in most cases not replace but act as complementary or assistive tool to the conventional analyses of volatile compounds by sensory methods and by traditional analytical techniques.

In 1982, Persaud and Dodd at university of Warwick, UK proposed a device in their seminal article published in Nature, which could reproducibly discriminate between a wide variety of odours, and its properties show that discrimination in an olfactory system could be achieved without the use of highly specific receptors [3]. At the beginning of the 1990s, more extended research began on the device. In 1993, Gardner and Bartlett defined the device as artificial nose or electronic nose (E-nose) - an instrument, which comprises of an array of electronic chemical sensors with partial specificity and an appropriate pattern-recognition system, capable of recognizing simple or complex odours [4-6]. Obviously, electronic nose system is an electronic device and inspired from biological nose but very far from the human nose. The functional concepts of electronic nose and biological nose are quite similar. Like the biological/mammalian nose, it detects odours by means of gas sensors which send signals to a recognition organ, which is the brain or a computer. Electronic nose devices have wide variations in their operating principles, the number of sensors, types of sensors as well as their sensitivity and selectivity. Therefore electronic nose systems have different names such as 'multi-sensor system', 'flavour sensor', 'aroma sensor', or 'odour sensing system' [7].

Electronic nose systems comprise of an array of sensors, associated electronics, pumps, odour conditioners, flow controllers, etc., and software for data pre-processing, statistical analysis, etc. As analytical instruments, these systems must be designed for long-term usage with high repeatability (the ability to obtain the same pattern for a sample on the same array over short intervals of time) and reproducibility (the ability of different sensor batches or different instruments to produce the same pattern for the same sample). The following sections will review different technologies and sensors used for development of electronic nose system.

15.2.1. E-nose Sensors

Sensors (Gas sensors) are the key components in the development of E-nose system. Different types of gas sensors are available. Some of the popular and widely available sensor types are metal oxide semiconductors (MOS); metal oxide semiconductor field effect transistors (MOSFET); conducting organic polymers (CP); piezoelectric crystals (bulk acoustic wave (BAW) & surface acoustic wave (SAW)), etc [8-11]. Few researchers have also reported other types of sensors such as fibre optic, electrochemical and bi-metal sensors [12-13]. Sensors such as MOS and MOSFET types work at higher temperature while CP, SAW and BAW work at room temperature. The sensors working at high temperature are considered to be less sensitive to moisture. The sensors used for E-nose system interact with the gas under investigation so that a series of physical and/or chemical interactions occur at the surface of sensors when volatile compounds pass over the sensor. A dynamic equilibrium develops as volatile compounds are constantly being adsorbed and desorbed at the sensor surface. The overview of few popular sensors and their working principles are described in the following sub-sections.

15.2.1.1. Metal Oxide Semiconductor Sensors

The most popular and commercially available MOS sensors also know as ceramic gas sensors were developed by Taguchi and became popular as gas alarming devices in the era of 1960s in Japan [14]. These sensors exhibit changes of conductivity induced by the adsorption of gases and subsequent surface reactions. The details of the various oxide materials and gas sensors based on metal oxides can be found in various textbooks and review articles [8, 13-15]. Most of the MOS sensors consist of a heater and a ceramic surface coated by a metal oxide semiconducting film. The metal oxide coating acts as oxidizing or reducing agent. It can be either of the n-type which responds to oxidizing compounds, or of the p-type which responds to reducing compounds [13]. Usually zinc oxide, tin dioxide, titanium dioxide, etc. are preferred materials in n-type semiconductors while p-type semiconductor materials are nickel oxide, cobalt oxide, etc. When n-type semiconductor is excited (via temperature or light), the resultant is an excess of electrons in its conduction band which increases the reactivity with oxidizing molecules, while P-type semiconductor when excited shows an electron deficiency in its valence band which helps reactions with reducing compounds.

The MOS sensors operate at higher temperature i.e. high power consumption, and sensitive to water vapors and highly sensitive towards compounds such as those of sulphur, weak acids and ethanol which normally 'blinds' them to any other volatile compound in the sample.

15.2.1.2. Conducting Organic Polymer Sensors

For the last one decade, other type of sensors in practice that gained popularity for developing E-nose system are conducting organic polymer (CP) sensors [16] and they are applied for characterization of foods [17-18]. The operating mechanism of conducting polymers is complex and not fully exploited but relies on change of resistance by the adsorption of gas. As far as the construction of these sensors is concerned, they consist of a substrate (e.g. fibre-glass or silicon), a pair of gold-plated electrodes, and a conducting organic polymer such as polypyrrole, polyaniline or polythiophene as a sensing element.

The polymer film is deposited by electrochemical deposition between both electrodes fabricated on the substrate. The conducting polymer film contains cation sites probably consisting of polarons or bipolarons which are small regions of positive charge in the polymer chain providing mobile holes for electron transport. When a potential gradient is applied across the electrodes, a current passes through the conducting polymer. When volatile compounds under investigation flow over the surface of the conducting polymer film, it alters the electron flow in the system and therefore the resistance of the sensor. Therefore, good selectivity in the CP sensors may be achieved by altering the polymers, counterion, solvent used or the electrical growth of the polymer coating [13].

15.2.1.3. Piezoelectric Crystal Sensors

Piezoelectricity phenomena was first discovered and demonstrated by the Curie brothers in 1880. They have reported that piezoelectricity phenomena as- certain anisotropic crystals, when subjected to mechanical stress, generate electric dipoles [19]. Piezoelectric crystals have great applications in the fields of optoelectronics, electronics, RF filters and liquid and gas sensing devices. The working principle of piezoelectric crystal sensors is change of mass transduced into a change in resonance frequency. The construction of piezoelectric crystals is simple. These are discs made of materials such as quartz, lithium niobate or lithium tantalate, coated with suitable materials such as lipids which have chemically and thermally stable properties. These sensors work at room temperature. When an alternating electrical potential is applied, the crystal vibrates at a fundamental/particular frequency, defined by its mechanical properties. When it is exposed to a vapour under investigation, the coating material adsorbs certain molecules. Due to adsorption of molecules, the mass of the sensing layer increases and hence decreases the resonance frequency of the crystal. The change in resonance frequency indicates the amount of the adsorbed molecules and hence the vapour under investigation [20-21].

Bulk acoustic wave (BAW) [22] and surface acoustic wave (SAW) [23] types of piezoelectric sensors are most popular among the other types of piezoelectric sensors like flexural plate wave (FPW) etc. [24]. The principle of BAW sensors was introduced by King in 1964 with his Piezoelectric Sorption Detector. These devices are also known as 'quartz crystal microbalance' (QCM or QMB) because, similar to a balance, their responses change in proportion to the amount of mass adsorbed [25]. The first SAW sensor was developed by Wohltjen and Dessy in 1979. The BAW and SAW sensors differ in their basic structure. In BAW sensors, waves (3-dimensional waves) travel through the crystal, while in SAW sensors, waves (2-dimensional waves: Rayleigh, Love and Bluestein-Gulyaev) propagate along the surface of the crystal. The operating frequencies of the SAW sensors are higher than that of the BAW sensors [26].

15.2.1.4. Metal Oxide Semiconductor Field-effect Transistor Sensors (MOSFET)

In 1975, Lundstrom and his co-researchers reported first metal oxide semiconductor field-effect transistor based hydrogen sensor which contains palladium as gate for MOSFET [27]. Since then several research works have been reported on MOSFET with different gate metals for different gases and their applications in various domains [28-30]. Different variants of MOSFET exist such as ISFET (Ion Sensitive), OGFET (Open Gate), and SGFET (Suspended Gate). However the standard MOSFET configuration is more commonly utilized for E-nose applications. A MOSFET sensor consists of three layers, i.e. a silicon semiconductor, a silicon oxide insulator and a catalytic metal (usually palladium, platinum, iridium or rhodium), also called the gate. A normal transistor operates by means of three contacts, two allow the current in (source) and out (drain), and the third acts as the gate contact that regulates the current through the transistor. When potential is applied across the gate terminal and drain terminal, this creates an electric field and hence causes conduction of the transistor. When the target molecules interact with the surface of metal gate, that alters the electric field, causes change in current flowing through the transistor. Gate metal film can be either thick layer of around 100–250 nm or thin around 5–30 nm. The selectivity and sensitivity of MOSFET sensors may be influenced by the operating temperature (50–200 °C), the composition of the metal gate, and the microstructure of the catalytic metal. MOSFET sensors usually have a relatively low sensitivity to moisture and are thought to be very robust. However, high levels of manufacturing facilities and expertise are essential to achieve good quality and reproducibility.

It is desired that the gas sensors used for E-nose should be of high sensitivity towards chemical compounds, that is, similar to that of the human nose and low sensitivity towards humidity and temperature. They should have medium selectivity. They must respond to different compounds present in the headspace of the sample. High stability and high reproducibility with linearity as well as reliability are other essential requirements. They should also have short reaction and recovery time, low power consumption; robustness and durability; easy calibration and easily processable data output facilities; and small dimensions.

15.2.2. E-nose Based on SAW Sensor (zNose)

15.2.2.1. Principle of zNose and its Operation

zNose (Model 7100, Electronic Sensor Technology, Newbury Park, CA, USA) system is based on the principle of high-speed gas chromatograph (GC) [31]. It consists of a six port valve and oven, a preconcentrating trap, a short GC capillary column (DB-5 1 m length, film thickness 0.25 mm, internal diameter 0.25 mm) and a sensitive surface acoustic wave (SAW) quartz microbalance detector, in which volatile analytes are condensed onto the surface of an oscillating crystal. The system is calibrated using a series of alkanes from C6 to C14. The retention time and retention index of the same are generated. The chromatography system consists of two sections. One section maintains a constant 3-ccm flow of helium gas passing through a column and impinging onto the surface of a temperature-controlled quartz crystal (Surface Acoustic Wave resonator). Capillary columns are stainless steel tubing with a polysiloxane coating on the inner surface. The column is resistively heated by passing an electrical current through it and temperature is measured by an RTD attached to the column. Direct heating and closed-loop control enables linear ramping of the column temperature at rates as high as 20 ^{o}C/second. The other section of the system is used to sample ambient air and consists of a heated inlet and a pump. Linking the two sections is a “loop” trap, which acts as a preconcentrator or absorber when placed in the air section and as an injector when placed in the helium section. The ‘loop’ trap is a 3-cm length of resistively heated metal capillary tubing filled with Tenax® absorbent.

System operation is a two-step process. Odor under investigation is first sampled and organic vapors are collected (preconcentrated) on the trap. After sampling, the trap is switched into the helium section where absorbed compounds are released as a narrow pulse by rapidly heating the trap to 250 ^{o}C. The individual compounds undergo an absorption interaction as they pass through the capillary column and exit separated in time with characteristic retention times. Retention time is a function of each compound’s volatility and absorption characteristics. In general, lighter and more volatile compounds exit the column first followed by the heavier and less volatile compounds. As they exit the column they are physically absorbed on the surface of the quartz crystal. This changes the frequency of quartz crystal [31-33].

Currently, the zNose technology is becoming popular for detection of ultra low concentrations of odors in various biological applications such as honey [34], sugars [35], oils [36] etc.

15.3. Electronic Tongue (E-tongue)

As described in previous sections, the overall quality of foods is assessed by different senses by human tasters. Taste of the foods is another critical parameter for assessment of foods. There are five basic taste experiences by biological tongue i.e. sour, salt, bitter, sweet and "umami". The biological tongue is the primary organ of gustation. The upper surface of the tongue is covered by papillae and taste buds. A pink tissue on the surface of tongue called mucosa keeps the tongue moist. Tiny bumps called papillae give rough structure. Several thousands of taste buds are present on the surfaces of the papillae. Taste buds are collection of nerve cells that connect to nerves running into the brain. Different lipid molecules present in the taste buds of tongue play a key role in sensing taste of the materials. The tongue has many nerves that help detect and transmit taste signals to the brain. An electronic tongue is a device, whose principle and functionality mimics that of the human gustation. It is also known as taste sensor and related to the sense of taste in similar ways as for the electronic nose to olfaction. Most of the foods have volatile as well as non volatile components. The E-nose is used to assess volatile constituents while non volatiles are assessed by E-tongue. In view of its functionality, the E-tongue functions in a similar way as the electronic nose. It consists of an array of non specific electrodes which produce signal pattern which is correlated to certain features or qualities of the sample by using pattern recognition algorithms. In general, E-tongue is aimed to produce qualitative as well as quantitative results.

15.3.1. Sensing Principles

There are different sensing principles - electrochemical, optical methods, measurements of mass changes- having the capabilities to be used in development of E-tongues that are reported in literature [37-41].

In electrochemical measurement of liquids, two basic electrochemical principles; potentiometric and voltammetric are often used for investigation of liquids. The Potentiometric measurement system contains two electrodes - one working electrode and other reference electrode. A pH electrode or ion sensor such as calcium, potassium, sodium, chloride etc. is one of the best examples of potentiometric sensors. The working electrode responds to the target molecule in the electrolyte solution while the reference electrode generates constant potential. This approach uses the potential difference between the working electrode and reference electrode as signal response of liquid under analysis. There is a wide scope of different types of membrane materials having different recognition capabilities that may be used to develop E-tongue [41].

The first E-tongue based on voltammetric measurement technique was developed by Winquist and his team in 1997, which consists of three electrodes, a working electrode, a counter electrode and a reference electrode [42]. In this method, potential is applied between working and reference electrode while current through the working electrode is measured. The size and shape of applied potential determines whether the target molecules shall lose or gain electrons. This technique is best suited for measurement of chemical compounds which are electroactive in nature. E-tongue based on voltammetry, in particular makes use of non-specific metal electrodes for acquiring the information related to the redox active species through the measurement of the current through these electrodes when a potential having specific shape is applied, depending on the type of voltammetric technique employed. The voltammetric techniques offer many advantages such as simplicity, robustness, and a wide range of analytical possibilities according to several parameters such as the shape and material of the electrodes, the excitation signal applied, and the particular processing performed on the data [42-43].

In optical methods [44], it is well known that when a beam of light is incident in a medium, it causes the charges (electrons, atoms, or molecules) in the medium to oscillate and thus emit additional light waves that can travel in any direction. The oscillating particles vibrate at the frequency of the incident light and re-emit energy as light of that frequency called scattering. If the emitted light is "out of phase" with the incident light then the two waves interfere destructively and the incident beam is attenuated. If the attenuation is nearly complete, the incident light is said to be "absorbed." Scattered light may interfere constructively with the incident light in certain directions, forming beams that have been reflected and/or transmitted. Therefore, each compound has its specific light absorption wavelength ranging from UV via VIS to NIR and IR and hence distinct absorption spectra. In optical techniques, by scanning a certain wavelength region, a specific spectrum or optical signature for the sample under investigation is obtained. These techniques offer advantages of high reproducibility and good long-term stability.

Apart from electrochemical and optical methods, mass sensitive devices [45] are also interesting and have special features making them useful for electronic tongues. The mass sensitive devices basically use piezo electric crystals. Piezoelectric crystal sensors are passive solid-state electronic devices, which can respond to changes in temperature, pressure, and most importantly, to changes in physical properties at the interface between the device surface and a foreign fluid or solid. Such changes in physical properties include variations in interfacial mass density, elasticity, viscosity, and layer thickness [46]. The incorporation of various chemically sensitive layers has enabled the transition from the microbalance to the mass sensor and resulted in the explosive growth of piezoelectric sensors in recent years. This type of sensor operates by observation of the propagation of an acoustic wave through the solid-state device. Sensing is achieved by correlating acoustic wave propagation variations to the amount of analyte captured at the surface and then to the amount or concentration of analyte present in the sample exposed to the sensor, or to the changes of physical properties of interfacial thin films. Measurement of the frequency shift and resonant resistance change of acoustic wave is one of the most accurate types of physical measurement. The frequency and resonant resistance of piezoelectric crystal is influenced by a large number of parameters mentioned above. A selective sensor is obtained when a sensor surface is coated with selectively interacting thin film [45-46].

15.3.2. E-tongue Based on Voltammetry

15.3.2.1. Measurement Setup

A block diagram of the basic experimental setup of an E-tongue system is shown in Fig. 15.1. It comprises of an array of working electrodes located at fixed distances from the reference electrode and counter electrode. Usually the working electrodes are made of noble metals or metal alloys such as Platinum (Pt), Gold (Au), Silver (Ag), Glassy carbon (GC) etc. An Ag/AgCl electrode is used as reference electrode while stainless steel or platinum is used as counter electrode. In this particular experimental setup, three working electrodes of Platinum (Pt), Gold (Au), and Glassy carbon (GC) and a stainless steel casing - engulfing a cylindrical structure in which the working and reference electrodes are embedded was used as the counter electrode. All these electrodes are connected to a standard potentiostat in a standard three electrode configuration. A relay module (developed here in the lab of Digital Systems Group, Central Electronics Engineering Research Institute, Pilani, India) for switching of the electrodes sequentially helps in connecting each of the working electrodes to the potentiostat circuitry so as to form a three electrode configuration every time a measurement is taken. The number of working electrodes, material of working electrodes and size of working electrodes are decided by the sample under investigation. In this setup, all the working electrodes are of diameter 2 mm and of 99 % purity. The potentiostat was from Gamry Instruments, USA.

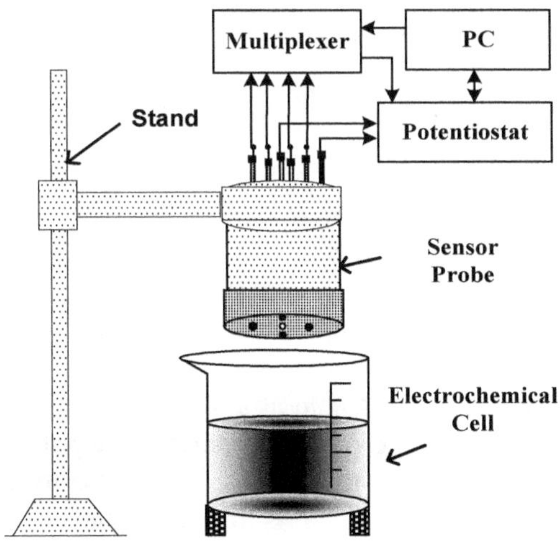

Fig. 15.1. Experimental setup for Electronic tongue system.

The potentiostat is interfaced to a Personal Computer. The voltammetric signal of interest is then imposed or impressed onto each of the working electrodes in sequence via potentiostat and the generated current response, which is indicative of the redox species present in the solution or sample, is acquired and sent to the PC via potentiostat for further processing and analysis [47].

15.3.2.2. Measurement Procedure

Depending on the shape of the signal used for excitation of the working electrodes, the voltammetry is of different types, the basic amongst them being pulse techniques or potential step voltammetric techniques. Pulse voltammetry is of special interest because of its advantages of greater sensitivity and resolution. The basis for all the potential step techniques is single potential step scan where in the potential is instantaneously changed from an initial value, $E_{initial}$ to a different final potential, E_{final}, and held there for a considerable amount of time as shown in Fig. 15.2.

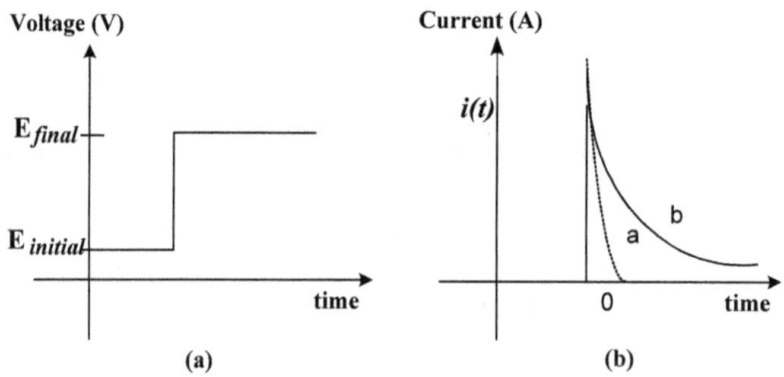

Fig. 15.2. (a) Waveform for single potential step imposed on the electrode; (b) Resultant current response on application of potential step.

Three types of pulse voltammetry are commonly used: large amplitude pulse voltammetry (LAPV), small-amplitude pulse voltammetry (SAPV), and differential pulse voltammetry (DPV). Moreover, the frequency of signal as well as combinations of different frequencies in a signal can also be used as excitation signal [48]. In the present case, large amplitude pulse voltammetry (LAPV) was used for the discriminating the beverages and will be described in the subsequent sections.

In LAPV [42-43, 49], the electrode is held at a base potential at which no or negligible electrode reactions occur. After a fixed waiting period, the potential is stepped to a final potential, the resultant that a current then flows to the electrode, with a sharp jump in current when the Helmholz double layer is formed. The current then decays as the double-layer capacitance is charged and electroactive compounds are consumed, until the diffusion-limited faradic current remains. The size and shape of the transient response reflect the amount and diffusion coefficients of both electroactive and charged compounds in the solution. When the electrode's potential is stepped back to its starting value, similar but opposite reactions occur. The instantaneous faradic current at the electrode is related to surface concentrations and charge transfer rate constants and depends exponentially on the difference of the electrode potential between the start value and the final potential.

The resulting current as a result of the applied potential step is measured, say *i(t)*. There might be more than one contribution to the total resultant current observed during an electrochemical measurement.

Faradaic current due to the redox reactions that occur due to the applied potential. This is due to the charge transfer across the electrode – solution interface.

Charging current (non-faradaic current) resulting due to the processes such as adsorption, desorption wherein the structure of the electrode - solution interface changes with changing potential or solution composition. This is the external current that flows transiently when the potential, electrode area or the solution composition changes.

The resulting current due to the reduction or oxidation of a target analyte is then a measure of its concentration [38, 49]. Thus, if a red-ox active species is reduced at the electrode surface, the reaction can be written as

$$Ox + ne^- = Red^-, \qquad (15.1)$$

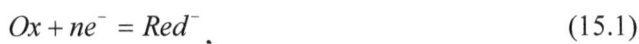

where Ox is the oxidized form and Red is the reduced form of the analyte. At standard conditions, this reaction has the standard potential E_0. The potential of the electrode at equilibrium, $E_{electrode}$ can be used to establish a correlation between the concentration of the oxidized form C_o and the reduced form C_r of the analyte, according to the Nernst relation

$$E_{electrode} = E_0 + \frac{RT}{nF}\ln\left(\frac{C_o}{C_r}\right) \qquad (15.2)$$

At the onset of a voltage pulse at the electrode, charged species and oriented dipoles will arrange next to the surface of the working electrode, forming a Helmholtz double layer. Thus, a charging current will initially flow as the layer builds up which then exponentially decays until reaching the level for the redox current. This current flow, i_c, is similar to the charging of a capacitor in series with a resistor and can be related as

$$i_c = \frac{E}{R_s} e^{\left(\frac{-t}{R_s C_d}\right)}, \qquad (15.3)$$

where R_s represents the solution resistance, t is the time, and C_d represents electrode equivalent capacitance representing the double layer capacitance. The redox current or the faradic current, i_f, from an electroactive species behaves in a similar way; initially a large current pulse is generated when compounds close to the electrode surface are oxidized or reduced, but decays with time. The current follows the Cottrell equation for a planar electrode given by

$$i_f = \frac{nFA\sqrt{D}C}{\sqrt{\pi t}}, \tag{15.4}$$

where, A is the area of the working electrode, D is the diffusion constant and C is the concentration of the analyte. Both faradaic and non-faradaic currents are inherent in the measured current response as shown in Fig. 15.2(b). As shown in Fig. 15.2 (b), the charging current, a (shown in dotted lines), for the potential step decays exponentially. The faradaic current, b (shown in solid line), also decays with time, but at a much slower rate as a function of $(\text{time})^{1/2}$ [38]. A single potential step may not be sufficient to provide information needed for the classification of samples. For this reason, a voltage waveform comprising of a sequence of potential pulses of varying amplitude is imposed onto the working electrodes and the resulting responses which we call the voltammograms are recorded.

The kind of potential waveform that was used to excite the electrodes is shown in Fig. 15.3 and the resulting response when each of the electrodes Au, GC and Pt are excited by the applied voltage waveform as in Fig. 15.3 for a particular sample of a juice is shown in Fig. 15.4.

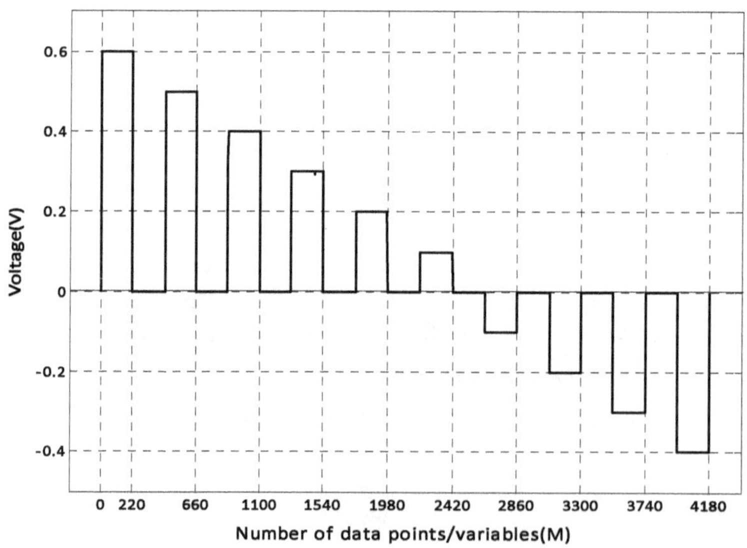

Fig. 15.3. Excitation signal for working electrodes.

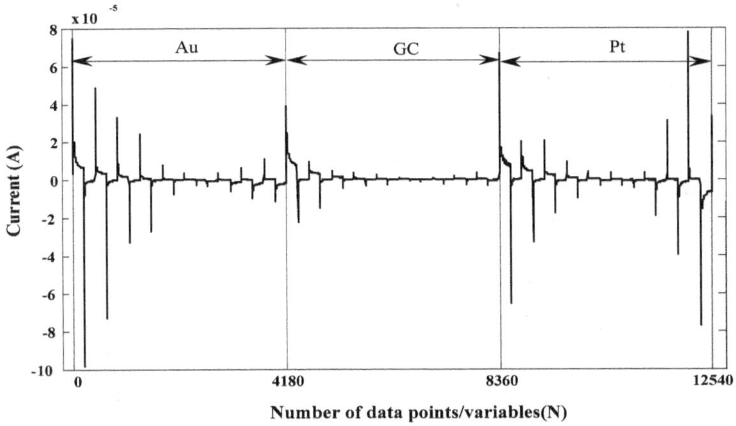

Fig. 15.4. A typical voltammetric response involving three electrodes Au, GC and Pt.

15.4. Data Analysis Methods

Data visualization of sensors' signals is known as basic data analysis technique in which different types of graphic analyses such as bar chart, profile, polar and offset polar plots are carried out. The operational method in biomimetic systems generates large amounts of data. In a complete system the data is collected, analyzed and presented in a predetermined way depending on its application, e.g., whether the milk is safe to drink (yes/no) or what sample tastes like (bitter/sweet). Moreover, raw signals generated by sensors contain noise. Thus, there is a need for pre-processing the raw sensor signals particularly in the domain of Biomimetic systems. Once raw signal is free from noise, then data compression or feature extraction comes into picture. These data analysis steps are often carried out using a multivariate analysis method. Several textbooks have been written by various authors in the field of multivariate analysis [50-51]. In multivariate analysis, various methods of data analysis such as principal components analysis (PCA), discriminant analysis (DA), feature weighting (FW) and cluster analysis (CA) are some of the popular methods in the field of biomimetic systems. In a broad sense, general multivariate data analysis methods are two step techniques- data compression or data reduction and pattern recognition. In first step of data compression or data reduction, it reduces high dimensionality in a multivariate problem where variables are partly correlated (e.g. sensors with overlapping sensitivities), allowing the information to be displayed in smaller dimensions. Many methods and techniques [52] have been developed for reducing the size of the data with insignificant loss of information. Wavelet transform [53], Hierarchical principal component analysis, exponential modeling [54], segmentation approach [55] and physical modeling have been used for variable reduction and data compression on the electronic tongue data. There are many multivariate techniques to choose from [56-57]. These techniques are used under the umbrella of pattern recognition (PARC) techniques which are routinely used in the analysis of data of electronic nose and electronic tongue.

Other methods for analysis also found in the literature are artificial neural networks (ANNs) [58]. Researchers have also analyzed E-nose and E-tongue data by different structure or networks methods starting from simple artificial neural network (ANNs) model, probabilistic neural networks (PNNs) to the radial basis function (RBF) [59-60]. These methods can be classified in number of ways like linear and nonlinear methods, supervised and non supervised methods etc. The choice of method depends on type and size of data and the type of outcome required of the analysis [61-62].

In simple words, in order to obtain final analytical results from the biomimetic systems, it is necessary to apply certain mathematical signal processing techniques. The PARC methods and multivariate calibration techniques are used to analyze the response of the sensor array because output of the sensor array in a multi-compound solution is complex in most cases and cannot be described by using theoretical equations. Usually, the processing of the data from the sensor array for biomimetic systems is usually performed in two ways- qualitative and quantitative. The overview of some of the frequently used data analysis methods such as PCA, LDA, ANNs and PNNs are described in the following subsections.

15.4.1. Principal Component Analysis (PCA)

PCA [50], a renowned application of linear algebra, is one of the most useful multivariate data analysis (MVDA) techniques for dealing with experimental data. It is a tool for reduction of dimensionality of a data set and also the redundancy of it. It is a linear transformation of the variables into a lower dimensional space which retains maximal amount of information about the variables. It uses an orthogonal transformation to convert a set of observations of possibly correlated variables into a set of values of linearly uncorrelated variables called principal components [51]. The number of principal components is less than or equal to the number of original variables. This transformation is defined in such a way that the first principal component has the largest possible variance (that is, accounts for as much of the variability in the data as possible), and each succeeding component in turn has the next highest variance possible under the constraint that it be orthogonal to (i.e., uncorrelated with) the preceding components. It extracts features from the observations for the users that are otherwise not apparent. Its operation can be thought of as revealing the internal structure of the data in a way that best explains the variance in the data. If a multivariate dataset is visualized as a set of coordinates in a high-dimensional data space (1 axis per variable), PCA can supply the user with a lower-dimensional picture, a "shadow" of this object when viewed from its (in some sense) most informative viewpoint. This is done by using only the first few principal components so that the dimensionality of the transformed data is reduced. This method maximizes the ratio between 'between class' variance and 'within class' variance. In most of the biomimetic systems, PCA is used to evaluate the sensor or as a pre-processing tool. The method is used because of its easy way to visualize the experimental data into groups with similar qualification and classify data. In mathematic form:

The goal is to map the vectors $x=[x_1,x_2,......x_d]^T$ onto vectors $z=[z_1,z_2,......z_M]^T$, where $M<d$. The vector x can be represented as a linear combination of a set of *d* orthonormal vectors u_i

$$x=\sum_{i=1}^{d} z_i u_i \quad (15.5)$$

where the coefficients z_i can be found from the following equation:

$$z_i = u_i^T x \quad (15.6)$$

This corresponds to a rotation of the coordinate system from the original x to a new set of coordinates given by z. To reduce the dimensions of the data only a subset (M < d) of the basis vectors u_i is kept. The remaining coefficients are replaced by constants bi and each vector x is then approximated as

$$\hat{x}=\sum_{i=1}^{M} z_i u_i + \sum_{i=M+1}^{d} b_i u_i \quad (15.7)$$

The basis vectors u_i, also called principal components are the eigenvectors of the covariance matrix of the data set. The coefficients b_i and the principal components should be chosen such that they best approximate the original vector x on average for the whole data set. However, the reduction of dimensionality from d to M causes an approximation error. The data points can be visualized using score plots wherein the data points are projected onto the new axes called the principal components. The score plot shows the relation between the observations or experiments. Groupings of observations in the score plot can be used for classification purpose. The data set is described in fewer dimensions using the principal components as new features. This shows that PCA is a very popular, unsupervised and widely used pre-processing tool.

15.4.2. Linear Discriminant Analysis (LDA)

Linear discriminant analysis (LDA) [56] and associated methods are predominately applied in statistics, pattern recognition and machine learning to find a linear combination of features which characterizes two or more classes of samples or observations. The LDA method is closely related to principal component analysis (PCA) where search for linear combinations of variables which best explain the data. LDA explicitly attempts to model the difference between the classes of data. PCA on the other hand does not take into account any difference in class. The LDA method easily handles the case where the within-class frequencies are unequal and their performances have been examined on randomly generated test data. This method maximizes the ratio of between-class variance to the within-class variance in any particular data set thereby guaranteeing maximal separability. In other words, the LDA finds a linear transformation ("discriminant function") of the two predictors, *X* and *Y* that yields a new set of transformed values that provides a more accurate discrimination than either predictor alone. A transformation function is found that maximizes the ratio of between-class variance to within-class variance. The transformation seeks to rotate the axes so that when the categories are projected on the new axes, the differences between the groups are maximized. The LDA method [57], finds the vectors in the underlying space that best discriminate among classes. For all samples of all classes the between-class scatter matrix S_B and the within-class scatter matrix S_W are defined by:

$$S_B = \sum_{i=1}^{c} M_i.(x_i - \mu).(x_i - \mu)^T , \qquad (15.8)$$

$$S_W = \sum_{i=1}^{c} \sum_{x_k \in X_i} (x_k - \mu_i).(x_k - \mu_i)^T , \qquad (15.9)$$

where M_i is the number of training samples in class i, c is the number of distinct classes, μ_i is the mean vector of samples belonging to class i and X_i represents the set of samples belonging to class i with x_k being the k-th sample of that class. S_W represents the scatter of features around the mean of each class and S_B represents the scatter of features around the overall mean for all given classes. The goal is to maximize S_B while minimizing S_W, in other words, maximize the ratio $\det|S_B|/\det|S_W|$. This ratio is maximized when the column vectors of the projection matrix (W_{LDA}) are the eigenvectors of $S_W^{-1}.S_B$. In order to prevent S_W becoming singular, PCA can be used as a preprocessing step and the final transformation is $W_{opt}^T = W_{LDA}^T W_{PCA}^T$ [51].

15.4.3. Artificial Neural Networks

Researchers have long felt that the neurons in the human brain are responsible for the human capacity to learn and adapt. Artificial Neural Networks (ANNs) are mathematical models for neuron behavior, inspired from the human brain and nervous system. On a very simplified and abstract level, ANNs models are based on the cognitive process of the human brain and of course much simpler than the human nervous system. A neural network consists of a set of interconnected processing algorithms functioning in parallel. Mathematical functions or neurons, link together, to build a network which mimics the human nervous system. In ANNs, neurons have weights which are assigned randomly. These weights are adjusted in such a way by means of an iterative or 'learning' process until the desired outputs are obtained. The final trained set of weights and functions are then saved as a 'neural network' [58-59].

In other words, the artificial neural networks (ANNs) are made of a set of elementary units also called perceptrons combined in a layered structure. The elements of the net (neurons) are generally disposed in layers and present a high degree of interconnection with each other. For each connection of the net we associate a weight that represents a sort of inside knowledge. A neural network, by altering the weights of connections across a learning rule, is able to learn a distinctive function from input/output (training set) of examples that are repeatedly presented to the model. From the ability to generalize results of learning, on patterns not present in the training set (test set), we deduce the goodness of the trained model.

A general structure of a network is shown in Fig. 15.5. The variables $x_1, x_2, ..., x_i, ..., x_n$ are the n inputs to the perceptron. These are analogous to inputs arriving from several different neurons to one neuron. The variables $w_1, w_2, ..., w_i, ..., w_n$ are the weights associated with the inputs. When w_i is positive, input x_i acts as an excitatory signal, it acts as inhibitory signal when input x_i is negative. The perceptron sums the product of these inputs and their associated weights $\left(\sum w_i x_i\right)$, compares it to a predefined threshold value also called bias and, if the summation is greater than the threshold value, computes an output using a nonlinear function (f). The signal output y is a nonlinear function (f) of the difference between the preceding computed summation and the threshold value and can be written as [58].

$$y = f\left(\sum w_i x_i - t\right), \tag{15.10}$$

where x_i is the input signals ($i = 1, 2, ..., n$), w_i are the weights associated with the signal x_i, t is the threshold value and $f(s)$ is the nonlinear function.

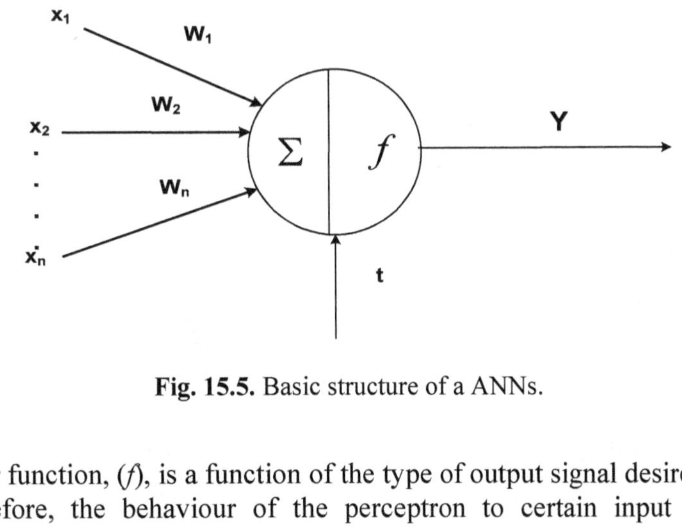

Fig. 15.5. Basic structure of a ANNs.

The nonlinear function, (f), is a function of the type of output signal desired in the neural model. Therefore, the behaviour of the perceptron to certain input information is determined by the weights of its input connections and by the level at which the threshold is set. The transfer function used will also define the shape of the transition step. Knowledge is stored as the values of adjustable parameters w_i, so initially the connection weights are set to small random values. Learning is then the process of adjusting the values in a way that roughly parallels the training of a biological system. The most common activation functions are linear, step, tanh and sigmoid.

A single perceptron has limited data processing capability but when connected with other perceptrons in a layered structure, then it transforms into a very powerful computational engine. Usually this is being done in two different stages: first giving the perceptrons neighbours to form a layer of units which receive inputs from the external source or sensors; secondly by introducing further layers, each taking, as their input, the output from the previous layer. In this way, the network of perceptrons is constructed and known as multilayer feedforward network.

In simple way, the basic structure of an ANN consists of three layers i.e. input layer, hidden layer and output layer. Each layer consists of a number of neurons. The input data is fed to the input neuron layer. The input data can be from any external source or from sensor. The output neuron layer represents the response from the network. In between the input and output layers one or several hidden layers can be used. ANN models solve problems by adapting to the nature of the data they receive. One of the ways to accomplish this is to use a training-data set and a testing-data set of input and output data. It starts with random assignments of weights to the paths joining the perceptrons in the different layers. Then an input x from the training-data set is passed through the artificial neural network. The neural network computes a value $\left(f(x)_{output}\right)$, which is compared with the actual value $\left(f(x)_{actual} = y\right)$. The error σ is computed from these two output values as:

$$\sigma = f(x)_{actual} - f(x)_{output} \tag{15.11}$$

This is the error measure associated with the last layer of the neural network. The technique of distribution of this error to the perceptrons in the hidden layers is called as back-propagation. The error associated with different perceptrons in the hidden layers is computed and different weights connecting different perceptrons in the networks are corrected so that they can approximate the final output more closely.

For a perceptron with an error σ associated with it, the associated weights may be updated as:

$$w_i(new) = w_i(old) + \alpha.\sigma.x_i, \tag{15.12}$$

where α is the learning constant, σ is associated error measure and x_i is the input to the perceptron. The input value x_i is passed through the network (now having updated weights) again, and if there are errors, the weights and errors are computed again. This process is iterated until the error value of the final output is within some user defined limits. The neural network then uses the next set of input-output data. This method is continued for all data in the training-data set. Finally the testing-data set is used to verify how well the neural network can simulate the relationship between the input and output. The learning rate factor α varies between 0 and 1 and accelerates or slows down the descent towards the global minimum of the system.

The backpropagation by gradient descent is generally a reliable procedure of learning networks; nevertheless, it has its limitations: it is not a fast training method, and it can be trapped in local minima.

To avoid the latter, a variant of the above algorithm called gradient-descent with momentum (GDM) introduces a third term, β, referred to as the momentum:

$$w_i(new) = w_i(old) + \alpha.\sigma.x_i + \beta.\Delta w_i(old), \tag{15.13}$$

where β takes a fixed value between 0 and 1 and serves to reduce to the probability of the system being trapped in a local minimum.

One of the problems that may occur during neural network training is called overfitting [59]. This situation occurs when the error on the training set is driven to a very small value, but the new data presented to the network shows a large error. The overfitting concept means that the network has memorized the training examples, but the generalization to new situations is incorrect. In practice, there are two methods that can be used to avoid overfitting, Bayesian regularization, and Early stopping. In the first case, the idea is to select the simplest network possible, and this is simplified by eliminating nodes whose weight connections are not significant enough. In the second case, three subsets of data are employed, A first subset is used for training the model, a second (validation set) is used to check if overfitting is taking place (detected as an increase in the validation error), and a third set (the external test set) is used only to compare performance between different models. All these precautions represent an increased experimental effort in generating data for the numerical model, which can be one of the drawbacks when working with ANNs [60-62].

15.4.4. Probabilistic Neural Networks

For classification purposes, a probabilistic neural network (PNN) has also been reported in literature [51, 62]. The PNN, introduced by Donald F. Specht in 1990 firstly, is relatively new compared to other neural networks like back-propagation artificial neural networks (BP-ANN) and radial basis function networks (RBFN) [63]. It takes its basic concept from the Bayesian statistical classifier which is the optimal statistical classifier. It maps the Bayes rule into a feed forward multilayer neural network. PNN associates an unknown pattern to the class to which this pattern is most likely to belong. PNN constitutes a classification methodology that combines the computational power and flexibility of ANNs. The main advantages of PNN over other ANNs include their simplified architecture which overcomes the difficulty of specifying an appropriate ANN model as well as their easy implementation during training and testing.

This type of network is composed of four layers. In the input layer there are *p* elements that correspond with the first *p* components of PCA that carry almost 100 % information of the old variables. The next layer is the pattern layer. This layer has a number of neurons equal to the training pattern vectors, grouped by classes, where the distance between the test vector and a learning pattern is assessed. The purpose of this layer is to measure and weigh with a radial function the distance of the input layer vector with each training set element. The third layer, the summation layer, contains one neuron for each class. This layer adds the outputs of the pattern neurons belonging to the same class. Finally, the "output layer" is simply a thresholder that seeks for the maximum value of the summation layer. Then the highest one is selected and takes that '1' as a result. The other outputs are set to '0'. In Fig. 15.6 a schematic diagram of the PNN network is shown [47].

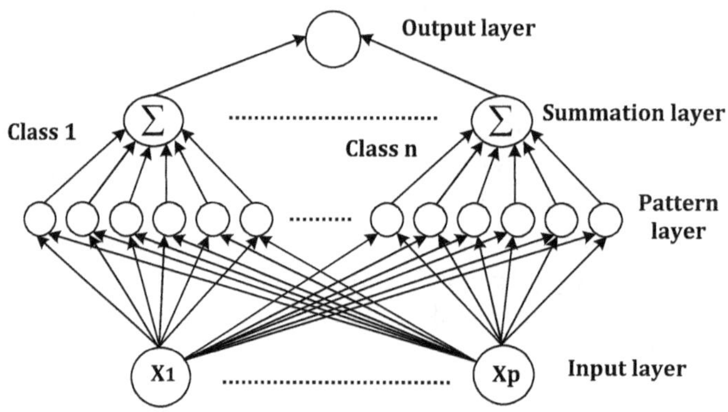

Fig. 15.6. Structure of basic probabilistic neuronal network (PNN).

A validation method was applied to the network in order to check the performance of the network. The method consisted of validating *N* distinct nets (in this case, *N* is the number of measurements) by using *N-t* training vectors and *N-(N-t)* testing vectors which were excluded from the training set for each group.

15.5. Applications of Biomimetic Systems

This section describes the applications of biomimetic systems (i.e. E-nose and E-tongue) as case study of two beverages namely Himalayan teas and Indian wines. First application involving electronic nose system relates to the analysis of Indian teas [64] and the later application involving E-tongue system relates to the classification of Indian wines [65].

15.5.1. E-nose (zNose) for Nondestructive Analysis of Indian Teas

Besides other beverages like coffee, cola, fruit juices, soft drinks etc., tea is one of the popular beverages not only for its taste but also for its beneficial medicinal properties [66]. There are many different tea species or clones among the world. Just like grapes and wines, the quality of tea is dependent on where and how it is grown along with the processing conditions. Different varieties of tea have different taste, flavor and color too. In Asia, the main tea-producing countries are India, China, Sri Lanka and Japan. Indian tea tastes quite different from that of China or Sri Lanka, and teas from Assam in Northern India differ in flavor of those from Nilgiri in the South and Kangra tea in the Northern India. Tea is processed from tender shoots (two and a bud) of *Camellia sinensis* (L.) O. Kuntze. The different classifications are dependent on the way the tea leaves are processed. The quality of tea can be attributed to the following three properties - physical appearance, infusion and flavor. Most of the quality parameters inherent in the tea shoots are determined by the germplasm characteristics and agro-climatic conditions [66-68]. Flavor is the most important parameter in the evaluation of the quality of made tea. Flavor of tea comprises of taste and aroma. Taste comes from non-volatile constituents, while aroma is due to volatile constituents. Over 500 volatile flavor components have been identified from tea. The flavor is dependent on the availability of precursors present in tea shoot, stimulation of conditions during tea manufacture for their liberation and retention of flavor components in the product [69]. Geographical and clonal variations affect the flavor constituents of black tea [68, 70]. Orthodox made tea from Kangra valley is based on China hybrid tea variety and has a unique flavor. Tea manufactured from different clones grown under Kangra conditions showed variation in quality when evaluated on the basis of taste and aroma. Crop yield is divided into three growth flushes in this region, early flush (April–mid June), rains flush (mid June–mid September) and backend flush (mid September–November). The previous study undertaken on manufactured orthodox tea over different growth flushes reported that early flush Kangra tea valued for high flavor undergoes a decline with progress in the season, reaching the trough in rains flush, followed by marginal recovery during backend flush [71].

Usually, the conventional assessment of a tea quality is performed by sensory evaluation where the human olfactory senses differentiating flavor, taste and color of the same commodity with little variation are used as important tools for quality judgment. Sensory evaluation of tea is one of the most difficult tasks in the overall tea attribute assessment. It relies on information provided by trained tasting panels, whose members may be influenced by physiological, psychological, and environmental factors. Therefore, the quality assessment of tea by human tasters is inevitably subjective with inconsistent results and expensive. Instrumental methods for the determination of odor, color and taste, such as gas chromatography, gas chromatography /mass spectrometry (GC-MS), high-performance liquid chromatography with mass spectroscopy (LC-MS), colorimeters, inductively coupled plasma atomic emission spectrometer, Infrared spectroscopy (NIR) [72-74] are costly, and they require trained personnel and are often of limited value because of destructive sampling and time-consuming procedures involved. However, all of the methods mentioned above are time-consuming in the identification of tea categories. Budinova et al. reported an application of NIR for the assessment of authenticity of tea [74-75]. In the works mentioned above, near infrared spectroscopy techniques have often been used to analyze quantitatively the valid tea composition. Recently, some researchers applied electronic tongue based on voltammetry and metal oxide based electronic nose for discrimination of different teas [76]. As it is known that the different tea varieties have chemical characters which are due to different tea processes and different origin of tea-leaves, therefore, the flavor features of each tea variety are reasonably different. Therefore, the aim of this chapter is to investigate flavor features of Kangra orthodox black tea of the same origin with different manufacturing seasons nondestructively within seconds using ultra fast GC technique coupled with chemommetric techniques. In this work, ultra fast GC frequency spectra coupled with the pattern recognition methods was used to identify seven varieties of orthodox black teas.

15.5.1.1. Experimental Details

15.5.1.1.1. Tea Samples

Tea samples from seven varieties of Kangra Orthodox black tea (Camellia sinensis (L.) O. Kuntze) were manufactured from tea shoots which comprising of a bud and subtending three leaves harvested during the three flushing seasons (Early, Rains, and Backend) at Banuri Tea Experimental Farm of the Institute of Himalayan Bio-Resource Technology (A constituent laboratory of CSIR, India), Palampur. The tea samples were subjected to undergo the unit process operation of withering under ambient air at constant flow; rolling using a peizy roller for 0.5 h, fermentation for 3 h, and drying with hot air at 95 °C in a miniature dryer. Six samples in triplicates of each variety namely K1, K2, K3, K4, K5, KB and KL have been used for investigation.

15.5.1.1.2. Tea Sample Preparation for the zNose

To study the effect of temperature on emission of aroma from tea samples, tea samples were prepared at two temperatures i.e. room temperature and at 50 °C. Tea samples (10 gm) were weighed and transferred into 60 mL brown glass vial sealed with a screw cap containing a silicon septum. The samples were allowed to equilibrate with the headspace in the vial for 24 hrs at room temperature (24 °C) so that maximum headspace could be generated in the vial. Eighteen sample vials of each variety were prepared the same way. In the case of samples at 50 °C, the sample bottles were kept at 50 °C in oven for 4 hours. The zNose was provided with a 5-cm stainless still needle at the inlet which was used for sampling through the septa of the vials. The sampling was set for 10 sec (sample flow 20 ml/min) after which the system switched to 20 sec of data acquisition mode.

15.5.1.1.3. zNose GC Parameters

The zNose parameters were selected after certain experiments carried out. The final inlet temperature was set to 200 °C, the valve temperature was set to 165 °C, and the initial column temperature was kept at 40 °C. During analysis the column temperature was increased at the rate of 10 °C per second to reach a final column temperature of 200 °C. The SAW sensor was operated at 60°C and the trap was operated at 250 °C. Helium flow was set at 3.00 ccm. After each data sampling period the system needed a 30-s baking period, in which the sensor was heated briefly (5–8 s) to 150 °C and after which the temperature conditions of the inlet, column, and sensor were reset to the initial conditions. In between each sample measurement at least one blank was run to ensure cleaning of the system and a stable baseline.

15.5.1.2. Data Analysis

The data samples were first preprocessed and analyzed using chemometric techniques. In this study, different data preprocessing methods were applied; standard normal transformation (SNV), Detrending, first order derivative and Multiplicative scatter correction (MSC). SNV is a mathematical transformation method of the log (1/R) spectra used to remove slope variation and to correct for scatter effects. Compared to SNV, first derivatives eliminate baseline drifts and small spectral differences are enhanced. To avoid enhancing the noise, which is a consequence of derivative, spectra are first smoothed. This smoothing is done by using the Savitzky–Golay algorithm, which is a moving window averaging method: a window is selected where the data are fitted by a polynomial of a certain degree. The central point in the window is replaced by the value of the polynomial.

Principal component analysis (PCA) was used for the extraction of the relevant features of the spectral response signals. Besides, Linear Discriminant Analysis (LDA) is used as classifier with the extracted features subspace of Principal components derived using PCA. The data obtained were evaluated using the following two commercial software package: Unscrambler (X, 2011, CAMO ASA, Trondheim, Norway) and Matlab (version 7.1, Mathworks, USA) for programming the feature extraction and the pattern recognition techniques.

15.5.1.3. Results and Discussion

15.5.1.3.1. Frequency Spectral Response of Tea Samples

Fig. 15.7 shows the raw frequency spectra of seven different varieties of orthodox tea samples. As shown in the figure, meaningful information or variations in the resultant spectra are only observable within the 2-10 sec time duration. The rest of the data (other than the data recorded during 2-10 sec) was discarded and not included in further data analysis.

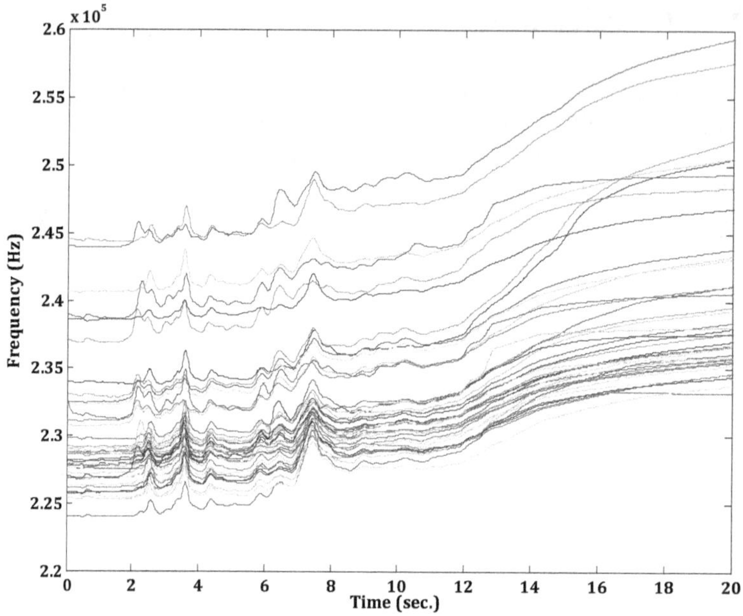

Fig. 15.7. Raw frequency spectra of different teas sampled at room temperature.

The spectral data in the 2-10 sec time duration was further processed using SNV, detrending and MSC transformations. The final spectra obtained after undergoing the transformations by the above preprocessing techniques is shown in Fig. 15.8. The trend of spectra was quite similar, but with a careful observation, some crossover between the different varieties was detected. In order to make good use of the spectra, further treatment with the chemometric methods described above were used to discriminate the different tea samples.

15.5.1.3.2. Principal Component Analysis (PCA) of Frequency Spectra

The frequency spectra obtained from the zNose for aroma profiles of seven classes of Himalayan orthodox black teas at two different temperatures (i.e. room temperature and 50 °C) are analyzed using PCA.

The two-dimensional score plot of PCA of the data set at room temperature for the seven tea varieties is shown in Fig. 15.9. The first PC contained 59.88 % of the information; second PC contained 22.38 % and the third PC contained 6.45 %. The first two PCs contained more than 82 % of information of tea samples. As shown in Fig. 15.9, seven distinct clusters of seven varieties (K1, K2, K3, K4, K5, KB and KL) are clearly visible. The clusters of KB and KL varieties are far away from K1, K2, K3, K4 and K5 varieties. There is small overlap between clusters of K2 and K3. In the case of headspace generated at 50 °C temperature, the overlap between K2 and K3 is larger as shown in Fig. 15.10. The first PC of the samples only contained 34.5 % of information while second PC and third PC contained 28.4 % and 19.7 % of information. The aroma generated at room temperature has better discrimination capability as compared to the headspace generated at 50 °C. Therefore, for further classification only data samples of aroma generated at room temperature were adopted for classifier design.

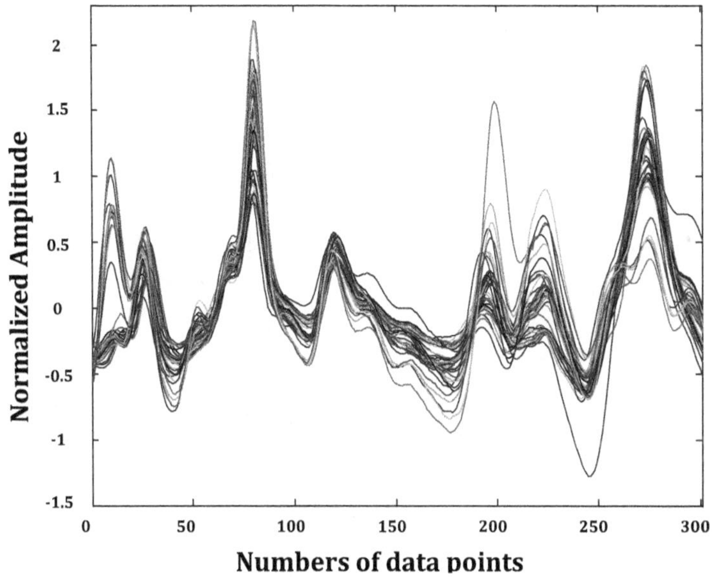

Fig. 15.8. Frequency spectra of different teas (after preprocessing) sampled at room temperature.

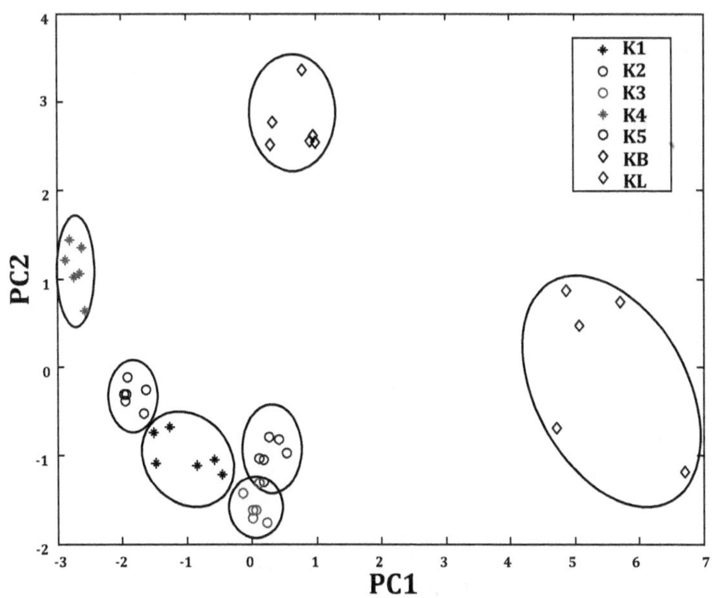

Fig. 15.9. Score plot of different tea samples stored at room temperature.

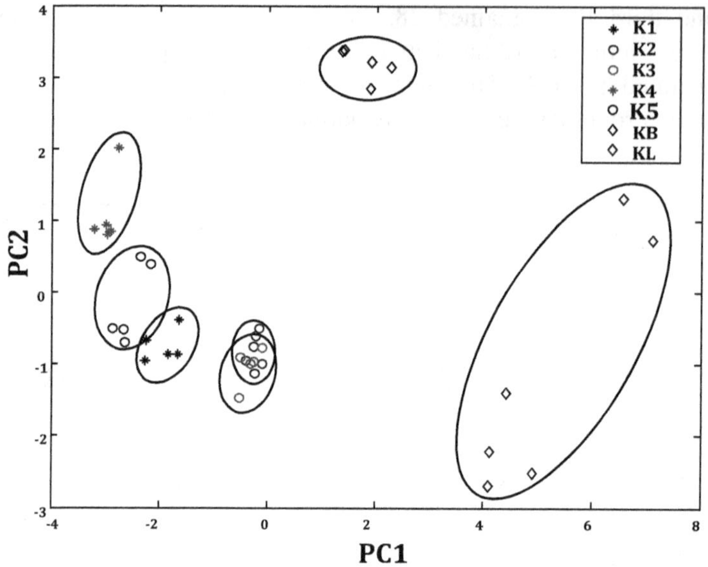

Fig. 15.10. Score plot of different tea samples stored as 50 °C.

15.5.1.4. Classification of Samples by PCA-LDA

In this study, LDA was applied on the feature subspace spanned by the principal component vectors extracted by PCA. In order to optimize the LDA model, PCA-LDA architectures with different numbers of inputs (PCs) were examined (1PC to 5PCs).

Finally, 4 PCs were selected as the inputs to the PCA-LDA model. Total 42 samples were selected for training the classifier. PCA-LDA analysis has shown an overall 97.5 % classification accuracy for the whole data set involving all seven varieties of tea samples. The confusion matrix for all the tea classes is shown in Table 15.1. A plot of the LDA classification results is shown in Fig. 15.11. Out of all the 42 samples of 7 classes - K1, K2, K3, K4, K5, KB, KL- one sample corresponding to 'K2- class' has been misclassified to 'K3- class' as seen in Table 15.1 and in Fig. 15.11.

Table 15.1. Confusion matrix of the classified tea samples.

Sample	**Predicted Samples**							
Actual Sample		k1	k2	k3	k4	k5	k6	k7
	k1	6	0	0	0	0	0	0
	k2	0	5	1	0	0	0	0
	k3	0	0	6	0	0	0	0
	k4	0	0	0	6	0	0	0
	k5	0	0	0	0	6	0	0
	k6	0	0	0	0	0	6	0
	k7	0	0	0	0	0	0	6

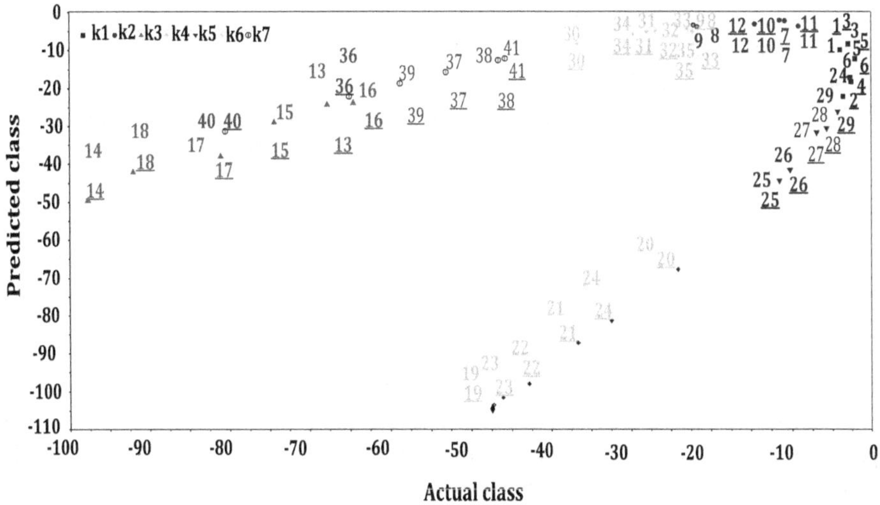

Fig. 15.11. A plot depicting the classification results (Actual class (denoted by plain text) vs. Predicted class (denoted by underlined text).

15.5.2. Authentication of Indian Wines Using E-tongue

According to the Rigveda (an ancient Indian sacred collection of Vedic Sanskrit hymns which is counted among the four canonical sacred texts of Hinduism known as the Vedas), wine is perhaps the oldest fermented product known to man. Wine refers to the fermented by-product of grapes (Vitis vinifera) but sometimes may include undistilled alcoholic fermented fruit product. In India, wine industry is in its infancy [77]. Recently, wines are gaining popularity in India but have yet to be characterized for successful marketability. Indian wine producers are anxious to market their products based on the quality attributes with comparison to international wines.

The quality of wines is normally estimated and controlled by human tasters (based on colour, aroma and taste profiles) and advanced analytical tools such as mass spectrophotometer (MS), liquid chromatography (LC), gas chromatography (GC), spectroscopy or combinations of plurality of these analytical tools etc. [78]. The analytical techniques are used for quantitative analysis and are often selective. Even though these analytical tools are highly reliable and accurate, they are quite expensive and require experienced operators. Furthermore, these analytical equipments are difficult to use in on-line applications of quality control. Therefore, rapid, reliable and low-cost methods for assessment of wines quality are of great interest to the wine industry. Over the past decade, there have been a number of attempts to apply electronic taste (E-tongue) and electronic odor (E-nose) technologies to wines, with varying degrees of success. Because of the wide variety of technologies and approaches, it is difficult to summarize the developments effectively or to make valid comparisons between the various studies [79-82]. The potentiometry type of E-tongue and its different variants (ISFET sensor type, silicon type, interdigitated with conducting polymers/lipids type and etc.) have been reported by several researchers for applications in wine industry; from wine classification to process control [83-90]. The potentiometric sensors are well known for their detection limitations towards charged species and also sensitive to noise [80]. As a result, there is a growing interest in voltammetry techniques and linear sweep voltammetry is of particular interest in the development of electronic tongues [91].

The voltammetric E-tongues using copper, copper and gold electrodes have been successfully applied for classification of Chinese yellow wines, red wines, rice wines, wine ages, and also for whiskies and have been reported in the literature [92-95]. Application of hybrid voltammetric E-tongue for detecting artificial taste enhancing adulterants in wines and discrimination of different Spanish red wines were reported by Parra et al. [96-98]. Similarly the hybrid tongue has been also used for the characterization and quantification of grape variety in red wines as reported by Gutierrez et al. [99]. There are no recorded applications of E-tongue technology to the analysis of Indian wines, thus the focus of the present study was to assess the ability of the electronic tongue based on voltammetry to distinguish between Indian wines of different variety and winery in India. In this study, a Voltammetry based Electronic tongue based on three working electrodes (gold, glassy carbon and platinum) was developed to classify eight brands of Indian wines wherein large amplitude pulse waveforms were used for excitation of each of the working electrodes. Different electrodes have different sensitivity to samples, and the signals obtained from different electrodes presented different information regarding the samples. Therefore, the composition of the sensor array is very important in voltammetric type of E-tongue. Furthermore, the Indian wines were first analyzed by the responses from each of the three electrodes, and the analysis results show that each of the electrodes was as important as the other and therefore required in discrimination. The main purpose of this study was to investigate whether the eight types of Indian wines of different brands could be classified by the E-tongue using principal component analysis (PCA) and artificial neural networks (ANN).

15.5.2.1. Material and Methods

15.5.2.1.1. Wine Samples

Forty bottles of Indian wines were obtained from local stores (Rajasthan, India) which included five bottles of each of the eight brands of Indian grape wine i.e. Vino (VN), Grover Cabernet Shiraz (GCS), Grover Shiraz Rose (GSR), Chenin Blanc (CB), Galaxy (GX), Sula Cabernet Shiraz (SCS), Sula Savignon Blanc (SSB), Vinsura Shiraz (VS). These eight brands of wine from five different vineyards as shown the Table 15.2 were examined in this study. There were four types of red wines, two white wines, one rose and one sparkling white wine samples. Each measurement was taken and recorded immediately after opening the bottle cap. The measurement procedure included no sample pretreatment. Initially, after opening the cap of the bottle, wine was filled in glass containers of 100 ml volume. A fresh portion of each wine was measured every time. The samples were analyzed in 3 replicas in a random order.

Table 15.2. Different brands of Indian wines.

S.N.	Vineyards	Brands	Type
1.	Chateau Indage Estate Vineyards	Vino	White sparkling
2.	Grover Vineyards	Cabernet Shiraz	Red
3.	Grover Vineyards	Shiraz Rose	Rose
4.	ND Vineyards	Chenin Blanc	White
5.	ND Vineyards	Galaxy-Red	Red
6.	Sula Vineyards	Sula Cabernet Shiraz	Red
7.	Sula Vineyards	Sula Savignon Blanc	White
8.	Vinsura Vineyards	Vinsura Shiraz	Red

15.5.2.1.2. Operation of the E-tongue System

In this study, current flowing through the working electrode and the counter electrode was measured when a potential pulse was applied over the working electrode with respect to the reference electrode. The applied potential waveform consists of three cycles i.e. electrochemical cleaning cycle, conditioning cycle and measurement cycle respectively. The Fig. 15.12 shows only the measurement cycle. In electrochemical cleaning cycle a positive potential of 2 V was applied for 500 ms, thereafter a negative potential of -2 V was applied for the same period. After the electrochemical cycle, the next cycle is conditioning cycle in which the potential is kept at 0 V for 10 sec.

The final cycle is the measurement cycle which is based on successive voltage pulses of gradually changing amplitude between which the base potential is applied, and the current is continuously measured. The frequency of the pulse was kept at 1 Hz for all these measurements. The potential range for all the measurements was varied from 1 V to -1 V in steps of 100 mV.

15.5.2.2. Data Analysis

In this paper, principal component analysis (PCA) and Artificial neural networks (ANN) were used to differentiate eight types of Indian wines. The data obtained were evaluated using the following two commercial software packages - Unscrambler (version 9.5, 2007, CAMO ASA, Trondheim, Norway) and Matlab (version 7.1, Mathworks, USA) - for feature extraction and pattern recognition.

15.5.2.3. Results and Discussion

15.5.2.3.1. Feature Extraction

In multivariate type of systems like voltammetric E-tongue, number of working electrodes plays a major role in resultant data set for analysis (three electrodes in this study). A large data set was obtained by the three working electrodes in the experiment;

however, most of the data were unusable or redundant. Thus, certain methods are required to be applied to obtain useful data out of the raw data set.

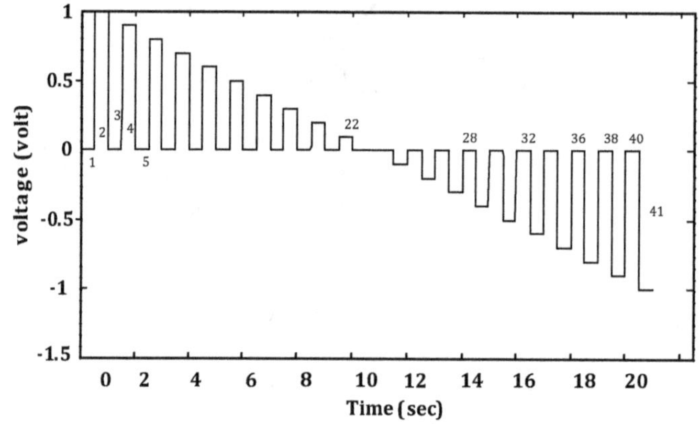

(a)

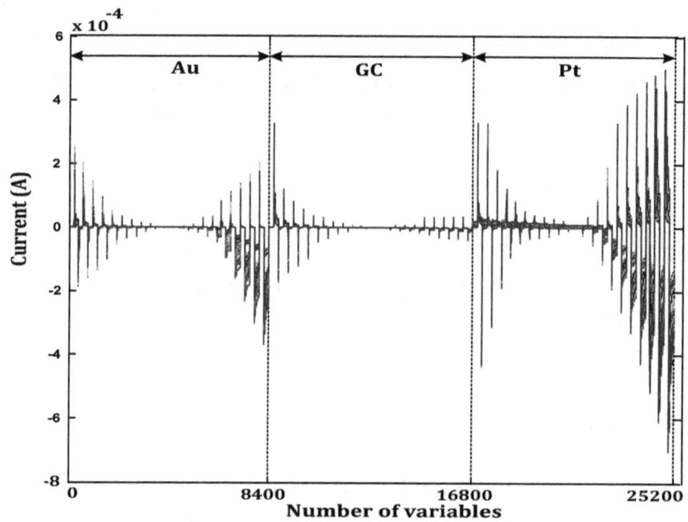

(b)

Fig. 15.12. (a) Excitation signal, and (b) response signals of all Indian wines using all electrodes.

Descriptive statistics are a way to summarize data into a few numbers that contain most of the relevant information. These descriptive statistics can be categorized in terms of measures of central tendency (location) of data by computation of parameters such as geometry mean, harmonic mean, mean, median, and trimmed mean of the data and also measures of dispersive parameters such as inter quartile range (IQR), standard deviation, variance etc. Some times higher order statistical parameters such as higher order moments, skewness and kurtosis provide important information about the data

[47]. In this work, two statistical parameters of the voltammetric signal were extracted i.e. difference between maximum value and minimum value, and mean value of the segment.

The obtained voltammetric signals using three working electrodes are arranged in series by appending them to form a single vector as shown in Fig. 15.12 (b). The size of response vector for each electrode is 8400 values/variables generated for a sample. In this paper, the response vector of each electrode is segmented into 41 steps; two statistical values (difference between the maximum value and the minimum value and mean value) were exacted from the each step cycle for analysis. A total of 246 [(2 data points per step × 41 steps) × 3 electrodes] features of the voltammetric signals were collected for each sample in the experiment.

15.5.2.3.2. PCA Analysis of Voltammetric E-tongue

In the work, initially, each of the individual wine varieties (four red wines, two white wines, one rose wine and one sparkling white wine) was analyzed by PCA to see if they could be discriminated on the basis of wine brands and also the wine type (Red, white, rose and sparkling white). The same data analysis procedure was also applied to the extracted four statistical features (maximum, minimum, mean and median) from these data sets so that fair comparison with full data set can be done. In voltammetric E-tongue, working electrodes play a key role in the discrimination of the samples. To investigate the effects of the electrodes (gold (Au), platinum (Pt), and glassy carbon (Gc)), the voltammetric responses obtained by the individual working electrodes for eight different wine samples were also analyzed by PCA separately. The number of effective electrodes that could classify Indian wines was chosen as an experimental object.

15.5.2.3.3. Results of PCA on Raw Data

The analysis for different wines with PCA score plots of the raw data (without preprocessing) obtained by Au, Gc, Pt and their appended combination (Au, Gc & Pt) are shown in Fig. 15.13 (a-c) and (d). The result showed that the Indian wines with different types could be fairly classified by individual working electrodes.

The score plot of gold electrode clearly separates white wines, sparkling white wine and red wines as shown in Fig. 15.13 (a). Six clusters of different wines are well separated however two varieties overlap. The rose wine cluster overlaps with one of the red wine cluster. In the case of glassy carbon electrode as shown in fig 13(b), six clusters of six varieties are well separated and two clusters overlap. Clusters for white wines are well distinct from red and sparkling white wines. The rose wine cluster overlaps with one of the red wines' cluster. In the case of Pt electrode, five clusters for different wines are well separated, however three clusters overlap. Here again, white wine clusters are well distinct from those of red wines and also far away from that of sparkling white wine.

Interestingly, rose wine cluster is also well distinct from that of other wines' groups. The overlapping clusters are all from red wines only. It is clear from the score plots for all the three electrodes that white wines, sparkling white wine and red wines have clearly separated clusters.

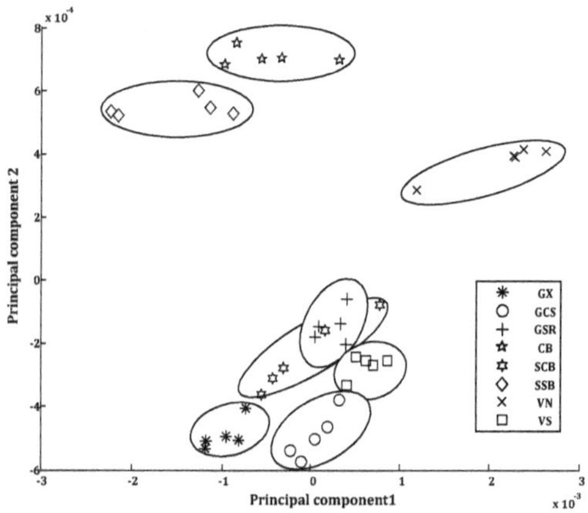

(a) Gold electrode

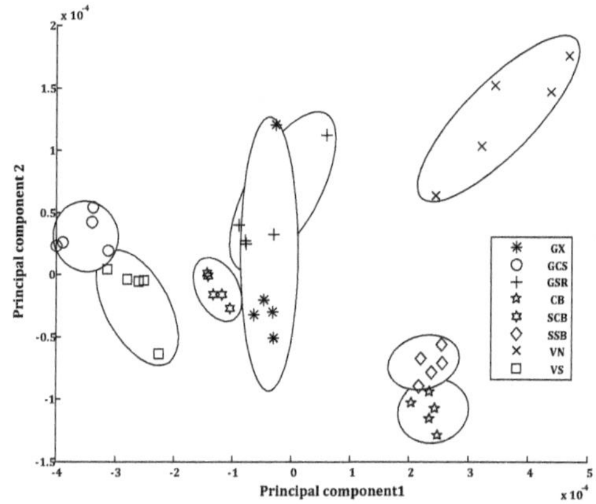

(b) Glassy carbon

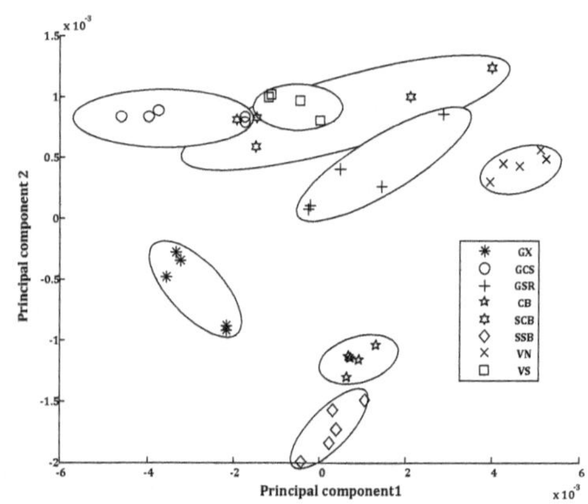

(c) Platinum electrode

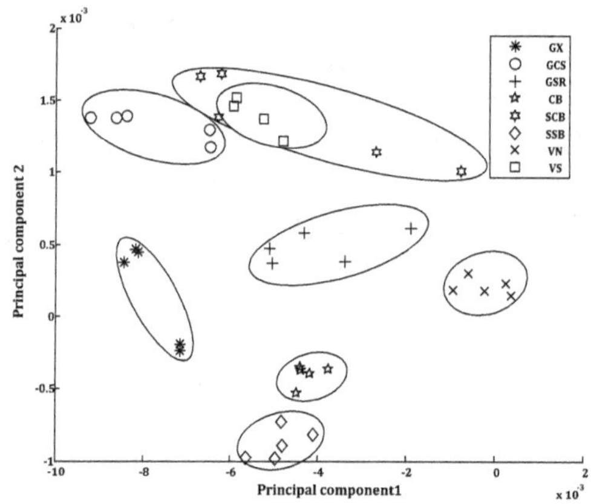

(d) E-tongue signal

Fig. 15.13. (a-c) Discrimination of PCA plots for eight types of Indian wines using Au, GC & Pt electrodes and (d) discrimination of PCA plots for eight types of Indian wines from the merged data of the three electrodes.

The PCA score plot of merged data from all three electrodes is shown in Fig. 15.13 (d). Again six of the clusters (two from red wines, two from white wines, one each from rose and sparkling white wines) are well separated, while the other two (from red wines) overlap (top-middle section of Fig. 15.13 (d)). All the white wine clusters are well separated from the red wines clusters so that cross-validation resulted 100 % of the samples to be categorized correct according to their brand. The sparkling white wine (middle right section of Fig. 15.13 (d)) is also distinct cluster but tends towards white wine clusters. The rose wine cluster is also between the red wines and white wines (middle section of Fig. 15.13 (d)). The explained variances by different principal components (PC1& PC2) are showed in Table 15.2.

The PC1 and PC2 together explain more then 90 % of the variance in the data for each of the electrodes individually and the appended data from all of the electrodes has shown highest explained variance i.e. 97.08 % as shown in Table 15.3. All the samples were classified reliably by PCA in two-dimensional principal components.

Table 15.3. Explained variance by PC1 & PC2 using different electrodes for Indian wine samples.

Electrode	Explained variance by PC1	Explained variance by PC2	Total explained variance
Gold	82.26	13.54	95.80
Glassy Carbon	85.32	08.21	93.53
Platinum	79.70	11.79	91.49
Combined (Au, Gc & Pt)	90.52	06.56	97.08

15.5.2.3.4. Results of PCA on Reduced Data

In voltammetric electronic tongues, the nature of the signals involves the recording of currents generated in the solutions under study related to an applied potential. Voltammetric signals contain hundreds of measures and usually overlapping regions with non-stationary characteristics, since all components in the solution which are electrochemically active below a specific potential, all contribute to the current measured.

In this sense, the voltammograms should be firstly analyzed in order to corroborate its analytical content. In order to reduce the huge amount of data generated in each measurement, a preprocessing stage employing extraction of statistical features was used. The compression of the original sensor information was achieved up to 97.34 % without any loss of relevant information using four statistical parameters. In this way, and using the proposed sensor array, the corresponding compressed data corresponding to the extracted statistical features were processed by applying PCA. To determine the discrimination capability of the extracted features based voltammetric E-tongue, PCA analysis of extracted features for each electrode as well as the merged or appended features of all electrodes was carried out. The resultant PCA score plots are shown in Fig. 15.14 (a) to (d). The explained variances by each of the three electrodes and their combination for the extracted features' data are shown in Table 15.4.

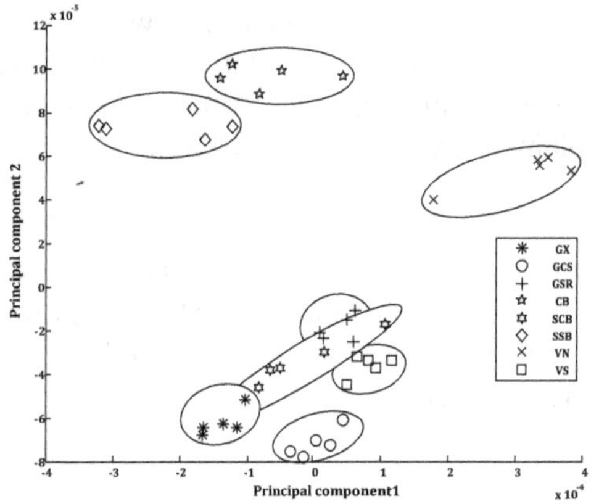

(a) Gold electrode

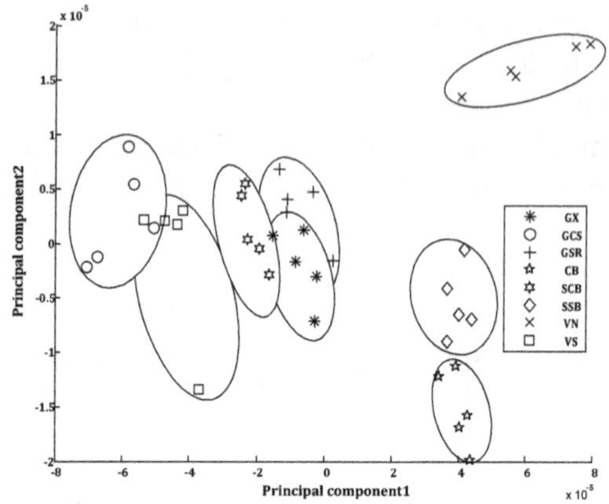

(b) Glassy carbon

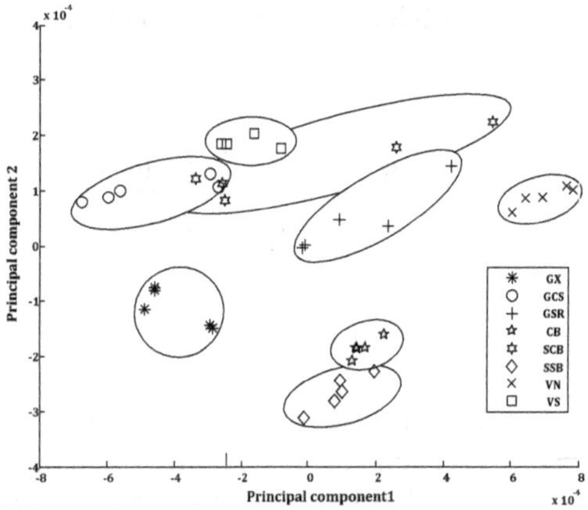

(c) Platinum electrode

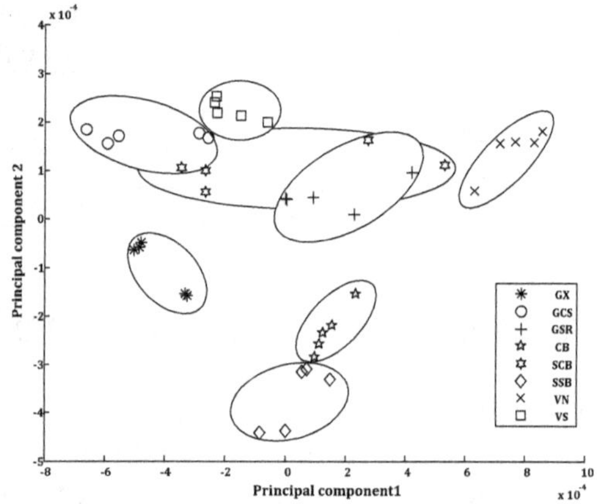

(d) E-tongue signal

Fig. 15.14. (a-c) Discrimination of PCA plots for eight types of Indian wines using extracted features from Au, GC & Pt electrodes and (d) discrimination of PCA plots for eight types of Indian wines extracted features from the merged data of the three electrodes.

Table 15.4. Explained variance by PC1 & PC2 using different electrodes for Indian wine samples (extracted features).

Electrode	Explained variance by PC1	Explained variance by PC2	Total explained variance
Gold	84.85	12.38	97.23
Glassy Carbon	94.02	4.54	98.56
Platinum	80.18	13.79	93.97
Combined (Au, Gc & Pt)	73.22	18.86	92.08

As shown in Fig. 15.14, the extracted features from all the electrodes have shown good discrimination capabilities for eight types of Indian wines. The PC1 has explained variance of more than 73 % while PC2 has explained about 19 % of the variance in the extracted features' data for each of the electrodes individually. The appended data of the extracted features from all of the electrodes has shown total explained variance i.e. 92.08 % as shown in Table 15.4. All the samples were classified reliably by PCA in two-dimensional principal components.

The PCA score plots of the extracted features' data corresponding to individual electrodes (Au, Gc & Pt) are more or less similar to those of the raw data. In the case of extracted features' data appended from each of the electrodes, six clusters are well separated while one red wine and one rose wine cluster overlap. The overall results of PCA are well acceptable. The reduced features' data has similar information as the raw data for the discrimination of the samples into their corresponding brands.

15.5.2.4. Classification Results of Indian Wines

As described in previous sections, the data compressed with statistical methods and next processed using PCA, only provides a visualization of the different grouping regions. In this sense, an actual classifier was built employing an ANN. This model was able to perform properly the qualitative classification of the eight main classes of Indian wines. In order to find the appropriate ANN model, significant effort is needed to optimize the configuration details that determine its operation. Normally, this is a trial-and-error process, where several network structural parameters like training algorithms, number of hidden layers, transfer functions, etc. are fine-tuned in order to find the best configuration to optimize the performance of the model. After some initial tests, we designated to adopt a typical three-layer simple feed-forward neural network trained by back-propagation algorithm. The final ANN architecture model had 235 input neurons, 8 neurons in the hidden layer and eight binary (1/0) output neurons (one for each Indian wine). The optimal number of neurons in the hidden layer was determined by varying the number of neurons in the hidden layer and comparing the performances. The hyperbolic tangent sigmoid function was used as the activation function for hidden layer as well as for output layer and the weights between layers were modified by backpropagating the error signals (computed by the output neurons), layer by layer, with the conjugate gradient algorithm [58]. The expected output error was programmed to reach a value of 0.025.

Nearly 75 % of the data was utilized to train the network while the remaining 25 % of the data was used to test the system. The network was trained with the scope of representing the 246 dimensional electrode array output in a suitable eight-dimensional output space assigning to the eight classes. The ability to recognize unknown samples was tested by submitting the remaining 25 % of data collected to the trained network. Since the result can be affected by initial weights in the network, we repeated each experiment for 10 times and the results were averaged. The averaged accuracy increases sharply at small numbers of neurons in the hidden layer and saturates with increasing number of neurons in that layer.

After training, a confusion matrix was built using the information of the testing subset (remaining 25 % of the data) in order to characterize the accuracy of the identification model and obtain unbiased data. Table 15.5 shows the results of the classification, expressed as the corresponding confusion matrix; its content displays expected vs. found result or response as calculated from the adopted ANN model. As it can be observed, correct classification for seven wines i.e. Galaxy, Cabernet Shiraz, Shiraz rose, Sula Cabernet Shiraz, Sula Savignon Blanc, Vino and Vinsura Shiraz was obtained, while only one sample of Chenin Blanc was misclassified as the Sula Savignon Blanc wine. The percentage of classifying correctly the samples during training of network involving 75 % of the data was 100 %. The overall classification accuracy obtained for eight types of wines was 97.5 %. Hence, the developed E-tongue based on voltammetry with three working electrodes with reduced feature extraction technique has great potential to discriminate Indian wines.

Table 15.5. Confusion matrix for classification of Indian wines.

Predicted	**Galaxy**	**Caber Shiraz**	**Shiraz Rose**	**Chenin Blanc**	**Sula Caber Shiraz**	**Sula Savig Blanc**	**Vino**	**Vinsura Shiraz**
Galaxy	100	0	0	0	0	0	0	0
Cabernet Shiraz	0	100	0	0	0	0	0	0
Shiraz Rose	0	0	100	0	0	0	0	0
Chenin Blanc	0	0	0	80	0	20	0	0
Sula Cabernet Shiraz	0	0	0	0	100	0	0	0
Sula Savignon Blanc	0	0	0	0	0	100	0	0
Vino	0	0	0	0	0	0	100	0
Vinsura Shiraz	0	0	0	0	0	0	0	100

15.6. Conclusions

The feasibility of the biomimetic systems, both E-nose and E-tongue in classification and authentication of beverages has been studied.

In the case of the case study involving E-nose, it can be concluded that zNose measurements coupled with suitable pattern recognition has a high potential to estimate aroma of tea which is one of the major components of tea quality. The technique can be further integrated with electronic taste and electronic eye for development of electronic panel for estimating the quality of teas. It also has potential to be applied to the tea quality control, tea manufacturing process monitoring, and rapid classification in the tea industry. In comparison to subjective sensory assessing methods and time-consuming chemical methods, the results obtained by the developed zNose model can represent a considerable improvement in estimating the tea aroma.

The results in the case of the case study involving E-tongue show that minor inter-winery variations can also be exploited to characterize a particular winery. Further experiments are required to establish standards to allow application of this approach in a commercial setting. The E-tongue can thereby distinguish differences in the non-volatile profiles of closely related wine types, and has established important relationships between wines from different grape varieties. Further efforts are needed to investigate application of voltammetric E-tongue for its ability to recognize relationships between some fruit wines and grape wines which can have some marketing potential. Thus, the E-tongue may have the ability to elucidate the relationship between wines, and ensure quality and uniformity based on non-volatile components. Further studies are proposed to investigate chemical compositions, correlating E-tongue signals with human perceptions of taste, and multi sensor fusion models with E-nose signals.

The advantages of both the systems can be exploited by developing a hybrid system which includes the E-nose and E-tongue (Electronic eye i.e. camera may also be included and embedded) for complete analysis of the samples wherein the information regarding the smell and taste (and even the colour) can be coupled and correlated for better analysis and results.

Acknowledgements

The study was carried out under a supra institutional research project (SIP-21) financed by the Council of Scientific and Industrial Research (CSIR), India.

References

[1]. A. Rehman, N. Iqbal, P. A. Lieberzeit, F. L. Dickert, Multisensor biomimetic systems with fully artificial recognition strategies in food analysis, *Monatsh Chem*, 140, 2009, pp. 931–939.

[2]. M. G. Varnamkhasti., S. S. Mohtasebi., M. Siadat, Biomimetic-based odor and taste sensing systems to food quality and safety characterization: An overview on basic principles and recent achievements, *Journal of Food Engineering*, 100, 3, 2010, pp. 377–387.

[3]. K. Persaud, G. H. Dodd, Analysis of discrimination mechanisms in the mammalian olfactory system using a model nose, *Nature*, 299, 1982, pp. 352–355.

[4]. P. N. Bartlett, N. Blair, J. W. Gardner, Electronic nose. Principles, applications and outlook, *ASIC, 15e Colloque*, Montpellier, 1993, pp. 478–486.

[5]. J. W. Gardner, P. N. Bartlett, A Brief History of Electronic Noses, *Sensors and Actuators B.,* 18, 1993, pp. 211–220.

[6]. P. N. Bartlett, J. M. Elliot, J. W. Gardner, Electronic noses and their application in the food industry, *Food Technology*, 51, 1997, pp. 44–48.

[7]. P. Mielle, 'Electronic noses': Towards the objective instrumental characterization of food aroma, *Trends in Food Science & Technology, Special Issue on Flavour Perception*, 7, 1996, pp. 432–438.

[8]. V. Demarne, S. R. Sanjine, Thin film semiconducting metal oxide gas sensors, in: Sberveglieri, G. (Ed), Gas Sensors, *Kluwer Academic Publishers*, Dordrecht, 1992, pp. 89–116.

[9]. D. Hodgins, The Electronic Nose: Sensor Array-Based Instruments that Emulate the Human Nose, in: Marsili, R. (Ed), Techniques for analyzing food aroma, *Marcel Dekker Inc.,* New York, 1991, pp. 331–371.

[10]. C. M. Mari, G. B. Barbi, Electrochemical gas sensors, in: Sberveglieri, G. (Ed), Gas Sensors, *Kluwer Academic Publishers*, Dordrecht, 1992, pp. 329–364.

[11]. T. A. Dickinston, J. White, J. S. Kauer, D. R. Walt, A chemical-detecting system based on a crossreactive optical sensor array, *Nature*, 382, 1996, pp. 697–700.

[12]. E. Schaller, J. O. Bosset, F. Escher, Electronic Noses' and Their Application to Food, *LWT - Food Science and Technology*., Vol. 31, Issue 4, April 1998, pp. 305–316.

[13]. D. James, M. Simon, S. M. Scott, Z. A. Zulfiqur, T. O. William, W. T. O O'Hare, Chemical Sensors for Electronic Nose Systems, *Microchim. Acta,* 149, 2005, pp. 1–17.

[14]. A. Chiba A, Chemical sensor technology. Vol. 4 In: Yamauchi S (Ed), Kodansha Ltd, *Elsevier*, Tokyo, 1992, pp. 1-18.

[15]. G. Eranna, B. C. Joshi, D. P. Runthala, R. P. Gupta, Oxide Materials for Development of Integrated Gas Sensors—A Comprehensive Review, *Critical Reviews in Solid State and Materials Sciences,* 29, 3, 2004, pp. 111-188.

[16]. D. Blackwood, M. Josowicz, Work function and spectroscopic studies of interaction between conducting polymers and organic vapor, *J Phys Chem* 95, 1995, pp. 493-502.

[17]. A. Guadarrama, J. A. Fernandez, M. Iniguez, J. Souto, J. A. de Saja, Array of conducting polymer sensors for the characterisation of wines, *Analytica Chimica Acta*, 411, 1–2, 2000, pp. 193–200.

[18]. B. C. Sisk, N. S. Lewis, Estimation of chemical and physical characteristics of analyte vapors through analysis of the response data of arrays of polymer-carbon black composite vapor detectors, *Sens. Actuators B*, 96, 2003, pp. 268-282.

[19]. J. Curie, and P. Curie, Development by pressure of polar electricity in hemihedral crystals with inclined faces, *Bull. Soc. Min. de France*, 3, 1880, pp. 90-93.

[20]. G. Lippmann, Principle de la conservation de l'électricité, *Ann. Chim. Phys*., Ser. 5, 24, 1881, pp. 145-178.

[21]. H. Makino, N. Kamiya, Effects of DC electric field on mechanical properties of piezoelectric ceramics, *Jpn. J. Appl. Phys*, 33, 1994, pp. 5323-5327.

[22]. Y. Bing-Sheng, N. Li-Hua, Y. Shou-Zhuo, Ion chromatographic determination of salicylate in human serum with a bulk acoustic wave sensor as detector, *Journal of High Resolution Chromatography*, 20, 4, 1997, pp. 227-230.

[23]. A. Venema, E. Nieuwkoop, M. J. Velleloop, M. S. Nieuwenhuizen, A. W. Bardendsz, Design aspects of SAW gas sensors, *Sensors and Actuators*, 10, 1–2, 1986, pp. 47–64.

[24]. A. W. Wang, R. Kiwan, R. M. White, R. L. Ceriani, R. L., A silicon-based ultrasonic immunoassay for detection of breast cancer antigens, *Sensors and Actuators B*, 49, 1998, pp. 13–21.

[25]. J. Hartmann, J. Auge, P. Hauptmann, Using the quartz-crystal-microbalance principle for gas detection with reversible and irreversible sensors, *Sens. Actuators B*, 18, 1994, pp. 429-433.

[26]. J. Ricco, R. M. Crooks, G. C. Osbourn, Surface acoustic wave chemical sensor arrays: new chemically sensitive interfaces combined with novel cluster analysis to detect volatile organic compounds and mixtures, *Acc. Chem. Res*., 31, 1998, pp. 289-296.

[27]. I. Lundstrom, S. Shivaraman, C. Svensson, L. Lundkvist, A hydrogen−sensitive MOS field−effect transistor, *Applied Physics. Letters*. 26, 2, 1975, pp. 55-57.
[28]. K. Dobos, R. Strotman, G. Zimmer, Performance of gas-sensitive Pd-gate mosfets with SiO_2 and Si_3N_4 gate insulators, *Sensors and Actuators*, 4, 1983, pp. 593–598.
[29]. H. Sundgren, I. Lundstrom, F. Winquist, I. Lukkari, R. Carlsson, S. Wold, Evaluation of a Multiple Gas-Mixture with a Simple Mosfet Gas Sensor Array and Pattern-Recognition, *Sensors and Actuators B-Chemical*, 2, 2, 1990, pp. 115-123.
[30]. F. Winquist, I. Lundstrom, H. Bergkvist, Ethylene Production from Fruits Measured by a Simple Field-Effect Structure and Compared with a Gas-Chromatographic Method, *Analytica Chimica Acta*, 231, 1, 1990, pp. 93-100.
[31]. E. J. Staples, G. W. Watson, A gas chromatograph incorporating an innovative new surface acoustic wave (SAW) detector, *Pittcon Conference*, New Orleans, Louisiana, 1-5 March, 1998.
[32]. E. J. Staples, A New Electronic Nose, *Sensors*, 16, 1999, pp. 33-40.
[33]. E. J. Staples, The zNose, a new electronic nose using acoustic technology, *J. Acoust. Soc.*, Am. 108, 5, 2000, pp. 2495-2495.
[34]. J. Lammertyn, E. A. Veraverbeke, J. Irudayaraj, zNose™ technology for the classification of honey based on rapid aroma profiling, *Sensors and Actuators B: Chemical*, 98, 1, 2004, pp. 54–62.
[35]. E. A. Veraverbeke, J. Irudayaraj, J. Lammertyn, Fast aroma profiling to detect invert sugar adulteration with zNose, *Journal of the Science of Food and Agriculture*, 85, 2, 2005, pp. 243-250.
[36]. S. Miresmailli, R. Bradbury, M. B. Isman, Qualitative assessment of an ultra-fast portable gas chromatograph (zNose™) for analyzing volatile organic chemicals and essential oils in laboratory and greenhouses, *Arthropod-Plant Interactions*, 4, 3, 2012, pp. 175-180.
[37]. A. J. Bard and L. R. Faulkner, Electrochemical Methods – Fundamentals and Applications, *John Wiley & Sons, Inc*., 1980.
[38]. J. Wang, Analytical Electrochemistry, *Wiley-VCH,* 1994.
[39]. P. T. Kissinger and W. R. Heineman, Laboratory Techniques in Electroanalytical Chemistry 2nd ed., *Marcel Dekker, Inc*., 1996.
[40]. F. Winquist, C. Krantz-Rülcker and I. Lundström, Electronic Tongues and Combinations of Artificial Senses (pages 267–291), in Handbook of Machine Olfaction: Electronic Nose Technology, Tim C. Pearce, Susan S. Schiffman, H. Troy Nagle, Julian W. Gardner (Eds), *Wiley-VCH Verlag GmbH & Co,* 2004.
[41]. G. Korotcenkov, Chemical Sensors Comprehensive Sensor Technologies: Volume 6 Chemical Sensors Applications, *Momentum Press*, 2011.
[42]. F. Winquist, P. Wide, I. Lundstrom, I. An electronic tongue based on voltammetry, *Analytica Chimica Acta*, 357, 1-2, 1997, pp. 21-31.
[43]. F. Winquist, C. Krantz-Rulcker, P. Wide, I. Lundstrom, I. Monitoring of freshness of milk by an electronic tongue on the basis of voltammetry, *Measurement Science & Technology*, 9, 12, 1998, pp. 1937-1946.
[44]. P. Ciosek. W. Wróblewski, Sensor arrays for liquid sensing – electronic tongue systems, *Analyst*, 132, 2007, pp. 963-978.
[45]. R. Lucklum and P. Hauptmann, The quartz crystal microbalance. Mass sensitivity, viscoelasticity and acoustic amplification, *Sensors and Actuators B*, 70, 2000, pp. 30-36.
[46]. T. Yamazaki, J. Kondoh, Y. Matsui, S. Shiokawa, Estimation of components in mixture solutions of electrolytes using a liquid flow system with SH-SAW sensor, *Sensors and Actuators*, 83, 2000, pp. 34-39.
[47]. P. C. Panchariya, A. H. Kiranmayee, Statistical feature extraction and recognition of beverages using electronic tongue, *Sensors and Transducers*, Vol. 112, Issue 1, January 2010, pp. 47-55.

[48]. T. Shi-Yi, D. Shao-Ping and C. Zhong-Xiu, Multifrequency large amplitude pulse voltammetry: a novel electrochemical method for electronic tongue, *Sensors and Actuators B*, 123, 2006, pp. 1049-1056.
[49]. F. Winquist, C. Krantz-Rülcker, and I. Lundström, Electronic Tongues, *MRS Bulletin*, Oct 2004, pp. 726–731.
[50]. H. Kim Esbensen, Multivariate Data Analysis – In Practice, , 5th Edition, *CAMO process AS,* 2004.
[51]. R. O. Duda, P. E. Hart and D. G. Stork, Pattern Classification, *Wiley–Interscience,* 2001.
[52]. S. Holmin, P. Spangeus, C. Krantz-Rulcker, F. Winquist, Compression of electronic tongue data based on voltammetry – a comparative study, *Sensors and Actuators B*, 76, 2001, pp. 455-464.
[53]. T. Artursson, M. Holmberg, Wavelet transform of electronic tongue data, *Sensors and Actuators B*, 87, 2002, pp. 379-391.
[54]. T. Artursson, P. Spangeus, M. Holmberg, Variable reduction on electronic tongue data, *Anal. Chim. Acta*, 452, 2002, pp. 255-264.
[55]. A. H. Kiranmyee, P. C. Panchariya, A. L. Sharma, New data reduction algorithm for voltammetric signals of electronic tongue for discrimination of liquids, *Sensors and Actuators A,* 187, 2012, pp. 154– 161.
[56]. P. Belhumeur, J. Hespanha, D. Kriegman, Eigenfaces vs. Fisherfaces: Recognition Using Class Specific Linear Projection, in *Proc. of the 4th European Conference on Computer Vision*, 1, 14-18 April 1996, Cambridge, UK, pp. 45-58.
[57]. W. Zhao, R. Chellappa, A. Krishnaswamy, Discriminant Analysis of Principal Components for Face Recognition, in *Proc. of the 3rd IEEE International Conference on Automatic Face and Gesture Recognition,* 14-16 April 1998, Nara, Japan, 336-341.
[58]. S. Haykin, Neural Networks: A Comprehensive Foundation, 2nd ed., *Prentice Hall*, 1999, pp. 156–255.
[59]. M. Bos, A. Bos, W. E. Van der Linden, Data processing by neuralnetworks in quantitative chemical analysis, *Analyst*, 118, 1993, pp. 323–328.
[60]. Verbeke, Handbook of Chemometrics and Qualimetrics: Part B, *Elsevier*, 1998, pp. 238-239.
[61]. F. Despagne, D. L. Massart, Neural networks in multivariate calibration, *Analyst*, 123, 1998, pp. 157R–178R.
[62]. C. M. Bishop, Neural Networks for Pattern Recognition, *Oxford University Press*, Oxford, 1999.
[63]. D. F. Specht, Probabilistic Neural Networks, *Neural Networks*, 3, 1990, pp. 109-118.
[64]. S. Kumar, P. C. Panchariya, P. Bhanu Prasad, A. L. Sharma, Non destructive classification of Himalayan orthodox black teas, *Sensors & Transducers*, Vol.145, Issue 10, October 2012, pp. 77-85.
[65]. H. K. Anga, P. C. Panchariya, A. L. Sharma, Authentication of Indian wines using voltammetric electronic tongue coupled with artificial neural networks, *Sensors & Transducers*, Vol.145, Issue 10, October 2012, pp. 65-76.
[66]. J. H. Weisburger, Tea and health: a historical perspective, *Cancer Lett*, 114, 1997, pp. 315–317.
[67]. M. E. Harbowy, D. A. Balentine, Tea chemistry, *Crit Rev Plant Sci*, 16, 1997, pp. 415–480.
[68]. P. L. Fernandez-Caceres, M. J. Martin, F. Pablos, A. G. Gonzalez, Differentiation of tea (Camellia sinensis) varieties and their geographical origin according to their metal content, *J Agric Food Chem*, 49, 2001, pp. 4775–4779.
[69]. B. B. Borse, R. L. Jagan Mohan, Nagalakshmi, S. Krishnamurthy, N., Fingerprint of black teas from India: identification of the regio-specific characteristics, *Food Chem*, 79, 2002, pp. 419–424.

[70]. A. Gulati, A. S. D. Ravindranath, Seasonal Variations in Quality of Kangra Tea (Camellia sinensis (L) O Kuntze) in Himachal Pradesh, *J. Sci. Food Agric.*, 71, 1996, pp. 231-236.

[71]. N. Togari, A. Kobayashi, T. Aishima, Pattern recognition applied to gas chromatographic profiles of volatile component in three tea categories, *Food Research International*, 28, 1995, pp. 495–502.

[72]. M. Frank, H. Ulmer, J. Ruiz, P. Visani, U. Weimar, Complementary analytical measurements based upon gas chromatography–mass spectrometry, sensor system and human sensory panel: a case study dealing with packaging materials, *Anal Chim Acta*, 431, 2001, pp. 11–29.

[73]. C. W. Huck, W. Guggenbichler, G. K. Bonn, Analysis of caffeine, theobromine and theophylline in coffee by near infrared spectroscopy compared to high-performance liquid chromatography (HPLC) coupled to mass spectrometry, *Journal of Pharmaceutical Biomedical Analysis*, 538, 2005, pp. 195–203.

[74]. G. Budinova, D. Vlacil, O. Mestek, K. Volka, Application of infrared spectroscopy to the assessment of authenticity of tea, *Talanta*, 47, 1998, pp. 255–260.

[75]. R. Dutta, E. L. Hines, J. W. Gardner, K. R. Kashwan, M. Bhuyan, Tea quality prediction using a tin oxide-based electronic nose: an artificial intelligence approach, *Sensor and Actuator B*, 94, 2003, pp. 228–237.

[76]. P. Ivarsson, S. Holmin, N. E. Hojer, C. Krantz-Rulcker, F. Winquist, Discrimination of tea by means of a voltammetric electronic tongue and different applied waveforms, *Sens Actuat B,* 76, 2001, pp. 449–454.

[77]. V. K. Joshi, D. Attri, Panorma of research and development of wines in India, *Journal of Scientific & Industrial Research*, 64, 2005, pp. 9-18.

[78]. M. Urbano, M. D. L. De Castro, P. M. Pérez, J. García-Olmo, M. A. Gómez-Nieto, Ultraviolet–visible spectroscopy and pattern recognition methods for differentiation and classification of wines, *Food Chemistry*, 97, 1, 2006, pp. 166.

[79]. C. Di Natale, R. Paolesse, M. Burgio, E. Martinelli, G. Pennazza, A. D'Amico, Application of metalloporphyrins-based gas and liquid sensor arrays to the analysis of red wine, *Analytica Chimica Acta*, 513, 2004, pp. 49-56.

[80]. F. Winquist, I. Lundstrom, P. Wide, The combination of an electronic tongue and an electronic nose, *Sensors and Actuators B, Chem.*, 58, 1999, pp. 512–517.

[81]. S. Buratti, S. Benedetti, M. Scampicchio, E. C. Pangerod, Characterization and classification of Italian Barbera wines by using an electronic nose and an amperometric electronic tongue, *Analytica Chimica Acta*, 525, 1, 2004, pp. 133–139.

[82]. S. Buratti, D. Ballabio, S. Benedetti, M. S. Cosio, Prediction of Italian red wine sensorial descriptors from electronic nose, electronic tongue and spectrophotometric measurements by means of Genetic Algorithm regression models, *Food Chemistry*, 100, 1, 2007, pp. 211–218.

[83]. J. Artigas, C. Jimenez, C. Dominguez, S. Minguez, A. Gonzalo, J. Alonso, Development of a multiparametric analyser based on ISFET sensors applied to process control in the wine industry, *Sensors and Actuators B*, 89, 2003, pp. 199-204.

[84]. A. Legin, A. Rudnitskaya, L. Lvova, Y. Vlasov, C. Di Natale, A. D'Amico, Evaluation of Italian wine by the electronic tongue: recognition, quantitative analysis and correlation with human sensory perception, *Analytica Chimica Acta*, 484, 1, 2003, pp. 33–44.

[85]. A. Riul, H. C. Jr., de Sousa, R. R. Malmegrim, D. S. dos Santos, A. C. P. L. F. Jr., Carvalho, F. J. Fonseca, O. N. Oliveira, L. H. C. Mattoso, Wine classification by taste sensor made from ultra-thin films and using neural networks, *Sensors and Actuators B*, 98, 2004, pp. 77-82.

[86]. A. Rudnitskaya, I. Delgadillo, A. Legin, S. M. Rocha, A. M. Costa, T. Simões, Prediction of the Port wine age using an electronic tongue, *Chemometrics and Intelligent Laboratory Systems,* 84, 2007, pp. 50–56.

[87]. G. Verrelli, L. Francioso, R. Paolesse, P. Siciliano, C. Di Natale, A. D'Amico, A. Logrieco, Development of silicon-based potentiometric sensors: Towards a miniaturized electronic tongue, *Sensors and Actuators B*, 123, 2007, pp. 191–197.
[88]. A. Rudnitskaya, L. M. Schmidtke, I. Delgadillo, A. Legin, G. Scollary, Study of the influence of micro-oxygenation and oak chip maceration on wine composition using an electronic tongue and chemical analysis, *Analytica Chimica Acta*, Vol. 642, 1-2, 2009, pp. 235–245
[89]. A. Rudnitskaya, S. M. Rocha, A. Legin, V. Pereira, J. C. Marques, Evaluation of the feasibility of the electronic tongue as a rapid analytical tool for wine age prediction and quantification of the organic acids and phenolic compounds. The case-study of Madeira wine, *Analytica Chimica Acta*, 662, 1, 2010, pp. 82–89.
[90]. D. Kirsanov, O. Mednova, V. Vietoris, P. A. Kilmartin, A. Legin, Towards reliable estimation of an "electronic tongue" predictive ability from PLS regression models in wine analysis, *Talanta*, 90, 2012, pp. 109–116.
[91]. P. Ivarsson, S. Holmin et al, Discrimination of tea by means of a voltammetric electronic tongue and different applied waveforms, *Sensors and Actuators, B, Chem*. 76, 2001, pp. 449–454.
[92]. J. Wu, J. Liu, M. Fu, G. Li, Z. Lou, Classification of Chinese Yellow Wines by Chemometric Analysis of Cyclic Voltammogram of Copper Electrodes, *Sensors*, 5, 2005, pp. 529-536.
[93]. S. Tian, S. Deng, C. Ding, C. Yin, H. Li, Discrimination of Red Wine Age Using Voltammetric Electronic Tongue Based on Multifrequency Large-Amplitude Voltammetry And Pattern Recognition Method, *Sensors and Materials*, 19, 5, 2007, pp. 287–298.
[94]. Z. Wei, J. Wang, L. Ye, Classification and prediction of rice wines with different marked ages by using a voltammetric electronic tongue, *Biosensors and Bioelectronics*, 26, 12, 2011, pp. 4767–4773.
[95]. W. Novakowski, M. Bertotti, T. R. L. C. Paixão, Use of copper and gold electrodes as sensitive elements for fabrication of an electronic tongue: Discrimination of wines and whiskies, *Microchemical Journal*, 99, 1, 2011, pp. 145–151.
[96]. V. Parra, T. Hernando, M. L. Rodriguez-Mendez, J. A. De Saja, Electrochemical sensor array made from bisphtalocyanine modified carbon paste electrodes for discrimination of red wines, *Electrochimica Acta*, 49, 2004, pp. 5177-5185.
[97]. V. Parra, A. Arrieta, J. A. Fernández-Escudero, M. Íñiguez, J. A. De Saja, M. L. Rodríguez-Méndez, Monitoring of the ageing of red wines in oak barrels by means of an hybrid electronic tongue, *Analytica Chimica Acta*, 563, 1–2, 2006, pp. 229–237.
[98]. V. Parra, A. Alvaro, A. Arrieta, J. A. Fernandez-Escudero, M. L. Rodrıguez-Mendez, J. A. De Saja, Electronic tongue based on chemically modified electrodes and voltammetry for the detection of adulterations in wines, *Sensors and Actuators B*, 118, 2006, pp. 448–453.
[99]. M. Gutiérrez, C. Domingo, J. Vila-Planas, A. Ipatov, F. Capdevila, S. Demming, S. Büttgenbach, A. Llobera, C. Jiménez-Jorquera, Hybrid electronic tongue for the characterization and quantification of grape variety in red wines, *Sensors and Actuators B: Chemical,* 156, 2, 2011, pp. 695–702.

Chapter 16
Magnetic Bead Based Biosensors: Design and Development

Wen-Yaw Chung, Kimberly Jane Uy, Yi Ying Yeh, Ting Ya Yang, Hao Chun Yang, Yaw-Jen Chang, T. Y. Chin, Kuang-Pin Hsiung, Dorota G. Pijanowska

16.1. Introduction

There is a growing popularity with the use of point of care technology (POCT). It is a tool that can be used for the diagnostics that will be used not only for detection of a disease but also for healthcare. There is a growing concern in the world right now for the growing number of obesity cases. Obesity is a condition in which the accumulation of body fat in a person has reached excessiveness. When there is excessive fat accumulation in the body, there is a high risk for the person to have health problems such as metabolic syndromes and diseases that can cause mortality. [1] Right now, to determine if a person is obese, his body mass index (BMI) is taken into consideration. The BMI will only take into consideration the height and weight of the person to do its determination. Studies have shown that it is actually the fat count of a person that determines if he is obese. The fat cells also known as adipocytes are composed of several proteins one of which is called adiponectin. Several studies show that the adiponectin count in the body is dysregulated when a person is obese. Further discussion of the protein will be included in this chapter.

The determination of the adiponectin is done using a diagnostic protocol called Enzyme Linked Immunosorbent Assay or ELISA for short. The ELISA protocol will measure the change in fluorescence to determine the amount of adiponectin in the sample. The enzyme attached to the detection antibody will chemically react to the substrate added into the processed sample. For this work, it is aimed to change the fluorescence detection with the magnetic field detection. This is done by replacing the enzyme with magnetic beads. Magnetic beads are ideal for such kind of bioassay application due to the following reasons: (a) cells exhibit few magnetic properties, (b) signals from

Wen-Yaw Chung
Electronic Engineering Department,
Chung Yuan Christian University, Zhong Li, Taiwan

magnetic beads are stable with time, (c) results from magnetic detection is independent of the colour and clarity of the sample, and (d) magnetic labelling has functionalities that may be used to improve bioassay performance such as magnetic filtration and manipulation [2]. Further discussion on the properties of magnetic beads is included in the following article.

In order to measure the magnetic field from the magnetic beads, a specific magnetic sensitive structure called magnetic MOSFET (MAGFET) is used and is described in this chapter. The physical structure of the MAGFET is a MOSFET with the drain having been split to two. The structure is able to detect the magnetic field perpendicular to the channel. Because the MAGFET structure is compatible to the CMOS technology, it can be implemented as a common library item.

A prototype was also developed as a proof of concept for this type of sensing. The prototype made use of commercially available components that mimic the operation of each component in the micro-level. A detailed description of this work is also included in this paper together with experimental results.

16.2. Materials and Methods

16.2.1. Adiponectin

The adipose tissue plays an important role in energy storage, fatty acid metabolism and glucose homeostasis. Adiponectin is one of the protein hormone secreted into the bloodstream exclusively by the adipose tissue. It was first described during 1995 [5]. It is also described in other literature under the following names: Acrp30, AdipoQ, apM1, and GBP28. It is composed of 247 amino acids and has a molecular mass of 28 kDA [5]. Freeze etch electron microscopy was performed in order to visualize the architecture of Adiponectin. Adiponectin trimer, hexamer and HMW isoforms are shown in Fig. 16.1. The structure of the trimer consists of a ball and a stick where the ball is composed of globular domain and the rigid stick is composed of collagen triple helix [6] (Fig. 16.1A). The image shows the trimer to have a length of approximately 18 nm. The hexameric architecture of the Adiponectin is composed of two trimers connected in parallel in a head to head fashion form with a resemblance to a letter Y [6] (Fig. 16.1B). For the HMW isoform, it only showed the tail end view [6] (Fig. 16.1C).

In an experiment done to prove that adiponectin is altered in obesity, adiponectin levels in adipose tissue samples from lean (ob/+) and obese (ob/ob) mice was tested. The experiment shows that the amount of adiponectin in an obese (ob/ob) mice is reduced by 70 to 90 % as compared to the lean (ob/+) mice. Also, in order to prove the correlation of the amount of adiponectin with regard to obesity, the same type of experiment was done for an adipose specific gene, aP2. The result shows that the amount of aP2 gene was not affected by the obesity of the mice. The results from the aforementioned experiment are illustrated in Fig. 16.2. Adipose tissues from human individuals were also examined. For this experiment, four obese individuals with a body mass index

(BMI) of 39 ± 1.4 and three normal individuals with a BMI of 21 ± 0.3 were tested. The test shows a reduction in the amount of adiponectin of about 50 to 80 % on obese individuals as compare to that of the normal individuals. Fig. 16.3 shows the results for samples one through seven. This experiment clearly proves that the adiponectin is dysregulated in both mouse and human that are obese.

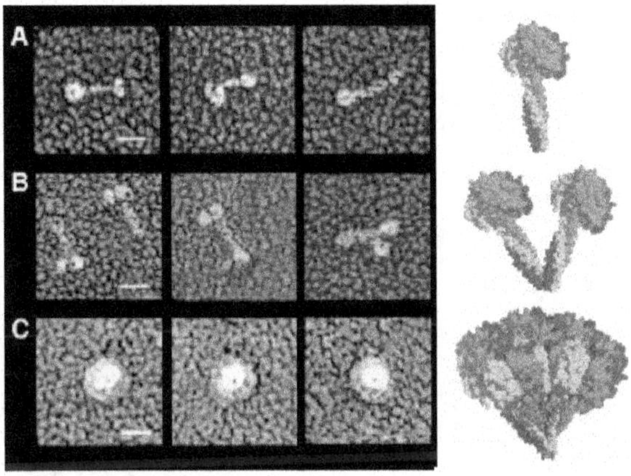

Fig. 16.1. Adiponectin (A) trimer, (B) hexamer and (C) HMW isoforms image [6].

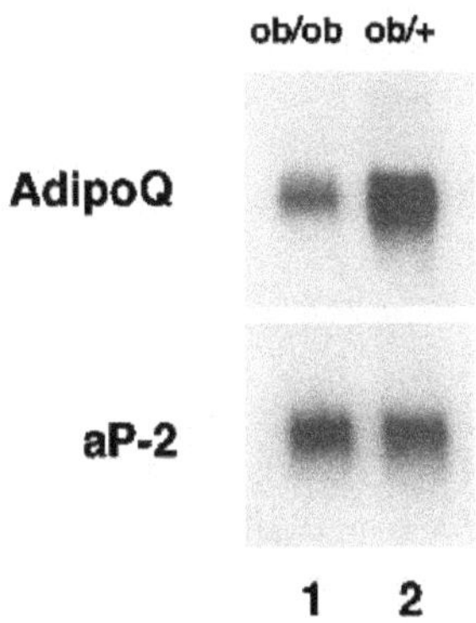

Fig. 16.2. Testing for the quantity of Adiponectin (aipoQ) in an (1) obese and (2) lean mice [7].

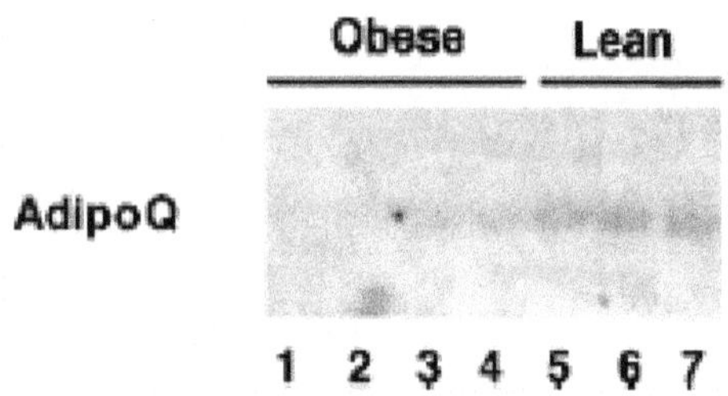

Fig. 16.3. Adiponectin (aipoQ) level testing result for human sample [7].

16.2.2. Superparamagnetic Beads

Magnetic beads are made, for the most part with, iron oxide together with some magnetic material. Since Iron oxide is a ferromagnetic material and that there is quite low magnetic content inside the beads, the saturation magnetic moment exhibited is quite small. On the outer layer, bio-reactive molecules are coated on. This will enable biological entities to be coated on the surface of the magnetic beads.

For this work, the magnetic beads exhibit a superparamagnetic behavior. This type of magnetic beads will only have magnetic moment when a source of magnetic field is present. Consequently, when the magnetic field is removed, the magnetic moment of the beads also disappears. Magnetic beads are composed of nano-size iron oxide crystals. When there is no magnetic field present, the magnetization of the crystals are randomly distributed. At this state, the magnetic beads do not exhibit any magnetization. When magnetic field is present, the magnetization of the crystals will align to the direction of the magnetic field. Fig. 16.4 is an illustration of the two states of magnetic beads as described in this chapter.

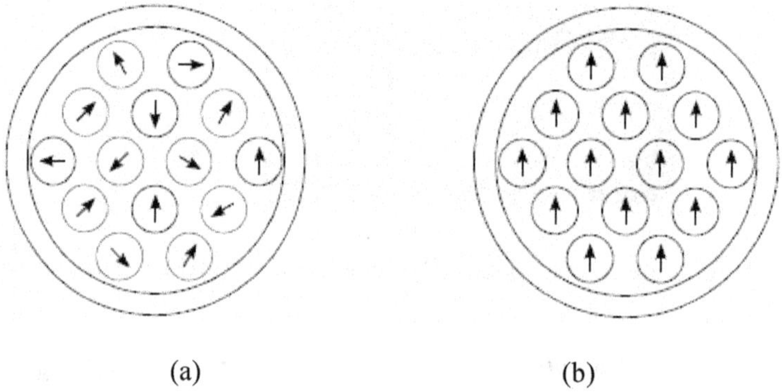

Fig. 16.4. State of superparamagnetic bead with (a) at the demagnetized state and (b) at the magnetized state [11].

16.2.3. Magnetic Particle Assay (MPA)

Before we can understand the magnetic particle assay (MPA), we must first understand enzyme linked immunosorbent assay (ELISA). The MPA is actually a protocol based on ELISA. Fig. 16.5 is an illustration of the difference between the MPA protocol and the ELISA protocol. ELISA can perform a broad range of both qualitative and quantitative diagnostics, thereby, making it a cost effective diagnosis technology. ELISA protocol requires multiple liquid washing steps that are dependent on spectrophotometer. The whole process is performed on a microplate. To perform a diagnostic test, a coating antibody that can bind to specific antigen is placed on to the microplate. The antigen is then introduced and bind to the coating antibody which immobilizes it. The excess antigen will then be washed away. The detection antibody is then introduced to bind to the target analyte. Again, the excess will be washed away. The streptavidin-HRP that catalyzes with the dye will then be conjugated with the second antibody. The presence of the analyte is indicated by the change in the liquid. Fig 5a illustrates how the antibodies and analyte is arranged on the microplate.

For the MPA protocol, the streptavidin-HRP is replaced with the magnetic bead. Instead of measuring the fluorescence of the sample liquid, the quantity of magnetic beads is taken as the representation of the concentration of analyte in the sample. The total magnetic moment of the magnetic beads when magnetized are measured and is converted to the relative value of the analyte concentration. Fig. 16.5 b is the structure of the antibodies and the analyte for the MPA protocol.

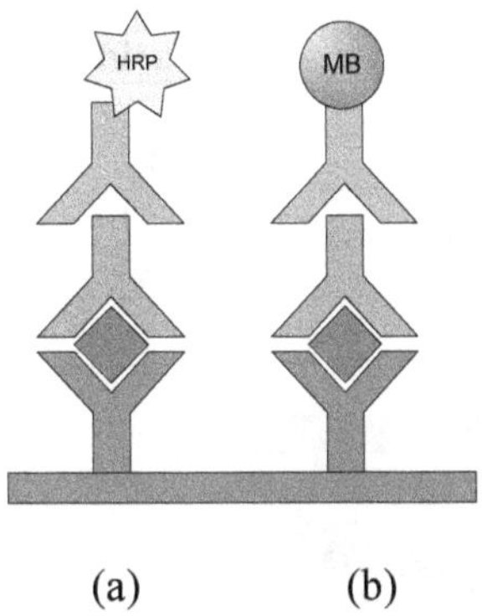

Fig. 16.5. Structure of biological entity for (a) ELISA protocol, and (b) MPA protocol.

16.3. Results and Discussion

16.3.1. Nitrocellulose Based Enzyme Linked Immunosorbent Assay (ELISA)

To POCT device that can perform diagnostics, we must be able to perform the ELISA protocol on top of the sensor. A way to do this is to perform the protocol on a nitrocellulose (NC) membrane. The NC membrane is a sticky membrane that can be used to immobilize biological materials.

For this work, the target analyte for the ELISA is the adiponectin. To make sure that the NC ELISA can work properly, this experiment was done to generate a standard curve for different levels of adiponectin. The work made use of plasma sample taken from a lean rat. Table 16.1 shows similarity of amino acid between species with the mouse as the base of comparison. The protocol was performed on a glass slide coated with NC membrane. The glass slide is divided into eight separate wells. The sample plasma was taken and divided into several samples having different concentration by means of dilution. The dilution factor for this experiment goes from 200x to 500x with intervals of 100x. Fig. 16.6 is a photo of the NC slide where in the ELISA protocol was performed. The experiment was performed on four identical NC slides. Each concentration level was tested twice on each slide having a total of eight wells for each dilution factor.

The reading of the result made use of a chromatograph machine that will show the level of fluorescence depending on the concentration of the sample. The experimental result shows linearity in fluorescence of the sample with regard to its dilution factor. The linearity of the experimental result is at 0.9843. Fig. 16.7 is a graph that shows the experimental results having the x-axis as the dilution factor and the y-axis as the fluorescence value.

Table 16.1. Percentage of similarity of adiponectin in different species having mouse adiponectin as reference sample.

Species	% Amino acid similarity
Rat	92
Human	83
Cow	91
Monkey	83

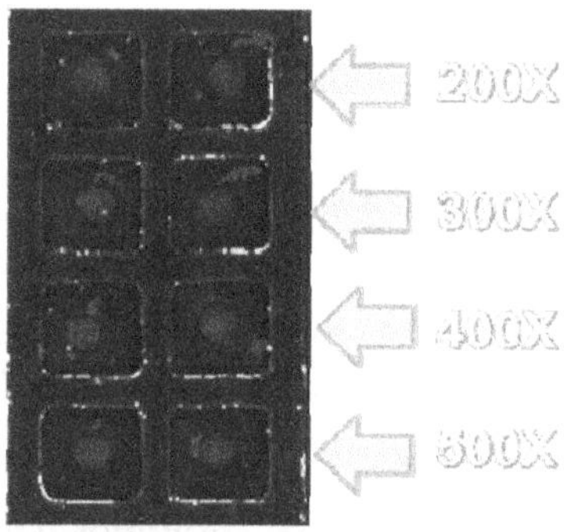

Fig. 16.6. Glass slide with NC membrane for concentration level analysis with the wells having dilution factor.

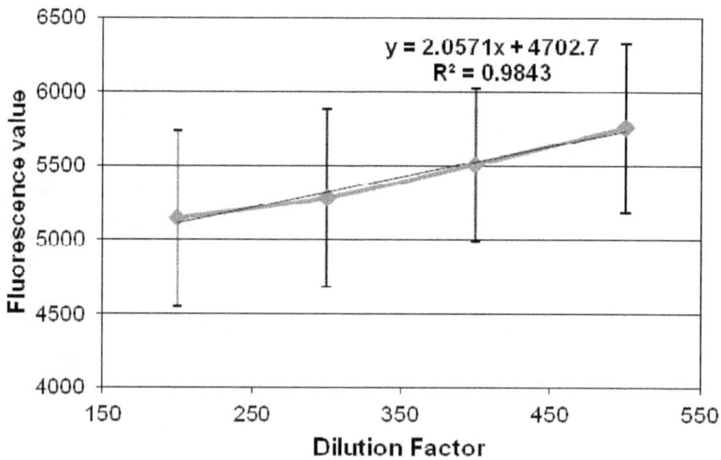

Fig. 16.7. Results for ELISA protocol implemented on NC membrane.

16.3.2. Sensor System Model

A way of sensing the magnetic beads is through the use of Hall sensors. For the preliminary study in this work, a system model for the microscale system was developed using commercially available components. Fig. 16.7 shows an illustration of the system model. The model made use of a hall sensor from Winson model number WSH136. The hall sensor has a sensitivity of 3.3 mV/gauss. On top of which, a magnet sheet is placed to serve as the source of magnetic field for the magnetization of the magnetic beads. In place of magnetic beads, steel balls are used. The steel balls are ideal for such applications since they themselves have no magnetic moment. They, however, are able to absorb the magnetic moment from the magnetic field and be magnetized. A porous plate was placed on top of the magnetic sheet to keep the steel balls in place during the experiment.

The experiment is done using nine steel balls to represent the magnetic beads (Fig. 16.8). Fig. 16.9 is the curve that shows the output voltage of the sensor with regard to the quantity of steel balls on the sensor. The output voltage ranges from 1.217 V to 1.728 with a sensitivity of 56.77 mV per steel ball. The characteristic curve from Fig. 16.9 shows a linearity of 0.98.

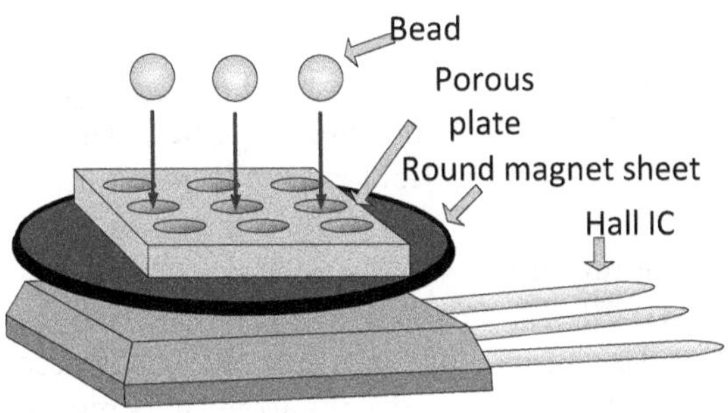

Fig. 16.8. Illustration of the system model implemented with commercially available component [10].

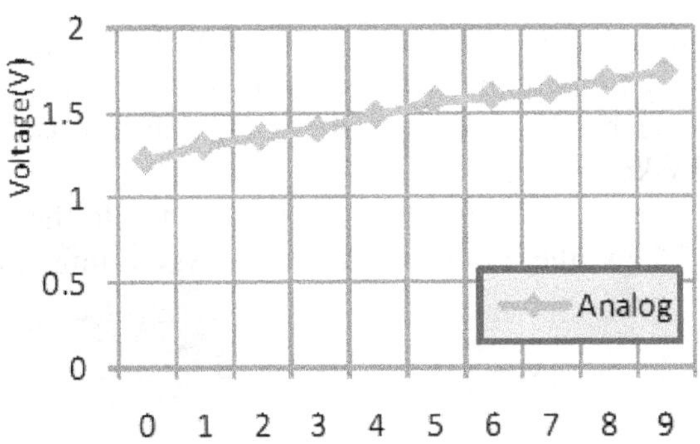

Fig. 16.9. Analog output value for the system generated from 0 to 9 numbers of steel balls placed on the sensor system [10].

16.3.3. Magnetic Field Effect Transistor

The hall sensor uses the Hall Effect as its basic principle. Another type of sensor that uses such principle is the Magnetic Field Effect Transistor (MAGFET). This type of sensor is ideal for POCT devices because of its compatibility with the CMOS technology. The operation of semiconductor is based on the formation of Lorentz force. Lorentz force FB is generated on the semiconductor during its operation. The is equivalent to the electrical charge qe moving in a stable magnetic field BZ with the speed *v*. Eq. (16.1) is the mathematical representation of these values to generate the Lorentz force. The Hall electrical field intensity is the combination of magnetic field and speed of electrical charge. Fig. 16.10a shows an illustration of the structure of the basic MAGFET. The basic structure of a MAGFET is the same as a MOSFET. There is a drain and a source. The main difference is that the MAGFET has two separate drains but only one source as illustrated. Under normal conditions, the electron charge flows equally between the two drains. But under a magnetic field, the electron is pushed to one side of the MAGFET depending on the direction of the applied magnetic field. To further understand the effect of magnetic field on the MAGFET, Fig. 16.10b is an illustration of the electron flow in the MAGFET when external magnetic field is applied. Using the right hand rule, we can see that when the magnetic field is applied towards the MAGFET, the electron charge is pushed by the Lorentz force to one side of the MAGFET.

The equivalent circuit of the MAGFET is given in Fig. 16.11. A MAGFET is composed of two MOSFET having the gates connected together. The current deviation Δi is controlled by the external magnetic field applied on to the MAGFET. It can be calculated using Eq. (16.2). In this equation the S represents the relative sensitivity of the MAGFET, I_d is the drain current of the MAGFET, and BZ is the magnetic field perpendicular to the MAGFET. Eq. (16.3) to Eq. (16.6) will explain the operation of the MagFET with respect to the source and drain current. The source current is the total of the two drain currents. The value of the drain current is composed of the magnitude of the drain current, the offset current and the noise current. The magnitude of each drain current is the half the value of the source current subtracted or added with half the value of the current change.

The microscale sensor device was developed under the specification in the system model. In place of the commercially available hall sensor is the MAGFET device. To magnetize the magnetic beads for the reading, an inductor is placed on top of the MAGFET. This will take the place of the magnetic sheet. The value of the magnetic field supplied to the system will depend on the current fed to the micro-coil. Fig. 16.12 is the illustration of how the system will work when implemented using CMOS technology.

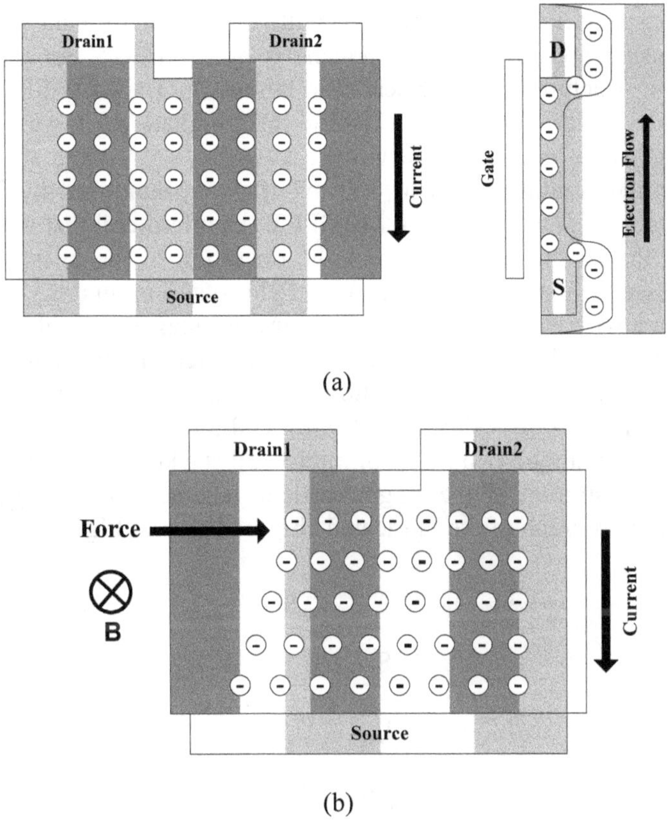

Fig. 16.10. Model and operation of MAGFET device with (a) at normal operation, and (b) under magnetic field.

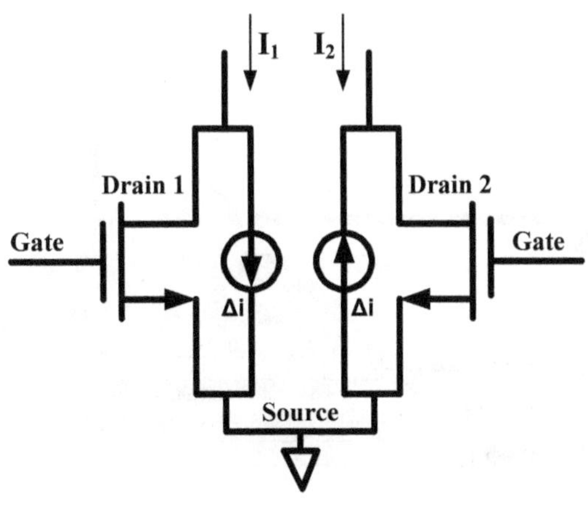

Fig. 16.11. Equivalent schematic diagram for a device.

The sensor was fabricated using the TSMC 0.35 um technology. The whole sensor was composed of 625 MAGFET interconnected with each other to make a singular sensor device. The sensing area of each component is at 5 um x 5 um. Fig. 16.13 is an illustration of the structure of the fabricated in this work. The MAGFET was simulated under the supply voltage of 3.3 V. During the simulation, it was found that the maximum output value with the gate voltage value at 0.5 V. The structure was also simulated under the condition of 0 to 100 mT. It was found that the structure has a sensitivity of 1 mV/mT. On top of each MAGFET is a micro-coil that will magnetize the magnetic beads that are to be sensed. For the operation of the micro-coil, When DC electricity is passed through a wire, a magnetic field rotates around the wire in a specific direction. When a wire is turned into a coil, an inward magnetic field is generated. And when there is multiple wires stacked together, the magnetic field generated by each will combine together to form a larger magnetic field strength. For this work, the micro-coil is composed of two stacks of wire made from metal 2 and metal. Fig. 16.14 is a cross section of the sensor with the two levels of micro-coil stacked on top of the MAGFET. It also includes an illustration of the magnetic field generated by the micro-coil and its directionality. A micrograph of the fabricated chip is shown in Fig. 16.15.

$$F_B = q_e \cdot (v \, x \, B_Z) = q_e \cdot E_H \tag{16.1}$$

$$\Delta i = S x I_d x B_Z \tag{16.2}$$

$$I_S = I_{D1} + I_{D2} \tag{16.3}$$

$$I_D = I_{Dmag} + I_{off} + I_{nA} \tag{16.4}$$

$$I_{D1mag} = I_S/2 - \Delta I/2 \tag{16.5}$$

$$I_{D2mag} = I_S/2 + \Delta I/2 \tag{16.6}$$

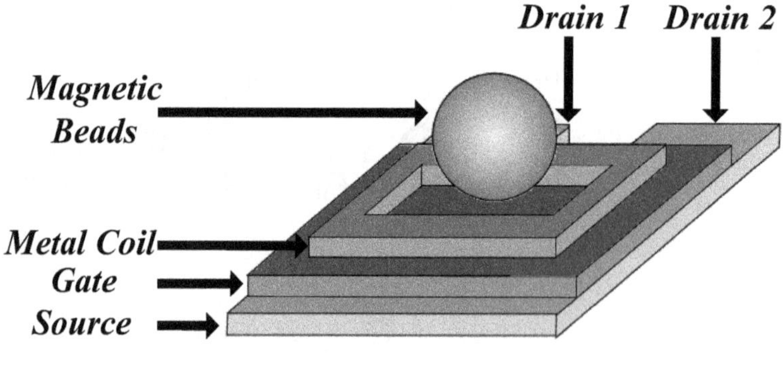

Fig. 16.12. System illustration for the development in the micro level.

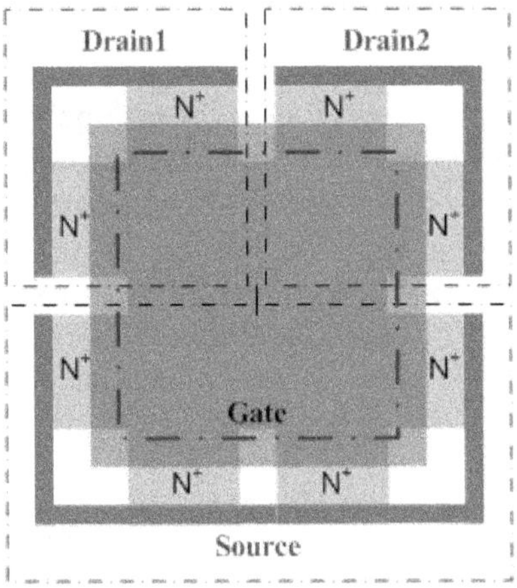

Fig. 16.13. Illustration of the structure of the fabricated MAGFET.

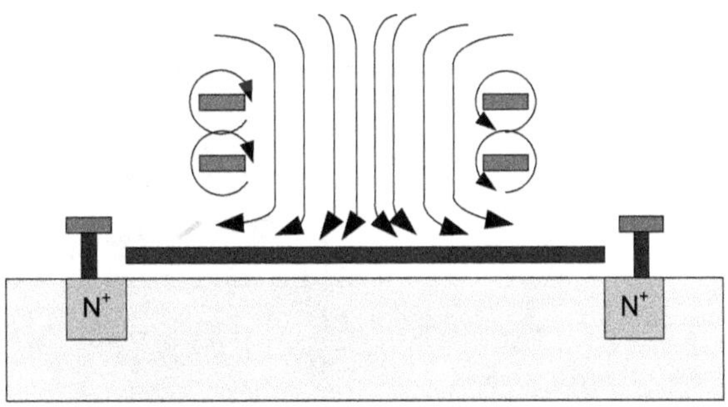

Fig. 16.14. Cross section of the fabricated MAGFET with the micro-coil on top having the directionality of the generated magnetic field illustrated.

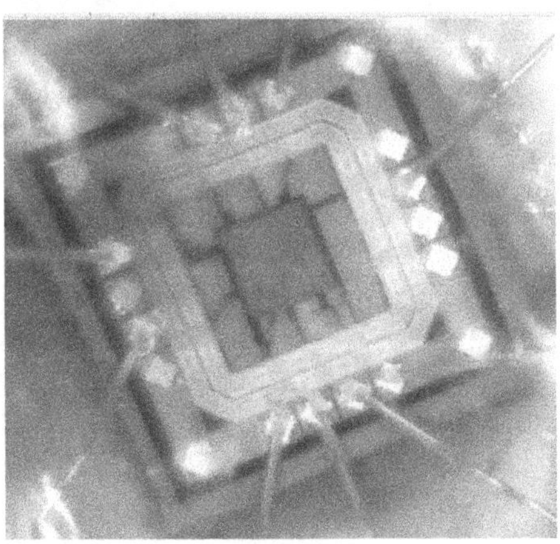

Fig. 16.15. Micrograph of the MagFET Sensing Device.

16.4. Testing and Result

The sensor has been tested using the semiconductor analyzer Analog B1500. Since the MagFET has four symmetric ports therefore it can be assumed to have four symmetric CMOS device. Initial test proved that the four devices are symmetric. The configuration of the said experiment is given in Fig. 16.16 where in simple current to voltage analysis has been performed. The results show each symmetric device have the same characteristic curves for the I_D-V_G curve shown in Fig. 16.17.

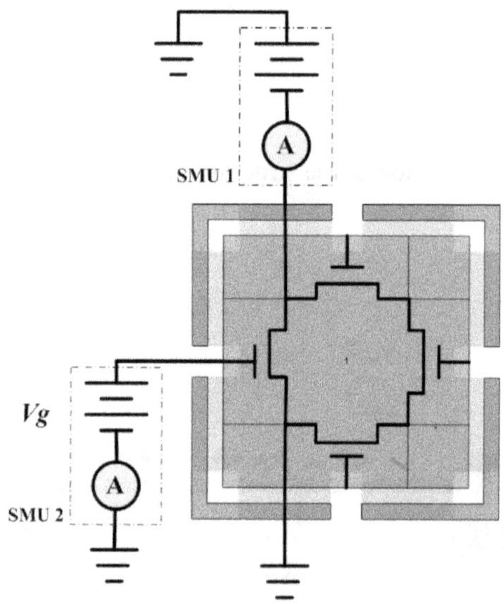

Fig. 16.16. Schematic for the Current to Voltage Testing for the Four Symmetric Devices.

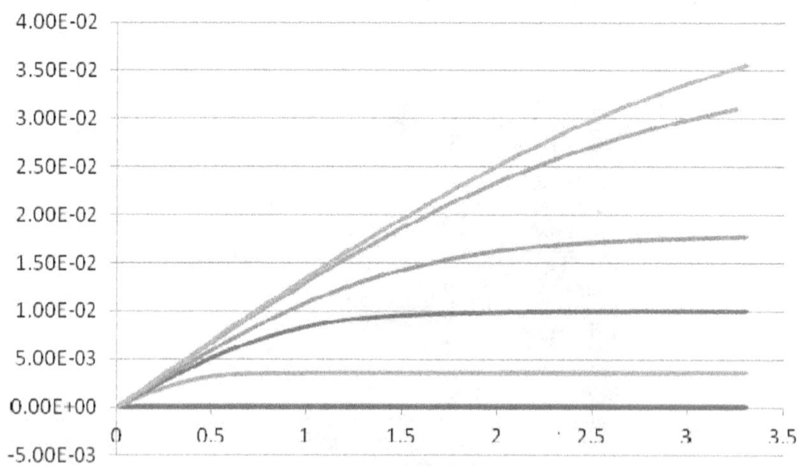

Fig. 16.17. ID-VG Curve for Four Symmetric Device.

For the operation of the MagFET to be used in detecting magnetic beads. The device was tested having to control the current that passes through the two drains. The schematic for the said test is given in Fig. 16.18. The operational gate voltage was extracted by feeding the two drains with different current and sweeping the gate voltage. The result is illustrated by a graph in Fig. 16.19. From the test, further experimentation made use of gate voltage valued at 2.5 V. having different current value fed through the drains, the work was able to extract the differential voltage with regard to the differential current of the between the two drains. The result is shown through a graph in Fig. 16.20. The linearity for the test is calculated to be at 99.93 %. All four plausible connections for the device have been tested and resulted to the same graph.

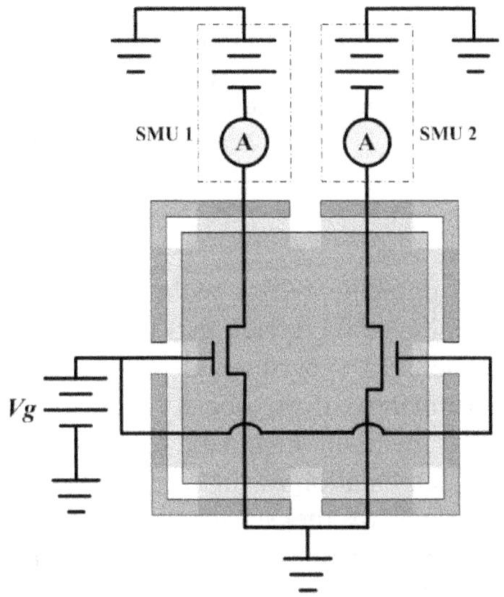

Fig. 16.18. Schematic for Testing MagFET Operation.

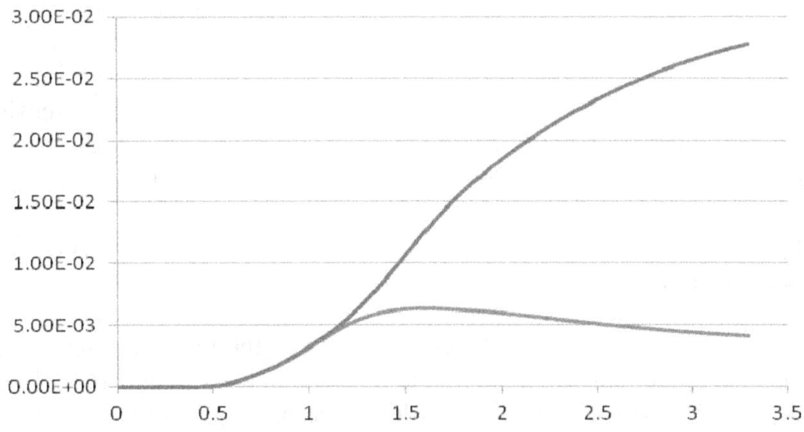

Fig. 16.19. Gate Voltage Sweep.

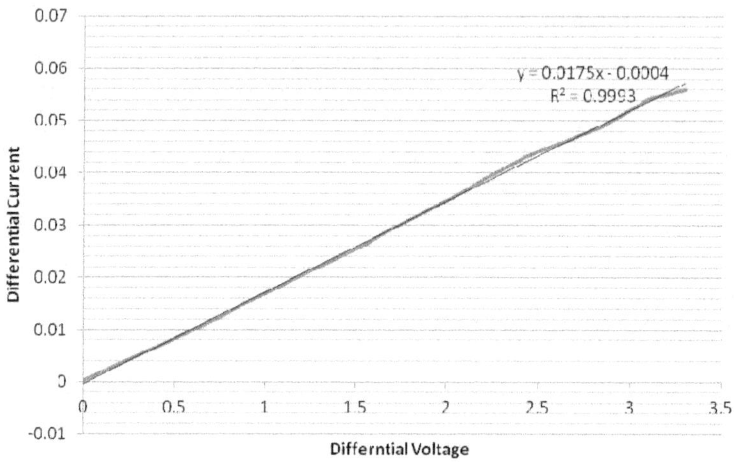

Fig. 16.20. Differential Voltage vs. Differential Current for the MagFET Device.

16.5. Conclusion

The aim of this work is to implement data gathering and experiment that would be needed in implementing a micro scale device that would be able to perform diagnostic experiments. This chapter includes experimental proof that quantifying of adiponectin can be a basis for the determination of the obesity or leanness of a person. It was also mentioned in this chapter that the main diagnostic tool for quantifying adiponectin is the ELISA protocol. For the implementation of the POCT device, the ELISA protocol must be implemented on the surface of the sensor. This is to be done by placing NC membrane on the surface of the sensor. To prove that this is possible, experiment was done using a glass slide covered with a layer of NC. Several concentration of adiponectin was tested and the results show that the output fluorescence level is a linear curve with regard to the dilution level of the sample adiponectin. A prototype of the target system was also implemented. This prototype was able to prove that the hall device is able to quantify the number of steel balls present on the surface of the sensor. This was taken as the basis for the implementation of the microscale sensing device.

The system is intended to detect magnetic particles. The MAGFET is a sensing device that is used for magnetic particle detection. This work included a discussion of the fabrication of the sensing device that made use of the MAGFET. The device was developed to include a micro-coil that will provide a magnetic field to the magnetic beads during the testing. Preliminary testing on the MAGFET has been done for its basic electronic characteristics.

The future work for this research includes the testing of the fabricated sensor. The work aims to characterize the sensor with regard to its reaction to the magnetic field applied. It also aimed that the ELISA protocol be implemented on the surface of the sensor. After the characterization, a readout sensor is to be developed for POCT device application.

Acknowledgement

This work would like to acknowledge the National Science Counsel, Taiwan, R.O.C. for funding this project (NSC 99-2221-E-033-064 and NSC-100-2221-E-033-051) and Hall sensors donated by Winson Semiconductor Corp, Taiwan. As well as National Chip Implementation Centre, Taiwan, R.O.C. for the chip fabrication Support.

References

[1]. Y. Arita, S. Kihara, N. Ouchi, M. Takahashi, K. Maeda, J. Miyagawa, K. Hotta, I. Shimomura, T. Nakamura, and K. Miyaoka, Paradoxical decrease of an adipose-specific protein, adiponectin, in obesity, *Biochemical and Biophysical Research Communications*, 257, 1, 1999, pp. 79-83.

[2]. O. Florescu, M. Mattmann, and B. Boser, Fully integrated detection of single magnetic beads in complementary metal-oxide-semiconductor, *Journal of Applied Physics*, 103, 2008, 046101.

[3]. M. Donoval, M. Daricek, V. Stopjakova, and D. Donoval, Magnetic FET-Based On-chip Current Sensor For Current Testing of Low-voltage Circuits, *Journal of Electrical Engineering,* 2008. 59, 3, pp. 122-130.

[4]. A. H. Berg, T. P. Combs and P. E. Scherer, ACRP30/adiponectin: an adipokine regulating glucose and lipid metabolism, *Trends in Endocrinology & Metabolism*, March, 2002. pp. 84-89.

[5]. P. E. Scherer, S. Williams, M. Fogliano, G. Baldini, and H. F. Lodish, A Novel Serum Protein Similar to C1q, Produced Exclusively in Adipocytes, *The Journal of Biological Chemistry*, November, 1995. pp. 26746-26749.

[6]. T. S. Tsao, E. Tomas, H. E. Murrey, C. Hug, D. H. Lee, N. B. Ruderman, J. E. Heuser, and H. F. Lodish, Role of Disulfied Bonds in Acrp30/Adiponectin Structure and Signaling Specificity, *The journal of Biological Chemistry*, December, 2003, pp. 50810-50817.

[7]. E. Hung, P. Liang, and M. Spiegelman, AdipoQ Is a Novel Adipose-Specific Gene Dysregulated in Obesity, *The Journal of Biological Chemistry*, May, 1996. pp. 10697-10703.

[8]. G. F. Santillan-Quinonez, R. S. Murphy, and V. H. Champac, Equivalent Rectangular Active Region and SPICE Macro Model for Split-Drain MAGFETs, *ECS Transactions*, 2010, pp. 393-400.

[9]. O. Florescu, M. Mattmann, and B. Boser, Fully integrated detection of single magnetic beads in complementary metal-oxide-semiconductor, *Journal of Applied Physics*, 2008.

[10]. W. Y. Chung, K. J. Uy, Y. Y. Yeh, T. Y. Yang, H. C. Yang, and H. W. Li, Building a Prototype for a Magnetic Nanoparticle Bead Based Biosensing Device, in Proceeding of the *2nd International Conference on Sensor Device Technologies and Applications*, 2011, pp. 155-159.

[11]. F. Colle, Hallsensor Based Detection of Magnetic Particles for Lab-on-a-chip, MSc Thesis, *Ghent University*, 2007.

[12]. S. I. Liu, J. F. Wei, and G. M. Sung, SPICE Macro Model for MAGFET And Its Applications, *IEEE Transactions on Circuits and Systems II: Analog and Digital Signal Processing*, 46, 4, 1999, pp. 370-375.

[13]. G. F. Santillan-Quinonez, R. S. Murphy, and V. H. Champac, Equivalent Rectangular Active Region and SPICE Macro Model for Split-Drain MAGFETs, *ECS Transactions*, 2010, pp. 393-400.

[14]. M. Daricek, M. Donoval, A. Satka, and T. Kosik, Characterization of MagFET structures, in *Proceedings of the 15th IEEE International Conference on Electronics, Circuits and Systems, (ICECS' 2008).*, 31 August - 3 September 2008, pp. 1233- 1236.
[15]. V. Frick, H. B. Nguyen, and L. Hebrard, A novel chopping-spinning MAGFET device, in *Proceedings of the 17th IEEE International Conference on Electronics, Circuits, and Systems (ICECS' 2010),* 12-15 December 2010, pp. 815-818.

Chapter 17
Human Blood Analytes Biochemical Sensors Based on Microsphere Stimulated Raman Spectroscopy

A. R. Bahrampour, N. Jahangiri, M. Saliminasab, M. Taraz, M. H. Zandi

17.1. Introduction

Blood analysis mostly carries out for medical diagnosis. For example, over 600 million blood cholesterol tests are performed annually worldwide to diagnose cardiovascular disease such as infarction or more than 180 million diabetics are advised to monitor their blood glucose levels several times each day [1]. Due to the spread of such diseases and their complications, the need for non-invasive determination of body analytes has attracted much interest in last decades. In this context, a wide range of optical technologies have been developed to design a non-invasive method for body analytes sensing [2-5].Among these methods, Raman spectroscopy, by generating a distinct spectrum for each analyte, can resolve the numerous low-concentration analytes of turbid media such as human skin or blood.

Raman spectroscopy such as spontaneous Raman spectroscopy and stimulated Raman spectroscopy is widely used in different branches of science, but very few applications of this technique for medical diagnostics have been reported. The most important disadvantage of spontaneous Raman scattering for diagnostic applications is the intrinsically weak cross section (on the order of 10^{-30} cm), which is about 14 orders of magnitude lower than the typical fluorescence cross section. Thus, traditionally Raman scattering has not been a useful technique for sensing molecules in low concentrations. So stimulated Raman spectroscopy(SRS) can be employed. Also in 1974, it was reported that the Raman scattering cross-section can be considerably increased if the molecules are adsorbed on roughened metal surfaces [7]. In the following decades SERS has been investigated as a useful technique that provides tremendously large Raman signal enhancement in the vicinity of metal nanoclusters.

A. R. Bahrampour
Department of Physics, Sharif University of Technology, Tehran, Iran

In the vicinity of metal nanoclusters, field enhancement occurs because of the resonant interaction between the optical field and surface plasmons in the metal. In essence, the light from a laser beam excites the surface plasmons, which are collective oscillations of conduction electrons. Those Plasmons then radiate a dipolar filed. The coherent interaction of the incoming electric filed with the dipolar filed leads to a redistribution of electric filed intensities in areas around the metal clusters. A molecule nearby or adsorbed on metal feels enhanced excitation intensity. So its Raman scattered filed is enhanced in the same way that the incident laser field is. Indeed, one can liken the metallic clusters to tiny antennas that enhance and transmit Raman scattered light. Under special conditions, a Raman enhancement factor due to the excitation of SPR of up to 10^{14} has been observed [8].

The SERS technique, by using the stimulated Raman signal as a fingerprint of the analyte, creates a spectroscopic device which can act as a highly sensitive molecular sensor.

Stimulated Raman scattering has be described by the third order nonlinearity and need high power laser pump fluency which is not suitable for the in-vitro and in-vivo diagnostic purpose. To overcome this problem silica microsphere resonator is chosen.

The long photon lifetimes and small volume of ultra-high-Q microspheres in the resonance condition, allow to significantly reducing the threshold for nonlinear phenomena such as stimulated Raman scattering [9]. The field enhancement due to the resonance condition of microspheres cause to reduce the stimulated Raman scattering thresholds of its surrounding media and hence observing stimulated Raman scattering with low power. So silica microspheres are good candidates for next-generation Raman sources. This device enables a large reduction in the necessary threshold pump power, while fiber- coupling notably improves overall efficiency and provides a convenient method of optical filed transport [10]. Based on ultra-threshold stimulated Raman lasing from silica microsphere, we have tried to design a novel non-invasive biochemical sensor for measurement of biological glucose levels with low power lasers, which is not above the safety limitation for human body applications [11]. Concentration of some body analytes such as troponin I is extremely low, and it is necessary to increase the accuracy of system. To detect of such analytes at biological levels, surface enhancement Raman scattering (SERS) can be used. So, our team has offered an optical microsphere coupled to an optical fiber based on SERS and SRS for an ultrahigh sensitivity troponin I sensor [12].

17.2. Structure and Optical Properties of Skin

Over the last decades, optical methods have obtained witnessed widespread and exciting applications in noninvasive biomedical diagnostics, as well as therapy and surgery. Optical methods for noninvasive clinical diagnostics are valuable due to their simplicity, safety and low cost.

Despite the numerous benefits of optical methods in medicine, there are some serious disadvantages. One of them is the transport of the light beam through the turbid tissues, such as human skin. The evident solution of this problem is the control of the optical properties of the biological tissue. Biological tissue can be considered as a scattering medium that shows all optical characteristics of turbid physical systems.

17.2.1. Structure, Physical and Optical Properties of Fibrous Tissues

Human skin is the largest organ of the body that serves as a protective and sensing barrier between the organism and the environment. It consists of three main visible layers from surface: epidermis (50–200 μm thick, the blood-free layer), dermis (1–4 mm thick, vascularized layer) and subcutaneous fat (from 1 to 6 mm thick depending on the body site) [13]. Inhomogeneous distribution of blood and pigment content in skin produces variations in average optical properties of skin layers.

Quantitatively, scattering and absorption coefficients and refractive index of the medium define the light-tissue interaction.

Both experimental and theoretical findings indicate that there is a reduction in absolute transmitted intensity through human skin [14]. Since absorption in biological tissue is substantially less than scattering in the NIR spectral range, the exponential profile of light attenuation in tissue is dependent mainly on the scattering coefficient [13, 15].

Changes of a special biochemical agent in a tissue vary refractive index mismatch and correspondingly, the scattering coefficient [15]. Measurement of scattering coefficient allows one to monitor the changes of biochemical agent in the tissue. The average scattering properties of the skin are defined by the scattering properties of the reticular dermis because of relatively big thickness of the layer(up to 4 mm [16]) and comparable scattering coefficients of the epidermis and the reticular dermis [13].

Physical and optical properties of the dermal layers are mainly defined by the fibrous structure of the tissue. These tissues consist mainly of collagen fibers packed in lamellar bundles [17,18] that are immersed in an amorphous ground (interstitial) substance, a colorless liquid containing protein, proteoglycans, glycoproteins and hyaluronic acid [19]. Collagen bundles show a wide range of widths and thickness. They cross each other in all directions but remain parallel to the tissue surface. All these inhomogeneities give a high scattering of a tissue in normal state [18].

The optical model of fibrous tissue in a local region can be presented as a slab with a thickness l containing scatterers(collagen fibrils). The average value of the refractive index of a fibrous tissue is:

$$n_{tissue} = n_{col} V_{col} + n_{gr} (1\text{-}V_{col}) , \qquad (17.1)$$

where n_{col}, n_{gr} and V_{col}, V_{gr} are the refractive indices and volume fractions of collagen and ground material, respectively. For fibrous tissues V_{col} is usually equal to 0.3 [17, 20].

Considerable refractive index mismatch between collagen fibers and a ground material causes the system to become turbid, i.e., causes multiple scattering and poor transmittance of propagating light. The refractive index of the background is a controlled parameter and may transit the system from multiple to low-step and even single-scattering mode [13].

The wavelength dependence of the refractive index of collagen fibrils(n_{col}) can be determined by [21]:

$$n_c(\lambda) = 1.439 + \frac{15880.4}{\lambda^2} - \frac{1.48 \times 10^9}{\lambda^4} + \frac{4.39 \times 10^{13}}{\lambda^{13}}, \tag{17.2}$$

where λ is the wavelength, nm.

We assume that the radius and density of the fibrils are unchangeable and only refractive index of interstitial fluid(ISF) as background substance is controlled by the biochemical agent administration.

The interstitial fluid (ISF) surrounds the collagen fibrous structures and cells. The refractive index mismatch between the ISF and cutaneous fibrous material controls skin scattering properties.

To describe dynamics of the refractive index and corresponding alterations of the scattering coefficient when the biochemical agent diffuses within the interstitial substance of the tissue we can consider it as diffusion through a partially permeable membrane.

The temporal dependence of the refractive index of the ISF caused by the clearing agent permeation into a tissue can be derived using the law of Gladstone and Dale as [21, 22]:

$$n_{ISF}(t) = [1 - C(t)]n_{ISF}(\lambda) + C(t)n_{osm}(\lambda) \tag{17.3}$$

$$n_{ISF}(\lambda) = 1.351 + \frac{2134.2}{\lambda^2} + \frac{5.79 \times 10^8}{\lambda^4} - \frac{8.15 \times 10^{13}}{\lambda^6} \tag{17.4}$$

$$n_{osm}(\lambda) = n_w(\lambda) + \alpha C \tag{17.5}$$

$$n_w(\lambda) = 1.3199 + \frac{6.878 \times 10^3}{\lambda^2} - \frac{1.132 \times 10^9}{\lambda^4} + \frac{1.11 \times 10^{14}}{\lambda^6}, \tag{17.6}$$

where $n_{base}(\lambda)$ is the refractive index of the ISF at the initial moment, $n_{osm}(\lambda)$ is the refractive index of an agent solution, $n_w(\lambda)$ is the wavelength dependence of the refractive index of water, C is the agent concentration(g/ml) and α is a constant that is defined as the increase in refractive index of the solution for every one percent, increase in agent concentration, and is known "specific refraction increment " [23]. C(t) is the time dependence of biochemical agent concentration.

The analysis of the mechanism of permeability of chemical agents through human skin bases on Fick's diffusion theory since a tissue slab free of agent is immersed in solution with the agent concentration of C_0, the time dependent concentration of the chemical agent C(t) within tissue can be expressed as [21]:

$$C(t) \approx C_0(1-\exp(-t\pi^2 D / l^2)) = C_0(1-\exp(-t/\tau)), \qquad (17.7)$$

where l is the tissue sample thickness, $\tau = l^2(\pi^2 D)$ is the characteristic diffusion time and D is the diffusion coefficient. The diffusion coefficient for the human skin is $2.56 \pm 0.13 \times 10^{-6}$ cm^2/s [24].

In the case of unchangeable scatterer size, all changes in the tissue scattering are connected with the changes of the refractive index of the ISF. The change of refractive index of ISF provides the change in relative refractive index of scattering particles and, consequently, the change in scattering coefficient. For noninteracting particles the scattering coefficient of a tissue is defined by the following equation [21, 25].

$$\mu_s(t) = \frac{\phi}{\pi a^2}\sigma_s(t)\frac{(1-\phi)^3}{1+\phi}, \qquad (17.8)$$

where a is the cylinder radius, ϕ is the volume fraction of the tissue scatters, and $\sigma_s(t)$ is the time-dependent cross-section of scattering.

For a thin dielectric cylinder in the Rayleigh-Gans approximation of the Mie scattering theory the scattering cross-section $\sigma_s(t)$ for un polarized incident light is given by [21, 26]:

$$\sigma_s = \frac{\pi^2 a x^3}{8}(m^2-1)^2\left(1+\frac{2}{(m^2+1)^2}\right), \qquad (17.9)$$

where $m=n_c/n_I$ is the relative refractive index of the scattering particle and surrounding medium and x is the dimensionless relative scatterers size which is determined as $x=2\pi a n_I/\lambda$, a is the cylinder radius.

Attenuation of light intensity for ballistic photons, in a medium with scattering and absorption is described by the Beer-Lambert law [15]:

$$I = I_0 e^{-\mu_t z} \tag{17.10}$$

where $\mu_t=\mu_s+\mu_a$ is the attenuation coefficient of ballistic photons, μ_s, μ_a are the scattering and absorption coefficients, respectively, and I_0 is the incident light intensity. Since absorption in tissues is substantially less than scattering ($\mu_a \ll \mu_s$) in the NIR spectral range, the exponential attenuation of ballistic photons in tissue is dependent mainly on the scattering coefficient:

$$I = I_0 e^{-\mu_s z} \tag{17.11}$$

Since the tissue scattering coefficient changes with biochemical agent concentration, the exponential profile of light attenuation in tissue is dependent on agent concentration. Therefore, one can monitor agent concentration by measuring the exponential slope of light attenuation in tissue.

However, the refractive index mismatch between the ISF and fibrous tissue controls skin scattering properties, but the equilibrium point between levels of a biochemical agent in ISF and blood is affected by many uncontrollable factors such as humidity or skin temperature. The relationship between changes of agent in blood and ISF, with respect to both time and concentration, is not well understood, especially during dynamic changes. Also, there are not all important biochemical agents in ISF.

As a result of so many unresolved issues related to the calibration of ISF readings, regulatory authorities and medical communities have become reluctant to accept ISF measurement as a full substitute of blood agent readings. Therefore, a noninvasive measurement of biochemical agent in blood is called to fulfill an unmet medical need where noninvasive ISF measuring has been found not sufficient.

17.2.2. Structure, Physical and Optical Properties of Whole Blood

The optics of whole blood at physiological conditions is mainly determined by the optical properties of red blood cells (RBCs) and plasma, whereas the contribution of remaining blood particles in optical properties can be neglected. Analysis of light propagation in such medium can be performed by description of absorption and scattering characteristics of individual blood particles.

The main scatterers in blood are RBCs. The major part of this cell is the hemoglobin solution (90 %) [27]. So the refractive index mismatch between hemoglobin solution in RBC cytoplasm and blood plasma provides strong blood scattering [28].

In radiative transfer theory, the absorption coefficient μ_a, scattering coefficient μ_s and anisotropy factor g of an investigated medium are determined by the size of RBCs, as well as real(n) and imaginary(χ) parts of the complex refractive index($n+i\chi$) of RBCs and their environment (blood plasma) [13]. The scattering properties of blood are also dependent on RBC volume, shape and orientation, which are defined by blood plasma osmolarity [29], aggregation and disaggregation capability and hematocrit [30].

In optical model of blood, particles can be represented as absorbing and scattering homogeneous spherical particles that volume of each particle is equal to the volume of a real RBC [31]. Contribution of the RBC membrane in scattering can be neglected due to its small thickness [32].

The concentration of hemoglobin is related to the RBC volume as [32]:

$$C_{Hb} = 0.72313 - 0.00451V, \qquad (17.12)$$

where C_{Hb} is the concentration of hemoglobin(g/ml), and V is the erythrocyte (RBC) volume (μm^3). Both the real and the imaginary parts of the refractive index of erythrocytes are directly proportional to the hemoglobin concentration as [29]:

$$n_e(\lambda) = n_0(\lambda) + \alpha(\lambda)C_{Hb} \qquad (17.13)$$

$$\chi_e(\lambda) = \beta(\lambda)C_{Hb}, \qquad (17.14)$$

where α and β are the constants and the spectral dependences of them are presented in Ref [33]. n_0 is the refractive index of the erythrocyte cytoplasm and is given by [32]:

$$n_0(\lambda) = n_w(\lambda) + 0.007 \qquad (17.15)$$

The blood plasma contains up to 91 % water, 6.5–8 % proteins(hemoglobin, albumin and globulin) and about 2 % low-molecular-weight compounds [34]. Since the blood plasma does not have pronounced absorption bands in NIR spectral range, the imaginary part of the refractive index of the plasma can be neglected. So the wavelength dependence of the real part of the refractive index of the blood plasma(n_p) can be determined by the expression [35]:

$$n_p(\lambda) = 1.3254 + \frac{8.4052 \times 10^3}{\lambda^2} - \frac{3.9572 \times 10^8}{\lambda^4} - \frac{2.3617 \times 10^{13}}{\lambda^6}, \qquad (17.16)$$

where λ is the wavelength (nm).

It should be noted that the blood plasma osmolarity is also an important factor in changes in the scattering properties of blood [34]. The change in osmolarity induces a variation of the RBC volume due to water exchange and therefore has an impact on the hemoglobin concentration within the RBC and consequently on their refractive index.

By analogy with equation(17.13), the real part of the refractive index of hemoglobin in plasma can be defined as:

$$n'_P(\lambda) = n_P(\lambda) + \alpha C_{Hb} \qquad (17.17)$$

C is the hemoglobin concentration (g/ml) and α is the "specific refraction increment "of hemoglobin [28].

The refractive index of the RBC surrounded with plasma and agent with concentration C can be estimated as [36]:

$$n_{RBC} = n_{Hb}.FR + (1 - FR)(n_w + \alpha C) \qquad (17.18)$$

where FR is the volume fraction of hemoglobin in RBC.

The scattering and absorption coefficients and the anisotropy factor of whole blood are given by [13, 25]:

$$\mu_s = (1 - H)\sum_{i=1}^{M} N_i \sigma_{s_i} \qquad (17.19)$$

$$\mu_a = \sum_{i=1}^{M} N_i \sigma_{a_i} \qquad (17.20)$$

$$g = \sum_{i=1}^{m} \mu_{s_i} g_i \Big/ \sum_{i=1}^{m} \mu_{s_i} \qquad (17.21)$$

Here H is the hematocrit value; N_i is the number of erythrocytes per unit volume of the medium.

In terms of the Mie theory, the scattering (σ_s) and the anisotropy factor of a homogeneous sphere are expressed by [37]:

$$\sigma_s = \left(\frac{\lambda^2}{2\pi n_P^2}\right)\sum_{n=1}^{\infty}(2n+1)\left(|a_n|^2 + |b_n|^2\right) \qquad (17.22)$$

$$g = \frac{\lambda^2}{\pi n_P^2 \sigma_s}\left[\sum_{n=1}^{\infty}\frac{n(n+2)}{n+1}\mathrm{Re}\{a_n a_{n+1}^* + b_n b_{n+1}^*\} + \sum_{n=1}^{\infty}\frac{n(n+2)}{n+1}\mathrm{Re}\{a_n b_{n+1}^*\}\right] \qquad (17.23)$$

where $m=n_{RBC}/n_P$ is the ratio of the refractive indices of the particlesn (n_{RBC}) and surrounding medium (n_P), $x=2\pi a n_P/\lambda$ is the size parameter, a_n and b_n are the Mie coefficients and a_n^* and b_n^* are their complex conjugates.

Agent administration in tissues and blood allows one to control effectively its optical properties. Such control leads to higher optical clearing and the appearance of a large amount of least-scattered and ballistic photons, allowing for successful applications of different imaging techniques for medicine. The kinetics of tissue optical clearing defined generally by both the kinetics of dehydration and refractive index matching that is characterized by different time intervals in dependence on tissue and used agents. Optical clearing can increase effectiveness of number of therapeutic and surgical methods using laser action on a target, in depth of tissue.

17.3. Optical Microsphere Resonators

In the past decades, the advances in micro-fabrication techniques have made it feasible to consider optical resonators having physical dimensions in order of optical wavelength. As a particular mode of microcavity resonances, the whispering-gallery-mode (WGM) is a morphology-dependent resonance [9]. WGM occurs when light travels in a dielectric medium of circular geometries. After repeated total internal reflections at the curved boundary, the electromagnetic field can close on itself and give rise to resonances [2]. Fig. 17.1(a) shows the geometric optics view of the WGM resonance condition in microsphere. The approximate condition for optical resonance or WGM is [38]:

$$2\pi R_0 n_1 = m\lambda \qquad m=1,2,3,... \tag{17.24}$$

where, R and n_1 are the sphere radius and refractive index, λ is the vacuum wavelength of light, and m is an integer.

Optical processes in microcavities are a very active field of research for fundamental as well as applied purposes [9, 39-42]. Various types of geometries are currently studied like microspheres [43], microrings [44], microtoroids [45], and microdisks [46]. Of all geometries studied for confining light, silica microspheres have attained the highest optical quality-factor (Q) about 10^9-10^{11}, and are of interest for a number of optical effects [47-54]. The beauty of silica microsphere is that they are easily produced in a laboratory setting. Upon heating the end of silica optical fiber, the end reflows to form a spherical volume under the influence of surface tension [9]. Due to the high viscosity of silica, the reflowed structure is both highly spherical and extremely uniform. The spherical surface has very low intrinsic roughness, and thus has a very small surface scattering loss.

A significant challenge in utilizing high-Q narrow-linewidth optical resonators is the need to excite resonant modes efficiently while, simultaneously, making sure that the Q is not compromised. Evanescent coupling, in which an exterior field tunnels into the sphere, appears to be the most promising approach. The ideal microsphere WGM coupling device includes the following characteristics: (A) efficient WGM excitation performance with little potential for Q-spoiling, (B) simple sphere-to-coupler alignment, (C) clearly defined ports,(D) robust and integrable structure, and (E) a consistent and inexpensive fabrication process [55].

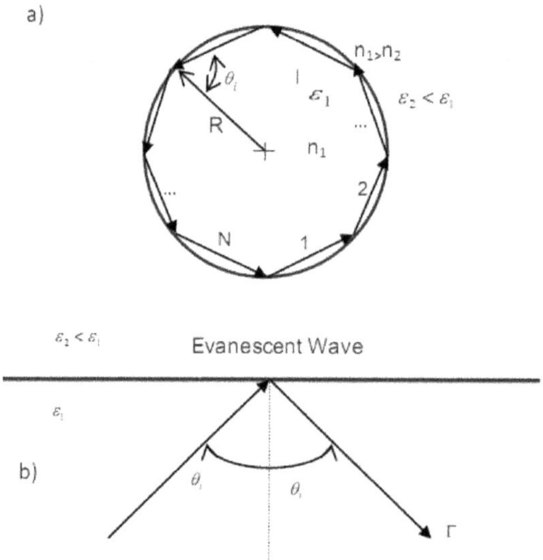

Fig. 17.1. (a) Ray optics model of light traveling inside sphere(N is the number of reflections for one round trip); (b) reflection of a plane wave from a dielectric interface [38].

Numerous coupling devices, such as half-block fiber couplers [56], tapered optical fibers [10], prism coupler [57] and angle-polished fiber tips [58] have been developed by several research groups. Among these, the tapered fibers have been shown to provide the most robust and efficient coupling to the WGMs without degrading [59]. So, trapped fiber coupling has been employed.

A dielectric sphere coupled to a tapered optical fiber is depicted in Fig. 17.2. The sphere has a radius of R_0, a uniform index of n_s, and is surrounded by a uniform background medium of index n_0.

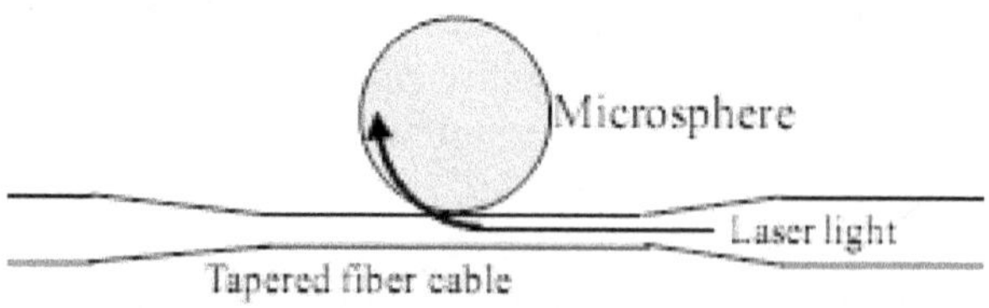

Fig. 17.2. Laser light is coupled into the microsphere's WGM through a tapered fiber cable.

The optical modes of a spherical dielectric particle can be calculated by solving Helmholtz equation in spherical coordinates, which has been treated by several authors [60]. The spherical coordinates are represented by the usual variables, r for the radial direction, ϕ for the azimuthal direction, and θ for the polar direction. Here θ is measured from the equatorial plane. The equatorial plane is defined as that great circle passing through the point on the sphere closest to the optical fiber (Fig. 17.3). It is along this plane that the modes are most strongly coupled to the fiber.

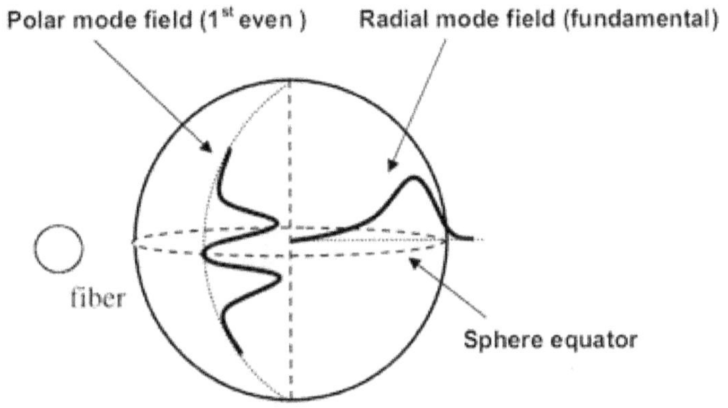

Fig. 17.3. Schematic of WGM field components in a microsphere coupled to a tapered optical fiber resonator (not to scale).

A dielectric sphere is an open cavity supporting tunneling leaky waves. The eigenvalues of the exact solution would need to be complex in order to satisfy the radiation condition. Such solutions, with fields that grow unbounded in the radial direction far from the sphere, are difficult to normalize [61]. On the other hand, it is only the near field which contributes to coupling with the external excitations. A significant simplification occurs if the sphere consists of a homogeneous dielectric, and if the optical modes reflect with grazing incidence upon the dielectric-air boundary, such that the polarization can be assumed to be constant along the optical trajectories. Under this assumption the optical modes can be solved by the scalar wave equation approximation and solutions fall into two classes, and are either electric in character(TM-case) or magnetic in character (TE-case). The field components can be expressed in terms of a single field components(E_ϕ for the TM-case or H_ϕ for the TE-case) and solutions are found by solving the scalar wave equation for either the E_ϕ or H_ϕ alone by the separation of variables approach, i.e. E_ϕ or $H_\phi=\Psi_{l,m,n}(r,\theta.\phi)=N_sR_r(r)\Theta_\theta(\theta)\Phi_\phi(\phi)$. TE modes possess an electric field is parallel to the surface of the sphere(i.e. $E_r=E_\phi=0$, $E=E_\theta\tilde{\theta}$); whereas the TM modes possess a magnetic field which is parallel to the surface of the sphere (i.e. $H_r=H_\phi= 0$, $H=H_\theta\tilde{\theta}$).The introduced eigenfunctions for the radial, azimuthal and polar fields can be associated with the radial mode number (n), the polar mode number (l) and the azimuthal mode number (m). So the component contributions of electromagnetic filed takes the form [9, 56]:

$$\Phi(\phi) = \exp[\pm im\phi] \tag{17.25}$$

$$\Theta(\theta) = \exp[-\frac{m}{2}\theta^2]H_N(\sqrt{m}\theta) \qquad m \gg 1 \gg \theta \tag{17.26}$$

$$R(r) = \begin{cases} j_l(kn_s r) & r \le R_0 \\ j_l(kn_s R_0)\exp[-\alpha_s(r - R_0)] & r > R_0 \end{cases} \tag{17.27}$$

And the coefficients are

$$N_s = \left\{\sqrt{\frac{\pi}{m}}2^{N-1}N!R_0^2[(1+\frac{1}{\alpha_s R_0})j_l^2(kn_s R_0) - j_{l-1}(kn_s R_0)j_{l+1}(kn_s R_0)]\right\}^{-1/2}$$

$$\alpha_s = \sqrt{\beta_l^2 - k^2 n_{sphere}^2} \qquad \beta_l = \frac{\sqrt{l(l+1)}}{R_0} \tag{17.28}$$

$$N = l - m \qquad k = \frac{2\pi}{\lambda}$$

The characteristic equation which describes the relationship between the wave vector k and the eigenvalues l and n is determined by matching tangential electric and magnetic fields across the surface $r=R_0$. The remaining $\vec{H}$ fields of the TE modes, or the $\vec{E}$ fields of the TM modes, are determined by Maxwell's equations. Matching tangential fields leads to the simple characteristic equation [56]:

$$(\eta_s\alpha_s + \frac{l}{R_0})j_l(kn_s R_0) = kn_s j_{l+1}(kn_s R_0)$$

$$\eta_s = \begin{cases} 1 & TE \text{ modes} \\ \frac{n_s^2}{n_0^2} & TM \text{ modes} \end{cases} \tag{17.29}$$

It is useful to invoke concepts from integrated optics to visualize the propagation of modes in a sphere. Parallel to surface of the sphere, in the direction of the zig-zag path, the mode has a propagation constant β_l which is shown to have the value $\beta_l = \sqrt{l(l+1)}/R_0$ [38, 57] (Fig. 17.4). The mode is confined to a belt around the equatorial plane by the curvature of the sphere in the polar direction. The projection of β_l onto the equator is commonly referred to as the "propagation constant" in integrated optics, because it is the wave vector in the net direction of propagation. The projection onto the equatorial plane turns out to have the value $\beta_m = m/R_0$. For any fixed value of l, m can range throughout $|m| \le l$. Different values of m imply that the modes travel in zig-zag paths with different inclinations with respect to the equatorial plane.

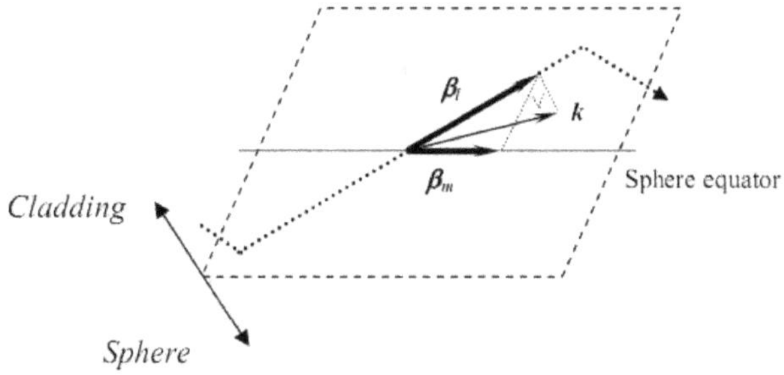

Fig. 17.4. Vector schematic for β_m , β_l and k, relative to the sphere surface and to a notional mode path projection [55].

Due to the presence of loss mechanisms such as material absorption, scattering losses or tunnel losses, the optical modes of a resonator are dissipative in character("leaky") and are referred to as "quasi-modes". Quasi-modes are distinct to their loss-less counterparts(modes) [62]. The extent to which dissipation is present in a resonant system is commonly expressed by the Quality-factor or Q-factor of the mode, which is defined by the energy storage time normalized with respect to the period of oscillation [63].

$$Q = \omega_0 \frac{E_{Stored}}{P_{loss}} = \omega_0 \tau_{ph} \tag{17.30}$$

In this equation, ω_0 is the resonance frequency, E_{stored} is the energy contained in the resonant system, and P_{loss} is the dissipated power. Equivalently, in the case of optical microcavities the optical Q-factor describes the photon lifetime of a mode τ_{ph}. In the case of a microsphere, the total Q-factor is comprised of several loss contributions: intrinsic material absorption, scattering losses (both intrinsic, as well as inherent to the surface of the cavity), surface absorption losses (e.g. due to the presence of adsorbed water), whispering gallery loss (or tunnel loss) and external coupling losses to a "useful" external mode (such as a prism or a waveguide) [9].

$$Q_{tot}^{-1} = Q_{mat}^{-1} + Q_{scatt}^{-1} + Q_{surf}^{-1} + Q_{ext}^{-1} + Q_{WGM}^{-1} \tag{17.31}$$

Our own interest in such microsphere resonators arises within the realm of Stimulated Raman laser, in which they have both small-cavity mode volumes (for large electric fields per photon) and ultralow resonator losses (for long photon-storage times).

In many applications, not only temporal confinement of light (i.e. the Q-factor), but also the extend to which the light is spatially confined is an important performance parameter. Several definitions of mode volume can be encountered in literature, and are discussed in this section. The most common definition of mode volume is related to the definition of the energy density of the optical mode [9].

$$w_e(r)+w_m(r)=\frac{1}{2}\varepsilon\tilde{E}\tilde{E}+\frac{1}{2\mu}\tilde{B}\tilde{B} \tag{17.32}$$

It is defined as the equivalent volume, the mode occupies if the energy density was distributed homogeneously throughout the mode volume, at the peak value:

$$V_{Mode}=\frac{\int(w_e(r)+w_m(r))dV}{\max(w_e(r)+w_m(r))}=\frac{\int\varepsilon(r)\left|\vec{E}(r)\right|^2 d^3r}{\max(\varepsilon(r)\left|\vec{E}(r)\right|^2)} \tag{17.33}$$

The integral is evaluated over all space, and also includes the regions where the field is evanescent. Two comments are in place. First, this definition of mode volume differs from the definition of effective mode volume which will be encountered when studying nonlinear optical processes. Secondarily, the precise definition of mode volume is strictly speaking not rigorous. It depends on the physical problem studied.

Our own interest in such microsphere resonators arises within the realm of Stimulated Raman laser, in which they have both small-cavity mode volumes (for large electric fields per photon) and ultralow resonator losses (for long photon-storage times).

In the microsphere stimulated Raman scattering can occur when both energy and momentum conservation among the pump photon, scattered photon and phonon are obeyed. Momentum conservation is intrinsically satisfied, due to the essentially flat dispersion relation of optical phonons. Due to the broad nature of the Raman gain spectrum, the resonance condition(i.e. energy conservation) for a Raman mode is strongly relaxed [52].

17.4. Raman Scattering

Detecting certain molecules in a solution with high sensitivity and molecular specificity is of great scientific and practical interest in many fields such as chemistry, biology, medicine, pharmacology, and environmental science [64]. The energy spectrum of molecular vibrations can serve as an unambiguous characteristic fingerprint for the chemical composition of a sample. Due to its sensitivity to molecular vibrations, Raman spectroscopy is a very important tool for the analysis of biomolecules.

The spectral information of the Raman signal can identify unique binding energies of molecules. The Raman spectroscopy has been employed in sensor technology for many years, as it provides many advantages over other spectroscopic techniques such as Fourier transform infrared (IR) spectroscopy, near-infrared (NIR) absorption, UV-visible absorption, fluorescence, nuclear magnetic resonance (NMR), x-ray diffraction, x-ray photoelectron spectroscopy or mass spectroscopy [65, 66]. Raman scattering is a non-destructive, non-invasive technique, because the threshold intensity of Raman lasing is ultralow.

17.4.1. Basic Principle

The spontaneous Raman scattering is an inelastic scattering between a photon and a molecule, arbitrated by a fundamental vibrational mode of any molecular medium. Raman scattering is named after Sir Chandrasekhara Venkata Raman who first observed the effect in 1928. During this process, the incoming photon with energy $h\nu_L$ will have a shift in its energy by the characteristic energy of vibration of $h\nu_M$. These shifts depend on whether the molecule is in its vibrational ground state or in its excited state. In the first case, the photon loses energy by excitation of a vibrational mode (Stokes scattering); in the second case, energy gains by de-excitation of such a mode (anti-Stokes scattering) are possible [67]. In this effect, incident light act as pump and generates the Stokes wave that serves as spectroscopic tool in biomedical application. Fig. 17.5 shows the energy levels for Stokes and anti-Stokes Raman scattering.

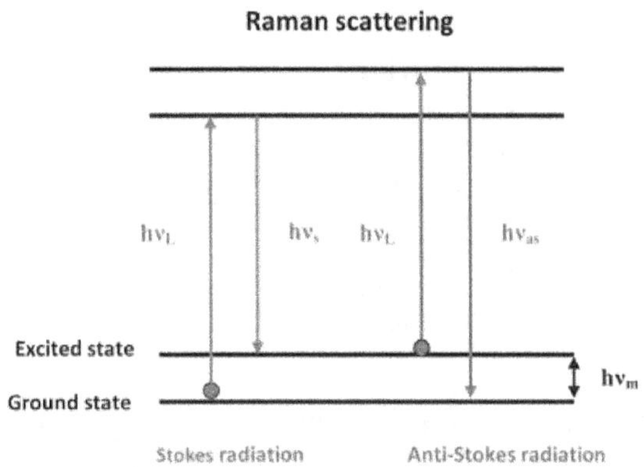

Fig. 17.5. Schematic diagram of Raman scattering.

The spontaneous Raman scattering process is typically a relatively weak process. However, under excitation by an intense laser beam, highly efficient scattering can occur as a result of the stimulated Raman scattering (SRS). SRS was discovered by Woodbury and Ng in 1962 and was described more fully by Eckhardt et al. in 1962. Since then, SRS has been studied extensively in a variety of molecular medium. SRS is an important nonlinear process that can turn optical fibers into broadband Raman amplifiers. With only a single input wave, the generation of Stokes frequencies begins with spontaneous scattering. As the scattered light grows in intensity, the process becomes stimulated, which can lead to Stokes power levels comparable to that of the pump input. To describe the effect, the earlier theory must be modified to include the effect of spontaneous scattering. Additionally, the effect of fiber loss on Stokes wave amplification must also be included. This section provides a theoretical approach for describing SRS in optical fibers.

17.4.2. Raman Gain

Optical fibers can be used to amplify a weak signal if that signal is launched together with a strong pump wave such that their frequency difference lies within the band width of Raman gain spectrum. Because Stimulated Raman Scattering is the physical mechanism behind amplification, such amplifiers are called Raman fiber amplifiers. As the scattered Stokes light increases in intensity, the stimulated scattering regime is eventually reached. Light at the Stokes wavelength can be input (along with the pump) and is amplified through the stimulated conversion process. This is the basis of a Raman amplifier, which in principle allows amplification for a signal light at any wavelength.

To obtain the Raman coupled equation, the Raman and pump plane waves must be satisfied in the wave propagation equation. Consider the simplest situation in which light at the Stokes wavelength along with the continuous-wave (CW) pump light is launched into an optical fiber.

When the nonlinear interaction between the pump and Stokes waves are occurred, the SRS process is governed by the following set of coupled equations:

$$\frac{dI_s(z)}{dz} = g_R I_p I_s - \alpha_S I_s \qquad (17.34)$$

$$\frac{dI_p(z)}{dz} = -\frac{\omega_P}{\omega_S} g_R I_p I_s - \alpha_P I_s, \qquad (17.35)$$

where I_s is the Stokes intensity, I_p is the pump intensity, and the g_R is the Raman gain coefficient, which is related to the cross section of spontaneous Raman scattering. α_s and α_p are the optical fiber losses at the Stokes and pump frequencies, respectively.

The Raman gain coefficient $g_R(\Delta\omega)$, where $\Delta\omega$ represents the frequency difference between the pump and Raman waves, is the most important quantity for describing SRS. It was measured for optical silica fibers in the early experiments on SRS, with improvement continuing in later years. Fig. 17.6 shows normalized Raman gain for fused silica as a function of the shift of frequency at the pump wavelength λ_p=1 μm. The most important feature of the Raman gain in optical silica fibers is that $g_R(\Delta\omega)$ extends over a large frequency range. As shown in Fig. 17.6, the Raman gain bandwidth is about 40 THz with a peak located near 13 THz. The Raman effect in silica fibers arises from a few basic vibrational modes that occur in fused silica and in structurally related glasses that are used as dopants.

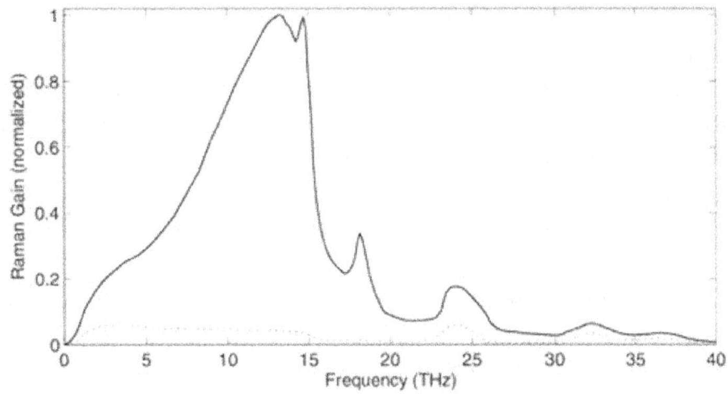

Fig. 17.6. Normalized Raman gain for fused silica. Pump and Stokes waves are co-polarized (solid curve). Pump and Stokes waves are orthogonally polarized (dotted curve) [68].

If Stokes light is launched with the pump at the optical fiber, it will be amplified due to the Raman gain. But if only the pump light is launched at the fiber input, spontaneous Raman scattering acts as a probe and is amplified with propagation. As a result, SRS leads to generation the Stokes waves whose frequency is determined by the peak of Raman gain. As shown in Fig. 17.2, The Stokes wave is co-polarized with the pump light because Raman gain is relatively small when they are orthogonally polarized. The most important disadvantage of normal Raman scattering for diagnostic applications is the intrinsically weak cross section (on the order of 10^{-30} cm^2), which is about 14 orders of magnitude lower than the typical fluorescence cross section. Thus, traditionally Raman scattering has not been a useful technique for sensing molecules in extremely low concentrations. In 1974, it was reported that the Raman scattering cross-section can be considerably increased if the molecules are adsorbed on roughened metal surfaces [71]. In the following decades SERS has been investigated as a useful technique that provides tremendously large Raman signal enhancement in the vicinity of metal nanostructures. The details of principle of SERS technique will be presented in the next section.

17.4.3. Principles of SERS

Plasmonic is one of the branch of nanophotonics, which explore the interaction between electromagnetic field and conduction electrons at metallic interface or in metal nanostructures on the order of or smaller than the wavelength. The history of using noble metal nanostructures (e.g., gold and silver nanoparticles) is less turbulent, with the application of metal nanoparticles for the staining of glass dating back to Roman times. The most spectacular applications of plasmonic to date are SERS, which utilizes the localized electric field enhancement in the near-field of metal nanostructures to enhance spontaneous Raman scattering of molecules. SERS has been an active area of research since its discovery in 1974 [69] and confirmation in 1977 [70].

Owning to the local electromagnetic field enhancements near the noble metal nanostructures, Raman signal could be considerably improved and novel optical labels based on surface enhanced Raman spectroscopy (SERS) instead of using fluorescence labels were proposed [71, 72]. SERS has been used as a signal transduction mechanism in biological and chemical sensing of trace analytes such as pesticides, anthrax spores [73], prostate-specific antigen [74], glucose [75], bacteria [76], and genetic materials [78]. A miniature inexpensive SERS device can be used in clinics and field.

The Raman signal enhancement originates from locally electromagnetic field enhancement near the surface of the metal nanostructures. Raman scattered power depends linearly on the excitation intensity. Thus, the total power of Stokes Raman light can be expressed as:

$$P_S(v_S) = N\sigma_{RS} I(v_L), \tag{17.36}$$

where N is the number of number of Stokes-active scatterers, σ_{RS} is the scattering cross section, and $I(v_L)$ is the intensity of excitation light. SERS can be used when the target analyte is brought to the surface of a noble metal nanostructure.

SERS utilizes a mutual electromagnetic enhancement due to surface plasmon resonance (SPR) and a chemical enhancement, both of which provided by gold or silver nanostructure. The electromagnetic enhancement results from the enhancement of the Raman light in a two-step process: firstly, the excitation of Raman-active molecules is proportional to the square of local electric field at the incident frequency $|E_{loc}(\upsilon_L)|^2/|E_0(\upsilon_L)|^2$, which can be very strong at the surface of metal structures. In the second step, the scattered Raman light is enhanced by the metal nanostructures which are proportional to the local electric field at the Raman frequency $|E_{loc}(\upsilon_s)|^2/|E_0(\upsilon_s)|^2$. The total power of the Stokes Raman light under SERS effect is:

$$P_{SERS} = N\sigma_{SERS} \frac{|E_{loc}(v_L)|^2}{|E_0(v_L)|^2} \frac{|E_{loc}(v_s)|^2}{|E_0(v_s)|^2} I(v_L) \tag{17.37}$$

In the most cases, the frequency of Raman scattered light is closed to the incident light. Thus, the frequency difference between them is in general much smaller than the linewidth of surface plasmon mode. So the commonly used expression for the enhancement of the power of the Stokes Raman light can be expressed as follows:

$$G_{SERS} = \frac{|E_{loc}|^2}{|E_0|^2} \tag{17.38}$$

When the molecules attach to the metallic nanostructure, the local electromagnetic field around the nanostructures improves RS by a factor of 10^6-10^8 for an ensemble of molecules, and by a magnitude of 10^{12}-10^{14} for a single molecule [78]. Under these conditions, the Raman cross section and, in turn, the signal intensity are extremely enhanced so that the level of detection down to a single molecule can be reached, while preserving all the structural information provided by Raman scattering. Therefore, advancement in SERS detection is linked to the progress in the synthesis and optimization of the optical characterization of nanostructures. SERS can be achieved and maximized by controlling both the electrical and chemical effects, mainly through careful design of the optical substrates and improving the absorption of the analytes of interest. Therefore, preparing optical substrate with optimized properties is of paramount importance.

17.4.4. The Effect of Nanostructures Morphology on SERS Enhancement

The most important SERS substrates are made of noble metals such as silver and gold. Although, silver is much more efficient than gold, collection of both metals yield the same electromagnetic enhancement factor (up to 10^{14}) at near-infrared(NIR) frequency [71]. This feature makes the gold nanoparticles (1–100 nm) very good candidate nanosensors for studying the internal organelles of the prokaryotic cells [70]. Other key factors affecting SERS intensity are the size and shape of the nanostructures. Several authors have studied the dependence of the SERS signal on the size of silver and gold nanoparticles. Smaller nanoparticles are advantageous in protein labeling and cellular imaging, because their small surface area reduces non-specific interactions and enables more targeted binding. In addition, smaller particles generate more confined electromagnetic fields so that they are more sensitive to single biomolecules, which take up a greater portion of the sensing area. However, the nanoparticle absorbance and scattering depend on nanoparticle size, scaling with nanoparticle volume for absorbance and with volume squared for scattering. Absorbance and scattering cross sections become comparable when the nanospheres are about 60 nm in diameter for silver nanospheres and 80 nm in diameter for gold nanospheres [79]. The third main factor affecting SERS intensity, which is regarded as the most important one, is the nanoparticle shape. Controlling the morphology provides a method to tune the optical and spectroscopic response of nanomaterials, which is an essential requirement for a wide range of applications such as genetic diagnostics, immunoassay labeling, and detection of trace amounts of drugs, biomolecules and pesticides. The most popular nanoparticle shapes are spheroids, triangular prisms, rods, and cubes (Fig. 17.3).

Due to the inhomogeneity of the distribution of SPRs throughout the whole particle surface, electromagnetic field concentrates in certain regions of the particle. Such an electromagnetic field concentration has been observed at the corner of triangular particles, the end of nanorods, and the edges and corners of nanobars and nanocubes. Complementary to these latest approaches for electromagnetic field concentration, remarkable progress has been directed to the controlled fabrication of the so-called "hot spots." Hot spots are defined as specific gaps between particles where the electromagnetic field intensity is extremely high due to coupling between their plasmon resonances. Carefully designed hot spots can become much more active and even enable the possibility of single molecule spectroscopy.

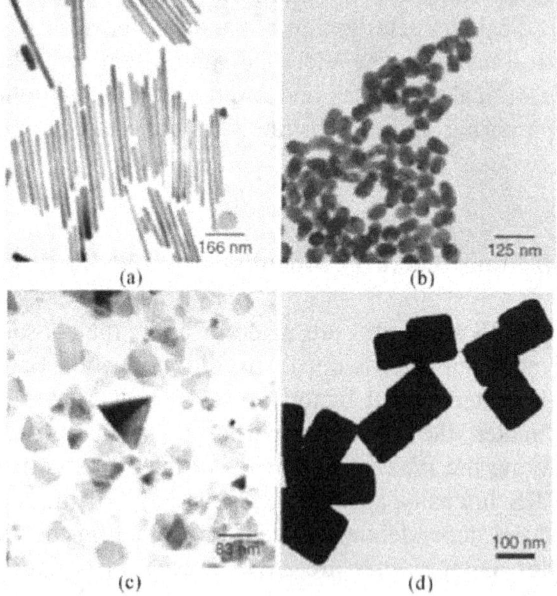

Fig. 17.7. Nanoparticle geometries: a) Gold nanorods, b) Gold colloids, (c) Silver triangular prisms, and d) Silver nanocubes [79].

17.5. Theoretical Development of a Non-invasive Micron Sized Blood Glucose Sensor Based on Microsphere Stimulated Raman Spectroscopy

Health and happiness of human have always been the main aim for scientists. To reach these top goals, many investigations have been done in thorough human physiology and pathology. In twenty-first century, human faced with an explosion in frequency of a dangerous disease called diabetes mellitus (Diabetes). Diabetes is a major world health problem. This disease is characterized by the presence of excess of glucose in the blood and tissues of the body. It is a metabolic disorder in which the pancreas underproduces (Type II diabetes) or does not produce (Type I diabetes) insulin. Insulin is a hormone produced by the pancreas that is needed by cells of the body in order to use glucose, the major source of energy for the human body [80]. It is estimated that the number of diabetics worldwide is 180 million and approximately 4 million deaths each year are caused by diabetes (9 % of deaths worldwide) [81]. The results of several studies on the prevalence of diabetes have projected that 366 million people will suffer from diabetes by 2030 [82]. Also the same study demonstrated that the most important demographic change to diabetes prevalence worldwide appears to be the increase in the proportion of people below the age of sixty-five years [83].

Diabetes is a chronic disease that without treatment can have very serious consequences. Since this disease is very widespread can lead to blindness, kidney failure and nerve damage. Also it is an important factor in accelerating hardening and narrowing of the arteries (atherosclerosis) leading to strokes, coronary heart disease and other blood vessel diseases [84]. The care and management of diabetes involves many costs. If predictions of diabetes prevalence are fulfilled, total direct healthcare expenditure on diabetes worldwide will be between 213 billion and 396 billion international dollars in 2025 [85]. It should be noted that a person with diabetes incurs medical costs that are two to five times higher than a person without diabetes. Healthcare professionals advise diabetics on the appropriate monitoring regime for their condition. Diabetics who use insulin (all Type I diabetes and many Type IIs) usually test their blood sugar more often (3 to 10 times per day). Assessment of the daily fluctuations of their blood glucose values gives a much greater understanding by patients and doctors for both to assess the effectiveness of their prior insulin dose and to help determine their next insulin dose. Indeed, home blood glucose measurement provides an important educational exercise for all seriously motivated diabetics as well as being the essential tool to achieve tight blood glucose control.

Reasonable insulin control is now relatively easily achieved in most cases with electrochemical glucose biosensors (glucometer) and the invention of this technology was one of the most important steps in managing diabetes, since it facilitates intensive therapy that reduces the risk of long-term complications. This is one of the major reasons that the glucometers occupy 85 % of the current world market for biosensors, which is estimated to be worth over 6.9 billion $ [86].

The method of blood glucose testing with a glucometer involves pricking the finger with a lancet (a small, sharp needle), putting a drop of blood on a test strip and then placing the strip into a meter that displays the blood glucose level [87] .Most current methods for self-blood-glucose-monitoring are invasive in that they require a blood sample for each test, usually obtained from a fingertip. Since, current blood (finger-stick) glucose tests are painful and inconvenient due to disruption of daily life, cause fear of hyoglycemia resulting from tighter glucose control and maybe difficult to perform in long term diabetic patients due to calluses on the fingers, a non-invasive method of blood glucose measurement would allow for a significant increase in the quality of life for the 180 million diabetics worldwide and a reduction of the yearly much higher health care costs [81]. In recent years many optical techniques are being investigated to reach this goal, e.g. near infrared(NIR) [88], Raman spectroscopy [89], fluorescent [90], polarimetric [91], photonic crystal [92], optoacoustic [93], and optical coherence tomography (OCT) [94].

We have explored the use of Raman spectroscopy to accomplish this end. This inelastic scattering, discovered by Raman and Krishna [95], is widely used in physics, chemistry, biology and medicine. Each molecule has its own distinct Raman vibrational frequencies. The Raman spectrum of solution glucose is shown in Fig. 17.7 [96]. The eight Raman modes of 436.4, 525.7, 854.9, 911.7, 1065.0, 1126.4, 1365.1, and 1456.2 are considered as the fingerprint Raman lines of blood glucose.

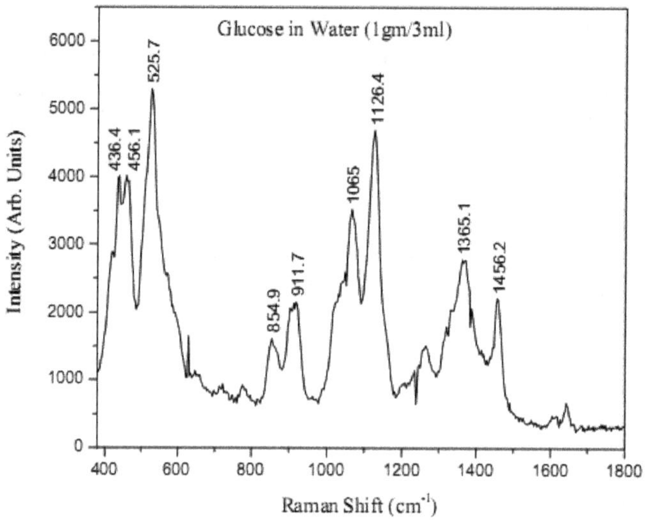

Fig. 17.7. Raman spectrum of aqueous glucose solution [96].

The concentration of human blood glucose is as low as ~0.7-1.1 $\times 10^{-3}$ g/ml for normal persons and >1.1$\times 10^{-3}$ g/ml for diabetes. The spontaneous Raman intensity from such low concentration glucose is too weak. A group at MIT showed that it is not possible to measure the blood glucose concentration with spontaneous Raman scattering [96]. Therefore stimulated Raman spectroscopy is employed. Stimulated Raman scattering can be described by the third order nonlinearity and need high power laser pump fluency which is not suitable for the in-vitro and in-vivo diagnostic purpose. To overcome this problem silica microsphere is chosen. By using ultralow-threshold stimulated Raman lasing in microspheres, we have tried to measure biological glucose levels with low power lasers, which is not above the safety limitation for human body applications. Fig. 17.8 shows the schematic of the glucose Raman microsphere sensor.

In this method two laser sources are employed. A low-power continuous pump and a low-power continuous probe beams. The frequency of the pump beam is changed, while the frequency of the probe beam is fixed. The pump beam is used to induce the Raman emission, while the probe beam serves to reveal the Raman modes. An adiabatic fiber taper is then used to couple laser beams into and out of the microsphere. The quality factor of microsphere is on the order of 10^9. The field enhancement due to the resonance condition of high quality factor microsphere causes to reduce stimulated Raman scattering thresholds of its surrounding media and hence observing stimulated Raman scattering with low power. When the difference between the pump and probe frequencies is coincident with a Raman vibrational mode frequency of glucose, the evanescent fields of the pump and probe that exist in the outside of microsphere cause stimulated Raman scattering from the glucose and the weak spontaneously Raman light will be amplified by several orders of magnitude (10 to 10^4) due to the pump photon flux (see Fig. 17.9).

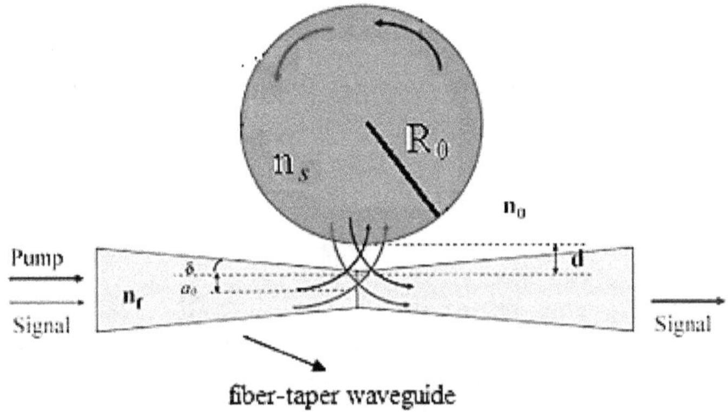

Fig. 17.8. Microsphere stimulated Raman spectroscopy system. The pump and probe signals are coupled to the microsphere through tapered optical fiber.

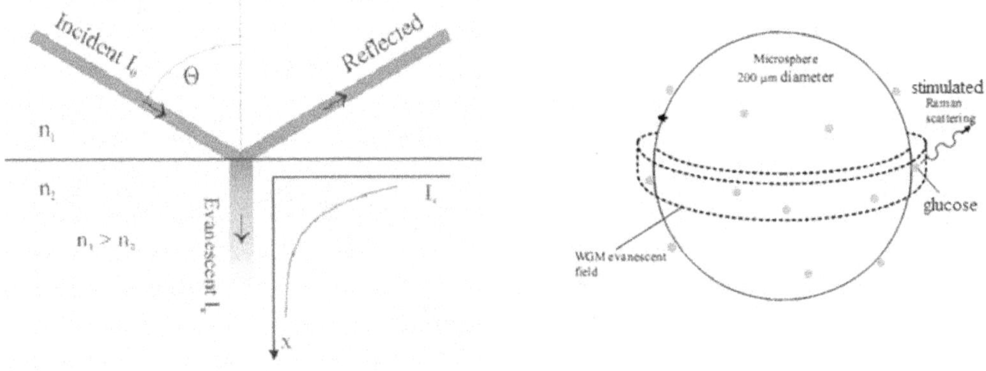

(a) (b)

Fig. 17.9. (a) Reflection of a plane wave from a microsphere interfaces, (b) Stimulate Raman scattering from glucose molecules on the surface of a microsphere.

Increased glucose in the blood is proportional to an increase in the stimulate Raman scattering from the glucose molecules and thus the signal intensity changes. Signal intensity will increase with increased glucose in the blood. The blood glucose concentration can be determined by analyzing the spectral information contained within the signal changes in the output of tapered fiber.

17.5.1. Governing Equations

17.5.1.1. Raman Gain

Raman gain is optical gain (amplification) arising from stimulated Raman scattering. It can occur in transparent solid media(e.g. optical fibers); liquids and gases under the influence of intense pump light, and is used in Raman amplifiers and Raman lasers. Its magnitude depends on the optical frequency offset between pump and signal wave, to some smaller extent on the pump wavelength, and on material properties [97]. Stimulated Raman scattering governed by the third-order susceptibility $\chi(3)$ is a nonlinear optical phenomenon in the sense that it occur when the response of a material system to an applied optical field depends in a nonlinear manner on the strength of the optical field [98].

A semiclassical model can be used to determine the gain, g_s, in which the electromagnetic fields are treated classically, and the medium is treated quantum mechanically [99]. Probe wave's equation can be written as:

$$\frac{\partial^2 E_s}{\partial z^2} + \mu_0 \varepsilon_0 n_g^2 \omega_s^2 E_s = -\mu_0 \omega_s^2 P_s^{(3)}, \tag{17.39}$$

where n_s is the refractive index, ω_p and ω_s are the frequency of pump and probe. Assuming plane wave solutions to the wave equation, the electric field at the probe frequency, ω_s is:

$$E_s = \frac{1}{2}(E_{0s} \exp\left[i(k_s z - \omega_s t)\right] + c.c.) \quad , \tag{17.40}$$

Here E_s is the vector constant and k_s is the propagation constant. By substituting E_s in equation (17.39) and solving for the propagation constant, k_s will yield:

$$k_s \approx \frac{n_s \omega_s}{c}\left(1 + \frac{\mu_0^2 c^3}{2 n_s^2 n_p} \chi(\omega_s)^{(3)} I_p\right), \qquad I_p = \frac{1}{\mu_0}\frac{n_g}{c}\left|E_p\right|^2 \tag{17.41}$$

where $\chi^3(\omega)$ is the third susceptibility and in the resonance condition can be broken down into its component parts, $\chi^3(\omega)=(\chi'+i\chi'')_R+\chi_{NR}$, χ'_R and χ''_R are the real and the imaginary part of the resonant contribution, χ_{NR} is a term accounting for the nonresonant background.

Now by substituting this expression in equation (17.41) and k_s in equation (17.40), E_s can be rewritten as:

$$E_s(z)=\frac{1}{2}E_{0z}\exp\left[-\kappa_s\frac{\mu_0^2c^3I_p}{2n_g^3}\chi_R{}''z\right]\times \exp\left[i\kappa_s\left(1+\frac{\mu_0^2c^2I_p}{2n_g^2}(\chi_R{}'+\chi_{NR})z\right)\right]\exp\left[-\omega_s t\right], \tag{17.42}$$

Now $I_s(z)$ can be written as:

$$I_s(z)=\frac{n_g}{\mu_0 c}\left|E_s\right|^2=I_s(0)e^{g_s/I_p z}, \tag{17.43}$$

where $I_s(0)$ is the initial intensity of the probe wave. On the other hands, in the resonance condition($\omega_0=\omega_p-\omega_s$), can be expressed in terms of the spontaneous Raman scattering cross section, $d\sigma/ d\Omega$, as:

$$\chi_R{}''=-\frac{\pi N\Delta c^4}{\omega_s^4 h\Gamma}\frac{d\sigma}{d\Omega}(4\pi)^3\varepsilon_0^2, \quad \Delta=1-\exp(\frac{hc(\Delta k)}{KT}) \; , \; \Gamma=2\pi c(30\text{cm}^{-1}) \tag{17.44}$$

where N is the density of molecules, and h is the Planck's constant. Substituting equation (17.44) into (17.42) for the stimulated Raman gain leads to:

$$g_s=\frac{16\pi N\Delta c^2}{\omega_s^3 n_{tissue}^2 h\Gamma}\frac{d\sigma}{d\Omega}, \tag{17.45}$$

But this is not the effective Raman gain because only the evanescent pump and probe fields that are located outside the cavity interact with glucose. Therefore it should be averaged on the radial component of filed.

In microsphere, the propagation equation for the pump and probe signal in the presence of Raleigh Scattering can be rewritten as follows [9]:

$$\frac{dI_s(z)}{dz}=-(\alpha-g_R I_p(z))I_s(z)+ 2G\sqrt{I_s(z)I_s(L-z)}\cos[nk(L-2z)], \tag{17.46}$$

$$\frac{dI_p(z)}{dz}=-(\alpha+g_R(I_s(z)+I_s(L-z)))I_p(z), \tag{17.47}$$

where I_s, I_p are the intensity of the signal and pump modes of the resonator and denotes the input wave. G and α are the scattering coefficient of both modes to each others and intrinsic loss respectively and g_R is Raman gain coefficient which is proportional to the glucose concentration.

The boundary condition in coupling region of fiber and microsphere for pump and probe beam can be written as [100]:

$$I_p(0)+r^2 I_p(L)-2r\sqrt{I_p(0)I_p(L)}=\chi^2 I_{in_p}, \tag{17.48}$$

$$I_s(0)+r^2 I_s(L)-2r\sqrt{I_s(0)I_s(L)}=\chi^2 I_{in_s}, \tag{17.49}$$

Here r and χ are the transmission and coupling coefficient and satisfy the relation $r^2+\chi^2=1$, I_{in} is incident field, I(0) is coupled intensity and I(L) is intensity after one circulation in the microsphere ($L=2\pi R_0$).

Outside of the sphere but very close to the surface, the fields decay exponentially in the radial direction. Because of the rapid decay in field amplitude, the pump and probe intensity for high azimuthal modes (m >> 1) can be approximated as [56]:

$$\begin{cases} I_s = I_s(a,z)\exp(-\alpha_s(r-R_0)) \\ I_p = I_p(a,z)\exp(-\alpha_p(r-R_0)) \end{cases}, \tag{17.50}$$

where

$$\alpha_s=\sqrt{\beta_l^2-k_s^2 n_{sphere}^2} \quad , \quad \alpha_p=\sqrt{\beta_l^2-k_p^2 n_{sphere}^2}$$

$$k_s=\frac{2\pi}{\lambda_s} \quad , \quad k_p=\frac{2\pi}{\lambda_p}$$

$$\beta_l=\sqrt{l(l+1)}/R_0 \quad , \quad |m|\le l$$

By substituting equation (17.50) into equation (17.47) and after simplification will be achieved:

$$\frac{dI_p(a,z)}{dz}= -\left(\begin{matrix} \alpha-g_s\dfrac{\left[\dfrac{1}{(\alpha_s+\alpha_p)^2}+\dfrac{R_0}{\alpha_s+\alpha_p}\right]}{\left[\dfrac{1}{\alpha_s^2}+\dfrac{R_0}{\alpha_s}\right]} \\ \times(I_{cw}(z)+I_{cw}(L-z)) \end{matrix}\right) I_p(a,z), \tag{17.51}$$

$$g_R=g_s\frac{\left[\dfrac{1}{(\alpha_s+\alpha_p)^2}+\dfrac{R_0}{\alpha_s+\alpha_p}\right]}{\left[\dfrac{1}{\alpha_s^2}+\dfrac{R_0}{\alpha_s}\right]}, \tag{17.52}$$

where g_R is the effective Raman gain in the presence of glucose and R_0 is the radius of microsphere.

17.5.1.2. Optimal Wavelength Region

Raman shifts are independent of excitation wavelength and thus Raman spectroscopy offers the flexibility to select a suitable excitation wavelength for a specific application. The NIR spectral region is commonly used in most reported methods. It has several spectral windows where hemoglobin, melanin, and water absorption band intensities are low enough to allow light to penetrate in the tissue, which enables noninvasive spectral measurements. The choice of NIR excitation for probing biological tissue is justified by three advantageous features: low-energy optical radiation, deep penetration, and reduced background fluorescence [101]. Fig. 17.10 illustrates the absorption spectra of major endogenous tissue absorbers [102]. As shown the "diagnostic window," in which a group of minima exists, is outlined over 830-1000 nm.

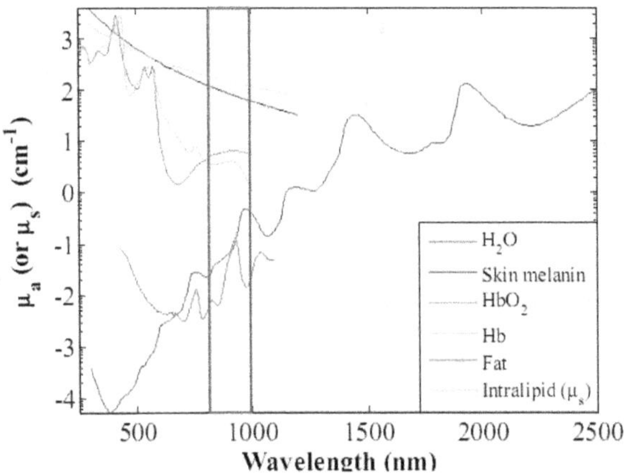

Fig. 17.10. Absorption spectra of water, skin melanin, hemoglobin, and fat. Also shown is the scattering spectrum of 10 % intralipid, a lipid emulsion often used to simulate tissue scattering [102].

17.5.2. Results and Discussion

For excitation of microsphere whispering-gallery modes, the probe and pump laser beam must circle the interior of the sphere through multiples of total internal reflections and returns in phase. The approximate condition for optical resonance of microsphere can be presented as [38]:

$$2\pi R_0 n_1 = m\lambda \qquad m=1,2,3,..., \tag{17.53}$$

where, R_0 and n_1 are the sphere radius and refraction index, λ is the wavelength of light, and m is an integer.

The microsphere in this work is approximately 200 nm in diameter and 1.5 in refractive index. When the pump and probe laser wavelengths are on resonance with the microsphere cavity and difference between the pump and probe frequencies is coincident with a Raman vibrational mode frequency of the glucose, the evanescent pump and probe fields illuminates nearby glucose and produces stimulated Raman signal.

As a result and with the respect of equation(17.53), our NIR Raman studies of biological tissue employ 834 nm as the excitation [13].

Pump wavelength can be achieved as:

$$\frac{1}{\lambda_p} - \frac{1}{\lambda_s} = \Delta k_{glucose}(cm^{-1}) , \tag{17.54}$$

pump wavelength must satisfy the resonance condition are shown with equation (17.53). By tuning the pump wavelengths from 740.5 nm to 800.9 nm, and measuring the Raman gain at the probe wavelength of 834 nm, the excitation Raman peaks from 436.4 cm to 1456.2 cm should be observed. The mode of 1126.4 cm-1 which has high Raman cross-section and narrow bandwidth are expected to be enhanced with lager gain factors. So in this mode, Stimulated Raman measurements were conducted with the following configurations (see Table 17.1).

Table 17.1. Parameters in microsphere stimulated Raman detector.

Symbol	Quantity	(SI)
λ_p	Pump wavelength	762.5 nm
λ_s	Probe wavelength	834 nm
I_p	Pump intensity	10^3 W/cm^2
I_s	Probe intensity	10^{-3} W/cm^2
R_0	Radius	100 μm
α	Interstice loss	10^{-4}
χ	Coupling coefficient	0.01

By numerical solving of equations (17.46), (17.47) and corresponding boundary conditions have been employed in equations (17.48), (17.49), for mode of 1126.4 cm^{-1} with respect of equation (17.53) and (17.54), the variation of pump and signal power are obtained and presented by Fig. 17.11 and Fig. 17.12 respectively.

These results show in the presence of glucose, pump frequency converts to signal frequency and the amplification of probe intensity will occur. Probe intensity changes are proportional to blood glucose levels. So blood glucose levels can be determined with measuring of probe intensity in the output of fiber after using a suitable calibration method.

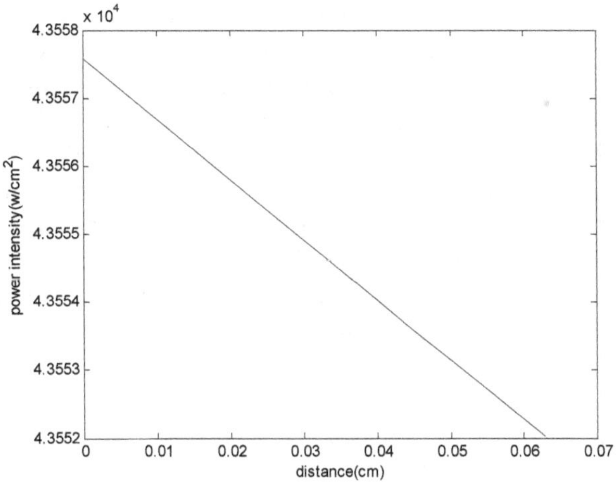

Fig. 17.11. Variation of the pump intensity versus the distance traveled by light on microsphere circumference [11].

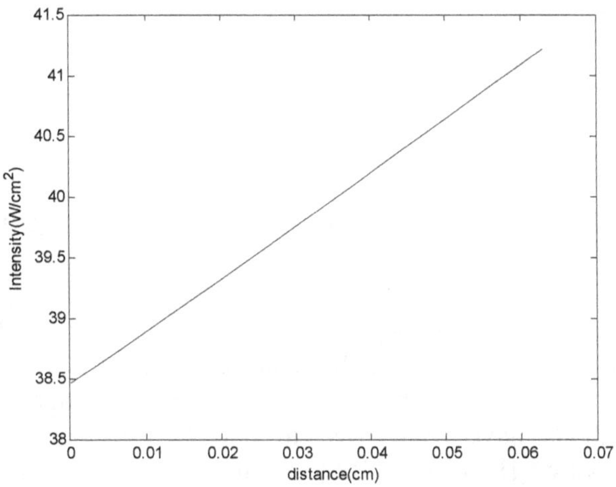

Fig. 17.12. Variation of the signal intensity versus the distance traveled by light on microsphere circumference [11].

17.6. A Novel Optical Sensor for Troponin I Enzyme Based on Surface-Enhanced Raman Spectroscopy in Microsphere

According to the World Health Organization, cardiovascular disease is the most common cause of death in the world. Detection of cardiac biomarkers can provide early identifications and diagnosis of acute coronary syndrome and yield information on the prognosis of the patient by assessing the risk of death. Multiple studies have expressed that cardiac Troponin I(cTnI) is an important prognostic marker in patients presenting with chest pain, even when creatine kinase (CK) and Myoglobin fraction are not elevated [103]. cTnI is one of the three subunits of the Cardiac Troponin complexes that are released into the blood stream upon injury to cardiac muscle, particularly acute myocardial infarction(AMI), with no overlap with skeletal muscle troponins under normal condition [104]. Comparing cTnI with CK-MG, MG, cTnc and cTnT, cTnI demonstrates superior sensitivity and specificity, with the ability to detect even minor amount of myocardial damage. cTnI concentration in patient blood surpasses the 0.3 ng/ml threshold in 3 to 4 hours after myocardial damage, with a peak usually at 12-24 hours. cTnI has a long half-life in the blood, remaining elevated for 3-5 days, and gradually returns to normal state within 7-10 days [105]. Fig. 17.13 shows the existence of cardiac biomarkers in patient blood after AMI.

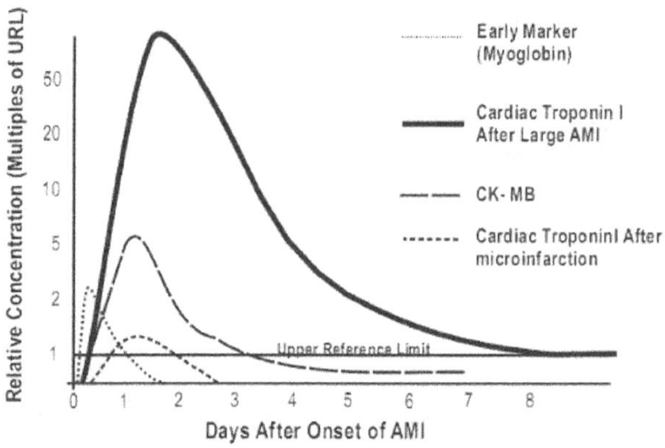

Fig. 17.13. Cardiac biomarkers concentration in blood after AMI [105].

Thus, European Society of Cardiology(ESC), American College of Cardiology (ACC) and American Heart Association (AHA) have stated that cTnI is the "gold standard" test for AMI diagnosis particularly in absence of clear electrocardiography(ECG) evidence [105]. Several methods have been used to detect cTnI, such as enzyme linked immounosorbent assay (ELISA) [106], radioimmunoassay (RIA) [107], and immunochromatographic [108] test. Limitations of these methods include a several complicated procedures, relatively slow process, and a highly reagent consumption. In addition, these methods require an expensive analyzer system for high sensitivity detection, because the elevated cTnI concentration in patient blood is in the range of 1-3 ng/ml. To detect the MG, cTnI and cTnT in biological level in patient blood after myocardial damage, the surface plasmon resonance (SPR) sensors are developed [109, 110]. The optical fiber based SPR sensor is employed to detect MG and cTnI in HEPES buffered saline (HBS) solution [109]. To immobilize the antigen of interest, corresponding antibodies are attached to a carboxymethylated dextran layer on the noble nanoparticles surfaces [109, 110]. A real time in vivo sensor for early detection of AMI will greatly enhance the patient care. The optical fiber based SPR sensor are suitable candidate for insertion into a vein for in vivo cTnI monitoring. The protein fouling and initiation of the clotting cascade on the optical fiber sensor are a serious problem for in vivo operation of this sensor.

The requirements for miniature, high sensitive sensors for the early detection of the onset of AMI, the ultra sensitive and label-free cTnI optical microcavity sensor is proposed. Optical microcavities such as microspheres, microdisk, microring and microtoroids have recently attracted considerable attention in the studies ranging from fundamental processes such as cavity quantum electrodynamics (QED) to nonlinear optics, microlasers, integrated electro-photonic micro-devices and bio/chemical sensing systems as applied areas [100]. Optical resonance modes, called whispering gallery modes (WGMs), are rotating around a circular orbit near the surface of such resonators via total internal reflection. In addition to temporal and spatial confinement of light in the microcavities, WGMs produce an evanescent field that extends into the surrounding medium. The strong optical confinement leads to great light-matter interaction between the photons of WGMs and the analyte. Field enhancement due to the resonance condition of high quality factor microsphere in cooperation with SPR effect, reduces the threshold intensity of nonlinear effects such as stimulated Raman scattering (SRS) of its surrounding medium [111]. The spectral information of Raman signal can identify unique binding energies of molecules. However, spontaneous Raman scattering cross section is on the order of 10^{-30} cm^2. Thus, Raman spectroscopy is not a reliable tool for sensing molecules in extremely low concentration. Surface enhanced Raman scattering (SERS) has been investigated as a useful technique that provides tremendously large Raman signal enhancement in the vicinity of metal nanostructures. Under special condition, the Raman enhancement factor due to the excitation of SPR, up to 10^{14} has been observed [112]. SERS technique by using Raman signal as a fingerprint of the analyte creates a spectroscopic device which can acts as highly sensitive molecular sensor. Non-invasive detection of cTnI by using SERS technique, could be a powerful tool in AMI diagnosis, but has not been studied extensively. The human skin presents a complex heterogeneous medium where blood and pigment are distributed variably in its depth. The skin is divided to three main layers from surface: epidermis (50-200 μm thick, the blood-free layers), dermis (1-4 mm thick, vascular layer) and subcutaneous fat (1-6 mm). According to the micron size of microsphere, there is a potential for inserting the microsphere on the skin groove. In other words, by installation microsphere sensor on the wrist watch, earring or ring, controlling the myocardial patients is provided permanently. However, due to complex boundary conditions of skin tissue and week signal generation, which result in noise increasing, the non-invasive detection method of cTnI is limited.

In this section to avoid from complex inhomogeneous boundary conditions of skin tissue, cTnI detection in HBS solution is proposed. Developing of optical microsphere coupled to the optical fiber based on SERS and SRS has offered a new opportunity for an ultrahigh sensitive cTnI sensor for the first time.

17.6.1. cTnI Sensor Structure

In the approach presented here, the design of microsphere resonator with silver nanoparticles grown on its surface in adjacent of optical fiber based on SERS technique is used to achieve cTnI detection. The polystyrene microsphere is placed in a well containing a HBS solution and cTnI biomarker. HBS at pH 7.4 has the same salt and pH environment similar to the human blood. The surrounding medium of an in vivo, non-invasive cTnI detector is the inhomogeneous skin tissue medium. The synthesis of the dextran layer based on the carboxymethylated dextran chemistry is demonstrated previously for protein immobilization on a noble metal surface [109, 110]. The cTnI biomarker is attached to a carboxymethylated dextran layer on a silver nanoparticle surface to increase signal to noise ratio. Fig. 17.14 shows a schematic of optical microsphere resonator that is coupled to the optical fiber in the presence of silver nanoparticles over the microsphere surface.

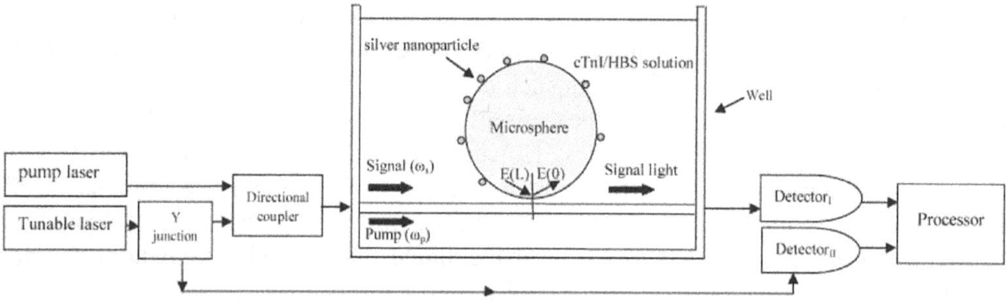

Fig. 17.14. Experimental setup of the composite sensing system for cTnI detection in HBS solution. The pump laser through a directional coupler is coupled to the optical fiber. Raman signal from a tunable laser is divided by Y junction coupler. The Raman signal output and reference signal are employed to obtain the cTnI concentration.

The pump laser at wavelength 565 nm through a directional coupler is coupled to the optical fiber as shown in Fig. 17.14. Raman signal from a tunable laser is divided by Y junction coupler. One of Y junction outputs is connected to the directional coupler and the other is connected to a measuring instrument to provide a reference signal. The pump output is filtered and the Raman signal is detected for the signal processing. The Raman signal output and reference signal are employed to obtain the cTnI concentration. The pump (ω_P) and signal (ω_S) frequencies are chosen as the resonance frequencies of the microsphere resonator. The pumping power is 1mW, while the signal power is 1 μW. The pump and signal light are coupled to the microsphere through the optical fiber simultaneously. The pump light with wavelength λ=565 nm, where it repeatedly circulates around the microsphere surface via total internal reflection, excites WGMs inside the microsphere. Due to the high quality factor of microsphere, evanescent field of WGMs produces a high intensity field capable of acting as an SERS excitation source. An important factor in the Raman signal power enhancement is the increased electromagnetic field due to the excitations of LSPR. The high field enhancement is concentrated around the silver nanoparticles, that so called "hot spots". As the microsphere sensor is dipped in HBS solution of cTnI, cTnI molecules can produce an SERS signal due to the electromagnetic field enhancement mechanism.

17.6.2. Electric Field Enhancement near the Silver Nanoparticle

To reach the ultrahigh sensitivity in enhancement mechanism for cTnI detection in HBS solution, combining of optical microsphere system with silver nanoparticles deposited on the microsphere surface is employed. To illustrate the local field enhancement more closely, Fig. 17.15 shows the near-field amplitude near the silver nanoparticle in the radial distance.

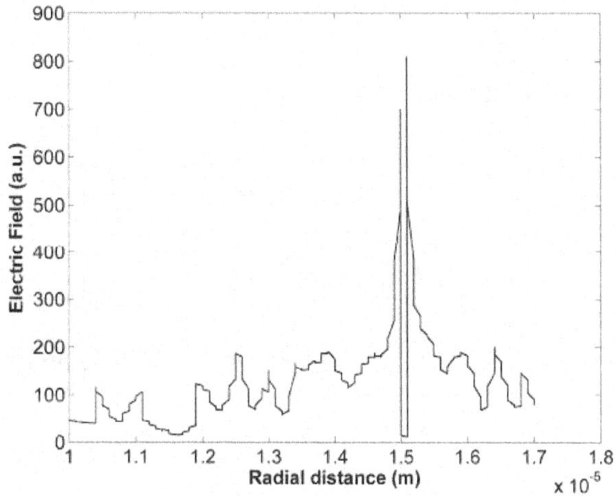

Fig. 17.15. Near-field enhancement near the 50 nm radius silver nanoparticle in the radial distance [12].

According to the curve, the electric field away from the silver nanoparticle due to the SPR is similar to the field of dipole radiation. However, the field amplitude inside the silver nanoparticle decayed due to the skin-depth effect of the metal and away from the surface of silver nanoparticle rapidly attenuates. The high intensity local field around the silver nanoparticle increases total power of the Stokes Raman light through SERS effect. Excitation of localized surface plasmon near the surface of silver nanoparticle enhances both the incoming and Stokes shifted light. By depositing one silver nanoparticle on the surface of microsphere system discussed above, high enhancement factor can hardly be achieved, but using a number of more than one silver nanoparticle the averaged electric field enhancement factor is in excess of 10^2 without significantly degrading the quality factor of the microsphere. Due to the small Stokes shift in Raman scattering, local field enhancement leading to Raman enhancement factor of 10^{10}.

17.6.3. cTnI Detection Based on SRS and SERS

As Raman light circulates the sphere along with the pump light, and encounters cTnI molecules near the microsphere surface, the potential exists for SRS. The coupled mode equation for the pump and Raman signal power are written as below [111]:

$$\frac{d\bar{P}_s(z)}{dt} = (-\alpha_s + g_R(\Delta\omega_s)\bar{P}_p)\bar{P}_s + k_s P_{s_i} \tag{17.55}$$

$$\frac{d\bar{P}_p(z)}{dt} = (-\alpha_p - g_R(\Delta\omega_s)\bar{P}_s)\bar{P}_p + k_s P_{p_i}, \tag{17.56}$$

where Ps, Pp, α_s, α_p, v_g, are the Raman signal power, pump power, the intrinsic losses of polystyrene microsphere in signal and pump frequency and group velocity of mode, respectively.

In the first order of approximation, the total losses of pump and signal (γ_p, γ_s) are constant:

$$\gamma_p = \alpha_p + g_R(\Delta\omega_s)P_p \tag{17.57}$$

$$\gamma_s = \alpha_s - g_R(\Delta\omega_s)P_s \tag{17.58}$$

So, the pump and signal power coupling (k_p, k_s) are defined by the following relations:

$$k_s = \frac{2v_g\chi^2(1+\exp(-\gamma_s L))}{L(1-\exp(-\gamma_s L))(1+r)} \tag{17.59}$$

$$k_p = \frac{2v_g\chi^2(1+\exp(-\gamma_p L))}{L(1-\exp(-\gamma_p L))(1+r)} \tag{17.60}$$

In an ultrahigh sensitive cTnI sensor, the microsphere material must be chosen such that the frequency of minimum intrinsic loss of microsphere is much closer to the frequency of maximum slope of the Raman spectrum of cTnI. The polystyrene intrinsic loss is shown in Fig. 17.16. The Raman gain spectrum for cTnI, $g_R(\Delta\omega)$ where $\Delta\omega$ represents the frequency difference between the pump and Raman waves, is the most important quantity for describing stimulated Raman scattering. Raman spectrum of cTnI is presented in Fig. 17.17.

Fig. 17.18 demonstrates the treatment of SRS in the microsphere. According to the diagram, the pulsed pump and Raman power variations as a function of time have a Gaussian shape (pulsed laser).

The pulsed pump and Raman light are coupled through the optical fiber to the microsphere. With the advent of pump frequency, light circulates around the sphere surface in the form of WGMs and produces a Raman scattering. Raman scattering is expected to occur first for the fundamental whispering gallery modes (n=1, l=m), because of the largest confinement of light in theses modes.

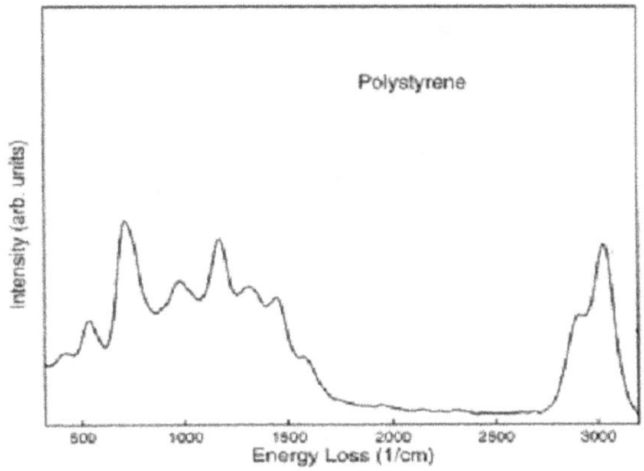

Fig. 17.16. Electron energy loss spectrum of a thick polystyrene sample [113].

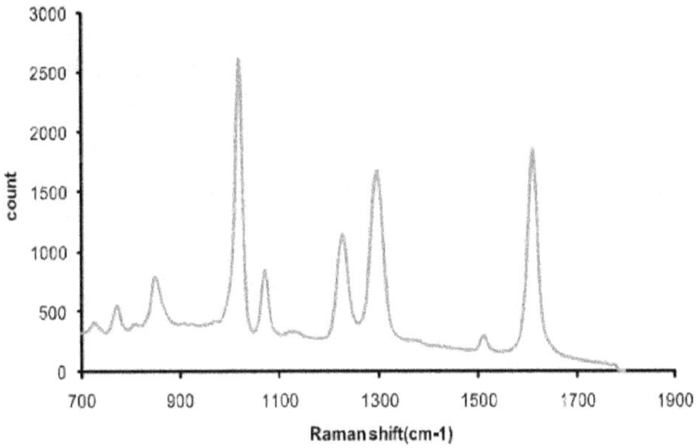

Fig. 17.17. Raman spectrum of cTnI [114].

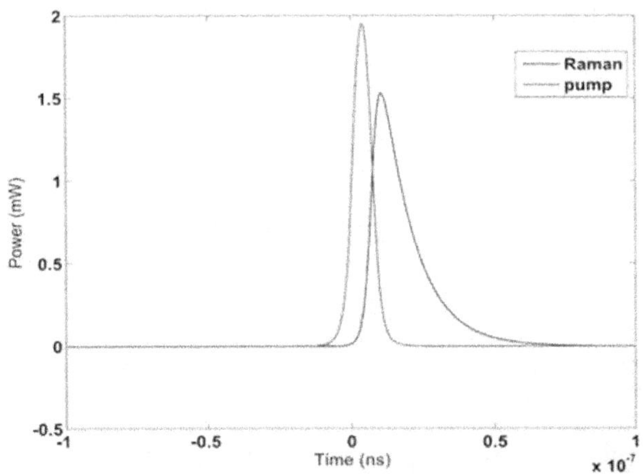

Fig. 17.18. Variations of Pump and Raman powers versus time for polystyrene [12] microsphere(r_m=15μm, n_m=1.59, χ=0.01, η=0.5).

The resonant WGMs excite the localized surface plasmon in the silver nanoparticles leading to enhanced pump light. Energy is transferred from pump frequency to Raman frequency. Under efficient conversion between pump and Raman frequencies, the Raman signal with a time difference (as shown in Fig. 17.6 reaches to its maximum magnitude. Therefore, the Raman gain is increased due to localized field enhancement through SERS effect.

Raman gain is calculated for a single cTnI molecule, and then the general Raman gain for average N molecules of cTnI is defined as follows:

$$g_R = Ng'_R \, , \tag{17.61}$$

where g'_R is the value of Raman gain for single cTnI molecule.

The cTnI molecules which are adsorbed on to the silver nanoparticles via dextran layer, can produce the enhanced Raman signal, due to the electromagnetic enhancement. The calculated enhanced Raman signal, which is provided by the silver nanoparticles, depends linearly on the cTnI molecule concentration. Fig. 17.19 shows that Stokes Raman power depends linearly on the cTnI molecules. The Raman power variations can be employed to detect cTnI molecules around the microsphere. As expected, as the microresonator sensor becomes more sensitive, the calculated Raman signal increases, implying that a stronger evanescent field is interacting with the HBS solution of cTnI.

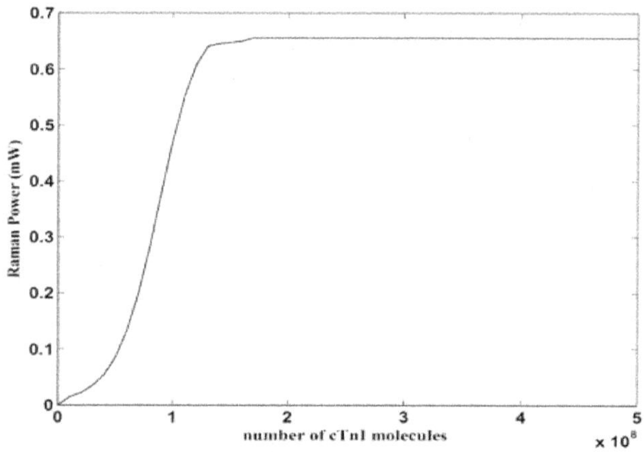

Fig. 17.19. Variation of Raman power versus cTnI molecules [12].

17.7. Conclusion

Biological Raman spectroscopy is a powerful technique for non-invasive tissue analysis and analyte concentration measurements. From its early development, many investigations have been done with the aid of more advanced instrumentation and calibration algorithms.

In this chapter, a new method has been devised to non-invasively measure the blood concentration. This approach is radically different from conventional non-invasive blood analytes measurement techniques. One of the important problems in determining blood glucose levels without injury based on stimulated Raman scattering technique is high pump power laser fluency which is above the safety limitation for human body application. In order to solve this problem, silica microsphere is employed.

Very high quality factor microsphere reduces the thresholds of nonlinear optical effects such as stimulated Raman scattering from its surrounding media. Also, Due to the high quality factor of the microsphere resonator and considerable light–matter interaction between the WGM evanescent field and the surrounding medium near the surface of the microsphere, it can be investigated for sensitive and label-free sensors.

We theoretically employed this sensor for non-invasive measuring of blood glucose levels. A low-power CW laser as pump and low-power CW laser as probe are employed. After laser start to work, the existence Pump and probe evanescent fields interact with glucose and cause stimulated Raman scattering. The results from these theoretical studies indicate that the stimulated Raman signal amplitude levels produced from the biological Glucose concentrations. By measuring the changes of the signal intensity in output of fiber, we can determine small amount of glucose with a low power pump in reasonable exposure time. Also the stimulated Raman lasing from optical microsphere resonator is used in combination with SERS to achieve an even higher electromagnetic field enhancement as compared to typical Raman scattering for cTnI detection in HBS solution. In this sensing system, localized electromagnetic field enhancement of the evanescent field, due to the resonant excitation of surface plasmons near the silver nanoparticles, enhanced the Raman intensity through the SERS effect. In simulation model presented here, an average electric field enhancement factor on silver nanoparticles greater than 10^2 (Raman enhancement factor of 10^{10}), without significantly degrading the quality factor of the microsphere, is achieved. Raman enhancement factors of 10^{12}–10^{14} are predicted for realistic experimental conditions. Our results show that the silver nanoparticle doped microsphere resonator presented a linear response for cTnI detection. The change of Raman intensity was employed to detect human cTnI in the HBS solution around the microsphere. The presence of silver nanoparticles on the resonator surface enhanced the sensitivity of the cTnI WGM sensor. This sensitivity will lead to a highly specific sensing system for extremely low concentrations of cTnI molecules in human blood samples. Therefore, this system can be used as a sensitive sensor to measure changes in small amount of analytes.

References

[1]. A. M. K. Enejder, T. W. Koo, and et al., Blood analysis by Raman spectroscopy, *Opt. Lett.*, Vol. 27, 2002, pp. 2004–2006.

[2]. A. J. Berger, Y. Wang, M. S. Feld, Rapid, noninvasive concentration measurements of aqueous biological analytes by near-infrared Raman spectroscopy, *Appl. Opt.*, Vol. 35, 1996, pp. 209–212.

[3]. I. Fine, Non-invasive method and system of optical measurements for determining the concentration of a substance in blood, *US Patent* No. 6400972, 2002.

[4]. O. S. Khalil, Non-invasive glucose measurement technologies: an update from 1999 to the dawn of the new millennium, *Diabetes Technol. Ther.*, Vol. 6, 2004, pp. 660-697.

[5]. S. A. Boppart, G. J. Tearney, B. E. Bouma, and et al., Noninvasive assessment of the developing xenopus cardiovascular system using optical coherence tomography, *Proc. Natl. Acad. Sci.*, Vol. 94, 1997, pp. 4256–4261.

[6]. M. A. Franceschini, D. A. Boas, A. Zourabian, and et al., Near-infrared spiroximetry: noninvasive measurement of venous saturation in piglets and human subjects, *J. Appl. Phys.*, Vol. 92, 2002, pp. 372–384.

[7]. M. Fleischmann, P. J. Hendra, A. J. McQuillan, Raman spectra of pyridine adsorbed at a silver electrode, *Chem. Phys. Lett.*, 1974, Vol. 26, pp. 163.

[8]. K. Kneipp, H. Kneipp, I. Itzkan, and et al., Surface-enhanced Raman scattering and biophysics, *J. Physics–Condensed Matter*, 2002, Vol. 14, pp. 597-624.

[9]. T. J. Kippenberg, Nonlinear Optics in Ultra-High-Q Whispering-Gallery Optical Microcavities, PhD Thesis, *California Institute of Technology*, California, USA, 2004.

[10]. J. C. Knight, G. Cheung, F. Jacques, T. A. Birks, Phase-matched excitation of whispering-gallery-mode resonances by a fiber taper, *Opt. Lett.*, Vol. 22, 1997, pp. 1129–1131.

[11]. A. R. Bahrampour, N. Jahangiri, M. Taraz, Development of a Non-invasive Micron Sized blood Glucose Sensor Based on microsphere Stimulated Raman Spectroscopy, *Sensors & Transducers,* Vol. 147, Issue 12, December 2012, pp. 129-142.

[12]. M. Saliminasab, A. R. Bahrampour, M. H. Zandi, Human cardiac troponin I sensor based on silver nanoparticle doped microsphere resonator, *J. Opt.*, Vol. 14, 2012.

[13]. V. V. Tuchin (Ed), Handbook of Optical Sensing of Glucose in Biological Fluids and Tissues, *Taylor & Francis*, Abingdon, UK, 2009.

[14]. J. Doutch, A. J. Quantock, V. A. Smith, K. M. Meek, Light Transmission in the Human Cornea as a Function of Position across the Ocular Surface: Theoretical and Experimental Aspects, *Biophys. J.*, Vol. 95, 2008, pp. 5092–5099.

[15]. K. V. Larin, M. Motamedi, R. O. Esenaliev, T. V. Ashitkov, Specificity of noninvasive blood glucose sensing using optical coherence tomography technique: a pilot study, *Phys. Med. Biol.,* Vol. 48, 2003, pp. 1371-1390.

[16]. A. R. Young, Chromophores in human skin, *Phys. Med. Biol.,* Vol. 42, 1997, pp. 789-802.

[17]. V. V. Tuchin, I. L. Maksimova, D. A. Zimnyakov, and et al., Light propagation in tissues with controlled optical properties, *J. Biomed. Opt.*, Vol. 2, 1997, pp. 401-417.

[18]. V. V. Tuchin, A N. Bashkatov, E. A. Genina, and et al, Optics of living tissues with controlled scattering properties, *Proc. SPIE*, Vol. 3863, 1999, pp. 10-21.

[19]. E. M. Culav, C. H. Clark, M. J. Merrilees, Connective tissue: matrix composition and its relevance to physical therapy, *Phys. Ther.*, Vol. 79, 1999, pp. 308-319.

[20]. P. O. Rol, Optics for Transscleral Laser Applications, D. N. Sci diss., *Institute of Biomedical Engineering,* Zurich, Switzerland, 1992.

[21]. A. N. Bashkatov, E. A. Genina, Yu. P. Sinichkin, and et al., Glucose and mannitol diffusion in human dura mater, *J. Biophys.*, Vol. 85, 2003, pp. 3310-3318

[22]. J. S. Maier, S. A. Walker, S. Fantini, et al., Possible correlation between blood glucose concentration and the reduced scattering coefficient of tissues in the near infrared, *Opt. Lett.*, Vol. 19, 1994, pp. 2062-2064.

[23]. R. Barer, Refractometry and interferometry of living cells: Part I. Basic Principles, *J. Opt. Soc. Am.,* Vol. 47, 1957, pp. 545–556.

[24]. V. V. Tuchin, A. N. Bashkatov, E. A. Genina, and et al., In vivo investigation of the immersion-liquid-induced human skin clearing dynamics, *Techn. Phys. Lett.,* Vol. 27, 2001, pp. 489-490.

[25]. J. M. Schmitt, G. Kumar, Optical scattering properties of soft tissue: a discrete particle model, *Appl. Opt.,* Vol. 37, 1998, pp. 2788-2797.
[26]. J. L. Cox, R. A. Farrell, R. W. Hart, and et al., The transparency of the mammalian cornea, *J. Physiol.*, Vol. 210, 1970, pp. 601-616.
[27]. R. F. Schmidt, G. Thews, Human Physiology, *Springer*, Berlin, Germany, 1989.
[28]. V. V. Tuchin, D. M. Zhestkov, A. N. Bashkatov, E. A. Genina, Theoretical study of immersion optical clearing of blood in vessels at local hemolysis, *J. Opt. Soc. Am.*, Vol. 12, 2004, pp. 2966-2971.
[29]. A. Roggan, M. Friebel, K. Dorschel, et al., Optical properties of circulating human blood in the wavelength range 400-2500 nm, *J. Biomed. Opt.*, Vol. 4, 1999, pp. 36-46.
[30]. A. V. Priezzhev, O. M. Ryaboshapka, N. N. Firsov, I. V. Sirko, Aggregation and disaggregation of erythrocytes in whole blood: study by backscattering technique, *J. Biomed. Opt.*, Vol. 4, 1999, pp. 76-84.
[31]. N. G. Khlebtsov and S. Yu. Shchegolev, Account of particles nonsphericity at determination of parameters of dispersion systems by a turbidity spectrum method. II. Characteristic functions of light scattering by systems of randomly oriented nonspherical particles in van de Hulst approximation, *Opt. Spectr.*, Vol. 42, 1977, pp. 663-668.
[32]. D. H. Tycko, M. H. Metz, E. A. Epstein, et al., Flow-cytometric light scattering measurement of red blood cell volume and hemoglobin concentration, *Appl. Opt.*, Vol. 24, 1985, pp. 1355-1365.
[33]. A. G. Borovoi, E. I. Naats, U. G. Oppel, Scattering of light by a red blood cell, *J. Biomed. Opt.*, Vol. 3, 1998, pp. 364-372.
[34]. V. V. Tuchin, X. Xu, R. K. Wang, Dynamic optical coherence tomography in studies of optical clearing, sedimentation, and aggregation of immersed blood, *Appl. Opt.,* Vol. 41, 2002, pp. 258-271.
[35]. A. N. Bashkatov, D. M. Zhestkov, E. A. Genina, et al., Immersion clearing of human blood in the visible and near-infrared spectral regions, *Opt. Spectr.*, Vol. 98, 2005, pp. 638-646.
[36]. V. V. Tuchin (Ed.), Handbook of Optical Biomedical Diagnostics, *SPIE Press*, Bellingham, WA, 2002.
[37]. C. F. Bohren, D. R. Huffman, Absorption and Scattering of Light by Small Particles, *Wiley*, New York, USA, 1983.
[38]. T. Ioppolo, N. Das, M. Volkan Otugen, Whispering gallery modes of microsphere in the presence of a changing surrounding medium: A new ray-tracing analysis and sensor experiment, *Appl. Phys. Lett*, Vol. 107, 2010, pp. 1031051-8.
[39]. S. Uetake, M. Katsuragawa, M. Suzuki, and K. Hakuta. Stimulated Raman scattering in a liquid-hydrogen droplet, *Physical Review A*, Vol. 6001, 1, 2000.
[40]. P. Michler, A. Kiraz, L. D. Zhang, et al., Laser emission from quantum dots in micro disk structures. *Appl. Phys. Lett.*, Vol. 77, 2, 2000, pp. 184-186.
[41]. D. K. Armani, T. J. Kippenberg, S. M. Spillane, K. J. Vahala. Ultra-high-Q toroid microcavity on a chip, *Nature*, Vol. 421, 6926, 2003, pp. 925-928.
[42]. P. Michler, A. Kiraz, L. D. Zhang, et al., Laser emission from quantum dots in micro disk structures, *Appl. Phys. Lett.,* Vol. 77, 2, 2000, pp. 184-186.
[43]. V. L. Seguin, Whispering-gallery mode lasers with doped silica microspheres, *Optical Materials*, Vol. 11, 1999, pp. 153-165.
[44]. A. Belarouci, K. B. Hill, Y. Liu, et. al., Design and modeling of waveguide-coupled microring resonator, *J. Luminescence*, Vol. 94-95, 2001, pp. 35-38.
[45]. D. K. Armani, T. J. Kippenberg, S. M. Spillane, K. J. Vahala, Ultra-high-Q toroid microcavity on a chip, *Nature*, Vol. 421, 2003, pp. 925-928.

[46]. T. Baba, P. Fujita, A. Sakai, et. al., Lasing characteristics of GaInAsP-InP strained quantum-well microdisk injection lasers with diameter of 2-10 /spl mu/m, *IEEE Photon. Tech. Lett.*, Vol. 9, 1997, pp. 878-880.

[47]. K. J. Vahala, Optical Microcavities, *Nature*, Vol. 424, 2003, pp. 839-846.

[48]. A. R. Bahrampour, A. Gholampour Ahir, A Novel Bio-Chemical Sensor Based on the Microsphere Raman Laser in the Presence of an External Non metallic Mirror, *Sensors & Transducers,* Vol. 135, Issue 12, December 2011, pp. 134-145.

[49]. M. Cai, O. Painter, K. J. Vahala, P. C. Sercel, Fiber-coupled microsphere laser, *Opt. Lett.,* Vol. 25, 2000, pp. 1430-1432.

[50]. D. W. Vernooy, A. Furusawa, N. P. Georgiades, V. S. Ilchenko, H. J. Kimble, Cavity QED with high-Q whispering gallery modes, *Phys. Rev. A*, Vol. 57, 4, 1998, pp. R2293-R2296.

[51]. F. Vollmer, D. Braun, A. Libchaber, et al., Protein Detection by Optical Shift of Resonant Microcavity, *Appl. Phys. Lett.*, Vol. 80, 2002, pp. 4057-4059.

[52]. V. Sandoghdar, F. Treussart, J. Hare, V. Lefevre Seguin, J. M. Raimond, S. Haroche, Very low threshold whispering-gallery-mode microsphere laser, *Phys. Rev. A*, Vol. 54, 3, 1996, pp. R1777-R1780.

[53]. C. Ming, G. Hunziker, K. Vahala, Fiber-optic add-drop device based on a silica microsphere whispering gallery mode system, *IEEE Photon. Technol. Lett.*, Vol. 11, 1999, pp. 686–687.

[54]. A. R. Bahrampour, A. Gholampour Azhir, R. Taghiabadi, K. Rahimi Yazdi, Sensitivity Enhancement of Biochemical Sensors Based on Er^{+3} Doped Microsphere Coupled to an External Mirror, *Sensors & Transducers,* Vol. 120, Issue 9, September 2010, pp. 152-161.

[55]. J. P. Laine, Design and Applications of Optical Microsphere Resonators, PhD Thesis, *Helsinki University of Technology*, Helsinki, Finland, 2003.

[56]. B. E. Little, J. P. Laine, H. A. Haus, Analytic Theory of Coupling From Tapered Fibers and Half Blocks into Microsphere Resonators, *Lightw. Tech.,* Vol. 17, 1999, pp. 704-715.

[57]. M. L. Gorodetskyd, V. S. Ilchenko, High-Q optical whisperinggallery microresonators: Precession approach for spherical mode analysis and emission patterns with prism couplers, *Opt. Comm.*, Vol. 113, 1994, pp. 133–143.

[58]. V. S. Ilchenko, X. S. Yao, L. Maleki, Pigtailing the high-*Q* microsphere cavity: a simple fiber coupler for optical whispering-gallery modes, *Opt. Lett.*, Vol. 24, 1999, pp. 723-725.

[59]. M. Cai, O. Painter, K. J. Vahala, Observation of critical coupling in a fiber taper to a silica-microsphere Whispering-Gallery Mode System, *Phys. Rev. Lett.* Vol. 85, 2000, pp. 74–77.

[60]. A. N. Oraevsky, Whispering-gallery waves, *Quantum Electron.*, Vol. 32, 2002, pp. 377-400.

[61]. H. M. Lai, P. T. Leung, K. Young, Time-independent perturbation for leaking electromagnetic modes in open systems with application to resonances in microdroplets, *Phys. Rev. A.*, Vol. 41, 1990, pp. 5187–5198.

[62]. R. K. Chang, A. J. Campillo, Optical Processes in Microcavities, *World Scientific*, Singapore, 1996.

[63]. J. Jackson, Classical Electrodynamics, Third ed., *John Wiley & Sons*, New York, USA, 1999.

[64]. S. P. Mohanty, E. Kougianos, Biosensors: A tutorial review, *IEEE Potentials*, Vol. 25, Issue 2, 2006, pp. 35-40.

[65]. R. Petry, M. Schmitt, J. Popp, Raman spectroscopy: a prospective tool in the life sciences, *Chem. Phys. Chem.*, 4, 1, 2003, pp. 14-30.

[66]. K. Kneipp, R. Aroca, H. Kniepp, E. Wentrup-Byrne, New approaches in biomedical spectroscopy, *Am. Chem. Soc*, 963, 2007, pp. 200-218.

[67]. K. Kneipp, H. Kneipp, I. Itzkan, and et al., Surface-enhanced Raman scattering and biophysics, *J Physics–Condensed Matter*, Vol. 14, 2002, pp. 597-624.
[68]. R. H. Stolen, E. P. Ippen, Raman gain in glass optical waveguides, *Appl. Phys. Lett.*, Vol. 22, 1973, pp. 276.
[69]. M. Fleischmann, P. J Hendra, A. J. McQuillan, Raman spectra of pyridine adsorbed at a silver electrode, *Chem. Phys. Lett.*, Vol. 26, 1974, pp. 163.
[70]. Y. C. Cao, R. C. Jin, J. M. Nam, C. S. Thaxton, C. A. Mirkin, Raman dye-labeled nanoparticle probes for proteins, *J. Am. Chem. Soc.*, Vol. 125, 2003, pp. 14676-7.
[71]. K. Faulds, W. E. Smith, D. Graham, Evaluation of surface-enhanced resonance Raman scattering for quantitative DNA analysis, *Anal Chem*, Vol. 76, 2004, pp. 412-417.
[72]. X. Zhang, M. A. Young, O. Lyandres, R. P. Van Duyne, Rapid detection of an anthrax biomarker by surface enhanced Raman spectroscopy, *Am. Chem. Soc.*, Vol. 127, 2005, pp. 4484-9.
[73]. D. S. Grubisha, R. J. Lipert, H. Y. Park, and et al., Femtomolar detection of prostate specific antigen: an immunoassay based on surf ace-enhanced Raman scattering and immunogold labels, *Anal. Chem.*, Vol. 75, 2003, pp. 5936-43.
[74]. K. E. Shafer-Peltier, Ch. L. Haynes, M. R. Glucksberg, R. P. Van Duyne., Toward a glucose biosensor based on surf ace-enhanced Raman scattering, *Am. Chem. Soc.*, Vol. 125, 2003, pp. 588-593.
[75]. R. M. Jarvis, A. Brooker, R. Goodacre, Surface -enhanced Raman spectroscopy for bacterial discrimination utilizing a scanning electron microscope with a Raman spectroscopy interface, *Am. Chem. Soc.*, Vol. 76, 2004, pp. 5198-5202.
[76]. M. Culha, D. Stokes, L. R. Allain, T. Vo-Dinh, Surface-enhanced Raman scattering substrate based on a self assembled monolayer for use in gene diagnostics, *Am. Chem. Soc.*, Vol. 75, 2003, pp. 6196-6201.
[77]. D. L. Jeanmaire, R. P. Van Duyne, Surface Raman spectro electrochemistry. Part I. Heterocyclic -aromatic, and aliphatic amines adsorbed on the anodized silver electrode, *J. Electroanal. Chem. Interface. Electrochem.*, Vol. 84, 1977, pp. 1-20.
[78]. J. A. Dieringer, R. B. L. Ii, K. A. Scheidt, R. P. Van Duyne, A frequency domain existence proof of single-molecule surface-enhanced Raman spectroscopy, *J. Am. Chem. Soc.*, Vol. 129, 2007, pp. 16249-56.
[79]. T. J. Davis, K. C. Veron, D. E. Gomez, Designing plasmonic systems using optical coupling between nanoparticles, *Phys. Rev. B*, Vol. 79, 2009, 79, pp. 155423-32.
[80]. R. Walkers, J. Rodgers, *Diabetes:* A Practical Guide to Managing your Health, *Dorling Kindersley Inc. Publishing*, London, UK, 2004.
[81]. http://www.livestrong.com
[82]. S. Wild, G. Roglic, A. Green, and et. al, Global Prevalence of Diabetes. Estimates for 2000 and projections for 2030, *Diabetes Care*, Vol. 27, 2003, pp. 1047–1053.
[83]. H. King, R. E. Aubert, W. H. Herman, Global burden of diabetes, 19952025: prevalence, numerical estimates and projections, *Diabetes Care*, Vol. 21, 1998, pp. 1414–1431.
[84]. P. J. Watkins, ABC of Diabetes, 5th, *BMJ Publishing*, London, UK, 2003.
[85]. D. Gun (Ed), Diabetes Atlas, *International Diabetes Federation*, 2th, Brussels, Belgium, 2003.
[86]. A. P. F. Turner, J. D. Newman, L. J. Tigwell, and et. al, Biosensors: A global view, *The Ninth World Congress on Biosensors*, 10-12 May 2006, Toronto, Canada.
[87]. E. Boland, Limitations of conventional methods of self-monitoring of blood glucose, *Diabetes Care*, Vol. 24, 2001, pp 1858–1862.
[88]. K. Maruo, M. Tsurugi, J. Chin, T. Ota, H. Arimoto, Y. Yamada, M. Tamura, M. Ishii, and Y. Ozaki, Noninvasive blood glucose assay using a newly developed near-infrared system, *IEEE J. Select. Top. Quant. Electr.*, Vol. 9, 2003, pp. 322–330.

[89]. A. M. K. Enejder, T. G. Scecina, J. Oh, M. Hunter, W. C. Shih, S. Sasic, G. L. Horowitz, M. S. Feld, Raman spectroscopy for noninvasive glucose measurements, *J. Biomed. Opt.*, Vol. 10, 2005.

[90]. K. M. Ye, J. S. Schultz, Genetic engineering of an allosterically based glucose indicator protein for continuous glucose monitoring by fluorescence resonance energy transfer, *Anal. Chem.*, Vol. 75, 2003, pp. 3451–3459.

[91]. G. L. Cotè, M. D. Fox, R. B. Northrop, Noninvasive optical polarimetric glucose sensing using a true phase technique, *IEEE J. Trans. Biomed. Eng.*, Vol. 39, 1992, pp. 752–756.

[92]. V. L. Alexeev, S. Das, D. N. Finegold, S. A. Asher, Photonic crystal glucose-sensing material for noninvasive monitoring of glucose in tear fluid, *Clin, Chem.*, Vol. 50, 2004, pp. 2353-2360.

[93]. R. O. Esenaliev, I. V. Larina, K. V. Larin, and et al., Optoacoustic technique for noninvasive monitoring of blood oxygenation: a feasibility study, *Appl. Opt.*, Vol. 41, 2002, pp. 4722–4731.

[94]. R. O. Esenaliev, K. V. Larin, I. V. Larina, and et al., Noninvasive monitoring of glucose concentration with optical coherence tomography, *Opt. Lett.*, Vol. 26, 2001, pp. 992–994.

[95]. C. V. Raman, K. S. Krishnan, A New Type of Secondary Radiation, *Nature*, Vol. 121, 1928, pp. 501-502.

[96]. R. R. Alfano, W. B. Wang, A. Doctor, Detection of Glucose Levels Using Excitation and Difference Raman Spectroscopy at IUSL, *New York State Center for Advanced Technology in Photonics Applications, Report 0904-1*, 2003, pp. 1-6.

[97]. http://www.rp-photonics.com/raman_gain.html

[98]. R. W. Boyd, Nonlinear Optics, 3th, *Academic Press*, New York, USA, 2007.

[99]. R. V. Tarr, P. G. Steffes, The Non-invasive Measure of D-Glucose in the Ocular Aqueous Humour Using Stimulated Raman Spectroscopy, *IEEE J. Lasers and Electro-Optics Society(LEOS) Newslett.*, Vol. 12, 1998, pp. 22-27.

[100]. J. Heebner, R. Grover, T. A. Ibrahim, Optical Microresonators, *Springer*, New York, USA, 2008.

[101]. W. C. Shih, K. L. Bechtel, M. S. Feld, Non-invasive Glucose Sensing with Raman Spectroscopy, *John Wiley & Sons, Inc*, USA, 2009.

[102]. http://www.omlc.ogi.edu/spectra/index.htm

[103]. M. D. McLaurin, F. S. Apple, E. M. Voss, C. A. Herzog, S. W. Sharkey, Cardiac troponin I, cardiac troponin T, and creatine kinase MB in dialysis patients without ischemic heart disease: evidence of cardiac troponin T expression in skeletal muscle, *Clin. Chem.*, Vol. 43, 1997, pp. 976-982.

[104]. J. E. Adams, G. S. Bodor, V. G. Davila-Roman, and et al., Cardiac troponin I: a marker with high specificity for cardiac injury, *Circul.*, Vol. 88, 1993, pp. 101-106.

[105]. J. Mair, and et al, Cardiac troponin I in the diagnosis of myocardial injury and infarction, *Clin. Chem. Acta.*, Vol. 245, 1996, pp. 19-38.

[106]. H. A. Katus, A. Remppis, S. Looser, and et al., Enzyme linked immuno assay of cardiac troponin T for the detection of acute myocardial infarction in patients, *J. Mol. Cell. Cardiol*, Vol. 21, 1989, pp. 1349-1353.

[107]. B. Cummins, M. L. Auckland, P. Cummins, Cardiac-specific troponin-I radioimmunoassay in the diagnosis of acute myocardial infarction, *Am. Heart J*, Vol. 113, 1987, pp. 1333-1344.

[108]. K. Penttil, H. Koukkunen, A. Kemppainen, and et. al., creatine kinase MB, troponin T, and troponin I - rapid bedside assays in patients with acute chest pain, *Intern. J. Clin. Lab. Res.*, Vol. 29, 1999, pp. 93-101.

[109]. J. F. Masson, L. Obando, S. Beaudoin, K. Booksh, sensitive and real-time fiber-optic-based surface plasmon resonance sensor for myoglobin and cardiac troponin I, *Talanta*, Vol. 62, 2004, pp. 86570.

[110]. R. Fireman Dutra, R. Kelly Mendes, V. Lins da Silva, L. Tatsuo Kubota, Surface plasmon resonance immunosensor for human cardiac troponin T based on self-assembled monolayer, *J. Pharmaceutical and Biomedical Analysis*, Vol. 43, 2007, pp. 1744-1750.
[111]. S. M. Spillane, T. J. Kippenberg, K. J. Vahala, Ultralow-threshold Raman laser using a spherical dielectric microcavity, *Nature*, Vol. 415, 2002, pp. 621-623.
[112]. K. Kneipp, Y. Wang, H. Kneipp, and et al., single molecule detection using surface-enhanced Raman scattering (SERS), *Phys. Rev. Lett.*, Vol. 78, 1997, pp. 1667-1670.
[113]. L. L. Kesmodel, S. Wild, G. Apai, High-resolution electron energy loss spectroscopy of insulating polymer surfaces, *Surface Science Letters*, Vol. 429, 1999, pp. 475-480.
[114]. F. S. Parker, Applications of Infrared, Raman, and Resonance Raman Spectroscopy in Biochemistry, *Springer*, Berlin, Germany, 1983.

Chapter 18
Simple and Robust Multipoint Data Acquisition Bus Built on Top of the Standard RS232 Interface

Alexey Pavluchenko, Alexander Kukla, Sergey Lozovoy

18.1. Introduction

Multichannel and multisensor systems are recently becoming increasingly popular in a variety of scientific and industrial applications [1]. Among other fields, this trend is particularly prominent in analytical chemistry [2, 3]. Integrating multiple data acquisition channels into a unified system with centralized control provides numerous benefits: simultaneous sensing of different physical quantities allows to improve measurement accuracy and extract otherwise hidden information by eliminating influencing factors, aggregating multiple measurement results, and employing multivariate statistical analysis in post-processing of obtained data [4–7]. Data management and processing is also greatly simplified with the use of centralized database.

Traditional structure of multisensor system usually includes single data converter that connects to multiple transducers via analog multiplexer from one side, and to the data collecting host through a single communication port on the other side (Fig. 18.1). This monolithic design has several drawbacks: lack of modularity, impossibility of true synchronous data acquisition because of multiplexer delays, impossibility to build a spatially distributed sensor network without introduction of additional signal repeaters and/or data converters, undesired redundancy in case when a fewer number of channels is required by application compared to that imposed by original design.

Alternative solution is represented by a modular system, where every sensor (or a set of small number of sensors) is equipped by its own data converter (correspondingly, a set of simultaneously operating data converters) and its own means of communication,

Alexey Pavluchenko
Institute of semiconductor physics of NASU, Kiev, pr-t Nauki 41, 03028, Ukraine

allowing to connect together as many devices as dictated by application. Since all data in such network is transmitted in digital form, spatial distribution of nodes can be arranged in much more flexible manner as compared to the above mentioned monolithic approach; thus either distributed, lumped, or mixed data acquisition facilities can be formed as needed. This flexibility comes in exchange for the increased cost of every single device, because each network node must include separate data converter, controller and communication hardware. This overhead, however, becomes negligible for devices produced in large quantities. Another case when added cost is outweighed by flexibility is a class of applications where single data acquisition channel is initially sufficient, but the need for additional expansion may occur in the future.

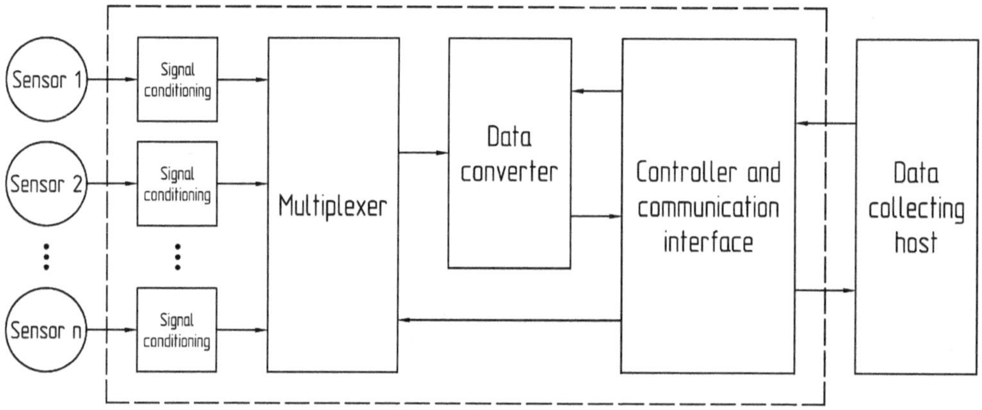

Fig. 18.1. Traditional multichannel data acquisition system architecture.

Selection of appropriate communication interface is an important part of sensory system design, since it affects, directly or indirectly, many aspects of system performance. These include maximum allowed number of channels, total bandwidth and bandwidth distribution, possible area coverage in case of distributed acquisition, data transmission reliability etc. On the other hand, design choices are governed by technical limitations and administrative restrictions, such as availability of certain infrastructure, long term availability of third-party provided components, possibility and complexity of future extensions and alterations of the designed system. Taking all this into account, we can formulate the list of desired properties for the sensor network communication layer:

- It must be a communication technology that is widespread, well specified and well supported by industry;
- It must provide possibility to build networks with sufficient number of nodes (dictated by application needs, primarily by required number of sensor channels), ideally with arbitrary number of nodes with reasonable performance degradation curve;
- It must provide sufficient bandwidth for selected application (dictated by required number of data acquisition channels, data sampling rate and associated protocol overheads);

- Availability of ready-made third-party solutions for necessary hardware and software components;
- Low cost of implementation and maintenance.

For wired networks, Ethernet is an obvious and primary candidate meeting most of these requirements in many situations. However, in order for multisensor system implemented over the underlying Ethernet local area network (LAN) to be cost-effective, certain prerequisites have to be met. Every device has to be connected to a separate Ethernet port, which makes it necessary to setup dedicated switching router, at least when number of devices in the network becomes sufficiently large. It might appear unfeasible to create all the required infrastructure for a small-scale sensor network only.

In other words, Ethernet is a good connectivity solution for a stationary sensor network, covering relatively large area, and seamlessly integrated into the existing on-site LAN infrastructure. Such seamless integration demands that the data server implemented in each device complies to a set of standard internet protocols, which leads to increased hardware and firmware complexity and correspondingly increased device cost.

For smaller scale applications, where only a few sensors have to be connected to a single host, and there are no demands for large area coverage and/or fast data sampling, more simplistic solutions can become applicable. Particularly, this covers the kind of applications where sensor network is used in place of traditional multichannel approach for the sake of flexibility and modularity. Such network usually connects small number (an order of 10) of sensors situated within the same laboratory, or in several adjacent rooms. In many scientific and industrial control applications performed measurements are quasi-static, in the sense that the measured quantity is not expected to change by amount exceeding system resolution between two subsequent measurements, that is, only stationary signal values are of interest and transitional parts of the signal are less important. These applications are thus not bandwidth-demanding and corresponding multi-sensor system can be implemented over a relatively low speed data transmission medium.

Below we will show how such network can be implemented with a very simple three-wire serial connection. Particularly, RS232 [8] is a viable alternative to other, more complex serial interfaces in this regard.

Despite its venerable age (50 years at the time of this writing), RS232 serial interface is still in the widespread use. Virtually all contemporary measurement and control instrumentation still comes equipped with RS232 compliant communication ports. Although nowadays the interface is rarely supported out-of-the-box by home and office use oriented equipment, adaptors for connecting RS232 capable devices to the modern extension interfaces, such as USB, PCI and PCI Express, are readily available [9–13].

The RS232 standard limits interface cable capacitance, which in its turn determines maximum network segment length. With the commonly available shielded cables network area coverage of about 100 m^2 can be expected [14, Chapter 3], and if longer network segments are necessary either special low-capacitance cables or active signal repeaters can be used, whichever appears more suitable and less expensive. The original standard covered transmission speeds up to 20 kbit/s, but modern implementations offer significantly faster communication capabilities. Most of the commercially available PCI and PCI Express RS232 adaptors are rated up to 115.2 kbit/s speeds, while industrial solutions provide transmission speeds up to 900 kbit/s [10, 13], and the rates as high as 3 Mbit/s are technically possible [15], albeit with lower signaling voltage levels and shorter link distances. The latter already comes close to 10 Mbit/s Ethernet, especially considering the associated protocol overheads.

Wide acceptance by industry and long life of RS232 are, of course, not accidental. The interface features sturdy mechanical attachment with D-subminiature connectors, small number of wires (three wires in minimal configuration), asynchronous transmission, high electromagnetic interference tolerance due to relatively high signaling voltage levels and dual polarity thereof, and very simple signaling logic. RS232 transceiver hardware is simple and easy to implement with either discrete components or with available integrated circuit solutions. Probably the most important factor for long-withstanding popularity is the ease of software control of RS232 attached devices, as the core software support for the serial interface is built-in into all popular operating system kernels.

At the first glance, it seems that multiple communication ports on the host side would be necessary to connect multiple devices. This may appear inconvenient, or even prohibitive to the system design, due to the connector bulkiness and limited number of both available ports and expansion bus slots for additional adaptors installation. However, it is in fact possible to connect virtually unlimited[1] number of data acquisition points to the host via one communication port, overcoming multiple port requirement by the use of ring network topology and duplex transmission peculiarities.

In this chapter we will describe the means of organizing the small-scale local area wired sensor network controlled by a single host, utilizing only a single communication port for data transmission. Although presented architecture employs RS232 as physical layer, proposed general principles can be applied to any serial interface that allows direct peer-to-peer unidirectional communication.

18.2. Network Topology

The simple unidirectional ring network with straightforward routing allows to connect theoretically unlimited number of data sourcing devices to the host equipped with only two – one incoming and one outgoing – data terminals. This network topology is shown schematically in Fig. 18.2.

[1] In practice, of course, it will be restricted by available bandwidth and physical segment length limitations.

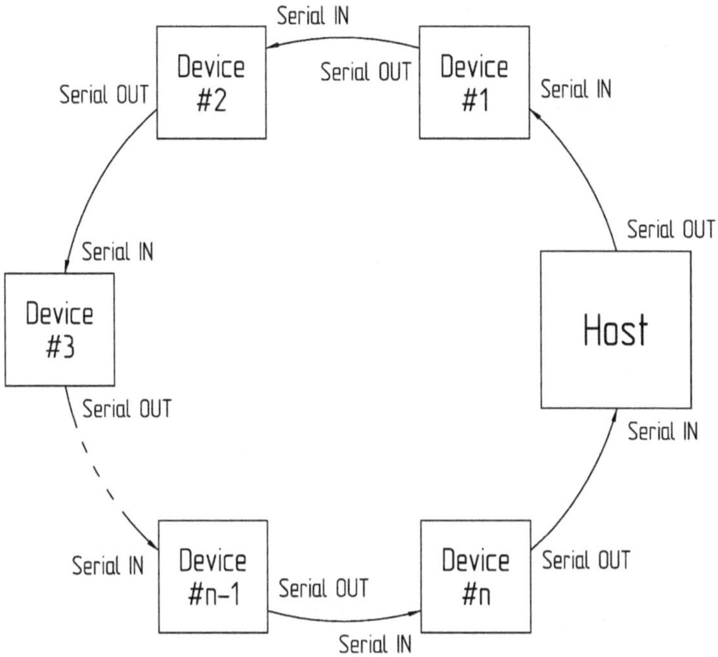

Fig. 18.2. Serial link ring network topology.

RS232 physical connection layer ideally suits this logical structure, as data transmission is asynchronous and it is possible to physically (spatially) separate the transmission and reception lines. Thus the only non-standard hardware needed to implement this ring network with RS232 is a set of specific cables (Fig. 18.3). All cables use standard DE-9 type connectors [16]. Corresponding internal structure of a network node is shown in Fig. 18.4, and schematic of the physical device attachment is shown in Fig. 18.5.

Notice that the loopback from TxD pin at the device output port to the TxD pin at the input port allows to use conventional 3-wire RS232 cable for device attachment when only a single device needs to be connected to the host. Of course, the special split cable (Fig. 18.3, top) can be used for this purpose as well. Also note that only one pair of connectors in the split cable connects the common (ground) wire. This is done to avoid formation of ground loops, since the ground pins of input and output ports of each device are already internally connected (see Fig. 18.4).

Devices designed in accordance with Fig. 18.4 schematic along with interconnections depicted in Fig. 18.5 form the physical layer of the multipoint serial bus. Note that in this chapter we use terms "bus" and "network" interchangeably, that is "bus" is interpreted in general sense as a combination of physical and logical means of digital data transmission between the endpoints (nodes) and the host, and not as a specific case of network topology.

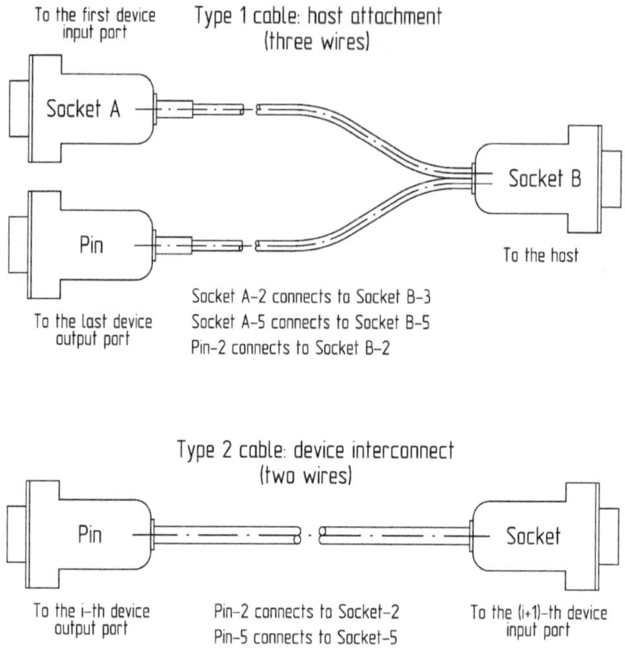

Fig. 18.3. Device attachment cables.

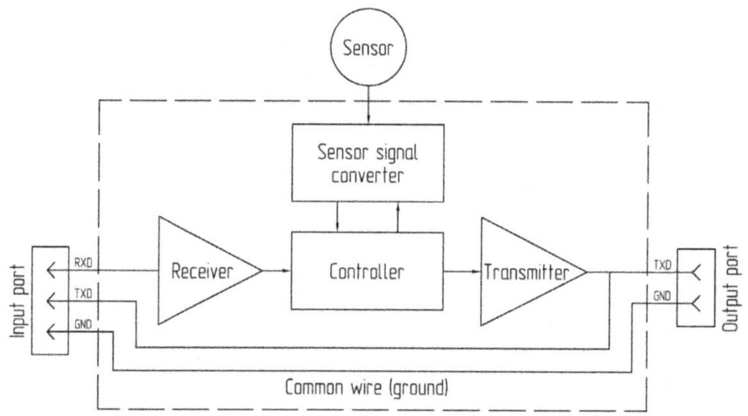

Fig. 18.4. Internal structure of the network node.

Since only three wires are used for physical connection, no hardware handshake as defined by standard [8] is utilized. The data flow is controlled through bus timeouts, status codes issued by connected devices in return to the host requests, and cyclic redundancy check (CRC) codes for data integrity verification.

During the normal bus operation, data packets originated at the host travel sequentially through each connected device. The devices act as active buffers electrically, and perform modifications to the data packet contents in accordance to the host requests on logical level. The packet containing modified or new data finally arrives to the host data reception terminal, where it is being interpreted and processed by the data registration and management software running on the host. Next section describes the means of logical data flow control in the proposed architecture in detail.

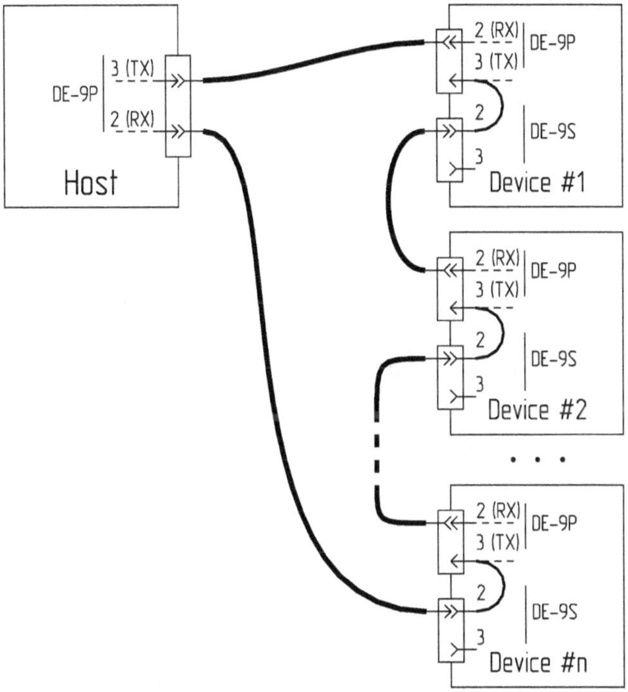

Fig. 18.5. Physical network layout. DE-9P designate pin connectors, DE-9S designate socket connectors; ground wires are not shown.

18.3. Protocol Outline

Variety of communication protocols can be applied on top of the described above physical topology to form the logical network layer. The protocol variant presented below is simple yet sufficient for the basic bus arbitration and payload transmission needs.

The logical bus structure is hierarchical, with the host acting as master client and all devices connected to the bus acting as slave servers with equal priority. All bus transactions are originated at the host. The host prepares and issues the data/control packet formatted in accordance with Fig. 18.6. Particular format of the packet fields is implementation dependent. It is convenient from the software standpoint to use fixed length packets for all transactions, but this is not mandatory.

Device address	Host command/ request code	Payload data	Device status/return code	Control code (checksum)

Fig. 18.6. Data packet structure.

Two types of transactions are possible on the bus: peer targeted and broadcast. Peer targeted transaction carries the request addressed to the particular device, like the data acquisition request. Broadcast transactions are used to deliver identical information or request to all devices connected to the bus, like the firmware or hardware reconfiguration requests, transmission mode change requests etc. Broadcast transactions are distinguished by specific value in the device address field of the data/control packet, which cannot be assigned as the actual valid device address. It is particularly convenient to use address zero for this purpose.

Initially, after the host power-up and the bus control software startup, the number of devices connected to the bus, as well as their respective addresses, is undetermined. To set up the bus configuration, the device enumeration transaction is broadcasted by the host. The payload field for this transaction initially contains zero. In response to enumeration request, each device reads the payload value from the received packet, increases this number by one and stores the result as its own address, then writes it to the payload field and sends the amended packet down to the next device on the ring.

Since the packet is being processed by all devices connected to the bus in sequence, when it finally arrives to the host the payload field contains the number of devices connected, and each device has its unique address assigned (assuming there were no transmission errors). After this procedure is complete, the host can exchange data with each device individually by targeting it with its specific address.

When the packet arrives to the server peer whose address is indicated in the corresponding packet field, the peer interprets the payload field in accordance to the code specified in the command/request field, modifies the payload field with new data if necessary, accordingly amends the control code field and transmits the modified packet down to the next device on bus. If an unrecoverable error condition arises during the packet reception or processing (this includes transmission errors detected by CRC), the device that encounters error condition puts corresponding error code to the status field and its own address to the address field. This way, all subsequent devices are effectively excluded from transaction and the host receives point of failure (POF) address along with the error code.

Particular set of command codes, status codes and control code computation method depend on the application requirements and constraints imposed by implementation specifics. The few general rules described above allow to implement arbitrarily complex schemes of interaction between the host and the peers. It has to be taken in consideration though, that due to the sequential structure of the bus the packet delivery time may depend on the targeted device address – generally, the further is the peer position on the ring relative to host, the longer it might take for a request packet to arrive to it, and the overall time required for a packet to return to the host may depend on the number of devices connected to the bus. These issues can become important in applications that demand strict timing and synchronization of data exchange. This will be discussed further in section 18.5.

18.4. Resilience Considerations

Network resilience is determined by its automatic reconfiguration ability and node fault tolerance. The resilient network must be able to automatically (that is, as a part of normal workflow) or semi-automatically (with a scheduled human operator intervention) reconfigure itself when the network composition changes due to external, as related to the host, circumstances. Such changes may include attachment of the new nodes, detachment or replacement of existing nodes, and malfunction of one or more of the existing nodes preventing them from communication with the host.

The well-known drawback of the simplistic ring topology is that a single node failure resulting in inability of the said node to receive or transmit data leads to the global failure of the network, as every node connectivity depends on every others, and there are no alternative routing paths. This is an architectural flaw and it cannot be worked around without modifying the topology itself. However, if the device detachment and reattachment is performed in a timely manner, the host can recognize the resulting temporary failure as a standard workflow situation and perform the network reconfiguration attempt as needed. By the same means, the non-scheduled accidental failures can be recognized as such, and the POF detection can be performed with a minor extension of the bus arbitration protocol described above.

18.4.1. Hot-swap and Hot-plug Capabilities

Since the bus does not contain power supply lines (devices are powered either individually or from dedicated common power supply source), physical detachment and attachment of devices can be trivially performed during the normal bus operation – virtually the only precaution to be taken is the properly implemented static discharge protection in transceivers. The resulting temporary interruptions in traffic flow can be detected by the host and acted upon accordingly, as described below.

If one of the nodes has to be excluded from the network, either the corresponding two segments must be merged into a singe one by cable replacement, or the corresponding device has to be replaced with a signal repeater. The first scenario may appear more time consuming as it might require physical re-arrangement of devices, depending on the original network layout and length of the segments being merged. The second method is more suitable for in-place reconfiguration, considering that the passive repeater can be formed by simply connecting together the socket and pin connectors of the input and output cables formerly used to attach the device being removed (type 2 cables shown in

Fig. 18.3). Care should be taken though not to exceed the maximum allowed segment length dictated by the cable capacitance. In the latter case, an active repeater should be used instead of the simple passive segments connection.

The network failures are detected by bus timeouts. When the host transmits data packet to the first device on bus, it expects to receive it back from the last device on bus during the predefined time interval. The exact amount of time to wait is determined by application needs and implementation specifics. At the very least, timeout interval should depend on transmission speed, number of devices on bus, and expected maximum delay caused by necessary packet processing within each device.

When the host detects bus timeout, it waits for the predefined period of time (retry interval) and then performs transaction retry. If the retry also results in bus timeout, the cycle can be repeated. Timeout interval $T_{bus\_timeout}$, retry interval T_{retry} and number of retries N_{retry} can be considered the network configuration parameters and should be individually optimized for every particular use case. If the transaction retry finally succeeds, the host is thus notified of reconfiguration necessity and issues device re-enumeration sequence, then performs other device-specific reconfiguration routines as needed. If maximum amount of retries is reached and no successful transactions could be performed, the host declares bus disruption and issues corresponding notification message to operator. Thus the scheduled temporary device detachment must not last longer than the sum of retry and timeout intervals multiplied by maximum allowed number of retries.

In the networks with low traffic, the temporary device detachment situation can "slip-through" between the regular transactions and pass unnoticed by the host. In this case, one or more of the three possible misconfiguration problems can occur: 1) At least one non-enumerated device is connected to the network in addition to the previously attached devices; 2) At least one of the previously attached devices is replaced by the new, non-enumerated device; 3) At least one of the devices was disconnected (detached and replaced with a repeater). All these situations can be automatically detected and resolved on the logical protocol level.

18.4.1.1. Resolving the Extra Device Attachment

Since an extra device does not have its address configured, it will only respond to broadcast transactions, and re-transmit all other data packets unaltered. As the rest of the devices connected to the bus still have their appropriate addresses assigned during initial enumeration, there is no way for the host to notice the appearance of additional peer unless it is explicitly queried with the specific broadcast transaction. Thus one possible way to resolve this misconfiguration problem is to program the host to perform broadcast re-enumeration or ping transactions periodically in accordance with predefined schedule. Note that this method can also be used to resolve any of the other mentioned misconfigurations. The drawback is that a certain, although likely relatively small, percent of bandwidth must be reserved for said periodic transactions.

Another, more economical bandwidth-wise, method can be implemented by extending the above described communication protocol with the addition of special beacon transactions initiated by devices, as opposed to usual data transfer and control transactions initiated by the host. If a device connected to the bus does not receive any packets with its target address for a predetermined period of time, it starts to periodically emit beacon packets to notify the host about its presence on the bus. The beacon packet payload field contains the originator address (it may also contain additional application-specific data, like e.g. the device type, serial number etc). If the beacon packet was issued by the non-enumerated device, the payload field contains zero, and the host thus becomes aware of misconfiguration, issues notification message to operator and initiates the re-enumeration routine.

The same beacon packet mechanism can be used for POF detection, as will be shown in the next subsection.

18.4.1.2. Resolving the Device Replacement

Replacement of one of the devices connected to the bus with a new non-enumerated one causes disappearance of one of the previously configured addresses from the node address list and appearance of the non-configured node responding only to the broadcast transactions in its place. This situation can be resolved as follows: when issuing the data packet targeted to specific peer, the host fills the status field with the special code indicating that the packed went unprocessed. When the packet is received and properly processed by the target peer, the latter replaces this code with the code indicating successful request completion (or respective error condition, if any arises). If the peer is not present on the bus, the packet returns to the host with the status field unaltered. The host thus becomes aware that the peer is no longer accessible by that address, notifies operator and initiates re-enumeration routine.

18.4.1.3. Resolving the Device Removal

From the host point of view this situation is identical to the device replacement situation discussed above, as the removed node simply stops processing the packets targeted to its previously configured address.

Alternatively to the automated misconfiguration detection methods described in this subsection, the network activity can be suspended by operator through the control software running on the host before the network reconfiguration procedures take place and resumed afterwards. In this case the host is notified of reconfiguration necessity directly by operator actions, and performs necessary reconfiguration routines as needed.

On a final note, it has to be kept in mind when implementing the host software, that after re-enumeration of the devices on bus, the host must keep track of the previous address assignments to maintain the logical link between the virtual device position in the network, which is related to its physical location, and its current network address, so that the data source channel numeration stays correct during the entire data acquisition session regardless of the possible network reconfigurations.

18.4.2. Failure Diagnostics

As has been mentioned above, for the considered network topology any node failure that results in inability of said node to receive or transmit data is fatal, as the entire network becomes inoperable. On the other hand, any sort of failure which keeps the nodes reception and transmission capabilities intact is non-critical in the sense that it does not affect the operation of other nodes. Besides, the situation when the packet reception/transmission unit within the device is still working while the packet processing unit has failed is very unlikely to happen, unless said units are implemented with two physically separate controllers. If, however, such situation emerges, both the fact of failure and its source point can be detected by the host either by persisting CRC code mismatch (the POF position in this case is deduced from the data packet address field as described in section 18.4), or by disappearance of the targeted peer from the bus (as described in subsection 18.4.1).

While it is impossible to achieve critical failure tolerance in a simple ring network without introduction of the additional physical connections between nodes, it is however possible to provide the automated means for POF detection by utilizing the same beacon transaction mechanism as described in previous subsection.

Like the bus timeout interval and retry interval, beacon timeout $T_{beacon\_timeout}$ and beacon repetition period T_{beacon} are the network configuration parameters that have to be optimized for every particular application. However, in order for proposed failure detection mechanisms to operate properly, certain relations must hold between the $T_{bus\_timeout}$, T_{retry}, $T_{beacon\_timeout}$ and T_{beacon}. One possible way to ensure the proper failure detection is to arrange the timeout and repetition intervals as follows:

$$\begin{gathered} T_{bus\_timeout} < T_{beacon\_timeout} < (T_{bus\_timeout} + T_{retry}), \\ T_{beacon} = T_{retry}. \end{gathered} \tag{18.1}$$

This arrangement is illustrated in Fig. 18.7.

It also has to be ensured that the $T_{beacon} = T_{retry}$ is greater than the maximum possible transaction time, that is, the time needed to cycle a single packet from the host serial port output to the host serial port input.

During the normal bus operation, all devices are periodically queried by the host for data collection. When the node receives a packet targeted to its address, it resets its internal beacon timeout timer. Thus normally beacon transactions are only initiated occasionally, if ever.

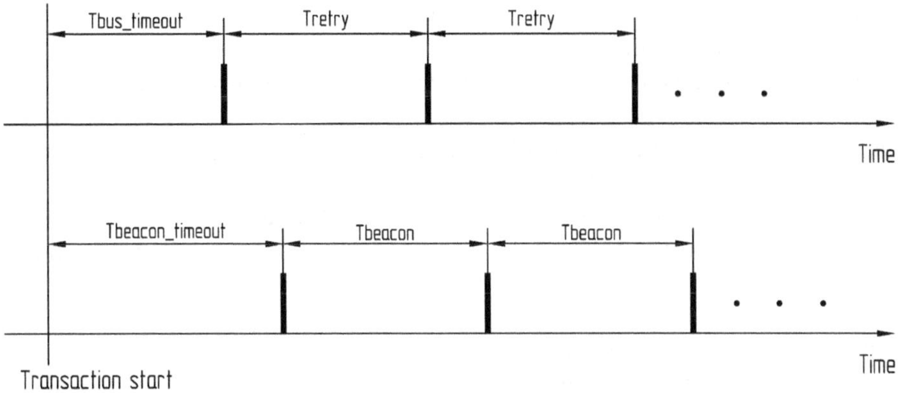

Fig. 18.7. Relation between the bus timing parameters.

If one of the network segments ceases to transmit, either because of the node (device) failure or the link (cable) failure, all devices situated further on the ring stop receiving packets. Their beacon timeout timers thus never reset, and eventually all of the nodes following the POF detect beacon timeout condition and start to transmit beacon packets.

The host meanwhile detects bus timeout since the ring is broken, and initiates the transaction retry cycle. Unless the failure was temporary and the network structure was restored before the passage of $T_{beacon\_timeout}$, the host should receive a burst of beacon packets from the devices connected to the bus below the POF location. By analyzing the origin addresses of the received beacon packets, the host is able to determine the POF location. It then continues to perform transaction retries until one of the retries succeed or until the predetermined number of unsuccessful retries N_{retry} is reached. If the transaction retry succeeds, the host initiates the device re-enumeration and associated re-configuration routines. If the maximum allowed number of retries is reached, the host reports failure condition along with the obtained POF location to operator. If neither successful transaction retry could be performed, nor any beacon packets were received by the host during the retry cycle, the host declares bus disruption and issues corresponding notification message to operator. The bus disruption condition thus can be caused either by unrecoverable failure at the last network segment (in the last device on bus, or in the link between the last device and the host), or by simultaneous failure of all network segments (e.g. due to the common power source failure).

Once the POF location has been determined, the exact nature of failure can be found out by on-site examination and corresponding recovery actions can be performed.

18.5. Bandwidth and Timing Considerations

The described protocol allows connection of virtually unlimited number of devices (limited only by the length of address field in the data packet). In practice the main limiting factor for the network size buildup is the increasing transaction delay, or in other words, the bus latency. The latter depends on a variety of factors: the physical layer data transmission rate, the packet processing time within each device, and the packet reception and transmission order – overlapped or non-overlapped.

Although data flow within the network ring is unidirectional, the communication must not necessarily be half-duplex; that is, the host does not have to wait for reception of a complete data packet before sending out the next one, if it has such capability. Likewise, any device connected to the network is allowed to simultaneously transmit and receive data.

The network therefore can operate in either of the two general packet transmission modes, hereby designated as pedantic mode and streaming mode. In pedantic mode, each node retransmits the data packet only after its reception is complete and control code verified, even if the packet is targeted towards a different peer. In streaming mode, the node only processes packets that are targeted to its address (or broadcast), and retransmits all other packets instantly without intermediate buffering. Note that the latter is possible because the address field is received before the rest of the packet data.

Pedantic mode provides maximum reliability and data integrity, while streaming mode provides higher data exchange rate at the cost of increased probability of data corruption. The time T required to transmit a single packet through the bus (that is, the bus latency) in pedantic mode depends, beside the other factors, on the number of connected devices, while in streaming mode it is determined primarily by the transmission baudrate. If we assume that the delay caused by packet processing within each device is negligibly small relatively to the data transmission speed, we can write the following expressions for transaction times:

$$T_p = (N+1)\frac{P}{B} \tag{18.2}$$

for pedantic mode and

$$T_s = 2\frac{P}{B} \tag{18.3}$$

for streaming mode, where N is the number of connected devices, P is the packet size in bits (including framing bits) and B is the data transmission baudrate.

The maximum allowed bus query rate (that is, the maximum frequency at which the host can retrieve the data from all connected devices) depends also on the packet scheduling algorithm implemented in the host software. This algorithm, similarly to the bus operation mode, can also be pedantic or streaming. If transmission and reception procedures are sequential and the packets are not queued, the minimum query period is determined by the number of devices on bus:

$$Q_p = N \cdot T. \tag{18.4}$$

If the transmission and reception procedures can run in parallel and the packets are queued in the FIFO buffers, the host can transmit the next packet without waiting for the reception of the previously issued one. In this case query period can be diminished to

$$Q_s = (N-1)\frac{P}{B} + T. \tag{18.5}$$

Thus in the slowest operation mode the minimum bus query period is

$$Q_p = N \cdot (N+1)\frac{P}{B} \tag{18.6}$$

and in the fastest mode it can be decreased by factor of N:

$$Q_s = (N+1)\frac{P}{B}. \tag{18.7}$$

Values (18.6) and (18.7) along with the internal buffering capabilities of connected devices determine the maximum possible data sampling rate.

In multisensor applications synchronous measurement (simultaneous reading from multiple channels) is often required. In the proposed architecture true direct synchronous sampling is not possible due to sequential nature of queries. The synchronous sampling still can be implemented if timing information is included in the payload along with the sampled data. The easiest way is for each device to perform sampling continuously with predetermined frequency, buffer the sampled data and then transfer the delay $T_{sampling\_delay}$ between the actual moment of sampling and the moment of sampled data deployment along with the data when acquisition query arrives. By knowing this delay, the device location on the bus, and associated transmission delays, the host can estimate the actual time of sampling. The pseudo-synchronous sampling then can be performed by extrapolating or interpolating the data values at the required time point.

If more precise synchronization is required, it can be achieved by equipping each device with an independent real time clock of desired resolution. The host then can schedule synchronous acquisition by setting the absolute moment of sampling with a broadcast transaction, and retrieving the stored sampled data from each device after the preset sampling moment passes.

Practical possibility of optimal bandwidth utilization closely depends on the data transmission modes available in physical network layer. The network can be built with half-duplex connections only, but this is not a requirement. A mixture of half-duplex and full-duplex connections can be used in different network segments – this can potentially degrade performance, but the network will remain functional, and no specific changes to the protocol are needed to reckon with such situation.

The same holds for data transmission rates – the baudrate must not necessarily be equal in all network segments, so the reception and transmission rates of a given node can be different. However, such heterogeneous network, while still functional, poses potential reliability problems, namely: the half-duplex segments can cause data loss, since the node can not receive data while transmitting, and in simple two-wire interconnection configuration there is no feedback route to indicate that the node is not ready to receive; different transmission rates on different segments can lead to bus congestion and, consequently, data loss if the node receives data at the higher rate that it can transmit, and there is not enough buffering space available within the receiving device. Particularly, the beacon transaction mechanism is likely to fail in the network containing half-duplex segments if several nodes start beacon packet transmission simultaneously.

To overcome these difficulties, certain precautions must be taken and corresponding restrictions applied to the data exchange protocol. First of all, the host has to determine if the bus is heterogeneous by querying each device of its capabilities – whether it is capable of full-duplex transmission, what are the maximum allowed reception and transmission rates and whether these rates can be configured or they are fixed. The enumeration and query transactions themselves must be performed in pedantic mode at the lowest predetermined rate with sufficiently long pauses between the packet transmissions. Obviously, each of the connected devices must accept incoming transmissions at a certain preconfigured minimum baudrate. If no *a priori* information is available on the capabilities of the attached devices, the host must negotiate appropriate transmission rate by means of trial and error.

Once the capabilities of each node are determined, the host can select appropriate data exchange mode. If the network appears to contain half-duplex segments, the pedantic mode should be used for all transactions. To ensure reliability of beacon transactions, the host then must configure the beacon transaction timing parameters $T_{beacon\_timeout}$ and T_{beacon} individually for each device, and the parameter values should be set so that no overlapping of beacon packets is ensured.

For maximum data exchange reliability and optimal bandwidth utilization, it is recommended to implement homogenous networks with full-duplex transmission at the equal baudrate in all segments whenever possible. In this case the possibility of data loss due to bus congestion is virtually excluded as long as each device has internal buffer capable to hold at least one incoming packet.

18.6. Application Example

The described network architecture was used in our implementation of the multichannel enzymatic biosensor system [17, 18], as a modular alternative to the previously designed monolithic version [19, 20].

General view of the working prototype is shown in Fig. 18.8. Each node consists of the two-channel single crystal ion-sensitive field-effect transistor, with one or both channels modified with the bioselective enzymatic membrane, connected to the secondary transducer containing the sensor signal measurement circuitry, analog to digital converters, integrated controller and RS232 transceivers. The network in current implementation operates at the baudrate of 19200 bits per second in half-duplex mode.

Devices provide effective data sampling rate of five samples per second (with internal oversampling and digital filtering). This is sufficient for the application purposes as the measurements are essentially static. Data exchange protocol uses packets of fixed size with the 8-bit address, request code, status code and control code fields and 64-bit payload field. Together with framing bits this amounts to the raw packet size of 120 bits, thus up to five devices can be connected to the bus without synchronization loss, and with an addition of the full-duplex streaming mode of operation this limit can be increased to 31 without changing the baudrate.

Although the current implementation does not incorporate advanced data flow control features described in section 18.4, they can be added in case of necessity since the secondary transducer controllers are in-system programmable.

The developed multichannel system was successfully used for sensors characterization.

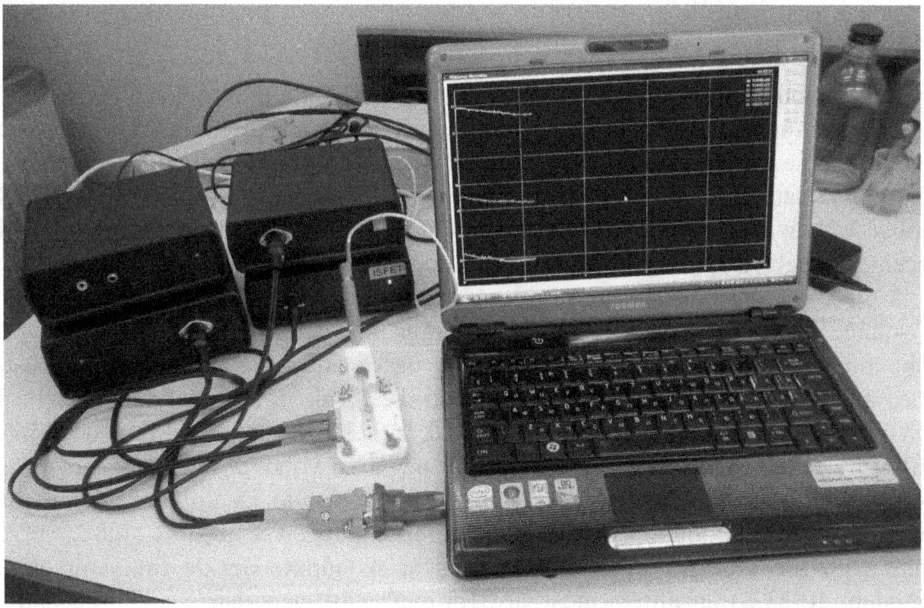

Fig. 18.8. Working prototype of the pH-sensor network with three physical points and six data acquisition channels, connected through a USB-to-RS232 adaptor to the PC-compatible laptop acting as a host.

18.7. Conclusion

Despite the relative simplicity of structure and implementation, described architecture is suitable as a replacement for both traditional multichannel systems and for sensor networks that utilize more complex data exchange protocols and data transmission media layers, and can be used to build multi-sensor setups of small to medium complexity. If no spatial separation of data acquisition points is required, the proposed architecture can also be used to form the local interconnection bus in a stacked board design. As has been shown, it is possible to partially overcome the lack of robustness inherent to simplistic ring network topology by means of the appropriate protocol conventions. This way, the system can be made scalable and, while not totally failure-tolerant, capable of semi-automated failure detection and diagnostics. The general principles expounded in this chapter can be as well applied to the sensor networks built upon different physical data link layers but having similar logical topology. Particularly, if higher transmission rates (up to 10 Mbaud) or larger area coverage (about a square kilometer) are required, the RS422 [21] compliant physical links can be used in a manner similar to the described here for RS232 links.

18.8. Appendix: Practical RS232 Transceiver Circuits

In this supplementary section we will consider practical aspects of interfacing the low-voltage CMOS or TTL instrumentation logic circuitry with the RS232 transmission lines.

Most of the modern integrated microcontrollers incorporate the universal asynchronous receiver/transmitter (UART) unit that automates logical formatting of the bitstream, providing the necessary clocking for selected baudrate and automatically generating the framing bits (start, stop and parity check signaling). In order to provide physical RS232 compliance, however, additional transformation of the electric signals being transmitted or received is necessary, which demands for the specific additional circuit elements.

The voltage levels at the microcontroller output are determined by the logic power supply voltage V_{dd}, usually 5 or 3.3 Volts. The mark state at the UART output is usually defined as logic high, that is close to V_{dd}, and the blank state is defined as logic low, which is close to the microcontroller ground. RS232 transmission line uses bipolar signaling with the mark state represented by negative voltage V_- and the blank state represented by positive voltage V_+, both in the range of 5 to 15 Volts at the transmitter output relative to the common wire, and the compliant receiver must be able to correctly operate with the voltages of 3 to 25 Volts at its input. Beside this voltage level conversion, RS232 transmitter must also provide sufficient current source and sink capabilities to drive the transmission line load, which can have static input impedance as low as 3 kOhm and parasitic capacitance as high as 2500 pF.

A variety of integrated RS232 transceivers is currently produced by most of the major semiconductor component manufacturers, including Maxim Integrated (www.maximintegrated.com), Analog Devices (www.analog.com) and Texas Instruments (www.ti.com) to name a few. While different parts can appear more or less tailored to specific application requirements, they are commonly intended for use in systems powered from a single low voltage power supply, and usually feature several line drivers and receivers on a single chip. The additional driver/receiver pairs are intended to drive the auxiliary data flow control signals. We will not try to cover peculiarities or provide selection guidelines for these devices here; if needed, this information can be found in datasheets, application notes and white papers provided on the manufacturers' web sites. Instead we will take a look at the transceiver internals and show how to build one with a small number of low-cost discrete components. This can prove useful in two ways: first, discrete component design may turn out to be more economical under certain circumstances; second, knowledge of the transceiver design principles can help developer to better identify potential problems posed by application requirements and to get better understanding of specific part characteristics and advantages when formulating selection criteria for the readymade third-party components.

Discrete component design by itself can be appealing for a minimalistic configuration which only uses data reception/transmission lines and does not employ hardware data flow control signals. In this case the net primary cost of components can become lower than that of integrated transceivers[2] This is likely to happen if the identical components are used in other parts of the device as well, e.g. in the analog signal conditioning circuits, and especially if the device design provides voltage sources suitable for RS232 line driving that are also used elsewhere in the circuit for different purposes. Use of discrete components can also potentially improve maintainability, since a wider range of drop-in replacement parts from a larger number of manufacturers can be expected.

All circuits presented in this section can be built with virtually any general purpose small signal transistors, such as 2N4401/2N4403, BC546/BC556 or BC846/BC856 complementary pairs for bipolar parts and 2N7002/BSS84 for field-effect parts. Commonly available fast switching diodes such as 1N4148 can be used where needed. The passive component value tolerances do not need to be specifically tight either, errors of about 10 % in resistances and 20 % in capacitances are acceptable.

18.8.1. Generating RS232 Line Voltages in a Single Power Supply System

As was already mentioned, RS232 transceivers use bipolar non-return to zero (NRZ) electrical signal encoding for transmission, with the typical voltage swing of 10 to 30 Volts between logical states, in order to improve electromagnetic interference tolerance. The modern trends in instrumentation design, however, lean towards the economical power saving solutions where single power supply is used for all parts of the

[2] It has to be taken into account though that discrete component design almost always demands for increased printed circuit board area which increases the board cost.

device circuit and the power supply voltage is kept as low as possible, usually at 5 Volts or even lower, relative to the circuit common point. This makes implementation of the RS232 compliant transceiver a challenging task. There are two interconnected problems to solve here: first, to provide bipolar line driving rails while having only unipolar power supply, and second, to provide sufficiently large voltage difference between the line driving rails while having only low voltage power supplies in the system being designed. In this subsection we will describe two common approaches (that can be also used in combination if desired) to resolution of these issues.

18.8.1.1. Rail Splitting

In some designs the high voltage power supply may be present in specific parts of the device circuitry, e.g. the 12 V supply can be used by analog signal conditioning circuits to increase signal to noise ratio and improve linearity, and 24 V supply is often used to power the industrial actuators. This voltage can be used to form the RS232 transmitter rails by splitting it with the voltage divider and connecting RS232 common line to the divider middle point.

The simplest implementation of rail splitter consists of the passive resistive voltage divider with capacitors added to decrease the dynamic output impedance (Fig. 18.9, *a*). The main problem of this circuit is that its static output impedance is determined by R1 and R2 values, and while it is desirable to minimize it, decreasing these resistances leads to increased power dissipation in voltage divider and thus degrades circuit efficiency. On the other hand, increasing R1 and R2 leads to higher sensitivity to the dynamic load variations which results in virtual ground (divider middle point) potential variation as the load on V_+ and V_- rails varies with time. Particularly, if the transmitter circuit design demands more power from one of the rails in either blank or mark state, the virtual ground potential will tend to shift toward the more loaded rail. If the transmitter power requirements are known beforehand, this can be accounted for by adjusting the R1 to R2 ratio correspondingly. In any case though, the R1 and R2 values in passive rail splitter should not exceed ~1 kOhm. If the resulting power loss from V_{cc} supply is unacceptable, an active rail splitter should be used instead. A simple addition of emitter follower at the virtual ground point (Fig. 18.9, *b*) dramatically improves the circuit performance, practically eliminating the virtual ground potential fluctuations as long as the load impedance on either rail is large enough so that VT1 remains in the active region of operation. The output impedance of this improved circuit stays very low even with the R1 and R2 resistances increased by an order of magnitude compared to the passive version.

If the supply voltage from analog part of the device is used for split rail design of RS232 transmitter, care should be taken not to couple the digital noise from the data transmission line to the precision analog circuitry. The current return path (the V_- virtual rail) should be separated from the analog ground in the printed circuit board layout, and proper filtering and possibly additional active regulation of V_{cc} power supply at the entry to the analog circuitry part of the device schematic should be employed.

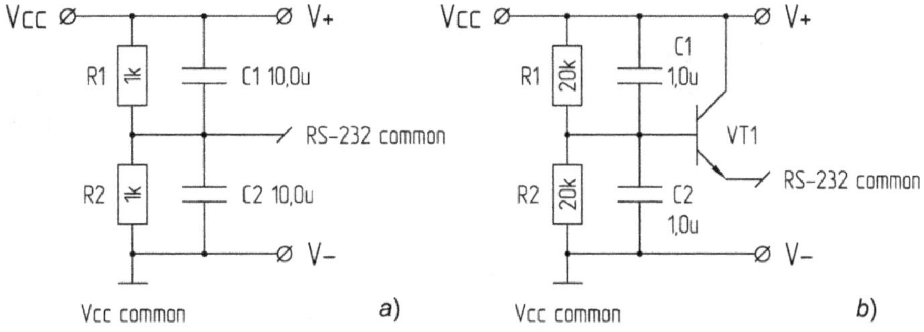

Fig. 18.9. The passive (a), and active (b) split-rail power supply schematic.

Another issue to keep in mind is that in the split rail design the RS232 common line becomes separated from the logic ground. Since receiver and transmitter must share the common line, a differential receiver design must be used (this will be discussed further in subsection 18.8.3). More radical solution is to use a dedicated controller for communication, powered from the same high-voltage rail through step-down regulator sharing the common point with the data transmission line, and isolate the entire data transmission part of the circuit from the rest of the device logic by means of optoelectronic coupling.

Extreme care must be exercised when building heterogeneous networks that may connect devices with different line voltage supply designs. If one of the devices shares logic ground with the RS232 common line, and the other uses split rail design with RS232 common line connected to virtual ground, such devices can still be connected together; however, it must be ensured that the logic grounds of the said devices stay separated. Particularly, this means that sharing of the logic power supply between devices is not allowed, neither is connection of the cables shielding to the logic ground on both sides.

18.8.1.2. Charge Pumping

Charge pumping is a very widespread technique used to obtain voltages of higher magnitudes and/or different polarity than that provided by the power supplies present in the system. Majority of integrated RS232 transceivers use this method internally to create the line driving voltages.

The principle of charge pump operation consists in alternating cycles of charging and discharging the capacitor whose leads are switched between the power supply and the circuit common point at one end, and between power supply and the load at another. In the first phase the first lead of the capacitor is connected to common point (ground), and the second one to the power supply. The capacitor is thus charged to the power supply voltage, V_{dd}. In the second phase, the first lead is disconnected from ground and connected to the power supply, and the second lead is disconnected from the power supply and connected to the load. The charge stored in the capacitor is being "pushed out" to the load by the electric field created by power supply voltage. Since the capacitor is still charged to V_{dd} from the previous phase, and the same voltage is now being applied to its first lead, the voltage magnitude applied to the load in second phase is effectively twice as high as the original power supply voltage, and approximately equals $2 \cdot V_{dd}$ (in practice the obtained voltage will always be slightly lower because of losses in the switches and other circuit nonidealities). The phases are then repeated continuously to provide the constant power supply to the load.

Voltage polarity reversal can be achieved in similar manner, by reversing the switching order to V_{dd} and to the ground between phases. Note that no voltage doubling occurs in this case. The charge pump circuit in RS232 transceivers usually consists of two cascaded stages, the doubled power supply voltage from the first stage output is fed into the second stage input for polarity reversal, thus both obtained RS232 line driving voltages V_+ and V_- have larger magnitude compared to the original power supply voltage V_{dd}.

Charge pump performance depends on switching frequency, charge reservoir capacitance, and switch implementation details. Designing optimal charge pump circuit is not a trivial task, and there are books dedicated entirely to the subject, see e.g. [22]. With a few guidelines, however, a circuit that performs well enough for a given practical application can be designed relatively easily. It has to be kept in mind though that even with an optimally designed circuit the load driving capability of the charge pump is quite limited, a few milliamperes of both source and sink can be expected at best. In regard to RS232 line driving it means that this technique is mostly suitable for powering the moderate speed (below 120 kbaud) transmitters.

Practical charge pump circuit is shown in Fig. 18.10. The voltage doubler stage is formed with VD1–VD3, C1 and C2 components. The role of the input switch is performed by a microcontroller (MCU) output; the role of the second switch is performed by VD1 and VD2 diodes. VD3 is optional and is used to provide an additional path for C2 charging and thus decrease the circuit startup time and improve dynamic load performance. C1 is the primary charge reservoir for the first stage; C2 is buffer charge storage necessary to compensate for the dynamic load current spikes. The polarity reversal stage is formed with R1–R4, VT1–VT4, VD4, VD5, C3 and C4 components. VT1–VT4 along with R1–R4 forms the input switch of the second stage. C3 acts as the primary charge reservoir for the second stage, VD4 and VD5 form the output switch, and C4 is the buffer charge storage, similarly to C2.

It is recommended to use Schottky barrier diodes (such as 1N5817 or 10BQ015) for VD1–VD5 to minimize the loss in voltage magnitude due to the voltage drop across forward biased diode junctions.

It may seem that R1 and R3 resistances could be increased to improve the circuit efficiency, however this is not the case. These resistances must be sufficiently low to guarantee that switching speed of VT3 and VT4 is high enough so that they always operate in counterphase, otherwise the situation when both VT3 and VT4 are conducting may occur, the resulting surge current then overloads the voltage doubler stage and the circuit performance significantly degrades.

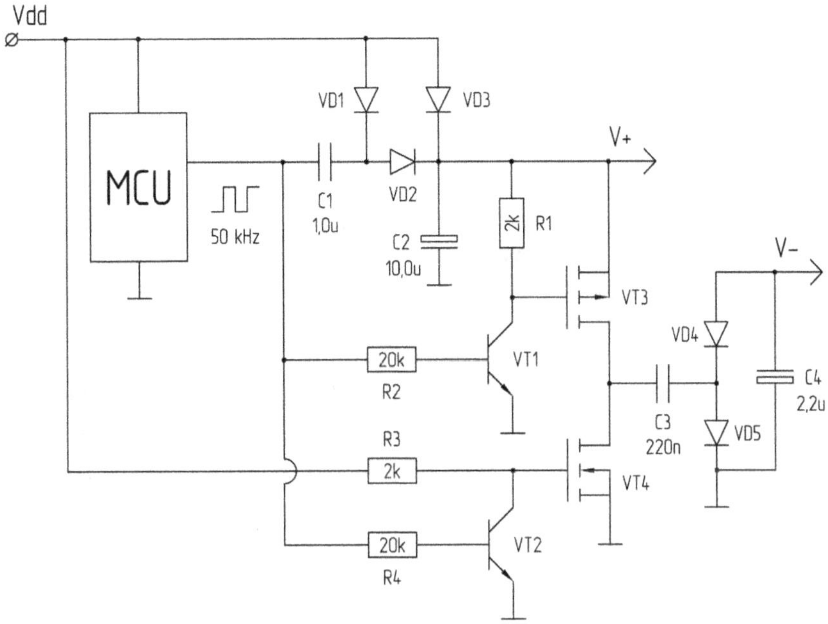

Fig. 18.10. Charge pump voltage converter schematic.

If the controller output does not provide sufficient current sourcing capability (about 250 microamperes), VT1 and VT2 can be replaced with field effect transistors identical to VT4. The R2 and R4 resistances should be decreased by about an order of magnitude in this case.

The actual values for charge pumping capacitors are selected in accordance with the following considerations. In order to maximize the load driving capability, we would like to transfer as much charge in each pumping cycle as possible, which means that we would like to maximize the C1 capacitance. The effective amount of charge stored in C1, however, is limited by the switching frequency, which determines the charging interval. Charge pumps implemented in integrated circuits usually operate at frequencies of several megahertz. In discrete design this is impractical for several reasons: first, the parasitic circuit elements (mounting capacitances, lead and track inductances) that limit the maximum switching speed are much more prominent; second, the use of specialized fast switching diodes and transistors may increase the design cost; finally, the square wave at the charge pump input is generated by microcontroller that by itself is usually clocked at several megahertz at most, and has to perform multiple tasks beside the charge pump clocking, which limits its ability to generate higher frequency signals.

On the other hand, we would not want to set the frequency too low, since this would not only increase the charging interval, but also equally increase the discharge interval, requiring larger buffer capacitances in order to keep sustained power supply to the load. Due to these limitations the suitable switching frequency range for discrete component design narrows down to several tens or hundreds of kHz.

The maximum allowed C1 capacitance can be estimated from the condition

$$5 \cdot R_{es} \cdot C_1 < T_{sw}/2$$

where R_{es} is the capacitor equivalent series resistance and T_{sw} is the switching period. The constant coefficient on the left side ensures that capacitor is charged to at least 99 % of the applied voltage. Assuming switching frequency of 50 kHz, and supposing that R_{es} does not exceed 1 Ohm[3], we obtain $C_{1max} = 2 \cdot 10^{-5}/10 = 2\ \mu F$. Allowing for some safety margin, we set C1 value to 1 μF.

Assuming $V_{dd} = 5$ V, and the voltage drop V_{d12} across forward biased VD1 and VD2 is approximately 1 V, we can determine the amount of charge transferred by the first stage in one pumping cycle:

$$Q = C_1 \cdot (V_{dd} - V_{d12}) = 4 \cdot 10^{-6}\ C.$$

Portion of this charge has to be transferred to the second stage to generate the negative supply V_-, and the rest is to be stored in C2 and sourced to the load. The maximum charging voltage of pumping capacitor C3 of the second stage is

$$V_{-max} = (2 \cdot V_{dd} - V_{d12} - V_{d5})$$

where V_{d5} is the voltage drop across forward biased VD5. Assuming $V_{d5} = 0.5$ V, and allowing for 50 % of the charge to be transferred from the first to second stage within each cycle, we can calculate the C3 capacitance:

$$C_3 = 0.5 \times Q/V_{-max} = 235.2\ nF.$$

Rounding to the nearest lower standard value, we set C3 to 220 nF. Buffer capacitors C2 and C4 are selected to be about 5..10 times larger than C1 and C3 respectively.

The circuit as is shown in Fig. 18.10 is able to drive 2 kOhm static loads on both positive and negative rails with $V_+ \approx 8.5$ V and $V_- \approx -7.0$ V (if V_{dd} equals 5 V), and is suitable for powering the transmitters designed in accordance with schematics presented in the next subsection (Fig. 18.11, *b* and Fig. 18.11, *c* in the low-speed variant of the

[3] Note that this implies that the V_{dd} supply is actually able to deliver at least 5 amperes of instantaneous current without significant voltage droop, in addition to any other load that is connected to it beside the charge pump. If this is not the case, it might be necessary to add small resistance in series with C1, effectively increasing R_{es}, which will result in lower maximum allowed C1 capacitance and, correspondingly, in weaker load driving capability of the designed circuit.

latter with resistance values respectively adjusted), for data transmission rates up to 120 kbaud. Startup time of the circuit (that is, the time from the moment when the 50 kHz square wave is applied to the circuit input until the V_+ and V_- reach their maximum magnitudes) is about 3 milliseconds. It must be ensured by system design that no data transactions are performed before the required startup interval elapses.

The charge pumping technique can be used to multiply logic supply voltage by factors higher than two, by combining more pumping stages. The load driving capability of the circuit, however, quickly degrades as the number of stages increases, and design becomes rather cumbersome. If the device logic is designed to operate from the voltage lower than 5 V it is advisable to either use integrated transceivers with built-in voltage converters, or add dedicated high voltage power supply for the data transmission part of the circuit. The latter may become mandatory if high data transmission rates and/or long transmission distances are required by application.

18.8.2. Transmitter Schematics

The very basic RS232 transmitter schematic includes a single transistor and four resistors (Fig. 18.11, *a*). When the controller UART output is in logic high state VT1 is closed and the transmitter output is pulled to the V_- voltage supply rail through the small resistance R2. When UART output goes logic low VT1 opens and pulls the transmitter output to the V_{dd} rail. R3 acts as a current limiter for short-circuit protection. R4 ensures that the output stays in mark (negative) state when microcontroller output switches to high impedance (e.g. when MCU goes into sleep mode or during controller reset). Some microcontrollers have built-in pullup resistors; in this case R4 can be omitted.

The simplicity of this design, however, comes at the cost of a few drawbacks. Since the V_{dd} power supply is used to drive the TX line, the circuit is only suitable for use in the devices where logic supply voltage equals 5 V. The microcontroller must be able to sink the current of about 1 milliampere to drive the switch; if this is not the case, VT1 can be replaced with a *p*-channel FET.

Since the passive pulldown is used to switch the output to the mark state, the output impedance of transmitter is asymmetric between the mark and blank states. More serious disadvantage of this design is that the substantial current flows through R2 when the line is in the blank state (the VT1 is conducting). This increases power dissipation and imposes additional strain on the V_- supply. This only happens during the blank bits transmission, and since the idle line stays in the mark state the redundant power dissipation should not be as severe an issue as it may seem, at least when no intense traffic is expected. However, there is another possible negative effect to consider: if the V_- power supply is not able to sink the required amount of current, the voltage magnitude on the V_- rail will droop during the blank bit transmission, and depending on the negative rail supply unit design it may not be able to recover fast enough when transmitter switches to the mark state, thus effectively limiting the maximum data transmission rate at which circuit is still operable.

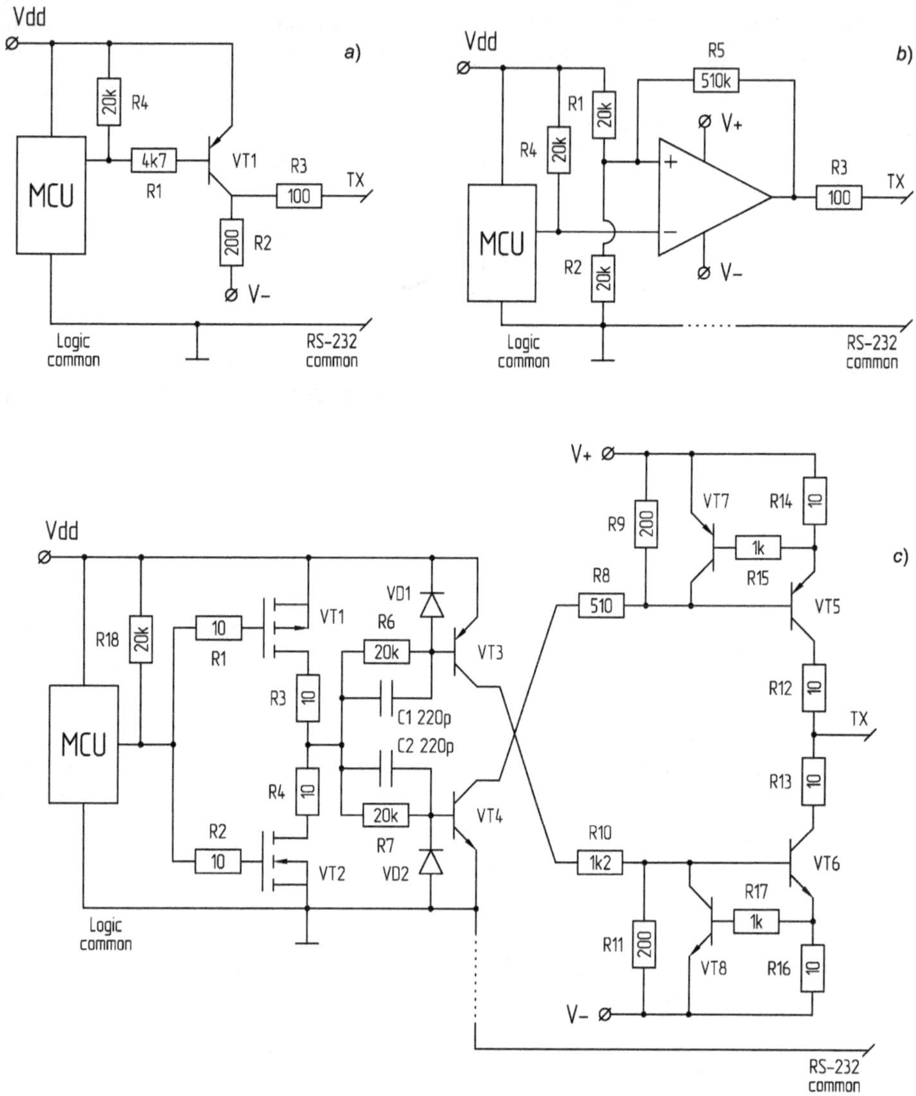

Fig. 18.11. RS232 transmitter schematics: (a) basic transmitter; (b) improved opamp-based transmitter; (c) high-speed transmitter.

If charge pumping is used to provide the negative supply for this variant of transmitter, it is advisable to adjust the charge pump design taking into account that, since the TX line is driven by logic supply voltage, the voltage doubler stage of the pump is now loaded only by the polarity inversion stage. To maximize the load driving capability of the latter C3 and C4 capacitances in the circuit shown in Fig. 18.10 can be increased to 1 uF and 10 μF respectively, and the C2 capacitor can be removed from the circuit. In this configuration the transmitter can operate with satisfactory performance at the rates up to 120 kbaud.

A better solution for the moderate data rates can be the use of an operational amplifier configured as comparator to drive the TX line (Fig. 18.11, *b*). The voltage divider formed by R1 and R2 provides the reference voltage for the non-inverting input, R3 acts as the output current limiter and isolates the amplifier output from the capacitive load. R4 serves the same purpose as in the design discussed above, and can be omitted if the microcontroller is equipped with the built-in pullup. R5 improves noise immunity and reduces the risk of ringing during the UART output state transitions. For low speed transmissions (below 20 kbaud) virtually any general purpose opamp, such as e.g. LM321, can be used. For higher rates up to 120 kbaud a wide bandwidth amplifier, e.g. LF351 or TL071, is more suitable.

The connection between RS232 common wire and logic ground is shown as dotted line in Fig. 18.11, *b* since the ground sharing is not mandatory for circuit operation, i.e. this transmitter design can be used with the split rail supply in which case the RS232 common point is lifted by several Volts above the logic ground, and V_- is directly connected to the latter.

If the data transmission rates over 120 kbaud are necessary, the circuit designs described above become unfeasible, as the requirements to the output voltage slew rate and the load driving capability become significantly more stringent. Either the dedicated discrete component design or the use of specifically selected integrated transceiver becomes necessary in applications demanding for high-speed data exchange.

It has to be noted though, that the systems implementing data transmission rates above 120 kbaud are technically not standard compliant. The EIA/TIA specification [8] limits the voltage slew rate at the transmitter output in order to ensure that the data being transmitted is not distorted due to crosstalk in the cable. In practice however data exchange rates close to 1 Mbaud are not uncommon, especially in the industrial applications. At such transmission rates the cable properties become crucial for the proper link operation, particularly the specific mutual capacitance between cable wires must be minimized. This can be done by using the cables with thicker isolation of each wire, and also by using the cables with higher number of twisted wires than actually required and connecting the unused wires to the common line (ground), effectively shielding the signal-carrying conductors from each other. Demands to the transmitter line driving capability also substantially increase in high-speed applications, and this has to be accounted for in the transmitter circuit design. These issues are discussed in more detail in [23], and general cable selection guidelines can be found in [24].

One possible variant of the high-speed transmitter circuit is shown in Fig. 18.11, *c*. The VT1 and VT2 act as a high-impedance buffer for the microcontroller output; R1 through R4 resistances are used to limit the switching current and prevent ringing during transitions of the UART output states. VT3 and VT4 act as the current drivers for the output stage switches VT5 and VT6. C1 and C2 are used to improve VT3 and VT4 switching performance and increase the slew rate of the VT5 and VT6 base driving signals, and VD1 and VD2 diodes are to protect the VT3 and VT4 base-emitter junctions from the resulting reverse voltage spikes. R12 and R13 are used to prevent the output waveform ringing when driving a highly capacitive load.

The subcircuits comprised of R14, R15, VT7 and R16, R17, VT8 form an active output current limiter (the output short-circuit protection). When emitter current of either VT5 or VT6 exceeds approximately 100 milliamperes, the resulting voltage drop across R14 or R16 causes VT7 or, correspondingly, VT8 to go into conduction and pull the base of VT5 or VT6 to the line supply rail thus decreasing its collector current.

The circuit performance primarily depends on proper selection of R6–R11 resistance values, and optimal values for R8 and R10 will also depend on the V_+ and V_- supply voltage magnitudes. The resistances shown in Fig. 18.11, *c* are selected for operation with V_+ and V_- magnitudes of 9 Volts. For different voltages, the appropriate R8 and R10 values can be estimated from the following heuristic relations:

$$R8 = 70 \cdot V_+ - 135;\ R10 = 70 \cdot (V_{dd} + V_-) + 270$$

where V_+ and V_- are taken by absolute values relative to the logic ground[4] in Volts, and the obtained resistance values are in ohms.

If the V_{dd} voltage is lower than 5 Volts, the R6 and R7 resistances should also be proportionally decreased.

Beside these selection guidelines, when designing a particular implementation of the high-speed transmitter it is always necessary to employ proper circuit simulation and prototyping, since at such transmission rates the circuit performance will depend not only on passive components selection, but also on the specific transistor properties, board layout and component mounting peculiarities.

Since relatively large currents are required to drive both the transmitter output stage and the load to ensure proper operation at high transmission rate, the power consumption substantially increases in comparison with the lower speed designs. The net average power consumed by the circuit as shown in Fig. 18.11, *c* with line driving voltages of 9 Volts when transmitting sustained bitstream of alternating ones and zeroes to the load of 5 kOhm || 1000 pF at 1 Mbaud can reach over one watt, although about one third of it is used to charge and discharge the load capacitance and thus is not dissipated into heat.

In applications where transmission rates up to 120 kbaud are sufficient the circuit should be adjusted by increasing the R1–R4, R8, R10, R12 and R13 resistances by an order of magnitude. R6 and R7 can be increased to 47 kOhm, and R9 and R11 set to 1 kOhm. Since R12 and R13 in the low-speed circuit variant are sufficiently large to serve as current limiters, the VT7, VT8 and R14–R17 components are no longer necessary. These modifications improve EIA/TIA specification compliance by limiting the output

---

[4] Note that this means that in a split-rail design V_+ and V_- in these expressions should be set equal to V_{cc} and zero, respectively.

voltage slew rate, and simultaneously significantly reduce power requirements. The low-speed version consumes less than 200 milliwatts when transmitting the bitstream at 120 kbaud all other conditions unchanged as compared to the high-speed example above.

18.8.3. Receiver Schematics

Basic RS232 receiver circuit can be built with a single transistor, three resistors and one diode (Fig. 18.12, *a*). Static input impedance of the circuit is determined by R1 resistance. Decreasing it improves switching characteristics of VT1, but it should not be made too small in order to not overload the transmitter. Practical values are from 3 to 10 kOhm, the lower end corresponding to higher data exchange rates. R2 is used to keep the receiver output in mark (logic high) state and prevent spurious false bit receptions when the receiver is detached from the line. VD1 protects the base-emitter junction of VT1 from breakdown during the mark (negative voltage) states of the RX line.

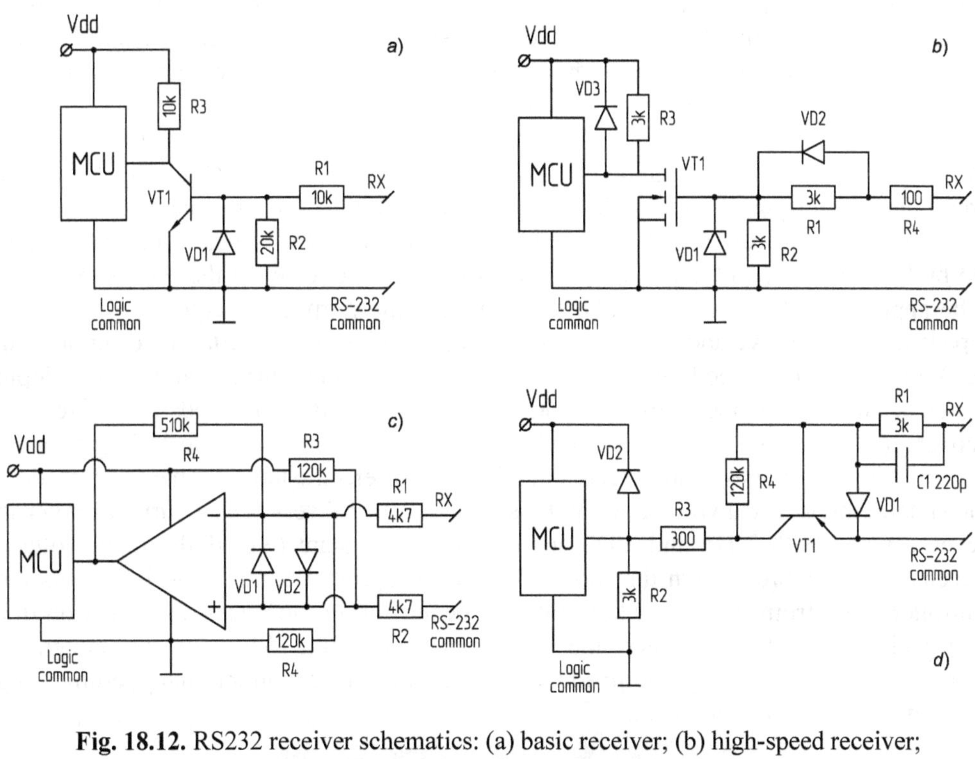

Fig. 18.12. RS232 receiver schematics: (a) basic receiver; (b) high-speed receiver; (c) opamp-based differential receiver; d) quasi-differential receiver for use with the split-rail line power supply design.

The R3 pullup resistor value can be selected quite arbitrarily, however for high-speed applications it should not be set too large to prevent VT1 from going into deep saturation during the blank (positive voltage) states of the RX line. Decreasing this resistor value improves switching performance, but increases the power dissipated by the circuit. Practical values for R3 are 1 to 10 kOhm, again the lower end corresponding to higher data rates.

The circuit as shown in Fig. 18.12, *a* can be used to reliably receive data at the rates up to 120 kbaud with any mark and blank voltage magnitudes within the range determined by the standard (i.e. starting from 3 Volts). With the R1 and R3 resistor values decreased to 3 kOhm the circuit can receive incoming bitstream at the rates of up to about 1 Mbaud. However, as the data rate increases, the circuit becomes more susceptible to asymmetry in the mark (V_-) and blank (V_+) voltages. Particularly, if the absolute magnitude of V_+ is significantly larger than that of V_- (say, 9 V against 3 V), the receiver may fail to operate properly. This happens because the high V_+ leads to increased forward base current of VT1 driving it into saturation, while the low V_- results in lower reverse recovery current during the switch-off stage; as a result, the transistor is unable to switch off in time.

One way to improve the circuit performance in this situation (beside lowering the R3 resistance even further) is to connect a Schottky barrier diode directly between the base and collector leads of VT1. This prevents the base-collector junction from forward biasing, thus keeping VT1 away from saturation regardless of the voltage applied to the receiver input during the blank state of the line.

The problem can also be remedied by replacing VT1 with a FET (Fig. 18.12, *b*). The circuit operation is similar to the described above, with a few discrepancies. Static input impedance of the receiver is determined by R2 when the RX line is in the blank state and by R1 when it is in the mark state, thus it is advisable to select these resistor values to be approximately equal. The VD2 diode serves the purpose of equalizing the input impedance for negative and positive input voltages, and, more importantly, of improving the VT1 switch-on speed by directing the gate charging current around the input resistance during the line transition from mark to blank. It also speeds-up the reverse recovery of VD1. The VD1 is replaced by Zener diode to protect the VT1 gate from excessive input voltages; it also serves the function of electrostatic discharge protection. The suitable breakdown voltage for VD1 is about 12 V. R4 serves as a current limiter in the case when both VD1 and VD2 are in conducting state (e.g. if the input voltage exceeds the VD1 breakdown threshold). VD3 is an optional component, it protects the controller input from voltage spikes that may occur during the transistor switch-on due to capacitive coupling between the VT1 gate and drain. Most of the integrated microcontrollers are already equipped with this kind of protection internally, eliminating the need in this extra component.

The circuit as shown in Fig. 18.12, *b* is able to receive the incoming bitstream at the rates up to 1 Mbaud with any V_+ and V_- magnitudes within the standard defined range.

Schematics shown in Fig. 18.12 *a* and *b* have the device logic ground and the RS232 common line directly connected. Sometimes this is unacceptable, for instance if the rail splitting is used to create V_+ and V_- voltages for the transmitter part of the transceiver under question, as described in subsection 18.8.1. In this case receiver circuit with differential or pseudo-differential input stage should be used.

For low data rates (below 20 kbaud) an operational amplifier can play the role of differential RS232 receiver, as shown in Fig. 18.12, *c*. Input impedance of the circuit is determined by the sum of R1 and R2 resistances. Resistors R3 and R4 ensure that the receiver output stays in mark (logic high) state when inputs are disconnected from the line. Diodes VD1 and VD2 provide the closed current path for the input signal and limit the differential voltage at the amplifier inputs. Since an amplifier is powered from the logic supply, a rail-to-rail output opamp such as OPA348 should be used[5]. Moreover, since the virtual ground potential in the split rail design is several Volts above the logic ground (assuming V_{cc} and V_{dd} supplies share common ground), and virtual ground is directly connected to the non-inverting receiver input, it must be ensured that the opamp can withstand corresponding common mode voltage at its inputs. For this purpose clamping diodes can be added between each amplifier input and power supply leads. The mentioned OPA348 has this protection built-in, so no external components are necessary. If a different opamp model is used, the developer should consult the documentation provided by the amplifier manufacturer.

Maximum data rate at which the opamp-based receiver is still operational is limited by the amplifier bandwidth. Although the high-speed amplifiers suitable for this application do exist, the design cost is likely to increase beyond the point of expedience. Instead, the circuit shown in Fig. 18.12, *d* can be used for high-speed reception in split rail design.

The circuit exploits the above mentioned fact that the virtual ground potential in a split-rail design is several Volts higher than the logic ground (usually 6 or 12 Volts). This allows to use virtual ground instead of V_{dd} supply voltage to drive the microcontroller input. When RX line is in the mark state its potential drops below virtual ground and VT1 goes into conduction, connecting virtual ground to the controller input. When the line goes into blank state, RX potential rises above virtual ground and VT1 closes. The MCU input is then tied to logic ground via R2.

Input impedance of the circuit is determined by R1 resistance. C2 capacitor improves switching speed of VT1 by providing the alternate path for VT1 base charge injection and withdrawal during the input voltage rise and fall. VD2 diode prevents the voltage at microcontroller input from going significantly higher than V_{dd}. Although microcontroller may have the built-in clamping diode, for high-speed operation it is still recommended to add an external fast recovery diode to improve switching speed. VD1 protects the base-emitter junction of VT1 from high reverse bias voltage when RX line is in the blank state. R4 ensures that the receiver output stays in mark (logic high) state when the input is disconnected from the line. R3 acts as a current limiter for VD2; its

[5] Alternatively, the amplifier can be powered from the transmitter V_+ and V_- rails, with a voltage limiter consisting of a resistor and two diodes clamping to V_{dd} and to logic ground added at the amplifier output. This configuration makes it possible to use a dual opamp in a single 8-pin package to build a complete transceiver, where transmitter part is implemented as shown in Fig. 18.11, *b*.

value should be selected to be about one tenth of R2. R2 resistance must be low enough to prevent VT1 from going into saturation if high-speed reception (at the rate of about 1 Mbaud) is required.

The circuit as shown in Fig. 18.12, *d* is able to receive the incoming bitstream at the rates up to 1 Mbaud with the virtual ground potentials in range from 6 through 12 Volts relative to logic ground and RX line input voltage absolute magnitudes within the standard defined range. For lower rates operation (up to 120 kbaud) resistances of R1 and R2 can be increased to ~10 kOhm to reduce power dissipation and reduce the strain on the transmitter; R3 can also be proportionally increased, and C1 can be omitted from the circuit. R4 must be increased proportionally to R2, i.e. up to ~360 kOhm.

18.8.4. Concluding Remarks

The purpose of this section was not only to present the design examples suitable for practical application, but also to demonstrate the engineering problems that arise in course of development of the serial communication interface within the restrictions posed by application demands.

From the presented material, it should have become clear to the reader that, like with virtually any other engineering task, there is no single universal solution that could equivalently effectively satisfy any combination of requirements in every possible range of the system characteristics. Although the existence of widely accepted industrial standard makes the development of interoperable equipment much easier, there is always an inevitable tradeoff between the maximum data transmission rate, maximum allowed link distance, interference tolerance, power consumption and design cost.

A skilled engineer must be able to evaluate all available approaches to the solutions of partial problems comprising the design task, determine whether a third party ready-made or newly developed original components would be more appropriate in each specific case, and find a combination of the said solutions that fits into a given set of technical and administrative restrictions while still providing the necessary functionality with guaranteed adequate performance. The authors hope that the contents of this chapter will help the interested developers to achieve this goal at least in regard to the one particular considered aspect of a sensory system design.

References

[1]. O. Kanoun, H.-R. Tränkler, Sensor Technology Advances and Future Trends, *IEEE Transactions on Instrumentation and Measurement*, Vol. 53, No. 6, 2004, pp. 1497–1501.

[2]. B. A. Snopok, I. V. Kruglenko, Multisensor systems for chemical analysis: state-of-the-art in Electronic Nose technology and new trends in machine olfaction, *Thin Solid Films*, Vol. 418, No. 1, 2002, pp. 21–41.

[3]. Yu. Vlasov, A. Legin, A. Rudnitskaya, C. Di Natalie, A. D'Amico. Nonspecific sensor arrays ("electronic tongue") for chemical analysis of liquids (IUPAC Technical Report), *Pure Appl. Chem.*, Vol. 77, No. 11, 2005, pp. 1965–1983.

[4]. D. L. Hall, An Introduction to Multisensor Data Fusion, *Proceedings of the IEEE*, Vol. 85, No. 1, 1997, pp. 6–23.
[5]. R. C. Luo, C.-C. Yih, K. L. Su, Multisensor Fusion and Integration: Approaches, Applications, and Future Research Directions, *IEEE Sensors Journal*, Vol. 2, No. 2, 2002, pp. 107–119.
[6]. R. G. Brereton, Introduction to multivariate calibration in analytical chemistry, *Analyst*, Vol. 125, 2000, pp. 2125–2154.
[7]. J. N. Miller, J. C. Miller, *Statistics and Chemometrics for Analytical Chemistry, 5th ed.*, Pearson Education, Harlow, UK, 2005 – 268 p.
[8]. *TIA-232-F: Interface between Data Terminal Equipment and Data Circuit-Terminating Equipment Employing Serial Binary Data Interchange*, Telecommunications Industry Association, 1997 – 51 p.
[9]. Moxa, Inc. (http://www.moxa.com)
[10]. MOXA Master Catalog: Industrial Networking, Computing and Automation Solution, *Moxa Inc.,* 2012, pp. 378-380, 454-455.
[11]. VS Vision Systems GmbH (http://www.visionsystems.de).
[12]. Advantech Co. (http://www.advantech.com).
[13]. Advantech Product Selection Guide, *Advantech Co.,* 2011, pp. 97–98.
[14]. S. Mackay, E. Wright, J. Park, Practical Data Communications for Instrumentation and Control, *Newnes*, Oxford, UK, 2003.
[15]. MAX13234E, MAX13235E, MAX13236E, MAX13237E: 3Mbps RS-232 Transceivers with Low-Voltage Interface, *Maxim Integrated* (http://www.maximintegrated.com).
[16]. *ITT Interconnect Solutions* (http://www.ittcannon.com).
[17]. A. L. Kukla, A. S. Pavluchenko, Yu. V. Goltvianskyi, Development of new variants of ion-sensitive electrodes based on the ISFETs, in *Abstracts of 7th Journees Maghreb-Europe "Materiaux et Applications aux Dispositifs et Capteurs" (MADICA 2010)*, Tabarka, Tunisie, 20-22 October 2010, p. 70.
[18]. A. L. Kukla, A. S. Pavluchenko, Yu. V. Goltvianskyi et al., Express Monitoring of Overall Toxicity and Separate Chemicals in Water Solutions with the ISFET Arrays, in *Report on the 34th AMOP Technical Seminar on Environmental Contamination and Response*, Alberta, Canada, October 4-6, 2011, pp. 966–975.
[19]. O. V. Demchenko, A. L. Kukla, A. S. Pavluchenko, Array of ISFETs for biochemical applications, in *Proceedings of the 2nd International Conference, Sensor Electronics and Microsystem Technologies, : Book of Abstracts*, Odessa, Ukraine, June 26–30 2006, p. 170.
[20]. O. O. Soldatkin, O. A. Nazarenko, S. V. Marchenko et al., Development of enzyme multibiosensor for determination of several toxic compounds, in *Proceedings of the International Conference, Functional Materials (ICFM' 07),* Partenit, Crimea, Ukraine, October 1–6 2007, p. 485.
[21]. TIA-422-B: Electrical Characteristics of Balanced Voltage Digital Interface Circuits, *Telecommunications Industry Association*, 1994.
[22]. F. Pan, T. Samaddar, Charge Pump Circuit Design, *McGraw Hill*, New York, USA, 2006.
[23]. Interface Circuits for TIA/EIA-232-F: Design Notes, *Texas Instruments, Inc.*, 2002.
[24]. D. Hess, J. Goldie, A Practical Guide To Cable Selection, *National Semiconductor Corp.*, 1993.

Index

D

E

F

G

H

L

M

N

O

P

U

V

W

Z

www.ingramcontent.com/pod-product-compliance
Lightning Source LLC
Chambersburg PA
CBHW060949210326
41598CB00031B/4771